Engineering Materials

This series provides topical information on innovative, structural and functional materials and composites with applications in optical, electrical, mechanical, civil, aeronautical, medical, bio- and nano-engineering. The individual volumes are complete, comprehensive monographs covering the structure, properties, manufacturing process and applications of these materials. This multidisciplinary series is devoted to professionals, students and all those interested in the latest developments in the Materials Science field, that look for a carefully selected collection of high quality review articles on their respective field of expertise.

Indexed at Compendex (2021) and Scopus (2022)

Srikanta Moharana · Bibhuti B. Sahu ·
Santosh Kumar Satpathy · Subhendu Chakroborty
Editors

Polymer Nanocomposites for 3D, 4D and 5D Printing

Fundamental to Applications

Springer

Editors
Srikanta Moharana
Department of Chemistry
School of Applied Sciences
Centurion University of Technology
and Management
Bhubaneswar, Odisha, India

Santosh Kumar Satpathy
School of Applied Sciences
Centurion University of Technology
and Management
Bhubaneswar, Odisha, India

Bibhuti B. Sahu
Basic Science and Humanities Department
Nalanda Institute of Technology (NIT)
Bhubaneswar, Odisha, India

Biju Patnaik University of Technology
Rourkela, Odisha, India

Subhendu Chakroborty
Department of Chemistry
School of Basic Sciences
Chandigarh University
Unnao, Uttar Pradesh, India

ISSN 1612-1317 ISSN 1868-1212 (electronic)
Engineering Materials
ISBN 978-981-96-4213-7 ISBN 978-981-96-4214-4 (eBook)
https://doi.org/10.1007/978-981-96-4214-4

This Springer imprint is published by the registered company Springer Nature Singapore Pte Ltd.
The registered company address is: 152 Beach Road, #21-01/04 Gateway East, Singapore 189721, Singapore

Preface

The rapid evolution of additive manufacturing technologies, colloquially known as 3D printing, has ushered in a new era of materials science and engineering. As we stand at the forefront of the Fourth Industrial Revolution, the convergence of nanotechnology, smart materials, and advanced manufacturing techniques is reshaping our approach to product design, fabrication, and functionality. Among the critical enablers of these technologies are polymer nanocomposites, which have opened new frontiers in material science by combining the versatility of polymers with the exceptional properties of nanoscale fillers. The integration of advanced materials, particularly polymer nanocomposites, with advanced printing technologies, is paving the way for unprecedented possibilities in the upcoming industrial revolution.

This book, *Polymer Nanocomposites for 3D, 4D and 5D Printing—Fundamental to Applications*, offers a comprehensive exploration of this exciting frontier, delving into the intricate world where polymer science meets cutting-edge printing technologies. The journey from 3D to 4D and now 5D printing represents a paradigm shift in manufacturing capabilities. While 3D printing has revolutionized prototyping and small-scale production, 4D printing introduces the dimension of time, allowing for the creation of dynamic, shape-changing structures. The emerging field of 5D printing pushes the boundaries even further, incorporating functionality as the fifth dimension to produce objects with embedded electronic or sensing capabilities. At the heart of these advancements lie polymer nanocomposites—materials that promise to unlock unprecedented possibilities in product performance and functionality.

This volume is structured to provide a holistic understanding of the field, from fundamental concepts to cutting-edge applications. We begin by examining the current state-of-the-art and emerging advancements in 3D, 4D, and 5D printing technologies, setting the stage for a deep dive into the materials that make these innovations possible. The role of nanotechnologies and nanomaterials in these printing techniques is explored in detail, highlighting how the manipulation of matter at the nanoscale can lead to macroscale improvements in printed objects.

This book comprises seventeen well-structured chapters on Polymer Nanocomposites for 3D, 4D, and 5D Printing. Chapter one "The Current State-of-the-Art and Emerging Advancements in 3D, 4D, and 5D Printing Technologies" provides a comprehensive overview of the latest advancements in these technologies. It highlights the significant trends shaping the future of additive manufacturing and introduces key concepts and innovations. Chapter two "Fundamental Concepts, Classification, and Materials Used for 3D, 4D, and 5D Printing" explores the fundamental principles of 3D, 4D, and 5D printing, focusing on polymers and their composites, providing a solid foundation for additive manufacturing materials science. Chapter three "Nanotechnologies and Nanomaterials in 3D, 4D, and 5D Printing Technologies" explores the integration of nanomaterials into 3D, 4D, and 5D printing, highlighting their potential to improve mechanical, thermal, and electrical properties while also discussing the challenges and challenges of nanotechnology in these advanced printing methods. Chapter four "Processing Techniques and Challenges of 3D-, 4D-, and 5D-Printed Polymer Composites" discusses the processing techniques used in fabricating polymer nanocomposites for 3D, 4D, and 5D printing, including extrusion, curing, and layer deposition and addresses the technical challenges and provides insights to overcome them. Chapter five "Additive Manufacturing Technologies for Polymer Nanocomposites" explores additive manufacturing technologies like FDM, SLA, and SLS for creating polymer nanocomposites, analyzing their applicability to various printing methods. Chapter six "3D, 4D, and 5D Printing Technology Towards Photoelectric Devices" offers 3D, 4D, and 5D printing's role in photoelectric device development, highlighting polymer nanocomposites' use in sensors, solar cells, and LEDs, enhancing electronics and energy harvesting. Chapter seven "Polymer-Graphene and Graphene Oxide Nanocomposites by Using 3D and 4D Printing for Multifunctional Applications" describes the use of graphene and graphene oxide in 3D and 4D printing to create multifunctional polymer nanocomposites, enhancing mechanical strength, electrical conductivity, and thermal stability. Chapter eight "Carbon-Nanotube-Reinforced 3D and 4D Printable Conductive Polymer Composites" explores the use of carbon nanotubes in reinforcing 3D and 4D printable polymer composites, highlighting their potential applications in flexible electronics and wearable devices. Chapter nine "Shape Memory Polymers and Its Composites for 3D-, 4D-, and 5D-Printed Applications" highlights the use of shape memory polymers (SMPs) in 3D, 4D, and 5D printing, focusing on their applications in adaptive and responsive systems like smart textiles, actuators, and biomedical devices. Chapter ten "Hydrogels and Hydrogel Nanocomposites for 3D- and 4D-Printed Applications" covers the use of hydrogels and hydrogel nanocomposites in 3D and 4D printing, highlighting their potential for creating biocompatible and bioactive structures. Chapter eleven "3D- and 4D-Printed Polymer Nanocomposites for Electronic Applications" explores the electronic applications of 3D- and 4D-printed polymer nanocomposites, focusing on their fabrication of circuits, antennas, and sensors for flexible, lightweight, and customizable electronics. Chapter twelve "Applications of 3D- and 4D-Printed Polymer Nanocomposites in the Medical and Biomedical Field" discusses the significant applications of 3D- and 4D-printed polymer nanocomposites in the medical and biomedical fields, highlighting their use in custom implants, prosthetics, tissue scaffolds, and drug delivery systems. Chapter thirteen "Elastomeric-Based 3D and 4D Printable Composite Materials for Engineering Applications" covers the development and applications of elastomeric-based composite materials in the field of engineering. These materials have unique properties that make them ideal for use in a variety of applications, including soft robotics, biomedical devices, and wearable technology. Chapter fourteen "Fundamentals and Recent Applications of 3D, 4D, and 5D Printing of Polymer and Its Nanocomposites: Challenges and Its Future Prospects" provides an in-depth analysis of the current challenges in 3D, 4D, and 5D printing using polymer nanocomposites, discussing recent applications, future prospects, research, and potential breakthroughs. Chapter fifteen "3D-Printed Polymer Nanocomposites for Water Treatment" explores the use of 3D printed polymer nanocomposites in water treatment, highlighting their potential for efficient filtration systems, adsorbents, and catalysts to remove contaminants. Chapter sixteen "5D-Printed Polymer Nanocomposites for Healthcare Applications" explores the use of 5D printed polymer nanocomposites in healthcare applications, focusing on the creation of complex medical devices, implants, and diagnostic tools for improved patient outcomes. Chapter seventeen "Multifunctional Polymer Nanocomposites for 3D and 4D Printing: Idea of 6D Printing" introduces 6D printing, a technology that aims to create dynamic, multifunctional systems. It explores its potential to produce advanced materials that can respond to multiple stimuli, offering new possibilities in various industries. Chapter eighteen "4D-Printed Polymer Nanocomposites for Smart Textiles" explores the use of 4D printing technology in creating polymer nanocomposites for smart textiles. The chapter discusses the potential applications of these materials in various industries, such as healthcare and aerospace. Additionally, it highlights the advantages of using 4D printing to enhance the functionality and performance of textiles. Chapter nineteen "Smart Materials for Next-Generation Manufacturing of 3D, 4D, and 5D Printing Technologies" explores the use of shape memory polymer by smart materials in advanced manufacturing processes, such as 3D, 4D, and 5D printing technologies. Shape memory polymers can change shape in response to external stimuli, making them ideal for creating complex structures with precise control. This chapter delves into the potential applications of these materials in various industries, highlighting their versatility and efficiency in next-generation manufacturing processes.

The book explores the processing techniques and challenges of 3D, 4D, and 5D printed polymer composites, focusing on key areas of nanocomposites such as polymer-graphene, graphene oxide, and carbon nanotube-reinforced composites. It also explores shape memory polymers and hydrogel nanocomposites, which can change form or properties in response to external stimuli. Moreover, this book explores the applications of advanced materials and printing techniques, including 3D and 4D printed polymer nanocomposites in medical and biomedical fields, electronic applications, water treatment, and smart textiles. It also introduces the concept of 6D printing, pushing the boundaries of additive manufacturing. The book addresses challenges and offers insights into future prospects, providing readers with a roadmap for further research and development in this rapidly evolving field.

This book is for a wide range of professionals, including materials and polymer scientists, engineers, researchers, and industry professionals. It offers valuable insights into the principles, processes, and potential of polymer nanocomposites in advanced printing technologies. The integration of polymer nanocomposites with 3D, 4D, and 5D printing technologies is transforming the way we print objects and possibilities. This fusion leads to a future where products are not just manufactured but grown, objects respond to their environment, and the line between digital and physical worlds blurs.

This book explores the transformative potential of polymer nanocomposites in advanced additive manufacturing, inviting readers to join us on an exciting journey. It showcases the collaborative nature of the scientific community and thanks all researchers and pioneers for their contributions. The expertise provided in this book has accelerated the progress of materials (polymer) science and engineering, paving the way for a new era.

Bhubaneswar, India — Srikanta Moharana, Ph.D.
Bhubaneswar, India — Bibhuti B. Sahu, Ph.D.
Bhubaneswar, India — Santosh Kumar Satpathy, Ph.D.
Unnao, India — Subhendu Chakroborty, Ph.D.

Contents

Editors and Contributors

About the Editors

Dr. Srikanta Moharana is presently working as Associate Professor in the Department of Chemistry, School of Applied Sciences, Centurion University of Technology and Management, Odisha, India. He obtained his Ph.D. and M.Phil. degrees in Chemistry from the School of Chemistry, Sambalpur University, India. He received his M.Sc. degree in Chemistry from the National Institute of Technology, Rourkela. He has published more than 67 research papers in international peer-reviewed journals, 75 book chapters, and many more, including renowned international publications such as the RSC, Elsevier, Wiley, Springer, and Taylor & Francis. Besides, he has edited 10 books under the banner of Elsevier, Springer Nature, RSC, and Taylor & Francis Publications. Under his guidance, two Ph.D. students have successfully received their Ph.D. degrees. Moreover, he has actively participated in and presented his research work at several international conferences and has been a reviewer for reputed international journals. He has been awarded *Prof. G. B. Behera's Best Ph.D. thesis award* under the banner of Orissa Chemical Society, Odisha, India. He has more than 5 years of teaching and research experience. His research experience and interests lie in carbon nanomaterials (graphene, carbon nanotubes) and functional ceramic-based polymer nanocomposite synthesis and characterization for advanced energy storage applications. He is a life member of the Orissa Chemical Society.

Dr. Bibhuti B. Sahu is presently working as Associate Professor, Department of Basic Science and Humanities, Nalanda Institute of Technology (NIT), Bhubaneswar, Biju Pattanaik University of Technology (BPUT), Odisha, India. He obtained his Ph.D. in experimental condensed matter Physics from the Department of Physics, Veer Surendra Sai University of Technology (VSSUT), Burla, Odisha, India, and M.Sc. and M.Phil. degrees in Physics from the School of Physics, Sambalpur University, Jyoti Vihar, Burla, Odisha, India. He has published more than 32 research articles (articles and book chapters) in peer-reviewed journals and 1 book under the banner of Springer Nature Publication. He has more than 18 years of teaching and 10 years of research experience. He has presented his work at various National and International conferences. His research interest is in Ferrites, Hexaferrites, Dielectric, Ferroelectric, and Multiferroic material synthesis, characterization, and application in energy storage and magnetic applications. Synthesis and characterization of ceramic-polymer nanocomposites for advanced energy conversion applications. He is a life member of the Orissa Physical Society.

Dr. Santosh Kumar Satpathy is Associate Professor in the Department of Physics, School of Applied Sciences, at Centurion University of Technology and Management, Odisha, India. He obtained his M.Sc., M.Phil., and Ph.D. in Physics from Sambalpur University, with his doctoral research conducted in collaboration with the Central Glass and Ceramic Research Institute, Kolkata. Dr. Satpathy's Ph.D. work focused on double perovskite materials. He has authored over 50 research papers in reputed national and international journals and contributed more than 15 book chapters published by prominent publishers. He has also edited a book with Elsevier and has two upcoming books in progress with RSC Advance and Taylor & Francis. He serves as a Guest Editor for Elsevier's Special Issue on Chemical Physics Impact. Under his supervision, six M.Phil. and two Ph.D. students have successfully completed their studies, while three Ph.D. candidates are continuing their research. With over 5 years of teaching and research experience, Dr. Satpathy's current research interests include ferroelectrics, advanced ceramic materials, multiferroics, polymers, ceramic composites, and perovskite materials. He has recently expanded his expertise to nanostructured perovskite systems and high-energy storage devices. Dr. Satpathy is a life member of the Orissa Physical Society.

Dr. Subhendu Chakroborty holds a Ph.D. in Chemistry and is the Assistant Dean-Research and Associate Professor at Chandigarh University, Uttar Pradesh, India. Recognized internationally, he was honored with the Best Keynote Speaker Award at the ISCMNB in Malaysia and won the International Picture Contest by the European Chemical Society for his illustration of "Everything is Chemistry." His research focuses on the synthesis of metal oxide, perovskite, and 2D nanomaterials for biomedical, sensor, biomaterial, functional, and environmental applications. With over 12 years of experience in teaching and research, Dr. Chakroborty has authored more than 90 scientific publications, including original research papers, review articles, textbooks, and invited book chapters in materials science and medicinal chemistry. He is a member of the Royal Society of Chemistry (MRSC) and the American Chemical Society (ACS) and serves as a peer reviewer for several international journals.

Contributors

Biswajeet Acharya School of Pharmacy and Life Sciences, Centurion University of Technology and Management, R.Sitapur, Paralakhemundi, Odisha, India

Monika Ahuja Department of Chemistry, Government Adarsh College, Affiliated to Barkatullah University Bhopal, Harda, Madhya Pradesh, India

Akshita Department of Chemistry, Amity Institute of Applied Sciences, Amity University Uttar Pradesh, Noida, India

Valentine Chikaodili Anadebe Department of Chemical Engineering, Alex Ekwueme Federal University Ndufu Alike, Abakakili, Ebonyi State, Nigeria

V. S. Aswany Department of Physics, Government College for Women (Affiliated to University of Kerala), Thiruvananthapuram, Kerala, India

Amulyaratna Behera School of Pharmacy, DRIEMS University, Tangi, Cuttack, Odisha, India

Subhendu Chakroborty School of Basic Sciences, Department of Chemistry, Chandigarh University Uttar Pradesh, Unnao, Uttar Pradesh, India

Jayesh Cherusseri Research Centre for Nanomaterials and Energy Technology (RCNMET), School of Engineering and Technology, Sunway University, Selangor Darul Ehsan, Malaysia

Barsha Choudhury School of Applied Sciences, Centurion University of Technology and Management, Bhubaneswar, Odisha, India

Omar Dagdag Department of Mechanical Engineering, Gachon University, Seongnam, Republic of Korea

N. K. Gaur Department of Physics, Barkatullah University, Bhopal, India

Sujit Kumar Guchhait Electrochemical Process Division, Sardar Swaran Singh National Institute of Bio-Energy, Kapurthala, India

Deepshikha Gupta Department of Chemistry, Amity Institute of Applied Sciences, Amity University Uttar Pradesh, Noida, India

Tejendra K. Gupta Department of Chemistry, Amity Institute of Applied Sciences, Amity University Uttar Pradesh, Noida, India

Tejendra Kumar Gupta Department of Chemistry, Amity Institute of Applied Sciences, Amity University Uttar Pradesh, Noida, India

Mohammad Harun-Ur-Rashid Department of Chemistry, International University of Business Agriculture and Technology (IUBAT), Dhaka, Bangladesh

Prashant Hitaishi Institute of Experimental and Applied Physics, Kiel University, Kiel, Germany

Mohammad Hossain Department of Chemistry, Jashore University of Science and Technology, Jashore, Bangladesh

Arezoo Hosseingholian Department of Biosciences, University of Milan, Milano, Italy

Shayan Hosseini Faculty of Chemical Engineering, Department of Polymer Engineering, Isfahan University of Technology, Isfahan, Iran

Abu Bin Imran Department of Chemistry, Bangladesh University of Engineering and Technology (BUET), Dhaka, Bangladesh

Pratik Kumar Jagtap Faculty of Science and Technology, The ICFAI University, Raipur, Chhattisgarh, India

Asheesh Kumar Department of Mechanical Engineering, Mahatma Gandhi Institute of Technology, Hyderabad, Telangana, India

Dinesh Kumar Department of Physics, JAIN (Deemed-to-be University), Bangalore, India

Debabrata Mishra Department of Physics and Astrophysics, University of Delhi, New Delhi, Delhi, India

Anchit Modi Department of Basic Sciences, IITM, IES University, Bhopal, India

Srikanta Moharana Department of Chemistry, School of Applied Sciences, Centurion University of Technology and Management, R.Sitapur, Paralakhemundi, Odisha, India

Drisya Mottammal Department of Physics, Government College for Women (Affiliated to University of Kerala), Thiruvananthapuram, Kerala, India

Seyyed Mohammad Amin Mousavi-Sagharchi Department of Microbiology, College of Basic Sciences, Shahr-E-Qods Branch, Islamic Azad University, Tehran, Iran

Alka Nair Department of Physics, Amity Institute of Applied Sciences, Amity University Uttar Pradesh, Noida, India

Hari Om Department of Chemistry, Bhagat Phool Singh Mahila Vishwavidyalay, Khanpur Kalan, Sonipat, Haryana, India

S. K. Parida Department of Physics, ITER, Siksha 'O' Anusandhan Deemed to Be University, Bhubaneswar, India

R. J. Parvathy Department of Physics, Government College for Women (Affiliated to University of Kerala), Thiruvananthapuram, Kerala, India

Deepthi N. Rajendran Department of Physics, Government College for Women (Affiliated to University of Kerala), Thiruvananthapuram, Kerala, India

Mohammadhossein Ramezanirad Student Research Committee, School of Medicine, Shahid Beheshti University of Medical Sciences, Tehran, Iran

Supriya Rana Institute of Nano Science and Technology, Knowledge City, Mohali, India

Mahdieh Ranjbar-Jamalabadi Faculty of Chemical Engineering, Department of Polymer Engineering, Isfahan University of Technology, Isfahan, Iran

Ashish Kumar Sahoo School of Applied Sciences, Centurion University of Technology and Management, Bhubaneswar, Odisha, India

Chimany Sahu Department of Physics, Barkatullah University, Bhopal, India

Deepak Sharma Deenbandhu Chhotu Ram University of Science & Technology, Murthal, Sonipat, Haryana, India;
Department of Chemistry, Bhagat Phool Singh Mahila Vishwavidyalay, Khanpur Kalan, Sonipat, Haryana, India

Chinmay Shau Department of Physics, Barkatullah University, Bhopal, India

Mohsen Sheykhhasan Cellular and Molecular Research Center, Qom University of Medical Sciences, Qom, Iran

S. P. Singh Department of Physics, Amity Institute of Applied Sciences, Amity University Uttar Pradesh, Noida, India

Devang Sukhla Department of Chemistry, Amity Institute of Applied Sciences, Amity University Uttar Pradesh, Noida, India

Md. Abu Bin Hasan Susan Department of Chemistry, University of Dhaka, Dhaka, Bangladesh;
Dhaka University Nanotechnology Center (DUNC), University of Dhaka, Dhaka, Bangladesh

Abhinay Thakur Division of Research and Development, Lovely Professional University, Phagwara, Punjab, India

Sheenam Thatai Department of Chemistry, Amity Institute of Applied Sciences, Amity University Uttar Pradesh, Noida, India

Susmi Anna Thomas Department of Physics, Government College for Women (Affiliated to University of Kerala), Thiruvananthapuram, Kerala, India

Mayank Tiwari Department of Physics and Astrophysics, University of Delhi, New Delhi, Delhi, India

Rohit Verma Department of Physics, Amity Institute of Applied Sciences, Amity University Uttar Pradesh, Noida, India

M. Vishnuvarthanan Department of Mechanical Engineering, Kalasalingam Academy of Research and Education, Srivilliputhur, Tamil Nadu, India

Krishna K. Yadav Department of Chemical Engineering, Sami Shamoon College of Engineering, Beer Sheva, Israel;
Nanotechnology Group, USAL–Nanolab, Departamento de Física Fundamental, University of Salamanca, Salamanca, Spain

The Current State-of-the-Art and Emerging Advancements in 3D, 4D, and 5D Printing Technologies

R. J. Parvathy, V. S. Aswany, Susmi Anna Thomas, Drisya Mottammal, Jayesh Cherusseri, Asheesh Kumar, Dinesh Kumar, and Deepthi N. Rajendran

Abstract Certain new materials have emerged that have the ability to modify their shape, strain, size and stiffness when they are exposed to external stimuli like water, electric field, heat, light, magnetic field, etc. Such materials show a wide range of applications that make them smart materials. One among them is the Shape Memory Polymers (SMPs). They are smart in the sense that they have the ability to regain their original shape and size once the deforming external stimuli is removed. By the addition of different varieties of nanofillers, the properties of these polymers can be modified, which makes them super important in various fields like biomedical, space, textiles, buildings, etc. This chapter emphasizes the importance of 3D, 4D and 5D printing technologies and how shape memory polymers are used for this purpose. The current state of art in printing technologies and their importance, and how it was widely used during COVID times were also explained in this chapter.

Keywords Printing technology · 3D printing · 4D printing · 5D printing · Polymers · COVID-19 · Mask

---

R. J. Parvathy · V. S. Aswany · S. A. Thomas · D. Mottammal · D. N. Rajendran (✉)
Department of Physics, Government College for Women (Affiliated to University of Kerala), Thiruvananthapuram, Kerala, India
e-mail: deepthinrphysics@gmail.com

J. Cherusseri
Research Centre for Nanomaterials and Energy Technology (RCNMET), School of Engineering and Technology, Sunway University, No. 5 Jalan Universiti, Bandar Sunway, 47500 Selangor Darul Ehsan, Malaysia

A. Kumar
Department of Mechanical Engineering, Mahatma Gandhi Institute of Technology, Hyderabad, Telangana, India

D. Kumar
Department of Physics, JAIN University, Bangalore, Karnataka, India

S. Moharana et al. (eds.), *Polymer Nanocomposites for 3D, 4D and 5D Printing*, Engineering Materials, https://doi.org/10.1007/978-981-96-4214-4_1

1 Introduction

In the modern world, digital revolution draws our attention to the fields of communication and computation. Designing and production of objects are easier nowadays due to the arrival of digital fabrication. Basic traditional models are challenged by these technologies, which directly influence the modelling of these structures [1]. The geometrical representation of physical objects by successive addition of materials embraces developments in digital fabrication technology. The life quality of the human population can be improved by introducing advancements in certain novel technologies like printing technology [2]. Material science also plays an important role here because of its use in manufacturing these structures [3]. The basic thing to know about in this field is to get an idea about how to manufacture a consumable thing from these materials. Regardless of the materials used, the primarily used manufacturing method is the 'subtractive' one. In 'subtractive' manufacturing, the raw materials are processed to make the desired product by manipulating, machining and reducing the raw materials. It helps to produce each and every single device, from basic plastic disposables to advanced surgical tools [4]. Then comes the method of 'forming' that involves changing the dimension of blocked material due to some forces. The third technology explains the method of 'casting' where the solid material is melted into liquid [5]. The difficulties faced by the subtractive and other manufacturing techniques have made the emergence of a new model in manufacturing called 'additive' manufacturing, in which the parts are formed by layer-by-layer addition of materials. The fourth technology explained above can also called 3D printing technology helps in discovering a wide range of processes for forming three-dimensional structures and prototypes [5]. The tactile reproduction of 3D digital models was achieved by enhancing the basic technology by digital fabrication. Digital fabrication is also effective in many other things, especially in printing applications. It helps to enhance the flexibility and quality of objects [6]. One by one layer of thin cross-sectional area is modeled which helps in the innovation of complex shapes and surfaces that are difficult to achieve [5]. Among these printing technologies, additive manufacturing or 3D printing is a fast-emerging one [3].

3D printing has a wide range of applications from small-scale industries like making clothes to large-scale industries like small rockets. It enables to produce limitless structure and shape, that is difficult to make using traditional technologies, via depositing materials in layer-by-layer form. The name additive manufacturing came from the successive addition of materials. However since the 3D material has a fixed characteristic, it cannot judge the state of a material in the future. Hence there arise certain products, which can alter their dimensions, measures, etc., called smart materials that started an era of 4D printed materials. Or otherwise called 'stimuli-responsive 3D'. Here, the change in shape with respect to time acts as the fourth dimension [2]. A 3D-printed object exposed to stimuli like heat, moisture, pressure, chemical reaction, etc. can exhibit self-assembling property, which is the main objective of 4D printing. The concept of 4D printing evolves after adding time dimension to 3D printing. This helps them to attribute shape deforming property that helps to evolve printing technology. Such 4D-printed transforming objects possess a large number of applications in various industrial sectors. Even though it is an advanced technology, 4D-printed objects with higher strength and complex shapes cannot be printed easily. Hence, we use 5D printing technology to print such structures having curved layers. Comparing 4D and 5D printing, 5D printing has two additional rotational axes-one is the rotation of the extruder head and the other is the rotation of the print bed [7]. The drawbacks faced by 3D and 4D printing can be overcame by the arrival of 5D printing technology, which has a variety of applications in the area of medical science and automobile fields.

With the variety of applications of printing technology in various fields like automobile industries, medical field, etc., 3D printing extends its application to medical fields such as drug delivery, fundamental research, non-biological implants, etc. A personalized manufacturing of devices is possible using it [8]. Design and manufacturing of organs and living tissues may be implied in the future in patients. There are many other applications too which include aerospace, electronics, automotive and other household products. Even if the printing technology has large applications nowadays, the limitation of raw materials for its production made its usage limited [9]. We can also ensure the emergence of printing technology in food fabrication due to the increased demands of consumers and their higher standards of living. Food printing can be done by formulating edible inks from printable biopolymers. Food decoration and customization using rheological principles encourage advancements in 2D/3D/4D/5D printing technologies [10]. The recent developments include incorporating state-of-the-art technologies like Augmented Reality (AR), Artificial Intelligence (AI), Printing technology and Internet of Things (IoT), to enhance product design and developments. Such an integrated approach will enhance the product efficiency and save the processing time. Hence, the research is ongoing to develop new innovations in reshaping this field in order to create a better future [11].

2 Types of Printing Technologies

The development of new materials using additive manufacturing technology came into being more than a quarter century ago. Materials are joined in layer-by-layer format to build novel complex geometric designs that show various applications in the fields of aerospace, biomedical, automotive, energy sectors, etc. [12].

2.1 3D Printing

3D printing uses bottom-up, layer-by-layer or top-down approach to make bulk materials with definite structure and shape [4]. It is always said to be a technology based on computer-aided design which combine numerical control systems and software for printing. Conventional manufacturing methods like removing raw materials, cutting, assembling, etc. cannot be practiced in 3D printing technology. Instead, it uses a bottom-up processing method that accumulates materials that make complex structural parts [13]. Basically, three-dimensional printing or 3D printing is a fabrication method that prints 3D objects using controlled layer deposition method. Among the two manufacturing methods, additive and subtractive manufacturing, 3D structure is made by grouping printable materials in a layered structure and by removing materials from solid materials. Which makes it an additive manufacturing method. 3D printing plays an important role in structure confinement, in which plastic or polymer-based fabrication explains additive manufacturing while wood, rock or metal-based ones pave the way for subtractive manufacturing. Both these manufacturing techniques have their advantages and disadvantages. Additive manufacturing technique is the most efficient one among the two, which produces large energy and least waste. So, it uses only a small number of machinery. The subtractive manufacturing method is more convenient and cost effective and also it need only a small time interval for the production of objects [14].

Even though 3D printing technology has been a rapidly emerging technology for the past 30 years, it is being underutilized. The time-efficient and cost-effective printing way of 3D printing technology has made its application in a wide range [14]. But recent years have made a drastic evolution in materials, machinery and software. This established the growth of using 3D printing technology in various fields like manufacturing, healthcare, education, entertainment, etc. The process of 3D printing starts with a computer-aided design (CAD) model that is forwarded to the concerned software for making a particular design. Then the 3D printer makes that structure by layer-by-layer deposition of the raw material mixture. The so-formed design is later on processed to prepare the final structure and hence it's also called rapid prototyping. There are mainly eight types of 3D printing technologies. Each method differs in its durability, material selection, cost and manufacturing speed.

2.1.1 Stereolithography (SLA)

In 1986, Charles. W. Hull invented the first known 3D printer based on the method of stereolithography. In this method, the structure is fabricated by making layers of resin (photo polymer) and solidifying it. Hence it is assumed to be the primary 3D printing method created and is the most used type also. A layer of liquid resin is solidified by incidenting it with an ultra violet laser. Then the platform containing the solidified layer is slightly moved downwards and another resin layer is added above it, which is then solidified to form the second layer. By repeating this process, we are able to fabricate the structure [15].

2.1.2 Fused Deposition Modelling (FDM)

It is one of the cost-effective 3D printing methods. Here the heated thermoplastic filaments in an extruder are deposited on a construction platform using a nozzle.

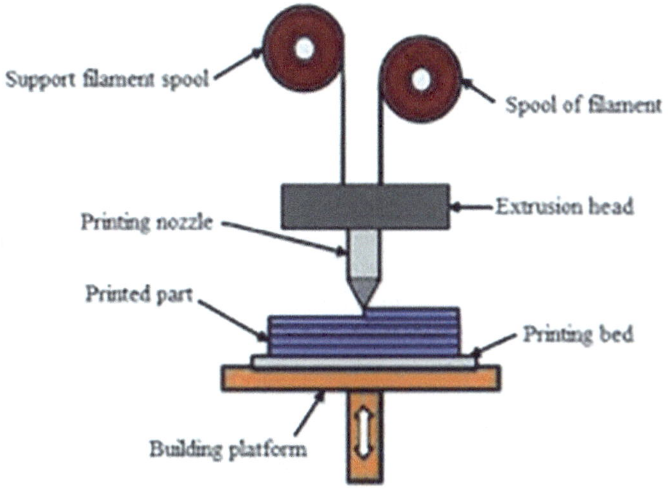

Fig. 1 Schematic diagram of fused deposition modelling. Reproduced from [17] under the terms of the CC BY 4.0 license

This platform is preserved at a low temperature as compared to the temperature of the deposited thermoplastic. Hence rapid solidification occurs and the above-mentioned process in stereolithography continuous. That is, the platform moves down and the second layer is deposited further. This process continues until the fabrication is completed [15]. The fusion mechanism in the FDM process depends mainly on solidification time. Fast freezing of molten fibers creates voids that form bonds lacking mechanical properties [16]. A schematic diagram showing the FDM process is mentioned in Fig. 1.

Even though high-quality machinable materials are generated by the FDM process, our craving for more effective 3D printed parts made us discuss polymer composite utilization in the FDM process. The outcome of a prototype depends on its dimensional accuracy [17]. The best multiple material usage, color and surface finish are the advantages of this method. Its cost is higher than SLA which makes its usage limited.

2.1.3 Digital Light Processing (DLP)

It uses photocurable resins to manufacture a single layer of 3D object by using spatially-controlled solidification. This solidification can be attained by using either UV or white light [18]. This 3D printing technology is similar to SLA with the main difference of using a digital light projector that flashes single images from each layer at once. For flashing light, a light-emitting diode (LED) or a UV light source is used. This light is further directed to a Digital micromirror device (DMD) that generates special light patterns and also controls where the light should be projected. The digital screen of the projector creates square pixels on each layer of image called voxels. Compared to SLA, DLP has fast print times, fine features and smooth surface finish. However, their brittleness made them unsuitable for using as mechanical parts [19].

2.1.4 Selective Laser Sintering (SLS)

This method is very important due to its suitability in selecting materials. The SLS method can use any materials like polymers, ceramics, metals, etc. or composites. The material used must be in the form of a powder which contains a polymer binder that must be removed later [20]. The basic process behind this method is power bed fusion. Here, a thermoplastic powder is heated below its melting point and is coated on a platform. Then the surface is scanned using a laser beam which solidifies the cross section of the structure. Once the entire cross-section is scanned, the platform will move down and the layer-by-layer deposition occurs until the complete making of that particular structure. This method is suitable for applications like functional part manufacturing, low-run production, complex ducting, etc. Higher cost and the need for longer lead times are the major disadvantages of SLS.

Similar to FDS, SLS also uses thermoplastic polymers because of their ability to get processed and reshaped on thermal variations. However not all polymers are suitable for SLS. The polymer should be a combination of certain intrinsic and extrinsic properties. The chemical structure of the polymer varies with respect to the change in its intrinsic properties, while the physical characteristics are related to extrinsic properties. A combination of these properties is needed for the SLS 3D printing method and its schematic representation is shown in Fig. 2 [21].

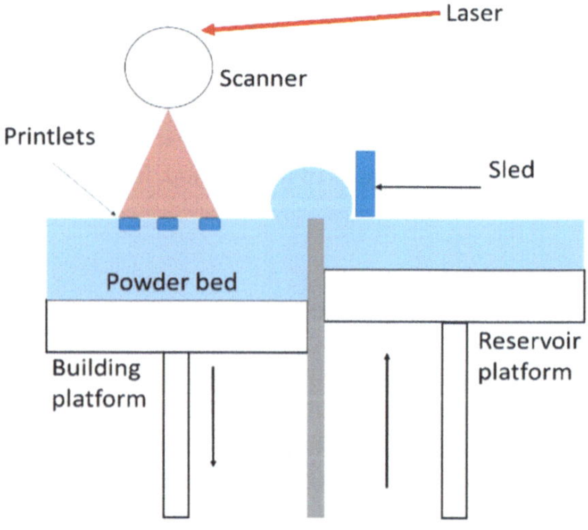

Fig. 2 Schematic diagram showing an SLS 3D printer. Reproduced from [21] under the terms of the CC BY 4.0 license

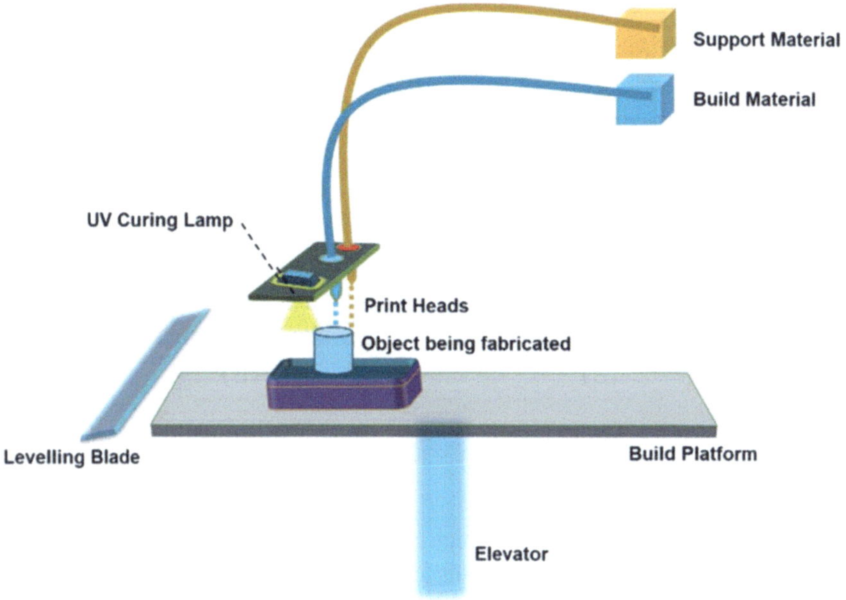

Fig. 3 A schematic diagram showing the material jetting process. Reproduced from [23] under the terms of the CC BY 4.0 license

2.1.5 Material Jetting (MJ)

As the name suggests it is similar to an inkjet printer that works using a photopolymer resin. This method uses a single process to produce multi-material parts by UV curing. Using this method, we can improve the mechanical properties of composites by spatially varying the composition of the photopolymers. A wide variety of materials and colors are prepared using a single process. Compared with other 3D printing technologies, MJ can build materials using point-wise deposition [22]. In this method, hundreds of photopolymer droplets are deposited on a platform and are solidified using UV light. After the making of a single layer, the platform is lowered by one layer thickness and this process repeats until the final structure is made.

The main advantage of this method is that it can fabricate multiple objects in a line without affecting the build speed. The material properties can be tuned by material jetting which helps to overcome the limitations of conventional manufacturing techniques. A schematic representation of the material jetting process is shown in Fig. 3. Thus MJ allows to print thin and high-quality parts [23].

2.1.6 Drop on Demand (DOD)

Thousands of nozzles are needed for the execution of this method, which is triggered by piezoelectric and thermal mechanisms. Heat is generated in the printing head by

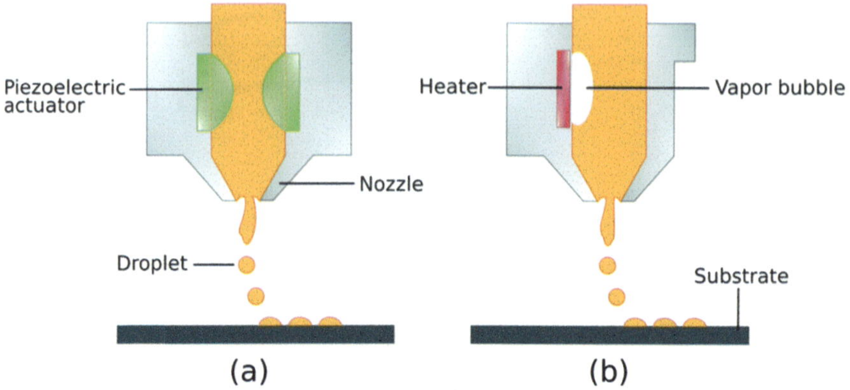

Fig. 4 Drop-on-demand 3D printing methods: **a**) piezoelectric DOD **b**) thermal DOD. Reproduced from [24] under the terms of the CC BY 4.0 license

the induction of an electric current, which creates bubbles in the material that evaporates. Then a slight volume drop is formed from the nozzle. The high vapor pressure solvents cause temperature elevation which deprives the organic compounds. In this method, piezoelectric printing heads are used [24]. A representation of both thermal DOD and piezoelectric DOD is shown in Fig. 4a and b, respectively.

DOD also produces colored prototypes which is one of its advantages. It also gives a good surface finish. However, the structures made using this manufacturing method are unsuitable for mechanical parts due to their brittleness. The high cost of production is also another limitation of this process [25].

2.1.7 Selective Laser Melting (SLM)

This additive manufacturing method is commonly used in industrial fields. There are several steps to complete this process which range from the preparation of computer data to the removal of invented structure from the platform [26]. The distinctive SLM configuration is shown in Fig. 5.

2.1.8 Electron Beam Melting (EBM)

This method uses a hot bed that has a temperature greater than 870 K. So, we cannot be able to produce a fine microstructure using this method. It has a working principle similar to that of SLM where layer-by-layer deposition of powder using an electron beam/laser occurs. Even though there exist some differences like when SLM operates in an inert atmosphere, EBM suggests a vacuum. This affects the properties and quality of the manufactured parts [27]. In this method, a computer-controlled electron gun is used to create a 3D structure from metal powder. Here, the energy

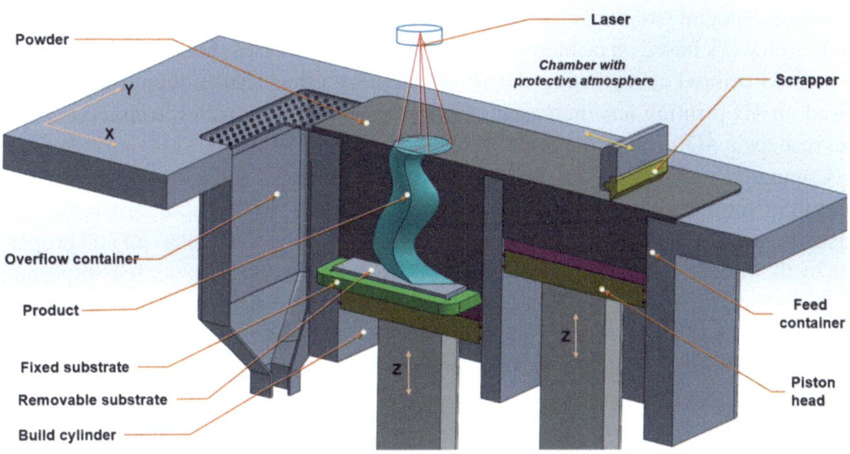

Fig. 5 Schematic diagram of SLM 3D printing process. Reproduced from [26] under the terms of the CC BY 4.0 license

source is an electron beam which helps in powder melting and also maintains the powder bed in a vacuum at high temperatures. After fabrication, it needs a cooling time [28].

The main advantages of this method are non-porous microstructure, non-oxidation of manufacturing parts and crack-free output structure. Using this method, we are able to produce brittle intermetallic materials. But this process also faces certain limitations like the use of selected materials only, greater processing time, size limitation on building parts and high cost [29]. Knowing all these methods in additive manufacturing, it is important to recognize the right process for manufacturing the specific materials. Based on some guidelines like material properties, post-processing requirements, surface quality, accuracy of material parts, etc. will help us to select a better AM process [30].

3 4D Printing

4D printing technology is based on 3D printing, but needs an extra stimulus and responsive materials. 4D printing is a time-dependent printing method that aids in creating a structure that varies its shape with respect to time. That means they exhibit intelligent behavior because of the interaction between stimulus and smart materials. Unlike 3D printing, this printing technology is predictable, printer independent and time dependent [31]. Initially, it is believed that 4D printing is what that is a time-combined 3D printing technology, where the property, functionality and shape of a 3D printed material varies with respect to time. But with an increase in studies in this technology, more definition arises for it. Since 4D printing can create shape-changing structures, special structures can be designed using mathematical modeling. In 2013, a research work based on printed active composites (PACs) was used to form complex structured printed sheets using the shape memory effect. Since then, the research based on 4D printing has increased. A preprogrammed command from a computer can reshape a 4D-printed material [32].

Compared to 3D printing, there are several advantages for 4D printing like fast growth of multi and smart materials, more deformable and flexible structures, etc. Also, the structures developed via 4D printing can enhance their behavior and properties by themselves. While performing 4D printing we need to consider five important factors which are the AM process, stimuli, modeling, mechanism of interaction and the material used for printing. There exist several AM processes, as explained in the previous session. The second factor stimuli can be of three types: physical, chemical and biological. Moisture, light, magnetic and electrical energy, temperature, etc. constitute the physical stimuli; while the chemical stimuli include pH level, oxidant, chemicals, etc. Then comes the biological stimuli like enzymes, glucose levels, etc. The third and fourth factors comprise mathematical modeling and interaction mechanisms. Not all materials undergo transformation by the applied stimulus. So better interaction mechanisms like physical manipulation or mechanical loading are applied that help in shape change. After supplying the interaction mechanism, the smart material is stimulated using a computer program that determines the time duration of the stimuli [32]. A schematic image showing the factors resulting from 4D printing is shown in Fig. 6.

The fabrication of intelligent materials using 4D printing put forth a wide range of applications such as making of self-folding and self-assembly materials.

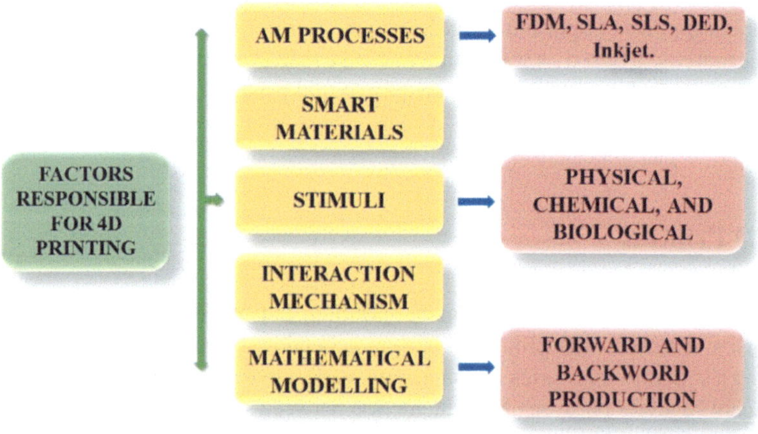

Fig. 6 A diagrammatic image showing the factors affecting 4D printing. Reproduced from [32] under the terms of the CC BY 4.0 license

4 5D Printing

The shape-shifting behavior of 4D printing technology gives an edge over 3D printing. However there are some limitations faced by 4D objects because they require high strength to make curved objects. There arises the 5D printing technology. The limitations faced by 3D and 4D printing due to their difficulty in making complex structures were overcomed by 5D printing. This method has two extra axes: the rotation of the print bed and the rotation of the extruder head. Hence 5D printed structures are widely applicable in the fields of healthcare and medical industry. It embraces a lot of promises to face contests in the healthcare, dentistry and engineering industries. This technology helps to create complicated and difficult shaped structures [7]. 5D is a combination of additive and subtractive techniques. Safety and industrial norm requirements make 5D printing an important technological advancement. A 3D-printed object may possess weak points in it due to its flat layers. But using a 5D printed technology will avoid this type of problem. A 5D printer also uses 25% less raw material as compared to 3D printing. This is one of the major benefits of a 5D printing technology. Only less materials are needed to make strong complex objects [33].

Thus, studies are still going on in this field to create advanced smart materials which can be used in various industrial purposes. The emergence of printing technology from 2D to 5D has marked a rapid growth in material objects in recent years. So, it can be concluded that a 2D printer is that which uses ink, so it can be called an inkjet printer or laser printer. While in a non-conventional printing method, called 3D printing, layered 3D objects are created. Then comes an extra dimension called time dimension that puts forth 4D printing in which the original size and shape of building parts get changed. And finally, the 5D printed objects that are similar to 3D, but are able to fabricate curved objects [34].

5 Materials Used in Printing Technology

A 3D printer uses different types of materials like resins, biological matters, concrete, nickel titanium, thermoplastics, metals, composite materials, etc. But 4D printing technology uses smart materials like photoreactive, piezo reactive, thermos reactive, hydrogels, etc. High-quality materials are needed for making 3D printed objects. This technology can produce fully functional parts using a variety of materials including polymers, ceramic, metallic and their combinations like composites, hybrids or functionally graded materials.

5.1 Metals

Metal 3D printed materials are widely used in automobile, medical, aerospace and manufacturing industries. The physical properties of such materials are excellent and hence they can be used to print complex structures from human organs to aerospace parts. Examples include aluminium alloys, nickel-based alloys, cobalt-based alloys, stainless steels and titanium alloys. Among them, cobalt-based alloys are mostly used in 3D printed structures in dental applications because of their resilience, elongation, high recovery capacity and high specific stiffness. Nickel-based alloys are mostly used for aerospace applications because their heat resistance is up to 1200 °C and have high corrosion resistance. Titanium alloys are also used in biomedical industries and for making aerospace components due to their high operating temperature and high stress [35, 36].

5.2 Polymers

Using fusion deposition modelling (FDM), 3D printed structures are formed by depositing successive layers of thermoplastic filaments such as polypropylene, poly-lactic acid, etc. Thermoplastics with low melting point are widely used in 3D printing technology. Such materials have a wide range of applications in the field of manufacturing medical devices, orthopaedic implants and biomaterials [3].

5.3 Ceramics

Ceramics are durable, strong and fire-resistant materials that show better mechanical properties. They are widely used in building and construction industries because they can be molded to any shape before the setting state. The main applications of ceramics came in aerospace and dental industries. Examples of such ceramic materials include alumina, zirconia and bioactive glasses. Here, alumina is a ceramic oxide that can act as a catalyst and absorbent used in high-technology industries[37]. Alumina has high density and curing complexity. Glass ceramics and bioactive glasses improve the bending strength of the 3D printed objects and this makes them useful in applications that need more mechanical strength. Zirconia is the main component used in nuclear power sectors. Hafnium-free zirconium is preferable for element tubing because it is less susceptible to radiation and possesses low absorption of thermal neurons [3].

6 Composites

Low weight, versatility and tailorable properties make it better for high-performing industries. Mainly two composite types are used in printing technology. They are polymer composites reinforced with glass fiber and that with carbon fiber. High strength, good fatigue resistance and corrosion performance make carbon fiber reinforced polymer composite applicable in aerospace industries. Similarly, glass fiber reinforced polymer composites have high performance and cost effectiveness. Also, fiber glass is burn resistant and has high thermal conductivity and low coefficient of thermal expansion. These properties make it suitable for manufacturing 3D-printed structures [38].

7 Smart Materials

Certain materials have the capability to change the shape and geometry of the object in the presence of external conditions like water and heat. Such materials are called smart materials [39]. The 3D-printed objects produced using these smart materials are soft robotic systems and self-evolving structures. Shape memory polymers and shape memory alloys are examples of smart materials that can be used as 4D printing materials. Shape memory polymers are basically functional materials that can change their nature with respect to external stimuli like heat, light, electricity, chemicals, etc. The evaluation of the material quality of shape memory polymers can be identified from surface roughness, part density and dimensional accuracy of the printed objects [40].

The above-mentioned intelligent/smart materials contribute to 4D-printed objects. This technology makes use of materials like liquid crystal elastomers, shape memory polymers, composite hydrogel, multifunctional materials, etc. 4D printing technology enables the design of sensitive and intelligent objects that can adapt to compound environmental factors. Incorporating engineering characteristics with internal systems creates a fascinating world of metamaterials. 4D metamaterials can be modified to match with environmental variations like changes in temperature. 4D printing based on material jetting technology creates multi-material printing. Hydrogels are biocompatible materials which have wide applications in the medical field as most of their structures constitute water. They also play a crucial role in the soft robotic industry and flexible electronics. Mathematical modelling is essential to know the functional assemblies of 4D printed objects. The smart materials used in 4D printing technology include shape memory alloys, electro-responsive polymers, temperature-responsive polymers, smart inorganic polymers, shape memory polymers, electroactive polymers, magnetic shape memory alloys and photo-responsive polymers [41]. Similar to 3D printed objects, 5D printing technology also uses various materials like metals, plastics, composites, etc. 5D printing has potential applications in the food industry. 5D printing technology is still in its early stage and the material options may expand in the future. Furthermore, a new printed technology called 6D printing has arrived which is an upgraded version of 5D printing technology using smart materials. By 6D printing, we are able to save time and cost by using appropriate processing parameters [42].

8 Current State-Of-Art in Printing Technology

The work [43] reported emphasizes how the 3D printing technology helps to face the issues created by the COVID -19 pandemic. Mainly the shortage in personal protective equipment (PPE) supplies due to its high demand. Another essential thing is the mechanical ventilator which also faced supply shortage. The work aims to bring the importance of in situ 3D printing practice in manufacturing parts of medical devices in health care. Also, a summary of the materials and ongoing research in the medical field are analyzed here. In such a crucial situation during the pandemic, additive manufacturing can make a difference. This technology helps to produce mask models or 3D-printed valves that replace normal masks with emergency respirators. The main thing to be noted in health care is that it is important to customize the tools and devices with respect to each patient. And these devices are too expensive. These limitations can be overcomed using 3D printing technology that manufactures customized products at affordable price.

The production of accessories, medical devices, or 3D printed parts to fight the COVID-19 pandemic is embarked upon by 3D printing companies. For making medical devices using additive manufacturing method, the materials used are classified into three: metals, polymer-based and ceramic-based. Among them maximum number of applications is for polymer-based materials which constitute polyamides, polycarbonates, rubber-like materials, acrylonitrile butadiene styrene, etc. While ceramic-filled epoxy resin and alumina–zirconia composites are the ceramic-based materials used for medical purposes, alloys of Ti and of cobalt and chromium are the metals. There are numerous companies that work in the 3D printing sector that can make different varieties of mask models. Copper3D team is one among them that made a powerful and airtight mono-block structure. This structure is 3D printed with the ceramic material polylactic acid (PLA) that helps to provide maximum protection from the external environment. The frame of the mask is sealed with 3D printed material MDFlex to make it watertight and antimicrobial. This review also points out the work by Swennen et al that proposed the idea and illustration of additively manufactured and reusable face masks. The face mask and filter membrane are made of polyamide composites, which are reusable. However the filter membrane and head fixation are disposable parts made of polyamide composites. Such a custom-made face mask is shown in Fig. 7A. The filter membrane support, the membrane, and its 3D prototype image is shown in Fig. 7B, C, D, respectively.

The 3D-printed surgical protective face mask called stopgap face mask, formed by Veterans Health Administration (VHA) is used for protection against liquid during

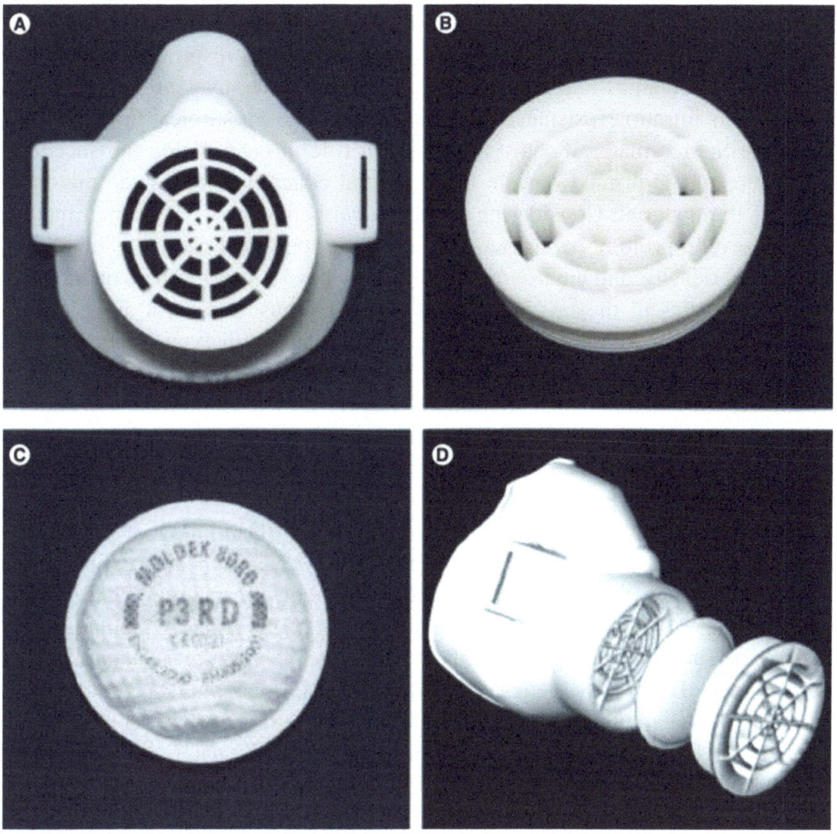

Fig. 7 A 3D printed face mask **A**) custom-made reusable face mask, **B**) filter membrane support, **C**) filter membrane and **D**) 3D image of the prototype. Reproduced from [44] under the terms of CC BY 4.0 license

COVID-19. This mask is composed of a 3D-printed mask with a filter cover, a rectangular patch of filter materials and two elastic strips. This reusable mask is made of medical-grade nylon which is compatible with disinfectant cleaners. Most face mask companies use PLA for 3D printing because of its low extrusion point and ease of printing. Another company named "Lowell Makes" had developed 3D printed COVID-19 masks that use less porosity filaments and offer the opportunity of sterilization without any harm. The material used here is PETG (a glycol-modified version of polyethylene terephthalate (PET)) which has good impact resistance. For the best operation of the mask, it should be perfectly printed without gaps between layers. However since this mask is not accepted by any monitoring agencies like that of FDA, the company did not guarantee that it can prevent COVID-19 transmission. Vital medical products related to COVID-19 can also be prepared using 3D printing technology. This method can be used to print definite accessories and components like personal protective equipment (PPE) that cannot be made with non-additive manufacturing techniques. Although 3D printing can be used to make PPE equipment like face shields, goggles, medical masks, respirators, etc., it also faces certain challenges like their inability to deliver an equivalent level of barrier protection, air filtration resistance, fluid protection, etc. for surgical mask and N95 respirators. Source materials have a significant role in determining the successful manufacture of 3D-printed objects. An essential concern to guarantee this is the control of the material. Considering the technical behavior of 3D printed medical products, several printing parameters like scanning speed, deposition velocity, laser beam energy density, and environmental humidity have a major impact on the physical characteristics of the last 3D printed product. Thus, this work aims to bring the importance of additive manufacturing in building complex engineering structures, especially during COVID times. We have explored and discussed research on different types of materials that are utilised in AM for medical purposes. In addition to the multiple benefits, this review also emphasizes its limitations. The biological products are additively manufactured and their biocompatibility and sterilization are explored.

This review [45] explained the state of the art of 3D printing of engineering materials and the recent advancements in the past 5 years. It also explained the different techniques, their applications and the varieties of materials that are used for processing it. Also, it explained different types of 3D printing techniques like binder jetting, material jetting, material extrusion, composites, smart materials, sheet lamination, etc. Different varieties of resources used for making 3D printing objects include ceramics, composites, metals, etc. 3D printing objects have certain advantages like rapid prototyping, repair, personal customization, materials, etc. It's clear that before making a product, a prototype of it needs to be made which helps to visualize the big object in a small version. 3D printing produces prototypes at a reduced rate compared to other conventional manufacturing processes. 3D printing also develops new materials for different applications. It also helps to change the dimensions of fabricated components and repair the damaged components. In addition to these advantages, 3D printing technology also faces certain limitations like size, time, surface finish, support material, etc. This technique helps to make objects with a small build volume compared to other techniques. Depending upon the size, it needs a large amount of time for fabrication. We need to maintain a balance between surface finish and the speed of production. During the fabrication, there may be unoccupied spaces which are reduced by using a support material usually called filler, that supports the structure. After fabrication, this filler needs to be removed from the whole object. The product formed has a poor surface finish and need a post processing so that it can give a better surface finish quality. All these applications make it a better and more efficient technology. A major area of research in this field is reducing the time of printing. Even though this current situation does not permit the standalone use of the 3D printing technique, it can be incorporated in different multistage manufacturing processes.

Verboven et al. in their work [46] explains the use of application techniques that are inexpensive and are capable of continuous manufacturing as of printing techniques. These applications are in the field of lighting and they are alternating current powder electroluminescent (ACPEL) devices, light-emitting electrochemical cell (LEC) and the organic light emitting diode (OLED). All of these are composed of thin printable and flexible layers. This review work emphasizes the importance of such devices and their principles of working, deposition techniques, materials, applications, outlook and state-of-the-art. A detailed review of light-emitting mechanisms, an awareness of the printing and coating techniques, device performance and current state of the art is explained here. Among these devices, ACPEL and LEC devices have simple steps with four and three layers stacking respectively. But OLED has more interlayers to increase its efficiency. An ACPEL device is like a light-emitting capacitor containing phosphor particles in a dielectric that is packed in between two electrodes. When high voltage alternating current is applied to this device, light is emitted by these embedded phosphor particles. It is constructed either in reverse or forward structures. For a forward structure ACPEL, the light emission occurs through a transparent substrate made of polyethylene terephthalate (PET) or polyethylene naphthalate (PEN). The electrodes used in it are transparent conductive oxides or transparent conductive polymers.

On the other hand, OLED is a multilayer structure containing a bottom electrode, a hole transport layer, hole injection layer, hole blocking layer, emission layer, electron transport layer, top electrode and electrode injection layer. All layers have a thickness that depends on selected material. OLED can be of two different configurations: bottom-emitting OLED (BE-OLED) and top-emitting OLED (TE-OLED). In BE-OLED, light is ejected through a transparent anode and substrate. While in TE-OLED, light transmits out through a transparent cathode. The structures of both these OLEDs are inverted. Degradation of OLED stack materials can be blocked by a high barrier encapsulation by indispensable for OLED. Similar to ACPEL devices, OLED has a flexible substrate. Now the third device called LEC is constructed from two electrodes where one is transparent and a solid electrolyte containing mobile ions. LEC can be of stacked type arrangement and surface or plasmon type. Several printing and coating methods are used to deposit layers of LEDs. Spin coating, blade coating, spray coating, inkjet printing, screen printing, etc. can be used to print the parts of these LEDs. Thus, the above three LED devices have certain advantages and limitations that make it useful for different applications. Most of the parts of these devices can be made by using printing technology.

A comprehensive review of 4D printing by F. K. Aldawood {Aldawood, 2023 #146} describes the challenges, opportunities and state of the art by 4D printing. In 4D printing, there is a change in time, but for 3D the fabricated structure is static. That means a 4D-printed design is able to change its shape with respect to environmental factors. This review paper explains the benefits, features, drawbacks and limitations of 4D printing technology. It also explains the applications in areas of electronics, aerospace, food, renewable energy, health care, fashion etc. A 4D printing operation is based on three rules. One is the majority of multi-material self-morphing behavior which involves relative expansion. The second rule said that all these behaviors are based on four physical properties: thermal expansion, mass diffusion, organic growth and molecular charge. Now the third rule which portrays the layers of 4D printed structures. The majority of 4D printed structures have two layers-an active and a passive layer. Similarly for materializing the 4D printing process, three principles must be followed. The first one is creating composite materials with stimulus response, the second is to offer stimulus for a printed material having a particular response. The third is to know the time needed to change and respond to a printed object. In order to fabricate a 4D-printed object, we can follow three approaches. In response to a stimulus, a smart material changes its behavior, which is the first approach. In the second approach, a 3D-printed device is implanted with polymer-based components and the third one initiates self-organization. Thus, the fabrication of a 4D printed material is influenced by five factors: materials, stimulus, manufacturing method, programming model and interaction between materials and stimuli.

Even though 4D printing has similar advantages as that of 3D printing, a 4D printing technology can produce way more items as compared to 3D printing. Instead of 3D printing, the benefit of using 4D printing technology is savings of energy, time, material and money. This new technology also helps to reduce product loss, errors and waste during the manufacturing process. 4D printing technology is also sustainable, rapid and energy efficient compared to other processes. 4D printing is basically a creative process that can modify the topology and stimulate the motion of a printed object. 4D printing technology plays an important role in fabricating biomedical scaffolds because of its capability to react to environmental variations. This technology is also important in producing microrobots with better stimulation and advanced design. As the 4D design is more complex than 3D printed objects, its main disadvantage is the lack of literature. Another one is the difficulty in optimizing the size and shape function parameters of 4D printed objects that show less accuracy. 4D printing has several application areas which include the fields of health care like dentistry, orthopedics, pharmaceutical, drug delivery, etc., in the fields of aerospace, electronics, food, fashion, renewable energy, etc. Since 4D printing technology is a developing area of research, a large research gap can be seen in it that discusses the scaling of manufacturing method, next generation and predictive modeling. The availability of materials for manufacturing 4D-printed objects is also another drawback. The advancing 4D printing technology can be assisted by materials with upgraded properties. In addition to these technologies, recent developments in this field proposed the arrival of new printing technologies like 5D and 6D. 5D printing helps to print curved surfaces without any supporting system by moving both the printed object and the print head.

In this report [42] Sundus Nida et al explained the emerging applications of 5D and 6D in the field of the food industry. But initially, this paper discussed 3D and 4D printing technologies used in making customized food products. The food products are manufactured in layered 3D structures. During printing, the consistency, viscosity and elasticity of the materials need to be known. The precision of a 3d printed object is affected by the rheological properties which are improved by adding additives such as lipids, salts, and carbohydrates. Even if 3D printing techniques are of different types, it is better to use extrusion-based printing in food industry. In this process, food is extruded out of the nozzle in the form of filament which is then deposited on the platform, and corresponding 3D structures are formed by selective sintering

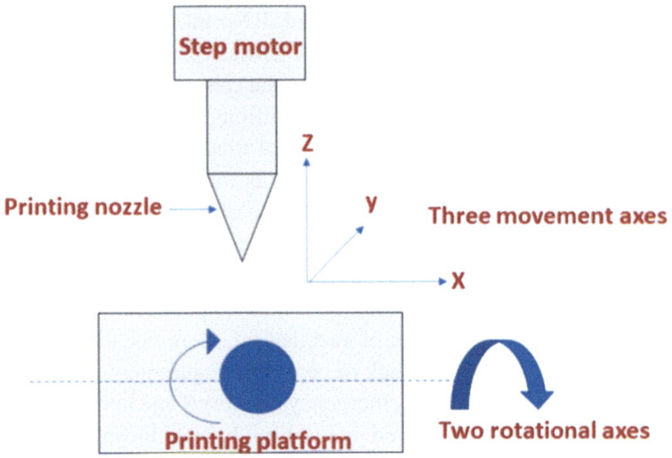

Fig. 8 Schematic representation of 5D printing. Reproduced from [42] under the terms of the CC BY 4.0 license

process. Starch and sugar mixtures are developed using binder jetting techniques while low-viscosity foods like that used in pizza base filling and cake decorations, can be made by inkjet printing. In various food applications, the main aim of 4D printing is to change the shape of 3D printed objects. It also helps to change the flavour and colour of food products and hence changes their nutritional value. A new hybrid additive/subtractive manufacturing technology called 5D printing helps to print complex structures using two rotational axes and motion of the print head in X, Y and Z directions, as shown in Fig. 8.

5D printing can be easily applied in food industry like food packing and food printing which include concave shaped snacks and ball-shaped confectionary items. Food packaging section includes developing complex box structures, cups, bowls, etc. These 5D printed objects can be prepared using some smart materials using a novel technology called 6D printing technology. It is similar to 5D printing and possesses several advantages as compared to 3D and 4D printing technologies. A mechanical characteristic feature of 6D and 4D printed objects is called bending deformation [47].

Both samples are formed by comparing the peaks. The 4D sample is printed in a flat horizontal shape while the 6D is printed in V-shaped layers. This V shape causes better bending angles and deformation. The selection of better processing parameters for 6D printing helps to save cost and time. That is the reason why we use 6D printing technology in the food industry for producing highly sensitive strong materials. Thus, 5D and 6D printing technology helps to fabricate printed objects using only 25% less materials with high mechanical strength. So, it is concluded that many open doors are still underexplored in this area of research that should be studied and used in various commercial applications.

In the review paper [48], the importance of 3D printing technology in medical applications is explained, especially during covid times. The perspectives and current state of the art of 3D printing technology during the crisis by COVID-19 emphasized the importance of this technology in saving millions of people worldwide. Since it was a worldwide pandemic, a global supply of medical equipment is necessary. Among them, the most important thing needed is the personal protective kits for health care workers. In this review, the importance of 3D printing in lung and heart surgeries and how this technology explores its use before and after the COVID pandemic, is portrayed. 3D printed models were very useful in treating long-term issues or after-effects faced by the patients. It also helped with printing respirators and protective shields. Due to a lack of vaccination, there occurs a wide spread of the COVID-19 virus and we face a lack of healthcare equipment and PPE kits. This shortage made us to use these kits at emergency situations and hence may face a lack of ventilators. This problem was solved up to a limit by the innovation of 3D printing in making health care products. 3D printing can efficiently and quickly make more PPE kits and ventilator valves. This technology can also produce tools and supplies that must be useful to face post-pandemic world. This review paper emphasized the recent challenges faced by 3D printing technology in the medical field.

3D modeling has a wide range of applications in medical field like pre-operation planning with patient-specific and virtual 3D-printed implants. It is also applicable for preparing the patient for surgery. The 3D printing technology helps the surgeon to plan and practice the surgery before the actual one. This helps to reduce the time for surgery and enhances its easiness. Such a virtual 3D model can give provide a direct picture of the object of interest, which makes it more advantageous over 2D printed images or scans. This technology is also important in making patient-specific medical implants that are needed for specialized treatments like bone anatomy. The prototype of such a model gives a dimensional accuracy which confirms that there is no mechanical failure present here. Some other 3D models like the cardiovascular one are used for treating heart abnormalities. The 3D model of a hyper obstructive cardiomyopathy is shown in Fig. 9. Similar is the case of lung surgery. Figure 9A, B represents the presurgical condition while Fig. 9C, D are that for post-surgical.

The breathing complications due to the presence of fluid in the lungs can be overcome by treating it with 3D-printed ventilator parts like ventilator splitter and also by bio-printed scaffolds. Unlike 2D structures, 3D bio-printed scaffolds can show better porosity in order to stimulate the structure of the lungs effectively. 3D bioprinting also helps to develop fully functional lung tissue that can be used for treating lung complications caused by COVID-19. By incorporating various cells with decellularized scaffolds or biological or synthetic scaffolds we are able to develop human lung tissues. The importance of vaccines and therapeutic needles for the treatment of serious diseases like COVID-19 has been a global endeavor. Many researches are going on in these areas, especially in knowing the features of microneedles. A group of small patches with a collection of micro-scale dissolvable or hollow needles that are used for vaccine administration is termed as a microneedle. Such patches are stable at room temperature and this may help to reduce vaccine hesitancy by phobias

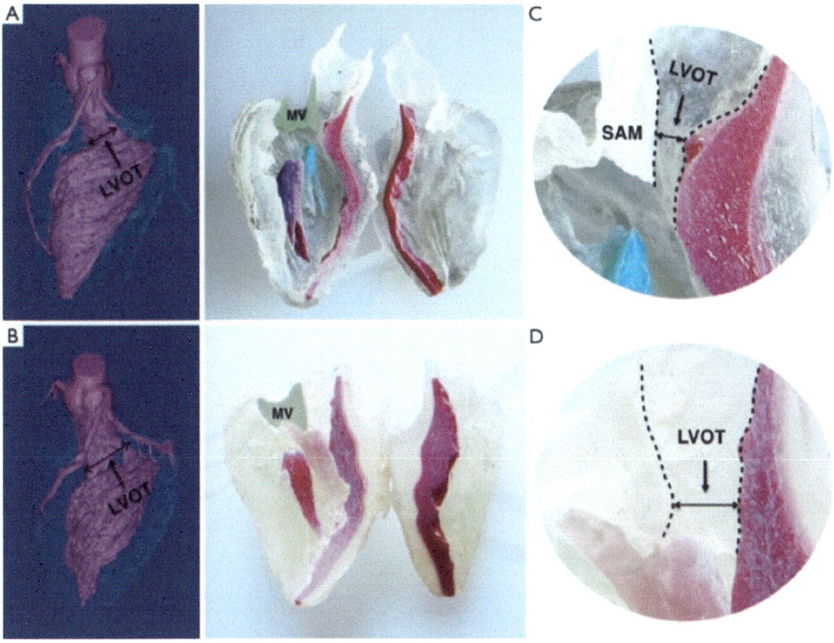

Fig. 9 **A**), **B**) Presurgical and **C**), **D**) post-surgical 3D models representing hypertrophic obstructive cardiomyopathy. Reproduced from [48] under the terms of the CC BY 4.0 license

of needles. 3D printing technology gained quick popularity in fabricating microneedles for the purpose of vaccine or drug delivery. Another strategy for delivering drugs or vaccines is to make drug-loaded particles with 3D printing technology. An innovative 3D printing method supported by multi-photon lithography can be used to manufacture antigen nanoparticles. We are able to control the characteristic features of a 3D printed object by direct laser writing or multiphoton lithography, which helps in the development of more and more 3D printed objects for the future.

During the COVID-19 pandemic, 3D printing technology has developed to great heights due to its efficiency in producing better protective objects like PPE. It was common practice to see 3D-printed face shields during early pandemic days. Various countries have relaxed their Food and Drug Administration guidelines in order to pave the way for making 3D-printed shields reliable and easy. This helped the local communities to fabricate high-quality 3D printed models instead of importing the kits. Hence several models of face shields were also introduced. The main purpose of a face shield is to create a physical barrier between the external environment and the user's face. It is basically a transparent plastic. A 3D-printed face shield in assembled and disassembled form is shown in Fig. 10a, c and b, respectively.

In this review, the author also put forward a comparison between the properties of different types of face shields including printing time, protection, comfort to wear and fitting. Among all the varieties present, the Easy 3D printed shield shows better

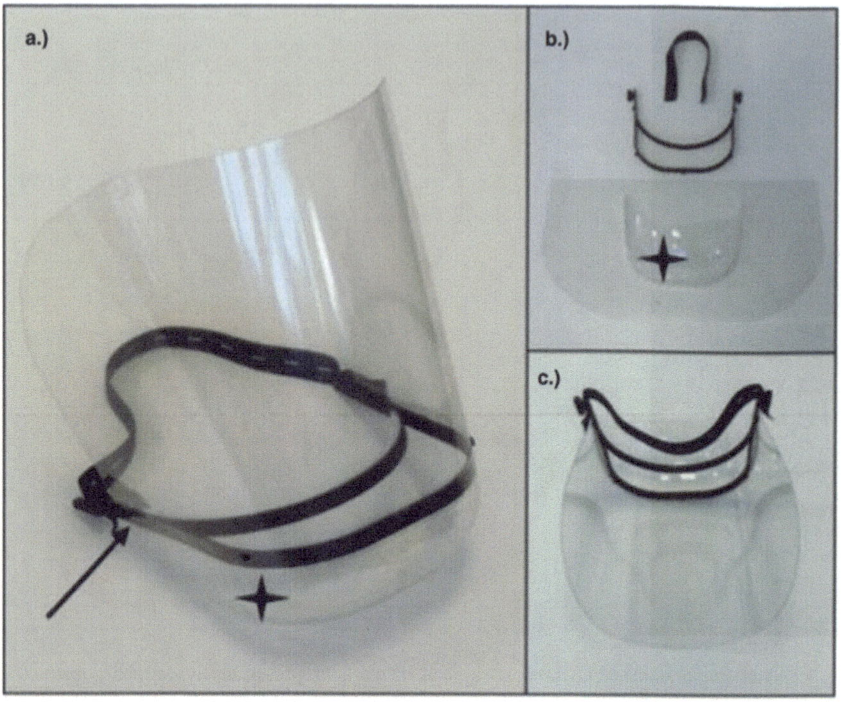

Fig. 10 Models of a 3D printed face shield **a** in assembled form **b** in disassembled form and **c** in reverse assembled form respectively. Reproduced from [48] under the terms of the CC BY 4.0 license

protection and its design is also user-friendly and comfortable. All the different types of 3D printed face shields like PRUSA RC1, RC2, Easy 3D, Budmen V3, etc. are created using materials like polyethylene terephthalate (PETG), acrylonitrile butadiene styrene (ABS), PLA, etc. Compared to other materials, PLA is better because of its flexible nature. Hence a face shield made of PLA is highly durable since it does not break easily. A face shield must be re-sterilized in order to reduce waste and preserve supplies. However, there are studies that show the fact that a 3D printed PETG shows less degradation when cleaned with hydrogen peroxide as compared to PLA 3D printed ones.

Another important application is respirator masks which are sealed helmets that pump air in and out through a filtration unit attached to the user's body. This mask offers easier and purified breathing as well as eye and ear protection. It has a rating higher as compared to N95 masks. However it is a strenuous process to manufacture a powered air purifying respirator (PAPR) and fully making it using 3D printing technology is very difficult. Hence a modified helmet with some of it's parts 3D printed, is used which enhance the safety parameters. Such a modified helmet attached with filters is shown in Fig. 11. Here two breathing filter circuits are attached to the anaesthesia tube that pumps the filtered air back into the environment. Another application of 3D printing technology is using a 3D printed adapter to fabricate an air purifying unit. They are much cheaper to make than a PAPR unit, since a PAPR requires a lot of components. The post-filter efficiency of the mask is also checked by changing the rate of airflow through it. Still, it shows better efficiency with a low cost of production. Hence it is confirmed that 3D printing can also be used for cost-effective production of PPE masks especially during the verge of a pandemic.

It is concluded that it is very difficult to meet all the needs of people during a pandemic if we only prefer to use traditional manufacturing methods. Even the traditional manufacturing techniques can also produce protective equipment, we face difficulty in producing more of them in numbers using these methods only. Also, we find less efficiency for products prepared using traditional techniques. There comes the innovative treatment of producing health care protective equipment using 3D printing technology. The lungs and cardiovascular complications that occur due to the COVID-19 pandemic have made us to think of extra long-term care for us to survive. The post-surgical treatment of the heart and lung surgeries was taken good care of by the 3D bio-printed tissues. 3D printing technology is also useful so that the

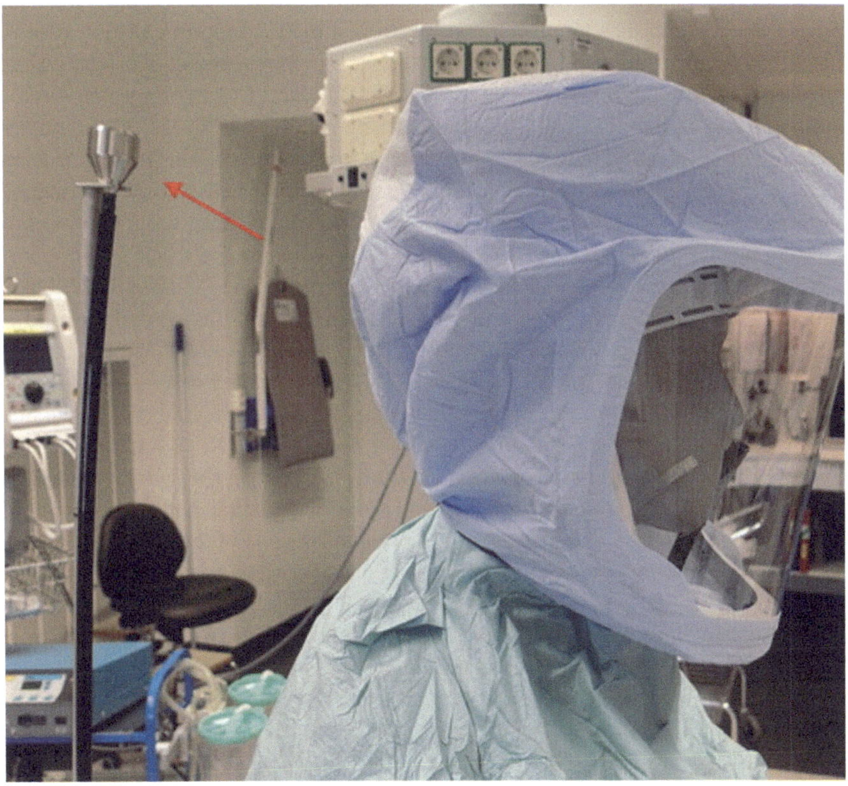

Fig. 11 Modified 3D-printed ventilated helmet incorporated with filter units. Reproduced from [48] under the terms of the CC BY 4.0 license

parts printed can be easily interchangeable. Hence large manufacturers can use it for making more versatile products. Another important aspect of 3D printing is that it can make multiple products from the same machine even at the same time. 3D printing faces many challenges while printing biomaterials since the printing depends on biomaterial properties like biodegradability, biocompatibility and mechanical properties. All these properties mainly depend on the type of 3D printer used. Each country's government should organise genuine material supplies for 3D printing their own PPE kits needed for their country. Thus, in the medical field a future direction or path is open to fulfil the surgical and other needs of patients. Also, it may improve the functionality of medical devices using robotics and bioelectronics. Thus, we are able to see a widespread technological progress in the medical field in the coming years.

The importance of printing technology in tissue engineering is explained in the review [49]. An auto-healing regenerative mechanism of tissue engineering, by the development of new tissues instead of damaged ones, by 4D printing technology is explained here. Tissue engineering will increase the speed and feasibility of tissue healing mechanisms. For replacing damaged tissues new tissues are either cultured in vitro and implanted or enhanced tissue growth by implanting stimulants. In either of these methods, tissues are developed by using a matrix support called scaffolds. This scaffold fabrication using 4D printing technology is researched and studied and its future in tissue engineering is investigated. The use of technology in evolving biofluidics and organoids using tissue engineering may lead to the successful implementation of 4D printing in tissue engineering in medical fields. This review concentrates on the design considerations, biological materials used and the fabrication techniques used for making scaffolds. Tissue engineering for the production of scaffolds can be classified into two types namely ex vivo and in situ tissue engineering. In ex vivo, the stem cells from a donor are isolated and seeded on a scaffold that is exposed to a suitable environment in bioreactors. Later on, this seeded scaffold is implanted on a target tissue and is then fabricated with a similar shape and size as that of the affected tissue. The use of more biomaterials and low efficiency of printed parts reduces its applications. The immune compatibility between the recipient and donor immune systems is also another drawback faced by ex vivo tissue engineering. All these drawbacks can be cleared by in situ tissue engineering which is used to fabricate scaffolds of required size, shape and properties. In this method, the scaffold is implanted in the body and it attracts the endogenous cells of the host to the injury site that causes tissue regeneration. The differences between ex vivo and in situ are illustrated in Fig. 12.

Scaffolds are designed and structured based on the function needed and the target site. The special features of these scaffolds are classified mainly into three types: biological, chemical, structural and physical aspects. A summary of these features is explained in Fig. 13.

Basically, it is clear that a 3D-printed structure does not show any dynamic variations in its behaviour after the completion of the printing process. But in 4D printing, the printed 3D structures are assisted by computer-aided design (CAD) and hence it shows variations with respect to external stimuli. A 4D-printed object is able to copy

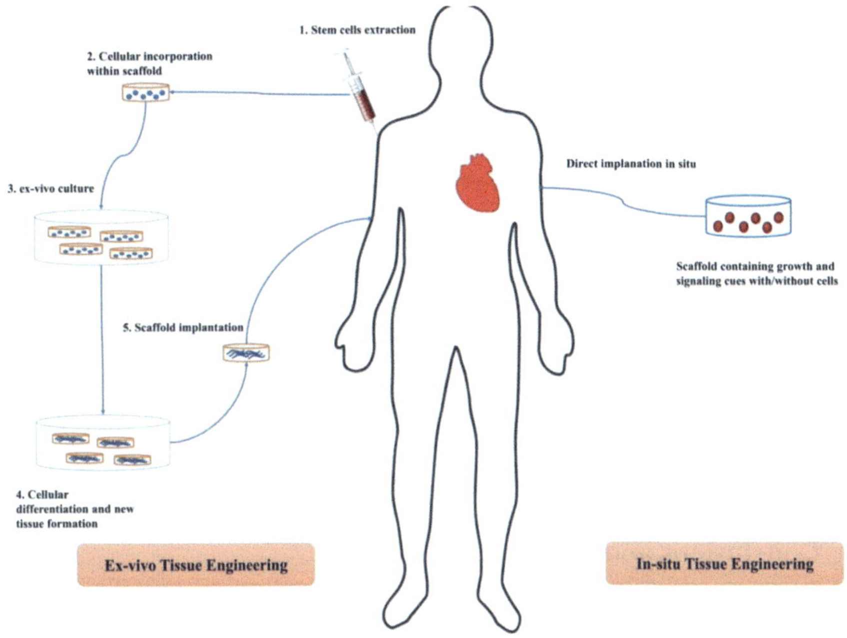

Fig. 12 A schematic illustration of the basic steps followed by ex vivo and in situ tissue engineering approaches. Reproduced from [49] under the terms of the CC BY 4.0 license

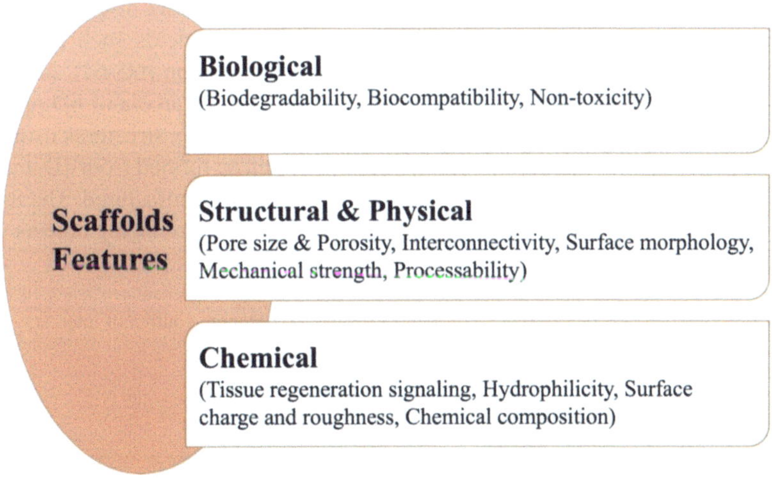

Fig. 13 Principal features for the design and fabrication of scaffolds. Reproduced from [49] under the terms of the CC BY 4.0 license

the structure of an organ and change its dynamic functions with respect to changes in pH, light, temperature, moisture, magnetic field, etc. This is the main advantage of 4D printing over 3D printing. Studies on the research of 4D printing technology particularly during in vivo studies is managed by using shape memory polymers with PCL-based 4D scaffolds. Since it is an upcoming area of research, 4D printing is one of the promising areas of research in the field of tissue engineering. Manipulating microscale-level fluids causes microfluidic design. Considering the challenges faced by 4D printing technology, the main problem is the migration of scaffolds away from the implanted site. Stabilisation of scaffolds in their target site is also another challenge faced by it. The large-scale production of scaffolds causes stability issues and their safety for a long-term run is also important. We are also doubtful about the cost-effectiveness of scaffolds. Thus, this study holds the importance of 4D printing technology in fabricating scaffolds that cause reduced failures with higher success rates than risk in organ transplantation.

9 Conclusions and Future Outlook

In this book chapter, we explained about different types of printing technologies and their importance in day-to-day life. The importance of promising technologies using additive manufacturing to create potential objects in industrial revolution is explained. The working principle and applications of 3D, 4D and 5D printing technologies and their emergence in various area of research are discussed. These advanced printing technologies can also be used in other fields such as energy conversion [50–53], energy storage [54–59], water purification [60–64], medicine [65], etc. After all these developments in printing technology an era of 6D printing is on the verge of technology. It is difficult to fabricate complex structures using 3D printing technology. Hence these complex structures can be formed from 4D and 5D printing and a variety of innovative printed products can be manufactured. The future of printing technology holds great promise by creating timely changing objects with respect to external stimuli.

Declaration of Conflicts of Interest The authors declare no potential conflict of interest.

References

1. Gershenfeld, N.: How to make almost anything: The digital fabrication revolution. Foreign Aff. **91**, 43 (2012)
2. Mallakpour, S., Tabesh, F., Hussain, C.M.: 3D and 4D printing: From innovation to evolution. Adv. Coll. Interface. Sci. **294**, 102482 (2021)
3. Shahrubudin, N., Lee, T.C., Ramlan, R.: An overview on 3D printing technology: Technological, materials, and applications. Procedia Manufacturing **35**, 1286–1296 (2019)

4. Jakus, A.E.: An introduction to 3D printing—past, present, and future promise. In: 3D printing in orthopaedic surgery, pp. 1–15. Elsevier (2019)
5. Jandyal, A., et al.: 3D printing–A review of processes, materials and applications in industry 4.0. Sustain. Oper. Comput., **3**: p. 33–42 (2022)
6. Scopigno, R., et al.: Digital fabrication techniques for cultural heritage: a survey. in Computer graphics forum. Wiley Online Library (2017)
7. Anas, S., et al.: Concept of 5D printing technology and its applicability in the healthcare industry. Materials Today: Proceedings **56**, 1726–1732 (2022)
8. Shafiee, A., Atala, A.: Printing technologies for medical applications. Trends Mol. Med. **22**(3), 254–265 (2016)
9. Kumaresan, R., et al.: 3D printing technology for thermal application: a brief review. J. Adv. Res. Fluid Mech. Therm. Sci. **83**(2), 84–97 (2021)
10. Cheng, Y., et al.: Rheology of edible food inks from 2D/3D/4D printing, and its role in future 5D/6D printing. Food Hydrocolloids **132**, 107855 (2022)
11. Rane, N., Choudhary, S., Rane, J.: Enhanced product design and development using Artificial Intelligence (AI), Virtual Reality (VR), Augmented Reality (AR), 4D/5D/6D Printing, Internet of Things (IoT), and blockchain: A review. Virtual Reality (VR), Augmented Reality (AR) D, **4** (2023)
12. Prakash, K.S., Nancharaih, T., Rao, V.S.: Additive manufacturing techniques in manufacturing-an overview. Materials Today: Proceedings **5**(2), 3873–3882 (2018)
13. Chen, Y., et al.: Improving 3D/4D printing characteristics of natural food gels by novel additives: A review. Food Hydrocoll. **123**, 107160 (2022)
14. Deshmukh, K., et al.: Introduction to 3D and 4D printing technology: State of the art and recent trends. 3D and 4D printing of polymer nanocomposite materials, p. 1–24 (2020)
15. Pereira, G.R., Gasi, F., Lourenço, S.R.: Review, analysis, and classification of 3D printing literature: types of research and technology benefits. IJAERS **6**, 167–187 (2019)
16. Vyavahare, S., et al.: Fused deposition modelling: a review. Rapid Prototyp. J. **26**(1), 176–201 (2020)
17. Mogan, J., et al.: Fused deposition modelling of polymer composite: a progress. Polymers **15**(1), 28 (2022)
18. Mu, Q., et al.: Digital light processing 3D printing of conductive complex structures. Addit. Manuf. **18**, 74–83 (2017)
19. Luongo, A., et al.: Microstructure control in 3D printing with digital light processing. In Comput. Graph. Forum. Wiley Online Library (2020)
20. Kruth, J.-P., et al.: Lasers and materials in selective laser sintering. Assem. Autom. **23**(4), 357–371 (2003)
21. Gueche, Y.A., et al.: Selective laser sintering (SLS), a new chapter in the production of solid oral forms (SOFs) by 3D printing. Pharm. **13**(8), 1212 (2021)
22. Tee, Y.L., et al.: 3D Printing of polymer composites with material jetting: Mechanical and fractographic analysis. Addit. Manuf. **36**, 101558 (2020)
23. Gülcan, O., Günaydın, K., Tamer, A.: The state of the art of material jetting—a critical review. Polym. **13**(16), 2829 (2021)
24. Mahmood, M.A.: 3D printing in drug delivery and biomedical applications: a state-of-the-art review. Compd. **1**(3), 94–115 (2021)
25. Khalil, E., et al.: Innovative Textile Printing Technology. In: Advanced Technology in Textiles: Fibre to Apparel, pp. 139–159. Springer (2023)
26. Nguyen, D.-S., Park, H.-S., Lee, C.-M.: Applying selective laser melting to join Al and Fe: An investigation of dissimilar materials. Appl. Sci. **9**(15), 3031 (2019)
27. Konda Gokuldoss, P., Kolla, S., Eckert, J.: Additive manufacturing processes: Selective laser melting, electron beam melting and binder jetting—Selection guidelines. Mater., **10**(6): p. 672 (2017)
28. Alojaly, H.M., Hammouda, A., Benyounis, K.Y.: Review of recent developments on metal matrix composites with particulate reinforcement. (2023)

29. Calderon, M.A., et al.: Extrapolating evidence-based medicine of AIT into clinical practice in the United States. J. Allergy Clin. Immunol.: Pract., **11**(4): p. 1100–1115 (2023)
30. Kumbhar, N.N., Mulay, A.: Post processing methods used to improve surface finish of products which are manufactured by additive manufacturing technologies: a review. J. Inst. Eng. (India): Ser. C, **99**: p. 481–487 (2018)
31. Momeni, F., Liu, X., Ni, J.: A review of 4D printing. Mater. Des. **122**, 42–79 (2017)
32. Ahmed, A., et al.: 4D printing: Fundamentals, materials, applications and challenges. Polymer **228**, 123926 (2021)
33. Reddy, P.R., Devi, P.A.: Review on the advancements of additive manufacturing-4D and 5D printing. Int. J. Mech. Prod. Eng. Res. Dev. **8**(4), 397–402 (2018)
34. Mishra, S.B., et al.: Journey from 2D to 5D Printing: A Brief Review. Adv. Mech. Ind. Eng., p. 95–101 (2022)
35. Trevisan, F., et al.: Additive manufacturing of titanium alloys in the biomedical field: processes, properties and applications. J. Appl. Biomater. & Funct. Mater. **16**(2), 57–67 (2018)
36. DebRoy, T., et al.: Additive manufacturing of metallic components–process, structure and properties. Prog. Mater. Sci. **92**, 112–224 (2018)
37. Tang, X., Yu, Y.: Electrospinning preparation and characterization of alumina nanofibers with high aspect ratio. Ceram. Int. **41**(8), 9232–9238 (2015)
38. Liu, Z., et al.: Modification of glass fiber surface and glass fiber reinforced polymer composites challenges and opportunities: from organic chemistry perspective. Curr. Org. Chem. **19**(11), 991–1010 (2015)
39. Lee, J.-Y., An, J., Chua, C.K.: Fundamentals and applications of 3D printing for novel materials. Appl. Mater. Today **7**, 120–133 (2017)
40. Yang, Y., et al.: 3D printing of shape memory polymer for functional part fabrication. Int. J. Adv. Manuf. Technol. **84**, 2079–2095 (2016)
41. Haleem, A., et al.: Significant roles of 4D printing using smart materials in the field of manufacturing. Adv. Ind. Eng. Polym. Res. **4**(4), 301–311 (2021)
42. Nida, S., Moses, J.A., Anandharamakrishnan, C.: Emerging applications of 5D and 6D printing in the food industry. J. Agric. Food Res. **10**, 100392 (2022)
43. Tarfaoui, M., et al.: 3D printing to support the shortage in personal protective equipment caused by COVID-19 pandemic. Mater. **13**(15), 3339 (2020)
44. Abbas, M.Z.: Industrial applications of 3D printing to scale-up production of COVID-19-related medical equipment. J. 3D Print. Med., **5**(2): p. 97–110 (2021)
45. Singh, T., Kumar, S., Sehgal, S.: 3D printing of engineering materials: A state of the art review. Materials today: proceedings **28**, 1927–1931 (2020)
46. Verboven, I., Deferme, W.: Printing of flexible light emitting devices: A review on different technologies and devices, printing technologies and state-of-the-art applications and future prospects. Prog. Mater. Sci. **118**, 100760 (2021)
47. Arikan, N.A.: A short review of production with 5D and 6D printing
48. Hagen, A., et al.: 3D printing for medical applications: Current state of the art and perspectives during the COVID-19 crisis. Surgeries **2**(3), 244–259 (2021)
49. Adel, I.M., ElMeligy, M.F., Elkasabgy, N.A.: Conventional and recent trends of scaffolds fabrication: a superior mode for tissue engineering. Pharm. **14**(2), 306 (2022)
50. Kumar, D., et al.: Nanocarbon assisted green hydrogen production: development and recent trends. Int. J. Hydrogen. Energy **50**, 118–141 (2024)
51. Cherusseri, J., et al.: Polymer-based composite materials: characterizations. In: Composite materials: processing, applications, characterizations, pp. 37–77 (2017)
52. Thomas, S.A., et al.: A comprehensive review on tungsten oxide nanostructures-based electrochromic supercapacitors and machine learning models for design and process parameter optimization. Energy Storage **5**, e499 (2023)
53. Thomas, S.A., et al.: Rapid synthesis of hierarchical tin disulfide (SnS2) nanostructures by a microwave-assisted hydrothermal method for high-performance supercapatteries. ACS Appl. Electron. Mater. **6**, 3346–3361 (2024)

54. Thomas, S.A., Cherusseri, J.: Recent advances in synthesis and properties of zirconium-based MXenes for application in rechargeable batteries. Energy Storage **5**, e475 (2023)
55. Sadanandan, A., et al.: A critical review on two-dimensional $Ti_3C_2T_x$ MXenes for anti-corrosion coatings. Prog. Org. Coat. **183**, 107757 (2023)
56. Thomas, S.A., Cherusseri, J.: Boron carbon nitride (BCN): an emerging two-dimensional nanomaterial for supercapacitors. J. Mater. Chem. A. **11**, 23148–23187 (2023)
57. Pramanik, S., et al.: Metal matrix composites: theory, techniques, and applications. In: Composite materials: processing, applications, characterizations, pp. 369–411 (2017)
58. Cherusseri, J., Kar, K.K.: Recent progress in nanocomposites based on carbon nanomaterials and electronically conducting polymers. In: Polymer nanocomposites based on inorganic and organic nanomaterials, pp. 229–256 (2015)
59. Richard, B., et al.: Minireview on fluid manipulation techniques for the synthesis and energy applications of two-dimensional MXenes: advances, challenges, and perspectives. Energy Fuels **37**, 6999–7013 (2023)
60. Zaed, M.A., et al.: Low-cost synthesis of $Ti_3C_2T_x$ MXene-based sponge for solar steam generation and clean water production. Ceram. Int. **50**, 27910–27922 (2024)
61. Zaed, M.A., et al.: Utilization of recycled materials for low-cost MXene synthesis and fabrication of graphite/MXene composite for enhanced water desalination performance. Sep. Purif. Technol. **354**, 129055 (2025)
62. Zaed, M.A., et al.: Synthesis and characterization of hierarchical $Ti_3C_2T_x$ MXene/graphitic-carbon nitride/activated carbon@ luffa sponge composite for enhanced water desalination. Open Ceram. **19**, 100645 (2024)
63. Zaed, M.A., et al.: Hierarchical $Ti_3C_2T_x$ MXene@ Honeycomb nanocomposite with high energy efficiency for solar water desalination. Chemosphere **366**, 143459 (2024)
64. Zaed, M.A., et al.: Unlocking desalination's potential: harnessing MXene composite for sustainable desalination. Chem. Eng. J. **500**, 156910 (2024)
65. Cherusseri, J., et al.: SARS-CoV-2-on-chip for long COVID management. Biosensors **12**, 890 (2022)

Fundamental Concepts, Classification, and Materials Used for 3D, 4D, and 5D Printing

S. K. Parida

Abstract In this chapter, the fundamental concept of the 3D, 4D and 5D printing technology has been discussed briefly. Unlike traditional manufacturing procedures, 3D printing is an additive process that has become a practical way to build technical components. Here, 3D, 4D and 5D printings are the modern advanced emerging technologies for a wide range of applications. Since every material has different features and applications, it is critical to comprehend the advantages and disadvantages of each choice. From a fundamental perspective, available materials, fabrication speed, each unique application and resolution are the characteristics of these advanced techniques. Once again, due to their low post-processing waste, ease of manufacture, low human involvement, minimum material waste, and energy efficiency, 3D, 4D, and 5D printing are sustainable technologies for industrial applications. The merits and downsides of various 3D, 4D and 5D printing technologies are covered in this study. A thorough explanation of the various materials suitable for every kind of 3D, 4D and 5D printing methods are discussed along with numerous application areas are also presented in the study. The results of this collective study would be useful to upcoming scholars and industry professionals working in 3D, 4D and 5D printing.

Keywords 3D, 4D and 5D printing · Additive process · Manufacturing processes · Fabrication speed · Material waste

1 Introduction

The manufacturing process of physical objects from three-dimensional printing by laying down successive thin layers of materials is called 3D printing. The additive manufacturing can be known as 3D printing. A fascinating advancement in additive manufacturing, 3D printing transforms the way we manufacture everything from basic tools to intricate components. The materials employed in this technique are

S. K. Parida (✉)
Department of Physics, ITER, Siksha 'O' Anusandhan Deemed to be University, Bhubaneswar, India
e-mail: santoshparida@soa.ac.in

S. Moharana et al. (eds.), *Polymer Nanocomposites for 3D, 4D and 5D Printing*, Engineering Materials, https://doi.org/10.1007/978-981-96-4214-4_2

crucial. These are necessary for 3D printers to assemble designs into tangible items. Japanese researcher Hideo Kodama invented 3D printing in the year 1981. In the 1st step, the digital object is converted to a computer-aided design (CAD) model and then sliced into different layers. The principle of the 3D printing process focuses on the stereolithographic process which has the x-, y- and z-axis. 3D printing has high speed, low cost, high accuracy, diverse and robust material properties [1]. Figure 1 represents a 3D printing technology flow diagram.

Here, 3D printing has many potential applications in the real world based upon engineering and manufacturing, medical, aerospace, fashion, design, and consumer products. The negative impact of 3D technology is that 3D printing takes a lot of time to print the components and the process is very slow. The 3D printer has printing weapons, violation of copyrights, fraud, and harms authenticity. Interestingly, materials like ceramics, polymers, smart materials, concentre, metal and alloys and biomaterials are often used in 3D printing technologies. However it does not have enough strength, expensive cost of raw materials and misuse of technology.

The drawbacks of 3D printing can be overcome by 4D printing technology. The 4D printing technology was invented by Skylar Tibbits in February 2013. The additive manufacturing technology known as 4D printing allows objects to self-assemble or reorganize over time. The form of 4D printed material is affected by environmental factors such as temperature, humidity, and other variables. Smart materials and intricate patterns that are "programmed" to cause your 3D print to alter shape are the basis of 4D printing. By including a temporal dimension, 4D printing allows to produce dynamically controlled forms on demand. 4D printing involves four dimensions. The x, y, and z axes, as well as the fourth dimension, time, can save up to 90% of time and material. The 4D printing flow diagram is given in Fig. 2 [2].

4D printing has different applications due to its self-changing properties and it can also change shape. 4D printing is applicable in all fields like medical field, polymer industry aerospace and fashion. In aerospace, 4D printing can be employed in future wars and antenna arrays. The US Army has previously expressed considerable interest

Fig. 1 3D printing technology flow diagram

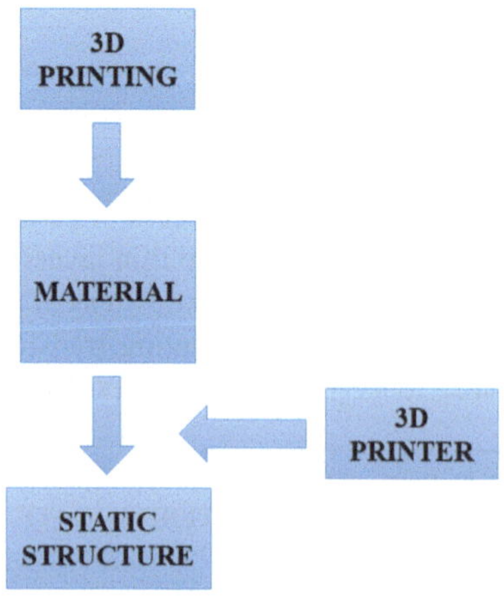

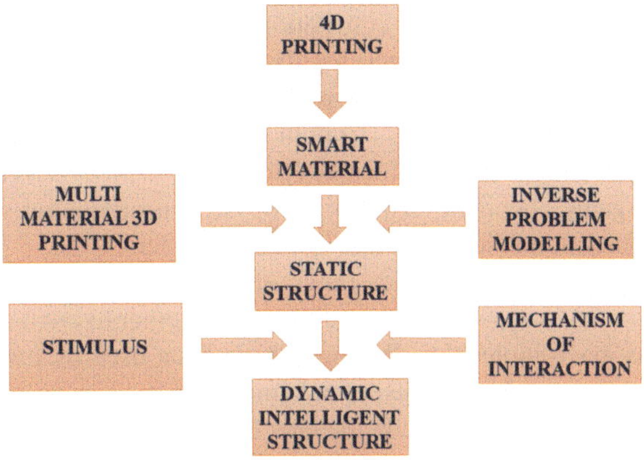

Fig. 2 4D printing technology flow diagram

in 4D printing [1]. The next generation of medical gadgets will use 4D printing. Shipping volume will be in a flat pack format, which may be activated after delivery to begin 3D printing. 4D printing can help tackle difficulties in the medical, dental, and other related industries.

The negative impact of 4D printing is that it is not able to print complex shapes that have curved surfaces. This can be overcome by 5D printing technology as it is stronger than 3D and 4D printing. 5D printing is a revolutionary additive manufacturing process in which the print head and printable item rotate along the x, y, and z axes, resulting in five degrees of freedom. It makes exceptionally precise curved layers or concave forms under design limits. Using this method, the printer head prints in five axes while the printed portion moves. The object can be produced from all five axes rather than just one by having the print bed move forward and backward along the x, y, and z axes simultaneously [3]. As an application 5D printing technology is mostly applied in the medical and automobile sectors. In these printing technologies materials like thermoplastic polymer, powder, metal, and UV curable resin, etc. are used. But in comparison to metals and ceramics, polymeric materials are suitable for printing technology due to their properties like resolution, high efficiency, and customization.

2 Type of Polymer Composites

There has been increasing demand to utilize polymer composites in research [4]. Polymer composites provide freedom with regards to ultimate product qualities, design, possibilities for starting from raw material, and ever-expanding progressive breakthroughs in synthesis procedures [5], apart from their extensive commercial availability and cost-effectiveness in comparison to other materials like metals and ceramics for budget-friendly purpose [6]. The superiority of polymer-based nanocomposites over conventional materials is considerable because of their excellent corrosion resistance, highly configurable product properties, flexible production processes, and greater corrosion resistance properties. The term "polymer" corresponds to "many parts" from the Greek word "poly" and "mero". Polymers are large molecules with moral masses that range from dozens to millions. About 80% of the organic chemical business is focused on producing synthetic polymers, including plastics, textile fibres, and synthetic rubbers.

Polymers are a wide range of materials made up of small molecules which are known as monomers that are linked together to create long chains which are known as macromolecules. Polymers have been used by humans for ages in many forms including oils, tars, resins and gums. The modern polymer industry emerged with the industrial revolution. In the later 1830s, Charles Goodyear developed vulcanization, a method that produced a usable type of natural rubber [7]. Polymers are at the core of many significant industrial products. Rapid expansion in manufacturing is being fuelled by social factors and the need to replace traditional resources. The term "polymer" was introduced by Swedish scientist J. J. Berzelius. He viewed benzene (C_6H_6) as a polymer of ethyne (C_2H_2). Later this definition was modified [8]. Henri Braconnot's pioneering work on derivative cellulose compounds, dating back to 1811, is considered one of the earliest significant contributions to polymer research. In the late 1800s, vulcanization increased the durability of natural rubber, making it the first semisynthetic polymer to gain popularity. By combining formaldehyde and phenol at precisely the right temperature and pressure in 1907, Leo Baekeland created Bakelite, the first synthetic polymer. In 1909, Bakelite made its official debut [9]. The Ziegler–Natta catalyst, created by German chemist Karl Ziegler and Italian chemist Giulio Natta, was important to the study of synthetic polymers and brought them the 1963 Noble Prize in Chemistry. 1974. Paul Flory's contributions to the field of polymer science earned him the Chemistry Noble Prize.

Artificial organic polymers have been used in the form of wax and shellac for ages. Cellulose, a plant polymer that is responsible for the strength of natural fibres and ropes. By the early nineteenth century, natural rubber derived from rubber trees was widely used for polymer and medicine granules [10]. Goodyear's vulcanized rubber outperformed untreated natural rubber in terms of strength, abrasion resistance, elasticity, temperature sensitivity, gas impermeability, chemical and electric current resistance, and polymer composition [11]. Plastic, rubber, coatings, adhesives, sealants, natural and synthetic fibers, and other materials that are commonly employed in modern life are all included in polymer science. One of the most significant concepts of the twentieth century is the idea of polymers. This idea was first established in the 1920s amid discussion, and it is intimately associated with H. Staudinger, who was awarded the Nobel Prize in 1953.

Synthetic polymers include common materials like polyesters and nylons, as well as those employed in medical applications such as organs and biodegradable sutures. Plastic materials are classified as commodity (high volume, cheap cost) or engineering (higher cost, low volume) based on economic and application factors. The first category includes polyethylene (PE), polypropylene (PP), and polyvinyl chloride (PVC), while the second group includes polycarbonate (PC), poly(ether ether ketone) (PEEK), and polyimide (PI), among others. Natural, manufactured (modified natural), and synthetic fibres have high aspect ratios, strength, and modulus, among other qualities that vary by application. Elastomers have the capacity to stretch and retract quickly [12].

Polymer characteristics are classified according to their scale and physical properties. A polymer's fundamental attribute is the identity of its component monomers. Microstructure refers to the arrangement of monomers inside a polymer at the chain level [13]. At the level of the nanoscale, the chemical properties explain how chains interact with physical forces. They explain the bulk polymer's interaction with solvents and other compounds at the macroscale [14, 15].

- **Chemical Property**

The properties of a polymer are determined by the attractive forces that occur between its chains. These interchain interactions are much stronger than the attractions between ordinary molecules because of lengthy polymer chains. Dipoles in the monomer units of polymers can influence the intermolecular forces. Adjacent chains of polymers containing amide or carbonyl groups have the potential to form hydrogen bonds due to the strong attraction between the partially positively charged hydrogen atoms in N=H groups on one chain and the partially negatively charged oxygen atoms in C=O groups on the other.

- **Mechanical Property**

A polymer's bulk characteristics are often of relevance to end users. These parameters determine the polymer's macroscopic behaviour. A material's tensile strength indicates its ability to withstand stress without permanent deformation [16].

- **Transport Property**

Diffusion refers to the rate at which molecules travel across the polymer matrix. Polymers have a significant role in film and membrane applications. When referring to polymers, the phrase "melting point" refers to the transition from a crystalline or semi-crystalline phase to a solid amorphous phase, not from solid to liquid [17, 18].

- **Elasticity**

Bond distortions are responsible for elastic behaviour in most elastic materials, including metals often employed in springs. Bond lengths deviate from the lowest energy equilibrium when force is applied, and strain energy is electrostatically stored.

3 Comments and Classification on Polymers

There are different types of classification of polymers. The most common classification is based on the origin of polymers [19]. Due to their diverse properties, complicated structure and many uses polymers cannot be classified in a single category. Depending on the raw materials available polymers are divided into three different types. These three different types of polymers are natural, synthetic and semi-synthetic polymers. Natural polymers exist naturally and are found in both plants and animals. Examples are rubber, cellulose, starch, and proteins. Biopolymers are another name for biodegradable polymers. Naturally occurring polymers, such as cellulose nitrate and cellulose acetate, are converted chemically to create polymers that are semi-synthetic. Synthetic polymers are those created by humans. Plastic is the most widely used and common manmade polymer. It is used in many different industries and dairy products, such as nylon-6 and polyether [20]. Here, the different types of polymers are discussed and presented in Fig. 3.

- **Natural Polymers**

Polymers which are extracted from plants or animals or any living beings are known as natural polymers [21]. There are kinds of polymers inorganic and organic. Because they contribute to essential life processes and serve as fundamental structural elements, organic polymers are significant to living things. For instance, the solid parts of every plant are composed of polymers. Nature also contains a large number of inorganic polymers, including diamond and graphite [20].

- **Synthetic Polymers**

Synthetic polymers are man-made polymers. Numerous procedures are used to create synthetic polymers. Polymers can be made from a range of simple hydrocarbons, including ethylene and propylene, by adding monomers to the developing chain in a sequential order. Repeated ethylene monomers combine to form polyethylene, an additive polymer [20].

- **Semi-Synthetic Polymers**

Semisynthetic fibers are polymers made from natural fibers that have undergone straightforward chemical processing to enhance their tensile strength and durability. For example, acetate rayon, cupra ammonium silk, and viscous rayon [21].

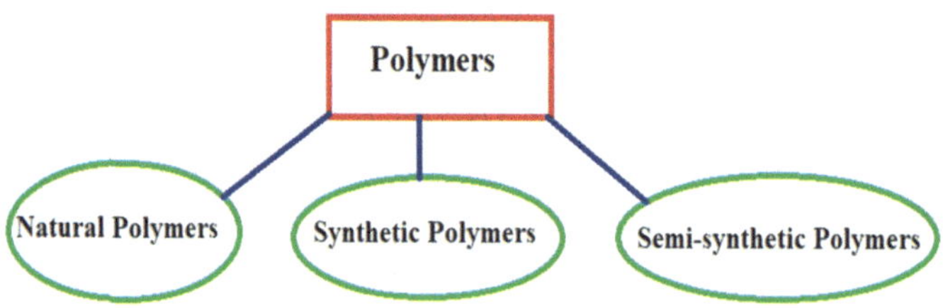

Fig. 3 Types of polymers

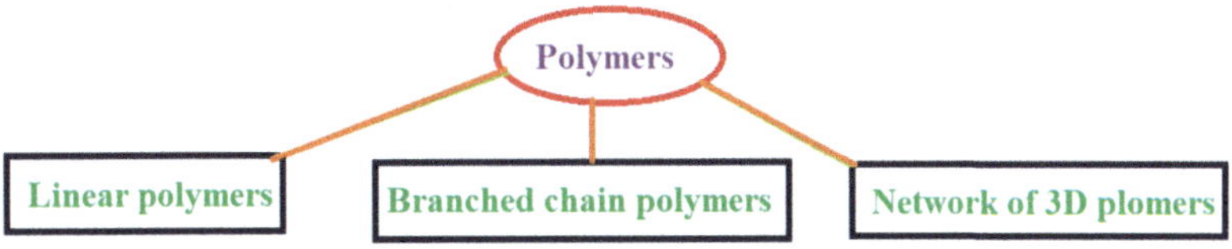

Fig. 4 Types of polymers on the basics of structure

4 Classification of Polymers According to Their Structure

In this section, the type of the polymers on the basics of structure has been discussed and presented in Fig. 4.

- **Linear Polymers**

Linear polymers are composed of long, straight chains of monomeric units linked together. For example, high-density polythene (HDPE), nylon, and polyester [22].

- **Polymers With Branched Chains**

Branched-chain polymers have branched all the way along the key chain in addition to their untwisted nature. For example, amylopectin, glycogen, and low-density polythene (LDPE) [22].

- **Network of Three-Dimensional Polymers**

The network of three-dimensional polymers is made up of monomer molecules that are connected by only covalent bonds. Heavy cross-links in these massive molecules prevent the mobility of individual monomeric components. For example, Bakelite, urea–formaldehyde, and melamine formaldehyde [22].

5 Classification of Polymers According to Their Methods of Synthesis

Here, in this section, different types of polymers on the basics of their synthesis techniques have been discussed and presented in Fig. 5.

Fig. 5 Types of polymers on the basics of structure synthesis methods

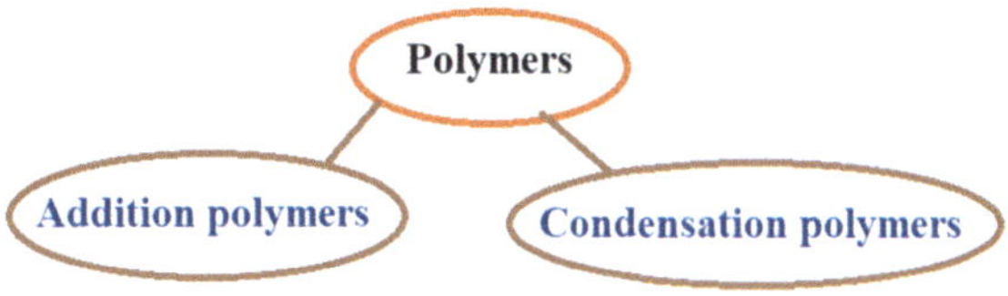

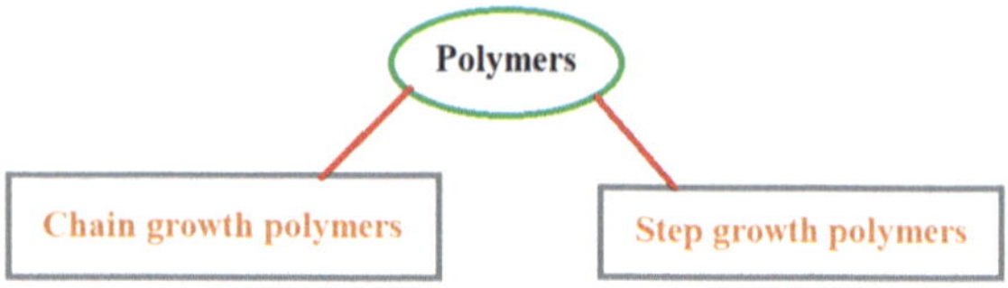

Fig. 6 Types of polymers on the basics of growth polymer chain

Addition Polymer: These polymers are obtained from another method of polymerization that includes repeatedly adding monomers to lengthy chains to yields polymers. Examples include polystyrene, polypropylene, and polyethylene.

Condensation polymer: Usually comprising two monomers, a series of condensation processes yields condensation polymers. For example, polyester, Nylon-6, and Bakerite.

6 Classification of Polymers According to Their Growth Polymer Chain

In this section, the classification of the polymer on the basics of their growth of polymer chain has been discussed briefly. Figure 6 shows the types of polymers on the basics of the polymer growth chain that are present.

Chain growth polymers: Polymers with chain growth are created by adding monomer units one after the other to a reactive intermediate developing chain. For example, polystyrene, polypropylene, and polyethene.

Step-growth polymers: Polymers with step growth are created through a sequence of separate responses that result in the loss based on tiny molecules like NH_3, HCl, and H_2O as well as the development of bonds between two distinct monomers.

7 Classification of Polymers According to Molecular Forces

In this section, it has been discussed briefly about the types of polymers that are divided into four different parts. Figure 7 shows four types of polymers, which come due to the strength of the molecular forces present.

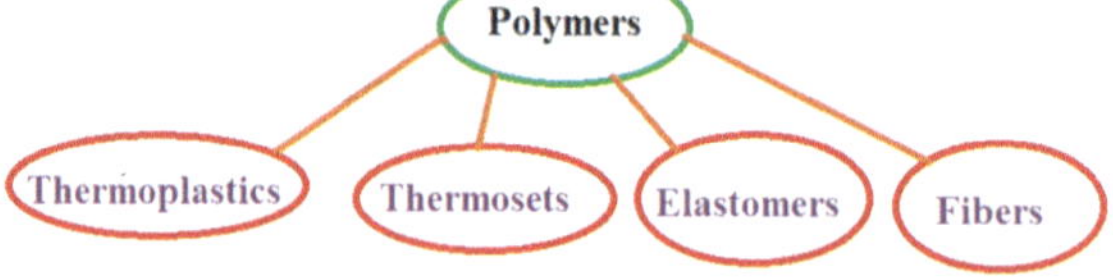

Fig. 7 Types of polymers on the basics of molecular forces

Thermoplastics: The molecules in a thermoplastic are held together by relatively weak intermolecular interactions, which causes the material to soften when heated and return to its former state when cooled. They are bulk of linear and slightly branched polymers. Every major thermoplastic is produced by chain polymerization. Examples include credit cards, vehicle bumpers, food packaging, and insulation.

Thermosets: Thermosets cannot be reshaped by heating. Thermosets are often three-dimensional networks of polymers with strong crosslinking between the polymer chains. The chains are limited by cross-connecting movements and add to the solid substance that gives thermosets their strength and robustness. They are primarily employed in the building and automobile sectors. They are also utilized in the production of glues, toys, boat hulls and varnishes.

Elastomers: Elastomers are rubbery polymers that are repeatedly stretched to their unstretched length before returning to their original size when pressure is released. Elastomers, despite their cross-linking destiny, are cross-linked materials. Low crystallinity and a glass transition temperature over T_g are requirements for an elastomer. Rubber bands and other elastic materials are used to make elastomers [22].

Fibers: A large number of the polymers used in plastics are also the same as in synthetic fibers. It's common to say that a fiber has an aspect ratio of at least 100. Denier units are used to express the thickness of the fiber. A unit of measurement for fiber linear mass density is the denier. It is referred to as the weight in grams per 9000 m. Natural fibers include silk, cotton, wool, and wood pulp; synthetic fiber includes nylon, carbon, PE, PTFE, and Kevlar [23].

8 3D Printing

Additive manufacturing, or 3D printing, has made it possible to create supplies on demand with precise electrical and mechanical properties and complex geometries. AM methods have been developing in a range of industrial and new applications, and they often involve polymers and composites. Generally, layer upon layer of materials is added throughout the 3D printing process to create the desired shape of an object. It is easy and can produce complicated designs straight from a computer model. With 3D printing, once the object is designed in software, no physical modifications to the machine or equipment are necessary. Thus, labor- and time-intensive production processes are being transformed by 3D printing technology. They additionally minimise the waste of resources. Different materials, including composites, metal, ceramic, and polymer, are used by different kinds of 3D printers. Using CAD software, a 3D model is first created before being converted into an STL (surface tessellation language) format for 3D printing. In order to plan the printing path of 3D items, the STL file will be divided into 2D layers. An other file type, such an OBJ file, can have texture and material colour information, while a PLY file can hold three-dimensional scanned objects. For simplicity of management, the data are saved in XML, a human-readable format. Polymers are a common choice for 3D printing materials because of their cost-effectiveness, convenience of usage, and flexible design. Several manufacturing parameters, namely temperature, cooling and heating rate, can affect the degree of crystallinity and the size of the material microstructures during the printing process. Studying the materials and adopting the best procedure are the first steps towards a successful 3D printing project. Next, the manufacturing process needs to be optimised, and lastly, the print quality must be reviewed. Some 3D painting techniques are developed such as Multiple Exposure (ME), Vanishing Point (VP) and so on [24].

- **Multiple Exposure**

The printing process employed in ME differs according to whether the material is liquid or solid. Layer by layer, the solid filament is placed on the platform or previously printed layers using FFF, a 3D printing technique in which the filament is melted into a semi-liquid state at the hot end and extruded through the nozzle. Materials made of thermoplastics are utilized as feedstock. Controlling the feed rate that pushes the molten filament out of the nozzle and keeping the compression pressure between the roller and the hot end below a crucial limit are essential for an excellent print because they prevent the filament from buckling [25]. The degree of orientation of polymer chains is influenced by the shear stress generated in the melted state of a polymer, which is dependent on printing speed. The polymer strength in the chain direction is greater than that in the transverse direction, therefore the polymer chains prefer to align along the printing direction. The temperature of the 3D printing platform's surroundings should also be considered, as it affects the rate of cooling of the melted polymer and the subsequent crystallinity of the re-solidified polymer, both of which influence the mechanical performance of the printed object in the end.

- **Vanishing Point**

VP, also referred to as SLA, employs UV-solidified photopolymers, a resin based on acrylic. The two primary machine configurations are top-down and bottom-up. UV light travels in the desired direction above the tank in top-down VP, hardening the photopolymer in the liquid bath by forming crosslinked networks. The platform descends from the reaction zone by a certain layer thickness after a layer is produced to make room for newly cured resin to repopulate the zone. The printer then prints the subsequent layer, and so on, until the product is finished. Until the product is completed, the printer continues printing each layer after that. The top-down the light source is set up beneath the resin bath by VP printers, and the moving platform is constructed with an upside-down layout. The bottom of the bath has a transparent coating that lets laser light through but stops the cured portion from adhering to it. Furthermore, VP can produce 3D objects with great resolution in the submicron range if the polymerization excitation technique is altered [26]. To ensure a good photo-initiation process and high-quality products, diluent and chain transfer compounds can be added to photopolymers, which typically consist of monomer/oligomer and photo-initiators [27]. The viscosity, wetting behaviour of the resin, and laser intensity are three curing reaction parameters that need to be effectively managed for an effective polymer curing process [27, 28].

- **Powder Bed Fusion-Laser Beam (PBF-LB)**

Currently gaining popularity in mass production, PBF is one of the most adaptable technologies. By carefully injecting thermo-energy, such as lasers, it prints three-dimensional forms out of powder. Selectively sintering and fusing the powders together to generate the required shape for each 2D layer, the laser passes over the powder material. The powder bed is heated during printing to be slightly above the crystallisation temperature to impede crystallisation and slightly below the melting temperature to minimise the heat required to melt the powder. An operator pulls out more powder from the supply platform after printing one layer, recoats the layer, and then continues the process. In practical application, polyamide and polycaprolactone (PCL) powders are the major materials due to their consolidation behaviour and large sintering window, even if any thermoplastic polymer powder can be used as a feedstock in theory. Key printing parameters in PBF are scanning speed, layer thickness, laser power, and powder size [29, 30]. In order to achieve low porosity and high geometric accuracy, suitable powder recoating, and suitable mobility of the melted powder, these elements should be taken into account. PBF can manufacture complex objects [31]. Product size depend upon the volume of PBF machine. The machine is very costly, it's resolution is high and the size is also large. The product needs post processing as it has rough surface.

- **Sheet Lamination**

For polymer composites, sheet lamination (SL), one of the common additive manufacturing techniques, can also be utilised by adhering material sheets to create three-dimensional structures. Through the use of adhesive and thin material sheets rather than welding, laminated item manufacturing technology may print polymer composites. To melt and fuse the sheets together during the printing process, pressure and heat are applied in tandem. The extra material in the present piece is cut to create a planned layer for the 3D sculpture once layer-by-layer bonding is finished. SL is well suited for quick production because material handling is simple. Heat treatment to improve the interlayer bonding and post-processing to smooth the surface can be necessary. Using this SL technique, adhesive is applied to the sheets, followed by the dispensing of polymer powder on each sheet and the bonding of the sheets. The printed item is crushed to consolidate the part after being heated in an oven to melt the polymer. SL has been successfully printed 3D objects using composites. A structure having maximum volume of 12 × 12 × 4 inches can be create using SL machine.

9 Applications of 3D Printing

3D printing is a versatile, efficient, and profitable method of production, but it requires more than just a machine and material. Generally, 3D printing budget must account for additional expenses such as storage, ventilation, handling materials, and equipment setups, especially if intended to take advantage of its exceptional adaptability. 3D printing allows us to swiftly prototype and build items for a wide range of applications at a low cost. However, selecting the appropriate 3D printing procedure is only one side of the equation. Ultimately, the materials will determine whether we can build pieces with the necessary mechanical properties, functional attributes, or appearance. There are dozens of plastic materials available for 3D printing, each with distinct properties that make it best suited to specific applications. To make it easier to identify the right material for a certain part or product, let us start with an overview of the many types of plastics and 3D printing procedures.

There are two main types of plastics used.

- The thermoplastic is the most often used type of plastic. The ability to go through multiple melt and solidification cycles distinguishes them from thermosets. Thermoplastics may be heated and formed into the desired shape. Because no chemical bonds form throughout the process, thermoplastics can be recycled, melted, and reused. The properties change slightly between melting cycles.
- Thermosetting polymers (also known as thermosets) are permanent solids once cured. Polymers in thermosetting materials crosslink during the curing process, which is initiated by heat, light, or suitable radiation. Thermosetting plastics degrade when heated rather than melted, and do not reform after cooling. Thermosets cannot be recycled or reconfigured into their original components.

Inevitably 3D printing starts with a printer. Many different types of 3D printers are available in the market [1], some of the most commercial ones are as follows:

A. **Stereolithography (SLA) 3D printers**: SLA is one of the most popular 3D printing processes, available to both hobbyists and professionals. These printers focus one or two lasers on a vat of resin [2], curing a small portion of it. They create the product in tiers. They can employ extremely powerful lasers designed for engineering-grade resins. Some of the popular 3D printing materials of SLC are given in Table 1.

B. **Fused deposition modelling (FDM) 3D printers**: FDM printers can be consumer or industrial grade. A nozzle [3] warms and softens filament that has been wrapped around a spool. The heated filament is subsequently deposited in layers by the nozzle. Some of the popular 3D printing materials of FDM are given in Table 2.

C. **Digital light printing (DLP) 3D printers**: Many users choose DLP printing to make larger components or larger volumes of parts since it can flash a full layer at once rather than just a small section [4].

Table 1 Shows popular 3D printing materials in SLA

Materials	Description	Applications
Alumina 4N resin	Technical ceramic made of 99.99% pure alumina that has exceptional mechanical, thermal, and conductive properties	Features include heavy-duty tools, wear-and chemical-resistant components, and insulators against heat and electricity
ESD resin	ESD-safe material to enhance operations in electronics manufacturing	Fixture and tooling for the production of electronics Custom trays for handling and storing components, anti-static prototypes, and end-use components
Flame retardant (FR) resin	Material that is stiff, creep-resistant, heat-resistant, and flame-resistant for use in indoor and industrial settings with high temperatures or ignition sources	Interior components found in cars, trains, and airplanes Custom fixtures, jigs, and spare parts for use in industrial settings, as well as internal and protective components for consumer or medical electronics
Draft resin	3D printing materials can be used as one of the fastest materials for printing 4 times faster than standard resins and can go to 10 times faster than FDM	May be used for initial prototypes application and rapid iterations
Standard resins	Features include high resolution and a smooth matte finish	Used as initial prototypes and follow the concept models
Clear resin	The only totally clear material for plastic 3D printing; polishes to near optical transparency	Parts with optical transparency and Mill fluidics
Polyurethane resins	Superior durability over an extended period of time. Stability in UV, temperature, and humidity; flame retardancy; sterilizability; and resilience to chemicals and abrasions	High-performance components for automobiles, airplanes, and machinery end-use parts are robust and tough. Durable, long-lasting functional prototypes
Tough and durable resins	Durable, functional, dynamic, and strong materials, able to withstand impacts, stretching, bending, and compression without breaking. Many substances with characteristics resembling those of PE or ABS	Enclosures and dwellings fittings and jigs, connectors Test-and-tear models
Rigid resins	Materials that are rigid, robust, and highly filled to withstand bending. Dimensionally stable under pressure, resistant to heat and chemicals	Fixtures, jigs, and tools Fan blades and turbines Components of fluid and airflow, automotive housings and electrical casings

(continued)

Table 1 (continued)

Materials	Description	Applications
Silicone 40A resin	The superior material qualities of cast silicone and the first readily available 100% silicone 3D printing material	Small batches of silicone parts, validation units, and functional prototypes. Personalized medical equipment Soft molds, masking tools, and flexible fixtures for resin or urethane casting
Flexible and elastic resins	Rubber, TPU, or silicone's flexibility is able to tolerate compression, bending, and flexing and withstands numerous cycles without tearing	Prototyping consumer items Features that comply with robotics. Anatomical models and medical equipment, props and models for special effects
High temp resin	Superior precision and endurance to high temperatures	Warm air, gas, and liquid movement fittings, housings, and mounts that withstand heat Inserts and molds
Jewellery resins	Supplies for vulcanized rubber molding and investment casting. Strong shape retention, fine details, and ease of casting	Custom jewellery, masters for reusable molds, and try-on pieces
Medical and dental resins	A large selection of biocompatible resins for making dentistry and medical devices	Dental and medical equipment, such as prosthetics, dentures, and surgical guidance

D. **Selective laser sintering (SLS) 3D printers**: SLS printers employ powdered materials [5]. The powder is heated just below its melting point before depositing a small layer of it onto the build platform. A laser scans the surface in the desired pattern, sintering and hardening the powder. The platform travels down a layer, and another layer of powder is deposited. Table 3 shows the popular SLS 3D printing materials. Some of the popular 3D printing materials of SLS are given in Table 3.

10 4D Printing

The development of three-dimensional (3D) printing technology is the source of four-dimensional (4D) printing. At a TED conference held at MIT in 2012, Skylar Tibbits gave a demonstration of how changes made to a static printed item using 3D printing might cause it to alter shape over time [32]. The initial research on 4D printing was published in 2013. It was built on the concept of printed active composites. Because of this, a printed sheet might be made into a complex arrangement by utilising the shape memory effect [33]. Right now, 4D printing can produce a large variety of objects that 3D printing is unable to. Using 3D printing production tools and environmental stimuli, 4D printing changes the material's shape [34].

Table 2 Shows popular 3D printing materials of FDM

Materials	Description	Applications
Composites (carbon fiber, Kevlar, fiberglass)	Sturdy, powerful, or incredibly resilient, only a few pricey industrial FDM 3D printers are compatible	Working prototypes Fixtures, jigs, and tools
PVA (polyvinyl alcohol)	Soluble supporting substance and disappears in water	Supporting documentation
HIPS (high-impact polystyrene)	Most frequently used soluble support material with ABS, it dissolves in limonene chemical	Supporting documentation
Nylon	Most frequently used soluble support material with ABS, it dissolves in limonene chemical	Working prototypes Parts that withstand wear
TPU (thermoplastic polyurethane)	Stretchable, elastic, resistant to impact, and highly effective at reducing vibration	Adaptable prototypes
ABS (acrylonitrile butadiene styrene)	Robust and long-lasting, resistant to heat and impact, needs air, and a heated bed to print	Working prototypes
PETG (polyethylene terephthalate glycol)	Suitable for quicker production at lower printing temperatures, humidity and resistance to chemicals, very transparent, perhaps safe for food	Applications that are waterproof and Snap-fit parts
PLA (polylactic acid)	The FDM materials that print the quickest, strong and stiff yet easily broken, less resilient to heat and chemicals, organic and odorless	Conceptual models resemble prototypes

Shapes, colours, volumes, features, or functions of 3D printing materials may change significantly in 4D printing in response to external stimuli or environmental factors [35, 36]. Materials for 4D printing that are now available include hydrogels, liquid crystal elastomers (LCEs), shape memory polymers (SMPs) and their composites, shape memory alloys (SMAs), and dielectric elastomers [37, 38]. Together with other technical advancements like intelligent robotics, quantum computing, and human augmentation, the potential for 4D printing has only increased [39]. By fusing new materials with established printing techniques, 4D printing enables the direct manufacturing of sensors, electrical devices, programmable structures, soft actuators, and other functional devices [40]. When it comes to printability and design freedom, 4D printing offers greater flexibility than traditional SM manufacturing methods [41].

Different 4D printing methods are discussed below.

- **Material Jetting**

Table 3 Shows popular 3D printing materials of SLS

Materials	Description	Applications
Nylon composites	To make nylon materials stronger and more rigid, glass, aluminum, or carbon fiber reinforcement is added	Prototyping with functionality and structural final components
Nylon 12	Robust, resilient, strong, and long-lasting; resistant to UV rays, heat, moisture, solvents, temperature, and water; impact resistant and able to bear repeated wear and tear	Final components of functional prototyping for pharmaceuticals
Nylon 11	Though less stiff, it is comparable to Nylon 12 in terms of impact resistance, elongation at break, and strength	Final components of functional prototyping for pharmaceuticals
TPU	Rubbery, elastic, and flexible, adaptable to deformation, superior stress absorption and high UV stability	Flexible, rubber-like final components, Medical equipment Prototyping with functionality
Polypropylene	Robust and pliable Resilient to chemicals Waterproof and handleable	Final components to functional prototyping for pharmaceuticals

Suitable for fluid samples, material jetting is a different kind of 3D printing technology that can create 4D structures [42]. The process of material jetting involves continuously or dropwise spraying the materials onto a prefabricated platform [43]. When compared to other 3D printing methods, material-jetting printers have the advantage of being able to print one or more materials at once. 4D structures have been created using an inkjet printer, which is one kind of material-jetting 3D printing method, in a number of earlier research. Inkjet 3D printing was used by Ding et al. to produce active composite materials directly. Heat could be applied to the 3D-printed shape to immediately change its shape. The pre-strained elastomer changed form when heated because it softened and released strain. As it cooled to room temperature, the altered shape persisted. The inset of illustrates the shape transformation procedure. By using heating and cooling cycles, the shape-transformation process was reversible [44].

- **Powder Bed Fusion**

Another kind of 3D printing method used for 4D printing is powder-bed fusion. SLS is a potential 3D printing method that uses laser sintering to process powdered materials. It is a subtype of powder bed fusion techniques [45]. In order to produce a 3D structure with excellent mechanical qualities, SLS melts and fuses powder layer by layer using a high-power laser [46]. A magnetic gripper exercise is completed and a powder is 4D printed that responds to magnetism using polyurethane powder. Under a magnetic field, the 4D-printed gripper demonstrated reversible gripping [45].

SLS 3D printing was used by Gan et al. to create a piezoresistive sensor in a different investigation. Piezoresistive sensors were created by utilising a polyurethane/CNT nanocomposite's exceptional electrical conductivity.

- **Vat Photopolymerization**

Using a process called vat photopolymerization, photosensitive resins are polymerized layer by layer in a vat under UV light to create three-dimensional structures. Following material extrusion, vat photopolymerization is the second most often used 3D printing method for 4D printing. SLA and DLP are the two subgroups of vat photopolymerization [47]. A UV laser source is utilised in SLA 3D printing processes to polymerize (cure) a photosensitive resin point-by-point (using a moveable photon source) [48] to complete one layer. The previously cured layer is cured before the subsequent layer is applied. In DLP, a digital screen is used to expose UV light one layer at a time. DLP offers superior high resolution and is comparatively faster. By projecting the sliced image of a 3D model onto a screen using a projector, the DLP approach exposes photocurable resin to projected UV light, which causes it to polymerize [49].

- **Material Extrusion**

A slurry of photo- or thermo-curable resin, or thermoplastic filaments, is dispensed selectively via a nozzle or orifice layer by layer to create a three-dimensional structure using the material extrusion 3D printing technology. Fused Filament Manufacturing, or FDM for short, is an extrusion-based 3D printing process. In Fused Deposition Modelling (FDM) 3D printing, thermoplastic filaments are heated and fed through a printer extruder head before being deposited on a heated constructed platform [50]. The nozzle forces the molten filament through, where it is then placed on the plate. G-codes are used by computer-controlled motors to direct the motion of the print head. Another material extrusion-based 3D printing method is called DIW. In this method, a liquid material that may be photo- or thermo-cured is extruded through a nozzle under air or physical pressure, and it is then cured on a specially designed platform [51].

- **Materials Used in 4D Printing**

There are some materials which preferably used in the 4D printing method. 4D printing primarily involves materials that can change shape or properties over time in response to environmental stimuli. Here are some common materials used in 4D printing—(i) **Shape Memory Polymers (SMPs)**: These materials can "remember" a predefined shape and return to it when exposed to specific stimuli, such as heat or light, (ii) **Hydrogels**: Water-absorbing polymers that can swell or shrink in response to changes in temperature or pH, making them suitable for applications like soft robotics and biomedical devices, (iii) **Thermo-responsive Materials**: Polymers that change properties when heated, often used for actuators and smart textiles, (iv) **Electroactive Polymers (EAPs)**: These materials change shape or size when an electric field is applied, useful for sensors and actuators, (v) **Composites**: Combinations of materials that can exhibit multi-responsive behavior, enhancing functionality and

(vi) **Bio-materials**: Organic materials that can respond to biological stimuli, useful in medical applications.

These materials enable the creation of structures that can adapt and change over time, making 4D printing a promising technology for various innovative applications.

- **Applications of 4D Printing**

4D printing has several innovative applications across various fields, including: (i) **Medical Devices**: Smart implants and prosthetics that adapt to the body's needs, enhancing comfort and functionality, (ii) **Self-Assembly Structures**: Building materials that can change shape and assemble themselves in response to environmental conditions, (iii) **Robotics**: Soft robots that can transform their shape for improved movement and flexibility in various tasks, (iv) **Textiles**: Clothing that adapts to temperature or moisture levels, providing enhanced comfort and performance, (v) **Aerospace**: Components that can change shape for improved aerodynamics and performance during flight, (vi) **Architecture**: Structures that can adapt to environmental changes, improving energy efficiency and comfort and (vii) **Consumer Products**: Items like furniture that can change shape for different uses, enhancing functionality and saving space.

These applications highlight the potential of 4D printing to create dynamic and adaptable solutions across multiple industries.

11 5D Printing

The term 5D printing refers to additive manufacturing that uses five axes. Utilizing the three axes of 3D printing in conjunction with the reciprocating movement of the print bed and extruder head, 5D printing is made possible. In comparison to 3D and 4D printing, 5D printing allows for the manufacture of parts that are three to four times more durable. A part generated by 3D and 4D printing can endure up to 0.1 MPa, but the identical part produced by 5D printing may withstand up to 3.7 MPa, according to an experimental study done as an indicator of this condition [52, 53]. 5D printing is interesting in the automotive, aerospace, and biomedical industries because it can create curved surfaces [52]. Stacking flat layers atop one another during 3D printing can create flat layers. However intricate curved layers can be created with 5D printing [54].

- **Materials used in 5D Printing**

5D printing can utilize a variety of materials, including: (i) **Thermoplastics**: Commonly used in additive manufacturing, materials like PLA, ABS, and PETG are popular for their versatility, (ii) **Metals**: Metal powders such as titanium, aluminum, and steel can be used for robust applications, particularly in aerospace and automotive. (iii) **Composites**: Reinforced materials that combine polymers with fibers (like carbon or glass) offer enhanced strength and durability, (iv) **Ceramics**: Used for high-temperature applications and for creating intricate, durable parts. (v) **Smart Materials**: These can change properties in response to environmental conditions, allowing for adaptive structures, (vi) **Bio-based Materials**: Environmentally friendly options made from renewable resources are becoming more popular and (vii) **Elastomers**: Flexible materials that can be used for applications requiring rubber-like properties.

These materials enable 5D printing to achieve complex geometries and functional designs across various industries.

- **Advantage of 5D printing**

Some of the interesting and versatile potential advantages of the 5D printing technology are given below—(i) It enables the creation of intricate shapes that traditional methods can't achieve, allowing for more innovative designs, (ii) 5D printing often uses less material, reducing waste and costs, (iii) It can combine different materials in a single print, enhancing functionality and performance, (iv) This technology allows for highly personalized products tailored to specific needs and (v) Parts can be printed in one piece, decreasing assembly time and potential points of failure.

- **Disadvantage of 5D printing**

Besides many advantages of the 5D printing technology, there are also a few disadvantages, including: (i) **High Costs**: The initial investment in 5D printing technology and materials can be significant, (ii) **Complexity**: The technology and processes involved can be complicated, requiring specialized knowledge and training, (iii) **Limited Material Options**: Not all materials are suitable for 5D printing, which can limit design choices, (iv) **Speed**: Depending on the complexity of the design, 5D printing can be slower than traditional manufacturing methods. (v) **Quality Control**: Ensuring consistent quality across prints can be challenging, especially with intricate designs and (vi) **Equipment Size and Space**: Advanced 5D printers may require more space and infrastructure than traditional printers.

- **Applications of 5D printing**

5D printing has a range of exciting applications across various industries, including: (i) **Aerospace**: Manufacturing lightweight, complex components that enhance fuel efficiency and performance, (ii) **Automotive**: Creating parts with intricate designs and multimaterial properties, improving safety and functionality, (iii) **Health care**: Producing custom prosthetics and implants tailored to individual patient needs, (iv) **Architecture**: Constructing innovative building materials and structures that can adapt to environmental conditions, (v) **Consumer Products**: Designing unique, customized items like footwear and accessories that combine different materials for enhanced performance, (vi) **Military**: Developing advanced equipment and components that can withstand extreme conditions and (vii) **Robotics**: Fabricating lightweight, complex robotic parts that improve agility and functionality.

These applications demonstrate 5D printing's potential to revolutionize manufacturing and product design.

12 Conclusion

In conclusion, 3D, 4D, and 5D printing represent significant advancements in additive manufacturing technologies, each with unique capabilities and applications. 3D printing creates three-dimensional objects layer by layer; 4D printing introduces the dimension of time, enabling materials to change shape or properties in response to stimuli; while 5D printing incorporates advanced geometries and multi-material capabilities for even more complex designs. Each printing method is classified based on technology (e.g., FDM, SLA for 3D printing) and the specific functions they serve, from static objects in 3D to dynamic, adaptive structures in 4D and 5D. The choice of materials varies significantly, with 3D printing utilizing plastics and metals, 4D printing leveraging smart materials like shape memory polymers, and 5D printing focusing on composites and advanced polymers for enhanced functionality. Together, these technologies offer innovative solutions across diverse industries, paving the way for future advancements in manufacturing, design, and engineering.

Acknowledgements The author would like to express his sincere gratitude to the host Institution for providing the digital library facility for collection of the information.

Conflict of Interest No conflict of interest to declare.

References

1. Jing-Jun, W., Huang, L.-M., Zhao, Q., Xie, T.: 4D printing: history and recent progress. Chin. J. Polym. Sci. **36**, 563–575 (2017)
2. Sadiq Sha, H.A.J., Patil, P.P.: Review on 4D and 5D printing technology. **07**, 744–751 (2020)
3. Haleem, A., Javaid, M., Vaishya, R.: 5D printing and its expected applications in orthopaedics. J. Clin. Orthop. Trauma (2018)
4. Al Rashid, A., Khalid, M.Y., Imran, R., Ali, U., Koc, M.: Utilization of banana fiber-reinforced hybrid composites in the sports industry. Materials (Basel) **13**, 1–10 (2020)
5. Khalid, M.Y., Al Rashid, A., Arif, Z.U., Akram, N., Arshad, H., Garcia Marquez, F.P.: Characterization of failure strain in fiber reinforced composites: under on-axis and off-axis loading. Crystals **11**, 1–11 (2021)
6. Al Rashid, A., Koc, M.: Creep and recovery behaviour of continuous fiber-reinforced 3DP composites. Polymers (Basel) **13**, 1–14 (2021)
7. Brown, W.H., Poon, T.: Introduction to Organic Chemistry, 3rd edn. (2005)
8. Nicholson, J.W.: Educ. Chem. **28**, 70 (1991)
9. Bari, H.A.A., Suali, E., Hassan, Z.: Glycolic acid ethoxylate lauryl ether performance as drag reducing agent in aqueous media flow in pipelines. **8**, 4410–4415 (2008)
10. Al-Sarkhi, A.: Drag reduction with polymers in gas-liquid/liquid-liquid flows in pipes **2**, 41–48 (2010)
11. Singh, B.N., Kim, K.H.: Floating drug delivery systems: an approach to oral controlled drug delivery via gastric retention. J. Control **63**, 235–259 (2000)
12. Feldman, D.: Polymer history. Des. Monomers Polym. **11**, 1–15 (2008)
13. Gjullin, C.M.: Effect of clothing color on the rate of attack of Aedes mosquitoes. J. Econ. Entomol. **40**, 326–7 (1947)

14. Gillies, M.T.: The role of carbon dioxide in host-finding by mosquitoes (Diptera: Culicidae): a review. Bull. Entomol. Res. **70**, 525–532 (1980)
15. Gillies, M.T., Wilkes, T.J.: The range of attraction of animal baits and carbon dioxide for mosquitoes, studies in a freshwater area of West Africa. Bull. Entomol. Res. **61**, 389–404 (1972)
16. Carpita, N.C.: Demonstration of cellulose synthesis in detached cotton fibers. Presented at 176th ACS National Meeting (Cellulose, Paper and Textile Chemistry Division), Miami, Florida (1978)
17. Muhlethaler, K.: Biochim. Biophys. Acta **3**, 527 (1949)
18. Forge, A.: Ann. Bot. **41**, 447 (1977)
19. Mishra, B., Khare, B.P.K.: A review on polymer. J. Emerg. Technol. Innov. Res. (JETIR) **2**, 258–264 (2015)
20. Leticia Perez, R.: Classification, types, and chemistry of polymers. Pharm. Anal. Chem. **7**, 1–2 (2022)
21. Mustafa, N.S., Ali Omer, M.A., Garlnabi, M.E.M., Ismail, H.A.: Reviewing of general polymer types, properties and application in medical field. Int. J. Sci. Res. (IJSR) **5**, 212–221 (2016)
22. Kumari, P., Kumar, A.: A review on properties, classification and applications of polymer. J. Adv. Sch. Res. Allied Educ. **16**, 1125–1129 (2019)
23. Bijker, W.E.: The fourth kingdom: the social construction of Bakelite. In: Of Bicycles, Bakelites, and Bulbs: Toward a Theory of Sociotechnical Change. MIT Press, Cambridge, MA (2015)
24. Kazmer, D.: Three-dimensional printing of plastics. In: Kutz, M. (ed.) Applied Plastics Engineering Handbook: Processing, Materials, and Applications, 2nd edn, pp. 617–634. Elsevier Inc. (2017)
25. Turner, B.N., Strong, R., Gold, S.A.: A review of melt extrusion additive manufacturing processes: I. Process design and modeling. Rapid Prototyp. J. **20**, 192–204 (2014)
26. Schmidleithner, C., Kalaskar, D.M.: Stereolithography. In: Cvetkovic, D. (ed.) 3D Printing, pp. 1–22. InTech (2018)
27. Manapat, J.Z., Chen, Q., Ye, P., Advincula, R.C.: 3D printing of polymer nanocomposites via stereolithography. Macromol. Mater. Eng. **302**, 1–13 (2017)
28. Cho, Y.H., Lee, I.H., Cho, D.W.: Laser scanning path generation considering photopolymer solidification in micro stereolithography. Microsyst. Technol. **11**, 158–167 (2005)
29. Goodridge, R.D., Shofner, M.L., Hague, R.J.M., McClelland, M., Schlea, M.R., Johnson, R.B., Tuck, C.J.: Processing of a polyamide-12/carbon nanofibre composite by laser sintering. Polym. Test. **30**, 94–100 (2011)
30. Dizon, J.R.C., Espera, A.H., Chen, Q., Advincula, R.C.: Mechanical characterization of 3D-printed polymers. Addit. Manuf. **20**, 44–67 (2018)
31. Chatham, C.A., Long, T.E., Williams, C.B.: A review of the process physics and material screening methods for polymer powder bed fusion additive manufacturing. Prog. Polym. Sci. **93**, 68–95 (2019)
32. Ahmed, A., Arya, S., Gupta, V., Furukawa, H., Khosla, A.: 4D printing: fundamentals, materials, applications and challenges. Polymer (Guildford) **228**, 123926 (2021). https://doi.org/10.1016/j.polymer.2021.123926
33. Ge, Q., Qi, H.J., Dunn, M.L.: Active materials by four-dimension printing. Appl. Phys. Lett. **103**, 131901 (2013). https://doi.org/10.1063/1.4819837
34. André, J.C.: From Additive Manufacturing to 3D/4D Printing 3: Breakthrough Innovations: Programmable Material, 4D Printing and bio-Printing. John Wiley & Sons (2017)
35. Bodaghi, M., Noroozi, R., Zolfagharian, A., Fotouhi, M., Norouzi, S.: 4D printing self-morphing structures. Materials **12**, 1353 (2019). https://doi.org/10.3390/ma12081353
36. Wang, J., Wang, Z., Song, Z., Ren, L., Liu, Q., Ren L.: Biomimetic shape–color double-responsive 4D printing. Adv. Mater. Technol. **4**, 1900293 (2019). https://doi.org/10.1002/admt.201900293
37. Zhang, Z., Corrigan, N., Bagheri, A., Jin, J., Boyer, C.: Frontispiece: a versatile 3D and 4D printing system through photocontrolled RAFT polymerization. Angew. Chem. Int. Ed. **58** (2019)

38. Zheng, N., Fang, Z., Zou, W., Zhao, Q., Xie, T.: Thermoset shape-memory polyurethane with intrinsic plasticity enabled by transcarbamoylation. Angew. Chem. Int. Ed. **55**, 11421–11425 (2016)
39. Kuang, X., Roach, D.J., Wu, J., Hamel, C.M., Ding, Z., Wang, T., Dunn, M.L., Qi, H.J.: Advances in 4D printing: materials and applications. Adv. Funct. Mater. **29**, 1805290 (2019). https://doi.org/10.1002/adfm.201805290
40. Yang, W.G., Lu, H., Huang, W.M., Qi, H.J., Wu, X.L., Sun, K.Y.: Advanced shape memory technology to reshape product design, manufacturing and recycling. Polymers (Basel) **6**, 2287–2308 (2014). https://doi.org/10.3390/polym6082287
41. Chen, A., Pegg, E., Chen, A., Jin, Z., Gu, G.: 4D printing of electroactive materials. Adv. Intel. Syst. **3** (2021). https://doi.org/10.1002/aisy.202100019
42. Singh, M., Haverinen, H.M., Dhagat, P., Jabbour, G.E.: Inkjet printing-process and its applications. Adv. Mater. **22**, 673–685 (2010)
43. Suriano, R., Bernasconi, R., Magagnin, L., Levi, M.: 4D printing of smart stimuli-responsive polymers. J. Electrochem. Soc. **166**, B3274–B3281 (2019)
44. Ding, Z., Yuan, C., Peng, X., Wang, T., Qi, H.J., Dunn, M.L.: Direct 4D printing via active composite materials. Sci. Adv. **3**, e1602890 (2022)
45. Wu, H., Wang, O., Tian, Y., Wang, M., Su, B., Yan, C., Zhou, K., Shi, Y.: Selective laser sintering-based 4D printing of magnetism-responsive grippers. ACS Appl. Mater. Interfaces **13**, 12679–12688 (2021)
46. Kafle, A., Luis, E., Silwal, R., Pan, H.M., Shrestha, P.L., Bastola, A.K.: 3D/4D printing of polymers: fused deposition modelling (FDM), selective laser sintering (SLS), and stereolithography (SLA). Polymers (Basel) **13** (2021)
47. Wang, Z., Wang, Z., Zheng, Y., He, Q., Wang, Y., Cai, S.: Three-dimensional printing of functionally graded liquid crystal elastomer
48. Bagheri, A., Jin, J.: Photopolymerization in 3D printing. ACS Appl. Polym. Mater. **1**, 593–611 (2019)
49. Zhang, F., Zhu, L., Li, Z., Wang, S., Shi, J., Tang, W., Li, N., Yang, J.: The recent development of vat photopolymerization: a review. Addit. Manuf. **48** (2021)
50. Solomon, I.J., Sevvel, P., Gunasekaran, J.: A review on the various processing parameters in FDM. Mater. Today: Proc. 509–514 (2020)
51. Wang, Z., Guo, Y., Cai, S., Yang, J.: Three-dimensional printing of liquid crystal elastomers and their applications. ACS Appl. Polym. Mater. **4**, 3153–3168 (2022)
52. Sadiq Sha, H.A.J., Sha, S.H.A.J., Patil, P.: Review on 4D and 5D printing technology composites view project additive manufacturing view project review on 4D and 5D printing technology. Int. Res. J. Eng. Technol. (2020)
53. Haleem, A., Javaid, M.: Expected applications of five-dimensional (5D) printing in the medical field. Curr. Med. Res., Pract. **9**(5), 208–209 (2019)
54. Eryildiz, M.: New Concepts of Additive Manufacturing Techniques: 5D and 6D Printing. Gece Publishing (2022)

Nanotechnologies and Nanomaterials in 3D, 4D, and 5D Printing Technologies

Mahdieh Ranjbar-Jamalabadi, Shayan Hosseini, Mohammadhossein Ramezanirad, Arezoo Hosseingholian, Seyyed Mohammad Amin Mousavi-Sagharchi, and Mohsen Sheykhhasan

Abstract The fusion of nanotechnologies and nanomaterials in 3D, 4D, and 5D printing has initiated a new era in advanced manufacturing with unparalleled precision, functionality, and efficiency. These state-of-the-art technologies facilitate the production of intricate structures with customized properties, paving the way for innovative applications in various fields such as healthcare, electronics, and aerospace. This piece delves into recent progressions in nanomaterial-based inks and their utilization in 3D, 4D, and 5D printing procedures. Nanomaterials such as carbon nanotubes, graphene, and quantum dots possess distinctive characteristics that can be exploited to craft new printing methods and materials. The article underscores the significance of optimizing printing parameters, post-processing techniques, and material design to achieve superior print quality, resolution, and functionality. Furthermore, the authors investigate the incorporation of stimuli-responsive components into printed structures, enabling shape-shifting and programmable behavior over time (4D printing) or in response to external stimuli (5D printing). This dynamic functionality opens fresh possibilities for designing intelligent devices, drug delivery systems, and self-repairing materials. Lastly, the article explores the obstacles and

M. Ranjbar-Jamalabadi and S. Hosseini—These authors have same contribution and share first authorship.

M. Ranjbar-Jamalabadi · S. Hosseini
Faculty of Chemical Engineering, Department of Polymer Engineering, Isfahan University of Technology, Isfahan, Iran

M. Ramezanirad
Student Research Committee, School of Medicine, Shahid Beheshti University of Medical Sciences, Tehran, Iran

A. Hosseingholian
Department of Biosciences, University of Milan, Milano, Italy

S. M. A. Mousavi-Sagharchi
Department of Microbiology, College of Basic Sciences, Shahr-E-Qods Branch, Islamic Azad University, Tehran, Iran

M. Sheykhhasan (✉)
Cellular and Molecular Research Center, Qom University of Medical Sciences, Qom, Iran
e-mail: msheykhhasan@muq.ac.ir

S. Moharana et al. (eds.), *Polymer Nanocomposites for 3D, 4D and 5D Printing*, Engineering Materials, https://doi.org/10.1007/978-981-96-4214-4_3

prospects of nanotechnology-driven 3D, 4D, and 5D printing, emphasizing its potential to transform manufacturing practices across various sectors. Concentrating on sustainable materials and scalable processes, these technologies offer the potential to revolutionize our methods for designing and producing functional structures with advanced properties.

Keywords Nanotechnologies · Nanomaterials · 3D printing · 4D printing · And 5D printing

1 Introduction

For many years, medicine has been used as a treatment, but it had an effect on the whole body. But nowadays, a new science is happening at the atomic and molecular levels. Nanotechnology is changing how we make things on a small scale. It can change how we diagnose and treat diseases. Scientists are making new tools and technologies by manipulating them at the nanoscale. These tools and technologies could help solve some of the biggest medical problems [108, 111]. Materials are used for different purposes, such as diagnostics, drug delivery, and regenerative medicine. Nanoparticles can be designed to act as highly specific contrast agents, improving the resolution and accuracy of imaging techniques such as magnetic resonance imaging (MRI) and computerized tomography (CT) scans [36]. Researchers are creating new ways to deliver drugs by making tiny particles that go to specific parts of cells or tissues. These carriers can deliver medications directly to areas that are affected by disease, decrease side effects, and increase the impact of treatment [55]. Also, nanomaterials have the potential to trigger tissue regeneration and repair. Nanomaterials can support tissue damage and can help treat injuries and chronic diseases [118]. Bioprinting and nanotechnology are coming together in new ways. Nanocomposite bioprinting has great promise for future tissue engineering [34, 66].

Bioprinting is a big improvement in regenerative medicine. It uses three-dimensional (3D) printing technology to make complicated 3D structures. The materials used are good for living cells and bones. This allows researchers to make scaffolds that look like the natural structure of tissues, which could help cells grow and heal damaged areas [7, 39]. However, traditional bioprinting materials may not be strong enough or have the right signals from nature to fully replicate the complicated functions of natural tissues. Here's where nanocomposites come in. Nanoparticles are used to make biocompatible materials used for bioprinting [13, 109] (Fig. 1).

These tiny particles can give the bioink many good qualities and can make bioprinted structures stronger and more able to mimic the mechanical properties of native tissues like bone or cartilage [13]. Another use of these particles is to carry drugs or growth factors and release them in a controlled way to help tissues grow back [22, 90]. The uses of bioinks differ in many types of tissue engineering. For example, Bioprinting scaffolds with better strength and tiny magnets can help bones grow and heal after broken bones or illness [47]. Nanocomposite bioinks can be

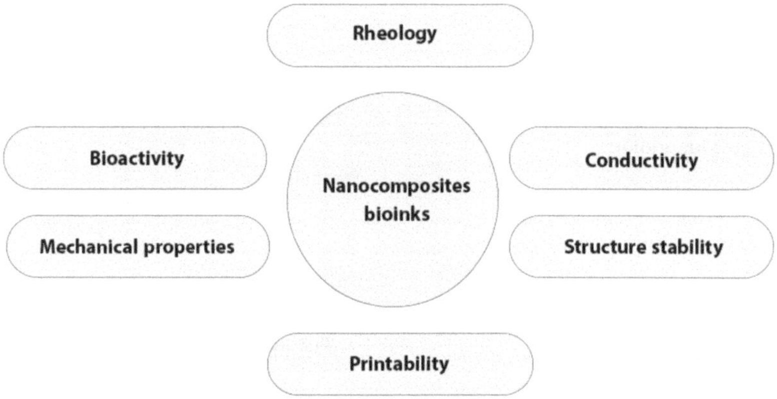

Fig. 1 Essential and required properties of nanocomposites bioinks

used to make scaffolds that look like cartilage and help heal damaged joints [50]. Bioprinting technology could be used to make skin grafts for patients with burns or other skin conditions [101].

However, there are still problems to solve, such as making sure these new materials stay safe and effective for a long time [74, 84]. Several challenges have remained in current research in 3D, 4D, and 5D bioprinting. Current bioprinting approaches still have technical challenges in terms of high-resolution cell deposition, controlled cell distributions, vascularization, and innervation within complex 3D tissues [74]. The successful application of bioprinting is still severely hampered by the limited number of biocompatible printable materials [75]. Vascularization is crucial for creating thicker tissues that require their own blood supply. Nanotech assistance in the formation of vascular networks within bioprinted tissues is a problem [60]. Further research about how nanotechnologies can facilitate these advancements in bioprinting is required because the transition to 4D and 5D printing is complex. In this chapter, we try to describe the use of nanotechnology science in bioprinting and how Nanomaterials can be used in 3D, 4D, and 5D printing technologies.

2 Bioprinting and Related Approaches

An inkjet printhead is used to spray bioink droplets onto a building platform in a technique called bioprinting. It has high-resolution printing capabilities and can be used to create complex structures. However, it is limited by the viscosity of the bioink and the viability of the cells during the jetting process [52]. Another one is laser-aided bioprinting. The localized gelation of a bioink solution is caused by a laser beam. A variety of biomaterials can be printed with high resolution. [96]. Laser-assisted bioprinting (LAB) provides a high-density cell distribution with consistent microwell geometries and high resolution (Fig. 2).

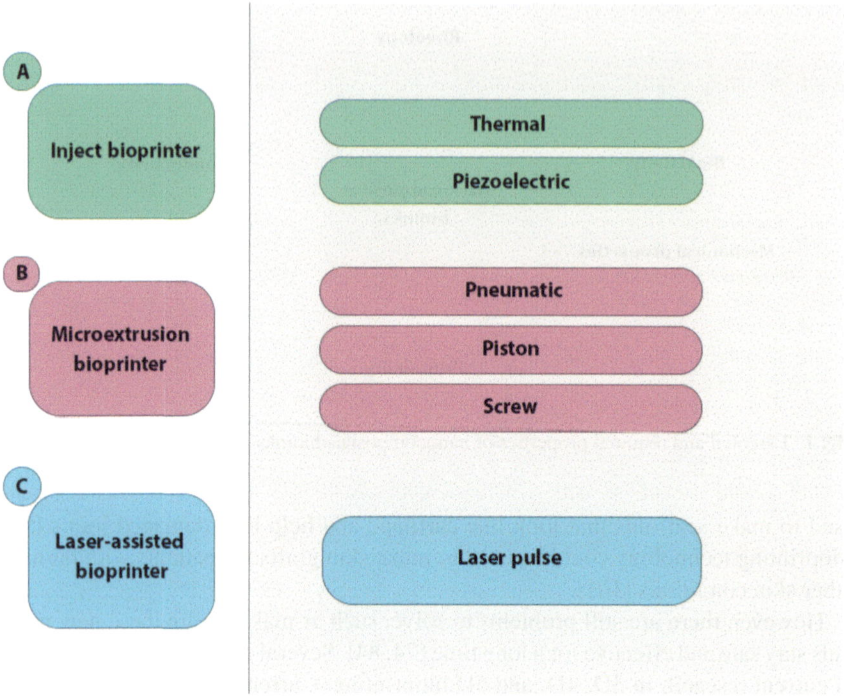

Fig. 2 Types of bioprinters

Traditional methods are problematic and do not provide consistent microwell geometries that affect cell function and proliferation as well as insufficient vascularization [117]. LAB is capable of printing cells with high viability in a single spatial arrangement that mimics the orientation of native tissues and organs. It can generate artificial cell niches. The advantages of using the LAB method in printing tissues are that it provides consistent high-density cell distribution, high resolution, and enhanced vascularization [117].

LAB is highly desired for printing viscous bioink with high cell viability and exceptional accuracy, but it is still limited to printing 2D designs due to its printing mechanism. Several studies have been fabricating a tubular 3D scaffold using the LAB printing process and have shown favorable results. LAB can be utilized in combination with other 3D bioprinting methods or fabrication techniques to produce 3D tissue constructs with higher mechanical properties for tissue engineering applications in the future [117] (Fig. 3). The stereolithographic bioprinting process employs a laser beam to cure a photosensitive bioink resin layer by layer. It's capable of delivering high-definition and constructing intricate structures [63]. For everyday users, fused deposition modeling (FDM) is the most popular technique for creating 3D objects. In an FDM printer, the thermoplastic filament is fed through a heated

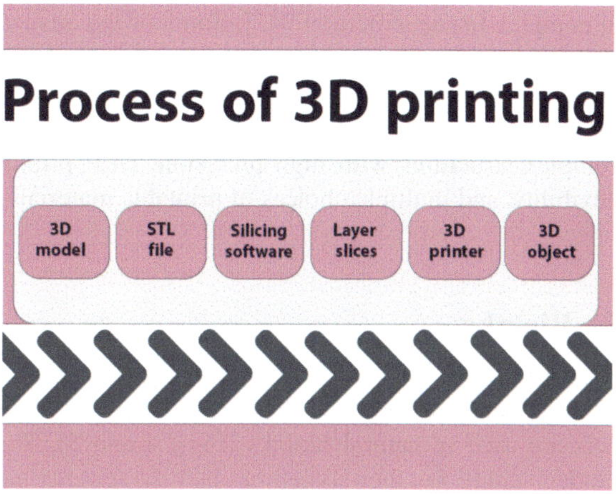

Fig. 3 Process of 3D printing

nozzle. The fibers are blown as they emerge and then placed on the production platform [88].

Every aspect of factor and input parameters in the FDM engineering process affects the quality and mechanical properties. The results have been influenced by several aspects, including the composition of the filament, the working parameters of the extruder, such as the speed and temperature, the specifications of the FDM machine, the type of filament polymer, and the parameters of the FDM process when printing the filament. Future work is important to determine the best results because there are still some relationships between factors and parameters that are not clear, making future work very important to be determined [62]. FDM techniques can create various complex shapes and structures while properly managing materials, resulting in less waste and various other advantages over conventional manufacturing. The advantage of the FDM technique is to produce a series of personalized items so that each product can be different while maintaining low prices. Custom designs do not require additional mold and equipment costs [62].

Selective laser sintering (SLS), a type of 3D printing that uses a laser to melt powders together, holds a variety of material possibilities produced into a powder [38]. The SLS printing method offers several advantages in chemistry. Some of these advantages include a solvent-free nature, availability of FDA-approved thermoplastic polymer/excipients, minimal to no post-processing step, and relatively high printing speed. Additionally, SLS printing permits the fabrication of personalized dosage forms of different shapes, sizes, and release characteristics [23]. Laser-based printers use a digital projector instead of a laser beam to cure the resin. DLP printers are capable of producing prints that are as precise and detailed as SLA prints, yet they are frequently more efficient and cost-effective [97].

The challenges associated with the limited size of the printable volume, resolution, and efficiency of the printing process. Additionally, there are significant limitations in the transparency of the printing materials used, which restricts the possible applications of DLP printing technology in biological tissue engineering, particularly in complex tissue structures [42]. Some of the advantages of using DLP-based 3D printing technology for tissue engineering include high resolution, high accuracy, and high production efficiency. Additionally, DLP printing can overcome some of the limitations of traditional manufacturing methods, such as the ability to produce complex structures with high precision. DLP printing also offers customization, flexibility, and multiple choices of printable materials [42].

3 Bioinks

Natural materials such as collagen, alginate, fibrin and hyaluronic acid and hydrogels are used in natural bioinks (Figs. 4 and 5). They're usually compatible and biodegradable, but they sometimes lack the tenacity needed for intricate organs [67]. Hydrogels are made of natural biological materials, synthetic materials, or combinations thereof. Hydrogels are either natural bioinks or synthetic bioinks. So technically, natural biomaterials are a popular choice for hydrogels used as bioinks because they are generally highly biocompatible [12]. The unique properties of hydrogels make them suitable for use as bioinks in bioprinting. They can provide an environment for cell growth. The capacity and biodegradability of the hydrogel is revealed. The gelation kinetics of hydrogels associated with the crosslinking process may contribute to preserving the structure of the construct [12].

Future research areas include improving the structural integrity, reproducibility and scalability of scaffold-free bioinks, exploring the potential of 4D bio-printing techniques in complex structure reconstruction, and bioink production and raising its standard [12]. Synthetic bioinks are made from synthetic materials such as poly(lactic-co-glycolic acid) and poly(ethylene glycol) They can be more durable than natural bioinks, but they may not be as biocompatible [67]. Drugs or other signaling molecules can be delivered to cells via nanoparticles, enhancing the mechanical properties of bioinks (Fig. 6). The long-term safety of nanobioinks is still being investigated [11]. For example, alginate is a brown algal polysaccharide used in bioprinting [87].

Collagen is a popular bioink because it is biocompatible, printable, and can be easily combined to form a gel [3]. Hyaluronic acid is a glycosaminoglycan that

Fig. 4 Two main sources of bioinks from nature (biotic and abiotic)

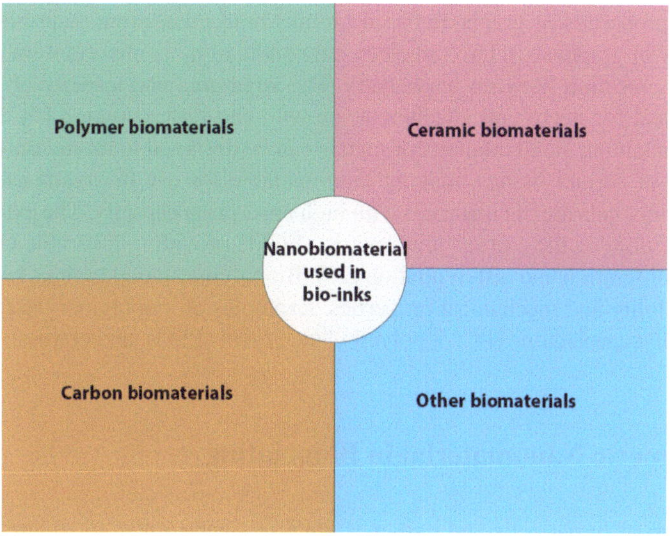

Fig. 5 Nanobiomaterials used in bioinks for bioprinting

Fig. 6 Nanobioinks and materials in 3D bioprinting

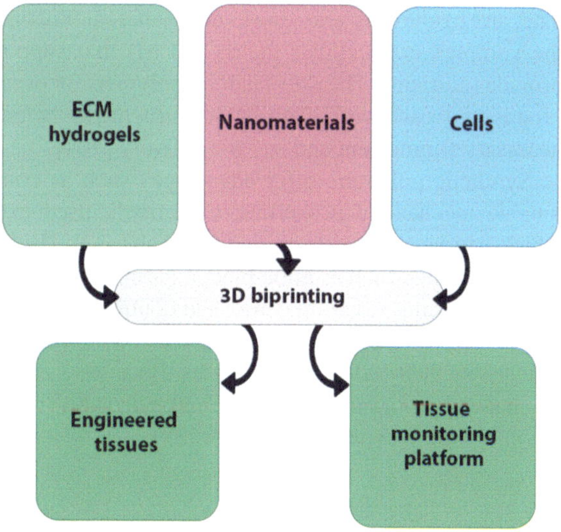

helps cells adhere together and grow [67]. Fibrin is a protein involved in blood coagulation [53]. Nano-based bioinks are made from biocompatible materials such as cellulose. These bioinks are highly biocompatible, making printing easier and more compatible with cells. The addition of nanoparticles to bioinks can enhance their properties and control the printing process. The bioink matrix contains nanoparticles that can promote cell adhesion, proliferation, and tissue regeneration [1]. The unique electrical properties of graphene-based bioinks hold great promise in bioprinting. The ability of graphene to have an electrical conductivity is important for signaling and communication between nerve cells. The large area and tenacity of graphene make it ideal for nerve cell attachment, growth, and differentiation [9]. Hydrogel bioinks containing small amounts of metals can be designed to mimic the electrical properties of certain tissues, making them suitable for use in organs such as the heart Polymer gels are 3D materials with high absorption capacity. The extracellular membrane mimics the extracellular matrix (ECM) providing favorable conditions for cell proliferation and differentiation [28]. Biopolymer-based bioinks have excellent printability and mechanical properties. Examples of biopolymer-based bioinks include polycaprolactone (PCL), polyethylene glycol (PEG), and chitosan [48].

4 Polymeric Nanomaterial in Bioprinting

3D printing of living cells and tissues is ideal for regenerative medicine and drug discovery. Polymers are used at the heart of this technology. Recent research has emphasized the ongoing development of natural polymers for optimal bioprinting techniques [7]. Bioprinting uses naturally occurring polymers such as collagen, alginate, and hyaluronic acid, due to their inherent biocompatibility and ability to mimic the complex extracellular matrix (ECM) that supports the cells in our bodies It provides nutrients. The use of natural polymers for organ 3D bioprinting is effective in producing tissues with well-defined structures, including intervascular connections necessary for nutrient and oxygen delivery [104].

Synthetic polymers offer advantages such as controlled degradation rates and tunable mechanical properties. Commonly used examples are PLGA and PCL, which can be tailored to tissue needs [10, 32]. High-quality bioprinting polymers have properties such as good biocompatibility, printability at physiological temperatures (reducing stress on cells), and controlled damage The combination would be [73]. One promising approach involves combining natural and synthetic polymers to increase their individual strengths. Because, synthetic polymers can enhance the mechanical properties of natural hydrogels, while natural polymers can improve cell viability within synthetic scaffolds [37].

5 Types of Nanomaterials Used in Bioprinting

Nanoparticles are commonly used as additives to enhance the mechanical properties, bioactivity, and performance of bio-inks [11] (Table 1). Metals (e.g., gold, silver), metal oxides (e.g., titanium dioxide, iron oxide), carbon-based materials, (e.g., carbon nanotubes, and graphene), and polymers (e.g., PLGA nanoparticles) [21] (Fig. 7). For example, a thin layer of titanium (Ti) was used as an absorbing laser in one research project. The thickness of the Ti films was between range from 25 to 400 nm. The experiments show that Ti films with thicknesses of 25 and 50 nm are better for the generation of small microdroplets, while if it was 100 nm it shows the splashing of hydrogel also thicknesses of 200 and 400 nm, the transferred particles can be seen on the donor slide [131]. According to the experimental result, the low amount of metal nanoparticles (0.5% to 2.5%) transferred within the microdroplets during laser bioprinting, and almost all nanoparticles (over 97%) remain in the hydrogel layer on the donor slide. so, the amount of transferred metal nanoparticles is considered too low and cannot produce any negative effect on the printed cells [131].

Carbon-based nanomaterials like graphene oxide and nanotubes have special electrical properties, so they are ideal for bioprinting tissues that have electrical signaling, such as nerves and muscles. Graphene nanoplatelets with a bioink can create a 3D

Table 1 Summarization of nanomaterial utilized in bioprinting

Nanomaterials	Example	Application	References
Ceramics	HAp nanocrystals Titanium dioxide	Incorporation of HA nanocrystals in hydrogel bioink improves bone cell growth and differentiation	[58]
Carbon-based	Graphene nanoplatelets Carbon nanotubes	Graphene nanoplatelets within bioink scaffold enhance neural stem cell proliferation and differentiation for nerve regeneration applications	[59]
Nanoparticles	Gold nanoparticles Silver nanoparticles	Addition of nanoparticles improves mechanical properties, bioactivity, and functionality of bioinks for bioprinting	[16]
Nanofibers	Polycaprolactone nanofibers Collagen nanofibers	Nanofibers fabricated from natural or synthetic polymers via electrospinning benefit cell behavior in bioprinted constructs	[60]
Nanocomposites	Hydrogel embedded with carbon nanotubes Gelatin hydrogel embedded with hydroxyapatite nanoparticles	Hydrogels embedded with nanoparticles/nanofibers improve overall properties for bioprinting applications	[61, 62]
Nanocarriers	Lipid-based nanoparticles Polymeric micelles	Nanocarriers improve controlled release kinetics and protect bioactive molecules within bioprinted constructs	[63, 64]
Nanogels	Crosslinked alginate nanogels Polyethylene glycol nanogels	Nanogels integrated into bioinks enhance rheology and provide conducive microenvironment for cell encapsulation	[65, 66]
Quantum Dots	Cadmium selenide quantum dots Cadmium telluride quantum dots	Quantum dots facilitate real-time monitoring of cellular behavior and tissue development in bioprinted constructs	[67]

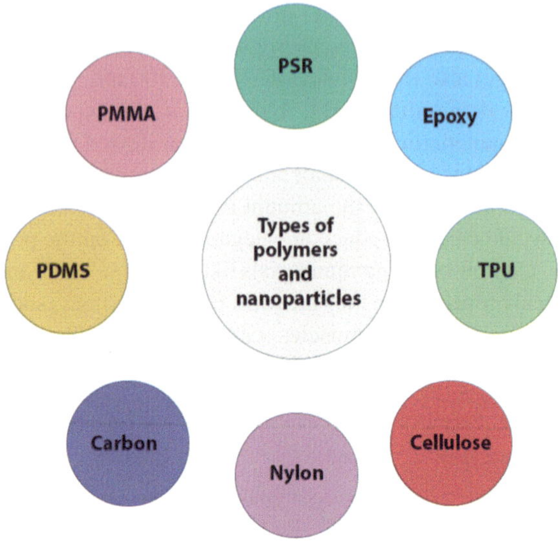

Fig. 7 Some important polymers used in nanoparticles

scaffold for neural stem cells [68]. Recent findings show that carbon materials (CMs) can combine with hydrogels an develop the materials for biomedical applications. 3D bioprinting technologies are used for the processing of hydrogels that contain CM to better mimic and show the cell behavior. Hydrogels with CMs can tune the rheological properties of biomaterial inks, also improve the mechanical properties of hydrogels and make new physical features for hydrogels, such as shape memory, roughness, and thermal and electrical conductivity [83].

Graphene can increase the proliferation and differentiation potential of the stem cells into functional neurons, it can be good for nerve regeneration applications [68, 128]. Graphene-based materials show unique mechanical and electrical properties, morphology, and permeability, making them the perfect choice for 3D bioprinting [86]. Graphene-based materials have shown perfect potential for the regeneration of both hard and soft tissues. Graphene has been used for 3D printing of bone, neural, and cardiac tissue. Studies also show that can combine graphene with anti-coagulant agents and growth factors for nerve cell growth and differentiation [86]. Hydroxyapatite (HAp) nanocrystals are a popular choice due to their similarity to the components of bones and teeth. HAp nanocrystals with a hydrogel bioink can improve printability, mechanical strength, and biocompatibility, promoting bone cell growth and differentiation (Fig. 8).

The most use of HAp is in the synthesis of biomedical materials such as bone, teeth, and cartilage. 3D printing, on the other hand, recently gained popularity as a fast, controllable process to produce HAp-based scaffolds [15]. 3D printing can be used to make HAp scaffolds with complex shapes. This is because HAp-based scaffolds are more complicated, which makes it easier to put different types of materials on top of each other. This can make tissues and organs look different. To make HAp-based scaffolds, we need to use 3D bioprinting to fix the problems with blood flow.

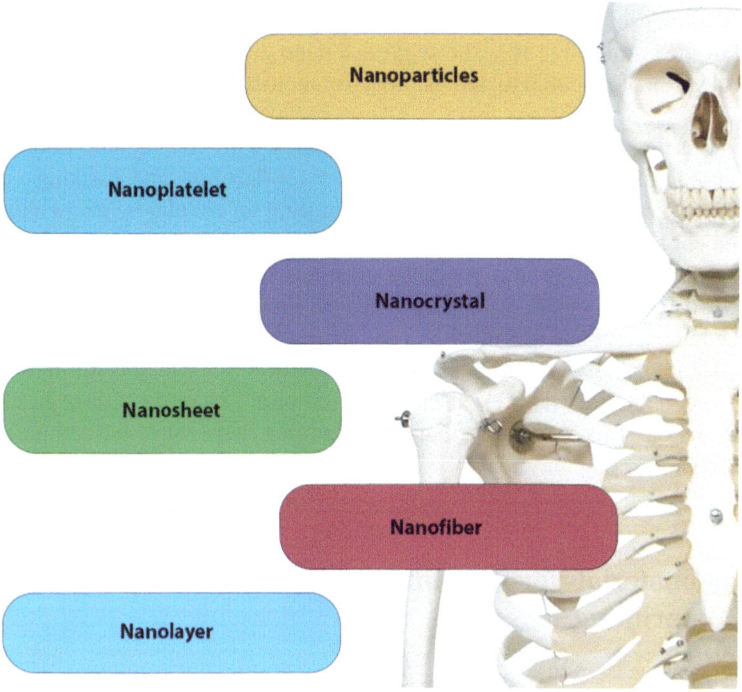

Fig. 8 Different forms of nanoparticles in bone and skeleton repairment

However, scaffolds using 3D printing technology also need to be improved to facilitate the growth and delivery of living cells [15]. Nanofibers have a large surface area and quite a lot of pores, they can mimic the structure of the ECM [51]. Nanofibers can be obtained from natural polymers (e.g., collagen, gelatin) or synthetic polymers (e.g., polycaprolactone, poly (lactic-co-glycolic acid)) by methods such as electrospinning [58]. Nanocomposites include high concentrations of nanoparticles or nanofibers and biomaterials together to improve their mechanical properties, electrical conductivity, Bioactivity is also improved [77, 116]. Nanocarriers are used to deliver bioactive molecules (e.g., growth factors, drugs, genes) to specific cells or tissues in a bioprinted way. These nanocarriers can protect the something that they carry from degradation and provide controlled release kinetics, as a result, can improve treatment goals [31, 134].

Nanogels are 3D networks of crosslinked polymers, that can encapsulate bioactive molecules or cells. They can be integrated into bioinks to improve their properties and provide a suitable microenvironment for cell encapsulation and proliferation [129, 135]. In a study, scientists used a combination of nanogels and gels called GelMA. It's made by adding methacryloyl groups to N-isopropylacrylamide-based nanogels to make them work such as GelMA. It is a stable hydrogel system with nanogels covalently embedded in the micropore network of GelMA hydrogel. This allows for specific functions that are not found in a single gel. The shape of the composite hydrogel shows the nanogel wrapped within the GelMA network and covering the surface of the pore wall [136]. The suggested nano gel-GelMA composite hydrogel system has a lot of potential to deliver bioactive molecules in 3D cell culture systems [136].

Quantum dots are semiconductor nanoparticles with size-tunable fluorescence optical properties [61]. They can be utilized for labeling and tracking cells in bioprinted structure and can do real-time monitoring for cellular behavior and tissue development [25, 61, 76]. Research discussed the efficient differentiation of hepatocytes using bioprinting and graphene quantum dots. The goal is to establish a 3D in vitro hepatic model for toxicological evaluation and improve drug testing accuracy [54]. Graphene quantum dots can increase the differentiation efficiency of stem cells into hepatocytes because they have an affinity for the insulin-like growth factor 1 receptor (IGF1R) present in the phospholipid membrane through electrostatic attraction. This could contribute to promoting hepatoblast differentiation via calcium channels [54].

6 Nanomaterials for Reinforcing Mechanical Properties in Bioprinting

Hydrogels are gaining attention as printing mediums because they offer properties such as viscosity, self-healing, and shear thinning, which are vital for the bioprinting process. Among the important properties that hydrogels must possess is viscosity. Viscosity measures the resistance of a hydrogel to deformation and displacement. This property is crucial because suitable viscosity can directly impact the printing process [20]. Additionally, self-healing capability is important as hydrogels, capable of returning to their original shape after deformation, can benefit from the significance of shear thinning in the printing process. Strengthening hydrogels with nanoparticles is an effective solution to enhance the performance and efficiency of bio-inks [29].

These nanoparticles can improve the mechanical, rheological, and biological properties of the inks. For example, adding nanoparticles to hydrogels can reduce their viscosity and assist in shear thinning. Furthermore, mineral nanoparticles can enhance the biological properties of the inks, such as osteogenic properties [43].

However, for the effective use of these techniques in 3D bioprinting, it is necessary to carefully examine the effect of adding nanoparticles on the various properties of the inks. This includes investigating the impact of nanoparticles on viscosity, rheology, mechanical properties, and biological properties of the inks [56]. In this way, it is possible to design and manufacture bio-inks with optimal properties for biomedical applications. Hydrogels made from natural polymers for biological applications may break under harsh mechanical conditions due to their natural properties. To address this issue, adding nanoparticles to hydrogels can improve their mechanical properties. Nanoparticles such as nanorods or graphene sheets, with electrostatic interactions with the polymer chains of hydrogels, enhance their mechanical properties. The use of these combinations can aid in tissue engineering applications, including tissue reconstruction and wound healing [80].

Biomaterial inks for 3D bioprinting of bone are typically based on hydrogels. These inks are soft and preferred to maintain cell functionality against shear forces. Shape fidelity in structures is reinforced, especially in 3D bioprinted bone, by using nanofibers or other nanomaterials to strengthen hydrogel structures [91].

Nanofibers

The use of nanofibers in hydrogels can enhance their mechanical properties. For example, alginate-based bioink hydrogel containing human stem cells has been reinforced using PLA nanofibers, resulting in increased shape fidelity and improved cell proliferation. Additionally, adding PCL nanofibers to PCL hydrogel through extrusion has enhanced shape fidelity and mechanical strength. Combining nanofibers with gelatin hydrogels can provide better mechanical and morphological properties. Furthermore, composite hydrogels of gelatin and cellulose nanofibrils have been produced through 3D printing, increasing their tensile strength and improving mechanical properties [95].

Nanoparticles

In the new era, the use of various nanoparticles in 3D printing of bone structures is common and important. For instance, creating coating layers of nano-sized HAp on alginate/gelatin-based 3D printed structures acts as a barrier to mineralization and improves mechanical properties. Additionally, adding Titanium dioxide or β-tricalcium phosphate to these structures helps enhance their mechanical and physical properties. The use of nanoparticles such as HAp in 3D printed scaffolds via extrusion or laser processing can create networks similar to native bone, increasing mechanical resistance, promoting bone tissue formation, and providing adequate mechanical support for cell attachment and proliferation. Gold nanoparticles and other nanomaterials are also used in 3D printing to strengthen bone tissue and create structures similar to native bone tissue, aiding in cell proliferation and osteogenic differentiation [115].

Nanoplatelets/2D Sheets

Adding nanomaterials to 3D printed scaffolds, including graphene nanoplatelets and graphene oxides, can improve their biological and mechanical properties. For example, in one study, adding graphene nanoplatelets to a liquid resin ink increased the resin viscosity, flexural modulus, and fracture toughness in 3D printed structures. Furthermore, significant improvements in tensile strength and elongation were observed in other composites using graphene oxides. These studies demonstrate that adding nanomaterials can aid in bone growth and enhance the biological and mechanical properties of 3D-printed scaffolds [69, 130].

Nanocrystals

Cellulose nanocrystals (CNC) and silver nanoparticles were used to improve the biological and mechanical properties of printing materials in the 3D printing process.

Adding CNC to a Polyethylene Glycol Diacrylate matrix and silver nanoparticles to a poly methyl methacrylate matrix created significant enhancements in the mechanical properties and antibacterial activity of the composite material [82]. This approach could be useful for producing dental repair materials and bone implants. Additionally, incorporating modified nanostructures into biopolymer materials improves the mechanical and biological properties of printing materials, which can be used in tissue engineering for structuring both soft and hard tissues [70].

Nanomaterials for Biopolymer Composites in 3D Printing

Biopolymer materials in 3D printing, including natural biopolymers like alginate and collagen, are used in both solid and liquid forms. However, there is a need for mechanical and biological compatibility for clinical applications [92]. Incorporating various nanomaterials, such as polymers, ceramics, carbon, and metals, into biopolymer matrices is proposed as a solution to this problem. The use of naturally derived biopolymers combined with nanomaterials and processed through 3D printing techniques is of particular interest due to their mastery of such applications [78].

Ceramic and Semiconductor Nanomaterials

Ceramic and semiconductor nanomaterials in 3D printing are used for different applications, including biomedical, energy, and electronics. These nanoparticles offer superior mechanical durability and electrical conductivity. In the field of biomedicine, ceramics are used to produce customized implants, while nanoscale ceramics in tissue engineering enhance strength and expand applications. Adding these nanomaterials to polymer matrices, such as PLA, improves the mechanical and biochemical properties of polymer scaffolds. The use of 3D printing also effectively enhances the distribution of these nanomaterials, particularly by adding HAp to scaffolds, which increases bone strength and reinforces orthopedic applications [80, 91].

Carbon Nanomaterials

Carbon nanomaterials such as CNTs, graphene, carbon dots, fullerenes, and carbon nanofibers are combined with polymer matrices for biomedical applications. These composites are used, for example, in 3D printing. For instance, graphene in PLGA is used to enhance cell growth in neural tissues. Additionally, CNTs and carbon nanofibers considerably improve the mechanical and biological properties of 3D-printed polymer composites. Overall, carbon nanomaterials improve the biological and physical properties of polymer composites, making them highly valuable in tissue engineering [56].

Polymeric Nanomaterials

Polymeric nanomaterials and cellulose nanofibrils are added to polymer matrices to create better tissue structures in 3D printing, while derivatives of cartilage and Polyhedral oligomeric silsesquioxane nanoparticles are used to improve the mechanical properties of scaffolds and medical implants. The use of these compounds creates new opportunities for the development of medical products and drug delivery systems in 3D printing [49].

Composite Nanomaterials

In 3D printing of nanomaterials, metallic nanoparticles such as gold, copper, titanium, and silver are used to improve the mechanical, thermal, optical, and electrical properties of printed components. These nanoparticles are used in various structures such as bone scaffolds and electric capacitors. Additionally, the use of hydrogels and nanomaterials such as nanosilicates and cellulose nanocrystals helps improve the stiffness and biocompatibility of printed structures [43, 65].

Carbon-Based Nanomaterials

Carbon-based nanomaterials, including graphene and carbon nanotubes, are of interest in the 3D printing process. These materials are used to enhance the performance and strength of printed parts due to their unrivalled properties, such as structural and electrical characteristics. The use of CNTs and graphene in printing materials aims to increase mechanical strength and electrical conductivity. Additionally, multi-walled graphene and carbon nanotubes are used in the fabrication of scaffolds and polymer composites, improving their electrical behaviors and dimensional stability. The use of these materials as fillers in composites and printed structures facilitates the improvement of mechanical and electrical properties [2].

Metallic Nanomaterials

Metallic nanoparticles such as gold, silver, copper, and titanium are of interest in additive manufacturing (AM) processes as components with improved mechanical, optical, thermal, and electrical properties (Fig. 9). These nanomaterials are used in the fabrication of tissue engineering scaffolds, electric composites, capacitors, sensors, and flexible electronic devices. Silver nanoparticles and modified silver nanoparticles can be used in sensors and capacitors to enhance electrochemical and electrical properties. Additionally, copper nanoparticles can improve mechanical and antimicrobial properties, while titanium nanoparticles and iron oxides offer therapeutic potential in anti-cancer treatments [16].

7 Nano-based Bioprinting in Cell Culture

Biomaterial inks need to be injectable, and biodegradable. Hydrogels are one type of substance that is utilized in tissue engineering called bioinks. Due to their biocompatible properties, these hydrogels are a super choice for biomedical applications [4]. Biocompatibility of a substance refers to its ability to support cell survival. Cell viability assessment is used to evaluate the biocompatibility of materials. Size, structure, shape, surface charge, and nanoparticle concentration are also examined. Surface chemical modifications can improve the biocompatibility of nanoparticles. In

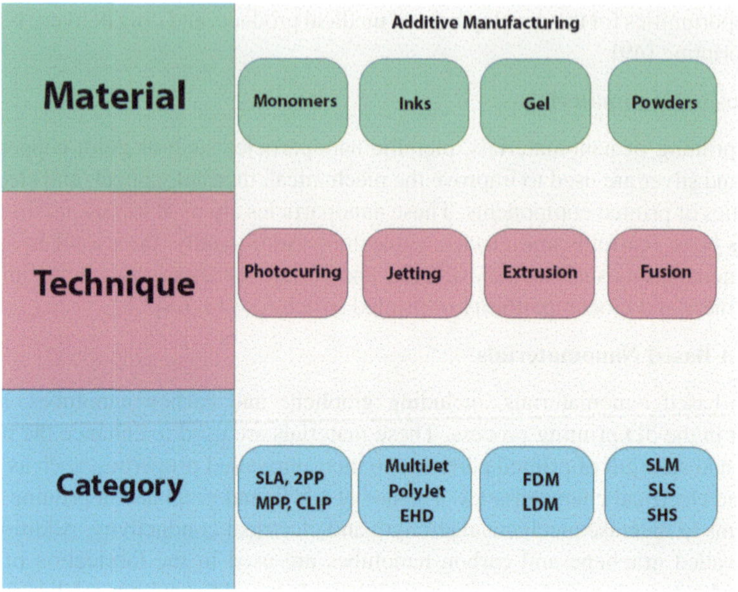

Fig. 9 AM and its materials and techniques

bio-printing applications, attention must be paid to the biocompatibility and biological performance of materials [8]. Cell viability assays are commonly used to evaluate the biocompatibility of common materials. Additionally, surface chemical modifications can enhance the biocompatibility of nanoparticles. In bio-printing, bioinks with nanoparticles can be used to combine functionalities and properties [19, 103].

8 Nano-Materials for Bone and Vascular Differentiation

Nanotechnology and nano-materials in 3D bioprinting bone fabrication enhance osteogenic differentiation by creating bone-mimicking microenvironments and improving cellular performance and mineralization [23].

Nano-Topography

Utilizing nano-topography in 3D bioprinting, using hierarchical scaffoldings of PCL, enhances protein absorption and cellular interaction. These methods improve cellular activities, cell growth, and myotube formation [24]. Moreover, surface modification of scaffolds with nanofibers improves cellular adhesion and proliferation of pre-osteoblasts. Implementing nano-topography in 3D scaffolds enhances new bone formation capability and improves the Alkaline phosphatase activity of mesenchymal stem cells. Overall, these modifications can assist in designing 3D scaffolds with better biological properties and enhance cellular-nano-topography interactions [27, 57].

Nano-Particles

The use of nanoparticles in bone tissue implants can enhance their biological and mechanical performance. Adding nHAp to materials such as PLLA and PLGA increases the growth and differentiation of bone and vascular cells without the need for additional growth factors. Additionally, nHAp can act as a carrier for important genes to encourage bone formation, and 3D scaffolds containing nHAp exhibit useful antibacterial properties. Overall, the use of nanoparticles in bone tissue implants can facilitate bone healing processes and prevent infections [30].

Nano-Crystals

Nano-crystals are used in composite structures to promote vascularization and bone formation. Recent studies have shown that adding nano-crystals to bone structures in 3D printing methods can promote cell viability, angiogenesis, and bone formation [79]. The use of bioactive molecules such as resveratrol and strontium ranelate can also aid in increasing angiogenesis and bone formation. Additionally, the release of D-foresamine helps accelerate angiogenesis and bone formation in bone defect sites. In general, the use of nano-crystals in composite structures can aid in improving vascularization and bone formation, leading to significant advancements in bone repair and strengthening [45, 59, 100].

Nano-Sheets

The use of black phosphorus (BP) nano-sheets in 3D scaffolds significantly improves the treatment of photothermal osteosarcoma and bone regeneration. The photothermal properties of BP enable the execution of photothermal therapies for surgical bone defects, and the phosphorus characteristics enhance cellular proliferation, osteogenic differentiation, angiogenesis, and bone formation in 3D scaffolds. Utilizing BP-modified BG scaffolds is degraded, leading to the formation of new bone tissue in bone defect sites. Overall, the use of these nano-materials aids in bone repair and strengthening and facilitates the production of new bone tissue [72].

Bioactive Molecules and Drugs

In 3D printed scaffolds, bioactive molecules such as Icariin and drugs like Bone Morphogenetic Protein 2 (BMP-2) are utilized to facilitate bone differentiation, angiogenesis, and tissue regeneration processes. Icariin improves the mechanical properties and angiogenesis of printed porous titanium scaffolds, while BMP-2 provides a significant enhancement in water affinity and cellular adhesion of the scaffolds. These scaffolds have controlled release capabilities of bioactive molecules and can assist in the repair of complex body tissues, including bone-tendon interfaces [81, 99].

9 Nano-Biomaterials Used in Practical Bioprinting Inks and Their Classification

The use of nano-biomaterials in nano-composite bioprinting enhances structural stability and cellular efficiency. These nano-biomaterials are derived from various materials such as ceramics, polymers, carbons, and biologics and are used to prepare nano-composite bioprinting inks [114]. These materials include calcium phosphates, graphene, carbon nanotubes, collagen nanofibers, and metallic nanoparticles such as silver and gold, which have unique properties and are used for applications such as tissue regeneration [106].

Nano-Ceramic Biomaterials

In skeletal tissue engineering, HAp nanoparticles, especially in 3D bioprinting, are used. These nanoparticles can improve the biological properties of bioink and enhance the differentiation of stem cells. Moreover, the use of nano-hydroxyapatite and core–shell PLGA nanoparticles with TGF-β1 growth factor is performed to increase human mesenchymal Stem Cells (hMSC) proliferation and osteogenic differentiation [110]. Laponite is also used in bioprinting as a non-toxic and natural material, improving the printability and mechanical properties of bioinks. Additionally, surface-modified bio-glass nanoparticles and iron oxide nanoparticles can accelerate mineralization and osteogenic differentiation processes and increase the mechanical hardness of the bio-gel [112, 113].

Carbon-Based Nanomaterials

Graphene and single and multi-walled CNTs are used in tissue engineering as materials with outstanding electrical and mechanical properties. The use of graphene in stimuli-responsive tissues such as neural and nerve tissues is noteworthy for enhancing stem cell adhesion and differentiation [40]. Graphene-based inks for 3D printing of cell culture scaffolds with excellent electrical and mechanical properties are among them. Carbonaceous nanofillers such as graphene, Graphene oxide, and CNTs are also used in tissue engineering to improve electrical conductivity and mechanical strength and can enhance the electrical and mechanical properties of bone [119].

Polymeric Nano-Biomaterials

The use of nano-scale cells, including nanofibrils and bacterial cells, in tissue engineering, is common. These nanocelluloses are used as hydrogel matrices and nanomaterial reinforcements but may induce inflammatory responses and cellular toxicity in the long term. Collagen, the primary protein in the ECM fibrous matrix, is used to produce bio-mimetic and reinforce hydrogel matrices. Various nanoparticles in the 3D bioprinting process, including PLA, gold, and silver, are used for purposes such as cell reinforcement and stimulation and can serve as nano-carriers for controlled delivery of bio-molecules [123].

10 Designing Functional Bioinks for Tissue Engineering

Cellular Interactions and Physiology of Nanomaterials

Biological nanomaterials are divided into two categories: biodegradable and non-biodegradable. Materials such as carbon and cellulose belong to the non-biodegradable category and usually accumulate in the body and then are excreted. On the other hand, biodegradable nanoparticles such as silicates and calcium phosphates release various ions that participate in cellular activities. However, excessive consumption of metallic ions and degradation products from polymeric nanoparticles may have adverse effects. While the use of nanomaterials in biological systems is feasible, the characteristics of nanobiomaterials and their biodegradability need to be carefully evaluated [94, 125] (Fig. 10).

The use of biological nanomaterials can enhance the properties of biomaterials, but loading them into printable biomaterials poses challenges, including limitations on biological activities and interference with printability. To address these challenges, various methods are used to prepare nanocomposite bioinks, including direct mixing prior to crosslinking, in situ nanocomposite formation, biomimetic mineralization, and melt mixing [71]. Cells are usually added to biomaterials in the final stage before printing to minimize stresses on them and effectively distribute them within the biomaterial structure. The use of nanocomposite bioinks can provide many advantages for improving the properties of biomaterials, including increased

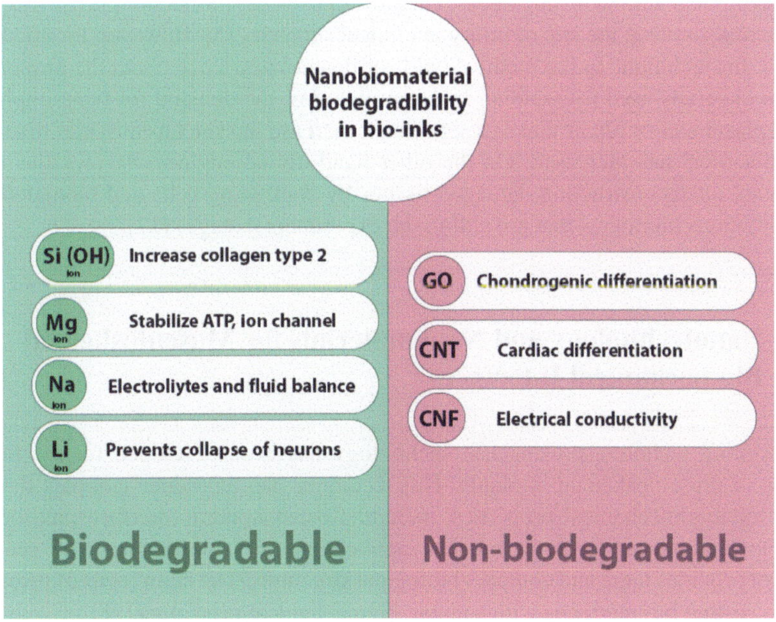

Fig. 10 Synthesis of nanocomposite bioinks for tissue engineering

printability, mechanical reinforcement, structural stability after printing, and stimulation of responsive cells such as cardiac, neural, and muscular cells. Research in 3D bioprinting focuses on developing functional organs with neural networks. The use of nanoparticles and hydrogels in scaffolds can improve cell adhesion and differentiation [126].

Natural ECM Hydrogels for 3D Printing Technology

Collagen, the main protein in the human body, is used to form ECM fibers, and chemical modifications such as methacrylation of collagen have been proposed to enhance its mechanical strength. Gelatin, a polymer derived from collagen, is of interest due to its biocompatibility and cellular adhesion properties, and GelMA improves mechanical properties and thermal stability [102]. Chitosan, a derivative of chitin, is attractive for its biodegradability, biocompatibility, and antimicrobial properties, and it can facilitate bone, skin, and nerve regeneration. Alginate, a polysaccharide from brown algae, is used in tissue engineering and biomedical applications and can protect bioactive molecules and be used in 3D bioprinting. Hyaluronic acid, a natural polysaccharide in ECM tissues, is used in HAp hydrogels for wound healing and drug delivery applications [132].

3D printing is a revolution in biomedical engineering. This method enables the fabrication of complex, multicellular tissues, vascular tissues, and multi-tissue structures. By integrating nanotechnology and 3D printing, it is possible to enhance the fabrication of customized 3D tissue structures. This technique can aid in addressing the shortage of organs and provide laboratory tissue models for drug candidate testing and preclinical animal testing needs. On the clinical side, this technique can directly create new tissue at the site of injury in clinical surgeries [5]. However, for effective use of this technique in the medical field, sufficient knowledge about the properties of the materials used is essential. In recent decades, the demand for bone implants for replacement or repair due to a significant increase in skeletal-muscular diseases and disorders has increased. On the other hand, tissue engineering provides new facilities for reconstructing damaged tissues by combining cells and biomaterials, using nanotechnologies that are still in the experimental stages [17].

11 Nanotechnology and Nanomaterials for Musculoskeletal Biomechanical Interfaces

3D printing technology using nanotechnology has been developed for the fabrication of musculoskeletal implants. This technique can reproduce complex tissues with suitable mechanical properties, including osteochondral interfaces and bone-muscle–tendon interfaces. In medical applications, 3D bioprinting with precise control provides large and complex biological structures. The main focus of research is on creating biointerfaces with suitable biomechanical properties [18].

Bioprinting

3D bioprinting techniques, such as inkjet, laser-based, microvalve, and extrusion, have specific advantages and drawbacks. For example, inkjet printing creates small defects and increases thermal stresses on cells [26]. The use of lasers may damage cells, but creates less stress on them. Extrusion enables adaptation to the selection of biomaterials and printing of larger organs, but may create shear stress on cells. The selection of biomaterials for 3D printing should support cells and ensure their survival. Bioprinting inks are usually classified based on hydrogels and natural or synthetic molecular compounds. Challenges such as long-term shape retention and gel swelling can hinder the capabilities of bioprinting. Combinations of polymers and crosslinking technologies are being investigated to solve these issues [33].

Nanobiomaterials for Practical Bioink Design

Recent research in tissue engineering has focused on nano-composite hydrogels with bio-nanomaterials. The use of these hydrogels in tissue engineering improves thermal, optical, electrical, and magnetic properties. 3D printing technology also has applications in several fields, including medicine and drug industries. Biopolymers with inherent thermoplasticity and wide thermal processing capabilities enable diverse inputs for 3D printing. Incorporating nanomaterials into tissue engineering scaffolds allows for the development of 3D printing methods with better characteristics [46].

Cell-Free Scaffold Printing Techniques

3D printing methods for fabricating tissue scaffolds have garnered attention. FDM uses layer-by-layer deposition of biopolymers, but blending nanofillers with polymer matrices can enhance structure clarity. Stereolithography utilizes UV laser beams for resin structuring and offers high resolution. SLS employs programmed laser beams to fuse polymers or ceramic granules, creating porous and lightweight structures. Adding nanomaterials can improve mechanical properties and address issues like weakness against mechanical loads [85].

Cellular Scaffold Printing Techniques

Using biopolymer hydrogels containing cells as bioinks for producing 3D structural additives, such as biomimetic tissue scaffolds, improves cell distribution capability. Currently, various approaches exist for bioprinting, including droplet-based, extrusion-based, and laser-assisted methods. These methods enable the production of high-resolution and accurate scaffolds but face challenges such as cell viability and printer processability [89, 122]. Droplet-based bioprinting (DBB) and laser-based bioprinting (LBB) methods are used for fabricating 3D biointerfaces. DBB utilizes various energy sources to form droplets, and nanomaterials like nanocellulose can enhance its performance. In LAB, lasers are used for high-precision structure fabrication, and nanoparticles suspended in collagen hydrogels can create cellular patterns without the need for light initiators. In extrusion-based bioprinting, using compatible bio-nanomaterial compositions can improve the process and alleviate material mixing issues [93].

Rapid advancements in nanomaterials and AM technologies have brought about remarkable changes in the field of tissue engineering. From the development of bionic ears and neural stimulating structures to the ability to manufacture custom biological tissues at lower costs and time, these advancements have had wide-ranging impacts on both medical and tissue engineering fields. Nanomaterials, composites, and 3D printing technologies, including stereolithography, powder bed techniques, and extrusion-based methods, have supported these advancements. However, challenges such as biocompatibility, material availability, composite processing, and personalized structures require appropriate solutions [98] (Fig. 11).

Regenerative medicine and tissue engineering, through the combination of cells, biomaterials, and 3D bioprinting technology, focus on constructing 3D tissue structures for use in biomedical applications. These methods enable the creation of substitute biological tissues with better performance in the human body. Hydrogels based on ECM using biomaterials such as collagen and gelatin are created, and integration with nanomaterials like graphene and carbon nanotubes improves mechanical and biochemical properties. These methods hold promise for the development of engineered tissues and the execution of regenerative medical projects. These materials can be composed of a single constituent or a combination of various components. The primary goal in designing these materials is to demonstrate their inimitable

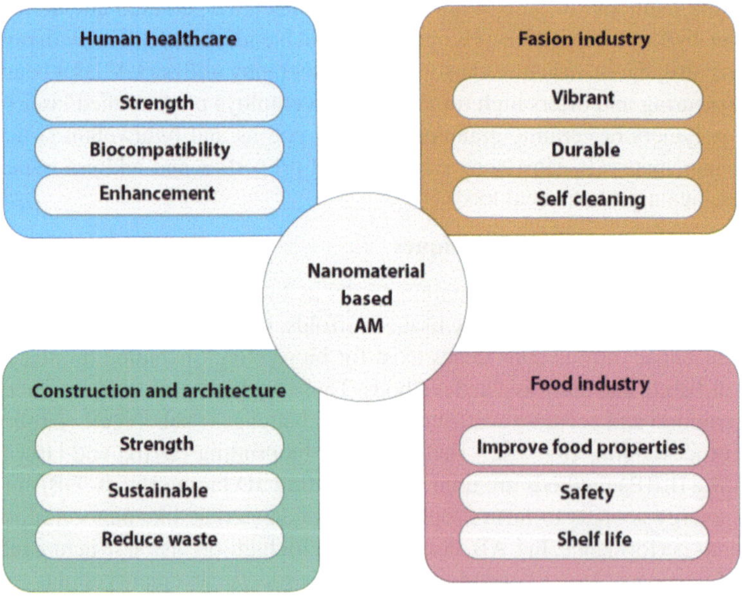

Fig. 11 Nanomaterial-based AM and properties

properties such as superior resistance, conductivity, or chemical reactivity compared to macro-scale materials. Surface and quantum properties, including surface energy and volume ratio, in nanomaterials, are important factors influencing their behavior compared to bulk materials [107]. The volume restriction and increased surface-to-volume ratio in nanometer-scale materials enhance reactivity and surface energy, providing various characteristics. For example, electron confinement and energy level quantization in nanometer-scale materials create changes in electronic, optical, and magnetic properties distinct from bulk materials [133]. These characteristics and unique phenomena demonstrated by nanomaterials present new approaches for their integration into the 3D printing process [121].

Combining nanotechnology with 3D printing offers new opportunities for producing innovative nanocomposites. Nanomaterials at the nanoscale possess unique properties such as friction, optical, electrochemical, thermal, and antibacterial properties. Integrating these nanomaterials with 3D printing technology can improve product performance and enhance mechanical and magnetic properties. For example, adding metal-filled nanocomposites to polymer matrices creates superior magnetic properties. These advancements present new opportunities for various applications such as aerospace, defense, and static and dynamic loading applications. However, the presence of photothermal sensitizers in printing materials causes challenges in terms of appearance and color, which can be addressed by using gold nanowires in polymer matrices [122]. These combinations can produce 3D-printed objects with a more uniform and vibrant white appearance and improve optical and color properties. Additionally, the biocompatibility, biodegradability, and antibacterial properties of various nanoparticle materials make them applicable for use in biomedical implants using 3D printing techniques. These advancements enable the creation of new facilities for 3D tissue replacements, biomedical implants, and smart wearable sensors [18, 127].

12 Present Challenges and Future Perspectives

Despite all the advantages of 3D, 4D, and 5D printing technology, these technologies still have challenges. This section examines the challenges in knowledge and facilities and potential avenues for future investigations and clinical applications. Due to the progress of research, opportunities and potential applications of AM Additive manufacturing products using nanotechnology have emerged. By combining NM Nanomaterials with AM Additive manufacturing, various applications have been used, especially in biology and medicine [120]. The development of drugs and drug delivery, types of implants and scaffolds, as well as customization in design and materials used in this field, are medicine research advances in research in the field of medicine [35].

3D printing gives the possibility to making of biomimetic scaffolds with a wide range of different materials 3D printing has been a suitable solution for making biomimetic scaffolds with a wide range of different materials. Today, the use of

3D printing techniques is active in a variety of manufacturing artificial body organs [14]. In biomimicry, there is a need to replicate the shape, microenvironment, and framework of natural organs or tissues. Also, this approach focuses on the development of printed biomaterials. Printed organs have limitations in terms of having vital elements. Still, with the approach of biological 3D printing, these limitations have been partially resolved. With the help of the 3D printing technique Through the use of 3D printing technology of biological inks containing cells cellulose, it has become possible to provide these organs. Bio-inks are a suitable environment for the growth and adhesion of cells, and they also benefit from ideal biocompatibility and biodegradability properties [122].

However, they have defects in their mechanical and physical properties. In order to improve these properties, the use of nanocomposites is considered an efficient approach. A wide range of nanomaterials, from metals to non-metals, along with biocompatible polymers that enable exceptional nanocomposites, have opened perspectives that have been unattainable so far Numerous nanomaterials, ranging from metals to non-metals, and biocompatible polymers that allow for remarkable nanocomposites have opened up hitherto unreachable possibilities. However, there are still significant gaps in understanding the interactions between nanohybrids in printed media and the need for standardized process parameters and synthesis techniques to ensure consistency in the production of different nanohybrids To guarantee consistency in the synthesis of various nanohybrids, standardised process parameters and synthesis procedures are necessary, as there are currently large gaps in our knowledge of the interactions of nanohybrids in printed media [124].

Recently, 3D printing technology has evolved to the next generation, 4D printing [6]. 4D bio-printed scaffolds are 3D biologically known structural in which their properties, shape and function can be changed over time by exposure to a defined external stimulus, e.g. electric/magnetic field, temperature, pH, light and ions change [44]. Although new bioinks such as hydrogels and hydrogel nanocomposites are very cell-friendly and optimized with shear thinning behavior, they must have stiffness and responsiveness to multiple external stimuli, thus providing a new level of Adds complexity to biological materials [105] (Fig. 12).

Currently, proposed additive manufacturing technologies such as 5D printing can enhance 4D printing technology. For this purpose, speeding up the manufacturing process and strengthening the printed output, the 5D printer can be done in more axes (containing five degrees of freedom) [64], where the movement of the search engine in five different angles and the printed body for printing parts It uses curved surfaces without requiring a structural support. This technique improves the strength of the final product, which leads to better-designed products. One of the areas where the need for 5D printing technology is obvious is the construction of artificial cartilages and bones, and scaffolds due to their curved shape. Therefore, the 5D printing technique is an efficient method to help repair large and complex bone defects.

In order to develop the current printing techniques, it is required to develop smart nanomaterials that can provide full performance with a multipurpose system and have optimal compatibility. 3D and 4D printed products have been able to offer potential applications in systems of the human body and clinical applications, such as

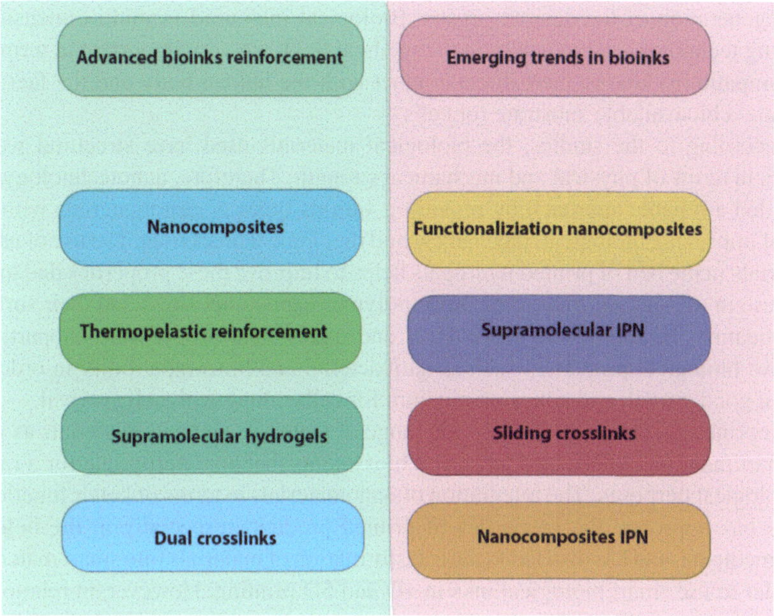

Fig. 12 Bioinks reinforcement and trends in bioinks

artificial muscles, implantable biomedical devices, and wearable sensors. However, due to the complexities of the process and nanomaterials, in order to achieve the best results and make parts with desirable and desired properties, extensive studies are being conducted, and there is a need to develop and improve these techniques with 5D and 6D printing methods. It is needed. Since 6D printing is a new concept of combining 4DD and 5D printing technologies, it leads to the development of additive manufacturing and the formation of a new generation of advanced smart materials that include nanoparticles in their structure [41]. Such progress in the field of multidimensional printing technology will improve and accelerate clinical treatment methods.

13 Conclusion

This chapter examines the technology of 3D, 4D, and 5D printing in nanotechnology and nanomaterials, which describes bioprinting, its process method, and the types of materials in them. Also, the investigation of nanomaterials, their function in enhancing the mechanical properties of composites during printing, as well as their mechanisms of action, are covered. Nano-based bioprinting is widely used in the fields of regenerative medicine and tissue engineering, which improves the behavior of cells in the culture environment and in the scaffold, which ultimately helps in the regeneration of the desired organs. Biological inks used in multi-dimensional printing technologies, especially in the medical field, are very important in terms of biocompatibility due to their direct contact with the human body and the fact that they are a bioavailable substrate for cells.

According to the studies, the biological materials used have structural weaknesses in terms of physical and mechanical strength. Therefore, nanotechnology has provided a reliable approach by providing various types of nanomaterials with the aim of improving biological, mechanical and rheological properties. The use of nanomaterials in the bed of printed hydrogels helps to improve these properties due to the increase in the surface-to-volume ratio (polymer nanocomposites) and their surface modification. In addition to rheological and mechanical properties, nanoparticles achieve biological properties such as antibacterial and biocompatibility in order to offer a good growth and adhesion medium for cells, along with cell survival.

Bioprinting technology uses a wide range of materials and nanomaterials as well as smart materials to produce advanced biosystems that are appropriate for a range of biological purposes. The integration of nanomaterials in terms of better functional levels has improved the properties of printed products, especially in the field of nanomedicine, and has also led to the transformation of materials into smart materials in order to use smart biological inks in 4D and 5D printing. However, in relation to the deformation of bioinks after 3D printing and their effect on the behavior and viability of cells, a thorough investigation is needed to become a controllable model for inducing the development of biological tissue.

The continuous development of nanotechnology and nanomaterials also enables new strategies in regenerative medicine and tissue engineering, such as 4D and 5D printing of stimuli-responsive materials, microfluidic (bio) printing, in vivo and in vivo printing (bio), and organs on chips. The 3D printing approach needs to be improved in the field of drug delivery, but it is hoped that this technology will emerge as a viable one soon and reach a commercial level to enable customizable release profiles. Advanced approaches needed in tissue engineering, regenerative medicine, and drug delivery are developed by the effective fusion of nanotechnology and nanomaterials with 3D printing and bioprinting.

References

1. Abdelhamid, H.N., Mathew, A.P.: Cellulose-based nanomaterials advance biomedicine: a review. Int. J. Molec. Sci., 23 (2022)
2. Afewerki, S., Magalhães, L., Silva, A.D.R., Stocco, T.D., Silva Filho, E.C., Marciano, F.R., Lobo, A.O.: Bioprinting a synthetic smectic clay for orthopedic applications. Adv Healthc Mater **8**(13), e1900158 (2019). https://doi.org/10.1002/adhm.201900158
3. Aigner, T., Stöve, J.: Collagens–major component of the physiological cartilage matrix, major target of cartilage degeneration, major tool in cartilage repair. Adv. Drug Deliv. Rev. **55**(12), 1569–1593 (2003)
4. Aizarna-Lopetegui, U., García-Astrain, C., Renero-Lecuna, C., González-Callejo, P., Villaluenga, I., Del Pozo, M.A., Sánchez-Álvarez, M., Henriksen-Lacey, M., Jimenez de Aberasturi,

D.: Remodeling arteries: studying the mechanical properties of 3D-bioprinted hybrid photoresponsive materials. J Mater Chem B **11**(39), 9431–9442 (2023). https://doi.org/10.1039/d3tb01480k

5. Al-Harbi, N., Hussein, M.A., Al-Hadeethi, Y., Felimban, R.I., Tayeb, H.H., Bedaiwi, N.M.H., Alosaimi, A.M., Bekyarova, E., Chen, M.: Bioactive hybrid membrane-based cellulose acetate/bioactive glass/hydroxyapatite/carbon nanotubes nanocomposite for dental applications. J. Mech. Behav. Biomed. Mater. **141**, 105795 (2023). https://doi.org/10.1016/j.jmbbm.2023.105795
6. Aldawood, F.K.: A Comprehensive Review of 4d printing: state of the arts, opportunities, and challenges. Actuators **12**(3), 101 (2023). https://www.mdpi.com/2076-0825/12/3/101
7. Aljohani, W., Ullah, M.W., Zhang, X., Yang, G.: Bioprinting and its applications in tissue engineering and regenerative medicine. Int. J. Biol. Macromolec. **107** Pt A, 261–275 (2018)
8. An, J., Chua, C.K., Yu, T., Li, H., Tan, L.P.: Advanced nanobiomaterial strategies for the development of organized tissue engineering constructs. Nanomedicine (Lond.) **8**(4), 591–602 (2013). https://doi.org/10.2217/nnm.13.46
9. Aydın, T., Gurcan, C., Taheri, H., Yilmazer, A.: Graphene based materials in neural tissue regeneration. Adv. Exp. Med. Biol. **1107**, 129–142 (2018)
10. Backes, E.H., Harb, S.V., Beatrice, C.A.G., Shimomura, K.M.B., Passador, F.R., Costa, L.C., Pessan, L.A.: Polycaprolactone usage in additive manufacturing strategies for tissue engineering applications: a review. J. Biomed. Mater. Res. Part B Appl. Biomater. (2021)
11. Bakht, S.M., Pardo, A., Gómez-Florit, M., Reis, R.L., Domingues, R.M.A., Gomes, M.E.: Engineering next-generation bioinks with nanoparticles: moving from reinforcement fillers to multifunctional nanoelements. J. Mater. Chem. B (2021)
12. Benwood, C., Chrenek, J., Kirsch, R.L., Masri, N.Z., Richards, H., Teetzen, K., Willerth, S.M.: Natural biomaterials and their use as bioinks for printing tissues. Bioengineering **8**(2), 27 (2021)
13. Bhattacharyya, A., Janarthanan, G., Noh, I.: Nano-biomaterials for designing functional bioinks towards complex tissue and organ regeneration in 3D bioprinting. Additive Manuf., 101639 (2020a)
14. Bhattacharyya, A., Janarthanan, G., Noh, I.: Nano-biomaterials for designing functional bioinks towards complex tissue and organ regeneration in 3D bioprinting. Addit. Manuf. **37**, 101639 (2020). https://doi.org/10.1016/j.addma.2020.101639
15. Bogala, M.R.: Three-dimensional (3D) printing of hydroxyapatite-based scaffolds: a review. Bioprinting **28**, e00244 (2022)
16. Bouguéon, G., Kauss, T., Dessane, B., Barthélémy, P., Crauste-Manciet, S.: Micro- and nano-formulations for bioprinting and additive manufacturing. Drug Discov. Today **24**(1), 163–178 (2019). https://doi.org/10.1016/j.drudis.2018.10.013
17. Bovone, G., Guzzi, E.A., Bernhard, S., Weber, T., Dranseikiene, D., Tibbitt, M.W.: Supramolecular reinforcement of polymer-nanoparticle hydrogels for modular materials design. Adv. Mater. **34**(9), e2106941 (2022). https://doi.org/10.1002/adma.202106941
18. Brusentsev, Y., Yang, P., King, A.W.T., Cheng, F., Cortes Ruiz, M.F., Eriksson, J.E., Kilpeläinen, I., Willför, S., Xu, C., Wågberg, L., Wang, X.: Photocross-linkable and shape-memory biomaterial hydrogel based on methacrylated cellulose nanofibres. Biomacromol **24**(8), 3835–3845 (2023). https://doi.org/10.1021/acs.biomac.3c00476
19. Cai, Y., Chang, S.Y., Gan, S.W., Ma, S., Lu, W.F., Yen, C.C.: Nanocomposite bioinks for 3D bioprinting. Acta Biomater. **151**, 45–69 (2022). https://doi.org/10.1016/j.actbio.2022.08.014
20. Cernencu, A.I., Dinu, A.I., Stancu, I.C., Lungu, A., Iovu, H.: Nanoengineered biomimetic hydrogels: a major advancement to fabricate 3D-printed constructs for regenerative medicine. Biotechnol. Bioeng. **119**(3), 762–783 (2022). https://doi.org/10.1002/bit.28020
21. Chakraborty, A., Roy, A., Ravi, S.P., Paul, A.: Exploiting the role of nanoparticles for use in hydrogel-based bioprinting applications: concept, design, and recent advances. Biomater. Sci. (2021)
22. Chakraborty, A., Roy, A., Ravi, S.P., Paul, A.: Exploiting the role of nanoparticles for use in hydrogel-based bioprinting applications: concept, design, and recent advances. Biomater Sci **9**(19), 6337–6354 (2021). https://doi.org/10.1039/d1bm00605c

23. Charoo, N.A., Barakh Ali, S.F., Mohamed, E.M., Kuttolamadom, M.A., Ozkan, T., Khan, M.A., Rahman, Z.: Selective laser sintering 3D printing–an overview of the technology and pharmaceutical applications. Drug Dev. Ind. Pharm. **46**(6), 869–877 (2020)
24. Chen, J., Zou, X.: Self-assemble peptide biomaterials and their biomedical applications. Bioact Mater **4**, 120–131 (2019). https://doi.org/10.1016/j.bioactmat.2019.01.002
25. Chen, Y.: Quantum dots for labeling live cells. Methods Mol. Biol. **906**, 193–198 (2012)
26. Chimene, D., Miller, L., Cross, L.M., Jaiswal, M.K., Singh, I., Gaharwar, A.K.: Nanoengineered osteoinductive bioink for 3D bioprinting bone tissue. ACS Appl. Mater. Interfaces **12**(14), 15976–15988 (2020). https://doi.org/10.1021/acsami.9b19037
27. Chinga-Carrasco, G., Rosendahl, J., Catalán, J.: Nanocelluloses—nanotoxicology, safety aspects and 3D bioprinting. Adv. Exp. Med. Biol. **1357**, 155–177 (2022). https://doi.org/10.1007/978-3-030-88071-2_7
28. Choi, Y.-J., Park, H., Ha, D.-H., Yun, H.-S., Yi, H.-G., Lee, H.: 3D bioprinting of in vitro models using hydrogel-based bioinks. Polymers **13**(3), 366 (2021)
29. Cidonio, G., Glinka, M., Kim, Y.H., Dawson, J.I., Oreffo, R.O.C.: Nanocomposite clay-based bioinks for skeletal tissue engineering. Methods Mol. Biol. **2147**, 63–72 (2021). https://doi.org/10.1007/978-1-0716-0611-7_6
30. Condi Mainardi, J., Rezwan, K., Maas, M.: Genipin-crosslinked chitosan/alginate/alumina nanocomposite gels for 3D bioprinting. Bioprocess Biosyst. Eng. **45**(1), 171–185 (2022). https://doi.org/10.1007/s00449-021-02650-3
31. Das, S.S., Bharadwaj, P., Bilal, M., Barani, M., Rahdar, A., Taboada, P., Bungău, S.G., Kyzas, G.Z.: Stimuli-responsive polymeric nanocarriers for drug delivery, imaging, and theragnosis. Polymers, 12 (2020)
32. Dash, T.K., Konkimalla, V.S.B.: Poly-ϵ-caprolactone based formulations for drug delivery and tissue engineering: a review. J. Controlled Release Official J. Controlled Release Soc. **158**(1), 15–33 (2012)
33. Deng, Y., Zhang, F., Jiang, M., Liu, Y., Yuan, H., Leng, J.: Programmable 4D printing of photoactive shape memory composite structures. ACS Appl. Mater. Interfaces **14**(37), 42568–42577 (2022). https://doi.org/10.1021/acsami.2c13982
34. Dvir, T., Timko, B.P., Kohane, D.S., Langer, R.S.: Nanotechnological strategies for engineering complex tissues. Nat. Nanotechnol. **6**(1), 13–22 (2020)
35. Eldeeb, A.E., Salah, S., Elkasabgy, N.A.: Biomaterials for tissue engineering applications and current updates in the field: a comprehensive review. AAPS PharmSciTech **23**(7), 267 (2022). https://doi.org/10.1208/s12249-022-02419-1
36. Emerich, D.F.: Nanomedicine—prospective therapeutic and diagnostic applications. Expert Opin. Biol. Ther. **5**, 1–5 (2005)
37. Flégeau, K., Pace, R., Gautier, H., Réthoré, G., Guicheux, J., Le Visage, C., Weiss, P.: Toward the development of biomimetic injectable and macroporous biohydrogels for regenerative medicine. Adv. Coll. Interface. Sci. **247**, 589–609 (2017)
38. Gao, B.-T., Zhao, H., Peng, L., Sun, Z.: A review of research progress in selective laser melting (SLM). Micromachines, 14 (2022)
39. Gao, G., Cui, X.: Three-dimensional bioprinting in tissue engineering and regenerative medicine. Biotech. Lett. **38**, 203–211 (2015)
40. Geetha Bai, R., Muthoosamy, K., Manickam, S., Hilal-Alnaqbi, A.: Graphene-based 3D scaffolds in tissue engineering: fabrication, applications, and future scope in liver tissue engineering. Int. J. Nanomedicine **14**, 5753–5783 (2019). https://doi.org/10.2147/ijn.S192779
41. Georgantzinos, S.K., Giannopoulos, G.I., Bakalis, P.A.: Additive Manufacturing for Effective Smart Structures: The Idea of 6D Printing. J. Compos. Sci. **5**(5), 119 (2021). https://www.mdpi.com/2504-477X/5/5/119
42. Gong, J., Qian, Y., Lu, K., Zhu, Z., Siow, L., Zhang, C., Zhou, S., Gu, T., Yin, J., Yu, M.: Digital light processing (DLP) in tissue engineering: From promise to reality, and perspectives. Biomed. Mater. **17**(6), 062004 (2022)
43. Hao, Z., Li, H., Wang, Y., Hu, Y., Chen, T., Zhang, S., Guo, X., Cai, L., Li, J.: Supramolecular peptide nanofiber hydrogels for bone tissue engineering: from multihierarchical fabrications

to comprehensive applications. Adv Sci (Weinh) **9**(11), e2103820 (2022). https://doi.org/10.1002/advs.202103820
44. Hata, S., Ihara, S., Saito, H., Murayama, M.: In-situ heating-and-electron tomography for materials research: from 3D (in-situ 2D) to 4D (in-situ 3D). Microscopy **73**(2), 133–144 (2024). https://doi.org/10.1093/jmicro/dfae008
45. He, J., Zhang, B., Li, Z., Mao, M., Li, J., Han, K., Li, D.: High-resolution electrohydrodynamic bioprinting: a new biofabrication strategy for biomimetic micro/nanoscale architectures and living tissue constructs. Biofabrication **12**(4), 042002 (2020). https://doi.org/10.1088/1758-5090/aba1fa
46. Hirsch, C., Roesslein, M., Krug, H.F., Wick, P.: Nanomaterial cell interactions: are current in vitro tests reliable? Nanomedicine (Lond.) **6**(5), 837–847 (2011). https://doi.org/10.2217/nnm.11.88
47. Holzwarth, J.M., Ma, P.X.: Biomimetic nanofibrous scaffolds for bone tissue engineering. Biomaterials **32**(36), 9622–9629 (2011)
48. Hospodiuk, M., Dey, M., Sosnoski, D.M., Ozbolat, I.T.: The bioink: a comprehensive review on bioprintable materials. Biotechnol. Adv. **35**(2), 217–239 (2017)
49. Hosseini, S., Shahrousvand, M., Mohammadi-Rovshandeh, J., Jahanbakhshi, M., Javadi, A., Soleimani, M., Jalalian, H., Hajikhani, M.: Fabrication of pH-responsive amphiphilic poly(Vinyl Alcohol-Methyl Methacrylate) copolymer nanoparticles for application in cancer drug delivery systems. Iranian J. Sci. **48**(1), 99–111 (2024). https://doi.org/10.1007/s40995-023-01573-w
50. Hutmacher, D.W.: Scaffolds in tissue engineering bone and cartilage. Biomaterials **21**(24), 2529–2543 (2000)
51. Ingavle, G.C., Leach, J.K.: Advancements in electrospinning of polymeric nanofibrous scaffolds for tissue engineering. Tissue Eng. Part B Rev. **20**(4), 277–293 (2014)
52. Jiang, T., Munguia-Lopez, J.G., Flores-Torres, S., Kort-Mascort, J., Kinsella, J.M.: Extrusion bioprinting of soft materials: an emerging technique for biological model fabrication. Appl. Phys. Rev. (2019)
53. Jing, P., Rudra, J.S., Herr, A.B., Collier, J.H.: Self-assembling peptide-polymer hydrogels designed from the coiled coil region of fibrin. Biomacromol **9**(9), 2438–2446 (2008)
54. Kang, H.K.: Studies on efficient hepatocyte differentiation using bioprinting and graphene quantum dots (2022)
55. Kargozar, S., Mozafari, M.: Nanotechnology and nanomedicine: start small, think big. Mater. Today Proc. **5**, 15492–15500 (2018)
56. Karmakar, R., Dey, S., Alam, A., Khandelwal, M., Pati, F., Rengan, A.K.: Attributes of nanomaterials and nanotopographies for improved bone tissue engineering and regeneration. ACS Appl. Bio Mater. **6**(10), 4020–4041 (2023). https://doi.org/10.1021/acsabm.3c00549
57. Kenry, Yeo, T., She, D.T., Nai, M.H., Marcelo Valerio, V.L., Pan, Y., Middha, E., Lim, C.T., Liu, B.: Differential collective cell migratory behaviors modulated by phospholipid nanocarriers ACS Nano, **15**(11), 17412–17425 (2021).https://doi.org/10.1021/acsnano.1c03060
58. Keshvardoostchokami, M., Majidi, S.S., Huo, P., Ramachandran, R., Chen, M., Liu, B.: Electrospun nanofibers of natural and synthetic polymers as artificial extracellular matrix for tissue engineering. Nanomaterials (Basel) **11**(1) (2020). https://doi.org/10.3390/nano11010021
59. Kim, J., Choi, H.S., Kim, Y.M., Song, S.C.: Thermo-responsive nanocomposite bioink with growth-factor holding and its application to bone regeneration. Small **19**(9), e2203464 (2023). https://doi.org/10.1002/smll.202203464
60. Ko, H., Milthorpe, B.K., McFarland, C.D.: Engineering thick tissues--the vascularisation problem. Europ. Cells Mater. **14**, 1–18; discussion 18–19 (2007)
61. Krátká, Z., Blažková, I., Vaculovičová, M., Adam, V., Kizek, R.: Modern imaging techniques (2014)
62. Kristiawan, R.B., Imaduddin, F., Ariawan, D., Ubaidillah, Arifin, Z.: A review on the fused deposition modeling (FDM) 3D printing: filament processing, materials, and printing parameters. Open Eng. **11**(1), 639–649 (2021)

63. Kumar, H., Kim, K.: Stereolithography 3D bioprinting. Methods Mol. Biol. **2140**, 93–108 (2020)
64. Lai, J., Wang, M.: Developments of additive manufacturing and 5D printing in tissue engineering. J. Mater. Res. **38**(21), 4692–4725 (2023). https://doi.org/10.1557/s43578-023-01193-5
65. Laird, N.Z., Acri, T.M., Chakka, J.L., Quarterman, J.C., Malkawi, W.I., Elangovan, S., Salem, A.K.: Applications of nanotechnology in 3D printed tissue engineering scaffolds. Eur. J. Pharm. Biopharm. **161**, 15–28 (2021). https://doi.org/10.1016/j.ejpb.2021.01.018
66. Langer, R., Vacanti, J.P.: Tissue engineering. Science **260**(5110), 920–926 (1993)
67. Lee, J., Lee, S.-H., Kim, B.S., Cho, Y.-S., Park, Y.: Development and evaluation of hyaluronic acid-based hybrid bio-ink for tissue regeneration. Tissue Eng. Regenerative Med. **15**(6), 761–769 (2018). https://doi.org/10.1007/s13770-018-0144-8
68. Li, N., Zhang, Q., Gao, S., Song, Q., Huang, R., Wang, L., Liu, L., Dai, J., Tang, M., Cheng, G.: Three-dimensional graphene foam as a biocompatible and conductive scaffold for neural stem cells. Scientific Reports, 3 (2013)
69. Li, Y., Zhu, K., Liu, X., Zhang, Y.S.: Blood-vessel-on-a-chip platforms for evaluating nanoparticle drug delivery systems. Curr. Drug Metab. **19**(2), 100–109 (2018). https://doi.org/10.2174/1389200218666170925114636
70. Loo, Y., Hauser, C.A.: Bioprinting synthetic self-assembling peptide hydrogels for biomedical applications. Biomed. Mater. **11**(1), 014103 (2015). https://doi.org/10.1088/1748-6041/11/1/014103
71. Loukelis, K., Helal, Z.A., Mikos, A.G., Chatzinikolaidou, M.: Nanocomposite bioprinting for tissue engineering applications. Gels **9**(2) (2023). https://doi.org/10.3390/gels9020103
72. Luo, Y., Wei, X., Huang, P.: 3D bioprinting of hydrogel-based biomimetic microenvironments. J. Biomed. Mater. Res. B Appl. Biomater. **107**(5), 1695–1705 (2019). https://doi.org/10.1002/jbm.b.34262
73. Lupuleasa, D.: Biocompatible Polymers For 3D Printing. Farmacia (2018)
74. Mandrycky, C.J., Wang, Z., Kim, K., Kim, D.H.: 3D bioprinting for engineering complex tissues. Biotechnol. Adv. **34**(4), 422–434 (2016)
75. Mao, H., Yang, L., Zhu, H., Wu, L., Ji, P., Yang, J., Gu, Z.: Recent advances and challenges in materials for 3D bioprinting. Progress Nat. Sci. Mater. Int. (2020)
76. Matea, C.T., Mocan, T., Flaviu, T., Pop, T.A., Moşteanu, O., Puia, C.I., Iancu, C., Mocan, L.: Quantum dots in imaging, drug delivery and sensor applications. Int. J. Nanomed. **12**, 5421–5431 (2017)
77. Meesorn, W., Shirole, A., Vanhecke, D., Espinosa, L.M.S., Weder, C.: A simple and versatile strategy to improve the mechanical properties of polymer nanocomposites with cellulose nanocrystals. Macromolecules **50,** 2364–2374 (2017)
78. Mei, Q., Rao, J., Bei, H.P., Liu, Y., Zhao, X.: 3D Bioprinting photo-crosslinkable hydrogels for bone and cartilage repair. Int. J. Bioprint **7**(3), 367 (2021). https://doi.org/10.18063/ijb.v7i3.367
79. Mo, X., Zhang, D., Liu, K., Zhao, X., Li, X., Wang, W.: Nano-hydroxyapatite composite scaffolds loaded with bioactive factors and drugs for bone tissue engineering. Int. J. Mol. Sci. **24**(2) (2023). https://doi.org/10.3390/ijms24021291
80. Moghaddam, A.S., Khonakdar, H.A., Arjmand, M., Jafari, S.H., Bagher, Z., Moghaddam, Z.S., Chimerad, M., Sisakht, M.M., Shojaei, S.: Review of bioprinting in regenerative medicine: naturally derived bioinks and stem cells. ACS Appl. Bio Mater. **4**(5), 4049–4070 (2021). https://doi.org/10.1021/acsabm.1c00219
81. Mohammadzadeh, M., Zarei, M., Abbasi, H., Webster, T.J., Beheshtizadeh, N.: Promoting osteogenesis and bone regeneration employing icariin-loaded nanoplatforms. J. Biol. Eng. **18**(1), 29 (2024). https://doi.org/10.1186/s13036-024-00425-4
82. Mohan, D., Teong, Z.K., Bakir, A.N., Sajab, M.S., Kaco, H.: Extending cellulose-based polymers application in additive manufacturing technology: a review of recent approaches. Polymers (Basel) **12**(9) (2020). https://doi.org/10.3390/polym12091876

83. Moura, D., Pereira, R.F., Goncalves, I.C.: Recent advances on bioprinting of hydrogels containing carbon materials. Mater. Today Chem. **23**, 100617 (2022)
84. Murphy, S.V., Atala, A.J.: 3D bioprinting of tissues and organs. Nat. Biotechnol. **32**, 773–785 (2014)
85. Olmos-Juste, R., Alonso-Lerma, B., Pérez-Jiménez, R., Gabilondo, N., Eceiza, A.: 3D printed alginate-cellulose nanofibers based patches for local curcumin administration. Carbohydr. Polym. **264**, 118026 (2021). https://doi.org/10.1016/j.carbpol.2021.118026
86. Patil, R., Alimperti, S.: Graphene in 3D bioprinting. J. Funct. Biomater. **15**(4), 82 (2024)
87. Piras, C.C., Smith, D.K.: Multicomponent polysaccharide alginate-based bioinks. J. Mater. Chem. B (2020)
88. Przekop, R.E., Gabriel, E., Pakuła, D., Sztorch, B.: Liquid for fused deposition modeling technique (L-FDM)—a revolution in application chemicals to 3D printing technology: color and elements. App. Sci. **13**(13), 7393 (2023). https://www.mdpi.com/2076-3417/13/13/7393
89. Pugliese, R., Montuori, M., Gelain, F.: Bioinspired photo-crosslinkable self-assembling peptides with pH-switchable "on-off" luminescence. Nanoscale Adv. **4**(2), 447–456 (2022). https://doi.org/10.1039/d1na00688f
90. Riehemann, K., Schneider, S.W., Luger, T.A., Godin, B., Ferrari, M., Fuchs, H.: Nanomedicine—challenge and perspectives. Angew. Chem. Int. Ed. **48**(5), 872–897 (2009)
91. Roato, I., Masante, B., Putame, G., Massai, D., Mussano, F.: Challenges of periodontal tissue engineering: increasing biomimicry through 3d printing and controlled dynamic environment. Nanomaterials (Basel), **12**(21) (2022). https://doi.org/10.3390/nano12213878
92. S, S., R, G. A., Bajaj, G., John, A.E., Chandran, S., Kumar, V.V., Ramakrishna, S.: A review on the recent applications of synthetic biopolymers in 3D printing for biomedical applications. J. Mater Sci. Mater Med. **34**(12), 62 (2023). https://doi.org/10.1007/s10856-023-06765-9
93. Safir, F., Vu, N., Tadesse, L.F., Firouzi, K., Banaei, N., Jeffrey, S.S., Saleh, A.A.E., Khuri-Yakub, B.P.T., Dionne, J.A.: Combining acoustic bioprinting with AI-assisted raman spectroscopy for high-throughput identification of bacteria in blood. Nano Lett. **23**(6), 2065–2073 (2023). https://doi.org/10.1021/acs.nanolett.2c03015
94. Sahay, G., Alakhova, D.Y., Kabanov, A.V.: Endocytosis of nanomedicines. J. Control. Release **145**(3), 182–195 (2010). https://doi.org/10.1016/j.jconrel.2010.01.036
95. Sarker, M., Naghieh, S., McInnes, A.D., Schreyer, D.J., Chen, X.: Strategic design and fabrication of nerve guidance conduits for peripheral nerve regeneration. Biotechnol. J. **13**(7), e1700635 (2018). https://doi.org/10.1002/biot.201700635
96. Saunders, R.E., Derby, B.: Inkjet printing biomaterials for tissue engineering: bioprinting. Int. Mater. Rev. **59**, 430–448 (2014)
97. Sekmen, K., Rehbein, T., Johlitz, M., Lion, A., Constantinescu, A.: Thermal analysis and shrinkage characterization of the photopolymers for DLP additive manufacturing processes. Continuum Mech. Thermodyn. (2022).https://doi.org/10.1007/s00161-022-01137-0
98. Shahid, N., Erum, A., Hanif, S., Malik, N.S., Tulain, U.R., Syed, M.A.: Nanocomposite hydrogels-a promising approach towards enhanced bioavailability and controlled drug delivery. Curr. Pharm. Des. **30**(1), 48–62 (2024). https://doi.org/10.2174/0113816128283466231219071151
99. Shavandi, A., Hosseini, S., Okoro, O.V., Nie, L., Eghbali, F., Melchels, F.: 3D bioprinting of lignocellulosic biomaterials. Adv Healthc Mater **9**(24), e2001472 (2020). https://doi.org/10.1002/adhm.202001472
100. Sheng, X., Li, C., Wang, Z., Xu, Y., Sun, Y., Zhang, W., Liu, H., Wang, J.: Advanced applications of strontium-containing biomaterials in bone tissue engineering. Mater Today Bio **20**, 100636 (2023). https://doi.org/10.1016/j.mtbio.2023.100636
101. Shevchenko, R.V., James, S.L., James, S.E.: A review of tissue-engineered skin bioconstructs available for skin reconstruction. J. R. Soc. Interface **7**, 229–258 (2010)
102. Sohaebuddin, S.K., Thevenot, P.T., Baker, D., Eaton, J.W., Tang, L.: Nanomaterial cytotoxicity is composition, size, and cell type dependent. Part. Fibre Toxicol. **7**, 22 (2010). https://doi.org/10.1186/1743-8977-7-22

103. Stern, S.T., Adiseshaiah, P.P., Crist, R.M.: Autophagy and lysosomal dysfunction as emerging mechanisms of nanomaterial toxicity. Part. Fibre Toxicol. **9**, 20 (2012). https://doi.org/10.1186/1743-8977-9-20
104. Stoppel, W.L., Ghezzi, C.E., McNamara, S.L., Black, L.D., Kaplan, D.L.: Erratum to: clinical applications of naturally derived biopolymer-based scaffolds for regenerative medicine. Ann. Biomed. Eng. **43**, 2023–2023 (2015)
105. Subeshan, B., Baddam, Y., Asmatulu, E.: Current progress of 4D-printing technology. Progress in Additive Manufacturing **6**(3), 495–516 (2021). https://doi.org/10.1007/s40964-021-00182-6
106. Szymański, T., Mieloch, A.A., Richter, M., Trzeciak, T., Florek, E., Rybka, J.D., Giersig, M.: Utilization of carbon nanotubes in manufacturing of 3D cartilage and bone scaffolds. Materials (Basel), **13**(18) (2020). https://doi.org/10.3390/ma13184039
107. Tavares, M.T., Gaspar, V.M., Monteiro, M.V., JP, S.F., Baleizão, C., Mano, J.F.: GelMA/bioactive silica nanocomposite bioinks for stem cell osteogenic differentiation. Biofabrication **13**(3) (2021). https://doi.org/10.1088/1758-5090/abdc86
108. Tekade, R.K., Maheshwari, R., Soni, N.D., Tekade, M., Chougule, M.B.: Nanotechnology for the Development of Nanomedicine (2017)
109. Theus, A.S., Ning, L., Jin, L., Roeder, R.K., Zhang, J., Serpooshan, V.: Nanomaterials for bioprinting: functionalization of tissue-specific bioinks. Essays Biochem. **65**(3), 429–439 (2021). https://doi.org/10.1042/ebc20200095
110. Theus, A.S., Ning, L., Jin, L., Roeder, R.K., Zhang, J., Serpooshan, V.: Nanomaterials for bioprinting: functionalization of tissue-specific bioinks. Essays Biochem. (2021)
111. Tong, S., Fine, E.J., Lin, Y., Cradick, T.J., Bao, G.: Nanomedicine: tiny particles and machines give huge gains. Ann. Biomed. Eng. **42**, 243–259 (2013)
112. Topuz, F., Nadernezhad, A., Caliskan, O.S., Menceloglu, Y.Z., Koc, B.: Nanosilicate embedded agarose hydrogels with improved bioactivity. Carbohydr. Polym. **201**, 105–112 (2018). https://doi.org/10.1016/j.carbpol.2018.08.032
113. Tortorella, S., Vetri Buratti, V., Maturi, M., Sambri, L., Comes Franchini, M., Locatelli, E.: Surface-modified nanocellulose for application in biomedical engineering and nanomedicine: a review. Int. J. Nanomedicine **15**, 9909–9937 (2020). https://doi.org/10.2147/ijn.S266103
114. Tripathi, S., Mandal, S.S., Bauri, S., Maiti, P.: 3D bioprinting and its innovative approach for biomedical applications. MedComm **4**(1), e194 (2023). https://doi.org/10.1002/mco2.194
115. Tsegay, F., Elsherif, M., Butt, H.: Smart 3D printed hydrogel skin wound bandages: a review. Polymers (Basel) **14**(5) (2022). https://doi.org/10.3390/polym14051012
116. Vengatesan, M.R., Mittal, V.: Nanoparticle- and Nanofiber-Based Polymer Nanocomposites: An Overview (2016)
117. Ventura, R.D.: An overview of laser-assisted bioprinting (LAB) in tissue engineering applications. Med. Lasers Eng. Basic Res. Clin. Appl. **10**(2), 76–81 (2021)
118. Verma, S., Domb, A.J., Kumar, N.: Nanomaterials for regenerative medicine. Nanomedicine **6**(1), 157–181 (2011)
119. Wang, X., Wang, Q., Xu, C.: Nanocellulose-based inks for 3D bioprinting: key aspects in research development and challenging perspectives in applications-a mini review. Bioengineering (Basel), **7**(2) (2020). https://doi.org/10.3390/bioengineering7020040
120. Wang, Z., Agrawal, P., Zhang, Y.S.: Nanotechnologies and nanomaterials in 3D (Bio)printing toward bone regeneration. Adv. NanoBiomed Res. **1**(11), 2100035 (2021), Article 2100035. https://doi.org/10.1002/anbr.202100035
121. Warwar, C., Martin, P., Vasilyev, G., Vilensky, R., Fadul, R., Redenski, I., Srouji, S., Zussman, E.: Injectable hydrogels based on inter-polyelectrolyte interactions between hyaluronic acid, gelatin, and cationic cellulose nanocrystals. Biomacromol **23**(8), 3222–3234 (2022). https://doi.org/10.1021/acs.biomac.2c00316
122. Wei, L., Li, Z., Li, J., Zhang, Y., Yao, B., Liu, Y., Song, W., Fu, X., Wu, X., Huang, S.: An approach for mechanical property optimization of cell-laden alginate-gelatin composite bioink with bioactive glass nanoparticles. J. Mater. Sci. Mater. Med. **31**(11), 103 (2020). https://doi.org/10.1007/s10856-020-06440-3

123. Yoon, J., Han, H., Jang, J.: Nanomaterials-incorporated hydrogels for 3D bioprinting technology. Nano Converg **10**(1), 52 (2023). https://doi.org/10.1186/s40580-023-00402-5
124. Yoon, J., Han, H., Jang, J.: Nanomaterials-incorporated hydrogels for 3D bioprinting technology. Nano Convergence **10**(1), 52 (2023). https://doi.org/10.1186/s40580-023-00402-5
125. Yue, K., Trujillo-de Santiago, G., Alvarez, M.M., Tamayol, A., Annabi, N., Khademhosseini, A.: Synthesis, properties, and biomedical applications of gelatin methacryloyl (GelMA) hydrogels. Biomaterials **73**, 254–271 (2015). https://doi.org/10.1016/j.biomaterials.2015.08.045
126. Zhang, H.B., Xing, T.L., Yin, R.X., Shi, Y., Yang, S.M., Zhang, W.J.: Three-dimensional bioprinting is not only about cell-laden structures. Chin. J. Traumatol. **19**(4), 187–192 (2016). https://doi.org/10.1016/j.cjtee.2016.06.007
127. Zhang, P., Teng, Z., Zhou, M., Yu, X., Wen, H., Niu, J., Liu, Z., Zhang, Z., Liu, Y., Qiu, J., Xu, X.: Upconversion 3D bioprinting for noninvasive in vivo molding. Adv. Mater. **36**(14), e2310617 (2024). https://doi.org/10.1002/adma.202310617
128. Zhang, S., Chen, X., Shan, M., Hao, Z., Zhang, X., Meng, L., Zhai, Z., Zhang, L., Liu, X., Wang, X.: Convergence of 3D bioprinting and nanotechnology in tissue engineering scaffolds. Biomimetics **8**(1), 94 (2023)
129. Zhang, X., Malhotra, S., Molina, M.A., Haag, R.: Micro- and nanogels with labile crosslinks—from synthesis to biomedical applications. Chem. Soc. Rev. **44**(7), 1948–1973 (2015)
130. Zhang, Y., Zhu, L., Tian, J., Zhu, L., Ma, X., He, X., Huang, K., Ren, F., Xu, W.: Smart and functionalized development of nucleic acid-based hydrogels: assembly strategies, recent advances, and challenges. Adv Sci (Weinh) **8**(14), 2100216 (2021). https://doi.org/10.1002/advs.202100216
131. Zhigarkov, V., Volchkov, I., Yusupov, V., Chichkov, B.: Metal nanoparticles in laser bioprinting. Nanomaterials **11**(10), 2584 (2021)
132. Zhou, B., Jiang, X., Zhou, X., Tan, W., Luo, H., Lei, S., Yang, Y.: GelMA-based bioactive hydrogel scaffolds with multiple bone defect repair functions: therapeutic strategies and recent advances. Biomater Res **27**(1), 86 (2023). https://doi.org/10.1186/s40824-023-00422-6
133. Zhou, K., Ding, R., Tao, X., Cui, Y., Yang, J., Mao, H., Gu, Z.: Peptide-dendrimer-reinforced bioinks for 3D bioprinting of heterogeneous and biomimetic in vitro models. Acta Biomater. **169**, 243–255 (2023). https://doi.org/10.1016/j.actbio.2023.08.008
134. Zhou, Q.-L., Chen, Z.-Y., Wang, Y.-X.J., Yang, F., Lin, Y., Liao, Y.-Y.: Ultrasound-mediated local drug and gene delivery using nanocarriers. BioMed Res. Int. (2014)
135. Zou, Y., Li, D., Shen, M., & Shi, X. (2019). Polyethylenimine-based nanogels for biomedical applications. *Macromolecular bioscience*, e1900272.
136. Zu, G., Meijer, M., Mergel, O., Zhang, H., van Rijn, P.: 3D-Printable hierarchical nanogel-GelMA composite hydrogel system. Polymers **13**(15), 2508 (2021)

Processing Techniques and Challenges of 3D-, 4D-, and 5D-Printed Polymer Composites

S. K. Parida

Abstract This chapter explores the evolving landscape of additive manufacturing technologies- 3D, 4D, and 5D printing and their application in polymer composites. It examines the processing techniques employed in each technology, highlighting their respective advantages and challenges. Key topics include material selection, design considerations, mechanical properties, and the impact of environmental stimuli on shape-changing capabilities. Challenges such as material compatibility, process control, and scalability are discussed, along with future directions and opportunities for innovation in this rapidly advancing field.

Keywords 3D 4D and 5D printing · Polymer composites · Advantages and challenges · Environmental stimuli · Process control · Opportunities for innovation

1 Introduction

The construction of a 3-dimensional object from a digital 3D model or CAD model is known as 3D printing or additive manufacturing [1, 2]. This is done by different processes in which, under computer control, materials are deposited and other materials are added together layer by layer. In 1945, Murray Leinster explained the general concept of 3D printing and the procedures used in it in his short story "Things Pass by". David E. H. Jones arranged the concept of 3D printing in his regular column Ariadne in the journal New Scientist in 1974 [3, 4]. In the 1980s, materials and equipment that are used in additive manufacturing were developed. Hideo Kodama from Nagoya Municipal Industrial Research Institute invented in April 1980 two additive methods for the fabrication of three-dimensional plastic models with photo-hardening thermoset polymer, where the UV exposure area is controlled by a scanning fibre transmitter [5]. For XYZ Plotter, he filed a patent, which was published on November 10, 1981. The results of his research were published in journals in

S. K. Parida (✉)
Department of Physics, ITER, Siksha 'O' Anusandhan Deemed to Be University, Bhubaneswar, India
e-mail: santoshparida@soa.ac.in

S. Moharana et al. (eds.), *Polymer Nanocomposites for 3D, 4D and 5D Printing*, Engineering Materials, https://doi.org/10.1007/978-981-96-4214-4_4

April and November of 1981 [6, 7]. In the 1980s and 1990s, additive manufacturing processes for melting materials and metal sintering, like selective laser sintering, selective laser melting, and direct metal laser sintering, usually had their own names. Now, processes like fabrication, stamping, casting, and machining are known as non-additive, which were at that time known as the processes for metalworking. For material deposition, new techniques were developed at Standford and Carnegie Mellon University by the mid-1990s, which include micro-casting [8] and sprayed materials [9]. 3D printing technology was widely used in research industries and manufacturing in the early 2000s. These are mostly used in the architecture and medical industries; however, they are mostly used for low-accuracy modelling rather than the manufacturing of common items [10]. As of 2020, 3D printers will have great quality and be more affordable for more people to enter the world of 3D printing. Those affordable printers are fused deposition modelling printers. In November 2021, a British patient named Steve Verze received the world's first fully 3D-printed prosthetic eye. In April 2024, the world's largest 3D printer was revealed at the University of Marine. It can span 96 feet (29 m).

Additive manufacturing, often known as 3D printing, has quickly gained popularity in the world of engineering due to its numerous advantages. The goal of 3D printing is design flexibility, individualization [11], decentralization [12], and the ability to carry out procedures that were previously unachievable using conventional technologies [13]. Some of the advantages include faster prototyping, lower production costs, increased product customisation and higher product quality [14]. Furthermore, 3D printing's capabilities have expanded beyond traditional manufacturing, such as lightweight construction [15] or repair and maintenance [16], to include applications in prosthetics [17], bioprinting [18], food industry [19], rocket building [20], design and art [21], and renewable energy systems [22]. 3D printing technology may be utilised to create battery energy storage devices, which are critical for renewable energy generation and delivery. 8-dimensional printing also known as 4D printing creates a three-dimensional item using the same principles as 3D printing: computer-programmed material deposition in consecutive layers. However, with 4D printing, the final 3D shape may morph into many shapes in response to external stimuli, with the fourth dimension representing the time-dependent shape change after printing [23–25]. It is therefore a sort of programmable matter in which, following the production process, the printed object reacts to environmental conditions (humidity, temperature, voltage, etc.) and changes shape appropriately [26–28].

Stereolithography is a 3D printing technology that employs photopolymerization to connect substrates that have been placed down layer by layer, resulting in a polymeric network. Unlike fused-deposition modelling, where the extruded material hardens instantly to create layers, 4D printing is primarily based on stereolithography, in which ultraviolet radiation is typically employed to cure the stacked materials after the printing process is done [29]. Anisotropy is essential for designing the direction and amplitude of transformations under a particular circumstance, since it allows the micromaterials to be arranged in such a way that the end print has embedded directionality [30, 31]. 4D printing technology is mostly used in construction, defence, aerospace, health, automotive, textile etc. Using 4D printing objects can change properties or shape using external catalyst like temperature, pH, electricity, moisture, light, and magnetic field.

The concept of 3D printing was given by American universities in 2016. 5D printing is a new branch of additive manufacturing. 5D printing produces three-dimensional objects utilising five axes. A three-dimensional printer with five axes produces more durable parts than 3D and 4D printer output. In addition to the x, y, and z axes, the moving extruder head and the moving print bed constitute two additional axes. An FDM (melt deposition production) printer with these axes is known as a 5D printer, and the output produced by this printer is known as 5D printing production. In these printers, production is based on CAD (computer-aided design) data, as is the case with three-dimensional printers [32]. The main aim of production with 5D printing is to produce parts with better strength and prevent material waste. 5d printing technology are improving their applications for dentistry, orthopaedic, surgery and prosthetics.

2 Importance of 3D Printing

The importance of 3D printing can be summarized as follows: (i) **Customization:** Enables tailored products that meet specific needs, enhancing user satisfaction and functionality, (ii) **Rapid Prototyping:** Accelerates the design process, allowing for quick iterations and faster time-to-market, (iii) **Cost Efficiency:** Reduces material waste and production costs, especially for small-scale manufacturing, (iv) **Complex Geometries:** Facilitates the creation of intricate designs that are difficult or impossible with traditional methods, (v) **Supply Chain Simplification:** Local production reduces dependency on global supply chains and lowers transportation costs, (vi) **Sustainability:** Promotes eco-friendly practices by minimizing waste and enabling the use of recycled materials, (vii) **Innovation:** Drives advancements in various fields, from healthcare (custom implants) to aerospace (lightweight components), (viii) **Accessibility:** Makes manufacturing more accessible to individuals and small businesses, fostering entrepreneurship. Overall, 3D printing is transforming industries, enhancing efficiency, and enabling innovation across various sectors [33–38].

3 Processing Techniques of 3D Printing

Here are the main processing techniques used in 3D printing:

a. **Fused Deposition Modelling (FDM)**:

- Extrudes thermoplastic filament layer by layer.
- Common for prototyping and consumer applications.

b. **Stereolithography (SLA)**:

- Uses a laser to cure liquid resin into solid layers.
- Provides high precision and smooth finishes.

c. **Selective Laser Sintering (SLS)**:

- Utilizes a laser to fuse powdered materials (typically plastics or metals).
- Suitable for complex geometries and functional parts.

d. **Digital Light Processing (DLP)**:

- Similar to SLA but uses a digital light projector to cure resin.
- Faster than SLA due to simultaneous layer curing.

e. **Binder Jetting**:

- Applies a binding agent to layers of powder, then fuses them in a separate step.
- Can be used for metals, ceramics, and sand.

f. **Material Jetting**:

- Deposits droplets of material (usually photopolymer) layer by layer.
- Enables high-resolution multi-material parts.

g. **Directed Energy Deposition (DED)**:

- Melt material (usually metal) as it is deposited, often used for repairs and additive manufacturing in aerospace.

h. **Sheet Lamination**:

- Layers of material sheets are glued and cut into shape.
- Suitable for large, low-cost models.

Each technique has its unique advantages and applications, making 3D printing versatile for various industries [39–42]. But, due to some technical issues, people pay more attention to 4D printing technique. Figure 1 shows the difference between 3 and 4D printing techniques and their relevant applications.

4 Challenges of 3D-Printed Polymer Composites

3D-printed polymer composites offer exciting opportunities for creating materials with tailored properties, but they also come with several challenges. Here are some of the key challenges associated with 3D printing polymer composites:

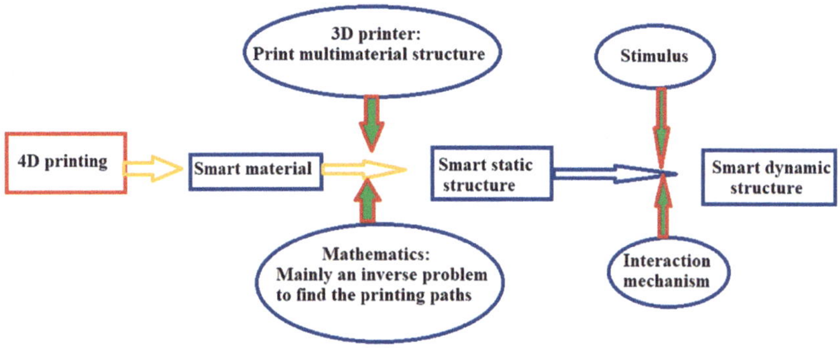

Fig. 1 Difference between 3 and 4D printing techniques

a. **Material Compatibility**: Combining polymers with other materials, such as fibers or nanoparticles, can lead to compatibility issues. Ensuring proper adhesion between the polymer matrix and the reinforcement is crucial for achieving desired mechanical properties.
b. **Printability**: Not all polymer composites are easily printable. The viscosity and flow characteristics of the composite material can affect the quality of the print. Finding the right balance between material composition and printer settings is often challenging.
c. **Processing Parameters**: The processing parameters, such as temperature, speed, and layer height, need to be optimized for each specific composite material. These parameters can significantly impact the final properties of the printed part.
d. **Anisotropy**: 3D-printed parts often exhibit anisotropic properties, meaning their strength and other characteristics can vary depending on the direction of the print layers. This is particularly true for composites where the orientation of fibers or fillers affects the overall behavior of the material.
e. **Interlayer Adhesion**: Achieving strong interlayer adhesion is crucial for the mechanical integrity of 3D-printed parts. In composites, this can be more challenging due to the presence of different materials, which might not bond well with each other across layers.
f. **Quality Control**: Ensuring consistent quality in 3D-printed polymer composites is difficult. Variations in material composition, printer settings, and environmental conditions can lead to inconsistencies in the final product.
g. **Post-Processing**: Many polymer composites require post-processing steps, such as curing or annealing, to achieve their full potential. These steps can add complexity and time to the production process.
h. **Cost**: High-quality composite materials and the equipment required for 3D printing can be expensive. This can limit the accessibility and scalability of the technology for certain applications.
i. **Environmental and Health Concerns**: The use of certain composite materials may raise environmental and health concerns, particularly if they involve nanoparticles or other potentially hazardous substances.

j. **Design Complexity**: While 3D printing allows for complex geometries, designing for 3D-printed composites can be complex due to the interactions between the polymer matrix and the reinforcement materials.

Addressing these challenges requires ongoing research and development in materials science, printer technology, and process optimization to fully harness the potential of 3D-printed polymer composites [43–54].

5 Importance of 4D Printing

4D printing is an advanced form of 3D printing that involves the use of materials and designs that enable printed objects to change their shape or properties over time in response to environmental stimuli such as heat, light, humidity, or other external factors. The "fourth dimension" refers to time, as these printed objects are not static but dynamic, allowing them to adapt and transform post-production. Therefore, 3D printing + time = 4D printing. Here, Fig. 2 shows the concept of 1D, 2D, 3D and 4D printing and predicts their relevant applications.

Here are some key points highlighting the importance of 4D printing:

a. **Adaptive Functionality**: 4D printing enables the creation of objects that can adapt their shape and functionality according to environmental conditions. This capability is particularly valuable in fields like aerospace, automotive, and construction, where adaptive structures can lead to enhanced performance and efficiency.
b. **Biomedical Applications**: In the medical field, 4D printing holds the potential for creating implants and devices that can change shape to accommodate changes in the human body, such as swelling or movement. This adaptability can lead to more personalized and effective medical treatments.
c. **Self-Repairing Materials**: 4D printing can produce materials that have self-healing properties, enabling them to repair themselves after damage. This feature is crucial for extending the lifespan and reliability of materials used in various industries.

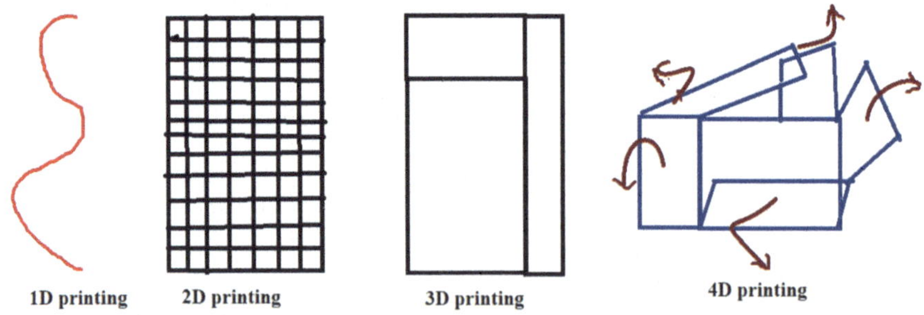

Fig. 2 1D, 2D, 3D, and 4D printing concepts

d. **Sustainability**: By creating materials that can adapt to changing conditions, 4D printing can contribute to sustainability efforts. For example, structures that automatically adjust to environmental changes can reduce energy consumption in buildings or transportation systems.
e. **Complex Design Possibilities**: 4D printing opens up new possibilities for complex designs that would be difficult or impossible to achieve with traditional manufacturing methods. This allows for innovative product designs and engineering solutions that can respond dynamically to their environment.
f. **Reduced Maintenance**: The ability of 4D-printed materials to adapt or self-repair can reduce the need for maintenance and replacement, leading to cost savings and increased efficiency over the lifecycle of a product.
g. **Smart Textiles**: In the fashion and textile industries, 4D printing can create fabrics that change color, shape, or texture in response to stimuli, leading to innovative clothing designs that can adapt to weather conditions or user preferences.
h. **Improved Performance**: In sectors such as sports and wearable technology, 4D printing can be used to develop equipment and devices that adjust to optimize performance and comfort for the user.
i. **Customizable Products**: 4D printing allows for the creation of highly customizable products that can be tailored to meet specific user needs or environmental conditions, enhancing user experience and satisfaction.
j. **Innovation and Research**: The exploration of 4D printing technology drives innovation and research in materials science, leading to the development of new smart materials and expanding the possibilities for future applications.

Overall, 4D printing represents a significant advancement in manufacturing and design, offering numerous benefits across various industries by enabling dynamic and responsive products. As research and development in this field continue, the potential applications and impact of 4D printing are expected to grow [55–62].

6 Processing Techniques of 4D-Printed Polymer Composites

The processing techniques for 3D-printed polymer composites involve various methods to incorporate reinforcing materials into the polymer matrix, enhance material properties, and optimize the printing process. Here are some key processing techniques used in the 3D printing of polymer composites:

a. **Material Extrusion (Fused Filament Fabrication—FFF/FDM)**

- **Description**: This is the most common method for 3D printing polymer composites, where a composite filament is fed through a heated nozzle and deposited layer by layer.
- **Reinforcement**: Typically involves short fibers, nanoparticles, or continuous fibers embedded in the polymer matrix.

- **Advantages**: Simple and cost-effective, allowing for the use of a wide range of composite materials.
- **Challenges**: Achieving uniform dispersion of reinforcement and optimizing print parameters to ensure good adhesion and mechanical properties.

b. **Vat Photopolymerization (SLA/DLP)**

- **Description**: Involves curing liquid resin composites layer by layer using a light source (laser or projector).
- **Reinforcement**: Incorporation of nanoparticles, such as ceramics or carbon nanotubes, to enhance mechanical and thermal properties.
- **Advantages**: High resolution and smooth surface finish.
- **Challenges**: Limited to light-curable resins, potential for settlement of particles, and challenges with achieving uniform reinforcement distribution.

c. **Powder Bed Fusion (Selective Laser Sintering—SLS)**

- **Description**: Uses a laser to sinter powdered composite material layer by layer.
- **Reinforcement**: Commonly uses polymer powders mixed with ceramic or metal particles.
- **Advantages**: Good for producing complex geometries and functional parts.
- **Challenges**: Controlling particle size distribution and achieving consistent density and mechanical properties.

d. **Direct Ink Writing (DIW)**

- **Description**: A syringe-based extrusion method that prints viscoelastic inks containing polymer composites.
- **Reinforcement**: Allows for high loadings of short fibers, nanoparticles, or continuous fibers.
- **Advantages**: Suitable for multi-material printing and complex architectures.
- **Challenges**: Requires precise control of ink rheology and curing processes.

e. **Stereolithography with Composite Slurries**

- **Description**: Involves the use of a slurry containing a photopolymer resin mixed with reinforcement particles.
- **Reinforcement**: Typically includes ceramic or glass fibers/particles.
- **Advantages**: High precision and ability to create detailed features.
- **Challenges**: Settling of particles and ensuring homogenous distribution throughout the slurry.

f. **Laminated Object Manufacturing (LOM)**

- **Description**: Involves the layer-by-layer lamination of sheets of composite material, which are bonded and cut to shape.
- **Reinforcement**: Incorporates fibers or fabrics into polymer sheets.
- **Advantages**: Good for large parts and allows for incorporation of pre-fabricated composite layers.

- **Challenges**: Limited to specific geometries and requires additional post-processing for complex shapes.

g. **Hybrid Manufacturing**

- **Description**: Combines traditional manufacturing techniques (e.g., injection molding) with 3D printing to create composite parts.
- **Reinforcement**: Utilizes various forms of reinforcement depending on the hybrid approach.
- **Advantages**: Allows for complex parts with enhanced properties and reduced manufacturing time.
- **Challenges**: Requires careful integration of processes to ensure compatibility and performance.

By leveraging these techniques, manufacturers can produce polymer composites with enhanced mechanical, thermal, and functional properties, expanding the potential applications of 3D printing in various industries [63–74].

7 Challenges of 4D-Printed Polymer Composites

4D printing of polymer composites involves creating materials that can change shape or properties over time in response to external stimuli, such as temperature, moisture, or light. While this technology holds great promise for applications in fields like aerospace, biomedical devices, and smart textiles, it also presents several challenges. Here are some of the key challenges associated with 4D printing polymer composites:

a. **Material Selection and Compatibility**

- **Complex Material Requirements**: Developing materials that respond predictably to stimuli is challenging. The materials must exhibit significant and reversible changes in shape or properties without degrading over time.
- **Compatibility**: Ensuring compatibility between the polymer matrix and the responsive materials (e.g., shape-memory alloys, hydrogels) is crucial for effective functionality.

b. **Process Complexity**

- **Printing Precision**: Achieving the precise control needed to print materials that can later change shape requires advanced printing techniques and precise calibration.
- **Multi-material Integration**: Integrating multiple materials with different properties and responses into a single printed object is complex and often requires specialized printing equipment and processes.

c. **Stimulus Control**

- **Environmental Sensitivity**: The responsiveness of 4D-printed materials depends on specific environmental conditions, which can be difficult to control in real-world applications.
- **Stimuli Limitations**: Finding appropriate stimuli that can reliably trigger changes without causing material degradation or failure is challenging.

d. **Mechanical Properties**

- **Structural Integrity**: Ensuring that 4D-printed objects maintain structural integrity throughout shape changes is crucial. The transformation should not compromise the strength or durability of the material.
- **Anisotropy**: As with 3D-printed composites, anisotropy can be an issue, where mechanical properties vary based on the direction of printed layers.

e. **Design and Simulation**

- **Complex Design Requirements**: Designing objects that can undergo specific transformations involves complex modeling and simulation, often requiring advanced computational tools.
- **Predictive Modeling**: Accurately predicting the behavior of 4D-printed composites under various conditions is challenging and requires sophisticated simulation models.

f. **Cost and Scalability**

- **High Costs**: The materials and equipment needed for 4D printing are often expensive, which can limit widespread adoption.
- **Scalability**: Scaling up production for commercial applications while maintaining quality and performance is a significant challenge.

g. **Long-Term Stability and Durability**

- **Material Fatigue**: Repeated transformations can lead to material fatigue and eventual failure. Ensuring long-term stability and durability is critical for practical applications.
- **Degradation**: Exposure to environmental factors such as UV light, moisture, and temperature variations can lead to degradation over time.

h. **Post-Processing and Assembly**

- **Additional Steps**: Some 4D-printed composites require post-processing, such as curing or thermal treatment, to activate their responsive properties, adding complexity to the production process.
- **Assembly Challenges**: Incorporating 4D-printed components into larger systems can be challenging due to the dynamic nature of these materials.

i. **Limited Range of Stimuli**

- **Stimulus Range**: Currently, the range of stimuli that can be effectively used to trigger transformations in 4D-printed materials is limited, which can restrict potential applications.

j. **Environmental Impact**

- **Sustainability Concerns**: The development of new materials for 4D printing must consider environmental impact and sustainability, particularly regarding the disposal and recyclability of these materials.

Addressing these challenges requires ongoing research and development in material science, printing technologies, and computational modelling. As the technology advances, solutions to these challenges will likely emerge, enabling broader applications and more robust implementations of 4D-printed polymer composites [75–86].

8 Importance of 5D Printing

5D printing is an advanced form of 3D printing that adds complexity and functionality to the printed objects by utilizing additional axes of movement in the printing process. While traditional 3D printing typically involves layer-by-layer construction using three linear axes (X, Y, and Z), 5D printing incorporates two additional rotational axes, allowing for more intricate designs and improved material properties. The detail of the 5 axes of the 5D printing technique is given in Fig. 3. The two newly added axes are called movable print bed and movable print head along with the X-, Y-, and Z- axes.

Here are some key points highlighting the importance of 5D printing:

a. **Enhanced Structural Integrity**

- **Continuous Printing**: Unlike traditional 3D printing, which relies on layer-by-layer deposition, 5D printing can produce structures with continuous paths, enhancing mechanical strength and reducing weak points.
- **Reduced Anisotropy**: By printing along curved surfaces and optimizing the orientation of fibers or layers, 5D printing can reduce anisotropic properties, resulting in parts that are stronger and more uniform in all directions.

b. **Complex Geometry and Design Flexibility**

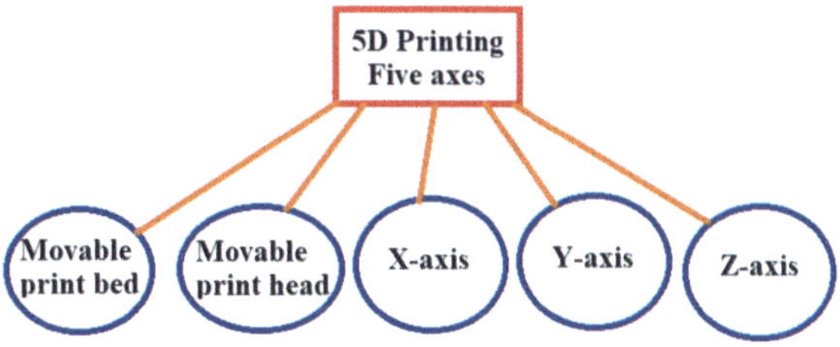

Fig. 3 Shows 5D printing process

- **Intricate Designs**: 5D printing enables the creation of complex geometries that are difficult or impossible to achieve with traditional 3D printing, allowing for more innovative designs and applications.
- **Design Optimization**: The additional axes allow for more freedom in design optimization, enabling the creation of parts that are lighter, stronger, and more efficient.

c. **Material Efficiency**

- **Reduced Material Waste**: By optimizing the printing path and minimizing the need for support structures, 5D printing can reduce material waste, making the process more sustainable and cost-effective.
- **Optimized Material Usage**: The ability to control material orientation and deposition more precisely leads to efficient use of materials, potentially reducing costs and environmental impact.

d. **Improved Surface Finish**

- **Smoother Surfaces**: The use of additional axes allows for smoother and more continuous surface finishes, reducing the need for post-processing and improving the aesthetic quality of printed parts.

e. **Multifunctional Parts**

- **Integrated Functions**: 5D printing can incorporate multiple functions into a single part, such as varying stiffness, tailored thermal conductivity, or embedded electronics, leading to more versatile and functional components.

f. **Reduced Production Time**

- **Efficient Manufacturing**: By optimizing the printing process and reducing the need for support and post-processing, 5D printing can decrease production times, accelerating the manufacturing cycle.

g. **Applications in High-Performance Industries**

- **Aerospace and Automotive**: The ability to produce lightweight, strong, and complex components is particularly valuable in aerospace and automotive industries, where performance and efficiency are critical.
- **Medical Devices**: In the medical field, 5D printing can enable the production of customized implants and devices with enhanced mechanical properties and biocompatibility.

h. **Innovation and Research**

- **Advanced Research Opportunities**: The capabilities of 5D printing open up new avenues for research in materials science, engineering, and design, driving innovation and technological advancement.

i. **Sustainable Manufacturing**

- **Eco-Friendly Production**: By minimizing waste and optimizing material usage, 5D printing contributes to more sustainable manufacturing practices, aligning with global efforts to reduce environmental impact.

j. **Competitive Advantage**

- **Market Differentiation**: Companies that adopt 5D printing technology can gain a competitive edge by offering superior products with enhanced properties and functionality.

Overall, 5D printing represents a significant advancement in additive manufacturing, offering numerous benefits in terms of design flexibility, material efficiency, and product performance. As the technology continues to develop, it is expected to play a transformative role in various industries, enabling the creation of innovative solutions and high-performance products [87–90].

9 Processing Techniques of 5D-Printed Polymer Composites

5D printing is an advanced extension of 3D printing technology that incorporates additional degrees of movement to enhance the manufacturing process. This approach allows for more complex geometries and improved mechanical properties by utilizing not only the three linear axes (X, Y, Z) but also two additional rotational axes. When applied to polymer composites, 5D printing can offer significant advantages in terms of structural integrity, material efficiency, and design flexibility. Here are some key processing techniques for 5D-printed polymer composites:

a. **Multi-Axial Printing**

- **Description**: 5D printing employs multiple axes of movement to orient the print head or build platform, allowing for dynamic positioning during the printing process. This enables the printer to build structures that follow curved surfaces and optimized paths.
- **Advantages**: This technique can significantly enhance the strength and durability of the printed parts by aligning fibers or material layers along the directions of maximum stress.
- **Applications**: Ideal for creating complex geometries with optimized load-bearing capabilities, such as aerospace components and automotive parts.

b. **Continuous Fiber Reinforcement**

- **Description**: In 5D printing, continuous fibers can be embedded in the polymer matrix, with the printing paths designed to follow complex 3D curves and surfaces.

- **Advantages**: This results in significantly improved mechanical properties, such as increased tensile strength and impact resistance, by ensuring fibers are aligned with the expected load directions.
- **Applications**: Useful in industries requiring high-strength, lightweight materials, such as sports equipment and structural components.

c. **Dynamic Layering**

- **Description**: Unlike traditional layer-by-layer deposition, dynamic layering adjusts the layer thickness and orientation during the printing process based on the part geometry and stress requirements.
- **Advantages**: Reduces anisotropy in printed parts and improves interlayer adhesion, resulting in more homogeneous material properties.
- **Applications**: Beneficial for parts that require consistent strength and performance across different orientations.

d. **Adaptive Support Structures**

- **Description**: 5D printing allows for the creation of adaptive support structures that can be dynamically generated and adjusted based on the part geometry and printing angle.
- **Advantages**: Minimizes material waste and post-processing by optimizing support structures only where necessary.
- **Applications**: Ideal for complex, intricate designs where traditional support structures would be cumbersome or difficult to remove.

e. **Rotational Deposition**

- **Description**: The use of rotational axes allows for the deposition of material onto non-planar surfaces, enabling more efficient material distribution and the ability to print directly onto existing objects.
- **Advantages**: Facilitates the creation of more complex shapes and the integration of additional features into existing components.
- **Applications**: Suitable for hybrid manufacturing approaches, where existing parts are enhanced with additional features or functions.

f. **Enhanced Composite Formulations**

- **Description**: 5D printing leverages advanced composite formulations, incorporating nanoparticles, microfibers, or other additives that respond to the dynamic printing process.
- **Advantages**: Allows for the development of materials with tailored properties, such as increased toughness, thermal resistance, or electrical conductivity.
- **Applications**: Beneficial for creating multifunctional materials for use in electronics, aerospace, and medical devices.

g. **Synchronized Process Control**

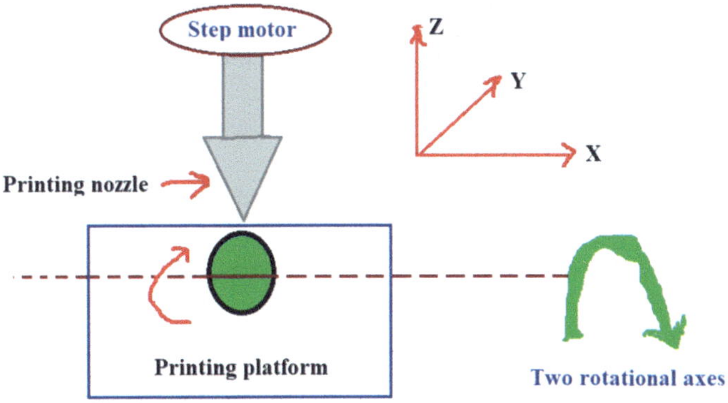

Fig. 4 Detailed diagram of the 5D printing technique

- **Description**: Precise control over printing parameters, such as speed, temperature, and pressure, is synchronized with the additional axes to ensure optimal deposition and bonding.
- **Advantages**: Enhances the precision and quality of the printed parts, reducing defects and improving performance.
- **Applications**: Critical for high-performance components requiring tight tolerances and consistent quality.

In conclusion, 5D printing of polymer composites presents numerous opportunities to create advanced materials with superior properties and functionalities. By leveraging the additional degrees of freedom, manufacturers can optimize the structural and mechanical characteristics of parts, paving the way for innovative applications across various industries [91–94]. The detail of the 5D printing technique is given in Fig. 4.

10 Challenges of 5D-Printed Polymer Composites

5D printing, which introduces additional axes of movement during the printing process, offers exciting opportunities to create polymer composites with enhanced properties and complex geometries. However, this technology also presents several challenges that need to be addressed to realize its full potential. Here are some of the key challenges associated with 5D-printed polymer composites:

a. **Complexity of Printing Systems**

- **Advanced Equipment Requirements**: The incorporation of additional rotational axes requires more complex and sophisticated printing systems. This complexity can lead to increased costs and maintenance requirements.

- **Precision and Calibration**: Ensuring precise control and calibration of the additional axes is crucial for producing high-quality prints, which can be technically challenging and time-consuming.

b. **Material Development**

- **Material Compatibility**: Developing materials that are compatible with the 5D printing process and can benefit from the additional degrees of freedom is a significant challenge. This includes ensuring uniform dispersion of reinforcement materials like fibers or nanoparticles.
- **Dynamic Material Properties**: Creating polymer composites that can withstand the stresses and strains introduced by the additional axes of movement requires advanced material formulations.

c. **Design and Simulation**

- **Complex Design Tools**: Designing for 5D printing involves creating parts that take full advantage of the additional axes. This requires sophisticated design tools and software capable of simulating complex movements and predicting material behavior.
- **Predictive Modeling**: Accurate modeling of the interactions between materials and the printing process is essential to optimize designs and ensure the final product meets performance requirements.

d. **Process Optimization**

- **Control of Layer Deposition**: Managing the deposition of layers with additional axes is more complex and requires optimized printing strategies to ensure uniformity and integrity.
- **Temperature and Speed Management**: Adjusting printing parameters, such as temperature and speed, to accommodate the additional complexity of 5D printing is crucial for achieving consistent quality.

e. **Mechanical Properties**

- **Anisotropy Management**: While 5D printing can reduce anisotropy, achieving uniform mechanical properties across different orientations still requires careful control of the printing process and material selection.
- **Structural Integrity**: Ensuring that printed parts maintain structural integrity, especially at points where multiple materials or directions intersect, is challenging.

f. **Cost and Scalability**

- **High Costs**: The sophisticated machinery and materials required for 5D printing can lead to higher production costs, potentially limiting its commercial viability.
- **Scalability Challenges**: Scaling up 5D printing for mass production while maintaining quality and consistency is a significant challenge that requires further development.

g. **Post-Processing**

- **Complex Post-Processing**: The intricate geometries possible with 5D printing may require additional post-processing steps, such as surface finishing or curing, to achieve the desired final properties.
- **Assembly and Integration**: Integrating 5D-printed components into larger systems can be complicated due to their complex shapes and unique properties.

h. **Quality Assurance**

- **Inspection and Testing**: Ensuring the quality and performance of 5D-printed composites requires advanced inspection and testing techniques, which can be costly and time-consuming.
- **Standardization**: Developing industry standards for 5D printing processes and materials is necessary to ensure reliability and consistency.

i. **Environmental Impact**

- **Sustainability Concerns**: As with other advanced manufacturing technologies, the environmental impact of 5D printing, including material waste and energy consumption, must be carefully managed.

j. **Limited Material Options**

- **Material Availability:** The range of materials that can be used effectively in 5D printing is currently limited, necessitating ongoing research to expand material options.

Addressing these challenges requires continued research and innovation in materials science, design methodologies, and printing technologies. As these challenges are overcome, 5D printing has the potential to transform the manufacturing of polymer composites, offering new possibilities for performance and design across various industries [95–98].

11 Conclusion

The exploration of 3D, 4D, and 5D printing technologies for polymer composites reveals a dynamic and transformative landscape in manufacturing and materials science. Each of these technologies offers unique advantages and faces distinct challenges that must be addressed to unlock their full potential.

- 3D printing has already revolutionized how we design and manufacture products, offering unprecedented design freedom, customization, and material efficiency. Despite its widespread adoption, challenges such as anisotropy, surface finish, and material limitations continue to drive research and innovation. Ongoing advancements in materials and printing techniques promise to expand the applicability and improve the performance of 3D-printed composites.

- 4D printing extends the capabilities of 3D printing by introducing time-dependent transformations, enabling the creation of smart materials and structures that can adapt to their environments. While this technology opens new possibilities in fields such as biomedical devices, aerospace, and soft robotics, challenges remain in developing responsive materials, ensuring structural integrity, and controlling stimuli responsiveness. As research progresses, 4D printing is poised to enable new functionalities and applications that were previously unimaginable.
- 5D printing represents a further leap forward, incorporating additional axes of movement to enhance mechanical properties and design complexity. By optimizing fiber orientation and reducing anisotropy, 5D printing can produce stronger and more efficient composite structures. However, the complexity of the printing systems and the need for advanced design and simulation tools present significant hurdles. Continued research and development are crucial to overcoming these challenges and realizing the full benefits of 5D printing in high-performance industries.

Across all these technologies, several overarching challenges persist, including material development, process optimization, cost, and scalability. Addressing these challenges will require a multidisciplinary approach, combining advances in materials science, engineering, and computer-aided design. Looking forward, the integration of artificial intelligence, machine learning, and data-driven design methodologies holds the promise of accelerating innovation in the field of additive manufacturing. By harnessing these technologies, researchers and engineers can develop smarter, more efficient processing techniques and unlock new applications for polymer composites.

In conclusion, 3D, 4D, and 5D printing of polymer composites are at the forefront of a manufacturing revolution, offering transformative capabilities across various industries. While challenges remain, the potential benefits are immense, promising to redefine how we design, produce, and utilize materials in the future. Continued research, collaboration, and innovation will be essential to overcoming these challenges and fully realizing the promise of these groundbreaking technologies.

Acknowledgements The author would like to thank the host institute for providing a digital library facility for the collection of the required data and information.

Conflict of Interest Statement: The author declares that he has no known financial or personal conflicts of interest that could have appeared to influence the work reported in this chapter.

References

1. Wei Gao, Yunbo Zhang, Devarajan Ramanujan, Karthik Ramani, Yong Chen, Christopher B. Williams, Charlie C.L. Wang, Yung C. Shin, Song Zhang, Pablo D. Zavattieri.: The status, challenges, and future of additive manufacturing in engineering. Comput.-Aided Des., **69**, 65–89 (2015)

2. Tuan D. Ngo, Alireza Kashani, Gabriele Imbalzano, Kate T.Q. Nguyen, David Hui, Additive manufacturing (3D printing): A review of materials, methods, applications and challenges. Compos. Part B: Eng.. **143**, 172–196 (2018)
3. Ariadne.: New Sci. **64**(917): 80. (1974). ISSN 0262–4079
4. Ellam, Richard.: 3D printing: you read it here first. New Sci. Retrieved 23 August 2019 (2019)
5. Kodama, H.: Background of my invention of 3D printer and its spread. Pat. Mag. Jpn. Pat. Atty. Assoc. **67**, 109–118 (2014)
6. Hideo Kodama.: A scheme for three-dimensional display by automatic fabrication of three-dimensional model. IEICE Trans. Electron. (Jpn. Ed.), J64-C, 237–241 (1981)
7. Kodama, H.: Automatic method for fabricating a three-dimensional plastic model with photo-hardening polymer. Rev. Sci. Instrum. **52**, 1770–1773 (1981)
8. Amon, C.H., Beuth, J.L., Weiss, L.E., Merz, R., Prinz, F.B.: Shape deposition manufacturing with micro casting: processing, thermal and mechanical issues. J. Manuf. Sci. Eng. **120**, 656–665 (1998)
9. Beck, J.E., Fritz, B., Daniel Siewiorek, Lee Weiss. Manufacturing mechatronics using thermal spray shape deposition. In: Proceedings of the 1992 Solid Freeform Fabrication Symposium, (272–279) (1992)
10. Peng Wu, Jun Wang, Xiangyu Wang.: A critical review of the use of 3-D printing in the construction industry. Autom. Constr., **68**, 21–31 (2016)
11. Weller, C., Kleer, R., Piller, F.T.: Economic implications of 3D printing: Market structure models in light of additive manufacturing revisited. Int. J. Prod. Econ. **164**, 43–56 (2015)
12. Ben-Ner, A., Siemsen, E.: Decentralization and localization of production: the organizational and economic consequences of additive manufacturing (3D Printing). Calif. Manage. Rev. **59**, 5–23 (2017)
13. Zhaolong Li, Qinghai Wang, Guangdong Liu.: A review of 3D printed bone implants, Micromachines, **13**, 1–25 (2022)
14. Sivasankaran, P., Radjaram, B.: 3D printing and its importance in engineering—a review. In: 2020 International Conference on System, Computation, Automation and Networking (ICSCAN), Pondicherry, India, 1–3 (2020)
15. Zhi Zhang, Lei Zhang, Bo Song, Yonggang Yao, Yusheng Shi, Bamboo-inspired, simulation-guided design and 3D printing of light-weight and high-strength mechanical metamaterials. Appl. Mater. Today, **26**, 101268 (2022)
16. Westerweel, B., Basten, R.: Jelmar denBoer, Geert-Jan vanHoutum printing spare parts at remote locations: fulfilling the promise of additive manufacturing. Prod. Oper. Manag. **30**, 1615–1632 (2021)
17. Manero, A., Smith, P., Sparkman, J., Dombrowski, M., Courbin, D., Kester, A., Womack, I., Chi, A.: Implementation of 3D printing technology in the field of prosthetics: past, present, and future. Int. J. Environ. Res. Public Health **16**, 1–15 (2019)
18. Caprioli, M., Roppolo, I., Chiappone, A., Larush, L.: Candido Fabrizio Pirri, Shlomo Magdassi, 3D-printed self-healing hydrogels via Digital Light Processing. Nat. Commun. **12**, 1–9 (2021)
19. Nachal, N., Moses, J.A., Karthik, P., Anandharamakrishnan, C.: Applications of 3D printing in food processing. Food Eng. Rev. **11**, 123–141 (2019)
20. Zastrow, M.: 3D printing gets bigger, faster and stronger. Nature **578**, 20–23 (2020)
21. Schubert, C., van Langeveld, M.C., Donoso, L.A.: Innovations in 3D printing: a 3D overview from optics to organs. Br. J. Ophthalmol. **98**, 159–161 (2013)
22. Al Ahbabi, K.J.A., Alrashdi, M.M.S., Ahmed, W.K.: The capabilities of 3D printing technology in the production of battery energy storage system. In: 2021 6th International Conference on Renewable Energy: Generation and Applications (ICREGA), Al Ain, United Arab Emirates, 211–216 (2021)
23. Tibbits, S.: 4D printing: multi-material shape change. Archit. Des. **84**, 116–121 (2014)
24. Jin Choi, O-Chang Kwon, Wonjin Jo, Heon Ju Lee, Myoung-Woon Moon.: 4D printing technology: a review, 3D printing and additive manufacturing, **2**, 159–167 (2016)
25. Navaf, M.: Kappat Valiyapeediyekkal Sunooj, Basheer Aaliya, Plachikkattu Parambil Akhila, Cherakkathodi Sudheesh, Shabir Ahmad Mir, Johnsy George, 4D printing: a new approach for

food printing; effect of various stimuli on 4D printed food properties. A Compr. Rev., Appl. Food Res. **2**, 1–12 (2022)
26. Qi Ge, Conner K. Dunn, Jerry Qi, H., Martin L. Dunn.: Active origami by 4D printing, Smart Materials and Structures, **23**, 094007 (2014)
27. Pietro Cataldi, Mufeng Liu, Mark Bissett, Ian A. Kinloch.: A review on printing of responsive smart and 4D structures using 2D materials, advanced materials technologies, **7** (2022)
28. Eujin Pei, Giselle Hsiang Loh, Technological considerations for 4D printing: an overview progress in additive manufacturing, **3**, 95–107 (2018)
29. Qi Ge, Amir Hosein Sakhaei, Howon Lee, Conner K. Dunn, Nicholas X. Fang, Martin L. Dunn.: Multimaterial 4D printing with tailorable shape memory polymers. Sci. Rep., **6**, 1–11 (2016)
30. Yi-Chen Li, Yu Shrike Zhang, Ali Akpek, Su Ryon Shin, Ali Khademhosseini.: 4D bioprinting: the next-generation technology for biofabrication enabled by stimuli-responsive materials, Biofabrication, **9** (2016)
31. Sydney Gladman, A., Elisabetta A. Matsumoto, Ralph G. Nuzzo, L. Mahadevan, Jennifer A. Lewis.: Biomimetic 4D printing. Nat. Mater., **15**, 413–418 (2016)
32. Haleem, A., Javaid, M.: Expected applications of five-dimensional (5D) printing in the medical field. Curr. Med. Res. Pract. **9**, 208–209 (2019)
33. Kan, W.H., Chiu, L.N.S., Lim, C.V.S., Zhu, Y., Tian, Y., Jiang, D., Huang, A.: A critical review on the effects of process-induced porosity on the mechanical properties of alloys fabricated by laser powder bed fusion. J. Mater. Sci. **57**(21), 9818–9865 (2022)
34. Krishnan, M., Atzeni, E., Canali, R., Calignano, F., Manfredi, D., Ambrosio, E.P., Iuliano, L.: On the effect of process parameters on properties of AlSi10Mg parts produced by DMLS. Rapid Prototyp. J. **20**(6), 449–458 (2014)
35. Oliveira, J.P., LaLonde, A.D., Ma, J.: Processing parameters in laser powder bed fusion metal additive manufacturing. Mater. Des. **193**, 108762 (2020)
36. Scipioni Bertoli, U., Wolfer, A.J., Matthews, M.J., Delplanque, J.P.R., Schoenung, J.M.: On the limitations of volumetric energy density as a design parameter for selective laser melting. Mater. Des. **113**, 331–340 (2017)
37. Vafadar, A., Guzzomi, F., Rassau, A., Hayward, K.: Advances in metal additive manufacturing: A review of common processes, industrial applications, and current challenges. Appl. Sci. (Sci.) **11**(3), 1213 (2021)
38. Trenfield, S., Madla, C., Basit, A., Gaisford, S.: Binder jet printing in pharmaceutical manufacturing, 3D Print. Pharmaceut. 41–54 (2018)
39. Williams, J., Adewunmi, A., Schek, R., Flanagan, C., Krebsbach, P., Feinberg, S., Hol-lister, S., Das, S.: Bone tissue engineering using polycaprolactone scaffolds fabricated via selective laser sintering. Biomater. **26**(23) 4817–4827 (2005)
40. Bennett, J., Garcia, D., Kendrick, M., Hartman, T., Hyatt, G., Ehmann, K., You, F., Cao, J.: Repairing automotive dies with directed energy deposition: industrial application and life cycle analysis. J. Manuf. Sci. Eng. **141**(2), (2018)
41. Liu, G., Zhang, X., Chen, X., He, Y., Cheng, L., Huo, M., Yin, J., Hao, F., Chen, S., Wang, P., Yi, S., Wan, L., Mao, Z., Chen, Z., Wang, X., Cao, Z., Lu, J.: Additive manufacturing of structural materials, Mater. Sci. Eng. 100596 (2021)
42. Yap, C., Chua, C., Dong, Z., Liu, Z., Zhang, D., Loh, L., Sing, S.: Review of selective laser melting: materials and applications. Appl. Phys. Rev. **2**(4), 041101 (2015)
43. Bock, T.: The future of construction automation: technological disruption and the upcoming ubiquity of robotics. Autom. Constr. **59**, 113–21 (2015)
44. Noorani, R.: 3D printing: technology, applications, and selection. CRC Press, Boca Raton (2018)
45. Kietzmann, J., Pitt, L., Berthon, P.: Disruptions, decisions, and destinations: enter the age of 3D printing and additive manufacturing. Bus. Horiz. **58**(2), 209–15 (2015)
46. Yossef, M., Chen, A.: Applicability and limitations of 3D printing for civil structures. Civ Constr Environ Eng Conf Present Proc. **35**, 87–99 (2015)

47. Jung, K.-Y., Roh, M.-S.: A study for an appropriate risk management of new technology deployment in nuclear power plants. Ann. Nucl. Energy **99**, 157–64 (2016)
48. Uppalla, S., Tadikamalla, M.: A review on 3D printing of concrete—the future of sustainable construction. I-Manager's J. Civ. Eng. **7**(3), 49–62 (2017)
49. Xu, J., Ding, L., Love, P.: Digital reproduction of historical building ornamental components: from 3D scanning to 3D printing. Autom. Constr. **76**, 85–96 (2017)
50. Buswell, R.A., Leal de Silva, W.R., Jones, S.Z.: Dirrenberger J. 3D printing using concrete extrusion: a roadmap for research. Cem. Concr. Res. 112:37–49 (2018)
51. Hambach, M., Volkmer, D.: Properties of 3D-printed fiber-reinforced portland cement paste. Cem. Concr. Compos. **79**, 62–70 (2017)
52. Perkins, I., Skitmore, M.: Three-dimensional printing in the construction industry: a review. Int. J. Constr. Manag. **15**(1), 1–9 (2015)
53. Shakor, P., Nejadi, S., Paul, G.: A study into the effect of different nozzles shapes and fibre-reinforcement in 3D printed mortar. Materials. **12**, 1708 (2019)
54. Panda, B., Paul, S.C., Tan, M.J.: Anisotropic mechanical performance of 3D printed fiber reinforced sustainable construction material. Mater. Lett. **209**, 146–9 (2017)
55. Li, Y.C., Zhang, Y.S., Akpek, A., Shin, S.R., Khademhosseini, A.: Biofabrication **9**, 012001 (2016)
56. Castro, N.J., Meinert, C., Levett, P., Hutmacher, D.W.: Curr. Opin. Biomed. Eng. **2**, 67 (2017)
57. Gao, B., Yang, Q., Zhao, X., Jin, G., Ma, Y., Xu, F.: Trends Biotechnol. **34**, 746 (2016)
58. Zopf, D.A., Mitsak, A.G., Flanagan, C.L., Wheeler, M., Green, G.E., Hollister, S.: Otolaryngol. Head Neck Surg. **152**, 57 (2015)
59. Yi, H.G., Lee, H., Cho, D.W.: Bioeng. **1**, 10 (2017)
60. Awasthi, S., Verma, T., Agarwal, M., Singh, J.V., Srivastava, N.M., Nichterd, M.: Clin. Epidemiol. Glob. Health **3**, 107 (2017)
61. Cohen, Y.B., Cardoso, V.F., Ribeiro, C., Lanceros-M´endez, S.: Advanced piezoelectric materials, Woodhead Publishing, (2017)
62. Monzon, M.D., Paz, R., Pei, E., Ortega, F., Suarez, L.A., Ortega, Z., Aleman, M.E., Plucinski, T., Clow, N.: Int. J. Adv. Manuf. Technol. **89**, 1827 (2017)
63. Kim, S., Laschi, C., Trimmer, B.: Soft robotics: a bioinspired evolution in robotics. Trends Biotechnol. **31**, 287–294 (2013)
64. Wang, X., Guo, R., Liu, J.: Liquid metal based soft robotics: materials, designs, and applications. Adv. Mater. Technol **4**, 1800549 (2019)
65. Lee, S.M., Jung, W.K., Park, J., et al.: Development of a 4D hand gripping aid using a knitted shape memory alloy and evaluation of finger-bending angles in elderly women. Fashion Text **9**, 11 (2022)
66. Langford, T., Mohammed, A., Essa, K., et al.: 4D printing of origami structures for minimally invasive surgeries using functional scaffold. Appl. Sci. **11**, 332 (2020)
67. Ahmed, A., Arya, S., Gupta, V., et al.: 4D printing: Fundamentals, materials, applications and challenges. Polymer **228**, 123926 (2021)
68. Koch, H.C., Schmelzeisen, D., Gries, T.: 4D textiles made by additive manufacturing on Pre-Stressed Textiles—An overview. Actuators **10**, 31 (2021)
69. Gladman, A.S., Matsumoto, E.A., Nuzzo, R.G., et al.: Biomimetic 4D printing. Nat. Mater. **15**, 413–418 (2016)
70. Tang, G.P., Liu, Q.P., Song, B.: Study on 4D printing of liquid crystal elastomers process and properties. J. Mech. Eng **57**, 234–242 (2021)
71. Naficy, S., Gately, R., Gorkin, R., et al.: 4D printing of reversible shape morphing hydrogel structures. Macromol. Mater. Eng. **302**, 1600212 (2017)
72. Guo, J., Zhang, R., Zhang, L., et al.: 4D printing of robust hydrogels consisted of agarose nanofibers and polyacrylamide. ACS Macro Lett. **7**, 442–446 (2018)
73. Joshi, S., Rawat, K., Karunakaran, C., et al.: 4D printing of materials for the future: Opportunities and challenges. Appl. Mater. Today **18**, 100490 (2020)
74. Mitchell, A., Lafont, U., Hołyńska, M., et al.: Additive manufacturing — A review of 4D printing and future applications. Addit. Manuf. **24**, 606–626 (2018)

75. Bodaghi, M., Damanpack, A.R., Liao, W.H.: Self-expanding/ shrinking structures by 4D printing. Smart Mater. Struct. **25**, 105034 (2016)
76. Invernizzi, M., Turri, S., Levi, M., et al.: 4D printed thermally activated self-healing and shape memory polycaprolactonebased polymers. Eur. Polym. J. **101**, 169–176 (2018)
77. Choong, Y.Y.C., Maleksaeedi, S., Eng, H., et al.: 4D printing of high performance shape memory polymer using stereolithography. Mater. Des. **126**, 219–225 (2017)
78. Choong, Y.Y.C., Maleksaeedi, S., Eng, H., et al.: High speed 4D printing of shape memory polymers with nanosilica. Appl. Mater. Today **18**, 100515 (2020)
79. Zhao, T., Yu, R., Li, X., et al.: 4D printing of shape memory polyurethane via stereolithography. Eur. Polym. J. **101**, 120–126 (2018)
80. Ren, L., Li, B., He, Y., et al.: Programming Shape-morphing behavior of liquid crystal elastomers via parameter-encoded 4D printing. ACS Appl. Mater. Interfaces **12**, 15562–15572 (2020)
81. Su, J.W., Gao, W., Trinh, K., et al.: 4D printing of polyurethane paint-based composites. Int. J. Smart Nano Mater. **10**, 237–248 (2019)
82. Monzón, M.D., Paz, R., Pei, E., et al.: 4D printing: processability and measurement of recovery force in shape memory polymers. Int. J. Adv. Manuf. Technol. **89**, 1827–1836 (2017)
83. Chu, H., Yang, W., Sun, L., et al.: 4D printing: a review on recent progresses. Micromachines **11**, 796 (2020)
84. Piedade, A.P.: 4D printing: the shape-morphing in additive manufacturing. J. Funct. Biomater **10**, 1–7 (2019)
85. Haleen, A., Javaid, M., Vaishya, R.: 4D printing and its applications in orthopaedics. J. Clin. Orthop. Trauma **9**, 275–276 (2018)
86. Biswas, M.C., Chakraborty, S., Bhattacharjee, A., et al.: 4D printing of shape memory materials for textiles: mechanism, mathematical modeling, and challenges. Adv. Funct. Mater. **31**, 2100257 (2021)
87. Ravinder Reddy, P., Anjani Devi, P.: (2018) Review on the advancements to additive manufacturing-4d and 5d printing. Int J Mech Prod Eng Res Dev (IJMPERD) 8(4):397–402. Top of Form
88. Foresti, R., Rossi, S., Pinelli, S., et al.: In-vivo vascular application via ultra-fast bioprinting for future 5D personalised nanomedicine. Sci. Rep. **10**, 3205 (2020)
89. Sadiq Sha, H.A.J., Patil, P.P.: Review on 4D and 5D printing technology. Int. Res. J. Eng. Technol. (IRJET) **7**(6), 2395–0056 (2020)
90. Sadiq Sha, H.A.J., Patil, P.P.: Review on 4D and 5D printing technology, Int. Res. J. Engg. Technol. **7**(6) 744–751 (2020)
91. Reddy, P.R., Devi, P.A.: Review on the advancements of additive manufacturing-4D and 5D printing. Int. J. Mech. Prod. Eng. Res. Dev. **8**(4), 397–402 (2018)
92. He, C., Zhang, M., Devahastin, S.: Investigation on spontaneous shape change of 4D printed starch-based purees from purple sweet potatoes as induced by microwave dehydration. ACS Appl. Mater. Interfaces **34**(12), 37896–37905 (2020)
93. Haleem, A., Javaid, M., Vaishya, R.: 5D printing and its expected applications in Orthopaedics, J. Clin. Orthopaedics Trauma **10**(4) 809–810 (2019)
94. Chen, Y., Zhang, M., Sun, Y., Phuhongsung, P.: Improving 3D/4D printing characteristics of natural food gels by novel additives: a review. Food Hydrocoll. **123**, 107160 (2022)
95. Sathe, G.: Meet ethereal machine's Halo, a "5D Printer" That's Just Won the best of innovation CES 2018 Award, (2017)
96. Liu, Z., He, C., Guo, C., Chen, F., Bhandari, B., Zhang, M.: Dehydration-triggered shape transformation of 4D printed edible gel structure affected by material property and heating mechanism. Food Hydrocolloids **115**, 106608 (2021)
97. Aruna, R., Mohamed Abbas, S., Vivek, S., Suresh, G., Meenakshi, C.M., Srinivasan, T., Selva Ganapathy, K.: Evolution of 5D printing and its vast applications: a review. Recent Advances in Materials and Modern Manufacturing: Select Proceedings of ICAMMM **2021**, 691–714 (2022)
98. Anas, S., Khan, M.Y., Rafey, M., Faheem, K.: Concept of 5D printing technology and its applicability in the healthcare industry. Materials Today: Proceedings **56**, 1726–1732 (2022)

Additive Manufacturing Technologies for Polymer Nanocomposites

Ashish Kumar Sahoo and Barsha Choudhury

Abstract The advancement and application of Additive Manufacturing (AM) technologies, particularly 3D and 4D printing, signify a transformative period in manufacturing and product development. Additive manufacturing (3D printing) has become integral to the aerospace, automotive, and healthcare sectors, enabling complex designs, personalization, and sustainability. The capability for rapid prototyping, customization, and the production of complex geometries, which traditional manufacturing methods struggle to achieve efficiently, is the foremost and core benefit of AM technology. This chapter provides a comprehensive overview of the advancements in 3D, 4D, and 5D printing technologies, highlighting their transformative impact on various industries. In this regard, PolyJet printing, known for its precision and speed, produces high-detail objects using UV-hardened polymers. However, material costs remain a limitation. 4D printing introduces time as a dimension, allowing shape changes in response to environmental cues. Applications span biomedicine, aerospace, and dental restorations. 5D printing combines additive and subtractive methods, promising intricate designs for jewellery, automotive parts, and food packaging. Overall, these technologies revolutionize manufacturing processes.

Keywords Additive manufacturing · 3D · 4D and 5D printing · Polymer nanocomposite

1 Introduction

Additive manufacturing has existed for over thirty years and has drawn a significant amount of interest among researchers because of its ability to create complex designs seamlessly without requiring assembly. Continuous technological enhancements have added ingredients for its substantial growth for its versatility with diverse materials. The push for additive manufacturing comes from a need for personalized

A. K. Sahoo (✉) · B. Choudhury
School of Applied Sciences, Centurion University of Technology and Management, Bhubaneswar, Odisha, India
e-mail: ashish.sahoo@cutm.ac.in

S. Moharana et al. (eds.), *Polymer Nanocomposites for 3D, 4D and 5D Printing*, Engineering Materials, https://doi.org/10.1007/978-981-96-4214-4_5

products, swifter development timelines, a stronger commitment to environmental sustainability, and the pursuit of cost-effective production and shorter lead times. It also paves the way for innovative economic models [1]. Additive manufacturing encapsulates the key traits of groundbreaking innovations, as it's adaptable, effective, and resourceful [2]. As a result, traditional production techniques are increasingly making room for these advanced methods [3].

In this regard, 3D printing technology is quickly becoming popular across various industries, including aerospace, automotive, and healthcare. It's particularly gaining a grip on the production of medical equipment and personal health devices [4]. This field is an evolution of 3D printing, with the fourth dimension being time. It incorporates time-saving designs and materials that can respond automatically to changes in their surroundings. The x, y, and z axes represent the three dimensions of space in traditional printing, and the addition of time as the fourth dimension allows for a significant reduction in material usage—up to 90%. This innovative approach is becoming increasingly prevalent in modern industrial practices [5]. Subsequently, the emergence of 4D printing technology expands upon the concept of 3D printing by introducing the ability to create objects that can change shape or self-assemble over time. This process is driven by environmental factors such as temperature and humidity [6]. At the core of 4D printing are smart materials that react to these external stimuli, enabling the printed object to transform. On the other hand, 5D printing merges both additive and subtractive manufacturing processes to fabricate intricate designs. This method utilizes a dynamic print head that operates along the traditional X, Y, and Z axes, complemented by two additional rotational axes integrated into the print bed, enhancing the structure's complexity [6]. Pioneered by the team at Mitsubishi Electric Research Laboratories in 2016, this advanced form of printing was put to the test with the creation of pressure caps. The remarkable results revealed that caps crafted with a 5D printer exhibited superior mechanical strength (measured at 3.7 MPa) compared to those produced by conventional 3D printers (0.1 MPa). This groundbreaking discovery was further validated by Ethereal Machines in Bengaluru, India, with their development of the "Halo 5D printer [6]." This versatile machine has proven its utility in manufacturing intricate jewellery pieces and automotive components, achieving greater structural integrity while utilizing 25% less material than its 3D counterparts [6]. The promising applications of 5D printing extend into the culinary and packaging industries. For instance, the technology holds the potential to revolutionize the production of food items with elaborate geometries, such as spherical sweets or uniquely shaped crisps. Moreover, the packaging sector could witness a transformation with the ability to construct complex containers, cups, and bowls that would typically necessitate additional support when using 3D printing methods [6].

2 3D Printing Technology

3D printing enables the creation of physical objects from geometric designs through successive material addition [7]. Since its commercialization by Charles Hull in 1980, this technology has experienced remarkable growth, finding applications in diverse sectors such as artificial heart pumps [8], jewellery [9], corneas [10], rocket engines [11], and even a steel bridge in Amsterdam [12] and the aviation and food industries. Originating from the layer-by-layer fabrication of 3D structures directly from computer-aided design (CAD) drawings [12], 3D printing is widely recognized for its innovative potential and versatility, offering new possibilities for enhancing manufacturing efficiency. It supports printing various materials, including thermoplastics, ceramics, graphene-based materials, and metals [13], promising to revolutionize industries by increasing production speed and reducing costs while also empowering consumer influence over product customization. By locating manufacturing facilities closer to consumers, 3D printing facilitates a more flexible and responsive production process with enhanced quality control and reduced global transportation needs, utilizing fleet tracking technology for efficient distribution [14]. However, despite its advantages, the technology raises concerns about potential job displacement in labor-intensive economies and the unauthorized production of weapons and counterfeit goods due to the accessibility and simplicity of the printing process [15].

3 Applications of 3D Printing Technology

3D printing is rapidly transforming the medical field with its ability to tailor-make medical devices and implants. This technology is particularly beneficial for producing items like ear prostheses and jaw replacements that perfectly fit an individual's unique anatomy. All these are possible due to the integration of 3D imaging software. In the field(realm) of surgery, 3D printing is invaluable. For instance, in the delicate procedure of living-donor liver transplants, 3D-printed models of the organ are instrumental in reducing surgery duration and minimizing risks to the donor, thereby enhancing the overall success of the operation. The field of bioprinting is also advancing, providing more precise models for both non-biological and biological research. This is especially useful in cancer studies, where it helps simulate the complex spatial and temporal aspects of tumor development [1]. In addition, the scope of 3D printing extends to the fabrication of living tissues and organs, which holds the promise of future implantation into human patients. For over a decade and a half, 3D printers have been increasingly adopted for a wide array of biomedical uses, spanning from tissue engineering to medical research. A variety of 3D printing instruments are being developed for applications ranging from surgical preparation to the creation of anatomical models, such as liver mimics and temporary splints. Researchers are actively employing this technology to arrange cells and construct diverse tissues and organs, including vascular systems and heart patches.

Ongoing research in bioprinting is leading to significant progress. As one of the most rapidly growing sectors, 3D additive manufacturing is set to revolutionize practices throughout the biomedical field, encompassing everything from research models and surgical planning to the production of devices and the potential replacement of tissues or organs [1].

3.1 Production of Humerus Bone with 3D Printer

The process of creating a 3D-printed model of the humerus bone involves several steps. First, a digital representation of the bone is created using computer-aided design (CAD) software. This model defines the surface geometry of the bone, capturing its intricate features and contours. After this, the CAD model is converted into an STL (stereolithography) file format. In STL, the bone's surface is discretized into numerous small triangles. These triangles collectively represent the bone's shape. Think of it as assembling a jigsaw puzzle using simple geometric pieces; each triangle corresponds to a piece that fits seamlessly with its neighbours. Once we have the STL file, it can be fabricated to a 3D printer. The printer follows a layer-by-layer approach, depositing material (such as plastic filament or resin) to build up the bone structure. Such models are valuable tools for education, medical training, and research [16]. Figure 1a–d depicts the different production stages of humerus bone.

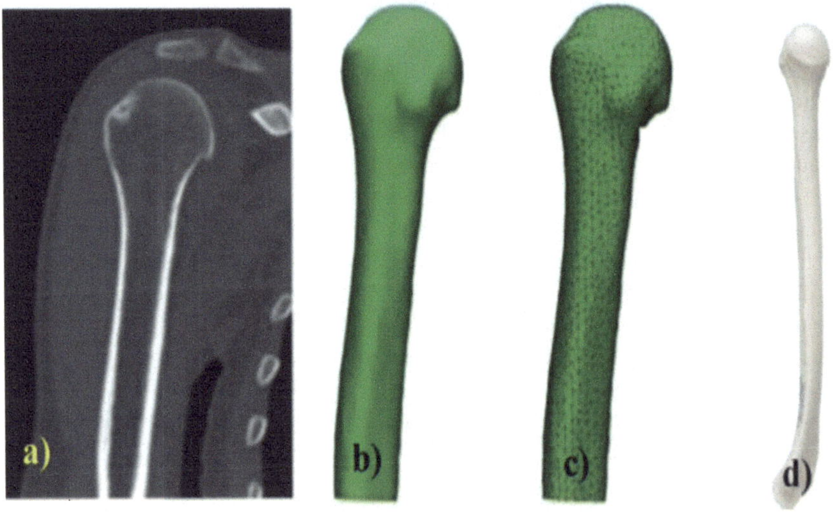

Fig. 1 Production stages of humerus bone with 3D printer; **a** imaging of humerus bone with the help of computerized tomography, **b** modelling, **c** slicing the model for printing, **d** printed humerus bone [16]. Copyright 2021. Adapted from Elsevier

3.2 Patient-Specific Implants

The utilization of 3D printing technology for manufacturing patient-specific prostheses holds significant importance in the medical field. Personalized implant applications, especially in dental procedures, offer convenience for healthcare professionals and enhance patient comfort. However, traditional implant production methods remain the gold standard due to their biological compatibility and superior mechanical properties. Despite these advantages, traditional implants often fall short in terms of aesthetic appearance, ideal angulation during placement, and hygiene maintenance. To address these limitations, personalized implants can now be created using 3D printing technology.

One key advantage of this approach is the elimination of traditional measurement methods, leading to shorter waiting times. 3D printing is commonly preferred for knee joints, tibia, femur, fibula bone implants, and dental implants. For instance, Fig. 2 illustrates hip and knee implants produced through 3D printing. The tibial tray, femoral component, acetabular cup, and hip stem fabricated using selective laser melting (SLM) or additive manufacturing (AM) methods demonstrate the precision achievable with patient-specific designs. By obtaining anatomical information through successful surgical imaging and creating digital models, complications and placement errors can be minimized. Ultimately, patient-specific implants enhance the likelihood of successful outcomes [16].

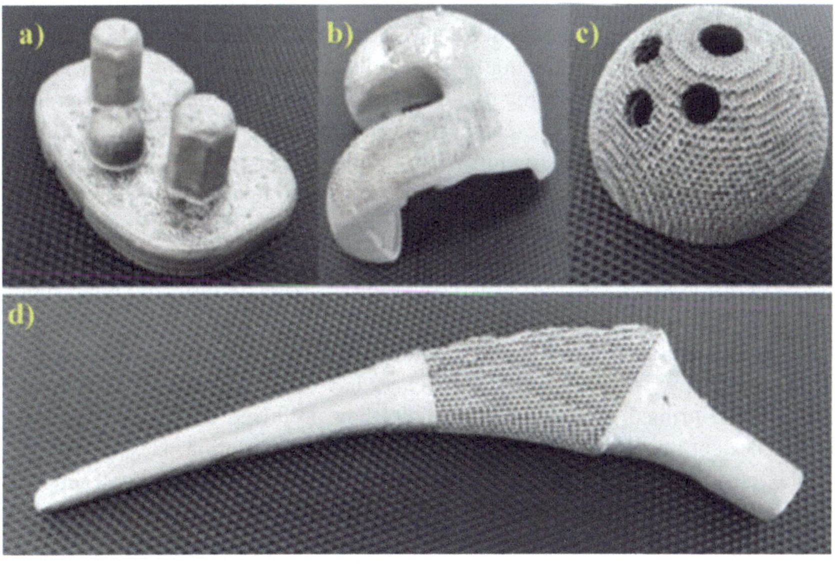

Fig. 2 Hip and knee implants manufactured by 3D printer technology; **a** tibial component, **b** femoral component, **c** acetabular cup, **d** hip stem [16] Copyright 2021. Adapted from Elsevier

3.3 3D Printing in Veterinary Medicine

3D printing holds promise not only for human medicine but also for veterinary applications. Veterinarians use this technology in various areas, such as preoperative planning and anatomical visualization. By creating 3D models of animal anatomy, veterinarians can better understand differences and anticipate challenges before surgery, thereby reducing errors and ensuring patient safety [17, 18]. Successful cases include using 3D printing to improve a dog's lameness through surgical intervention. In another instance, a 3D printer aided in tumor removal surgery for a dog, allowing precise planning and successful treatment [19]. Additionally, 3D-printed prostheses offer hope for disabled animals. These custom implants, tailored to an animal's specific anatomy, can replace damaged bones, tissues, or muscles. Notably, a hornbill diagnosed with cancer received a 3D-printed prosthesis, demonstrating the technology's effectiveness in both preoperative imaging and surgical planning.

Figures 3 illustrate examples of 3D-printed prostheses for animals. These include a secretary bird with a broken leg, an acute penguin with an amputated foot, and a congenital front leg prosthesis for a kitten [16]. Patient-specific implants, often complex in shape, are feasible through 3D printing, making it a valuable tool in veterinary orthopedics. As the field continues to evolve, we can expect increased adoption of 3D printing in veterinary medicine [16]. Researchers are actively exploring the field of organ printing using 3D printers. Although current technology cannot yet produce organs that function identically to natural ones, it has made significant progress. Scientists have successfully created organs that closely resemble the originals by closely examining a patient's anatomical structure. For instance, Zein et al. achieved a milestone by producing the first fully printed liver. This liver was used in preoperative planning, allowing for better surgical preparation. By studying livers from both patients and donors, researchers developed synthetic fibers with 3D-printed networks of vessels and bile structures. The resulting transparent 3D liver provided a detailed visualization of these anatomical features. Comparing the printed liver to the natural one confirmed the study's success, despite existing limitations [6].

Organ printing technology serves not only for transplantation but also for creating replicas of diseased organs (Fig. 4). Consider the case of a 3D-printed kidney model based on tomography images from a patient with kidney cancer [16].

This model facilitated preoperative planning and guided successful tumor removal during surgery [16]. Additionally, researchers have explored the functional capabilities of 3D-printed organs. In one study, they produced an auricle (ear) with embedded electronics, providing an alternative hearing solution. Despite the ear's complex curved structure, the bionic ear closely adhered to anatomical geometry [20]. These advancements hold great promise for the future of organ printing and medical applications.

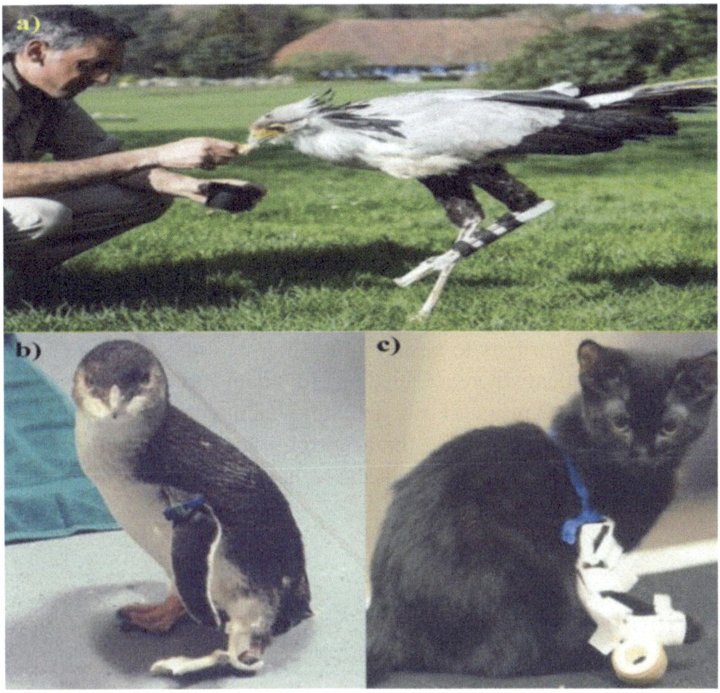

Fig. 3 3D printed prostheses for animals: **a** Leg prosthesis for secretary bird, **b** A prosthesis for a penguin whose foot has been amputated, **c** Prosthesis produced for a kitten who is not one of the congenital front leg bones [16] Copyright 2021. Adapted from Elsevier

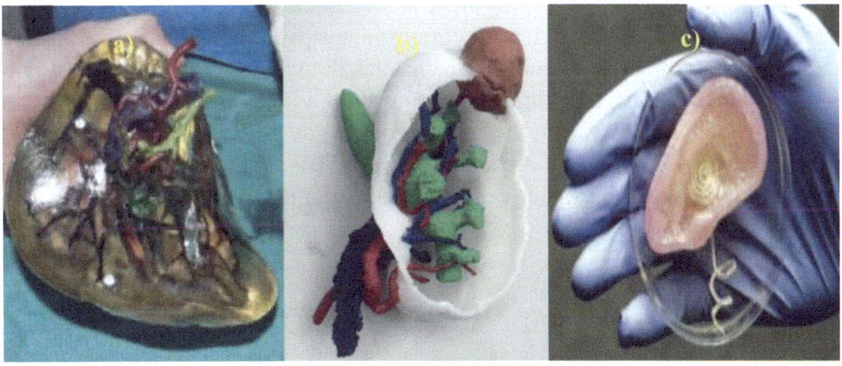

Fig. 4 3D printing organs: **a** The first fully 3D printed liver known, **b** Kidney model produced by additive manufacturing for surgical planning according to tomography image data, **c** Bionic ear that has hearing manufactured with a 3D printer [16]. Copyright 2021. Adapted from Elsevier

3.4 Limitations of 3D Printing

The speed of 3D printing is often perceived as slow, and the materials produced can lack the necessary strength, which is compounded by the high cost of raw materials. Additionally, there are significant ethical issues associated with the technology. For instance, the potential for copyright infringement is high, as individuals with access to 3D printers could replicate objects that are legally protected, thereby undermining copyright laws. Furthermore, the capability of 3D printers to replicate keys, identification cards, and credit cards poses a risk of fraud and identity theft. Another pressing concern is the ability to manufacture weapons, such as plastic guns, which could be used by criminals or terrorists to easily smuggle arms into secure areas. This was highlighted by an incident where a Canadian individual successfully created a functional firearm using a 3D printer. From an employment perspective, the automation associated with 3D printing reduces the need for human labor, leading to potential job losses in the manufacturing sector as the technology becomes more widespread [21].

4 4D Printing Technology

4D printing expands upon the concept of 3D printing by introducing the ability to create objects that can change shape or self-assemble over time. This process is driven by environmental factors such as temperature and humidity. At the core of 4D printing are smart materials that react to these external stimuli, enabling the printed object to transform. This emerging field is an evolution of 3D printing that adds time as the fourth dimension. It incorporates innovative designs and materials that can adapt and respond automatically to changes in their surroundings, resulting in significant time savings. In traditional printing, the x, y, and z axes represent the three dimensions of space. However, the addition of time as the fourth dimension allows for a substantial reduction in material usage, up to 90%. This innovative approach is becoming increasingly prevalent in modern industrial practices [21].

4.1 4D Printing of Shape Memory Polymers

The rapid advancements in engineering have led to the creation of innovative materials designed to enhance performance and functionality. Among these, materials that respond to specific stimuli by altering their physicochemical properties have garnered significant interest. These Stimulus-Responsive Materials (SRM) have promising applications ranging from functional nanocomposites to precise drug and gene delivery systems [22]. A subset of SRMs, known as Shape-Changing Materials (SCM), undergoes transformation under light exposure. Electroactive Polymers

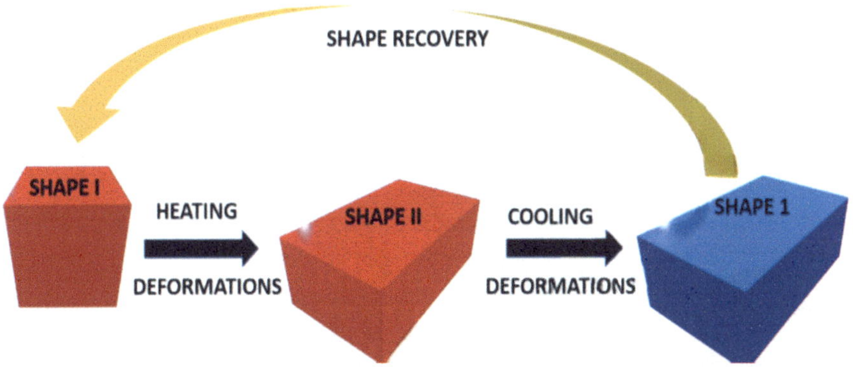

Fig. 5 Schematic delineation of the shape-memory effect in polymers [26]. Copyright 2020. Adapted from Elsevier

(EAP) and piezoelectric materials are prime examples of SCMs [23, 24]. Another category, Shape Memory Materials (SMM), characterized by the Shape Memory Effect (SME), can revert to their original form upon the application of an appropriate stimulus, even after being significantly and quasi-plastically deformed [25].

The discovery of shape memory materials dates back to Chang and Read's observation of the Smart Memory Effect (SME) in a Gold-Cadmium alloy in 1951, followed by a similar discovery in an Indium-Titanium alloy in 1953 [25, 26]. However, it wasn't until 1963 that the practical applications of these materials were reported [25]. SMMs respond to external stimuli and have a profound impact on everyday life, commonly found in diverse areas such as smart robotics, drug delivery, tissue engineering scaffolds, and self-healing systems [27]. These materials exhibit extraordinary thermomechanical behaviors (illustrated in Fig. 5) regaining their initial shape through SME, and are referred to as superelastic, pseudoelastic, or viscoelastic [26].

4.2 Applications of 4D Printing

The field of 4D printing is emerging as a novel area of study. In the realm of aerospace, this technology shows promise for future warfare applications and the development of antenna arrays. The U.S. military has expressed significant interest in the potential of 4D printing. This innovative printing technique is poised to revolutionize medical device manufacturing. Items shipped in a compact, flat-packed state can be expanded post-delivery, resuming their intended three-dimensional form. Furthermore, 4D printing offers solutions to challenges within medical, dental, and related sectors. This technology adapts its form in response to specific environmental triggers, making it ideal for creating dental restorative materials. Key qualities of dental fillings, such as durability, color fidelity, adhesion, lifespan, and resistance to failure,

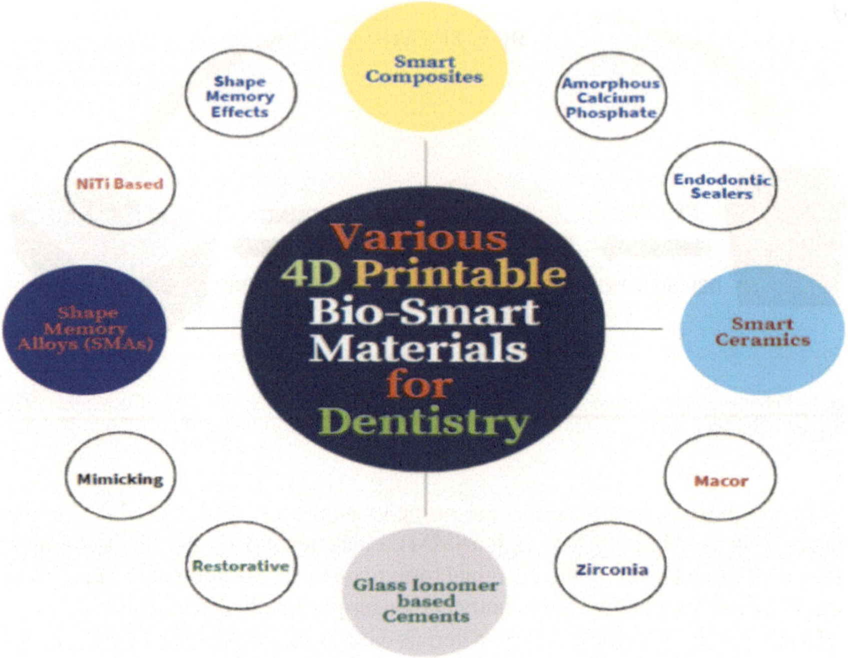

Fig. 6 Illustration of various 4D printable bio-smart materials for dentistry (reproduced with permission [28] Copyright 2022. Adapted from Elsevier

are critical. 4D printing introduces new possibilities in removable prosthetic dentistry, creating materials that mimic natural tissues, thereby minimizing the risk of damage to complex anatomical structures, like the maxillary sinus, during implantation. The precision and adaptability of 4D printing stand to enhance medical accuracy [21]. The different applications of 4D printing technology have been illustrated in Fig. 6.

4.3 Advantages of 4D Printing

4D printing exhibits unique self-altering characteristics, enabling it to transform, contract, and expand as needed. This allows for the compression of larger objects, facilitating innovative designs, such as a heart model that can alter its shape using appropriate materials. The technology's ability to modify both shape and function to meet specific demands is groundbreaking. For instance, it enables the precise fabrication of heart valves without the limitations of traditional design constraints. Organs like hearts, kidneys, and livers can be printed using intelligent materials. Moreover, 4D printing is adept at creating smart valves capable of regulating blood flow by adjusting their diameter. This dynamic adaptability extends to all facets of 3D printing applications, with objects capable of evolving in response to temperature variations. Remarkably, 4D printing can reduce time and material usage by up to 90%, representing a significant advancement in manufacturing efficiency [5].

4.4 Limitations of 4D Printing

Compared to 5D printing technology, 4D printing exhibits a notably lower modulus. The stability of 4D-printed objects is more susceptible to changes in environmental temperatures. Additionally, the process of incorporating smart materials into the print head of 4D printers presents significant challenges. In terms of strength, objects created with 4D printing are inferior to those produced by 5D printing methods. Moreover, 4D printing is limited in its ability to fabricate integrated parts with curved surfaces or intricate geometries. Consequently, the development of 5D printing has been propelled by the need to address these shortcomings [5].

5 5D Printing Technology

5D printing merges both additive and subtractive manufacturing processes to fabricate intricate designs. This method utilizes a dynamic print head that operates along the traditional X, Y, and Z axes, complemented by two additional rotational axes integrated into the print bed, enhancing the complexity of the structures produced. Pioneered by the team at Mitsubishi Electric Research Laboratories in 2016, this advanced form of printing was put to the test with the creation of pressure caps. The results were remarkable, revealing that caps crafted with a 5D printer exhibited superior mechanical strength (measured at 3.7 MPa) compared to those produced by conventional 3D printers (0.1 MPa). This groundbreaking discovery was further validated by Ethereal Machines in Bengaluru, India, with their development of the "Halo 5D printer." This versatile machine has proven its utility in manufacturing intricate jewellery pieces and automotive components, achieving greater structural integrity while utilizing 25% less material than its 3D counterparts. The promising applications of 5D printing extend into the culinary and packaging industries. For instance, the technology holds the potential to revolutionize the production of food items with elaborate geometries, such as spherical sweets or uniquely shaped crisps. Moreover, the packaging sector could witness a transformation with the ability to construct complex containers, cups, and bowls that would typically necessitate additional support when using 3D printing methods [6].

6 Additive Manufacturing

6.1 What is Additive Manufacturing?

Additive Manufacturing (AM), previously known as Rapid Prototyping and commonly referred to as 3D Printing, is a term that encompasses a range of technologies used to construct physical objects directly from digital data. Historically, the term 'Rapid Prototyping' was employed across various sectors to denote a swift method of crafting a preliminary model or part before its final production or market launch. The aim was to create a foundational model quickly to enable future iterations and produce the final product. Rapid Prototyping is a process used in business and software development where the product is incrementally developed to allow early testing and feedback from clients and stakeholders. However, the term "Rapid Prototyping" no longer captures the full scope of current applications where the output often serves as the end product itself. As a result, standardization bodies like ASTM International have adopted the term "Additive Manufacturing" to better reflect the essence of these technologies, which involve the additive layer-by-layer fabrication of parts.

The main idea behind additive manufacturing (AM) is to create complex three-dimensional objects directly from computer-aided design (CAD) data, without the need for traditional process planning. Although this simplifies the process, it still requires a basic understanding of the AM machinery and the materials used. What sets AM apart from other manufacturing methods is its layering technique, where each layer represents a cross-section of the CAD model. The accuracy of the final object depends on the thickness of these layers, with thinner layers resulting in a closer match to the original design.

Commercial additive manufacturing (AM) machines all use a layered approach to create products, but they differ in the materials, methods of layer formation, and techniques for layer adhesion. These variations can impact the accuracy, material and mechanical properties, production speed, necessary post-processing, machine size, and overall cost of the final product. This section has introduced the basic principles of Additive Manufacturing and provided an overview of the standard AM workflow from conception to application. It has also examined the impact of AM on design and manufacturing processes and concluded with a discussion on its role as a crucial tool in product development, alongside other related tools [29].

6.2 Benefits of Additive Manufacturing

Additive Manufacturing, also known as AM, is considered a revolutionary technology in the field of product creation. It has the potential to replace traditional manufacturing methods and usher in a new era of industrial revolution. AM is often considered a disruptive innovation because it changes the way we approach product design and business establishment. The question arises, what makes AM so fascinating and inspiring?

Let's discuss the rapid production of components through Additive Manufacturing (AM). Its benefits go beyond just quick production. The entire product development cycle is accelerated due to the extensive use of digital tools. With 3D CAD models as the foundation and seamless transition to AM, concerns about data conversion or misinterpretation of design intentions are minimized. The advantages of using AM (additive manufacturing) over traditional manufacturing methods are evident in the reduced number of steps required to construct parts, regardless of their complexity, as it can be done in a singular operation. On the other hand, traditional manufacturing methods typically involve several iterative procedures. Introducing additional features into a design could significantly increase these steps. However, the time taken to produce prototypes with AM remains relatively unaffected by minor design modifications during the early stages of product development.

Additive manufacturing (AM) is a process that can help reduce the number of steps and resources required to create a prototype from CAD drawings. Traditional methods such as hand carving, molding, forming, and CNC machining can be labor-intensive, prone to errors, and require meticulous planning, including the creation of molds or fixtures. In contrast, AM simplifies or even eliminates these complex stages. When combined with auxiliary technologies like silicone rubber molding and various finishing tools, AM can produce a wide range of parts with unique features. Workspaces that incorporate AM can become cleaner, more efficient, and adaptable [30].

6.3 *Overview of the Standard AM Procedure*

The process of transforming a digital Computer-Aided Design (CAD) blueprint into a physical product involves several stages. The level of utilization of Additive Manufacturing (AM) varies based on the complexity and size of the product. For example, smaller items with simple designs may only use AM for creating visual prototypes, while larger items with significant engineering requirements may integrate AM at multiple points during their development cycle. In the early stages, the focus might be on swiftly producing basic prototypes, while later stages may require detailed finishing work, such as smoothing, priming, and painting, to prepare the parts for their final use. The ability of AM to create complex shapes without the need for traditional tooling considerations is particularly advantageous at this stage.

We will discuss the various AM stages in more detail later, but for now, let us summarize the typical eight-step process:

1. Step One: CAD Modeling—Every AM project starts with a software model that accurately depicts the item's external shape. This model can be created using various professional CAD software, resulting in a three-dimensional solid

or surface model. Alternatively, reverse engineering tools like laser or optical scanners can be used to generate this digital representation.
2. Step Two: STL Conversion—The STL file format is the AM industry standard and is widely accepted by AM machines. Most CAD platforms can export models in this format, which details the object's outer surfaces and lays the groundwork for slice calculations.
3. Step Three: Preparing the AM Machine—The STL file must be transferred to the AM machine, where it may require adjustments to ensure that it is correctly scaled and oriented for the build.
4. Step Four: Machine Configuration—Before starting the build, the AM machine settings must be fine-tuned. These settings include material specifications, energy sources, layer thickness, and timing parameters.
5. Step Five: Construction—The actual construction of the part is mostly automated, requiring minimal oversight. Monitoring is necessary to prevent issues such as material shortages or technical malfunctions.
6. Step Six: Extraction—After the build is complete, the parts need to be carefully removed from the machine, which may involve safety protocols to ensure that conditions like temperature are safe for handling.
7. Step Seven: Post-Construction Finishing—After removal, the parts often require further cleaning and finishing. This stage can include the removal of support structures, surface treatment, and possibly heat treatments. The finishing process can be extensive and labor-intensive, depending on the desired quality.
8. Step Eight: Application—The parts are now prepared for their intended use, which may involve assembly with other components to create the final product or model.

While we have covered the standard steps in the AM process, it is important to note that additional or alternative steps may be necessary for specific machines or methods. For example, some setups may bypass the STL file stage if the machine is compatible with native CAD data. Regular maintenance checks are crucial for the longevity and proper functioning of AM machines, which can be sensitive to environmental conditions. Material handling also requires attention, as some materials used in AM have a limited shelf life and must be stored to prevent degradation or unwanted reactions [30].

7 Application Areas for AM-Enabled Product Development with Polymer Nanocomposites

Additive Manufacturing (AM) is revolutionizing the way products are developed, offering unique solutions that diverge from conventional development pathways. The integration of AM with diverse data sources facilitates swift and cost-effective responses. Below are some areas where AM's design methodologies shine.:

7.1 Medical Modeling

AM's growing role in health care involves creating components from personal medical scans, such as CT, MRI, and 3D ultrasound. These scans undergo extensive processing to isolate pertinent details for modeling or integration into product designs. Notable software platforms enable this data transformation, leading to innovations like custom-fitted hearing aids, tailored from individual ear canal scans.

7.2 Reverse Engineering Data

Beyond healthcare, AM excels in capturing and processing intricate surface data. Laser scanning is a prevalent method for gathering detailed surface information from geometrically complex objects. The goal often extends beyond mere replication; it's about incorporating this data into innovative product designs. The spectrum of laser scanning technology ranges from premium, high-precision systems to more affordable, mobile phone camera-based solutions.

7.3 Architectural Modeling

In architecture, models are crafted to highlight specific design elements, often altering textures, colors, and forms to accentuate certain aspects. Consequently, architectural software may need to adapt features to align with AM capabilities.

7.4 Automotive

Traditional manufacturing methods can struggle with complex geometries due to cost or feasibility constraints. AM, however, excels in creating lightweight, hollow structures that reduce costs and enhance performance. The automotive sector leverages AM to produce intricate designs like honeycomb structures, optimizing the strength-to-weight ratio for safer, lighter vehicles.

7.5 Aerospace

The aerospace industry faces a promising future with AM, amidst rising global demand and a shift towards more efficient aircraft. AM enables the crafting of components with elaborate geometries and hollow interiors, significantly reducing weight.

With the potential to apply AM to a vast number of aircraft parts, the industry is exploring ways to maintain surface quality, especially for parts with internal features [30].

8 Application of AM in the Aerospace Industry

Additive manufacturing (AM) has emerged as a pivotal technology in the aerospace sector, contributing significantly to its revenue streams across production and maintenance services. Over the past decade, this industry has consistently been at the forefront of AM adoption. Currently, the aerospace industry accounts for 18.2% of AM's total market revenue. Notably, in 2016, the aerospace industry exhibited the highest growth rate at 1.6% annually, outpacing the motor vehicle sector's 1.0% growth. AM's revenue was recorded at $2.7 billion in 2016, marking a 12.9% increase from the previous year, and it is projected to exceed $100 billion in the next 20 years, predominantly within the aerospace domain.

The spectrum of AM components in aerospace is categorized into metallic and non-metallic parts, typically corresponding to critical and non-critical elements of aircraft, respectively. Since the mid-1990s, Boeing and Bell Helicopter have integrated polymer-based AM components for non-essential parts. As of March 2015, Boeing had produced over 200 distinct parts for a variety of aircraft models utilizing AM techniques. At that juncture, aircraft were equipped with upwards of 20,000 polymer AM components. Presently, Boeing has incorporated an extensive array of AM components into 16 commercial and defense aircraft models. In 2017, Boeing commenced incorporating a minimum of four titanium-alloy AM components in the manufacturing of the 787 Dreamliner, with plans to fabricate nearly 1,000 parts using AM, potentially saving between $2 and $3 million per aircraft [31]. Airbus stands as another prominent entity in the AM landscape. It has implemented AM-produced metal brackets and bleed pipes in the Airbus A320neo and A350 XWB test planes. Furthermore, Airbus has engaged in a long-term research collaboration with Arconic to develop sizable AM structural components measuring up to 1 m. Space agencies such as NASA and the European Space Agency and companies like SpaceX are investigating AM applications for rocket engine components like igniters, injectors, and combustion chambers. Other key players in the aerospace industry utilizing AM include Honeywell Aerospace, Lockheed Martin, and Northrop Grumman, driven by the unique design demands of the aerospace field and the advanced capabilities offered by AM technologies [32].

9 Why Manufacturers Adopt Additive Manufacturing Technologies: The Role of Sustainability

Additive Manufacturing (AM) technologies are recognized for their substantial contributions to sustainability. The application of AM benefits sustainable development, which encompasses the pursuit of economic well-being, environmental preservation, and social fairness. Sustainable manufacturing involves processes that avoid pollution, conserve energy and natural resources, are economically beneficial, and ensure the safety of workers, communities, and consumers.

9.1 Distinct Advantages of AM in Promoting Sustainability

AM technologies endow manufacturers and designers with distinct fabrication strengths that positively influence sustainability. These strengths are derived from AM's unique nature, including its additive, layer-by-layer production method, absence of traditional tooling, and reduced resource usage.

9.2 The Advantage of 'Complexity-For-Free'

AM's capability to produce complex geometric designs effortlessly, which would be challenging or expensive with conventional methods, is a notable advantage. This 'complexity-for-free' characteristic means that the cost of production is independent of the design's complexity [33]. The additive process supports the creation of lighter structures, which can lead to significant reductions in energy consumption and emissions, as demonstrated by Tang et al. (2016) [33]. This approach can reduce lifecycle costs, with savings in the production phase, use phase, and decommissioning [34]. For example, in the aviation sector, reducing material weight can lead to substantial fuel cost savings [35]. The 'complexity-for-free' benefit also enables the integration of multiple product components into a single, multifunctional part, thus removing the need for additional assembly [23, 36]. This integration can eliminate costs, time, and quality issues associated with assembly operations. Additionally, the freedom of design inherent in AM fosters creativity and innovation, leading to products with enhanced quality, functionality, durability, and easier manufacturing and maintenance [23].

9.3 Tool-Free Fabrication

Additive Manufacturing (AM) enables the transformation of 3D CAD designs into physical objects without the need for traditional tooling, molds, or setup time, thus offering a highly adaptable production environment. This unique characteristic of AM allows for the creation of tailor-made products in an eco-friendly way, as highlighted by Khorram Niaki et al. [24]. While customization is not solely the domain of AM, it excels in this area by avoiding additional costs or delays typically associated with custom work, as noted by a research group [37]. The 'economy-of-one' approach [38], which is described as facilitated by AM, allowing for cost-effective manufacturing of individual items, in contrast to the 'economies-of-scale' model. Conventional manufacturing achieves cost-efficiency by spreading fixed costs across large production volumes, like tooling. However, AM's lack of such fixed costs makes even single-item production financially viable, as Weller et al. [39] have observed. The absence of large-scale production requirements also supports just-in-time manufacturing, reducing inventory expenses, as suggested by Berman [40].

9.4 Resource-Efficient Production

AM is known for its minimal resource usage, including materials, energy, labor, time, and space. By building objects layer by layer from digital data, AM minimizes material flow cuts out many middlemen in the production process, thus streamlining the supply chain, as per Holmstrom et al. [41], Ford and and Ma et al. [42]. These efficiencies offer multiple sustainability advantages economically, environmentally, and socially. AM's ability to recycle unused raw materials, such as powders and resins, is significant, with Ford and Despeisse (2016) estimating a recyclability rate of 95–98% for metal powders [23]. Energy consumption and environmental impact assessments of AM reveal that its Specific Energy Consumption (SEC) is generally higher compared to traditional methods have been found [33]. However, AM can be more energy-efficient for small production runs, optimized designs, and certain processes like Binder Jetting, as shown in different studies [29, 36, 38]. A holistic view that includes the entire supply chain is essential for evaluating AM's environmental benefits. AM can enhance a company's environmental footprint through reduced transportation needs via distribution [33]. Therefore, the energy efficiency and environmental impact of AM are contingent upon factors like design optimization, production scale, process parameter choices, and supply chain structure [37, 43].

The reduced resource intensity of AM also has social ramifications. It simplifies the manufacturing process, enabling emerging economies to close the technological gap and enter the burgeoning high-tech manufacturing sector alongside developed nations [44]. Additionally, AM offers health advantages by reducing workers' exposure to harsh and dangerous conditions [45]. Nonetheless, there are potential risks during the handling and disposal of AM materials [23]. Moreover, the streamlined material flow and design flexibility foster collaborative creation and consumer participation in the design and production stages [44].

10 Polymer Composite Materials in Additive Manufacturing

Composite materials can be conceptualized as a synergistic combination of multiple constituents, where the resultant material exhibits enhanced characteristics that surpass the individual qualities of each component [46]. These materials are often referred to as reinforced systems, with fillers integrated within a binding matrix [47, 48]. The primary function of the matrix is to maintain structural integrity and direct the applied stresses throughout the composite, which are inherently diverse and frequently exhibit direction-dependent properties [48, 49]. The composition and nature of both the matrix and the reinforcements, along with their shapes, volumes, interface quality, and the manufacturing techniques employed, significantly dictate the final attributes of the composite [10] [50]. Composites can be engineered with metallic, ceramic, or polymeric matrices, offering a vast array of potential combinations [51]. Typically, a composite consists of one or more dispersed phases within a continuous phase. When multiple dispersed phases of distinct types are present, the material is classified as a hybrid composite [52]. The matrix serves as the continuous phase, while the dispersed phase, known as the reinforcement, enhances the composite. By amalgamating components with complementary physical and mechanical properties, composites leverage the strengths of each [53]. Embedding high-strength, high-modulus reinforcements into a polymer matrix can significantly elevate the material's mechanical and thermal performance [54]. Polymer matrix composites offer advantages over metals, such as intricate shape fabrication capabilities and reduced density, leading to benefits like decreased fuel consumption in transportation and aviation, increased speed in competitive sports, extended missile range, and augmented transport payload.

Composite materials are categorized based on their matrix type into organic, mineral, and metallic groups. Organic composites encompass materials like cardboard (comprising resins and cellulose fibers), laminated tires (a mix of rubber, steel, organic resins, glass fibers, carbon, and boron), and reinforced plastics (resins combined with short fibers) [52]. Mineral composites include substances like concrete (a blend of cement, sand, and additives), carbon–carbon composites (carbon matrix with carbon fibers), and ceramic composites (ceramic matrices with ceramic fibers) [55]. Lastly, metallic composites are represented by materials such as aluminum alloyed with boron or carbon fibers [56, 93–95]. These diverse composite materials find applications across various domains, including packaging, the automotive industry, lightweight construction, civil engineering, aviation, sports, biomedicine, thermomechanical parts, and the aerospace sector [46].

Material jetting technology is one of the most important AM technologies, and it incorporates a wide range of colors as well as materials. Here, both hard and soft polymeric materials can be fabricated in a single process with localized tuning material properties. Here, 3D-printed parts can be tailored to different stiffness without the incorporation of any complex designs. Among the various AM technologies, material jetting incorporates a wide variety of colors and materials, allowing hard and soft polymeric materials to be fabricated in a single process [56]. Material jetting allows localized tuning of material properties. 3D printed parts with different stiffness can be tailored without compromising the complexity of the design [56] which is discussed in the following caption.

Commercial AM polymers Although there are many choices of available polymers, a user must carefully consider which materials should be used to manufacture parts because the mechanisms of the four AM technologies discussed above distinctly differ from one another. There are two different types of techniques viz fused filament fabrication (FFF) and selective laser sintering (SLS). FFF involves melt depositing of polymer composites through filament extrusion as per the requirement of the input design. The machine consists of a spool of filament, which is fed through an array of rollers and gears linked to the extruder head. Here, the heated head melts the thermoplastic polymer and deposits it in the build platform. The schematic working of FFF has been depicted in Fig. 7. On the other hand, SLS is a powder-based technique that allows the freeform fabrication of 3D parts by successively layer-by-layer solidifying powder materials. The solidification is achieved by selectively fusing or sintering on specified materials from a laser emission system (as depicted in Fig. 8). To initiate printing using the above-mentioned techniques, the FFF material must be available in filament form and the SLA material should be in liquid resin form [57]. To suit different applications, the materials are sometimes required to be chemically and fire resistant, flexible, corrosion resistant, minimum water absorption, and optically transparent [57]. The Multijet Fusion (MJF) is a recently developed AM that is similar to SLS and uses powdered polymers. However, instead of using a laser to sinter or melt the material, MJF uses a fusing agent and detailing agent to bond the powders by IR radiation. Stereolithography (SLA) is another commonly used method of 3D printing and was one of the first methods of AM development. Various methods of SLA include digital light processing (DLP), micro stereolithography (MSL), digital light synthesis (DLS), and traditional laser SLA. SLA provides the highest printing resolution in comparison to the previously mentioned AM methods [57].

11 PolyJet Printing

PolyJet printing is a technique that utilizes UV light to harden polymers sprayed onto a build platform. A schematic diagram for PolyJet printing has been depicted in Fig. 9. A gantry system equipped with multiple nozzles dispenses the polymer in

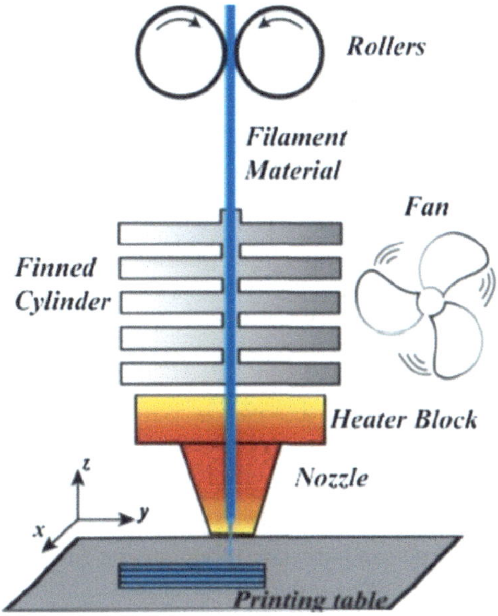

Fig. 7 Schematic showing how a typical FFF machine operates [57]. Copyright 2020. Adapted from Elsevier

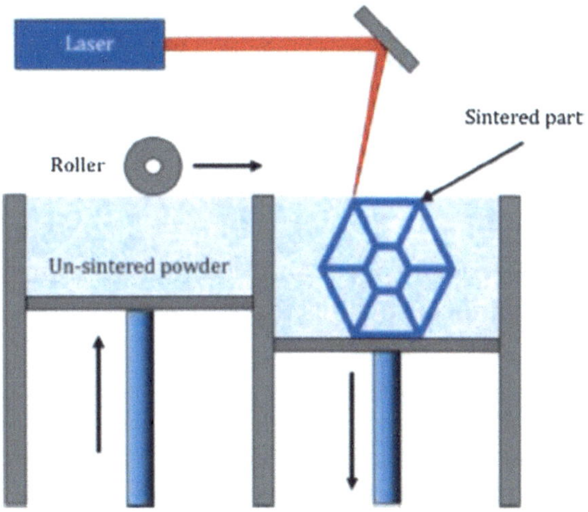

Fig. 8 Schematic diagram depicting the operation of a typical SLS machine [57]. Copyright 2020. Adapted from Elsevier

a liquid or semi-liquid state, which is then solidified by UV radiation. The gantry, which carries both the nozzles and the UV light source, moves along the x–y axis, while the build platform adjusts downward after each layer is completed, preparing for the subsequent layer's application. This method is versatile, allowing for the use of various photopolymers, provided they are in the appropriate physical state for spraying. The fidelity of the final print is influenced by several factors, including the height of each layer, the orientation of the print, the intensity of the UV light, and

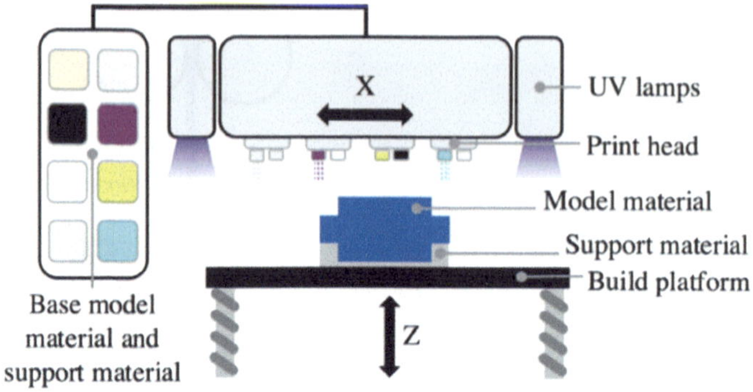

Fig. 9 Illustration of polyjet 3D printing [62]. Copyright 2022. Adapted from Elsevier

the rate at which the polymer is dispensed [58]. Although the UV light's exposure time is fixed due to its attachment to the moving gantry, adjustments to the polymer's spray rate can indirectly modulate the curing process.

The speed of PolyJet printing sets it apart from other printing methods, as it can deploy multiple nozzles simultaneously to apply polymer layers. This capability not only accelerates the printing process but also enables the creation of objects composed of different materials, with a high degree of detail [59]. The technology produces parts in various colors and with smooth surface finishes [60], eliminating the need for post-processing and making it an excellent choice for rapid prototyping. However, the cost associated with using a single material remains a limitation of this technology [61].

12 Recent Progress on Polymer Composites and Polymer Nanocomposites for AM

As the material spectrum for AM is limited to certain applications due to some specific requirements. Therefore, traditional AM polymers are limited to high-performance applications. Therefore, it has become a fact of major importance to develop high-performance materials to expand its capability. This can be achieved by preparing polymer composite materials, which are fabricated by a combination of two different materials with different properties. Though composite materials have been used since ancient times, it was not until the twentieth century that the composite material field gained significant momentum, owing to the explosive development of fiber-reinforced polymer composite [57]. These types of composites have extraordinary properties like lightweight and corrosion resistance. Subsequently, polymer nanocomposite materials with nanofillers dispersed in a polymer matrix have attracted considerable research attention due to the nano-effect. These show extraordinary mechanical, thermal, electrical and chemical properties with very minute doses of nanofillers. These are also less prone to anisotropy issues where extra laminar strength is decreased owing to a lack of fiber reinforcements. Moreover, the fabrication process in AM for PNC is exactly the same without any type of modifications to the machine [57].

13 Conclusion

The recent advancements in Additive Manufacturing (AM) technologies, in 3D and 4D printing, pose a revolutionary effect on manufacturing and product development. The core benefits of AM include its capability for rapid prototyping, customization, and the production of complex geometries, which traditional manufacturing methods struggle to achieve efficiently. This 'complexity-for-free' aspect allows for innovative designs and more efficient material usage, leading to substantial cost and energy savings across various industries, such as aerospace and healthcare.3D printing has been a game-changer in fields requiring high precision and customization, notably in medical applications where it facilitates the production of tailored implants and surgical models. The seamless transition from digital designs to physical objects without the need for traditional tooling also underscores the flexibility and adaptability of AM, paving the way for more personalized and efficient production processes.4D printing further enhances these capabilities by introducing smart materials that respond to environmental stimuli, allowing for the creation of dynamic structures that can change shape or function over time. This technology holds significant promise in areas such as aerospace, where adaptive components can improve performance and reduce maintenance needs, and in medical fields, where self-transforming implants and devices can offer better integration with biological tissues. Despite these advantages, there are challenges to address. The relatively slow speed of 3D printing, the high cost of materials, and potential ethical issues, such as copyright infringement and the illicit production of weapons, pose significant concerns. Furthermore, the shift towards automated manufacturing processes might lead to job losses in traditional manufacturing sectors. In conclusion, while the journey of additive manufacturing is marked by impressive technological advancements and promising applications, it also necessitates careful consideration of ethical, economic, and social impacts. Balancing innovation with responsibility will be crucial in harnessing the full potential of AM technologies in a way that benefits society as a whole.

References

1. Kathryn Thompson, M., Moroni, G., Vaneker, T., et al.: Design for additive manufacturing: trends, opportunities, considerations, and constraints. CIRP Ann. Manuf. Technol. **65**, 737–760 (2016). https://doi.org/10.1016/j.cirp.2016.05.004i

2. Bandyopadhyay, A., Heer, B., Keck, W.M.: Additive manufacturing of multi-material structures (2018)
3. Oropallo, W., Piegl, L.A.: Ten challenges in 3D printing. Eng. Comput. **32**, 135–148 (2016). https://doi.org/10.1007/s00366-015-0407-0
4. Pérez, M., Carou, D., Rubio, E.M., Teti, R.: Current advances in additive manufacturing. In: Procedia CIRP. Elsevier B.V., pp. 439–444 (2020)
5. Sha, A.J.S.H., Patil, P.: Review on 4D and 5D printing technology. Int. Res. J. Eng. Technol. (2020)
6. Fan, G., Meng, Y., Zhu, S., et al.: Three-dimensional printing for laparoscopic partial nephrectomy in patients with renal tumors. J. Int. Med. Res. **47**, 4324–4332 (2019). https://doi.org/10.1177/0300060519862058
7. Dawood, A., Marti, B.M., Sauret-Jackson, V., Darwood, A.: 3D printing in dentistry. Br. Dent. J. **219**, 521–529 (2015). https://doi.org/10.1038/sj.bdj.2015.914
8. Mao, X., Wang, Z.: Research progress of three-dimensional bioprinting artificial cardiac tissue. Tissue Eng Regen Med **20**, 1–9 (2023). https://doi.org/10.1007/s13770-022-00495-9
9. Yap, Y.L., Yeong, W.Y.: Additive manufacture of fashion and jewellery products: a mini review. Virtual Phys. Prototyp. **9**, 195–201 (2014). https://doi.org/10.1080/17452759.2014.938993
10. Nagarajan, B., Arshad, M., Ullah, A., et al.: Additive manufacturing ferromagnetic polymers using stereolithography—Materials and process development. Manuf. Lett. **21**, 12–16 (2019). https://doi.org/10.1016/j.mfglet.2019.06.003
11. Thomas, D.J.: Advanced active-gas 3D printing of 436 stainless steel for future rocket engine structure manufacture. J. Manuf. Process. **74**, 256–265 (2022). https://doi.org/10.1016/j.jmapro.2021.12.037
12. Tofail, S.A.M., Koumoulos, E.P., Bandyopadhyay, A., et al.: Additive manufacturing: scientific and technological challenges, market uptake and opportunities. Mater. Today **21**, 22–37 (2018)
13. Low, Z.X., Chua, Y.T., Ray, B.M., et al.: Perspective on 3D printing of separation membranes and comparison to related unconventional fabrication techniques. J Memb Sci **523**, 596–613 (2017)
14. Sniderman, B., Baum, P., Rajan, V.: Issue 19 I 2016 3D opportunity for life Additive manufacturing takes humanitarian action (2016)
15. Shahrubudin, N., Lee, T.C., Ramlan, R.: An overview on 3D printing technology: Technological, materials, and applications. In: Procedia Manufacturing. Elsevier B.V., pp. 1286–1296 (2019)
16. Bozkurt, Y., Karayel, E.: 3D printing technology; methods, biomedical applications, future opportunities and trends. J. Market. Res. **14**, 1430–1450 (2021)
17. Javaid, M., Haleem, A.: Additive manufacturing applications in orthopaedics: a review. J Clin Orthop Trauma **9**, 202–206 (2018)
18. O'Reilly, M.K., Reese, S., Herlihy, T., et al.: Fabrication and assessment of 3D printed anatomical models of the lower limb for anatomical teaching and femoral vessel access training in medicine. Anat. Sci. Educ. **9**, 71–79 (2016). https://doi.org/10.1002/ase.1538
19. Kim, S.E., Shim, K.M., Jang, K. et al.: Three-dimensional printing-based reconstruction of a maxillary bone defect in a dog following tumor removal. Vivo (Brooklyn) **32**, 63–70 (2018). https://doi.org/10.21873/invivo.11205
20. Livingston, E., Desai, A., Berkwits, M.: Sourcing personal protective equipment during the COVID-19 pandemic. JAMA **323**, 1912 (2020). https://doi.org/10.1001/jama.2020.5317
21. J Sadiq Sha, H.A., Sha, A.J.S.H., Patil, P.: Review on 4D and 5D printing technology. Int. Res. J. Eng. Technol. (2020)
22. Schmaljohann, D.: Thermo- and pH-responsive polymers in drug delivery. Adv. Drug Deliv. Rev. **58**, 1655–1670 (2006)
23. Ford, S., Despeisse, M.: Additive manufacturing and sustainability: an exploratory study of the advantages and challenges. J. Clean. Prod. **137**, 1573–1587 (2016). https://doi.org/10.1016/j.jclepro.2016.04.150
24. Khorram Niaki, M., Nonino, F.: Additive manufacturing management: a review and future research agenda. Int. J. Prod. Res. **55**, 1419–1439 (2017). https://doi.org/10.1080/00207543.2016.1229064

25. Feninat, F.El., Laroche, G., Fiset, M., Mantovani, D.: Shape Memory Mater. Biomed. Appl.
26. Subash, A., Kandasubramanian, B.: 4D printing of shape memory polymers. Eur. Polym. J., 134 (2020)
27. Wang, L.C., Song, W.L., Fang, D.: Twistable origami and Kirigami: from structure-guided smartness to mechanical energy storage. ACS Appl. Mater. Interfaces **11**, 3450–3458 (2019). https://doi.org/10.1021/acsami.8b17776
28. Javaid, M., Haleem, A., Singh, R.P., et al.: Significance of 4D printing for dentistry: materials, process, and potentials. J. Oral. Biol. Craniofac Res. **12**, 388–395 (2022). https://doi.org/10.1016/j.jobcr.2022.05.002
29. Tang, Y., Mak, K., Zhao, Y.F.: A framework to reduce product environmental impact through design optimization for additive manufacturing. J. Clean. Prod. **137**, 1560–1572 (2016). https://doi.org/10.1016/j.jclepro.2016.06.037
30. Gibson, I., Rosen, D., Stucker, B., Khorasani, M.: Additive Manufacturing Technologies
31. Alami, A.H., Ghani Olabi, A., Alashkar, A. et al.: Additive manufacturing in the aerospace and automotive industries: recent trends and role in achieving sustainable development goals. Ain Shams Eng. J., 14 (2023)
32. Najmon, J.C., Raeisi, S., Tovar, A.: Review of additive manufacturing technologies and applications in the aerospace industry. In: Additive Manufacturing for the Aerospace Industry. Elsevier Inc., pp. 7–31 (2019)
33. Niaki, M.K., Torabi, S.A., Nonino, F.: Why manufacturers adopt additive manufacturing technologies: the role of sustainability. J. Clean. Prod. **222**, 381–392 (2019). https://doi.org/10.1016/j.jclepro.2019.03.019
34. Gebler, M., Schoot Uiterkamp, A.J.M., Visser, C.: A global sustainability perspective on 3D printing technologies. Energy Policy **74**, 158–167 (2014). https://doi.org/10.1016/j.enpol.2014.08.033
35. Hopkinson, N., Hague, R.J.M., Dickens, P.M.: Rapid Manufacturing. Wiley (2005)
36. Yang, S., Zhao, Y.F.: Additive manufacturing-enabled design theory and methodology: a critical review. Int. J. Adv. Manuf. Technol. **80**, 327–342 (2015)
37. Priarone, P.C., Ingarao, G.: Towards criteria for sustainable process selection: On the modelling of pure subtractive versus additive/subtractive integrated manufacturing approaches. J. Clean. Prod. **144**, 57–68 (2017). https://doi.org/10.1016/j.jclepro.2016.12.165
38. Baumers, M., Tuck, C., Wildman, R. et al.: Transparency built-in: energy consumption and cost estimation for additive manufacturing Baumers et al. energy and cost estimation for additive manufacturing. J. Ind. Ecol. **17**, 418–431 (2013). https://doi.org/10.1111/j.1530-9290.2012.00512.x
39. Weller, C., Kleer, R., Piller, F.T.: Economic implications of 3D printing: market structure models in light of additive manufacturing revisited. Int. J. Prod. Econ. **164**, 43–56 (2015). https://doi.org/10.1016/j.ijpe.2015.02.020
40. Berman, B.: 3-D printing: the new industrial revolution. Bus. Horiz. **55**, 155–162 (2012). https://doi.org/10.1016/j.bushor.2011.11.003
41. Holmström, J., Liotta, G., Chaudhuri, A.: Sustainability outcomes through direct digital manufacturing-based operational practices: a design theory approach. J. Clean. Prod. **167**, 951–961 (2017). https://doi.org/10.1016/j.jclepro.2017.03.092
42. Ma, J., Harstvedt, J.D., Dunaway, D., et al.: An exploratory investigation of additively manufactured product life cycle sustainability assessment. J. Clean. Prod. **192**, 55–70 (2018). https://doi.org/10.1016/j.jclepro.2018.04.249
43. Bekker, A.C.M., Verlinden, J.C.: Life cycle assessment of wire + arc additive manufacturing compared to green sand casting and CNC milling in stainless steel. J. Clean. Prod. **177**, 438–447 (2018). https://doi.org/10.1016/j.jclepro.2017.12.148
44. Niaki, M.K., Nonino, F.: Springer Series in Advanced Manufacturing The Management of Additive Manufacturing Enhancing Business Value
45. Huang, S.H., Liu, P., Mokasdar, A., Hou, L.: Additive manufacturing and its societal impact: A literature review. Int. J. Adv. Manuf. Technol. **67**, 1191–1203 (2013)

46. Hsissou, R., Bekhta, A., Harfi, A. El.: Synthesis and characterization of a new epoxy resin homologous of DGEBA: DIGLYCIDYL BIS DISULFIDE carbon ether of bisphenol a (2018)
47. Asim, M., Jawaid, M., Nasir, M., Saba, N.: Effect of fiber loadings and treatment on dynamic mechanical, thermal and flammability properties of pineapple leaf fiber and kenaf phenolic composites. J. Renew. Mater **6**, 383–393 (2018). https://doi.org/10.7569/JRM.2017.634162
48. Amrollahi, S., Ramezanzadeh, B., Yari, H., et al.: Synthesis of polyaniline-modified graphene oxide for obtaining a high performance epoxy nanocomposite film with excellent UV blocking/anti-oxidant/anti-corrosion capabilities. Compos. B Eng. **173**,(2019). https://doi.org/10.1016/j.compositesb.2019.05.015
49. Rod, K.A., Nguyen, M.T., Elbakhshwan, M., et al.: Insights into the physical and chemical properties of a cement-polymer composite developed for geothermal wellbore applications. Cem. Concr. Compos. **97**, 279–287 (2019). https://doi.org/10.1016/j.cemconcomp.2018.12.022
50. Nobile, M.R.: Rheology of polymer-carbon nanotube composites melts. In: Polymer-Carbon Nanotube Composites: Preparation, Properties and Applications. Elsevier Ltd, pp. 428–481 (2011)
51. Devnani, G.L., Sinha, S.: Effect of nanofillers on the properties of natural fiber reinforced polymer composites (2019)
52. Ahmadijokani, F., Shojaei, A., Dordanihaghighi, S., et al.: Effects of hybrid carbon-aramid fiber on performance of non-asbestos organic brake friction composites. Wear **452–453**, 203280 (2020). https://doi.org/10.1016/j.wear.2020.203280
53. Baek, Y.M., Shin, P.S., Kim, J.H., et al.: Thermal transfer, interfacial, and mechanical properties of carbon fiber/polycarbonate-CNT composites using infrared thermography. Polym. Test. **81**,(2020). https://doi.org/10.1016/j.polymertesting.2019.106247
54. Das, P., Tiwari, P.: Thermal degradation study of waste polyethylene terephthalate (PET) under inert and oxidative environments. Thermochim. Acta **679**,(2019). https://doi.org/10.1016/j.tca.2019.178340
55. Bołtryk, M., Krupa, A., Pawluczuk, E.: Modification of the properties of the cement composites with the organic filler. Constr. Build. Mater. **167**, 143–153 (2018). https://doi.org/10.1016/j.conbuildmat.2018.02.025
56. Tee, Y.L., Tran, P., Leary, M., et al.: 3D Printing of polymer composites with material jetting: Mechanical and fractographic analysis. Addit. Manuf. **36**,(2020). https://doi.org/10.1016/j.addma.2020.101558
57. Wu, H., Fahy, W.P., Kim, S. et al.: Recent developments in polymers/polymer nanocomposites for additive manufacturing. Prog. Mater Sci., 111 (2020)
58. Singh, R.: Process capability study of polyjet printing for plastic components. Evol. Ecol. **25**, 1011–1015 (2011). https://doi.org/10.1007/s12206-011-0203-8
59. Ibrahim, D., Broilo, T.L., Heitz, C., et al.: Dimensional error of selective laser sintering, three-dimensional printing and PolyJet^{TM} models in the reproduction of mandibular anatomy. J. Cranio-Maxillofacial Surg. **37**, 167–173 (2009). https://doi.org/10.1016/j.jcms.2008.10.008
60. Jasiuk, I., Abueidda, D.W., Kozuch, C., et al.: An overview on additive manufacturing of polymers. JOM **70**, 275–283 (2018). https://doi.org/10.1007/s11837-017-2730-y
61. Nath, S.D., Nilufar, S.: An overview of additive manufacturing of polymers and associated composites. Polymers (Basel) **12**, 1–33 (2020)
62. Sanders, J., Wei, X., Pei, Z.: Experimental investigation of PolyJet 3D printing: effects of sample location and volume on power consumption. Manuf Lett **31**, 83–86 (2022). https://doi.org/10.1016/j.mfglet.2021.07.013

3D, 4D, and 5D Printing Technology Towards Photoelectric Devices

M. Vishnuvarthanan

Abstract In recent years, the evolution of additive manufacturing technologies has revolutionized various industries, particularly in the area of photoelectric devices. This chapter explores the transformative potential of 3D, 4D, and 5D printing technologies in developing and fabricating advanced photoelectric devices. The principles of layer-by-layer deposition in additive manufacturing techniques offer unparalleled flexibility, precision, and scalability in the production of intricate structures with tailored functionalities. Initially, this chapter provides a comprehensive overview of traditional photoelectric device fabrication techniques and their inherent limitations, giving way to the introduction of additive manufacturing as a disruptive alternative. Subsequently, it investigates the fundamentals of 3D printing, explaining the underlying processes of additive manufacturing. Through precise control of material deposition, layer thickness, and geometry, 3D printing enables the creation of complex photoelectric device architectures with enhanced efficiency and performance. Furthermore, the narrative progresses to discuss the emerging frontier of 4D printing, wherein materials possess the ability to alter their shape, properties or functionality in response to external stimuli over time. Moreover, the exploration extends to the promising area of 5D printing which transcends the physical dimensions of space and time by embedding the additional information within the printed objects. With the advances in nanotechnology and material science, 5D printing fills the photoelectric devices with unprecedented functionalities such as multi-spectral absorption, tuneable bandgaps, and enhanced light-trapping capabilities. The encoding of intricate nanostructures and metamaterial architectures, 5D printing enables the realization of next-generation photonic devices with unparalleled performance metrics.

Keywords 3D Printing · 4D Printing · 5D Printing · Additive manufacturing · Photoelectric devices · Smart materials · Nanotechnology

---

M. Vishnuvarthanan (✉)
Department of Mechanical Engineering, Kalasalingam Academy of Research and Education, Srivilliputhur, Tamil Nadu, India
e-mail: vishnuvarthanan.india@gmail.com

S. Moharana et al. (eds.), *Polymer Nanocomposites for 3D, 4D and 5D Printing*, Engineering Materials, https://doi.org/10.1007/978-981-96-4214-4_6

1 Introduction

1.1 Overview of Photoelectric Device

Photoelectric devices play a pivotal role in converting light energy into electrical energy or vice versa [1], supporting a broad spectrum of applications ranging from solar cells and photodetectors to light-emitting diodes (LEDs) and photonic sensors. Photoelectric devices operate based on the principles of the photoelectric effect, wherein incident photons interact with semiconductor materials to generate electron–hole pairs, initiating the electrical conduction or emission of light [2]. This phenomenon forms the basis for various optoelectronic processes, including photovoltaic conversion, photoconductivity, photoluminescence, and photodetection. The performance of photoelectric devices is governed by parameters such as absorption spectra, carrier mobility, recombination rates, and external quantum efficiency, which decide their efficiency, sensitivity, and spectral response characteristics [3].

The core components of photoelectric devices include semiconductor materials, electrodes, and encapsulation layers, each fulfilling distinct roles in device operation and performance. Semiconductor materials such as silicon, gallium arsenide, and organic polymers serve as the active medium for light absorption and charge carrier transport [4]. The electrodes facilitate the electrical contact with external circuits, enabling the extraction of generated charges or injection of electrical currents. The encapsulation layers provide mechanical protection, environmental stability, and optical management, enhancing the device's durability and performance in harsh operating conditions. The photoelectric devices encompass a wide array of functionalities tailored to specific applications and performance requirements. Solar cells also known as photovoltaic cells, harness the sunlight to generate electricity through the photovoltaic effect, offering a sustainable and renewable energy source for power generation [5]. Photodetectors including photodiodes and phototransistors, detect and convert incident light into electrical signals, enabling applications such as optical communication, imaging and sensing [6]. The Light-emitting diodes (LEDs) emit light when forward-biased, serving as essential light sources in displays, illumination and optoelectronic devices. The photoelectric devices find applications in spectroscopy, imaging, optical interconnect and biomedical sensing, demonstrating their versatility and utility across diverse fields [7].

Despite the significant advancements, photoelectric device fabrication faces challenges related to material compatibility, manufacturing scalability, device integration and performance optimization. The traditional fabrication techniques often entail complex processing steps, material wastage and limited design flexibility, hindering the realization of advanced photonic systems with tailored functionalities [8]. However, the emergence of additive manufacturing technologies including 3D, 4D and 5D printing presents unprecedented opportunities for overcoming these challenges. Photoelectric devices represent a cornerstone of modern optoelectronics, enabling in many of applications spanning energy conversion, light detection, emission and manipulation [9]. An understanding of the fundamental principles, key components and diverse functionalities of photoelectric devices is essential for advancing their fabrication processes and unlocking new frontiers in optoelectronic design and technology.

2 Overview of Traditional Photoelectric Device Fabrication Techniques

The traditional methods of fabricating photoelectric devices have been pivotal in laying the groundwork for advancements in optoelectronics and the methods are explained and it is shown in Fig. 1.

2.1 *Photolithography*

Photolithography is a fundamental technique extensively employed in the semiconductor industry for the production of photoelectric devices such as solar cells and photodetectors [10]. The process involves transferring a pattern from a photomask onto a photosensitive substrate, typically using light exposure and chemical etching. The photolithography enables high-resolution patterning and precise feature control,

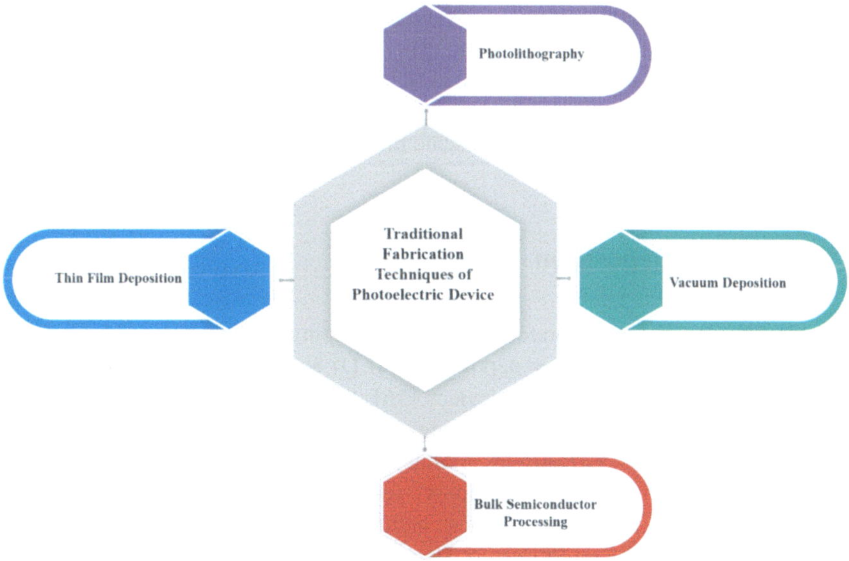

Fig. 1 Traditional fabrication techniques of photoelectric devices

it is constrained by expensive cleanroom facilities, limited material compatibility and sequential processing steps which impede scalability and customization [11].

2.2 *Vacuum Deposition*

The vacuum deposition techniques including physical vapor deposition (PVD) and chemical vapor deposition (CVD) are widely utilized for thin film deposition in photoelectric device fabrication [12]. PVD methods such as sputtering and evaporation involve the deposition of materials onto a substrate surface under vacuum conditions. On the other hand, CVD relies on chemical reactions to form thin films from gaseous precursors [13]. Although vacuum deposition offers excellent film uniformity and purity, it necessitates specialized equipment, stringent environmental controls and restricted substrate choices, thereby restricting its applicability in diverse device architectures.

2.3 *Bulk Semiconductor Processing*

Bulk semiconductor processing techniques such as crystal growth and wafer dicing are essential for the fabrication of crystalline photoelectric devices like solar cells and light-emitting diodes (LEDs) [14]. These methods entail the synthesis of semiconductor materials with controlled crystalline structures followed by slicing into individual wafers for device integration. The bulk semiconductor processing facilitates the production of high-quality materials with tailored optoelectronic properties, it suffers from material wastage, limited design flexibility and batch-to-batch variations posing challenges for large-scale manufacturing and customization [15].

3 Thin Film Deposition

The thin film deposition methodologies including spin coating, chemical bath deposition (CBD) and atomic layer deposition (ALD) are instrumental in fabricating functional layers in photoelectric devices [16]. Spin coating involves the application of a liquid precursor onto a rotating substrate, forming a thin film through centrifugal force. The CBD utilizes chemical reactions in solution to deposit thin films onto substrates, while ALD employs sequential surface reactions to achieve atomic-scale control over film thickness and composition. Despite offering versatility and cost-effectiveness, thin film deposition techniques often suffer from low throughput, limited thickness uniformity and material compatibility constraints [17]. The traditional photoelectric device fabrication techniques have played a pivotal role in advancing the field of optoelectronics, laying the foundation for the development of various photonic systems [18]. However, these conventional approaches have inherent limitations, including high production costs, limited design flexibility and material compatibility constraints which impede their scalability and customization. In light of these challenges, there is a growing imperative to explore additive manufacturing technologies.

4 Emergence of Additive Manufacturing Technologies

The emergence of additive manufacturing technologies has catalysed a paradigm shift in the fabrication of photoelectric devices, offering unprecedented opportunities for innovation, customization and scalability [19]. Additive manufacturing, commonly referred to as 3D printing, traces its origins back to the 1980s with the invention of stereolithography (SLA) by Chuck Hull [20]. SLA pioneered the concept of layer-by-layer fabrication, wherein a UV laser selectively solidifies photopolymer resin to build three-dimensional objects from digital models [20]. Over the past decades, additive manufacturing technologies have undergone rapid evolution, driven by advancements in materials, printing processes and equipment capabilities. Materials innovation has expanded the repertoire of printable materials to include semiconductors, conductive inks, quantum dots and metamaterials enabling the fabrication of functional photoelectric devices with tailored properties and functionalities. Printing processes have become more precise, scalable and versatile with improvements in resolution, speed and multi-material deposition capabilities, enhancing the reliability and complexity of printed structures. The development of hybrid additive manufacturing systems has facilitated the integration of complementary techniques such as subtractive machining, surface modification and post-processing, expanding the design possibilities and performance optimization avenues for photoelectric devices [21].

The advent of additive manufacturing technologies has revolutionized the landscape of optoelectronics offering novel solutions to traditional fabrication challenges and unlocking new frontiers in device design and performance [22]. The elimination of constraints of traditional manufacturing methodologies such as photolithography, vacuum deposition and bulk semiconductor processing, additive manufacturing enables the rapid prototyping, customization and on-demand production of photoelectric devices with intricate geometries, embedded functionalities and hierarchical architectures. Moreover, additive manufacturing facilitates the integration of advanced design paradigms such as topology optimization, lattice structures and conformal printing, enhancing the efficiency, durability and functionality of photoelectric devices across diverse applications [23]. Despite its transformative potential, additive manufacturing in photoelectric device fabrication faces challenges related to material selection, process scalability and printing resolution. The material compatibility, print quality and post-processing requirements vary depending on the specific additive manufacturing technique and application, continuous technological innovation to address these challenges [24]. Moreover, the scalability of additive manufacturing for mass production remains a key consideration, requiring advancements in printing speed, automation and material efficiency to achieve cost-effective and high-volume manufacturing of photoelectric devices. The emergence of additive manufacturing technologies has accompanied a new era of possibilities in the fabrication of photoelectric devices, offering unprecedented opportunities for innovation, customization and scalability [25].

5 Significance of 3D, 4D, and 5D Printing in Photoelectric Device Development

The integration of additive manufacturing technologies including 3D, 4D and 5D printing has emerged as a transformative approach in the development of photoelectric devices. One of the primary advantages of 3D, 4D and 5D printing in photoelectric device development lies in their inherent customization and design flexibility. It is unlike traditional fabrication methods which are often constrained by predefined processes and limited design options, additive manufacturing enables the creation of complex geometries, intricate structures and customized architectures tailored to specific performance requirements. Leveraging computer-aided design (CAD) software and advanced printing techniques can optimize the layout, morphology and composition of photoelectric devices, thereby enhancing their efficiency, reliability and functionality [26].

Furthermore, 3D, 4D and 5D printing facilitate the seamless integration of multiple functionalities within a single photoelectric device, offering enhanced versatility and performance optimization. Through the precise control of material deposition, layer-by-layer assembly and post-printing processing, additive manufacturing enables the incorporation of diverse components such as photovoltaic cells, light-emitting diodes (LEDs), sensors and energy storage elements into cohesive multifunctional systems [27]. This integration not only simplifies device assembly and reduces the manufacturing complexity but also enhances device functionality, enabling advanced functionalities such as self-powering sensors, adaptive photonic circuits and tuneable optoelectronic devices.

The additive manufacturing technologies facilitate rapid prototyping and iterative design cycles, accelerating the development timeline and reducing time-to-market for novel photoelectric devices. The elimination of the need for expensive tooling, specialized equipment, and extensive manual labour associated with traditional fabrication methods, 3D, 4D and 5D printing enable it to iterate designs quickly, test hypotheses and refine prototypes in a cost-effective and time-efficient manner. This iterative approach fosters innovation, creativity and collaboration, empowering researchers to explore novel concepts, optimize the device performance and address emerging challenges in optoelectronics with agility and precision. Additive manufacturing technologies offer scalability and mass production capabilities, enabling the efficient and cost-effective manufacturing of photoelectric devices in large quantities [28]. The traditional fabrication methods may face limitations in scaling production due to batch processing, material wastage and manufacturing complexity, additive manufacturing enables on-demand manufacturing, batch customization and distributed production models [29]. The leverage of automation, digital workflows and advanced materials, 3D, 4D and 5D printing facilitate the scalable manufacturing solutions for photoelectric devices, catering to diverse market demands, accelerating technology adoption and driving the commercialization efforts. The significance of 3D, 4D and 5D printing in photoelectric device development cannot be overstated. These advanced additive manufacturing technologies offer unparalleled opportunities for customization, multifunctional integration, rapid prototyping, scalability and mass production, revolutionizing the traditional type of device fabrication.

6 Fundamentals of 3D Printing

3D printing, also known as additive manufacturing, has emerged as a transformative technology with significant implications for the fabrication of photoelectric devices [30]. The steps involved in the 3D printing process are shown in Fig. 2.

6.1 Additive Manufacturing Process

3D printing operates on the principle of additive manufacturing, wherein objects are built layer by layer from digital design files [31]. It is unlike traditional subtractive manufacturing methods, which involve removing material from a solid block to create a desired shape, additive manufacturing adds material incrementally, enabling the fabrication of complex geometries with minimal waste.

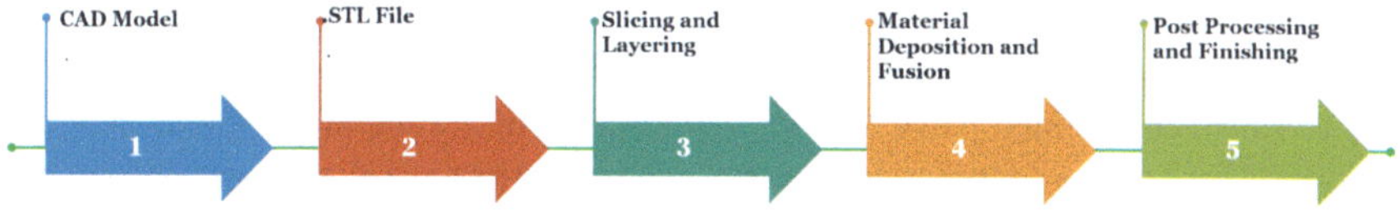

Fig. 2 Steps involved in 3D printing process

6.2 Digital Design and CAD Modelling

The 3D printing process begins with the creation of a digital design using computer-aided design (CAD) software. The CAD models define the geometry, dimensions and specifications of the object to be printed, providing a digital blueprint for the additive manufacturing process [32]. The CAD modelling allows for precise control over design parameters, enabling customization, optimization and iteration of photoelectric device prototypes.

6.3 Slicing and Layering

Once the digital design is finalized, it is sliced into thin cross-sectional layers using slicing software. Each layer represents a horizontal section of the object to be printed and is translated into a series of instructions for the 3D printer. The layer thickness determines the resolution and surface finish of the printed part with thinner layers yielding finer details and smoother surfaces [33].

6.4 Material Deposition and Fusion

The 3D printing process involves the deposition or fusion of material layer by layer to build up the desired object. Various material deposition techniques are employed in 3D printing including fused deposition modelling (FDM), stereolithography (SLA) and selective laser sintering (SLS). Each technique utilizes different materials and processes to achieve specific results offering versatility and flexibility in photoelectric device fabrication [34].

7 Post-Processing and Finishing

After the printing process is complete, post-processing steps may be required to clean, cure or finish the printed parts. It depends on the printing technology and material used, post-processing techniques may include support removal, surface smoothing, chemical treatment or heat curing. These steps help improve the mechanical, optical and aesthetic properties of the printed photoelectric devices.

8 Additive Manufacturing Techniques

The list of various additive manufacturing techniques is shown in Fig. 3.

8.1 VAT Photopolymerization

VAT photopolymerization also known as stereolithography (SLA) is a sophisticated 3D printing technique that holds immense promise in the fabrication of photoelectric devices [35]. VAT photopolymerization operates on the principle of photopolymerization, wherein liquid photopolymer resin is selectively cured or solidified by a UV laser to build up three-dimensional structures layer by layer. The process begins with the spreading of a thin layer of liquid photopolymer resin across a build platform within a vat or tank. A UV laser is then directed onto the surface of the resin, selectively curing or solidifying the resin according to the digital design data. Once a layer is cured, the build platform descends and a new layer of resin is spread over the cured layer. This layering process is repeated iteratively until the entire object is fabricated, with each layer adhering to the underlying layers to form a cohesive and intricate structure.

VAT photopolymerization finds diverse applications in the fabrication of photoelectric devices ranging from prototypes and functional demonstrators to custom-designed components and production parts [28]. The high resolution and surface finish offered by SLA makes it particularly suitable for producing intricate features,

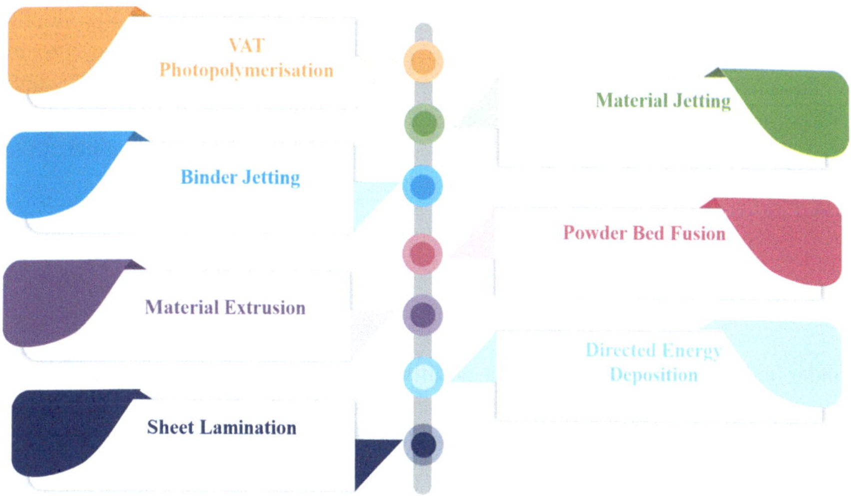

Fig. 3 List of additive manufacturing techniques

fine details and smooth surfaces in optoelectronic components such as lenses, waveguides and microstructures. Moreover, SLA enables the integration of multiple materials, additives and functional elements into printed parts, enhancing their optical, mechanical and electrical properties for specific applications. The SLA facilitates rapid prototyping, iterative design cycles and small-scale production of photoelectric devices, accelerating the development timeline and reducing time-to-market for new optoelectronic products [36].

8.2 Material Jetting

Material jetting is an innovative 3D printing technology that has garnered significant attention in the realm of photoelectric device fabrication [37]. Material jetting operates on the principle of depositing droplets of photopolymer resin or support material onto a build platform, layer by layer, using inkjet printheads. The process begins with the generation of a digital model using computer-aided design (CAD) software, which is then sliced into thin cross-sectional layers. Inkjet printheads selectively deposit droplets of photopolymer resin or support material onto the build platform, guided by the digital design data. The UV light or heat is then used to cure or solidify the deposited material, forming a layer. This layering process is repeated iteratively until the entire object is fabricated with each layer adhering to the underlying layers to create a cohesive and intricate structure.

Material jetting finds diverse applications in the fabrication of photoelectric devices, offering versatility and precision in producing complex geometries, customized features, and multi-material components [8]. The high resolution and fine feature detail offered by material jetting make it particularly suitable for producing lenses, waveguides, microstructures and other optical components with precise dimensions and smooth surfaces. Moreover, the material jetting enables the integration of multiple materials, additives and functional elements into printed parts, enhancing their optical, mechanical and electrical properties for specific applications [38].

8.3 Binder Jetting

Binder jetting is a groundbreaking additive manufacturing technique that holds immense potential in the realm of photoelectric device fabrication [37]. Binder jetting operates on the principle of selectively depositing a binding agent onto a powdered material bed, layer by layer to create a solid object. The process begins with the spreading of a thin layer of powdered material such as ceramics, metals or polymers across a build platform. A printhead selectively deposits droplets of a liquid binding agent onto the powdered material, binding the particles together to form a solid layer. The build platform then descends and a new layer of powdered material is spread over the previous layer. This layering process is repeated iteratively until the entire object is fabricated with each layer adhering to the underlying layers to create a cohesive and intricate structure.

Binder jetting finds diverse applications in the fabrication of photoelectric devices, offering versatility and efficiency in producing complex geometries, customized features and multi-material components [25]. The ability to print with a wide range of materials, including ceramics, metals and polymers makes binder jetting suitable for producing a variety of optoelectronic components such as lenses, waveguides, enclosures and structural supports.

8.4 Powder Bed Fusion

Powder bed fusion (PBF) is a transformative additive manufacturing technique that holds significant promise in the fabrication of photoelectric devices [39]. Powder bed fusion operates on the principle of selectively fusing powdered materials such as polymers or metals, layer by layer to create solid objects. The process begins with the spreading of a thin layer of powdered material across a build platform. A high-powered energy source such as a laser or electron beam, selectively fuses the powdered material, binding it together to form a solid layer. The build platform then descends and a new layer of powdered material is spread over the previous layer. This layering process is repeated iteratively until the entire object is fabricated with each layer adhering to the underlying layers to create a cohesive and intricate structure.

Powder bed fusion finds diverse applications in the fabrication of photoelectric devices offering versatility and precision in producing complex geometries, customized features and high-performance components [40]. The ability to print with a wide range of materials including polymers, metals and ceramics makes powder bed fusion suitable for producing a variety of optoelectronic components such as lenses, waveguides, enclosures and heat sinks. Moreover, powder bed fusion enables the integration of functional additives, dopants and coatings into printed parts, enhancing their optical, mechanical and electrical properties for specific applications.

8.5 Material Extrusion

Material extrusion also known as fused deposition modelling (FDM) is a leading additive manufacturing technique that holds immense potential in the fabrication of photoelectric devices. Material extrusion operates on the principle of selectively extruding thermoplastic filaments through a heated nozzle, which deposits layers of material onto a build platform according to a programmed digital design. The process begins with the generation of a digital model using computer-aided design (CAD) software which is then sliced into thin cross-sectional layers. The sliced model is then fed to the 3D printer, where the thermoplastic filament is heated to its melting point and extruded through the nozzle. The nozzle moves along the X, Y and Z axes depositing the melted material layer by layer to build up the desired object. Once a layer is deposited, it rapidly cools and solidifies, bonding to the underlying layers to form a cohesive and intricate structure.

Material extrusion finds diverse applications in the fabrication of photoelectric devices offering versatility and efficiency in producing complex geometries, customized features and functional prototypes [41]. The ability to print with a wide range of thermoplastics, including ABS, PLA, PETG and TPU makes material extrusion suitable for producing a variety of optoelectronic components such as housings, brackets, enclosures and support structures.

8.6 Directed Energy Deposition

Directed energy deposition (DED) is a cutting-edge additive manufacturing technique that holds immense promise in the fabrication of photoelectric devices. Directed energy deposition finds diverse applications in the fabrication of photoelectric devices offering versatility and efficiency in producing complex geometries, customized features and high-performance components. The ability to print with a wide range of metals and alloys, including aluminium, titanium and stainless steel makes DED suitable for producing a variety of optoelectronic components such as heat sinks, mounting brackets, structural supports and electrode arrays.

8.7 Sheet Lamination

Sheet lamination is an innovative additive manufacturing technique that holds significant promise in the fabrication of photoelectric devices. Sheet lamination operates on the principle of bonding layers of material together to create a solid object [42]. The process begins with the stacking of thin sheets or layers of material such as paper, plastic or metal onto a build platform. Each layer is precisely cut or shaped according to the digital design data using techniques such as laser cutting or CNC milling. Once all layers are stacked, they are bonded together using heat, pressure or adhesives to form a cohesive and intricate structure. The excess material may then be removed through machining or finishing processes to reveal the final part.

Sheet lamination finds diverse applications in the fabrication of photoelectric devices, offering versatility and efficiency in producing complex geometries, customized features and functional prototypes [8]. The ability to print with a wide range of materials including paper, plastics and metals makes sheet lamination suitable for producing a variety of optoelectronic components such as casings, enclosures, housings and substrates.

9 Applications of 3D Printing in Photoelectric Devices

3D printing technology has emerged as a transformative tool in the fabrication of photoelectric devices offering unprecedented flexibility, precision and efficiency in manufacturing processes [43]. The various applications of 3D printing in photoelectric devices are as follows.

9.1 Customized Optics and Lens Systems

One of the primary applications of 3D printing in photoelectric devices is the fabrication of customized optics and lens systems. 3D printing enables the rapid prototyping and production of complex optical components with intricate geometries and tailored properties. Using additive manufacturing techniques such as stereolithography (SLA) and digital light processing (DLP) the lenses, waveguides, diffusers and other optical elements with precise dimensions and optical characteristics can be created [44]. This capability allows for the customization of optics to meet specific performance requirements, such as focal length, aperture size and light transmission leading to enhanced efficiency and performance in photoelectric devices.

9.2 Functional Enclosures and Housings

The significant application of 3D printing in photoelectric devices is the fabrication of functional enclosures and housings. Additive manufacturing techniques such as fused deposition modelling (FDM) and selective laser sintering (SLS) enable the production of complex, lightweight and durable enclosures for electronic components, sensors and circuitry [45]. These 3D-printed enclosures can be customized to fit specific device configurations, accommodate various mounting options and provide protection against environmental factors such as moisture, dust and vibration. The 3D printing allows for the integration of features such as heat sinks, cooling channels, and electromagnetic shielding enhancing the overall performance and reliability of photoelectric devices in diverse operating environments.

9.3 Structural Supports and Mounting Components

3D printing technology offers an efficient solution for fabricating structural supports and mounting components in photoelectric devices [46]. Additive manufacturing techniques such as material extrusion and binder jetting enable the production of lightweight, high-strength structures with intricate geometries and tailored mechanical properties. These 3D-printed supports and mounts can be designed to precisely fit the contours of device components, providing stable and secure attachment points while minimizing weight and material usage. The 3D printing allows for the integration of features such as cable management channels, strain relief mechanisms and alignment guides, simplifying assembly and installation processes in photoelectric device manufacturing [47].

9.4 Customized Sensor Housings and Interfaces

The 3D printing technology enables the fabrication of customized sensor housings and interfaces for photoelectric devices. Additive manufacturing techniques such as powder bed fusion and directed energy deposition allow for the production of complex, multi-functional enclosures with integrated sensor mounts, connectors and cable routing features. These 3D-printed housings can be tailored to accommodate various sensor types, sizes and orientations, providing secure and reliable protection for sensitive electronic components [48]. Moreover, 3D printing enables the rapid iteration and optimization of sensor housing designs, facilitating the development of innovative sensing solutions for diverse applications in photovoltaics, spectroscopy, imaging and sensing.

10 Overview of 4D Printing in Photoelectric Devices

4D printing represents a groundbreaking advancement in additive manufacturing technology, offering the capability to create dynamic, shape-shifting structures that evolve over time in response to external stimuli. The concept of 4D Printing is shown in Fig. 4. 4D printing expands upon the principles of 3D printing by introducing the dimension of time as a dynamic element in the manufacturing process [49]. Traditional 3D printing involves the layer-by-layer deposition of materials to create static, predetermined structures and 4D printing adds an additional dimension of functionality by imbuing printed objects with the ability to change shape, adapt and respond to environmental stimuli autonomously or through external triggers [50]. This transformative capability is achieved through the use of smart materials, advanced design algorithms and programmable fabrication techniques that enable printed structures to undergo predefined shape transformations in a controlled manner.

The principles of 4D printing are rooted in the design, fabrication and activation of responsive materials that exhibit programmable shape-changing behaviours. Smart materials such as shape memory polymers, hydrogels and composites are integrated into the 3D printing process to impart dynamic properties to printed structures [51]. These materials undergo reversible changes in shape, size or stiffness in response to external stimuli such as temperature, humidity, light or magnetic fields. With

Fig. 4 Concept of 4D printing

the control of composition, geometry and activation parameters of printed objects, designers can engineer complex shape transformations that enable functionalities such as self-assembly, self-repair, adaptive optics, and reconfigurable circuits in photoelectric devices.

11 Applications of 4D Printing in Photoelectric Devices

4D printing holds immense promise for a wide range of applications in photoelectric device fabrication, offering unique opportunities for innovation and advancement in optoelectronic design and technology [52]. One of the key applications of 4D printing in this domain is the development of adaptive optics systems that can dynamically adjust their shape and focal properties to optimize light transmission, capture and manipulation in imaging, sensing and display applications. The 4D-printed structures can be utilized to create self-assembling photonic components such as lenses, gratings and waveguides that adapt to changing environmental conditions or user preferences. The 4D printing enables the fabrication of reconfigurable circuits and electronic interfaces in photoelectric devices allowing for dynamic control and optimization of electrical conductivity, signal routing and power distribution. The integration of shape-changing materials and mechanisms into printed circuits can create flexible, stretchable and self-healing electronic systems that adapt to mechanical stress, environmental fluctuations, and device malfunctions [53]. The concept and principles of 4D printing represent a paradigm shift in additive manufacturing technology opening new frontiers in the design and fabrication of photoelectric devices. The harnessing of dynamic capabilities of responsive materials and advanced fabrication techniques made 4D printing offer unique opportunities for creating adaptive, intelligent and self-regulating optoelectronic systems.

12 Smart Materials for Photoelectric Devices

Smart materials and responsive structures represent a transformative approach to photoelectric device fabrication offering dynamic functionalities and adaptive capabilities that enable enhanced performance and versatility [54]. Smart materials are a class of materials that possess inherent properties enabling them to respond to external stimuli with controlled and reversible changes in their physical, chemical or mechanical properties [55]. Smart materials exhibit a diverse range of properties and functionalities that make them ideal for applications in photoelectric devices. One key property of smart materials is shape memory which enables them to retain a pre-defined shape and recover their original form after deformation or exposure to external stimuli. Shape memory polymers and alloys can be used to create adaptive optics components such as lenses and mirrors that change shape in response to temperature or mechanical stress [56]. The important property of smart materials is self-healing which allows them to repair damage or restore functionality when subjected to mechanical or chemical stress. Self-healing polymers and composites can be incorporated into photoelectric device components to enhance durability, reliability and longevity reducing the need for maintenance and replacement [57].

Smart materials and responsive structures find diverse applications in the fabrication of photoelectric devices, offering unique opportunities for innovation and advancement in optoelectronic design and technology. One application is the development of adaptive optics systems that utilize shape-changing materials to dynamically adjust the focal properties, aberrations and alignment of optical components such as lenses, mirrors and prisms in response to changing environmental conditions or user requirements. Another application is the creation of self-assembling photonic components such as waveguides, gratings and filters that spontaneously reconfigure their structure and functionality in response to external stimuli. The smart materials and responsive structures can be used to create flexible, stretchable and wearable photoelectric devices that conform to irregular surfaces and withstand mechanical deformation [58]. The smart materials and responsive structures represent a transformative approach to photoelectric device fabrication offering dynamic functionalities and adaptive capabilities that enable enhanced performance, versatility and functionality.

13 Overview of 5D Printing in Photoelectric Devices

The evolution of additive manufacturing has led to remarkable advancements in various industries including the field of photoelectric devices. 3D printing has revolutionized the manufacturing processes and the emergence of 4D printing introduced dynamic capabilities to printed structures. Now, with the introduction of 5D printing, a new era of innovation is upon us. The concept of 5D printing is shown in Fig. 5.

5D printing represents a significant advancement beyond its predecessors, adding an additional dimension of functionality to the manufacturing process [59]. 3D printing involves the layer-by-layer deposition of materials to create static objects, 4D printing introduces dynamic capabilities by incorporating smart materials that respond to external stimuli. The 5D printing extends this concept further by integrating self-assembling, self-healing, or programmable components into the printed structures, enabling autonomous adaptation, evolution or optimization of the devices over time [60]. This additional dimension of functionality represents a paradigm

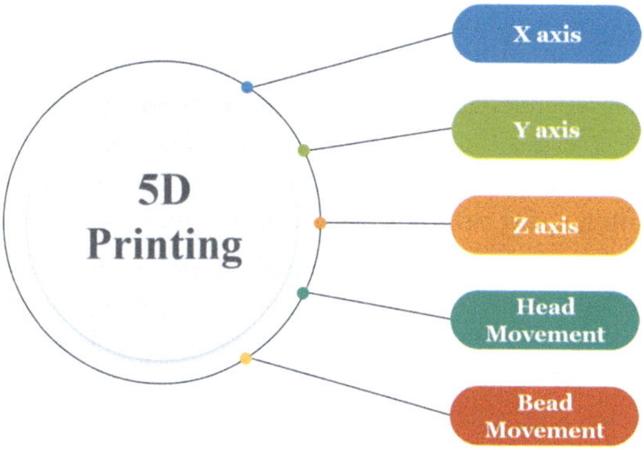

Fig. 5 Concept of 5D printing

shift in additive manufacturing offering unprecedented opportunities for innovation and customization in photoelectric device fabrication. The principles of 5D printing build upon those of its predecessors, emphasizing the integration of smart materials, advanced design algorithms and programmable fabrication techniques to create dynamic, self-regulating structures [61]. Unlike traditional manufacturing methods, which rely on static moulds or tooling, 5D printing enables the creation of complex, multi-functional devices with tailored properties and behaviours. With the unique capabilities of self-assembling materials, self-healing mechanisms or programmable components, the designers can create photoelectric devices that adapt to changing environmental conditions, user requirements or operational parameters autonomously or through external triggers.

14 Potential Applications of 5D Printing in Photoelectric Devices

The introduction of 5D printing opens up many potential applications in the fabrication of photoelectric devices, spanning from customized optics and adaptive sensors to self-repairing circuits and reconfigurable interfaces [62]. One application is the development of adaptive optics systems that dynamically adjust their shape, focal properties or irregularities in response to changing lighting conditions, user preferences or environmental factors. It can possible to create photoelectric devices that optimize light transmission, capture and manipulation in imaging, sensing and display applications autonomously by the integration of self-assembling materials or programmable actuators into the printed structures,

Another application is the creation of self-repairing circuits and electronic interfaces that detect and repair damage or degradation in real-time, ensuring continuous operation and reliability of the devices [63]. The incorporation of self-healing materials, embedded sensors and intelligent control systems into the printed structures can create photoelectric devices that adapt to mechanical stress, environmental fluctuations or component failures autonomously, extending their lifespan and reducing maintenance requirements. The introduction of 5D printing has far-reaching implications for the future of additive manufacturing and optoelectronics. Enabling the creation of dynamic, self-regulating structures with autonomous adaptation capabilities, 5D printing opens up new possibilities for innovation, customization and optimization in photoelectric device fabrication [64]. This transformative technology has the potential to revolutionize various industries including healthcare, aerospace, automotive and consumer electronics by unlocking new levels of performance, functionality and efficiency in printed devices. The introduction of 5D printing represents a significant milestone in additive manufacturing accompanying a new era of innovation and customization in photoelectric device fabrication.

15 Case Studies

15.1 3D Printing for Photoelectric Devices

In the applications of photoelectric devices, the integration of 3D printing technology has led to significant advancements, offering innovative solutions for design, prototyping and manufacturing. The exploration of various case studies can gain insights into the diverse applications and benefits of 3D printing in this field. One notable case study involves the development of customized solar panels using 3D printing technology. The traditional manufacturing processes for solar panels often involve complex and expensive fabrication techniques, limiting customization and scalability. However, researchers have leveraged 3D printing to overcome these challenges by creating intricate designs with precise geometries tailored to specific environments or applications. A research team at MIT developed a novel approach for 3D printing solar cells using perovskite materials. The utilisation of additive manufacturing techniques was able to fabricate solar panels with intricate microstructures that enhance light absorption and energy conversion efficiency. This breakthrough not only demonstrates the feasibility of 3D printing for producing high-performance photovoltaic devices but also opens up possibilities for cost-effective and customizable solar energy solutions. Another study revolves around the use of 3D printing in the production of light-emitting diodes (LEDs) and organic light-emitting diodes (OLEDs). These lighting technologies play a crucial role in various applications including displays, signage and general illumination. The traditional manufacturing processes for LEDs often involve complex assembly steps and material wastage. However, 3D printing offers a more streamlined and efficient approach allowing for the direct deposition of functional materials to create intricate LED structures.

The researchers at the University of California, Berkeley utilized 3D printing to fabricate high-resolution micro-LED arrays for display applications. The precise control of the deposition of LED materials was able to achieve sub-micron feature sizes and intricate geometries not feasible with conventional lithography techniques. This case study highlights the potential of 3D printing to revolutionize the production of LED devices enabling the development of next-generation displays with enhanced performance and functionality. The 3D printing technology has been instrumental in the development of novel photonic devices such as optical sensors and waveguides. These devices play a crucial role in various applications including telecommunications, biomedical imaging, and environmental monitoring. The traditional manufacturing techniques for photonic devices often involve complex fabrication processes and limited design flexibility. However, 3D printing offers a versatile and cost-effective solution for prototyping and producing customized photonic structures.

The researchers at the University of Cambridge utilized 3D printing to fabricate photonic crystal waveguides with precisely engineered bandgaps for controlling light propagation. The tailoring of the geometrical parameters of the waveguide structures was able to achieve unprecedented control over light confinement and dispersion properties. This case study demonstrates the potential of 3D printing to accelerate the development of advanced photonic devices with tailored functionalities for a wide range of applications. The case studies discussed above illustrate the transformative impact of 3D printing technology on the field of photoelectric devices. 3D printing offers unparalleled opportunities for innovation and advancement from customized solar panels to high-resolution LED arrays and advanced photonic structures,

15.2 4D Printing for Photoelectric Devices

The fusion of 4D printing technology with innovative materials has opened up new avenues for design, functionality and adaptability in photoelectric devices [65]. One study revolves around the development of self-assembling solar panels using 4D printing techniques. The traditional methods of manufacturing solar panels often involve complex assembly processes and rigid structures, limiting their adaptability to different environments and applications. However, researchers have harnessed the capabilities of 4D printing to create solar panels that can autonomously reconfigure their shape and orientation in response to changing sunlight conditions.

A research team at Harvard University utilized 4D printing to fabricate lightweight, foldable solar panels inspired by the natural folding mechanisms found in plants. The incorporation of shape-memory polymers (SMPs) into the panel's structure, they were able to achieve self-folding behaviour triggered by changes in temperature or light exposure. This innovative approach not only simplifies the manufacturing process but also enhances the efficiency and versatility of solar energy harvesting systems giving the way for applications in remote or dynamic environments. Another study focused on the development of photonic devices using 4D printing technology. Photonic devices such as optical filters, modulators and sensors play a crucial role in various applications including telecommunications, biomedical imaging and environmental monitoring. The traditional manufacturing techniques for photonic devices often involve labour-intensive processes and limited tunability.

However, 4D printing offers a promising solution for fabricating photonic structures with dynamically adjustable properties.

The researchers at the University of California, San Diego, utilized 4D printing to create photonic crystals capable of dynamically modulating light propagation and emission properties. The incorporation of responsive materials such as liquid crystal elastomers (LCEs) or hydrogels into the photonic crystal matrix, they were able to achieve reversible changes in optical bandgaps and refractive indices. This breakthrough enables the development of adaptive photonic devices with on-demand tunability, opening up new possibilities for reconfigurable optics and advanced sensing applications. The 4D printing has enabled the development of biomimetic photonic structures inspired by nature's intricate designs. With the mimicking of the hierarchical organization and responsive behaviours found in biological systems, researchers can create photonic devices with enhanced functionality and adaptability. A research team at MIT used 4D printing to fabricate bio-inspired photonic structures capable of dynamically adjusting their optical properties in response to environmental stimuli.

The incorporation of responsive materials such as shape-memory polymers and hydrogels, into the structure, they were able to achieve reversible changes in colour, reflectance and light absorption. This bio-inspired approach not only enhances the performance of photonic devices but also offers insights into the design principles underlying natural photonic systems giving the way for biomimetic engineering and adaptive optics. The case studies discussed above highlight the transformative impact of 4D printing on the field of photoelectric devices. From self-assembling solar panels to photonic devices and biomimetic photonic structures, 4D printing offers unprecedented opportunities for innovation and advancement.

15.3 5D Printing for Photoelectric Devices

The 5D printing holds tremendous potential for revolutionizing design, fabrication and performance in photoelectric devices. One case study involves the development of multifunctional photovoltaic devices using 5D printing technology. The traditional methods of manufacturing photovoltaic cells often rely on flat, rigid substrates and fixed material compositions limiting their efficiency and adaptability. However, researchers have used the capabilities of 5D printing to create solar cells with customizable microstructures, material gradients and energy harvesting functionalities.

A research team at Stanford University utilized 5D printing to fabricate solar cells with hierarchical architectures inspired by natural photosynthetic systems. The control of the composition and organization of light-absorbing materials at multiple length scales, they were able to enhance light absorption, charge separation and carrier transport within the device. This innovative approach not only improves the efficiency of solar energy conversion but also enables the integration of additional functionalities such as self-cleaning surfaces or energy storage components. Another study focused on the development of 5D-printed optical sensors with tailored spectral responses and sensitivity profiles. Optical sensors play a crucial role in various applications including environmental monitoring, biomedical diagnostics and industrial sensing. The traditional manufacturing techniques for optical sensors often involve complex lithographic processes and limited design flexibility. However, 5D printing offers a versatile and cost-effective solution for fabricating sensors with customizable geometries, material compositions and optical properties.

The researchers at MIT utilized 5D printing to create optical sensors capable of detecting specific wavelengths of light with high sensitivity and selectivity. With the incorporation of functional materials such as quantum dots or plasmonic nanoparticles into the sensor matrix, they were able to achieve spectral responses tailored to different sensing applications. This breakthrough enables the development of compact, multiplexed sensor arrays with enhanced performance and functionality opening up new opportunities for real-time monitoring and analysis. Furthermore, 5D printing has enabled the fabrication of biomimetic photonic structures inspired by nature's intricate designs. A research team at Harvard University leveraged 5D printing to fabricate bio-inspired photonic crystals capable of dynamically adjusting their optical properties in response to environmental stimuli. 5D printing offers unprecedented opportunities for innovation and advancement from multifunctional photovoltaic cells to customizable optical sensors and biomimetic photonic structures.

16 Conclusion

The advent of additive manufacturing technologies has catalysed a profound transformation across various industries, particularly within the area of photoelectric devices. This chapter has provided an in-depth exploration of the revolutionary potential of 3D, 4D and 5D printing technologies, enlightening their capacity to redefine the development and fabrication processes of advanced photoelectric devices. Beginning with a critical examination of traditional fabrication techniques and their inherent limitations, the narrative swiftly transitions to the disruptive emergence of additive manufacturing as a viable alternative. The explanation of the foundational principles of 3D printing, the chapter has underscored the unparalleled flexibility, precision and scalability offered by layer-by-layer deposition techniques. Moreover, the exploration into the frontier of 4D printing, where materials exhibit dynamic responses to external stimuli, transcending the confines of static structures. With the integration of smart materials and programmable design parameters, 4D printing unlocks new avenues for the fabrication of adaptive photoelectric devices capable of self-assembly, self-repair and shape transformation. This evolution promised to overcome traditional manufacturing constraints and usher in a new era of dynamic and responsive photonic systems. Furthermore, the research extended to the promising domain of 5D printing which exceeds the physical dimensions by embedding additional information within printed objects. 5D printing equipped with photoelectric devices with unprecedented functionalities including multi-spectral absorption, bandgaps and enhanced light-trapping capabilities. The encoding of intricate nanostructures and metamaterial architectures made by 5D printing gives the way for next-generation photonic devices boasting unparalleled performance metrics.

Finally, this chapter explained the paradigm shift catalysed by 3D, 4D and 5D printing technologies within the area of photoelectric devices. Surpassing of the limitations of traditional manufacturing methodologies, additive manufacturing offers unparalleled opportunities for innovation, customization and scalability in the development of advanced photonic systems. The future of 3D, 4D and 5D printing technology in photoelectric devices holds promise for enhanced efficiency, adaptability and functionality. The advancements may include intricate designs for solar panels, self-assembling photonic structures and embedding additional information for multifunctionality, driving innovation in renewable energy and optoelectronics.

References

1. Zhang, H., Lu, Y., Han, W., Zhu, J., Zhang, Y., Huang, W.: Solar energy conversion and utilization: towards the emerging photo-electrochemical devices based on perovskite photovoltaics. Chem. Eng. J. **393**, 124766 (2020)
2. Zou, W., Sastry, M., Gooding, J.J., Ramanathan, R., Bansal, V.: Recent advances and a roadmap to wearable UV sensor technologies. Adv. Mater. Technol. **5**(4), 1901036 (2020)
3. Yu, J., Liu, G., Chen, C., Li, Y., Xu, M., Wang, T., Zhao, G., Zhang, L.: Perovskite CsPbBr 3 crystals: growth and applications. J. Mater. Chem. C **8**(19), 6326–6341 (2020)
4. Morab, S., Sundaram, M.M., Pivrikas, A.: Review on charge carrier transport in inorganic and organic semiconductors. Coatings **13**(9), 1657 (2023)
5. Dambhare, M.V., Butey, B. and Moharil, S.V.: May. Solar photovoltaic technology: a review of different types of solar cells and its future trends. J. Phys. Conf. Ser. (Vol. 1913, No. 1, p. 012053). IOP Publishing (2021)
6. Wang, C., Zhang, X., Hu, W.: Organic photodiodes and phototransistors toward infrared detection: materials, devices, and applications. Chem. Soc. Rev. **49**(3), 653–670 (2020)
7. Babicheva, V.E.: Optical processes behind plasmonic applications. Nanomaterials **13**(7), 1270 (2023)
8. Zhu, Y., Tang, T., Zhao, S., Joralmon, D., Poit, Z., Ahire, B., Keshav, S., Raje, A.R., Blair, J., Zhang, Z., Li, X.: Recent advancements and applications in 3D printing of functional optics. Addit. Manuf. **52**, 102682 (2022)
9. Iqbal, M.A., Malik, M., Le, T.K., Anwar, N., Bakhsh, S., Shahid, W., Shahid, S., Irfan, S., Al-Bahrani, M., Morsy, K., Do, H.B.: Technological evolution of image sensing designed by nanostructured materials. ACS Materials Letters **5**(4), 1027–1060 (2023)
10. Fang, J., Zhou, Z., Xiao, M., Lou, Z., Wei, Z., Shen, G.: Recent advances in low-dimensional semiconductor nanomaterials and their applications in high-performance photodetectors. InfoMat **2**(2), 291–317 (2020)
11. Morbioli, G.G., Speller, N.C., Stockton, A.M.: A practical guide to rapid-prototyping of PDMS-based microfluidic devices: a tutorial. Anal. Chim. Acta **1135**, 150–174 (2020)
12. Oke, J.A., Jen, T.C.: Atomic layer deposition and other thin film deposition techniques: from principles to film properties. J. Market. Res. **21**, 2481–2514 (2022)
13. Sabzi, M., Mousavi Anijdan, S.H., Shamsodin, M., Farzam, M., Hojjati-Najafabadi, A., Feng, P., Park, N., Lee, U.: A review on sustainable manufacturing of ceramic-based thin films by chemical vapor deposition (CVD): reactions kinetics and the deposition mechanisms. Coatings **13**(1), 188 (2023)

14. Hossain, N., Mobarak, M.H., Mimona, M.A., Islam, M.A., Hossain, A., Zohur, F.T. and Chowdhury, M.A.: Advances and significances of nanoparticles in semiconductor applications–a review. Results Eng., 101347 (2023)
15. Rogdakis, K., Psaltakis, G., Fagas, G., Quinn, A., Martins, R., Kymakis, E.: Hybrid chips to enable a sustainable internet of things technology: opportunities and challenges. Discover Mater. **4**(1), 4 (2024)
16. Yaseen, M., Khattak, M.A.K., Khan, A., Bibi, S., Bououdina, M., Usman, M., Khan, N.A., Pirzado, A.A.A., Abumousa, R.A., Humayun, M.: State-of-the-art electrochromic thin films devices, fabrication techniques and applications: a review. Nanocomposites **10**(1), 1–40 (2024)
17. Agresti, A., Di Giacomo, F., Pescetelli, S. and Di Carlo, A.: Scalable deposition techniques for large-area perovskite photovoltaic technology: a multi-perspective review. Nano Energy, p. 109317 (2024)
18. Cheng, Z., Cao, R., Wei, K., Yao, Y., Liu, X., Kang, J., Dong, J., Shi, Z., Zhang, H., Zhang, X.: 2D materials enabled next-generation integrated optoelectronics: from fabrication to applications. Adv. Sci. **8**(11), 2003834 (2021)
19. Mehla, S., Selvakannan, P.R., Mazur, M., Bhargava, S.K.: Emerging technological applications of additive manufacturing. In: Additive Manufacturing for Chemical Sciences and Engineering, pp. 169–238. Springer Nature Singapore, Singapore (2022)
20. Pelz, J.S., Ku, N., Meyers, M.A., Vargas-Gonzalez, L.R.: Additive manufacturing of structural ceramics: a historical perspective. J. Mater. Res. Technol., **15**, 670–695 (2021)
21. Rao, J., Leong Sing, S., Liu, P., Wang, J., Sohn, H.: Non-destructive testing of metal-based additively manufactured parts and processes: a review. Virtual Phys. Prototyping **18**(1), e2266658 (2023)
22. Alkunte, S., Fidan, I., Naikwadi, V., Gudavasov, S., Ali, M.A., Mahmudov, M., Hasanov, S., Cheepu, M.: Advancements and challenges in additively manufactured functionally graded materials: a comprehensive review. J. Manuf. Mater. Process. **8**(1), 23 (2024)
23. Renjith, S.C., Park, K., Okudan Kremer, G.E.: A design framework for additive manufacturing: Integration of additive manufacturing capabilities in the early design process. Int. J. Precis. Eng. Manuf. **21**, 329–345 (2020)
24. Kumar, M.B., Sathiya, P.: Methods and materials for additive manufacturing: a critical review on advancements and challenges. Thin-Walled Struct. **159**, 107228 (2021)
25. Li, Y., Feng, Z., Hao, L., Huang, L., Xin, C., Wang, Y., Bilotti, E., Essa, K., Zhang, H., Li, Z., Yan, F.: A review on functionally graded materials and structures via additive manufacturing: from multi-scale design to versatile functional properties. Adv. Mater. Technol. **5**(6), 1900981 (2020)
26. Tan, N.Y.J., Zhang, X., Neo, D.W.K., Huang, R., Liu, K., Kumar, A.S.: A review of recent advances in fabrication of optical Fresnel lenses. J. Manuf. Process. **71**, 113–133 (2021)
27. Wu, Y., An, C., Guo, Y.: 3D printed graphene and graphene/polymer composites for multifunctional applications. Materials **16**(16), 5681 (2023)
28. Huang, Z., Shao, G., Li, L.: Micro/nano functional devices fabricated by additive manufacturing. Prog. Mater Sci. **131**, 101020 (2023)
29. Mehrpouya, M., Dehghanghadikolaei, A., Fotovvati, B., Vosooghnia, A., Emamian, S.S., Gisario, A.: The potential of additive manufacturing in the smart factory industrial 4.0: a review. Appl. Sci. **9**(18), 3865 (2019)
30. Zhang, C., Li, Y., Kang, W., Liu, X., Wang, Q.: Current advances and future perspectives of additive manufacturing for functional polymeric materials and devices. SusMat **1**(1), 127–147 (2021)
31. Hwang, D., Blake Perez, K., Anderson, D., Jensen, D., Camburn, B., Wood, K.: Design principles for additive manufacturing: leveraging crowdsourced design repositories. J. Mech. Des. **143**(7), 072005 (2021)
32. Kumar, A., Kumar, P., Mittal, R.K., Singh, H.: Printing file formats for additive manufacturing technologies. Adv. Additive Manuf., 87–102 (2023)
33. Lou, S., Jiang, X., Sun, W., Zeng, W., Pagani, L., Scott, P.J.: Characterisation methods for powder bed fusion processed surface topography. Precis. Eng. **57**, 1–15 (2019)

34. Berglund, G., Wisniowiecki, A., Gawedzinski, J., Applegate, B., Tkaczyk, T.S.: Additive manufacturing for the development of optical/photonic systems and components. Optica **9**(6), 623–638 (2022)
35. Li, M., Zhou, S., Cheng, L., Mo, F., Chen, L., Yu, S., Wei, J.: 3D printed supercapacitor: techniques, materials, designs, and applications. Adv. Func. Mater. **33**(1), 2208034 (2023)
36. Touri, M., Kabirian, F., Saadati, M., Ramakrishna, S., Mozafari, M.: Additive manufacturing of biomaterials—the evolution of rapid prototyping. Adv. Eng. Mater. **21**(2), 1800511 (2019)
37. Rahman, M., Islam, K.S., Dip, T.M., Chowdhury, M.F.M., Debnath, S.R., Hasan, S.M.M., Sakib, M.S., Saha, T., Padhye, R., Houshyar, S.: A review on nanomaterial-based additive manufacturing: dynamics in properties, prospects, and challenges. Progress Additive Manuf., 1–28 (2023)
38. Elkaseer, A., Chen, K.J., Janhsen, J.C., Refle, O., Hagenmeyer, V., Scholz, S.G.: Material jetting for advanced applications: a state-of-the-art review, gaps and future directions. Addit. Manuf. **60**, 103270 (2022)
39. Groarke, R., Vijayaraghavan, R.K., Powell, D., Rennie, A., Brabazon, D.: Powder characterization– methods, standards, and state of the art. Fundamentals of Laser Powder Bed Fusion of Metals, pp. 491–527 (2021)
40. Jiang, P., Ji, Z., Zhang, X., Liu, Z., Wang, X.: Recent advances in direct ink writing of electronic components and functional devices. Progress Additive Manuf. **3**, 65–86 (2018)
41. Chen, C., Feng, J., Li, J., Guo, Y., Shi, X., Peng, H.: Functional fiber materials to smart fiber devices. Chem. Rev. **123**(2), 613–662 (2022)
42. Gibson, I., Rosen, D., Stucker, B., Khorasani, M., Gibson, I., Rosen, D., Stucker, B., Khorasani, M.: Sheet lamination. Additive Manufacturing Technologies, pp. 253–283 (2021)
43. Hassan, K., Nine, M.J., Tung, T.T., Stanley, N., Yap, P.L., Rastin, H., Yu, L., Losic, D.: Functional inks and extrusion-based 3D printing of 2D materials: a review of current research and applications. Nanoscale **12**(37), 19007–19042 (2020)
44. Stender, B., Mantei, W., Wiedenmann, J., Dupuis, Y., Hilbert, F., Houbertz, R., von Edlinger, M., Kistner, C., Koeth, J.: Industrial-scale fabrication of optical components using high-precision 3D printing: aspects-applications-perspectives. 3D Printing of Optical Components, pp. 185–237 (2021)
45. Wu, H., Fahy, W.P., Kim, S., Kim, H., Zhao, N., Pilato, L., Kafi, A., Bateman, S., Koo, J.H.: Recent developments in polymers/polymer nanocomposites for additive manufacturing. Prog. Mater Sci. **111**, 100638 (2020)
46. Goh, G.L., Zhang, H., Chong, T.H., Yeong, W.Y.: 3D printing of multilayered and multimaterial electronics: a review. Adv. Electron. Mater. **7**(10), 2100445 (2021)
47. Butt, J.: Exploring the interrelationship between additive manufacturing and Industry 4.0. Designs **4**(2), 13 (2020)
48. Jiang, Y., Islam, M.N., He, R., Huang, X., Cao, P.F., Advincula, R.C., Dahotre, N., Dong, P., Wu, H.F., Choi, W.: Recent advances in 3D printed sensors: materials, design, and manufacturing. Adv. Mater. Technolo. **8**(2), 2200492 (2023)
49. Momeni, F., Ni, J.: Laws of 4D printing. Engineering **6**(9), 1035–1055 (2020)
50. Leist, S.K., Zhou, J.: Current status of 4D printing technology and the potential of light-reactive smart materials as 4D printable materials. Virtual Phys. Prototyping **11**(4), 249–262 (2016)
51. Zhang, J., Yin, Z., Ren, L., Liu, Q., Ren, L., Yang, X., Zhou, X.: Advances in 4D printed shape memory polymers: from 3D printing, smart excitation, and response to applications. Adv. Mater. Technol. **7**(9), 2101568 (2022)
52. del Barrio, J., Sánchez-Somolinos, C.: Light to shape the future: from photolithography to 4D printing. Adv. Opt. Mater. **7**(16), 1900598 (2019)
53. Rich, S.I., Jiang, Z., Fukuda, K., Someya, T.: Well-rounded devices: the fabrication of electronics on curved surfaces–a review. Mater. Horiz. **8**(7), 1926–1958 (2021)
54. Abbas, M.: Smart materials for changing the electrical properties of nanostructures. Compos. Adv. Mater. **30**, 26349833211016480 (2021)
55. Amukarimi, S., Ramakrishna, S., Mozafari, M.: Smart biomaterials—a proposed definition and overview of the field. Current Opin. Biomed. Eng. **19**, 100311 (2021)

56. Pilate, F., Toncheva, A., Dubois, P., Raquez, J.M.: Shape-memory polymers for multiple applications in the materials world. Eur. Polymer J. **80**, 268–294 (2016)
57. Zhou, Y., Li, L., Han, Z., Li, Q., He, J., Wang, Q.: Self-healing polymers for electronics and energy devices. Chem. Rev. **123**(2), 558–612 (2022)
58. Gong, M., Zhang, L., Wan, P.: Polymer nanocomposite meshes for flexible electronic devices. Prog. Polym. Sci. **107**, 101279 (2020)
59. Li, D., He, J., Wang, L., Gao, L., Lu, B.: 5D printing—fabrication of biological function tissue. China Mech. Eng. **31**(01), 83 (2020)
60. Rane, N.: 3D, 4D, and 5D printing in architecture, engineering, and construction (AEC) industry: applications, challenges, and future scope. Engineering, and Construction (AEC) Industry: Applications, Challenges, and Future Scope (September 28, 2023) (2023)
61. Kumar, P.R.A.M.O.D., Roy, S., Hegde, H., Bharti, S. and Kumar, M.: 4D and 5D printing: Healthcare's new edge. 3D Printing Techn. Nanomed, pp.143–163 (2019)
62. Aruna, R., Mohamed Abbas, S., Vivek, S., Suresh, G., Meenakshi, C.M., Srinivasan, T., Selva Ganapathy, K.: Evolution of 5D printing and its vast applications: a review. Recent Adv. Mater. Modern Manuf. Sel. Proc. ICAMMM **2021**, 691–714 (2022)
63. Reddy, P.R., Devi, P.A.: Review on the advancements of additive manufacturing 4D and 5D printing. Int. J. Mech. Prod. Eng. Res. Dev. **8**(4), 397–402 (2018)
64. Singh, R., Holmukhe, R.M., Gandhar, A., Kumawat, K.: 5D Printing: A future beyond the scope of 4D printing with application of smart materials. J. Inf. Optim. Sci. **43**(1), 155–167 (2022)
65. Roach, D.J., Kuang, X., Hamel, C.M., Dunn, M.L., Qi, H.J.: 4D printing based on multi-material design. In: Manufacturing in the Era of 4th Industrial Revolution: A World Scientific Reference Volume 1: Recent Advances in Additive Manufacturing, pp. 163–194 (2020)

Polymer-Graphene and Graphene Oxide Nanocomposites by Using 3D and 4D Printing for Multifunctional Applications

Supriya Rana, Sujit Kumar Guchhait, and Krishna K. Yadav

Abstract Mankind's quest for progress has led to the constant evolution of technologies that shape our lives. Among these, 3D printing, or additive manufacturing, stands out. 3D printing, also known as additive manufacturing, has revolutionized the way objects are designed and manufactured. This technology enables the creation of three-dimensional objects by layering materials based on digital models. With applications spanning from prototyping and manufacturing to medicine and art, 3D printing offers unparalleled customization, efficiency, and innovation. Its impact is profound in fields such as aerospace, automotive, healthcare, and consumer goods, where it reduces costs, accelerates production timelines, and allows for complex geometries that traditional manufacturing methods cannot achieve. This innovation has given rise to 4D printing, an enhanced version that responds dynamically to stimuli. Applicable across industries, these techniques address challenges while opening new possibilities. This approach overcomes traditional production constraints by utilizing advanced materials like smart polymers, carbon nanotubes, and graphene, yielding products adaptable to changing conditions. Out of various materials, polymer-graphene nanocomposites have emerged as a transformative technology in the realm of 3D and 4D printing. This chapter enlarges and encapsulates the fundamental synergy between graphene and polymers, showcasing their potential in additive manufacturing. The integration of graphene, with its remarkable mechanical, thermal, and electrical properties, enhances the performance of printed structures. The chapter explores the design and technical possibilities inherent in the combination of these materials, with a focus on both 3D and advanced 4D printing

S. Rana
Institute of Nano Science and Technology, Knowledge City, Mohali, India

S. K. Guchhait
Electrochemical Process Division, Sardar Swaran Singh National Institute of Bio-Energy, Kapurthala, India

K. K. Yadav (✉)
Department of Chemical Engineering, Sami Shamoon College of Engineering, Beer Sheva, Israel
e-mail: krishnay447@gmail.com; krishphy25@gmail.com

Nanotechnology Group, USAL–Nanolab, Departamento de Física Fundamental, University of Salamanca, Plaza de la Merced, Edificio Trilingüe, 37008 Salamanca, Spain

S. Moharana et al. (eds.), *Polymer Nanocomposites for 3D, 4D and 5D Printing*, Engineering Materials, https://doi.org/10.1007/978-981-96-4214-4_7

techniques. This innovative approach paves the way for functional and adaptable materials, offering unprecedented opportunities in various fields, from materials science to biotechnology. These advancements promise to reshape manufacturing capabilities and unlock novel applications, revolutionizing the landscape of printed structures and devices.

Keywords Graphene · Polymers · 3D/4D printing · Heterostructure

1 Introduction

Technology and science are made to make human life easy, and lots of progress is being made to make technology easier and more economical, where the need for energy in a sustainable manner is very crucial [4, 17, 55, 61, 67]. In this prospect, 2D materials have shown their applications in almost every sector of science, which can benefit humankind, including energy, electronics, optoelectronics, water purification, corrosion, and so on, which arises due to their exceptional properties highlighted in Fig. 1 [35–36, 66, 67, 74–76].

Graphene and its polymer nanocomposite materials have been intensively researched in the present field over the last several decades [60, 73]. Graphene has 2D structures having hexagonal lattices and carbons were arranged in sp^2 fashion [5, 23, 57, 58, 78, 79] and can be also arranged in various forms but different names (Fig. 2). Although graphene was first introduced in 1962, Geim et al. in 2004 revealed the scotch tape technique to extract graphene made these materials very famous [23]. Additionally, graphene oxide (GO) and reduced graphene oxide (rGO) can be also developed from graphene [22, 52]. The most studied oxidized form is graphene oxide, which is often synthesized using Hummer's or modified Hummer's method [82]. Furthermore, reduced graphene oxide (r-GO) has been obtained through the reduction of graphene oxide. Graphene's charge carriers can be defined as a massless Dirac fermion, which makes high mobility and high thermal conductivity [10, 45, 51, 69]. Graphene's 2D structure allows a large specific surface area (~2600 m^2/g) as compared to other carbon forms [34]. However, the number of graphene layers, the location and type of defects present, and other factors all have an impact on graphene's thermal and mechanical properties. Suspended graphene exhibits high thermal conductivity, with values that range from $4.84 \times 10^3 - 5.30 \times 10^3$ Wm^{-1} K^{-1} [29, 52]. It has also been reported that Graphene produced by chemical vapour deposition (CVD) had almost the same range of thermal conductivity [6]. At the same time, the addition of polymeric filler improves the mechanical properties (native graphene layers have ~1 TPa Young's modulus) [38, 65]. The interaction of polymeric nanofiller has been enhanced using matrix-nanofiller interactions and graphene dispersion [18]. Graphene, in its functional forms, can produce improved interfaces and polymer dispersions. As a result, polymer-graphene nanocomposites display a range of improved electrical, thermal, wear, optical, and electromagnetic interference shielding of all polymeric matrices [2, 13, 31, 37, 72].

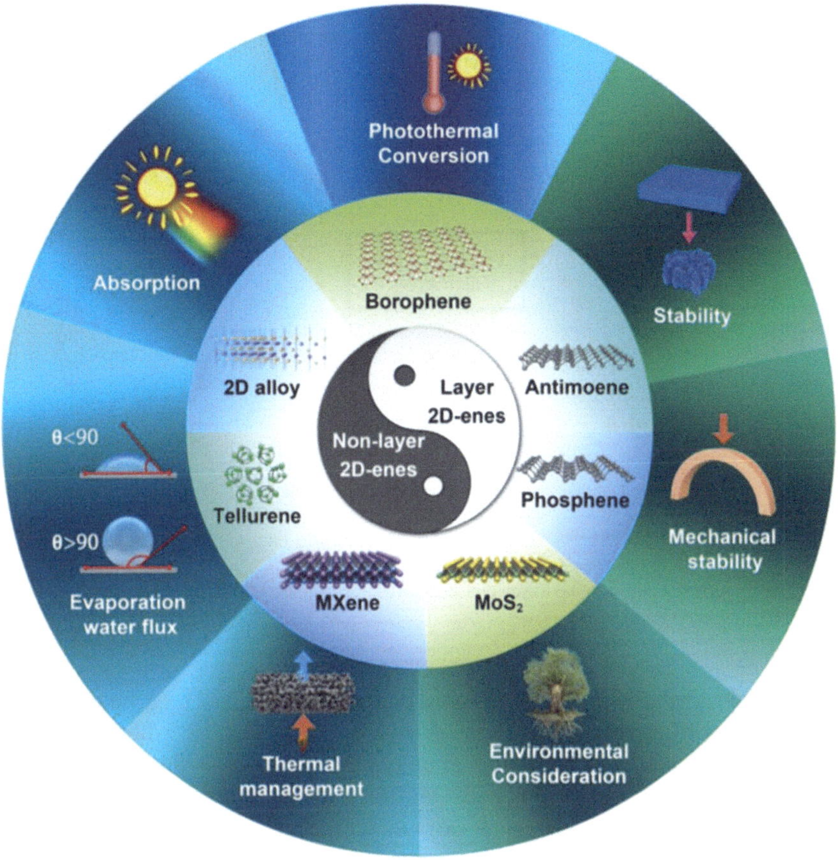

Fig. 1 Structure and their application of various 2D materials. Reproduced from Ref. [71] with permission from John Wiley & Sons, Copyrights © 2020 The Authors. Published by WILEY–VCH Verlag GmbH & Co. KGaA, Weinheim

The properties and application of end devices can be tuned by the pattering methods, where lithography-based methods were utilized most [26, 42, 77]. However, in recent times, three-dimensional (3D) and four-dimensional (4D) printing have emerged as next-generation manufacturing techniques, significantly influencing various fields such as materials science, computer science, biology, chemistry, and engineering [16, 25, 28, 59]. Three-dimensional printing allows the layer-by-layer addition of various materials to produce complex structures with high precision [80]. Despite its tremendous potential, 3D printing remains significantly underutilized, and there is a strong desire for the efficient, low-cost production of 3D materials. The latest breakthrough in 3D printing has garnered significant attention for its ability to develop items that can change their shapes after being removed from the printer [47]. Four-dimensional (4D) printing is a new era in printing that has

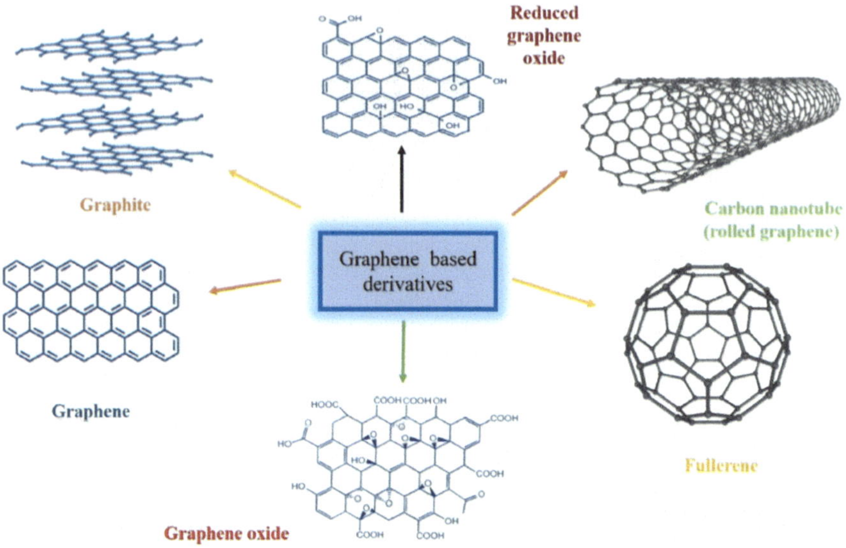

Fig. 2 Arrangements of graphene in various structures. Reproduced from Ref. [27] with permission from Elsevier, Copyright [2022]

been brought about by this technological advancement [62]. The primary goal of 4D printing is to enable the 3D-printed object to self-assemble when exposed to specific stimuli such as heat, pressure, humidity, chemical reactions, and so on [1]. In layman's terms, 4D printing is a revolutionary advancement that builds upon 3D printing by incorporating time as an extra dimension which implies that the shape or properties of 4D object can be changed with time. The cutting-edge technologies of 3D and 4D printing have enormous potential applications across various industries, including robotics, automotive, medical implants, electronics, and aerospace [16]. Dimensional accuracy makes these manufacturing processes suitable for use in such fields. To improve mechanical, thermal, and other properties like surface area, thermal stability, resilience strength, interfacial interaction, and desirable electrical properties, nanoparticles are being incorporated into the polymer matrix in 3D printing technology, which has made major advances in recent years [52]. Printability is an essential feature of 3D printing and is determined by the intricacy of the printed structure as well as the material's quality and manufacturing process. Reliability in 3D printing is primarily dependent on the proper viscosity and shear-thinning behavior of the material [56]. This chapter will deal with the technological and design potential that results from combining various materials, emphasizing both 3D and cutting-edge 4D printing techniques. This innovative technique opens previously unheard-of possibilities in a variety of sectors, engineering, materials science, energy, biomedical, and biotechnology, by paving the way for functional and flexible materials. These developments have the potential to transform the field of printed structures and devices by opening new applications and reshaping manufacturing capacities.

2 Polymer-Graphene-Based Nanocomposite Fabrication Techniques

Graphene-based nanocomposite materials such as pure graphene, functionalized graphene, graphene oxide, graphene nanoribbons, and doped graphene nanoribbons an important nanofillers for polymeric matrices. The polymer–graphene nanocomposites show advanced mechanical, electrical, thermal, and physical properties which are directly beneficial for the various industrial and technical applications. Various advanced fabrication techniques have been designed to fabricate next-generation 3D or 4D devices to process various materials such as polyamide, polycaprolactone, polyethylene, poly (lactic acid), etc. known as polymeric matrices in composite with graphene. These approaches include direct ink writing, selective laser sintering, stereolithography, and fused deposition modelling. Therefore, the selection of the appropriate method to achieve the desired superior and mechanical properties is highly beneficial. For example, to achieve the thermal stability of poly (lactic acid)–graphene oxide nanocomposites can be done by using direct ink printing technology and such materials used stereolithography and fused deposition help to achieve unique honeycomb morphology with significant advanced properties. Furthermore, the performance of such composite materials also depends on the type of polymer matrix, its functionalities, the type of nanofiller, and other associated printing parameters. For example, 3D-printed poly (methyl methacrylate)–graphene-based composite material has been done by using stereolithography whereas in 4D printing techniques acrylonitrile–butadiene–styrene, poly (lactic acid) and other thermosetting matrices in composite with graphene can be used. Consequently, the printing and design of such graphene-based nanofiller materials are highly important to achieve high-temperature stability and performance for various technical and industrial applications. Therefore, significant advancement should be needed when working with polymer–graphene nanocomposites for printing and various fabricating processes has been discussed in the following section.

2.1 *Direct Ink Writing*

In this technique, a viscous material is pressurized and used to create a three-dimensional shape of the desired material. During the processing, the platform used is stationary while the syringe (which contains pressurized materials) moves in all three-dimensional directions which causes the extruded material to undergo fusion layer by layer. Any solution, hydrogel, and paste form of the materials can be used and loaded for obtaining three-dimensional shapes. During the curing process, the different reactive materials are induced either using nozzle mixing or using heat/ UV light. The binders used during curing are three different kinds such as water-based binders, phosphoric and critic acid-based binders, and polymer-based binders. Therefore, depending on the type of binder and powder material used, it leads to chemical or physical bonds through adhesive forces or hydraulic setting reactions. The resolution of the final product depends on the binder viscosity, amount of added binder used, powder size, binder-powder interaction during curing, and speed of the deposition of the binder. The key advantage of this technique involves that desired material flexibility can be obtained, processing takes place at room temperature, and easy to remove from the support structure. However, this technique has some disadvantages, it requires the use of sacrificial materials that are needed to support the printing process, and sometimes such materials decrease the stiffness causing the collapse of the complex structure. The low efficiency and economic viability are major obstacle to this technique.

2.2 Selective Laser Sintering (SLS)

The SLS process uses powder materials instead of liquid binders. This method consists of a controlled laser beam with a specified controlled path. The powder is sintered by heating and placed in a scan path. During the sintering process, a high-powered laser is used, and the powder undergoes fusion by heating leading to molecular diffusion. This similar process is carried out for the next layer. The powder which is left unbounded is removed frequently to get the final product and this technique results in layer-to-layer sintering during the process followed by milling. The resolution of the final product depends on the particle size of the material, the power of the laser, scan speed, and spacing. The overall working technology of various methods of 3D printing is shown in Fig. 3. The SLS method can successfully process thermoplastic polymer and semi-crystalline materials of different thermal properties such as polycaprolactone (PCL), and polyamide (PA).

2.3 Stereolithography (SLA)

This method includes the use of photopolymers and UV lasers, where the photopolymers can be cured by using a UV laser in a controlled way. The photocurable resin results in polymerization during the curing process to form a 2D layer pattern. The overall setup includes layer-to-layer curing of the resin using the platform of the layered resin goes down which is already cured, and another layer of the uncured resin is available for curing. So, this method leads to layer-by-layer pattern deposition of the resin. Typical polymer materials such as epoxy resin and acrylic materials are used in this technique.

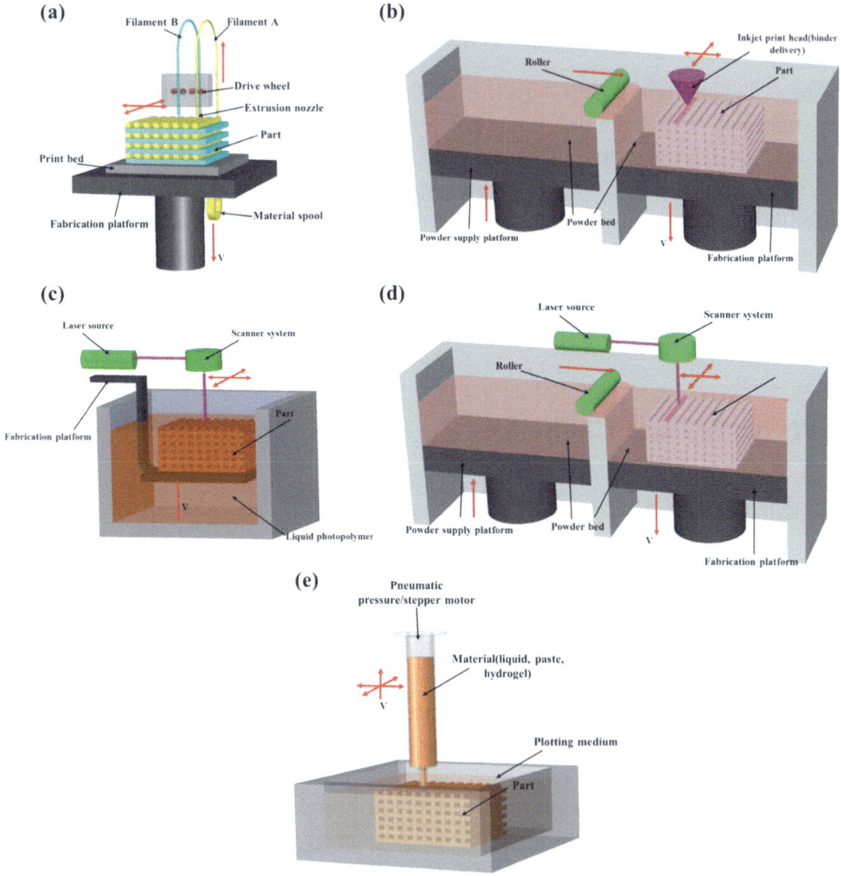

Fig. 3 Schematic representation of a typical **a** FDM setup **b** 3DP setup **c** SLA setup **d** SLS setup **e** 3D plotting setup. Reproduced from Ref. [70] with permission from Elsevier, Copyright [2022]

The curing reaction during polymerization is highly crucial to get an in-depth understanding of the quality of the final product. Therefore, the laser intensity, time of exposure of the laser, and speed of the scan, all these factors highly influence the printing resolution of the final product, depth of polymerization, and quality of the curing. Furthermore, during the polymerization process, the photoinitiators and UV absorbers play crucial roles in polymerization. The SLA is a nozzle-free technology where the clogging at nozzle can be avoided. Moreover, this technology can produce high-resolution printing and faster cycle processing. Despite huge applications, the high cost of the methods and the lack of being able to produce flexible photocurable resins limits its advancement in industrial applications. Cytotoxicity of the final product can be caused by photoinitiators or sometimes uncured resin remaining. These are the major issues still associated with this technique that need to be addressed.

2.4 Fused Deposition Modelling (FDM)

This model is an important 3D printing method and leads to the controlled extrusion of the material. In this method, low melting thermoplastic materials can be processed more efficiently, where the polymer-based material is melted at the nozzle and the material is extruded layer by layer followed by solidification. The quality of the desired product depends on the various printing parameters such as orientation, the thickness of the layer, the angle of orientation, raster width, and air gap between the layers. Moreover, this technique involves various advantages such as low cost, high speed, simple method, and potential for diverse material deposition.

3 Properties and Challenges

3.1 Enhanced Material Properties

The mechanical strength and electrical properties of the polymer can be increased by adding the graphene filler, where the property of GO influences the properties of such materials. Both these properties of polymer material are highly dependent on the filler concentration. Various factors influence the mechanical and electrical properties of the graphene-based nanocomposite material. The nanofiller uniform dispersion is highly influential in enhancing the mechanical properties. The nanoparticles show the agglomeration tendency which causes the negative feedback of the nanocomposite. The graphene-based composite improves the printing ability by improving uniformity. Therefore, during addition, a precise amount of filler was done which highly influenced the polymer chains and geometry of the composite. The solvent polarity and functionalization also influence the nature of dispersion. However, functionalized graphene nanomaterials are dispersed more in alkaline media and acidification of such materials can result in uncontrolled stacking. So, solvent polarization and mode of functionalization are highly influenced by the interfacial interaction of composite. On the other hand, the optimum amount of oxygen content and controlled interfacial interaction is beneficial for advanced mechanical, and electrical properties and good interfacial compatibility. The aspect ratio of nanofiller has an impact on composite properties, where a higher aspect ratio causes high rigidity and improved mechanical stiffness and strength. Therefore, optimizing the filler concentration and uniformity of distribution to obtain the desired properties. Moreover, the reduction process of graphene oxide, the morphology of the graphene-based composites, the rheological property of the nanofiller, and the optimum viscosity of the ink are highly influencing the printable, alignment, and other properties of the nanocomposite materials for wider applications.

3.2 Challenges in Nanocomposite

A wide area of research focuses on developing and enhancing mechanical stability in graphene polymer composite materials based on 3D and 4D structures. One of the difficult problems in the production processes is achieving the nanofillers to interface with the polymer consistently while maintaining a high level of interfacial interaction. Achieving uniformity and efficient nanofiller distribution is a challenging task. Nonetheless, there have been reports of improvements in filler dispersion using sonication and high-speed stirring. Exceptional mechanical and electrical performance of the materials are aided by in situ polymerization. Phase separation results from the polymerization process's lower entropy interaction between the original monomer and graphene filler. Selecting the appropriate monomer is therefore essential to resolving this issue. Because graphene is chemically unstable, large sheets precipitate in acidic environments. Additionally, the conventional technique to produce composite materials is layer-by-layer deposition. This technique forms assembly structures. Nonetheless, these fillers exhibit anisotropic behaviour in their mechanical characteristics. This technique has non-uniform distribution control and a low slow rate. Another challenge with employing 3D and 4D printing techniques is selecting the right printable materials. Therefore, further research is required to determine the precise flow rates, inexpensive manufacture, and sustainability of green material. An important issue with these materials is that most of the composites have poor mechanical strength. Consequently, further advancement is needed to attain the printable materials' efficacy. Additionally, the printing processes take a long time and are challenging to use to produce composite materials that are evenly distributed. Therefore, new printing methods must be developed to increase 3D and 4D printing resolution without sacrificing material properties or prolonging manufacturing times.

4 Fabrication Applications in Multifunctional Fields

4.1 Electronics and Conductive Materials

3D printing techniques can be effectively used for electronic device applications, and polymer-based composite materials can function as conductive electronic materials. Graphene-based polymers exhibit exceptionally advanced properties in the field of electronics and optoelectronics. R. D. Farahani et al. designed single-walled CNT nanocomposite microfabrication with three-dimensional patterns [19]. The nanocomposite network displays high capability compared to conventional technologies. The electrical conductivity of the materials was also measured, and this technique resulted in the formation of microelectronics materials having sensing components (Fig. 4a–c).

In their work, microspring networks were developed for freestanding strain sensors, which can detect out-of-plane strains. Figure 5 shows the electromechanical

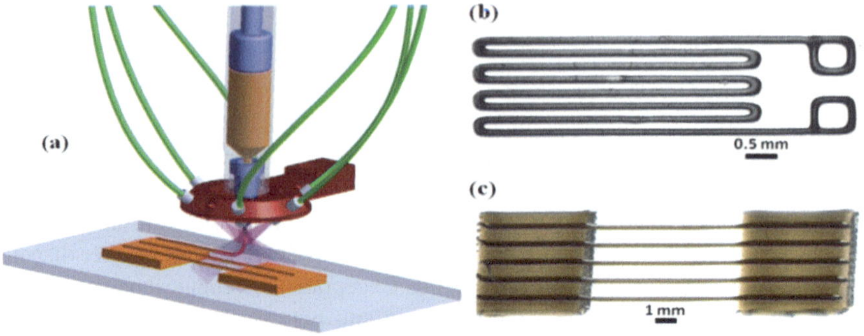

Fig. 4 **a** Direct-writing of nanocomposite microstructures using UV-assisted method. **b** Fabricated line network **c** microfiber coupon. Reproduced from Ref. [19] with permission from IOP Publishing, Copyright [2012]

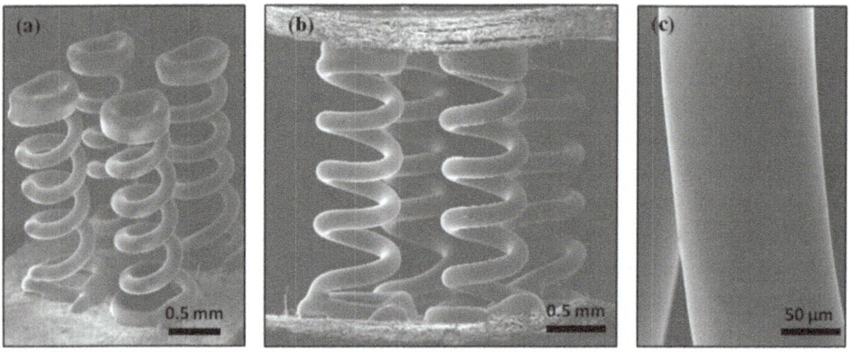

Fig. 5 **a** SEM of freestanding microsprings **b** final configuration of the sensor as can be seen from SEM **c** filament surface SEM image. Reproduced from Ref. [19] with permission from IOP Publishing, Copyright [2012]

testing of these networks. Figure 5a features an SEM image of four spring networks having five coils each, with the first and last coils flattened for attaching aluminium pads. The final sensor design in Fig. 5b arranges the springs in a rectangular array. The printed springs matched with the programmed path, which confirms the precision of the UV-DW technique. The nanocomposite filament surface with 130 μm has been shown in Fig. 5c.

A. J. Lopes et al. manufactured a hybrid system using direct print and stereolithography technology to design a 3D structure having electronic circuits [41]. To fabricate a 3D structural electronic device within a hybrid SL/DP machine without removing it, a process involving multiple steps (starts and stops) of the SL process. This method includes removing uncured resin, inserting electronic components, performing DP, and laser curing conductive traces. The general steps are designing the part with CAD, and preparing it with 3D Lightyear™ software and from Fig. 6 the process

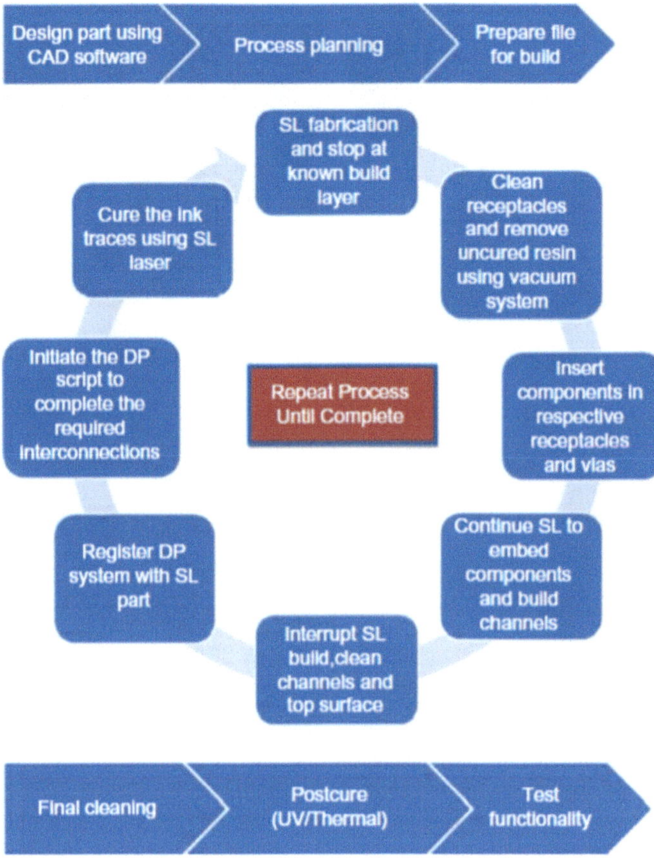

Fig. 6 The general embedded electronics fabrication process cycle. Reproduced from Ref. [41] with permission from Emerald Group Publishing Limited, Copyright © [2012]

can be easily understood. The findings proved that the 3D micro dispersing DP system provides control over the conductive trace deposition and the SL system provides the fabrication of 3D electronic structure.

X. Jin et al. designed Ag nanoparticles capsulated in PVP polymer developed on electrostatic pattering on the copper layer [32]. This is done by using the capping method to decorate PVP-capped Ag ions as catalytic precursors followed by reduction on the surface of polyimide film [32]. This material shows enhanced electrical conductivity which is highly beneficial for wearable electronic applications. Jakus et al. designed 3D printable graphene composite materials consisting of polylactide-co-glycoside and graphene-based materials [30]. This polymer matrix exhibits high mechanical robustness and flexibility for high electrical conductivity [30]. This system with modified graphene composition shows unique properties including mechanical, electrical, biological, and surgical. Figure 7a shows the performance

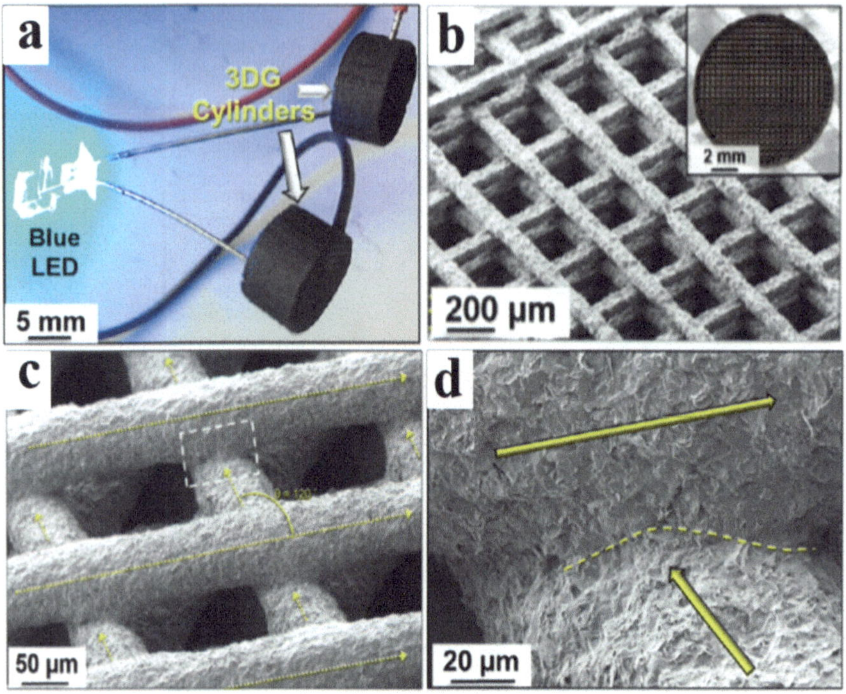

Fig. 7 **a** Electrical conductivity shown by turning blue LED from 3D printed, **b** SEM and optical image (inset) of 3D pattern, **c**–**d** Fiber–fiber junctions as shown in SEM (Arrows indicate fiber directions). Reproduced from Ref. [30] with permission from American Chemical Society, Copyright [2015]

of the 3D patterned model used for powering the blue LED, whereas SEM and optical images of fabricated interconnected fibers are shown in Fig. 7b–d. Moreover, magnetic nanostructured materials with graphene-based composite is another important area of electronic as well as electromagnetic application. Wajahat et al. synthesized functionalized graphene (FGP) nanocomposite with Fe_3O_4 and utilized them for 3D printing techniques [68]. FGP micro architectures exhibit high conductivity and magnetic properties that are highly beneficial to be used in magnet-guided cars, magnetic switches, EMI shielding and 3D connector of an electrical circuit [14].

Hushein et al. printed graphene with PVA via an extrusion-based 3D printing method and investigated their electrical conductivity [54]. Here they show two different 3D printing methods of graphene nanomaterial via the direct ink writing (DIW) method and the fused deposition modelling (FDM) method. For making graphene slurry, first, 15% polyvinyl alcohol (PVA) with 85% DI water mixed and 20% and 80% graphene slurry were prepared [15, 64]. In both 3D printed (DIW and FDM) models, as the voltage increases the resistivity also moves in an increased manner, however, both the DIW and FDM have almost similar resistivity levels.

5 Biomedical Applications

Nowadays, 3D and 4D printing is used in internal organs and tissue engineering. Patient-specific tissues naturally derived from polymers (gelatin, collagen, alginate, etc.) or synthetic polymer molecules (PEG, PLGA, PVA, etc.) are currently used for printing purposes in the biomedical field. The majority requirement is that their good mechanical, structural, biocompatible, and printable properties are essential for effective interactions with tissues and successful transplantation. Tissue engineering is one of the important fields of using graphene-based composite materials for biomedical applications. The development of scaffolds for tissue engineering having high mechanical properties, in vivo applications, and physical connectivity for cell infiltration is a critical and sophisticated manufacturing area. However, manufacturing technology has brought various processes for the fabrication of biocomposite scaffold materials with high mechanical strength, complex structure, controlled shape, and size that can effectively regenerate various tissues. Although, traditional technologies are available that have various loopholes such as a lack of ability to control the internal structure and porosity of scaffolds. The bio fabrication strategies have led to the development to minimize the positioning of the cells, structure, and function of various biological components are now achievable. Therefore, printing such highly biocompatible and biodegradable polymer composite scaffolds can be further strengthened by adding bioactive particles and these polymers maintain their mechanical properties. Bioprinting technology can be done by two different methods which include pre-seeding (direct) and post-seeding (indirect) method. The pre-seeding method causes the printing of the material and the cells simultaneously whereas the post-seeding method involves the printing of material first and further culturing it with proper cells.

Belaid et al. generated polylactic acid (PLA) scaffolds 3D printing material using the fused deposition modelling method for bone tissue engineering applications [9]. The schematic for printing a 3D using FDM has been shown in Fig. 8. SEM of 3D printing shows that when PLA (or with 0.2% GO) was used porous scaffolds obtained with pore size 405 ± 20 μm and 485 ± 30 μm, for PLA and 0.2% GO respectively (Fig. 9). The findings proved that GO incorporation results in enhancement in thermomechanical properties. Improved mechanical properties can be obtained by using advanced scaffolds where the hydrophobic surface transfers into a hydrophilic surface. Further, PLA/GO scaffolds act efficiently for cell proliferation and mineralization. The materials serve as biodegradable and biocompatible synthetic polymers for tissue engineering. The scaffold biocompatibility shows that 0.1–0.2% of GO has no significant difference in the viability of the cells. High viability and biocompatibility were observed in the presence of PLA and PLA/GO on MC3TC-E1 cells after 7 days of culture with 0.2–0.3% of GO. Therefore, the GO nanofiller shows enhanced biocompatibility for PLA and viability for cells on culturing.

Another work by Y Zhang et al. done detailed studies on the bone-forming activity using a 3D printing method [84]. In this study, 3D printing was employed to create β-tricalcium phosphate (TCP) bioceramic scaffold composites. These scaffolds were

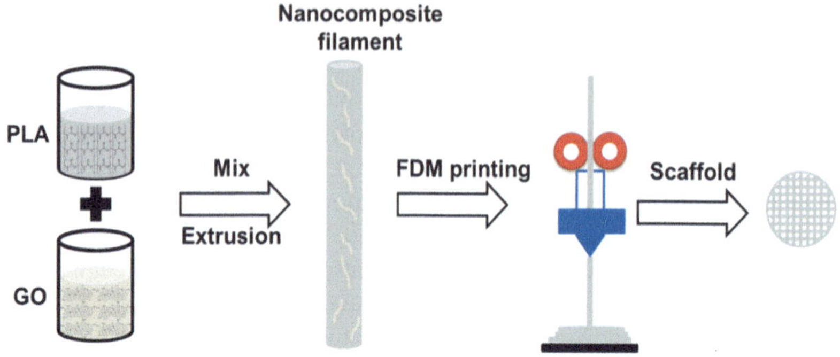

Fig. 8 Printing scheme using 3D FDM. Reproduced from Ref. [9] with permission from Elsevier, Copyright [2020]

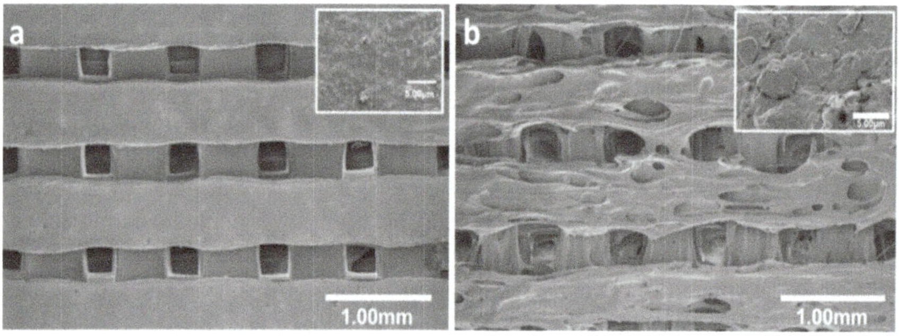

Fig. 9 SEM of FDM 3D printed **a** PLA and **b** PLA/GO (0.2%) scaffolds. Insets show a magnification of the surface. Reproduced from Ref. [9] with permission from Elsevier, Copyright [2020]

enhanced using a liquid chemical reduction method to disperse silver (Ag) nanoparticles on graphene oxide. The resulting composite was then integrated with the β-TCP scaffolds. The developed scaffolds exhibited both antibacterial and osteogenic activities when tested on rabbit bone marrow stromal cells. The findings highlight the potential of these bifunctional scaffolds in reconstructing large bone defects and treating infections.

P. Peng et al. blended polyetheretherketone (PEEK) with polyvinyl alcohol (PVA) to form PEEK/PVA scaffolds using 3D printing (Fig. 10) [21]. Therefore, the biomechanical properties of PEEK are combined with the remarkable degradability of PVA. Furthermore, the scaffolds exhibit enhanced interfacial bonding on combined with GO. This unique structural advancement leads to a decrease in interfacial free energy and forms a unique matrix where the strength of the PEEK/PVA scaffolds increases to ~150% on 1% GO loading. PEEK/PVA-GO scaffolds exhibit hydrophilic behaviour and in vitro, biodegradability promotes proliferation and osteogenic differentiation of MG63 cells and shows good bone regeneration ability. X. Zhou et al. investigated 3D bio-printed graphene oxide doped gelatin-based scaffolds for chondrogenic differentiation of human bone marrow mesenchymal stem cells [87]. They fabricate GO-gelatine methacrylate-PEGDA scaffolds with hierarchical structures using stereolithography. 3D printing GelMA-PEGDA-GO scaffolds serve as excellent material for chondrogenic differentiation and cartilage regenerative applications. Cabral et al. synthesized reduced graphene oxide nanoparticles to form scaffold structures for bone regeneration applications. 3D printing technique results in the formation of assembled 3D scaffolds incorporating rGO (Fig. 11) [12]. The synthesis was done by in situ reduction of L-Ascorbic and of GO present on tricalcium phosphate/gelatin/chitosan scaffold acid. The prepared functionalized rGO in situ scaffolds displaced enhanced wettability and mechanical properties. The scaffolds show antimicrobial performance against S. aureus, and E. coli and potential candidates for the bone tissue regeneration application.

E. M. Goncalves et al. presented three-phase composite 3D printing scaffolds comprising hydroxyapatite, CNT mixed with the polymeric matrix of PCL, where CNT content varies from 10 to 50 wt. % in the PCL matrix [24]. The scaffolds bring unique properties such as mechanical properties and electrical conductivity. The unique properties of 3D HA/PCL/CNTs composite scaffolds cause hydroxyapatite

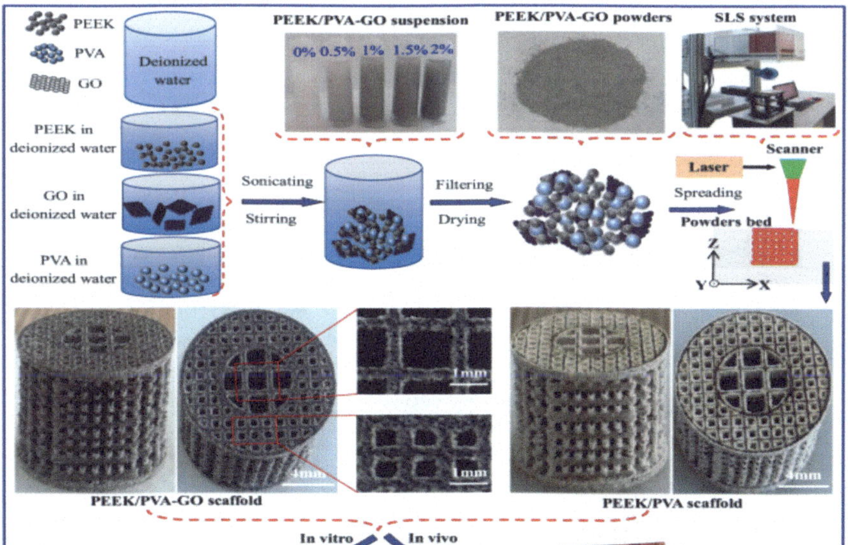

Fig. 10 Schematic showing the scaffold's fabrication along with cell culture and animal experimental design. Reproduced from Ref. [21]

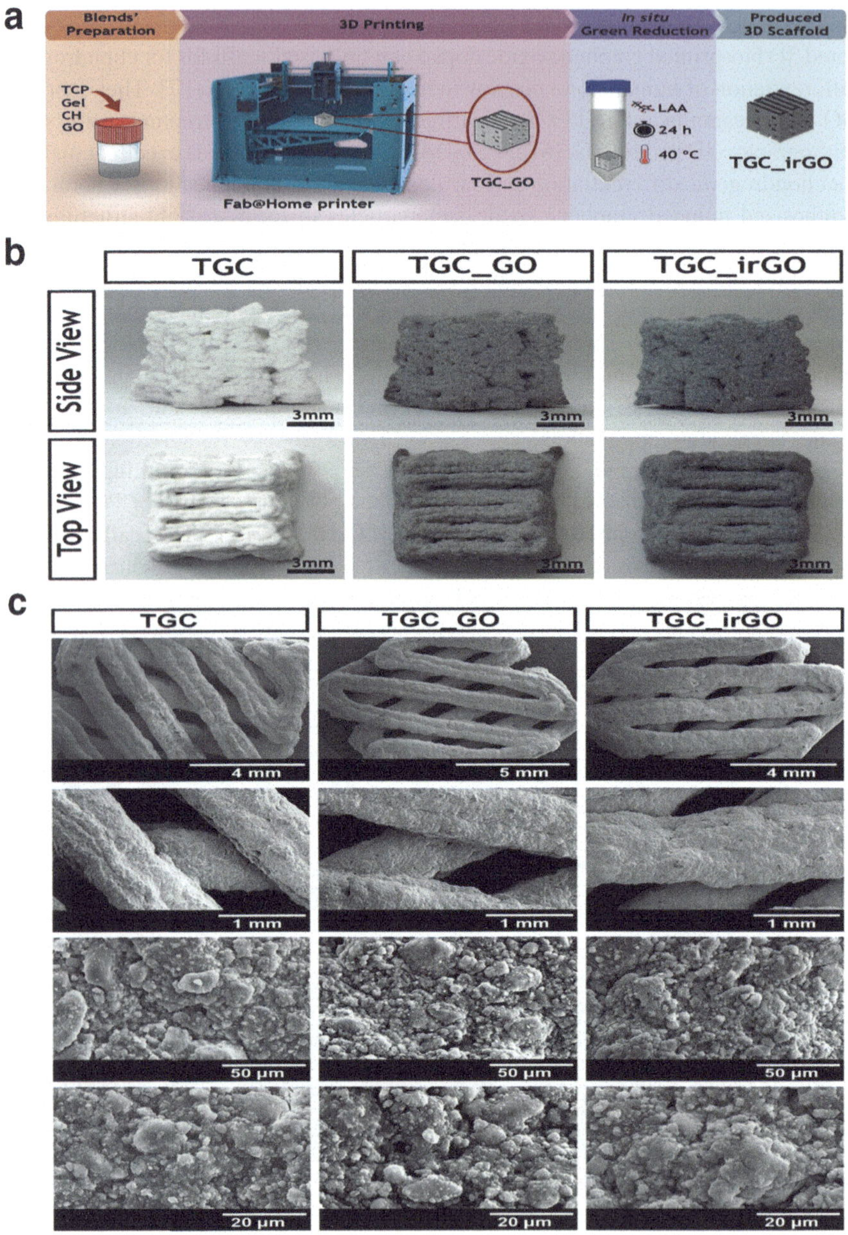

Fig. 11 Printing and SEM characterization of 3D scaffolds. **a** Experimental design for printing of TGC_irGO scaffolds; **b** Macroscopic view of scaffolds, **c** SEM images (at different magnifications) and surface topography of the produced 3D scaffolds. Reproduced from Ref. [12] with permission from Elsevier, Copyright [2019]

bioactivity, and cell adhesion in the field of bone regeneration. Y. Zhang et al. developed a novel bioactive poly (lactic-co-glycolic acid) PLGA/β-tricalcium phosphate (TCP) composite scaffold loaded with GO and bone morphogenetic protein-2-like peptide was loaded in situ (PTG/P) using 3D printing method [83]. The scaffolds repair bone defects to effective bond regulation where GO improves wettability and mechanical strength along with PTG/P scaffold shows excellent osteogenic ability. Pineda et al. fabricated polycaprolactone (PCL) scaffolds with thermally reduced GO by a 3D printing method. PCL/FrGO exhibits fourfold high viability for human mesenchymal stem cells and shows high potential for tissue engineering applications [3].

6 For Energy Applications

Supercapacitors emerge as a primary energy storage component due to their long-lasting lifetime, high power and energy density as well as fast charging rate [7, 49]. In supercapacitors, the capacitance can be enhanced by adding extra active materials, however, this can increase the size and cost of supercapacitor devices. Thus, energy density can be boosted by utilizing 3D cells, but ionic resistance might be increased, which can be optimized by 3D electrode architecture fabricated using 3D printing [48]. In 3D printing generally, polypyrrole (PPy) as a polymer was used with graphene where PPy acts as a pseudocapacitive material and boosts specific capacitance [81]. Zene et al. created polypyrrole–graphene aerogel (PPy-GA) composite electrodes which have 3D architectures and are highly compressible in nature (Fig. 12) [53]. Researchers developed resilient electrodes by combining super-elastic polypyrrole (PPy)-coated graphene aerogels (GA) through 3D printing and polymer self-assembly. The GA template, capable of withstanding up to 90% compression, can be molded into complex 3D structures with periodic macropores. These macropores ensure uniform PPy application, boosting the specific gravimetric capacitance of the GA electrode from 14 to 395 F g^{-1}, with 90% retention after 5000 charge–discharge cycles. The electrode maintains capacitance after 100 compression cycles and shows a significant increase in compressive strength from 0.14 to 2.4 MPa. Achieving an areal capacitance of 2 F cm^{-2} and an energy density of 0.78 mWh cm^{-2}, this method sets new benchmarks for compressible electrodes and paves the way for advanced, durable energy storage modules.

Teressa et al. in 2014 utilized the homogenization technique at high pressure to produce binder-free graphene oxide (GO) inks and used by micro-extrusion method for fabricating electrodes that have remarkable charge/discharge capabilities [46]. This approach holds significant promise for creating capacitors with elevated capacitive properties. Similarly, Sun et al. employed the micro-extrusion printing method where engineered micro-supercapacitors utilizing laminated graphene films [63]. These supercapacitors are fabricated in ways so that they can demonstrate outstanding capacitive performance, cycling stability, and flexibility. Apart from that, several

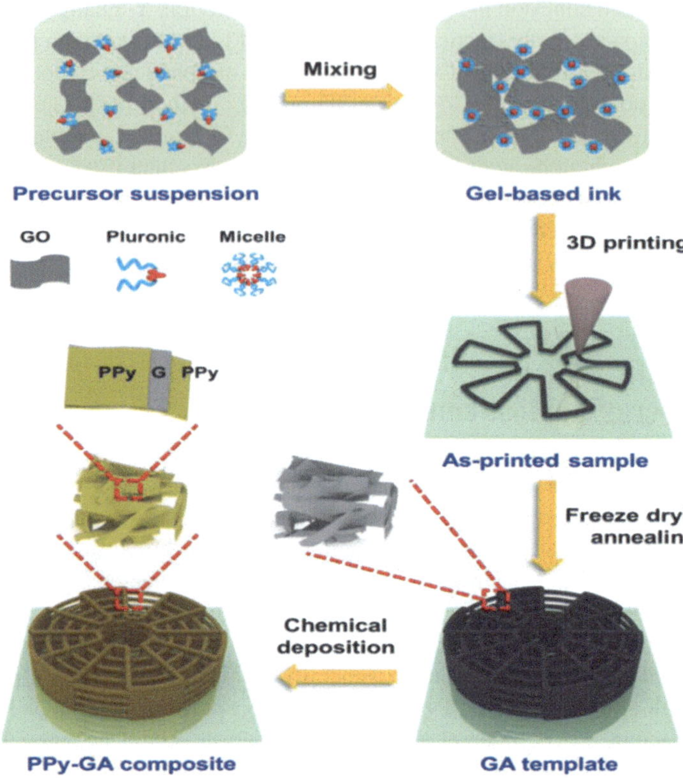

Fig. 12 A schematic diagram illustrating the fabrication process of 3D hierarchical PPy-GA composites. Reproduced from Ref. [53] with permission from John Wiley & Sons, Copyright [© 2018 WILEY–VCH Verlag GmbH & Co. KGaA, Weinheim]

other works were also done, and their Gravimetric, areal, and volumetric capacitive are represented in Tables 1 and 2.

Table 1 Gravimetric capacitive of various carbon-based electrodes

Electrodes	Thickness (mm)	Test Condition (A/g)	Cg (F/g)	Refs.
CNT-PPy	2	0.5	376	[39]
G-PPy	1	1.5	350	[86]
PANI-CNT	4	0.64	216	[50]
PPy-GA-o.1 M	2	6	395	[53]

Table 2 Volumetric and capacitive performance of various carbon-based devices

Electrodes	Thickness (μm)	Test Condition	C_v (F/cm^3)	C_a (F/cm^2)	Refs.
rGO-Au	0.45–13.7	1 V/s	0.56–17.1	7.7×10^{-4}	[40]
rGO-CNT	9	50 V/s	3.1	2.8×10^{-3}	[8]
PANI-NW	50	0.3 mA/cm2	105	4.52×10^{-2}	[43]
PPy-GA-o.1 M	2000	6 A/g	3.5	1.1	[53]

7 Challenges and Way Forward

In a nutshell, polymer-graphene and graphene oxide nanocomposites developed by using 3D and 4D printing technology show promising candidateship in various sectors like biomedical, energy, etc. Direct ink writing, SLA, and other methods are used in the 3D and 4D printing processes. However, when applied to different materials, these methods offer varying benefits and drawbacks. Consequently, it has been determined that selecting a suitable printing technique and optimizing its parameters are essential. The unique morphological features and design versatility of the polymer-graphene nanocomposites are integrated into the 3D and 4D printed items. Considering the printing technologies' remarkable features and technical advances, future research efforts are needed to build outstanding performance. In this section, significant challenges in the areas of fabrication process, design, methodology, cost, etc., regarding the 3D and 4D printing technologies are briefly discussed [22, 59]. To design 3D/4D printing, skilled professionals who can utilize the alternatives provided by additive manufacturing are needed. When contemplating the potential of these technologies, the present designing techniques do not provide 3D and 4D results to the utmost possible. As a result, adding geometric improvements to the designing techniques still needs a lot of work. After the design, future device manufacturers should get the required knowledge to meet the expanding architectural needs and optimize the product's manufacturing capacity by using the latest 3D and 4D printing technologies. The scalable and economical manufacturing of polymer-graphene and graphene oxide nanocomposite with consistent characteristics is another significant hurdle. The synthesis techniques used nowadays frequently entail intricate procedures and call for exact control over several elements, including size, morphology, and improved structural characteristics. Another challenge is to achieve a uniform and consistent dispersion of graphene in the matrix of polymers. While the fabrication processes always strive for a uniform distribution of graphene layers, it is challenging to develop a consistent process for the precise distribution of linked filler particles. Various proven methods such as forced convection flow, sonication, and high-speed stirring have been employed to enhance filler dispersion. While fatigue life increased for graphene-reinforced composites and this is explained by the fillers made of graphene dispersing effectively throughout the polymer matrix, improving the structural durability. In graphene nanocomposites, the distribution of networks is therefore extremely desirable since it provides greater space for the nanofillers to interface with the polymer chains and retain strong interfacial interaction. The vacuum that exists in the produced composites/print is the main reason for this reduction in mechanical strength. Consolidation or infiltration is thus used as an extra adjustment or refinement to raise the effectiveness of these printed goods even further. Furthermore, because of their delicate surface interaction with the matrix, the improved support may increase the permeability of the polymer composites when implemented. As a result, the reinforcement might be used to modify tempted permeability [22]. Nowadays, most printing techniques take a long time, and producing huge quantities of parts is challenging. Scalable and quick material processing should be the foundation for the development of new or enhanced printing methods. Machine learning (ML) models will grow more advanced and accurate as computing power increases and more data becomes available, which will help with the quick optimization of composite materials. Moreover, the advancement of explainable AI methods will improve our capacity to understand model estimates, providing an improved understanding of the complex connections between graphene properties and overall performance. To fully realize the promise of polymer-graphene and graphene oxide nanocomposites developed by using 3D and 4D printing technology and create cutting-edge materials that solve existing issues and open the door to a sustainable future, more study and coordination between scientists, engineers, and business experts are therefore required. When these challenges are overcome, polymer-graphene and graphene oxide nanocomposites will be widely used and economically viable in a variety of applications.

8 Conclusions and Future Prospects

In 3D and 4D printing technologies, polymer-graphene nanocomposites have emerged as a significant innovation. Due to its mechanical, thermal, and electrical capabilities, are integrated into printed structures to improve their performance. This chapter thoroughly explores the technological and design potential that results from combining various materials, emphasizing both 3D and cutting-edge 4D printing methods for multifunctional electromagnetic nature. Layer-by-layer synthesis of graphene/polymer composites using 3D-4D printing technology yields a wide range of structural designs. The effectiveness of various printing processes is determined by the material, structure, expenses, structural complexity, and advancement approach [85]. Moreover, the structural architecture in additive manufacturing is dependent on the various development approaches. 4D printing, in contrast to 3D printing, enables printed materials to self-assemble in response to outside stimuli [44]. Because of stimulus-sensitive programming, 4D printing technologies have been identified as innovative. The processes used in 3D and 4D printing include FDM, SLA, inkjet 3D printing, direct ink writing, and others [69]. When used in different materials, these procedures do have certain advantages and disadvantages. Future research must concentrate on expanding the diversity of printable materials since the limited material possibilities do not satisfy the standards set by different sectors. To increase the variety of printing materials, it will be necessary in the future to fabricate new and novel polymer materials in addition to new reinforcement agents with improved qualities. In this regard, graphene-based 3D and 4D-printed nanomaterials have been considered for multipurpose microscale devices [11]. Graphene nanocomposites printed in three and four dimensions have been explored thus far for applications related to energy, sensing, and high-temperature materials. Material performance is one of the most crucial factors in printing [20]. Printable polymer composites perform better when reinforcement agents are incorporated, but they still don't have the necessary mechanical strength and performance. However, several procedures can be used to improve the performance. Nevertheless, these extra stages come at an additional expense and extend the processing time [33]. More research is needed to develop commercial uses of 3D- and 4D-printed graphene nanocomposites for modern technological sectors. It is important to remember that the mechanical and geometric properties of the finished products are affected by the fabrication parameters used for each 3D and 4D printing process. Hence, the discovery of novel materials, the improvement of processing conditions, the scalability of processing methods, and lastly the performance of printed materials still needs special consideration.

References

1. Ahmed, A., Arya, S., Gupta, V., Furukawa, H., Khosla, A.: 4D printing: fundamentals, materials, applications and challenges. Polymer **228**, 123926 (2021). https://doi.org/10.1016/j.polymer.2021.123926
2. An, X., Butler, T.W., Washington, M., Nayak, S.K., Kar, S.: Optical and sensing properties of 1-pyrenecarboxylic acid-functionalized graphene films laminated on polydimethylsiloxane membranes. ACS Nano **5**(2), 1003–1011 (2011). https://doi.org/10.1021/nn102415c
3. Angulo-Pineda, C., Srirussamee, K., Palma, P., Fuenzalida, V.M., Cartmell, S.H., Palza, H.: Electroactive 3D printed scaffolds based on percolated composites of polycaprolactone with thermally reduced graphene oxide for antibacterial and tissue engineering applications. Nanomaterials **10**(3), 428 (2020). https://doi.org/10.3390/nano10030428
4. Yadav, K.K., Guchhait, S.K., Rana, S., Jha, M.: Excellent hydrogen generation from ultrathin nanosheets of cobalt cyclotetraphosphate. Mater. Sci. Eng. B: Solid-State Mater. Adv. Technol. **265**, 114983. https://doi.org/10.1016/j.mseb.2020.114983
5. Avouris, P., Dimitrakopoulos, C.: Graphene: synthesis and applications. Mater. Today **15**(3), 86–97 (2012). https://doi.org/10.1016/S1369-7021(12)70044-5
6. Azhari, A., Marzbanrad, E., Yilman, D., Toyserkani, E., Pope, M.A.: Binder-jet powder-bed additive manufacturing (3D printing) of thick graphene-based electrodes. Carbon **119**, 257–266 (2017). https://doi.org/10.1016/j.carbon.2017.04.028
7. Beidaghi, M., Gogotsi, Y.: Capacitive energy storage in micro-scale devices: recent advances in design and fabrication of micro-supercapacitors. Energy Environ. Sci. **7**(3), 867 (2014). https://doi.org/10.1039/c3ee43526a
8. Beidaghi, M., Wang, C.: Micro-supercapacitors based on interdigital electrodes of reduced graphene oxide and carbon nanotube composites with ultrahigh power handling performance. Adv. Func. Mater. **22**(21), 4501–4510 (2012). https://doi.org/10.1002/adfm.201201292
9. Belaid, H., Nagarajan, S., Teyssier, C., Barou, C., Barés, J., Balme, S., Garay, H., Huon, V., Cornu, D., Cavaillès, V., Bechelany, M.: Development of new biocompatible 3D printed graphene oxide-based scaffolds. Mater. Sci. Eng., C **110**, 110595 (2020). https://doi.org/10.1016/j.msec.2019.110595

10. Bolotin, K.I., Sikes, K.J., Jiang, Z., Klima, M., Fudenberg, G., Hone, J., Kim, P., Stormer, H.L.: Ultrahigh electron mobility in suspended graphene. Solid State Commun. **146**(9–10), 351–355 (2008). https://doi.org/10.1016/j.ssc.2008.02.024
11. Burela, R.G., Kamineni, J.N., Harursampath, D.: Multifunctional polymer composites for 3D and 4D printing. In: 3D and 4D Printing of Polymer Nanocomposite Materials, pp. 231–257. Elsevier. https://doi.org/10.1016/B978-0-12-816805-9.00008-9
12. Cabral, C.S.D., Miguel, S.P., de Melo-Diogo, D., Louro, R.O., Correia, I.J.: Green reduced graphene oxide functionalized 3D printed scaffolds for bone tissue regeneration. Carbon **146**, 513–523 (2019). https://doi.org/10.1016/j.carbon.2019.01.100
13. Chu, K., Li, W., Jia, C., Tang, F.: Thermal conductivity of composites with hybrid carbon nanotubes and graphene nanoplatelets. Appl. Phys. Lett. **101**(21). https://doi.org/10.1063/1.4767899
14. Cooperstein, I., Layani, M., Magdassi, S.: 3D printing of porous structures by UV-curable O/W emulsion for fabrication of conductive objects. J. Mater. Chem. C **3**(9), 2040–2044 (2015). https://doi.org/10.1039/C4TC02215G
15. Daniel, N., Liu, H.: Comparative study on tribological behavior of graphene/polyimide and carbon fibers/polyimide composites: a review. World J. Eng. Technol. **09**(01), 26–50 (2021). https://doi.org/10.4236/wjet.2021.91003
16. Deshmukh, K., Houkan, M.T., AlMaadeed, M.A., Sadasivuni, K.K.: Introduction to 3D and 4D printing technology: state of the art and recent trends. In: 3D and 4D Printing of Polymer Nanocomposite Materials, pp. 1–24. Elsevier (2020). https://doi.org/10.1016/B978-0-12-816805-9.00001-6
17. Devi, S., Wadhwa, R., Yadav, K.K., Jha, M.: Understanding the origin of ethanol oxidation from ultrafine nickel manganese oxide nanosheets derived from spent alkaline batteries J. Clean. Prod. **376**, 134147 (2022). https://doi.org/10.1016/J.JCLEPRO.2022.134147
18. Dhar, N., Kumar Sahu, S., Samanta, S., Rama Sreekanth, P.S.: An insight into mechanical properties of polymer nanocomposites reinforced with multidimensional filler system: a state of art review. Mater. Today: Proceed. **24**, 422–431 (2020). https://doi.org/10.1016/j.matpr.2020.04.294
19. Farahani, R.D., Dalir, H., Le Borgne, V., Gautier, L.A., El Khakani, M.A., Lévesque, M., Therriault, D.: Direct-write fabrication of freestanding nanocomposite strain sensors. Nanotechnology **23**(8), 085502 (2012). https://doi.org/10.1088/0957-4484/23/8/085502
20. Feng, C., Zhang, M., Bhandari, B.: Materials properties of printable edible inks and printing parameters optimization during 3D printing: a review. Crit. Rev. Food Sci. Nutr. **59**(19), 3074–3081 (2019). https://doi.org/10.1080/10408398.2018.1481823
21. Feng, P., Jia, J., Peng, S., Yang, W., Bin, S., Shuai, C.: Graphene oxide-driven interfacial coupling in laser 3D printed PEEK/PVA scaffolds for bone regeneration. Virtual Phys. Prototyping **15**(2), 211–226 (2020). https://doi.org/10.1080/17452759.2020.1719457
22. Fu, K., Wang, Y., Yan, C., Yao, Y., Chen, Y., Dai, J., Lacey, S., Wang, Y., Wan, J., Li, T., Wang, Z., Xu, Y., Hu, L.: Graphene oxide-based electrode inks for 3d-printed lithium-ion batteries. Adv. Mater. **28**(13), 2587–2594 (2016). https://doi.org/10.1002/adma.201505391
23. Geim, A.K., Novoselov, K.S.: The rise of graphene. Nat. Mater. **6**(3), 183–191 (2007). https://doi.org/10.1038/nmat1849
24. Gonçalves, E.M., Oliveira, F.J., Silva, R.F., Neto, M.A., Fernandes, M.H., Amaral, M., Vallet-Regí, M., Vila, M.: Three-dimensional printed PCL hydroxyapatite scaffolds filled with CNTs for bone cell growth stimulation. J. Biomed. Mater. Res. B Appl. Biomater. **104**(6), 1210–1219 (2016). https://doi.org/10.1002/jbm.b.33432
25. Guo, H., Lv, R., Bai, S.: Recent advances on 3D printing graphene-based composites. Nano Materials Science **1**(2), 101–115 (2019). https://doi.org/10.1016/j.nanoms.2019.03.003
26. Gupta, J., Juneja, S., Bhattacharya, J.: UV lithography-assisted fabrication of low-cost copper electrodes modified with gold nanostructures for improved analyte detection. ACS Omega **5**(7), 3172–3180 (2020). https://doi.org/10.1021/acsomega.9b03125
27. Hashmi, A., Nayak, V., Singh, K.R., Jain, B., Baid, M., Alexis, F., Singh, A.K.: Potentialities of graphene and its allied derivatives to combat against SARS-CoV-2 infection. Mater. Today Adv. **13**, 100208 (2022). https://doi.org/10.1016/j.mtadv.2022.100208

28. Huang, J., Xia, S., Li, Z., Wu, X., Ren, J.: Applications of four-dimensional printing in emerging directions: review and prospects. J. Mater. Sci. Technol. **91**, 105–120 (2021). https://doi.org/10.1016/j.jmst.2021.02.040
29. Ivanova, O., Williams, C., Campbell, T.: Additive manufacturing (AM) and nanotechnology: promises and challenges. Rapid Prototyping J. **19**(5), 353–364 (2013). https://doi.org/10.1108/RPJ-12-2011-0127
30. Jakus, A.E., Secor, E.B., Rutz, A.L., Jordan, S.W., Hersam, M.C., Shah, R.N.: Three-dimensional printing of high-content graphene scaffolds for electronic and biomedical applications. ACS Nano **9**(4), 4636–4648 (2015). https://doi.org/10.1021/acsnano.5b01179
31. Jeon, H., Kim, Y., Yu, W.-R., Lee, J.U.: Exfoliated graphene/thermoplastic elastomer nanocomposites with improved wear properties for 3D printing. Compos. B Eng. **189**, 107912 (2020). https://doi.org/10.1016/j.compositesb.2020.107912
32. Jin, X., Deng, Y., Cai, W., Lau, W., Hui, D., Yan, H., Mei, J., Liu, Y.: Material design and process development of electrostatically patterning silver capsuled composite particle for preparing conductive tracks on flexible substrate. Compos. B Eng. **105**, 111–115 (2016). https://doi.org/10.1016/j.compositesb.2016.07.010
33. Joshi, S., Rawat, K., Karunakaran, C., Rajamohan, V., Mathew, A.T., Koziol, K., Thakur, V.K., Balan, A.S.S.: 4D printing of materials for the future: opportunities and challenges. Appl. Mater. Today **18** (2020). 10049010.1016/j.apmt.2019.100490
34. Ke, Q., Wang, J.: Graphene-based materials for supercapacitor electrodes—a review. J. Materiomics, **2**(1), 37–54. Chinese Ceramic Society. https://doi.org/10.1016/j.jmat.2016.01.001
35. Khan, N., Yadav, K.K., Wadhwa, R., Ankush, Jha, M.: One-pot green method for fabrication of MoS2 decorated graphite rod and their application in H2 evolution. Mater. Lett. **313**, 131784 (2022). https://doi.org/10.1016/j.matlet.2022.131784
36. Khan, N., Yadav, K.K., Wadhwa, R., Sunaina, A., Jha, M.: Realizing ultralow overpotential during electrochemical hydrogen generation through photoexcitation of chromium disulphide. Int. J. Hydrogen Energy **56**, 1294–1300 (2024). https://doi.org/10.1016/j.ijhydene.2023.12.187
37. Kim, J., Lee, M., Jeon, S., Kim, M., Kim, S., Kim, K., Bien, F., Hong, S.Y., Park, J.: Highly transparent and stretchable field-effect transistor sensors using graphene-nanowire hybrid nanostructures. Adv. Mater. **27**(21), 3292–3297 (2015). https://doi.org/10.1002/adma.201500710
38. Lee, C., Wei, X., Kysar, J.W., Hone, J.: Measurement of the elastic properties and intrinsic strength of monolayer graphene. Science **321**(5887), 385–388 (2008). https://doi.org/10.1126/science.1157996
39. Li, P., Shi, E., Yang, Y., Shang, Y., Peng, Q., Wu, S., Wei, J., Wang, K., Zhu, H., Yuan, Q., Cao, A., Wu, D.: Carbon nanotube-polypyrrole core-shell sponge and its application as highly compressible supercapacitor electrode. Nano Res. **7**(2), 209–218 (2014). https://doi.org/10.1007/s12274-013-0388-5
40. Li, R.-Z., Peng, R., Kihm, K.D., Bai, S., Bridges, D., Tumuluri, U., Wu, Z., Zhang, T., Compagnini, G., Feng, Z., Hu, A.: High-rate in-plane micro-supercapacitors scribed onto photo paper using in situ femtolaser-reduced graphene oxide/Au nanoparticle microelectrodes. Energy Environ. Sci. **9**(4), 1458–1467 (2016). https://doi.org/10.1039/C5EE03637B
41. Lopes, A.J., MacDonald, E., Wicker, R.B.: Integrating stereolithography and direct print technologies for 3D structural electronics fabrication. Rapid Prototyping J. **18**(2), 129–143 (2012). https://doi.org/10.1108/13552541211212113
42. Ma, X., Li, F., Xie, Z., Xue, M., Zheng, Z., Zhang, X.: Size-tunable, highly sensitive microelectrode arrays enabled by polymer pen lithography. Soft Matter **13**(20), 3685–3689 (2017). https://doi.org/10.1039/C6SM02791A
43. Meng, C., Maeng, J., John, S.W.M., Irazoqui, P.P.: Ultrasmall integrated 3D micro-supercapacitors solve energy storage for miniature devices. Adv. Energy Mater. **4**(7) (2014). https://doi.org/10.1002/aenm.201301269
44. Momeni, F., Liu, X., Ni, J.: A review of 4D printing. Mater. Des. **122**, 42–79 (2017)

45. Morozov, S.V., Novoselov, K.S., Katsnelson, M.I., Schedin, F., Elias, D.C., Jaszczak, J.A., Geim, A.K.: Giant intrinsic carrier mobilities in graphene and its bilayer. Phys. Rev. Lett. **100**(1), 016602 (2008). https://doi.org/10.1103/PhysRevLett.100.016602
46. Nathan-Walleser, T., Lazar, I., Fabritius, M., Tölle, F.J., Xia, Q., Bruchmann, B., Venkataraman, S.S., Schwab, M.G., Mülhaupt, R.: 3D micro-extrusion of graphene-based active electrodes: towards high-rate AC line filtering performance electrochemical capacitors. Adv. Func. Mater. **24**(29), 4706–4716 (2014). https://doi.org/10.1002/adfm.201304151
47. Ngo, T.D., Kashani, A., Imbalzano, G., Nguyen, K.T.Q., Hui, D.: Additive manufacturing (3D printing): a review of materials, methods, applications and challenges. Compos. B Eng. **143**, 172–196 (2018). https://doi.org/10.1016/j.compositesb.2018.02.012
48. Nguyen, D.T., Meyers, C., Yee, T.D., Dudukovic, N.A., Destino, J.F., Zhu, C., Duoss, E.B., Baumann, T.F., Suratwala, T., Smay, J.E., Dylla-Spears, R.: 3D-printed transparent glass. Adv. Mater. **29**(26) (2017). https://doi.org/10.1002/adma.201701181
49. Nishanthi, S.T., Yadav, K.K., Baruah, A., Ganguli, A.K., Jha, M.: New sustainable and environmental friendly process of synthesis of highly porous Mo2S3 nanoflowers in cooking oil and their electrochemical properties. Electrochim. Acta (2019). https://doi.org/10.1016/j.electacta.2019.01.091
50. Niu, Z., Zhou, W., Chen, X., Chen, J., Xie, S.: Highly compressible and all-solid-state supercapacitors based on nanostructured composite sponge. Adv. Mater. **27**(39), 6002–6008 (2015). https://doi.org/10.1002/adma.201502263
51. Novoselov, K.S., Geim, A.K., Morozov, S.V., Jiang, D., Katsnelson, M.I., Grigorieva, I.V., Dubonos, S.V., Firsov, A.A.: Two-dimensional gas of massless Dirac fermions in graphene. Nature **438**(7065), 197–200 (2005). https://doi.org/10.1038/nature04233
52. Ponnamma, D., Yin, Y., Salim, N., Parameswaranpillai, J., Thomas, S., Hameed, N.: Recent progress and multifunctional applications of 3D printed graphene nanocomposites. Compos. B Eng. **204**, 108493 (2021). https://doi.org/10.1016/j.compositesb.2020.108493
53. Qi, Z., Ye, J., Chen, W., Biener, J., Duoss, E.B., Spadaccini, C.M., Worsley, M.A., Zhu, C.: 3D-printed, superelastic polypyrrole–graphene electrodes with ultrahigh areal capacitance for electrochemical energy storage. Adv. Mater. Technol. **3**(7) (2018). https://doi.org/10.1002/admt.201800053
54. R.H., Shajahan, M.I., Čep, R., Salunkhe, S., Murali, A.P., Sharad, G., Mohamed Abdelmoneam Hussein, H., Abouel Nasr, E.: Electrical conductivity analysis of extrusion-based 3D-printed graphene. Front. Mater. **11** (2024). https://doi.org/10.3389/fmats.2024.1328347
55. Rana, S., Yadav, K.K., Sunaina, Mehta, S.K., Jha, M.: Structurally engineered anisotropic cobalt-based nanostructures for efficient chlorine and oxygen evolution. Adv. Mater. Interfaces **9**(21), 2200740 (2022). https://doi.org/10.1002/ADMI.202200740
56. Rau, D.A., Bortner, M.J., Williams, C.B.: A rheology roadmap for evaluating the printability of material extrusion inks. Addit. Manuf. **75**, 103745 (2023). https://doi.org/10.1016/j.addma.2023.103745
57. Roscher, S., Hoffmann, R., Ambacher, O.: Determination of the graphene-graphite ratio of graphene powder by Raman 2D band symmetry analysis. Anal. Methods **11**(9), 1180–1191 (2019). https://doi.org/10.1039/c8ay02619j
58. Roscher, S., Hoffmann, R., Ambacher, O.: Determination of the graphene–graphite ratio of graphene powder by Raman 2D band symmetry analysis. Anal. Methods **11**(9), 1224–1228 (2019). https://doi.org/10.1039/C8AY02619J
59. Ryan, K.R., Down, M.P., Banks, C.E.: Future of additive manufacturing: Overview of 4D and 3D printed smart and advanced materials and their applications. Chem. Eng. J. **403**, 126162 (2021). https://doi.org/10.1016/j.cej.2020.126162
60. Silvestre, R., Garcia-Breijo, E., Ferri, J., Montava, I., Bou-Belda, E.: The influence of the structure of cotton fabrics on the adhesion of conductive polymer printed with 3D printing technology. Polymers **15**(3), 668 (2023). https://doi.org/10.3390/polym15030668
61. Sood, K., Yadav, K.K., Bhasin, K.K., Ghosh, T.K., Sarkar, S., Jha, M.: Unraveling the behavior of dual active sites of La2FeMnO6 for electrochemical oxygen generation in alkaline conditions. J. Electroanal. Chem. **938**, 117449 (2023). https://doi.org/10.1016/J.JELECHEM.2023.117449

62. Subeshan, B., Baddam, Y., Asmatulu, E.: Current progress of 4D-printing technology. Progress Add. Manuf. **6**(3), 495–516 (2021). https://doi.org/10.1007/s40964-021-00182-6
63. Sun, G., An, J., Chua, C.K., Pang, H., Zhang, J., Chen, P.: Layer-by-layer printing of laminated graphene-based interdigitated microelectrodes for flexible planar micro-supercapacitors. Electrochem. Commun. **51**, 33–36 (2015). https://doi.org/10.1016/j.elecom.2014.11.023
64. Sun, Z., Fang, S., Hu, Y.H.: 3D graphene materials: from understanding to design and synthesis control. Chem. Rev. **120**(18), 10336–10453 (2020). https://doi.org/10.1021/acs.chemrev.0c00083
65. Terrones, M., Martín, O., González, M., Pozuelo, J., Serrano, B., Cabanelas, J.C., Vega-Díaz, S.M., Baselga, J.: Interphases in graphene polymer-based nanocomposites: achievements and challenges. Adv. Mater. **23**(44), 5302–5310 (2011). https://doi.org/10.1002/adma.201102036
66. Usmani, A.B.S., Rana, S., Arora, A., Yadav, K.K., Sammi, H., Sardana, N., Jha, M.: Electrochemical oxygen generation from VO2 nanoflakes decorated onto graphite sheet. J. Alloy. Compd. **976**, 173058 (2024). https://doi.org/10.1016/j.jallcom.2023.173058
67. Wadhwa, R., Yadav, K.K., Rana, S., Jha, M.: Cobalt decorated graphitic carbon nitride photoanode for electrochemical ethanol oxidation: a sustainable way towards clean energy. Int. J. Hydrogen Energy (2023). https://doi.org/10.1016/J.IJHYDENE.2023.04.162
68. Wajahat, M., Kim, J.H., Ahn, J., Lee, S., Bae, J., Pyo, J., Seol, S.K.: 3D printing of Fe3O4 functionalized graphene-polymer (FGP) composite microarchitectures. Carbon **167**, 278–284 (2020). https://doi.org/10.1016/j.carbon.2020.05.045
69. Wang, F., Xu, Z.: Graphene and graphene oxide-reinforced 3D and 4D printable composites. In: 3D and 4D Printing of Polymer Nanocomposite Materials, pp. 259–296. Elsevier (2020). https://doi.org/10.1016/B978-0-12-816805-9.00009-0
70. Wang, X., Jiang, M., Zhou, Z., Gou, J., Hui, D.: 3D printing of polymer matrix composites: a review and prospective. Compos. B Eng. **110**, 442–458 (2017). https://doi.org/10.1016/j.compositesb.2016.11.034
71. Xie, Z., Duo, Y., Lin, Z., Fan, T., Xing, C., Yu, L., Wang, R., Qiu, M., Zhang, Y., Zhao, Y., Yan, X., Zhang, H.: The rise of 2D photothermal materials beyond graphene for clean water production. Adv. Sci. **7**(5). https://doi.org/10.1002/advs.201902236
72. Xu, Y., Liu, Z., Zhang, X., Wang, Y., Tian, J., Huang, Y., Ma, Y., Zhang, X., Chen, Y.: A graphene hybrid material covalently functionalized with porphyrin: synthesis and optical limiting property. Adv. Mater. **21**(12), 1275–1279 (2009). https://doi.org/10.1002/adma.200801617
73. Yadav, K.K., Arora, A., Jangra, S., Jha, M.: Chemically modified carbon nanotubes in 3D and 4D printing. Chem. Modif. Carbon Nanotubes Commer. Appl. 419–439 (2023). https://doi.org/10.1002/9783527838790.CH18
74. Yadav, K.K., Guchhait, S.K., Sunaina, A., Hussain, C.M., Ganguli, A.K., Jha, M.: Synthesis of zirconium diboride and its application in the protection of stainless steel surface in harsh environment. J. Solid State Electrochem. **23**(12), 3243–3253 (2019). https://doi.org/10.1007/s10008-019-04408-0
75. Yadav, K.K., Khan, N., Jha, M.: Application of surface modified carbon nanotubes in energy. ACS Symp. Ser. **1425**, 101–119 (2022). https://doi.org/10.1021/BK-2022-1425.CH005
76. Yadav, K.K., Kumar, G., Rana, S., Sunaina, Ghosh, S., Jha, M.: Understanding the role of sheet thickness on field emission from engineered hexagonal tin disulphide. Appl. Surf. Sci. **606**, 154816. https://doi.org/10.1016/J.APSUSC.2022.154816
77. Yadav, K.K., Shamir, D., Kornweitz, H., Peled, Y., Zohar, M., Burg, A.: Development of meta-chemical surface by dip-pen nanolithography for precise electrochemical lead sensing. Small Methods **8**(3), 2301118 (2023). https://doi.org/10.1002/smtd.202301118
78. Yadav, K.K., Singh, H., Rana, S., Sammi, H., Nishanthi, S.T., Wadhwa, R., Khan, N., Jha, M.: Utilization of waste coir fibre architecture to synthesize porous graphene oxide and their derivatives: an efficient energy storage material. J. Clean. Prod. 124240 (2020). https://doi.org/10.1016/j.jclepro.2020.124240
79. Yadav, K.K., Wadhwa, R., Khan, N., Jha, M.: Efficient metal-free supercapacitor based on graphene oxide derived from waste rice. Curr. Res. Green Sustain. Chem. **4**, 100075 (2021) https://doi.org/10.1016/j.crgsc.2021.100075

80. Yang, J., Zhou, B., Han, D., Cui, N., Li, B., Shen, J., Zhang, Z., Du, A.: High-precision three-dimensional printing in a flexible, low-cost and versatile way: a review. ES Mater. Manuf. (2021). https://doi.org/10.30919/esmm5f526
81. Yu, Z., Tetard, L., Zhai, L., Thomas, J.: Supercapacitor electrode materials: nanostructures from 0 to 3 dimensions. Energy Environ. Sci. **8**(3), 702–730 (2015). https://doi.org/10.1039/C4EE03229B
82. Zaaba, N.I., Foo, K.L., Hashim, U., Tan, S.J., Liu, W.-W., Voon, C.H.: Synthesis of graphene oxide using modified hummers method: solvent influence. Procedia Eng. **184**, 469–477 (2017). https://doi.org/10.1016/j.proeng.2017.04.118
83. Zhang, Y., Wang, C., Fu, L., Ye, S., Wang, M., Zhou, Y.: Fabrication and application of novel porous scaffold in situ-loaded graphene oxide and osteogenic peptide by cryogenic 3D printing for repairing critical-sized bone defect. Molecules **24**(9), 1669 (2019). https://doi.org/10.3390/molecules24091669
84. Zhang, Y., Zhai, D., Xu, M., Yao, Q., Zhu, H., Chang, J., Wu, C.: 3D-printed bioceramic scaffolds with antibacterial and osteogenic activity. Biofabrication **9**(2), 025037 (2017). https://doi.org/10.1088/1758-5090/aa6ed6
85. Zhang, Y., Zhang, F., Yan, Z., Ma, Q., Li, X., Huang, Y., Rogers, J.A.: Printing, folding and assembly methods for forming 3D mesostructures in advanced materials. Nat. Rev. Mater. **2**(4), 17019 (2017). https://doi.org/10.1038/natrevmats.2017.19
86. Zhao, Y., Liu, J., Hu, Y., Cheng, H., Hu, C., Jiang, C., Jiang, L., Cao, A., Qu, L.: Highly compression-tolerant supercapacitor based on polypyrrole-mediated graphene foam electrodes. Adv. Mater. **25**(4), 591–595 (2013). https://doi.org/10.1002/adma.201203578
87. Zhou, X., Nowicki, M., Cui, H., Zhu, W., Fang, X., Miao, S., Lee, S.-J., Keidar, M., Zhang, L.G.: 3D bioprinted graphene oxide-incorporated matrix for promoting chondrogenic differentiation of human bone marrow mesenchymal stem cells. Carbon **116**, 615–624 (2017). https://doi.org/10.1016/j.carbon.2017.02.049

Carbon-Nanotube-Reinforced 3D and 4D Printable Conductive Polymer Composites

Anchit Modi, Chimany Sahu, Monika Ahuja, Subhendu Chakroborty, and N. K. Gaur

Abstract The integration of carbon nanotubes (CNTs) into polymer matrices has led to the development of advanced composites with exceptional mechanical, electrical, and thermal properties, particularly suitable for 3D and 4D printing applications. This paper explores the synthesis, properties, and potential applications of carbon nanotube-reinforced 3D and 4D printable conductive polymer composites. CNTs, known for their high tensile strength, conductivity, and aspect ratio, enhance the performance of polymer composites, making them ideal for creating lightweight, strong, and conductive materials. The dispersion and alignment of CNTs within the polymer matrix are critical challenges addressed through advanced techniques such as in-situ polymerization, ultrasonication, and surface functionalization. Additionally, the scalability and cost of producing these composites are examined, highlighting recent advancements aimed at more economical production methods. Environmental and health implications are also considered, ensuring safe usage and disposal of CNT-infused materials. Innovations in 4D printing, which involve materials that can change shape or properties in response to external stimuli, are discussed, showcasing the expanding capabilities and applications of these composites in fields like aerospace, biomedical devices, and soft robotics. The ongoing research and technological advancements underscore the transformative potential of CNT-reinforced 3D and 4D printable conductive polymer composites in various industrial and technological domains.

A. Modi
Department of Basic Sciences, IITM, IES University, Bhopal, India

C. Sahu · N. K. Gaur
Department of Physics, Barkatullah University, Bhopal, India

M. Ahuja
Department of Chemistry, Government Adarsh College, Affiliated to Barkatullah University Bhopal, Harda, Madhya Pradesh, India

S. Chakroborty (✉)
School of Basic Sciences, Department of Chemistry, Chandigarh University Uttar Pradesh, Unnao, Uttar Pradesh, India
e-mail: subhendu.cy@gmail.com

S. Moharana et al. (eds.), *Polymer Nanocomposites for 3D, 4D and 5D Printing*, Engineering Materials, https://doi.org/10.1007/978-981-96-4214-4_8

Keywords CNT · Conductive polymer · Environmental

1 Introduction

In recent years, the advancement of additive manufacturing technologies, commonly known as 3D and 4D printing, has revolutionized the field of materials science. These techniques allow for the precise fabrication of complex structures with tailored properties, opening new horizons in various applications ranging from electronics to biomedicine. One of the most promising developments in this domain is the incorporation of carbon nanotubes (CNTs) into conductive polymer composites [1–4]. Carbon nanotubes, renowned for their exceptional mechanical strength, electrical conductivity, and thermal stability, have emerged as ideal reinforcements to enhance the properties of polymer matrices. These composites combine the versatility and adaptability of conductive polymers with the superior characteristics of CNTs, resulting in materials that are not only strong and conductive but also capable of responding dynamically to environmental stimuli in the case of 4D printing [5, 6]. This integration facilitates the creation of multifunctional materials that can be used in a variety of advanced applications such as flexible electronics, sensors, and smart textiles [7]. The development and optimization of these composites involve a multidisciplinary approach, encompassing material science, nanotechnology, and advanced manufacturing techniques. By fine-tuning the dispersion and alignment of CNTs within the polymer matrix, researchers can achieve enhanced electrical and mechanical performance. Furthermore, the ability to 3D and 4D print these materials allows for the creation of custom-designed structures with complex geometries, which are difficult to achieve with traditional manufacturing methods [8, 9].

In 3D printing, the addition of CNTs to polymer composites improves not only the conductivity and mechanical strength but also the thermal stability of the printed structures. This makes these composites particularly suitable for applications in electronics, where thermal management and electrical performance are crucial. Additionally, the mechanical robustness imparted by CNTs can lead to more durable and wear-resistant components, expanding their utility in harsh environments and mechanical applications [10]. The concept of 4D printing introduces a time-dependent dimension to the printed objects, enabling them to change shape or function in response to external stimuli such as temperature, light, moisture, or magnetic fields. The incorporation of CNTs in 4D printable materials further enhances these dynamic properties, allowing for more precise control over the stimuli-responsive behavior. This can lead to innovative applications such as self-healing materials, shape-memory devices, and adaptive structures that can transform their properties in real time [11].

Moreover, the environmental impact and sustainability of materials are becoming increasingly important considerations. CNT-reinforced polymer composites can be engineered to be more eco-friendly by selecting biodegradable or recyclable polymer matrices, thereby reducing their ecological footprint. The potential for using renewable resources and minimizing waste through additive manufacturing processes aligns well with global sustainability goals [12, 13]. The carbon nanotube-reinforced 3D and 4D printable conductive polymer composites mark a transformative development in materials science. These advanced materials offer unparalleled performance and functionality, paving the way for cutting-edge innovations across a broad spectrum of industries. As research continues to evolve, the potential applications and benefits of these composites are expected to expand, further driving technological progress and enhancing the quality of life [14].

2 Carbon Nanotubes (CNTs)

2.1 Structure and Properties of CNTs

Carbon nanotubes (CNTs) are cylindrical nanostructures composed of carbon atoms arranged in a hexagonal lattice, resembling a rolled-up sheet of graphene. They exhibit extraordinary mechanical, electrical, and thermal properties, making them one of the most studied materials in nanotechnology. The structure of CNTs can be categorized based on their configuration, which influences their properties and applications.

2.2 Basic Structure

The basic structure of carbon nanotubes (CNTs) is fundamentally derived from graphene, which is a single layer of carbon atoms arranged in a hexagonal lattice pattern. This arrangement provides the foundational building block for CNTs. Graphene is celebrated for its remarkable properties, including its exceptional strength, flexibility, and conductivity. When a graphene sheet is rolled into a seamless cylindrical shape, it forms what is known as a carbon nanotube. This cylindrical shape is characterized by diameters that typically fall within the nanometer range, while their lengths can extend up to several micrometers or even millimeters in some cases [15, 16]. The process of rolling the graphene sheet can occur in various ways, leading to different types of CNTs, such as single-walled carbon nanotubes (SWCNTs) and multi-walled carbon nanotubes (MWCNTs). In SWCNTs, a single graphene sheet forms a single cylindrical layer, whereas in MWCNTs, multiple concentric graphene cylinders are nested within each other. The seamless nature of the cylinder, combined with the atomic-scale dimensions, endows CNTs with unique properties. The precise method by which the graphene sheet is rolled (its chirality) determines whether the resulting CNT will have metallic or semiconducting electrical properties, which is crucial for various applications in nanoelectronics and materials science [17].

The cylindrical shape and nanoscale dimensions of CNTs result in an extraordinary aspect ratio, which is the ratio of length to diameter. This high aspect ratio is responsible for the exceptional mechanical properties of CNTs, including their high tensile strength and elasticity. Furthermore, the cylindrical structure allows for the possibility of electron transport along the tube axis with minimal scattering, contributing to their impressive electrical conductivity. This makes CNTs highly attractive for applications in fields such as nanocomposites, where they can reinforce materials, and in the development of miniaturized electronic devices, where their conductive properties are highly beneficial. Researchers are continually exploring ways to optimize the basic structure of CNTs to enhance their performance and expand their applications. This includes modifying the surface of CNTs through chemical functionalization to improve their interaction with other materials, thus broadening their applicability in areas like biomedical engineering, environmental sensing, and energy storage. The study of the basic structure of CNTs not only deepens our understanding of their intrinsic properties but also drives innovation in their practical uses, making CNTs a cornerstone of nanotechnology research and development [18, 19].

2.3 Electrical Properties

Carbon nanotubes (CNTs) display remarkable electrical properties, with their behaviour ranging from metallic to semiconducting depending on their chirality, described by the chiral indices (n, m). Metallic CNTs exhibit high electrical conductivity, comparable to metals, while semiconducting CNTs possess a bandgap suitable for electronic applications. With conductivity up to 10^6 S/m and a current carrying capacity exceeding traditional metals, CNTs find applications in nanoelectronics for creating faster transistors and efficient interconnects. Additionally, their sensitivity to environmental changes makes them valuable in sensors for detecting chemicals and biological substances. Furthermore, CNTs are utilized in energy storage devices like supercapacitors and batteries to enhance energy storage capacity and charge–discharge rates. Their ability to shield against electromagnetic interference also positions them in materials for Electromagnetic Interference (EMI) protection. Overall, the exceptional electrical properties of CNTs underpin their widespread application potential in diverse fields such as electronics, energy, and materials science [20].

2.4 Mechanical Properties

Carbon nanotubes (CNTs) possess extraordinary mechanical properties that stem from their unique atomic structure and cylindrical geometry. These properties, including high tensile strength, stiffness, resilience, and flexibility, make CNTs one of the strongest and most versatile materials known. With tensile strength exceeding that of steel by up to 100 times and Young's modulus surpassing 1 tera pascal (TPa),

CNTs are highly sought after as reinforcing agents in composite materials. In industries ranging from aerospace to automotive and sporting goods, CNT-reinforced composites offer unmatched mechanical performance, providing enhanced strength, stiffness, and durability. Moreover, the exceptional flexibility of CNTs allows for their incorporation into nanomechanical devices, including sensors and nanoelectromechanical systems (NEMS), enabling advancements in sensing, actuation, and data storage technologies. Beyond terrestrial applications, the lightweight nature and high specific strength of CNTs make them intriguing candidates for innovative concepts such as space elevators, where their remarkable mechanical properties could revolutionize space transportation. The ongoing exploration of CNTs' mechanical characteristics promises to unlock new possibilities in materials science and engineering, paving the way for groundbreaking advancements in various industries [21].

2.5 Thermal Properties

Carbon nanotubes (CNTs) possess exceptional thermal properties, stemming from their unique atomic structure and cylindrical geometry. With a high thermal conductivity exceeding that of most materials, including metals like copper and aluminum, CNTs efficiently conduct heat along their axis, making them highly sought after for thermal management applications. Moreover, their low coefficient of thermal expansion (CTE) and thermal stability render them ideal for use in environments with extreme temperatures. In electronic devices, CNT-based thermal interface materials (TIMs) are employed to enhance heat dissipation and prevent overheating by improving thermal contact between components. Furthermore, CNTs serve as effective heat spreaders and heat sinks in electronic packaging, ensuring efficient thermal management in compact devices. Beyond electronics, CNTs find applications in thermal barrier coatings, where they provide insulation and protect surfaces from high temperatures in aerospace, automotive, and industrial sectors. Additionally, CNT-based thermoelectric devices hold promise for energy harvesting and waste heat recovery applications. By dispersing CNTs in fluids and polymer matrices, nanofluids and nanocomposites with enhanced thermal properties are created, and used in various heat transfer and structural applications [22].

3 Types of Carbon Nanotubes

3.1 Single-Walled Carbon Nanotubes (SWCNTs)

Single-Walled Carbon Nanotubes (SWCNTs) are a distinct type of carbon nanotube that consists of a single graphene sheet rolled into a seamless cylindrical structure. These cylinders have a diameter typically ranging from 0.4 to 2 nm, making

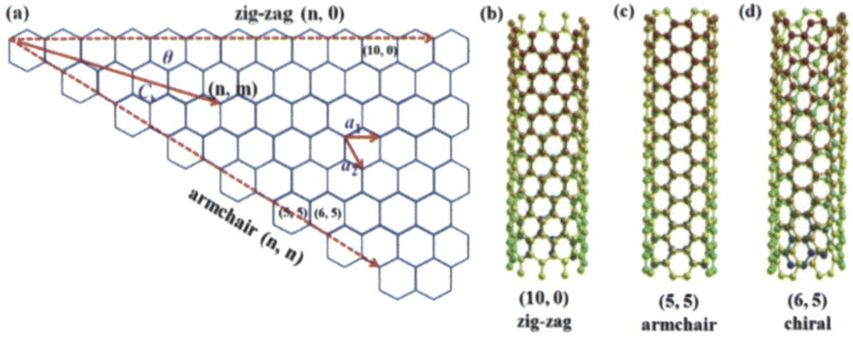

Fig. 1 **a** Unrolled single-layer graphene sheet showing the geometry of the SWCNT. (bed) Examples of the three types of nanotube sidewalls; zig-zag, armchair, and chiral [23]. Copyright 2016. Reproduced with permission from Elsevier

them exceptionally small and leading to a high aspect ratio. The unique electrical properties of SWCNTs are particularly noteworthy; they can be either metallic or semiconducting, a characteristic determined by their chirality, which is the specific angle and manner in which the graphene sheet is rolled. Figure 1 shows a single-layer graphene sheet with the geometry of a single-walled carbon nanotube (SWCNT) and the three types of chirality: zigzag, armchair, and chiral. This chirality dictates the electronic band structure of the nanotubes, thus influencing their conductivity and other electronic properties. SWCNTs are renowned for their exceptional mechanical strength, flexibility, and thermal conductivity, which make them highly desirable for a wide range of applications, including in nanoelectronics, photonics, and materials science.

Their ability to conduct electricity and their structural integrity offer significant advantages in the development of advanced composites, sensors, and nanoscale devices. Moreover, their surface can be chemically modified to enhance compatibility with other materials, expanding their utility in various fields such as drug delivery, energy storage, and environmental sensing. The ongoing research and development in SWCNTs aim to harness these unique properties to create innovative solutions in technology and industry [24].

3.2 Multi-walled Carbon Nanotubes (MWCNTs)

Multi-Walled Carbon Nanotubes (MWCNTs) are a type of carbon nanotube that consists of multiple concentric graphene cylinders nested within one another, forming a series of coaxial tubes. These structures can have diameters ranging from 2 to 100 nm, with each layer separated by an interlayer spacing of approximately 0.34 nm, which is similar to the distance between graphene layers in graphite. This unique multi-layered configuration endows MWCNTs with exceptional mechanical strength and durability, significantly surpassing the capabilities of single-walled carbon nanotubes (SWCNTs) in terms of structural integrity. Figure 2 shows the structures of different types of CNTs. The robustness of MWCNTs is attributed to the multiple graphene layers that contribute to their high tensile strength, resilience under stress, and resistance to damage. These properties make MWCNTs highly suitable for applications requiring enhanced durability, such as in composite materials used in aerospace, automotive, and construction industries.

Additionally, the nested structure of MWCNTs offers advantages in thermal and electrical conductivity, although their electrical properties are less tunable compared to SWCNTs due to the complex interactions between layers. MWCNTs also exhibit unique properties such as high aspect ratio, large surface area, and excellent chemical stability. These characteristics are leveraged in a variety of advanced applications including, but not limited to, reinforced polymer composites, conductive films, batteries, supercapacitors, and catalysts in chemical reactions [26]. The multi-layered structure allows for functionalization on both the inner and outer surfaces, which can be exploited to enhance their compatibility with different matrices and to introduce specific functionalities for targeted applications. Current research in the field of MWCNTs is focused on improving production methods to ensure uniformity and purity, reducing costs, and developing environmentally friendly synthesis processes. Scientists are also exploring novel applications in fields such as biotechnology, medicine, and environmental science, where the unique properties of MWCNTs can be harnessed for innovative solutions, such as drug delivery systems, biosensors, and water purification technologies. The ongoing advancements in MWCNT technology continue to expand their potential, paving the way for new industrial and technological breakthroughs [27].

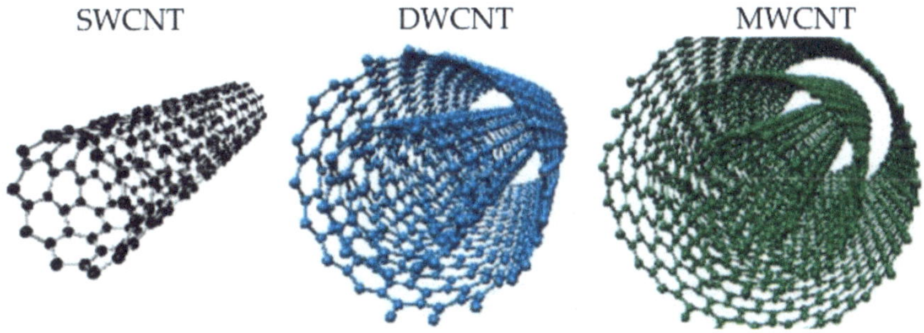

Fig. 2 Structures of carbon nanotubes [25]. Copyright 2022. Reproduced with permission from MDPI

4 Importance of Conductive Polymer Composites

Conductive polymer composites represent a significant advancement in materials science and engineering, offering a multifaceted combination of properties that make them indispensable across diverse industries. The structures of the different CNT-reinforced conducting polymers and their importance in different fields are presented in Fig. 3.

These composites are formulated by embedding conductive fillers, such as carbon nanotubes, graphene, or metal nanoparticles, within a polymer matrix. One of the primary advantages of conductive polymer composites lies in their electrical conductivity, which is achieved while retaining the mechanical flexibility and processability inherent to polymers. This unique characteristic enables their utilization in a myriad of applications, including flexible electronics, printed circuit boards, and electromagnetic interference (EMI) shielding, where traditional rigid conductive materials are impractical. Moreover, the mechanical flexibility of these composites opens up

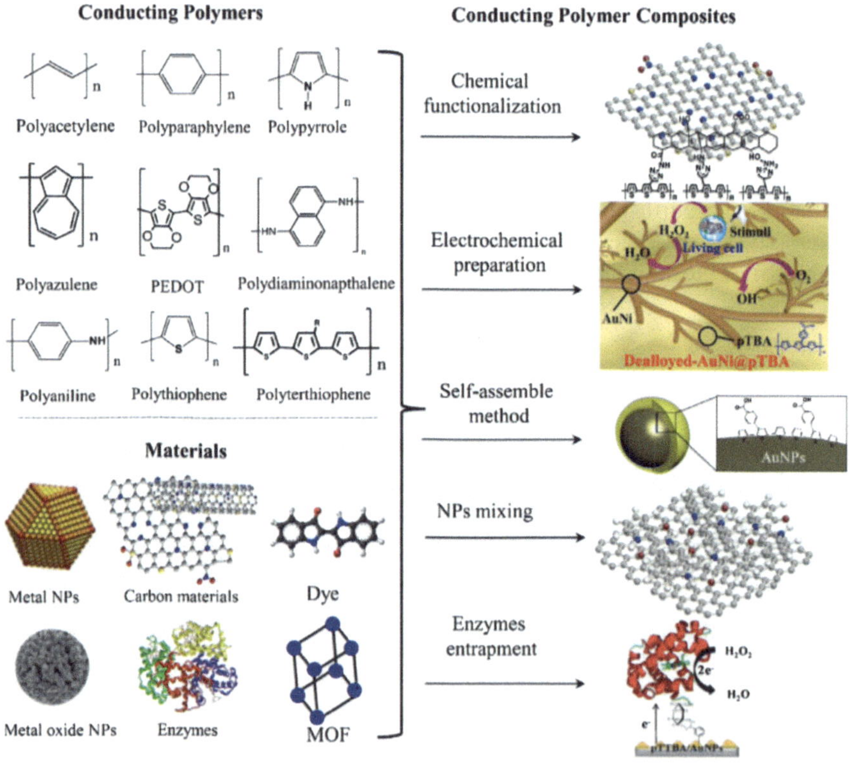

Fig. 3 Conducting polymers and their composite materials [28]. Copyright 2017. Reproduced with permission from Elsevier

avenues in emerging fields like wearable technology, where conformable and stretchable electronic devices are in high demand. The lightweight nature of conductive polymer composites further enhances their appeal, particularly in industries such as aerospace, automotive, and consumer electronics, where weight reduction is critical for improving fuel efficiency, reducing costs, and enhancing overall performance [29, 30]. Additionally, the inherent corrosion resistance of polymer matrices in these composites makes them well-suited for applications in harsh environments or outdoor settings, where traditional metallic conductors may degrade over time. Furthermore, the properties of conductive polymer composites can be tailored by adjusting the type, concentration, and dispersion of conductive fillers, allowing for optimization to meet specific application requirements. This versatility, coupled with their relative cost-effectiveness compared to pure metals, contributes to their widespread adoption in various industrial sectors. Overall, conductive polymer composites continue to drive innovation and enable the development of next-generation technologies and products that require lightweight, flexible, and electrically conductive materials. As research and development in this field progress, the potential applications and benefits of these composites are expected to expand even further, shaping the future of materials science and engineering [31, 32].

5 Fabrication of CNT-Reinforced Conductive Polymer Composites

5.1 Selection of Polymers

The selection of polymers for the fabrication of carbon nanotube (CNT) reinforced conductive polymer composites is a pivotal aspect influencing the properties and performance of the final material. In this chapter, we explore the considerations and criteria guiding the choice of polymer matrices, aiming to elucidate the key factors in optimizing the fabrication process and tailoring the properties of these advanced materials for diverse applications. Polymers play a crucial role as the matrix material in CNT-reinforced composites, providing a platform for dispersing and integrating CNTs while contributing essential properties such as mechanical strength, flexibility, and chemical resistance. Several polymers have demonstrated compatibility and synergy with CNTs, facilitating the creation of high-performance composites. Figure 4 shows the schematic diagram for different fabrication techniques for CNT-reinforced polymer composite materials. Among these, polyethylene (PE), known for its processability and mechanical robustness, finds application in flexible electronics and packaging materials [33].

Polycarbonate (PC), valued for its transparency and impact strength, serves in structural components across industries. Poly (methyl methacrylate) (PMMA), prized for its optical clarity, finds utility in optical sensors and displays. Polypropylene

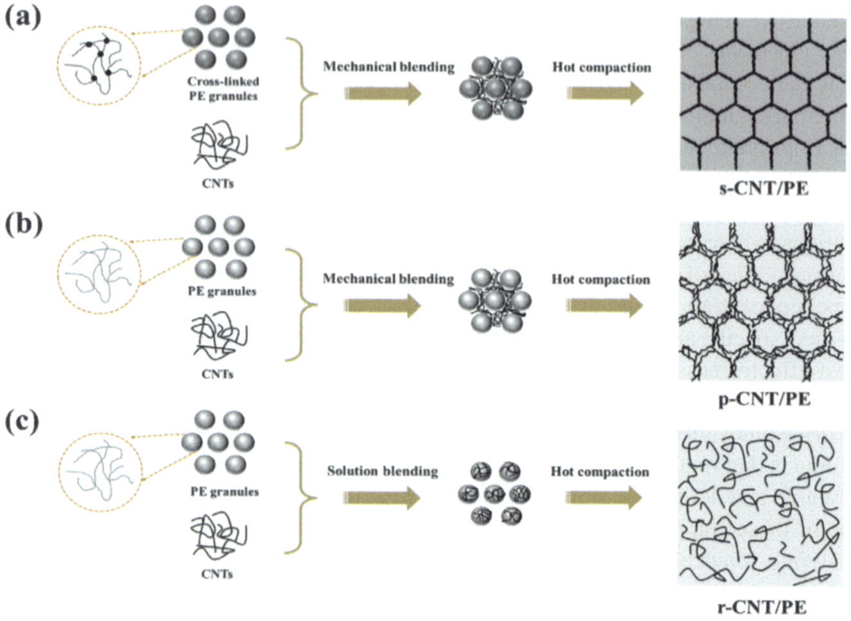

Fig. 4 Schematic for the fabrication of the s-CNT/PE (**a**), p-CNT/PE (**b**) and r-CNT/PE (**c**) composites [34]. Copyright 2021. Reproduced with permission from Elsevier

(PP), lauded for its chemical resistance and low density, is favored in automotive and consumer goods. Polyamide (PA), renowned for its strength and toughness, is essential in structural applications. Polyvinylidene fluoride (PVDF), with its chemical resistance and piezoelectric properties, excels in sensors and energy harvesting devices. Key considerations in polymer selection include compatibility with CNTs, mechanical properties, electrical conductivity, processing requirements, and targeted applications. Successful integration of CNTs into polymer matrices relies on achieving uniform dispersion, strong interfacial bonding, and optimal processing conditions. By addressing these considerations and leveraging the unique properties of polymers and CNTs, researchers and engineers can design and fabricate CNT-reinforced conductive polymer composites tailored to specific performance requirements across a spectrum of applications [35].

6 Dispersion Techniques

Achieving uniform dispersion of carbon nanotubes (CNTs) within the polymer matrix is crucial for harnessing their unique properties and maximizing the performance of CNT-reinforced conductive polymer composites. In this chapter, we explore various dispersion techniques employed to effectively disperse CNTs in polymer matrices, aiming to elucidate the key strategies and challenges associated with each method [36–38].

1. **Solution mixing**: Solution mixing involves dispersing CNTs in a solvent, followed by mixing with the polymer solution before solvent evaporation. This technique enables excellent dispersion control and homogeneity but requires a careful selection of solvents compatible with both CNTs and the polymer matrix.
2. **Melt mixing**: Melt mixing involves incorporating CNTs into the molten polymer using high-shear mixing techniques such as twin-screw extrusion or Banbury mixing. While melt mixing is suitable for large-scale production and offers good dispersion, it may result in CNT damage or agglomeration due to the high processing temperatures.
3. **In-situ polymerization**: In-situ polymerization involves dispersing CNTs in the monomer solution before initiating polymerization. This technique allows for intimate mixing at the molecular level, resulting in uniform dispersion and strong interfacial bonding between CNTs and the polymer matrix.
4. **Functionalization**: Functionalization of CNTs involves modifying their surface chemistry to enhance compatibility with the polymer matrix. Techniques such as covalent or non-covalent functionalization improve dispersibility and interfacial adhesion, leading to enhanced mechanical and electrical properties of the composite.
5. **Surfactant-assisted dispersion**: Surfactants can be used to facilitate CNT dispersion by reducing surface tension and promoting uniform distribution. However, surfactants may interfere with the electrical conductivity of the composite and require thorough removal post-processing.
6. **Ultrasonication**: Ultrasonication involves applying high-frequency ultrasound waves to break down CNT agglomerates and promote dispersion. While effective in achieving fine dispersion, ultrasonication may induce CNT damage or shorten length, necessitating optimization of parameters to balance dispersion and integrity.
7. **Electrospinning**: Electrospinning involves electrostatically spinning a polymer solution containing dispersed CNTs into nanofibers. This technique allows for precise control over fiber morphology and CNT alignment, offering opportunities for advanced applications in filtration, sensors, and tissue engineering.

Each dispersion technique has its advantages and challenges, requiring careful consideration of factors such as CNT properties, polymer compatibility, processing conditions, and desired composite properties. By understanding and optimizing dispersion techniques, researchers and engineers can develop CNT-reinforced conductive polymer composites with tailored properties for a wide range of applications in electronics, aerospace, automotive, and beyond [39].

7 3D Printing of CNT-Reinforced Composites

The advent of incorporating carbon nanotubes (CNTs) into various 3D printing methodologies has sparked a wave of excitement within the realm of materials science and engineering, unlocking unprecedented possibilities in the fabrication of composite structures with multifaceted properties. Figure 5 depicts the SEM images of 3D printed materials via different fabrication techniques.

Within the spectrum of 3D printing techniques such as Fused Deposition Modelling (FDM), Stereolithography (SLA), Digital Light Processing (DLP), Selective Laser Sintering (SLS), Selective Laser Melting (SLM), and Binder Jetting, the integration of CNTs introduces a wealth of advantages. For instance, in FDM, the utilization of thermoplastic filaments infused with CNTs enables the extrusion of layers, resulting in parts with heightened mechanical strength, electrical conductivity, and resistance to thermal deformation. Similarly, SLA and DLP techniques harness photopolymerizable resins mixed with CNTs to yield high-resolution parts boasting superior mechanical properties. Meanwhile, the powder bed fusion methods of SLS and SLM leverage CNT-reinforced powders to fabricate parts with customized properties, offering unparalleled geometric freedom and scalability [41, 42]. Furthermore, binder jetting presents a promising avenue for the production of large-scale parts, employing selective deposition of binding agents onto CNT-reinforced powdered beds. However, notwithstanding the tremendous potential of CNT-reinforced composites in 3D printing, a host of challenges loom, including the imperative of achieving uniform dispersion, meticulously controlling processing

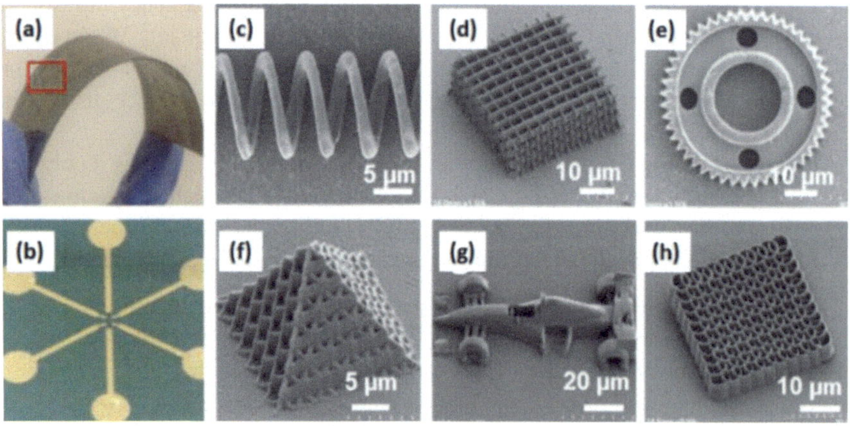

Fig. 5 Functional micro/nanostructures fabrication based on MTA composite resins by TPP lithography. Bent polyethylene terephthalate (PET) substrate (**a**) and SiO_2/Si substrate (**b**) used for TPP fabrication with Au electrode patterns; (**c**–**h**) Scanning electron microscopy (SEM) micrographs of various functional micro/nanostructures including a microcoil inductor (**c**); a woodpile (**d**); a microgear (**e**); a micro-pyramid (**f**); a microcar (**g**) and a spiral-like photonic crystal (**h**). The laser power and scanning speed used in the TPP fabrication were 15 mW and 0.5 mm/s, respectively [40]. Copyright 2017. Reproduced with permission from MDPI

parameters, and ensuring consistent part quality. Nonetheless, the amalgamation of CNTs into 3D printing processes heralds a new era of innovation across an array of industries, spanning aerospace, electronics, healthcare, and beyond, catalyzing groundbreaking advancements and pushing the frontiers of materials science ever further [43].

8 4D Printing with CNT-Reinforced Composites

The integration of carbon nanotube (CNT) reinforced composites into the realm of 4D printing represents a pioneering approach to fabricating dynamic structures capable of responding to external stimuli. In this chapter, we delve into the principles, techniques, challenges, and applications of 4D printing with CNT-reinforced composites, elucidating the transformative potential of this emerging field. 4D printing builds upon the principles of 3D printing by introducing an additional dimension-time, fabricating structures with the ability to change shape, properties, or functionality in response to environmental stimuli such as temperature, humidity, light, or pH. By integrating CNT-reinforced composites into 4D printing processes, researchers can engineer materials that exhibit dynamic behavior and adaptability, opening up new possibilities for applications in various industries. Several techniques have been explored for 4D printing with CNT-reinforced composites, including material extrusion, vat polymerization, powder bed fusion, and binder jetting. Each technique offers unique advantages and challenges in terms of achieving dynamic behavior, controlling stimuli-responsive properties, and ensuring uniform dispersion of CNTs within the printed structures [44]. CNT-reinforced composites can exhibit stimuli-responsive behavior by leveraging the unique properties of CNTs, such as thermal conductivity, electrical conductivity, and mechanical strength. By carefully tuning the composition, orientation, and distribution of CNTs within the composite matrix, researchers can design structures that undergo controlled shape changes or property variations in response to external stimuli. Figure 6 describes various controllable 4D printed materials for sensors, actuator and bio-inspired materials.

Despite the promising potential of 4D printing with CNT-reinforced composites, several challenges remain to be addressed, including achieving precise control over dynamic behavior, optimizing printing parameters, ensuring compatibility between CNTs and the polymer matrix, and scaling up production for industrial applications. However, the opportunities offered by this technology are immense, ranging from adaptive structures for aerospace and automotive applications to smart textiles, biomedical devices, and beyond [46].

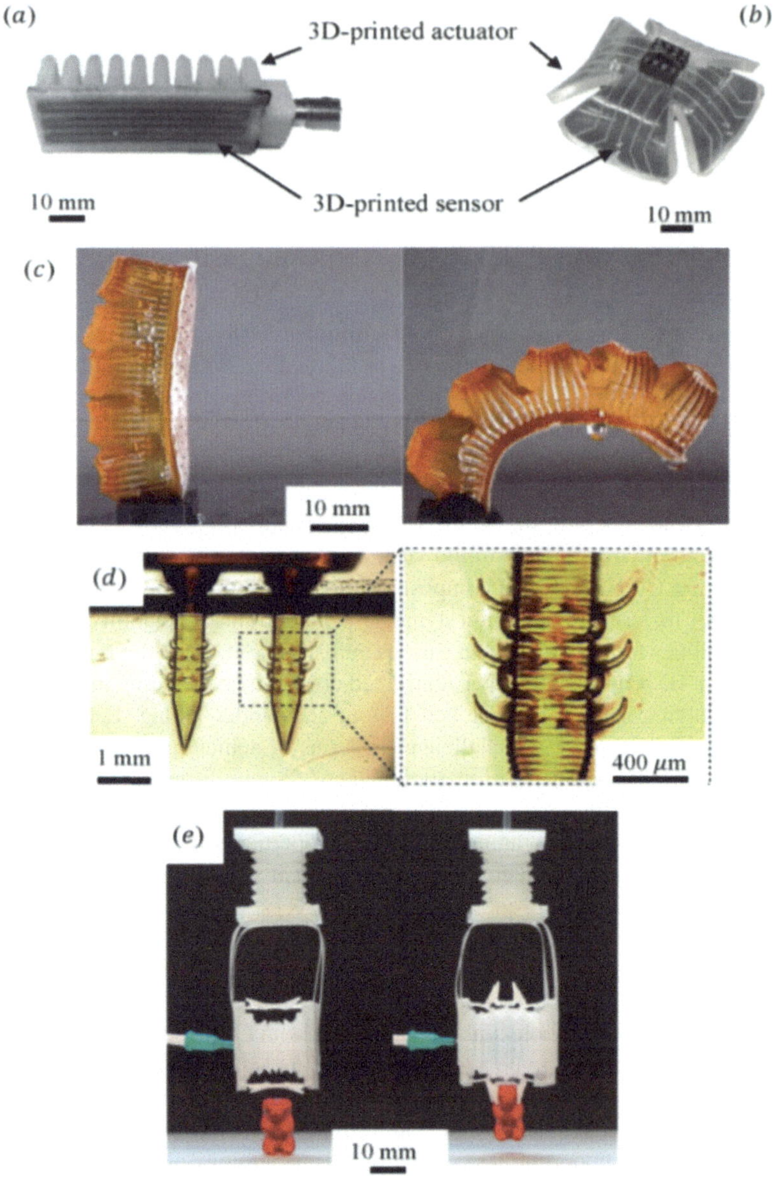

Fig. 6 Controllable 4D-printed systems **a** A combined sensor and 3D-printed pneumatic muscle **b** An embedded piezoelectric sensor in the hydrogel actuator **c** A sweating 4D-printed system with thermoregulation pores **d** The 4D-printed bioinspired microneedles with stiffness control **e** A buckling-induced 4D-printed active skin [45]. Copyright 2020. Reproduced with permission from MDPI

9 Applications of CNT -Reinforced 3D and 4D Printable Composites

The integration of carbon nanotube (CNT) reinforced composites into the realm of 4D printing represents a pioneering approach to fabricating dynamic structures capable of responding to external stimuli. In this chapter, we delve into the principles, techniques, challenges, and applications of 4D printing with CNT-reinforced composites, elucidating the transformative potential of this emerging field. 4D printing builds upon the principles of 3D printing by introducing an additional dimension-time, fabricating structures with the ability to change shape, properties, or functionality in response to environmental stimuli such as temperature, humidity, light, or pH. By integrating CNT-reinforced composites into 4D printing processes, researchers can engineer materials that exhibit dynamic behavior and adaptability, opening up new possibilities for applications in various industries [47]. Here are some key applications.

9.1 *Construction and Infrastructure*

One of the most significant contributions of CNT-reinforced composites in civil engineering is their ability to create smart materials. These materials are engineered to sense and respond to environmental changes, such as temperature fluctuations, stress, or damage. The inherent properties of carbon nanotubes, such as their exceptional strength, flexibility, and electrical conductivity, make them ideal for reinforcing composite materials. When integrated into concrete, polymers, or other building materials, CNTs enable these structures to monitor their own health in real time. For example, a CNT-reinforced concrete beam can detect cracks or stress points before they become critical, allowing for proactive maintenance and repair. This capability not only enhances the safety of structures but also extends their lifespan, reducing the need for frequent and costly repairs [48]. Furthermore, smart materials contribute to the sustainability of infrastructure by minimizing material waste and energy consumption over the life cycle of a structure Fig. 7 shows some 4D-printed structures of SWCNT/polymers along with the Raman spectrum taken from the box suspending the nanowire with an excitation wavelength of 785 nm.

In addition to smart materials, the advent of 4D printing technology has introduced the possibility of creating self-healing materials using CNT composites. Unlike traditional 3D printing, which produces static objects, 4D printing involves the use of materials that can change their properties over time in response to external stimuli. By incorporating CNTs into these materials, engineers can design infrastructure that automatically repairs itself when damaged. This self-healing capability is particularly valuable in areas prone to harsh environmental conditions or natural disasters, where infrastructure is constantly exposed to wear and tear. For instance, a bridge built with CNT-enhanced self-healing materials could autonomously close small cracks or

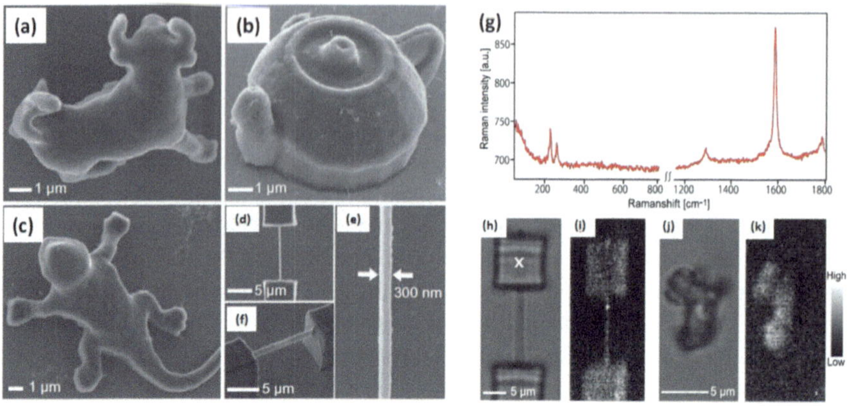

Fig. 7 3D micro/nano structural SWCNT/polymer composites are fabricated by using the TPP lithography. The structures in **a**–**f** are an 8-μm-long micro bull, a micro tea pod, a micro lizard, a nanowire suspended between two micro boxes, magnified image of (**d**), and a perspective view of the nanowire, respectively; **g** shows a Raman spectrum taken from the box suspending the nanowire with an excitation wavelength of 785 nm. Laser power and exposure time are 2.8 mW and 60 s, respectively; **h** A bright field image of the 300-nm-thick nanowire suspended between two boxes; **i** A G-band Raman image taken at the same area in (**h**). Laser power and exposure time are 1.4 mW and 10 s, respectively; **j** A bright field image of an 8-μm-long micro bull; **k** A G-band Raman image taken at the same area in (**h**). Laser power and exposure time are 2.8 mW and 10 s, respectively [40]. Copyright 2017. Reproduced with permission from MDPI

fractures caused by repeated stress, thereby preventing the formation of larger, more dangerous structural failures. This not only enhances the resilience of infrastructure but also significantly reduces maintenance costs and downtime [49]. The combination of smart and self-healing materials represents a paradigm shift in construction and infrastructure. As urbanization continues to accelerate, the demand for more durable, sustainable, and intelligent infrastructure is becoming increasingly urgent. CNT-reinforced composites offer a promising solution to these challenges, enabling the construction of buildings, bridges, and other structures that are not only stronger and longer-lasting but also more adaptive to their environments. This integration of nanotechnology into civil engineering marks a significant step towards the creation of infrastructure that can evolve with the changing needs of society, ultimately leading to safer, more efficient, and more sustainable urban landscapes. The application of CNT-reinforced composites in civil engineering is unlocking new possibilities in the construction and maintenance of infrastructure. By enabling the development of smart and self-healing materials, these advanced composites are helping to build a future where our cities are not only more resilient but also more responsive to the challenges posed by a dynamic and unpredictable world. As research and technology continue to advance, the potential for CNT-enhanced materials to revolutionize the construction industry will only grow, offering exciting prospects for the future of civil engineering [50].

9.2 Biomedical Applications

The integration of advanced materials like carbon nanotube (CNT)-reinforced composites into 3D and 4D printing technologies has opened new avenues in biomedical applications, particularly in the areas of customized implants and drug delivery systems. These innovations are revolutionizing the field of medical treatments, offering unprecedented levels of precision, efficiency, and patient-specific solutions. One of the most promising applications of CNT-reinforced composites in the biomedical field is the development of customized implants. Traditional implant manufacturing methods often struggle to meet the specific anatomical and mechanical needs of individual patients [21, 26]. However, 3D printing, when combined with CNT-reinforced composites, allows for the production of patient-specific implants with significantly enhanced mechanical properties. Carbon nanotubes are known for their exceptional strength, flexibility, and electrical conductivity. When these nanotubes are embedded in a polymer matrix and used in 3D printing, the resulting implants exhibit improved load-bearing capacity, durability, and compatibility with the human body. This is particularly beneficial in orthopedic applications, where implants need to withstand significant mechanical stress while promoting bone growth and integration. Moreover, the biocompatibility of CNT-reinforced composites reduces the risk of adverse reactions and improves the overall success rate of implant surgeries [6, 18, 36]. The ability to tailor the design of implants to match the patient's anatomy precisely not only improves the fit and function but also shortens recovery times and enhances the long-term outcomes for patients. Figure 8 shows some 3D/4D printed materials for different biomedical applications.

In addition to customized implants, the advent of 4D printing—a process where printed structures can change shape over time in response to external stimuli—has paved the way for innovative drug delivery systems. Unlike traditional drug delivery methods, which often rely on a one-time release of medication, 4D-printed structures can be designed to release drugs in a controlled manner, triggered by specific physiological conditions such as changes in temperature, pH, or moisture levels. This capability is particularly useful in targeted drug delivery, where it is crucial to deliver medication to a specific site within the body, such as a tumor, while minimizing exposure to surrounding healthy tissues. For instance, a 4D-printed drug delivery system could be engineered to remain dormant until it reaches a tumor site, where it would then change shape to release the therapeutic agent directly into the cancerous tissue [4, 19]. This level of precision in drug delivery not only enhances the effectiveness of the treatment but also reduces side effects and improves patient outcomes. Furthermore, the combination of CNT-reinforced composites and 4D printing offers the potential to create multifunctional biomedical devices that integrate structural support with therapeutic delivery. For example, a 4D-printed stent made from CNT-reinforced material could provide the necessary mechanical support to keep blood vessels open while simultaneously delivering drugs to prevent restenosis. This dual functionality represents a significant advancement over traditional stents, which lack the ability

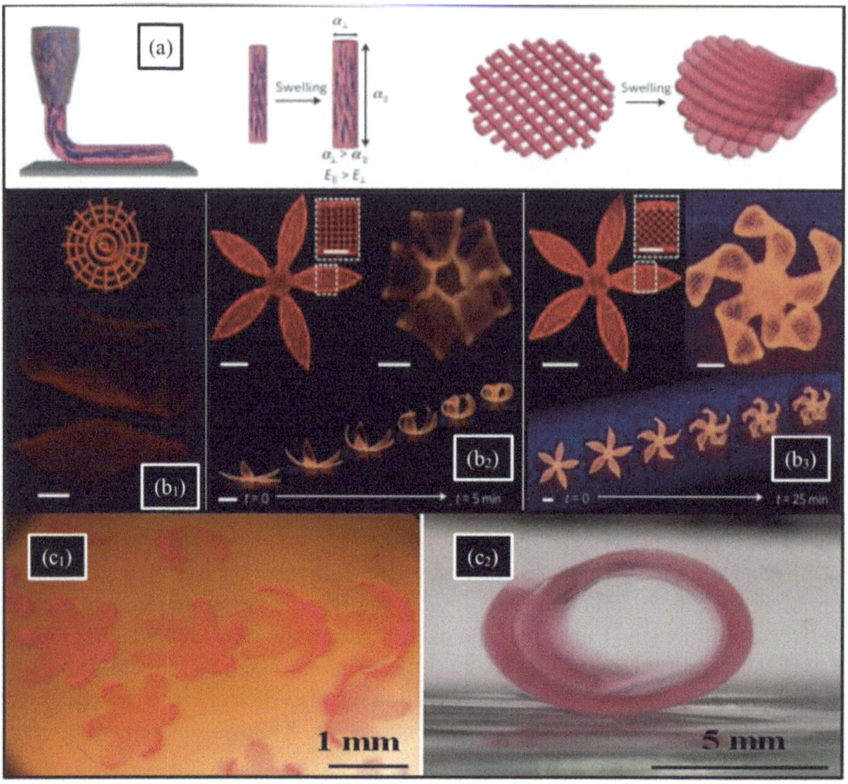

Fig. 8 **a** A schematic illustration of 4D-printed cellulosic fibrils manufactured through DIW and undergoes swelling deformation, (b_1–b_3) 4D-printed self-folding flower which swelled upon water stimulus, (c_1–c_2) Bone-like micro-developed 4D-printed tissue fabricated by using stimulus-responsive hydrogel [51]. Copyright 2022. Open access. Reproduced with permission from Elsevier

to adapt to changes in the body and deliver medication. The application of CNT-reinforced composites in 3D and 4D printing technologies is poised to transform the biomedical field. Customized implants and advanced drug delivery systems are just the beginning of what these innovations can achieve. As research and development continue, we can expect to see even more sophisticated and effective biomedical devices that leverage the unique properties of carbon nanotubes and the dynamic capabilities of 4D printing. These technologies promise to enhance patient care by providing more personalized, efficient, and successful treatment options [16, 32, 41].

9.3 Electronics and Sensors

The advancement of flexible electronics and sensors has gained significant momentum with the introduction of carbon nanotube (CNT)-reinforced composites. These materials are at the forefront of innovation in the electronics industry, where their unique properties are being harnessed to create next-generation devices that are not only more efficient but also adaptable to a wide range of applications. From wearable sensors to energy storage devices, CNT-reinforced composites are playing a pivotal role in shaping the future of electronics. One of the most promising areas where CNT-reinforced composites are making a substantial impact is in the development of flexible electronics. Unlike traditional rigid electronic components, flexible electronics can bend, stretch, and conform to different shapes, making them ideal for a variety of applications, particularly in wearable technology [8, 43]. Wearable sensors, for instance, are increasingly being used to monitor health metrics such as heart rate, glucose levels, and physical activity in real time. The incorporation of CNT-reinforced composites into these sensors enhances their performance by providing high levels of mechanical flexibility without compromising electrical conductivity. This ensures that the sensors remain functional even when subjected to repeated bending and stretching, which is essential for devices that are worn on the body. Additionally, CNTs contribute to the miniaturization of electronic components, allowing for the creation of thinner, lighter, and more comfortable wearable devices that can be seamlessly integrated into everyday life [25, 52]. Figure 9 shows the different techniques as well as the printed structures of composite materials for various sensing applications.

Beyond wearable sensors, CNT-reinforced composites are also being used to develop flexible displays, which are poised to revolutionize the consumer electronics market. These displays, which can be rolled, folded, or bent without damage, offer new possibilities for smartphones, tablets, and other portable devices. The exceptional electrical conductivity of CNTs enables the creation of thin, transparent conductive films that are essential for the operation of flexible displays. Moreover, the durability of CNT-reinforced materials ensures that these displays can withstand the mechanical stresses associated with daily use, such as frequent folding and unfolding, without degradation in performance. As a result, flexible displays are not only more resilient but also capable of delivering high-resolution images and videos, making them a desirable feature in the next generation of electronic devices. In addition to their applications in flexible electronics, CNT-reinforced composites are gaining traction as conductive materials in energy storage devices such as batteries and capacitors [9, 21]. The superior electrical conductivity of CNTs makes them an excellent choice for components that require efficient charge and discharge cycles. In batteries, CNT-reinforced electrodes can improve energy density and reduce charging times, addressing two of the most significant challenges in current energy storage technology. Furthermore, the structural integrity provided by CNTs enhances the longevity and stability of these devices, ensuring consistent performance over time. In capacitors, CNT-reinforced composites contribute to higher capacitance values,

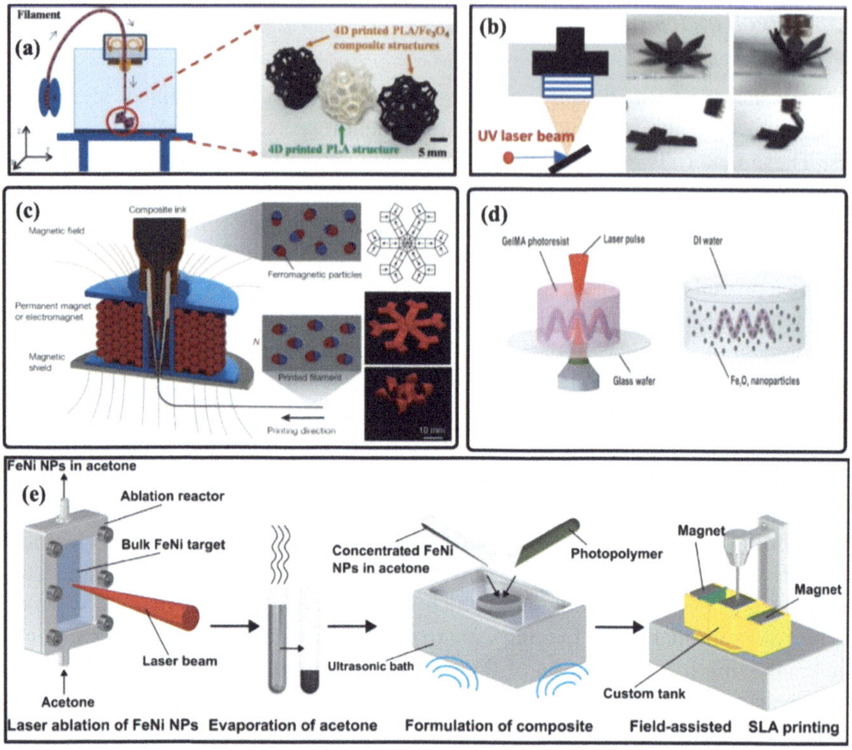

Fig. 9 Schematics of various 3D printing techniques; **a** fused filament fabrication of PLA-based magneto-active composites **b** DLP **c** Design of ferromagnetic domains in soft materials to develop magnetic composites using DIW **d** TPP used to develop MASMs **e** Masked type SLA technique used for the fabrication MAP structure [52]. Copyright 2024. Open access. Reproduced with permission from Elsevier

enabling the storage of larger amounts of energy in smaller volumes. This is particularly important in applications where space and weight are critical factors, such as in portable electronics and electric vehicles. Moreover, the integration of CNT-reinforced composites into energy storage devices opens up new possibilities for hybrid systems that combine the advantages of both batteries and capacitors. For example, a hybrid energy storage device could use CNT-reinforced electrodes to achieve the high energy density of batteries while maintaining the rapid charge and discharge capabilities of capacitors. This would result in a more versatile and efficient energy storage solution, capable of powering a wide range of electronic devices, from wearable sensors to electric cars. CNT-reinforced composites are revolutionizing the electronics and sensor industries by enabling the development of flexible, durable, and highly efficient devices. Their application in wearable sensors and displays is paving the way for more adaptable and user-friendly electronic products, while their role in energy storage devices is contributing to the advancement of more efficient and reliable power sources [53, 54].

In essence, the integration of CNT-reinforced composites into 4D printing processes not only promises innovative solutions to complex challenges but also opens doors to a wide array of applications across diverse sectors, propelling the advancement of technology and materials science.

10 Challenges and Future Directions

10.1 Dispersion and Alignment

Achieving uniform dispersion and proper alignment of carbon nanotubes (CNTs) within the polymer matrix remains a significant challenge. The inherent tendency of CNTs to agglomerate due to strong van der Waals forces leads to non-uniform dispersion, which adversely affects the mechanical, electrical, and thermal properties of the composites. Proper alignment is crucial for optimizing these properties, as the anisotropic nature of CNTs means their orientation can significantly influence the overall performance of the material. Future research is focused on developing advanced techniques to overcome these issues. Promising approaches include the use of surfactants, functionalization of CNT surfaces to improve compatibility with the polymer matrix, and the development of novel mixing methods such as high-shear mixing, ultrasonication, and electrospinning. Additionally, in-situ polymerization techniques, where the polymer is synthesized in the presence of well-dispersed CNTs, show potential for achieving better dispersion and alignment. Computational modeling and simulation also play a crucial role in understanding and optimizing the interactions between CNTs and polymer matrices at the molecular level [55].

10.2 Scalability and Cost

The high cost and scalability of producing CNT-reinforced composites are major barriers to their widespread industrial application. The production of high-quality CNTs is currently expensive, involving complex processes such as chemical vapor deposition (CVD), which are not easily scalable. Moreover, the integration of CNTs into polymer matrices in a cost-effective manner remains challenging. To address these issues, future research is aimed at developing more economical and scalable production methods for both CNTs and their composites. Advances in CVD techniques, such as the use of more affordable catalysts and more efficient synthesis methods, are essential. Additionally, the development of alternative methods, like liquid-phase exfoliation and large-scale roll-to-roll processing, holds promise for reducing costs. Researchers are also exploring the potential of using recycled materials and waste products as sources for CNT production, which could further lower costs and improve sustainability [56].

10.3 Environmental and Health Concerns

The potential environmental and health impacts of CNTs must be thoroughly investigated to ensure safe usage and disposal. CNTs can pose risks due to their small size, high aspect ratio, and potential to penetrate biological membranes, leading to toxicity concerns. Studies have shown that inhalation of CNTs can cause respiratory issues, and there are concerns about their long-term persistence in the environment. Future research should focus on comprehensive toxicity studies to better understand the health risks associated with CNTs. The development of safer synthesis methods and functionalization techniques that reduce toxicity without compromising performance is also crucial. Additionally, effective methods for the safe disposal and recycling of CNT-containing materials need to be established. Regulatory frameworks and guidelines for the handling, usage, and disposal of CNTs must be developed in parallel with technological advancements [57].

10.4 Innovations in 4D Printing

Continued advancements in 4D printing materials and techniques will expand the capabilities and applications of CNT-reinforced composites. 4D printing, which involves the use of smart materials that can change shape or properties over time in response to external stimuli, presents exciting opportunities for CNT composites. These materials can be programmed to respond to environmental conditions such as temperature, humidity, and light, leading to innovative applications in areas such as aerospace, biomedical devices, and soft robotics. Future research in this area will focus on the development of new smart polymers and the integration of CNTs to enhance the functionality and performance of 4D printed materials. Techniques for precisely controlling the orientation and distribution of CNTs within the printed structures will be crucial. Additionally, the development of multi-material printing technologies that can combine CNT-reinforced polymers with other functional materials will open up new possibilities for complex, multifunctional devices. The combination of 4D printing and CNT technology holds great potential for creating next-generation materials with unprecedented capabilities and performance [58, 59].

11 Conclusion

Carbon nanotube-reinforced 3D and 4D printable conductive polymer composites represent a significant stride forward in the realm of material science, marking the convergence of cutting-edge advancements in nanotechnology with the transformative capabilities of additive manufacturing technologies. This synthesis of innovation not only underscores the intrinsic potential of these materials but also signifies a paradigm shift in how we conceive and fabricate functional structures with tailored properties. Their unique combination of electrical conductivity, mechanical resilience, and thermal stability positions them as versatile candidates for a plethora of applications across diverse industries. Whether it's the creation of intricately patterned flexible circuits for next-generation electronics, the development of biocompatible scaffolds for tissue engineering, or the crafting of adaptive textiles for wearable technology, the utility of these materials knows no bounds. However, despite their immense potential, the journey towards realizing the full promise of carbon nanotube-reinforced composites in 3D and 4D printing is not without its challenges. Achieving uniform dispersion and proper alignment of CNTs within the polymer matrix remains a daunting task, necessitating ongoing research and development efforts to refine manufacturing techniques and optimize material properties. Additionally, scalability and cost considerations pose barriers to widespread industrial adoption, prompting the exploration of innovative solutions to enhance production efficiency and reduce manufacturing costs. Moreover, as with any emerging technology, there are legitimate concerns surrounding the environmental and health impacts of CNTs, underscoring the importance of comprehensive risk assessment and mitigation strategies to ensure safe usage and disposal.

Nevertheless, amidst these challenges lie boundless opportunities for continued innovation and advancement. Ongoing research and development endeavors are poised to push the boundaries of what is achievable, addressing current limitations and unlocking new potentials for these remarkable materials. With each breakthrough, the horizon of possibilities for carbon nanotube-reinforced composites in 3D and 4D printing technologies expands, ushering in a future where the boundaries of imagination are surpassed, and the frontiers of possibility are continuously redefined. In this journey of exploration and discovery, the synergy between material science and additive manufacturing technologies catalyzes innovation, driving us toward a future where the impossible becomes possible, and the unimaginable becomes reality.

References

1. Quanjin, M., Rejab, M.R.M., Idris, M.S., Kumar, N.M., Abdullah, M.H., Reddy, G.R.: Procedia Comput. Sci. **167**, 1210–1219 (2020). https://doi.org/10.1016/j.procs.2020.03.434
2. Alshahrani, H.A.: J. Sci. Adv. Mater. Dev. **6**(2), 167–185 (2021). https://doi.org/10.1016/j.jsamd.2021.03.006

3. Sanei, S.H.R., Popescu, D.: J. Compos. Sci. **4**(3), 98 (2020). https://doi.org/10.3390/jcs4030098
4. Medupin, R.O., Abubakre, O.K., Abdulkareem, A.S., Muriana, R.A., Abdulrahman, A.S.: Sci. Rep. **9**, 20146 (2019). https://doi.org/10.1038/s41598-019-56778-0
5. Kanoun, O., Bouhamed, A., Ramalingame, R., Bautista-Quijano, J.R., Rajendran, D., Al-Hamry, A.: Sensors **21**(2), 341 (2021). https://doi.org/10.3390/s21020341
6. Sezer, H.K., Eren, O.: J. Manuf. Process. **37**, 339–347 (2019). https://doi.org/10.1016/j.jmapro.2018.12.004
7. Nurazzi, N.M., Sabaruddin, F.A., Harussani, M.M., Kamarudin, S.H., Rayung, M., Asyraf, M.R.M., Aisyah, H.A., Norrrahim, M.N.F., Ilyas, R.A., Abdullah, N., Zainudin, E.S., Sapuan, S.M., Khalina, A.: Nanomaterials **11**(9), 2186 (2021). https://doi.org/10.3390/nano11092186
8. Kumar, A., Sharma, K., Dixit, A.R.: J. Mater. Sci. **55**, 2682–2724 (2020). https://doi.org/10.1007/s10853-019-04196-y
9. Mahmood, A., Perveen, F., Chen, S., Akram, T., Irfan, A.: Molecules **29**(2), 319 (2024). https://doi.org/10.3390/molecules29020319
10. Ye, W., Wu, W., Hu, X., Lin, G., Guo, J., Qu, H., Zhao, J.: Composite. Sci. Technol. **182**, 107671 (2019). https://doi.org/10.1016/j.compscitech.2019.05.028
11. Sajjad, R., Chauhdary, S.T., Anwar, M.T., Zahid, A., Khosa, A.A., Imran, M., Sajjad, M.H.: Adv. Ind. Eng. Polym. Res. **7**(1), 20–36 (2024). https://doi.org/10.1016/j.aiepr.2023.08.002
12. Islam, M.H., Afroj, S., Uddin, M.A., Andreeva, D.V., Novoselov, K.S., Karim, N.: Adv. Funct. Mater. **32**(40), 2205723 (2022). https://doi.org/10.1002/adfm.202205723
13. Mirabedini, A., Ang, A., Nikzad, M., Fox, B., Lau, K., Hameed, N.: Adv. Sci. **7**(11), 1903501 (2020). https://doi.org/10.1002/advs.201903501
14. Zhu, J.X., Wang, H., Zhu, Y. L.: J. Appl. Phys. **123**, 034505 (2018), https://doi.org/10.1063/1.5006523
15. Oladunni, J., Zain, J.H., Hai, A., Banat, F., Bharath, G., Alhseinat, E.: Sep. Purif. Technol. **207**, 291–320 (2018). https://doi.org/10.1016/j.seppur.2018.06.046
16. Selvaraj, M., Hai, A., Banat, F., Haija, M.A.: J. Water Process Eng. **33**, 100996 (2020). https://doi.org/10.1016/j.jwpe.2019.100996
17. Lee, J., Kim, H.C., Choi, J.W., Lee, I.H.: Int. J. Precis. Eng. Manuf.-Green Tech. **4**, 373–383 (2017). https://doi.org/10.1007/s40684-017-0042-x.
18. Volder, M., Tawfick, S.H., Baughman, R.H., Hart, A.J.: Science **339**(6119), 535–539 (2013). https://doi.org/10.1126/science.1222453
19. Lota, G., Fica, K., Frackowiak, E.: Energy Environ. Sci. **4**, 1592–1605 (2011).https://doi.org/10.1039/C0EE00470G
20. Huang, Y.Y., Terentjev, E.M.: Adv. Funct. Mater. **20**(23), 4062–4068 (2010). https://doi.org/10.1002/adfm.201000861
21. Tarfaoui, M., Lafdi, K., El Moumen, A.: Compos. B Eng. **103**, 113–121 (2016). https://doi.org/10.1016/j.compositesb.2016.08.016
22. Han, Z., Fina, A.: Prog. Polym. Sci. **36**(7), 914–944 (2011). https://doi.org/10.1016/j.progpolymsci.2010.11.004
23. Zhang, F., Hou, P.-X., Liu, C., Cheng, H.-M.: Carbon **102**, 181–197 (2016). https://doi.org/10.1016/j.carbon.2016.02.029
24. Kharlamova, M.V.: Prog. Mater. Sci. **77**, 125–211 (2016). https://doi.org/10.1016/j.pmatsci.2015.09.001
25. Espíndola, C., Correa, A.J., López-López, M., López-Cornejo, P., Bernal, E., Lebrón, J.A., Ostos, F.J., Benhnia, MR-E-I., Moyá, M.L.: Pharmaceutics **14**(12), 2806 (2022), https://doi.org/10.3390/pharmaceutics14122806
26. Pettes, M.T., Shi, L.: Adv. Funct. Mater. **19**(24), 3918–3925 (2009). https://doi.org/10.1002/adfm.200900932
27. Zhou, Y., Pervin, F., Lewis, L., Jeelani, S.: Mater. Sci. Eng. A **475**(1–2), 157–165 (2008). https://doi.org/10.1016/j.msea.2007.04.043
28. Naveen, M.H., Gurudatt, N.G., Shim, Y.-B.: Appl. Mater. Today **9**, 419–433 (2017). https://doi.org/10.1016/j.apmt.2017.09.001

29. Radzuan, N.A.M., Sulong, A.B., Sahari, J.: Int. J. Hydrogen Energy **42**(14), 9262–9273 (2017). https://doi.org/10.1016/j.ijhydene.2016.03.045
30. Zhang, W., Dehghani-Sanij, A.A., Blackburn, R.S.: J. Mater. Sci. **42**, 3408–3418 (2007). https://doi.org/10.1007/s10853-007-1688-5
31. Pang, H., Xu, L., Yan, D., Li, Z.: Prog. Polym. Sci. **39**(11), 1908–1933 (2014). https://doi.org/10.1016/j.progpolymsci.2014.07.007
32. Zhang, L., Deng, H., Fu, Q.: Compos. Commun. **8**, 74–82 (2018). https://doi.org/10.1016/j.coco.2017.11.004
33. Byrne, M.T., Gun'ko, Y.K.: Adv. Mater. **22**(15), 1672–1688 (2010). https://doi.org/10.1002/adma.200901545
34. Wang, M., Tang, X.-H., Cai, J.-H., Wu, H., Shen, J.-B., Guo, S.-Y.: Carbon **177**, 377–402 (2021). https://doi.org/10.1016/j.carbon.2021.02.047
35. Meer, S., Kausar, A., Iqbal, T.: Polym-Plast Technol. **55**(13), 1416–1440 (2016). https://doi.org/10.1080/03602559.2016.1163601
36. Kim, S.W., Kim, T., Kim, Y.S., Choi, H.S., Lim, H.J., Yang, S.J., Park, C.R.: Carbon **50**(1), 3–33 (2012). https://doi.org/10.1016/j.carbon.2011.08.011
37. Huang, Y.Y., Terentjev, E.M.: Polymers **4**(1), 275–295 (2012). https://doi.org/10.3390/polym4010275
38. Tijing, L.D., Park, C., Choi, W.L., Ruelo, M.T.G., Amarjargal, A., Pant, H.R., Im, I., Kim, C.S.: Compos. B Eng. **44**(1), 613–619 (2013). https://doi.org/10.1016/j.compositesb.2012.02.015
39. Lavagna, L., Nisticò, R., Musso, S., Pavese, M.: Mater. Today Chem. **20**, 100477 (2021). https://doi.org/10.1016/j.mtchem.2021.100477
40. Ghoshal, S.: Fibers **5**(4), 40 (2017). https://doi.org/10.3390/fib5040040
41. Cortés, A., Sánchez-Romate, X.F., Jiménez-Suárez, A., Campo, M., Ureña, A., Prolongo, S.G.: Polymers **12**(4), 975 (2020). https://doi.org/10.3390/polym12040975
42. Wang, X., Jiang, M., Zhou, Z., Gou, J., Hui, D.: Compos. B Eng. **110**, 442–458 (2017). https://doi.org/10.1016/j.compositesb.2016.11.034
43. Gnanasekaran, K., Heijmans, T., Bennekom, S., Woldhuis, H., Wijnia, S., With, G., Friedrich, H.: Appl. Mater. Today **9**, 21–28 (2017). https://doi.org/10.1016/j.apmt.2017.04.003
44. Zeng, C., Liu, L., Bian, W., Liu, Y., Leng, J.: Compos. B Eng. **194**, 108034 (2020). https://doi.org/10.1016/j.compositesb.2020.108034
45. Zolfagharian, A., Kaynak, A., Bodaghi, M., Kouzani, A.Z., Gharaie, S., Nahavandi, S.: Appl. Sci. **10**(9), 3020 (2020). https://doi.org/10.3390/app10093020
46. Khalid, M.Y., Arif, Z.U., Noroozi, R., Zolfagharian, A., Bodaghi, M.: J. Manuf. Process. **81**, 759–797 (2022). https://doi.org/10.1016/j.jmapro.2022.07.035
47. Falahati, M., Ahmadvand, P., Safaee, S., Chang, Y., Lyu, Z., Chen, R., Li, L., Lin, Y.: Mater. Today **40**, 215–245 (2020). https://doi.org/10.1016/j.mattod.2020.06.001
48. Ushiba, S., Shoji, S., Masui, K., Kuray, P., Kono, J., Kawata, S.: Carbon **59**, 283–288 (2013). https://doi.org/10.1016/j.carbon.2013.03.020
49. Li, Y., Qi, F.J., Yang, H., Gong, Q.H., Dong, X.Z., Duan, X.M.: Nanotechnology **19**, 055303 (2008). https://doi.org/10.1088/0957-4484/19/05/055303
50. Raza, S., Li, X., Soyekwo, F., Liao, D., Xiang, Y., Liu, C.: Eur. Polym. J. **160**, 110773 (2021). https://doi.org/10.1016/j.eurpolymj.2021.110773
51. Arif, Z.U., Khalid, M.Y., Zolfagharian, A., Bodaghi, M.: React. Funct. Polym. **179**, 105374 (2022). https://doi.org/10.1016/j.reactfunctpolym.2022.105374
52. Khalid, M.Y., Arif, Z.U., Tariq, A., Hossain, M., Khan, K.A., Umer, R.: Eur. Polym. J. **205**, 112718 (2024). https://doi.org/10.1016/j.eurpolymj.2023.112718
53. Liu, Y., Chou, T.W.: Mater. Today Adv. **5**, 100045 (2020). https://doi.org/10.1016/j.mtadv.2019.100045
54. Liu, Y., Zhang, W., Zhang, F., Leng, J., Pei, S., Wang, L., Jia, X., Cotton, C., Sun, B., Chou, T.: Compos. Sci. Technol. **181**, 107692 (2019). https://doi.org/10.1016/j.compscitech.2019.107692
55. Namathoti, S., Vakkalagadda, M.R.K.: Polymers **15**(6), 1371 (2023). https://doi.org/10.3390/polym15061371

56. Goh, P.S., Ismail, A.F., Ng, B.C.: Compos. A Appl. Sci. Manuf. **56**, 103–126 (2014). https://doi.org/10.1016/j.compositesa.2013.10.001
57. Mohan, V.B., Lau, K., Hui, D., Bhattacharyya, D.: Compos. B Eng. **142**, 200–220 (2018). https://doi.org/10.1016/j.compositesb.2018.01.013
58. Sousa, S.P.B., Peixoto, T., Santos, R.M., Lopes, A., Paiva, M.D.C., Marques, A.T.: J. Compos. Sci. **4**(3), 106 (2020), https://doi.org/10.3390/jcs4030106
59. Joshi, S., Rawat, K., Rajamohan, K.C.V., Mathew, A.T., Koziol, K., Thakur, V.K.: B. A.S.S, Appl. Mater. Today **18**, 100490 (2020), https://doi.org/10.1016/j.apmt.2019.100490

Shape Memory Polymers and Its Composites for 3D-, 4D-, and 5D-Printed Applications

V. S. Aswany, R. J. Parvathy, Drisya Mottammal, Susmi Anna Thomas, Jayesh Cherusseri, Dinesh Kumar, Asheesh Kumar, and Deepthi N. Rajendran

Abstract Polymers are a class of materials that have changed the lifestyle of humankind with fascinating functional applications. Among the various classes of polymers, shape memory polymers have attracted scientific interest due to their peculiar nature of stimuli-induced shape memory effect. The ability of shape memory polymers to regain their original state both at the microscopic and macroscopic levels is highly beneficial for their application in advanced 3D, 4D, and 5D printing technologies. In this chapter, we discuss the various aspects of shape memory polymers such as their intrinsic properties, classifications, and potential applications. The preparation of various composites with shape memory polymers is included with detailed discussion. The application of shape memory polymers in various advanced printing technologies such as 3D, 4D, and 5D printing are emphasized although a small description is involved in their other applications.

Keywords Shape memory polymers · Shape changing polymers · 3D printing · 4D printing · 5D printing

V. S. Aswany · R. J. Parvathy · D. Mottammal · S. A. Thomas · D. N. Rajendran (✉)
Department of Physics, Government College for Women (Affiliated to University of Kerala), Thiruvananthapuram, Kerala, India
e-mail: deepthinrphysics@gmail.com

J. Cherusseri
Research Center for Nanomaterials and Energy Technology (RCNMET), School of Engineering and Technology, Sunway University, No. 5 Jalan Universiti, Bandar Sunway, 47500 Selangor Darul Ehsan, Malaysia

D. Kumar
Department of Physics, JAIN (Deemed-to-be University), Bangalore, India

A. Kumar
Department of Mechanical Engineering, Mahatma Gandhi Institute of Technology, Hyderabad, India

S. Moharana et al. (eds.), *Polymer Nanocomposites for 3D, 4D and 5D Printing*, Engineering Materials, https://doi.org/10.1007/978-981-96-4214-4_9

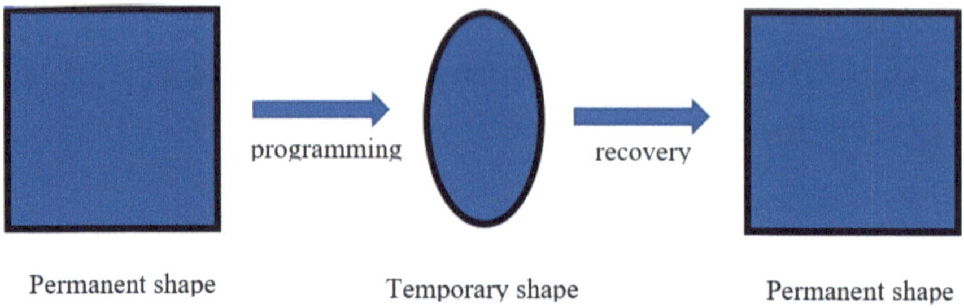

Fig. 1 Representation of the stimuli-induced shape memory effect in a shape memory composite [4]. Copyright 2002. Reproduced with permission from Wiley

1 Introduction

Materials that can recognize, analyze and respond to the external environment in a predicted or proposed manner can be considered as stimuli-responsive materials. Shape memory polymers (SMP) are one of the major classes of stimuli-responsive material for which shape-shifting effect can be installed [1]. Shape memory polymers (SMP) are stimuli-programmed materials that have the capability of changing their shape by applying an external trigger [2]. The external stimulus that offers active deformation in SMPs can be temperature, electrical field, magnetic field, solvent, ions, enzymes, pH, etc. [3]. After the removal of external stimulus, materials like shape memory polymers return to their original structure after deforming into a temporary shape [4]. Figure 1 represents the stimuli-induced shape memory effect in polymers.

2 Shape Memory Polymers

In 1941, L. B. Vernon discovered the very first shape memory effect in polymers. After the 1980s, studies were done to develop a shape memory polymer. In 1984, the CDF Chimie Company (France) developed the first SMP polynorbornene and commercialized by Japan Nippon Zeon Company [5]. However the applications of this polymer have been limited by its processibility. The second commercial SMP, Kurare TP-301, was developed by Kurare Corporation, Japan. It is a poly trans isoprene-based SMP with T_g −68 °C and melting temperature (T_m) of 67 °C. But this one was also limited in its applications due to heavy processibility. Later Asahi Company, Japan developed the third commercial SMP, Asmer. It is based on polystyrene butadiene with T_gs in the 60–90 °C range. In the late 1980s, Mitsubishi Heavy Industries developed diisocyanate/polyol-based polyurethane SMPs. The major advantage of polyurethane is the flexibility in T_gs temperature range which aids in designing SMP materials and promotes the shape memory effect in SMPs. Along with this, these are thermoplastic polymers which provide improvement in the processibility of the materials [6].

In the 1990s, advancements in SMPs were made in thermoplastic polyurethane-based SMPs. This is due to the attractive properties of these materials. One of the most important properties is the ability to control the structure of the material thereby governing the properties of the material. This results in the variation in transition temperature to a wide range thereby widening the applications of SMPs [6]. In 1997, Liang et al. proposed the preparation of polyurethane SMPs composites using glass, unidirectional Kevlar and woven fiberglass [7]. After the very first breakthrough, significant research occurred in the development of the field of SMPs in the 2000s. Shape Memory Polymer Composite developed by incorporating nanotechnology was one of the significant steps in the field of SMPs [5]. The hasty progress in shape memory polymers was endorsed by the advent of shape memory composites. The next important milestone was the utilization of heat-shrinkable polyethylene in SMPs. After 20 years, this application-driven industrial research along with academic research begin to investigate the underlying mechanisms and the design principles of these materials [7].

Shape memory polymers are a subclass of shape-changing polymers (SCP) refers to those which can change their shape either macroscopic or at microscopic levels. By incorporating shape memory behavior using a program or a non-program way, a polymer can be converted to a shape memory polymer up to a great extent. Shape memory polymers are stimuli-responsive SCPs and the shape-shifting behavior can be programmed. To be precise, SMP can be programmed to fix a temporary shape and then shift to its permanent shape upon stimulation. The shape memory effect illustrated by SMP can be classified into two; dual-shape memory effect and triple-shape memory effect. When SMP shows one temporary shape upon imposing an external stimulus and shows another permanent shape upon removing the external stimuli. Hence the associated behavior of SMP resulting in the involvement of two shapes is called as dual-shape memory effect. If SMPs are capable of memorizing more than one temporary shape and subsequently recovering to a permanent shape in a highly controllable manner, then this type of behavior in SMPs is called as triple shape memory effect. During the shape fluctuating process a temporary shape is usually reached on the application of an external environmental stimuli. This phase is called as programming step. This step implies the physical and exterior operation or stimulus that embarks in the process of shape-shifting reaction. The synthesis method of the material and the properties of the material do not influence the shape-shifting program in the polymer. In particular, when an external stimulus is applied to a SMP it changes its shape into a temporary phase upon the action of an external stimulus and returns to its original permanent shape after the removal of the applied stimulus. This stimulus can be reprogrammed to shift between the original shape A to temporary shape B in an alterable fashion [8].

A thermomechanical cycle of a shape memory polymer is shown in Fig. 2 [9]. Generally, shape memory polymers possess two phases; one a temporary phase and a permanent phase. The temporary phase is responsible for stabilizing the entire SMP and it helps to retain the shape memory effect of the material. The deformation in the

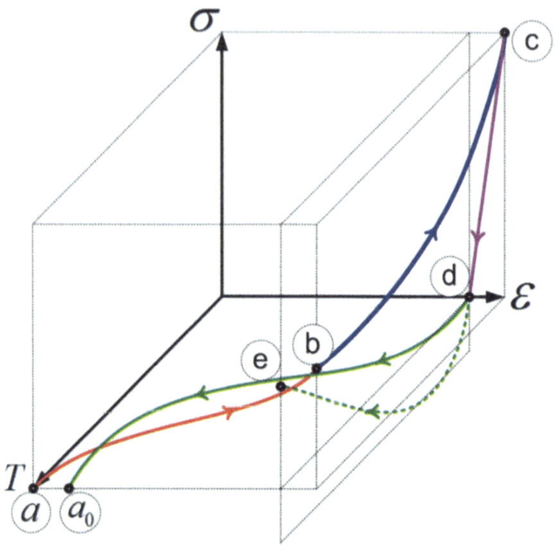

Fig. 2 Thermomechanical cycle of a shape memory polymer upon the application of external stimuli. Copyright 2012. Reproduced with permission from the Elsevier

permanent phase occurred due to the application of external stimulus is the driving force for the shape memory effect in SMP [10]. The shape-shifting effect in polymers can be obtained by applying different external stimuli. By applying different stimuli, the SMP shows different transitions between the temporary and permanent phases. Some of the transitions are glass transition, melting transition, reversible bonds, photo-induced shape memory, reversible transitions [11].

Figure 3 depicts the general structure of shape memory polymer which consists of both phases occurring while exhibiting the shape memory effect [11]. An SMP consists of two structures, a stable network or a hard phase and a second phase or soft phase. The first stable network stabilizes the whole SMP and it helps retain the permanent shape of the SMP after the application of the external trigger. The second phase is responsible for the shape memory effect. In the melting transition, when the temperature is higher than the switching temperature (minimum temperature required for the shape transition) is applied to the soft phase which has a low melting temperature compared to the hard phase. The soft phase melts and the hard phase crystallizes which results in the shape change in the material. As the soft phase melts, it creates a crosslinking bond between the molecules and the shape of the material shifts into new. As the temperature drops the molecules that melt and form crosslinking connections break and the soft phase crystallizes to bring back into the original permanent phase. Hence the soft block is the most crucial part of the SMP during the melting transition. Natural rubber, polyethers are some examples that show melting transitions for phase transitions in SMP [12]. A glass transition in SMP typically occurs below room temperature. When temperature is applied the soft phase semi-crystallizes and shows a temporary phase. The semi-crystallized phase is more prone to degradation due to the low crystallization in the soft phase [13]. Some shape memory polymers follow reversible covalent bonds in the temporary

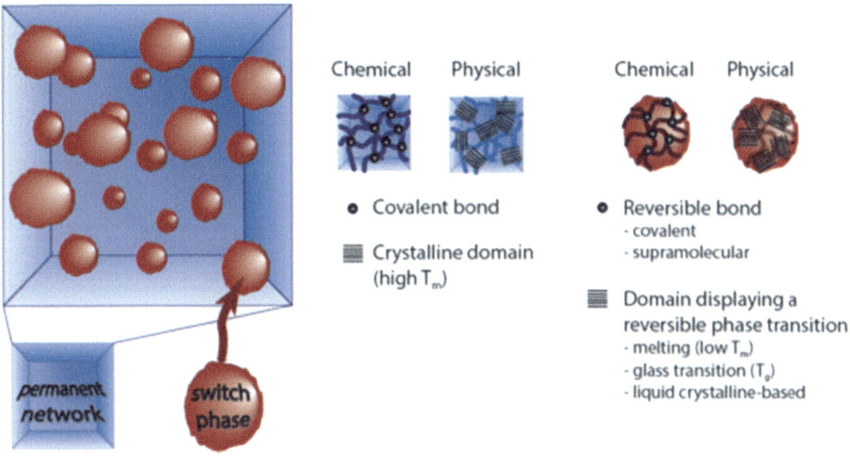

Fig. 3 General structure of SMPs during the shape memory effect [11]. Copyright 2015. Reproduced with permission from the Elsevier

phase during the application of an external stimulus. This shape memory effect has the power of self-healing by the application of functional moieties in the materials. Due to this even after damage to the material, the SMP shows an SME effect [14].

2.1 Types of Shape Memory Polymers

Depending on the thermomechanical responses, properties, mechanisms used for regaining the shape, elasticity, these polymers can be divided into varied clusters.

2.1.1 Shape Memory Polymers Based on Shape-Recovery Mechanisms

The first category of classification considered here depends on different shape-recovery mechanisms. Figure 4 shows the plot measured using oscillatory deformation at 1 Hz of tensile storage modulus versus temperature [15].

Glassy Thermosets Which Cross-Linked Chemically

A glassy polymer with a T_g at the temperature of interest and flexibility like rubber greater than T_g resulting from covalent cross-links is the most basic type of shape memory polymer [15]. Nevertheless, these materials are difficult to reconfigure following processing (casting or moulding) since the primary shape is a covalent set. A vinylidene random copolymer that is chemically cross-linked and composed

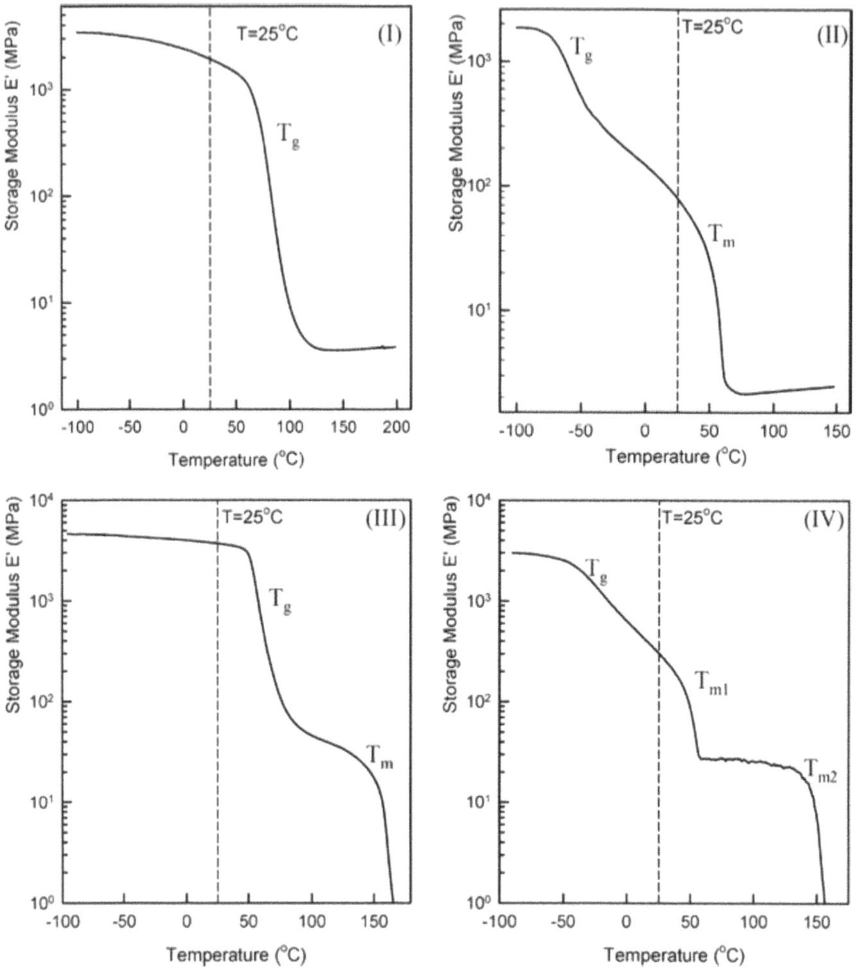

Fig. 4 The plot measured using oscillatory deformation at 1 Hz of tensile storage modulus versus temperature. (I) Glassy thermosets cross-linked chemically; (II) semi-crystalline rubbers cross-linked chemically; (III) thermoplastics; (IV) block copolymers cross-linked physically [15]. Copyright 1991. Reproduced with permission from the Royal Society of Chemistry

of two vinylidene monomers is an example. This type of thermoset shows complete shape recovery and quick shape fixing in hot water. This polymer also has the benefit of being optically transparent and castable. Polymers with T_g > room temperature in addition to chemically cross-linked polymers, may also fall into this category because of their good shape fixing by vitrification and lack of flow above T_g. These types of polymers exhibit an abundance of association per chain in a time scale of (1 s < t < 10 s) and these associations are physical cross-links. The physical cross-linking arises by this mechanism forms a three-dimensional network. This three-dimensional network possesses good flexibility and shows difficulty in thermal processing. Hence

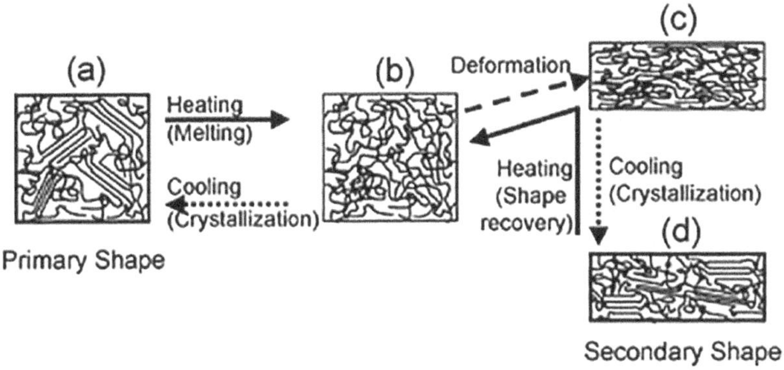

Fig. 5 Representation of shape recovery mechanisms of semi-crystalline rubbers cross-linked chemically [15]. Copyright 1991. Reproduced with permission from the Royal Society of Chemistry

this network follows solvent-based processing [15]. This polymer exhibits properties like thermoset polymers but shows some defects. Some of them are:

(1) difficulty in altering the transition temperature;
(2) trouble in changing the modulus plateau that controls the energy storage for deformation;
(3) due to the finite lifetime of association, the polymer jerks under stress at high temperatures;
(4) by the larger value of viscosity rise from heavy molecular weight, processing becomes ambitious.

To overcome these challenges the only thermal processing available for this type of polymers is solvent casting [16, 17].

Semi-crystalline Rubbers Cross-Linked Chemically

Figure 5 represents the shape recovery mechanism of a semi-crystalline rubber which is cross-linked chemically [15]. In semi-crystalline rubbers, glass transition temperature initiates the shape recovery mechanism. Along with this, the melting transition of semi-crystalline networks helps in the shape recovery process. Secondary forms are cemented here not by vitrification but by crystallisation. Permanent forms, which are unchangeable after processing, are created by chemical cross-linking, much like first class. Under critical temperature, this class of materials behaves more flexibly than glassy materials do, with their stiffness being directly correlated with the degree of crystallinity and, consequently, indirectly with the degree of cross-linking. The shape recovery mechanism of these polymers is faster compared to others due to the sharper transition zone in first-order transition. The semi-crystalline rubbers are considered shape memory materials due to their faster shape recovery mechanism along with superelastic properties and flexible modulus [18].

Thermoplastics

This group of shape memory polymers contains rigid amorphous or crystalline domains that can cross-link physically and thereby create significant temperature to exhibit elasticity which helps in the shape memory effect. Above the critical temperature T_g, this type of material is very easy to process. But below T_g, thermosets become soft and attain a secondary shape called polyurethane on cooling. This secondary structure is very beneficial since this structure shows a sharp glass transition that helps in the process of exhibiting the shape memory effect. These materials are comparable with chemically cross-linked polymers, but superior in processing in high T_g due to their physically cross-linked association. A miscible mixture of phenoxy resin and thermoplastic polyurethane, where PCL is the soft domain of the polyurethane, is an illustration of this substance. Shape memory fibres were created using electrospinning technology since this class of materials is simple to produce. In contrast to bulk polymers, a comparatively modest shape recovery was noted [19, 20].

Block Copolymers Cross-Linked Physically

Block copolymers contain a soft domain that crystallises and fixes a secondary shape in low temperature T_m. By supplying a suitable T_g temperature the original shape is regained. When temperature below T_m is provided to this type of material, they store the energy and use it to freeze the material in the secondary shape. Above the T_m temperature, the secondary shape melts and the original shape is regained. Usually, the temperature T_g of block copolymers lies between 68 and 90 °C. In between this temperature range, the materials attain their permanent stress-free structure. Above T_g when the material is heated it becomes flexible and rubbery. The advantage of the block copolymer with physical cross-link is when both domains flow, they have a permanent structure that can be reprogrammed by thermal treatment above 100 °C. One potential drawback could be that when the temporary shape is adjusted close to T_g, the "hard" domains could buckle within the stress, reducing the amount of recoverable strain. While the T_g temperature of the soft domains can be separately tailored to regulate the transition temperature, the hard/soft domain ratio can be adjusted to manage the rubbery qualities. The materials might also be made in a way that makes them biocompatible and biodegradable [15]. The incorporation of hydrophilic polyols, such as PEG, as soft domains, has expanded the possible applications of hydrophilic–hydrophobic balance to polymeric surfactants while permitting the possibility of strain recovery to be initiated by water vapour. Another benefit of this material is its low hard domain melting temperature, which is suitable for thermal processing and far lower than that of traditional TPUs. However, it is still big enough to provide a high application in shape-recovery mechanism application [10, 21, 22].

2.1.2 Shape Memory Polymers Classified by Shape Fixing Mechanisms

Depending on the stimulus or input that triggers the shape memory effect of material, shape memory polymer can be categorized into two. They are physically induced SMPs and chemically induced SMPs [23].

Physically Induced SMPs

When a shape memory polymer is induced by external physical triggers like temperature, electric field, magnetic field or UV light, then they are known as a physically induced shape memory polymer [24].

Thermally Induced SMPs: Thermally induced SMPs have been the subject of the most research. Direct thermal application is a method of activating these SMPs. A transitory form can be programmed when the applied temperature exceeds the polymer transition temperature (T_{trans}). The process of thermally induced SMPs is based on two thermal transitions: the glass transition temperature (T_g-based) and the melting temperature (T_m-based). The mechanism of thermally generated SMPs, which is based on T_m, is common to crosslinked polymers both chemically and physically. This kind of SMP has a copolymer with a low melting temperature phase and a high melting temperature phase representing the permanent network of the polymer. The most prevalent T_m-based SMPs are polyesters, polyolefins, and polyethers. These materials exhibit a crystalline hard phase that solidifies at high temperatures and a soft phase that melts at low temperatures [25]. For these materials, the crosslinking density and branching degree determine the switching temperature. SMPs can be utilised as T_g-based materials if their T_g is higher than 25 °C. Compared to T_m-based SMPs, T_g-based polymers have a slower form recovery because of their wider glass transition interval. Nonetheless, these polymers have applications in the orthodontic field due to their slow shape recovery effect and T_g near the body temperature which are desirable characteristics in preventing insertion-induced tissue damage. Thermally induced SMPs are very sensitive to temperature changes. Upon temperature shifting they first arrive at a rubbery state by distorting the primary shape to a temporary secondary shape on the application of an external stress. The polymer can then be cooled below its T_g to fix the induced distorted shape, and then heated above its T_g to regain the original shape [23, 26].

Light-induced SMPs: In order to create light-induced shape memory polymers, photosensitive units—such as photo isomerizable azobenzenes, triphenylmethane leuco derivatives, and photo-reversible cinnamon groups—must be incorporated into the polymeric structure. This creates a photo-responsive network and maintains the polymer's permanent shape. Applying an external tension and exposing the material to UV light fixes the temporary shape. In order to break the crosslinks of the polymer and promote the shape memory effect exposure to UV light at a particular wavelength is needed. But to recover the original permanent shape of the material the removal of UV rays and the application of a wavelength higher than the UV ray wavelength is adequate. According to Lendlein et al., UV light illumination can deform and fix grafted polymers with cinnamic groups into predefined geometries such (but not limited to) films, tubes, arches, or spirals. When subjected to UV light of a different wavelength, the newly formed structures can regain their original structure at room temperature and remain stable for extended periods of time [4, 23, 27].

Electric/magnetic SMPs: Indirect thermal-induced SMPs, also known as electric/magnetic-induced SMPs. When an external stimulus is applied the material shows a shape memory effect which is uncontrollable to a great extent. In order to bring the original structure back a magnetically induced or electrically induced stimulus is needed. Such type of SMPs is mentioned as electric or magnetic induced SMP. They follow the Joule heating effect for shape recovery. The heat produced by electricity promotes shape recovery of SMPs that are triggered by electricity. By the Joule effect, when a specific voltage is supplied, electricity is transformed into heat energy, which passes the temperature for transition and stimulates SMPs. The shape memory transition temperature of electrically stimulated SMPs can be quickly reached with good thermal conductivity, which speeds up the SMPs' recovery. This makes thermal conductivity a crucial component of electrically induced SMPs. Additionally, SMPs are separated into thermoset and thermoplastic varieties. Compared to thermoplastic SMPs, thermoset SMPs are more rigid, have superior dimensional stability, and are more durable in the environment. Epoxy resins are popular thermosetting polymers with a variety of great qualities, including strong acid and alkali resistance, high mechanical strength, and heat stability [23, 28].

Chemically Induced SMPs

When shape memory polymers are stimulated by any chemical activators then it is known as chemically stimulated or induced SMPs. Some of the activators as pH, water, solvents and biological or chemical agents.

pH-Induced SMPs: The physiological pH of the human body fluctuates according to the parts of the body and the diseases that are present. Consequently, pH-sensitive SMPs have discovered a fascinating use in the biological sciences. These SMPs are created by including sensitive or reversible crosslinks. pH groups that can trigger the inclusion or dissociation of complexes in response to varying pH environments. Hane and colleagues, for example, produced a pH SMP through the cross-linking of alginate modified with β-cyclodextrin (β-CD-Alg) and alginate modified with diethylenetriamine (DETA-Alg). The crosslinked alginate chains function as a fixing phase, whereas the pH-reversible β-cyclodextrin functions as a flexible phase. It is feasible to temporarily produce a shape in the complex when it is submerged in a pH of 11.5, and the original shape returns when the pH is shifted to 7. $95.7 \pm 0.9\%$ for the recovery ratio and $94.8 \pm 1.1\%$ for the fixity ratio, respectively. This material has a strong potential for use in medicine since its pH value is relatively similar to the pH of the body. It is also convenient and safe to use [23, 29].

Water-stimulated SMPs: The highly prevalent and safe liquid for use in biological applications is water. Bonded water can modify the hydrogen bonds in SMPs and make the chains more flexible, which lowers T_g. Water, for instance, when used as a plasticizer can lower the crosslinked polyvinyl alcohol's transition temperature and permit the material to return to its original shape at a lower temperature [30]. Indocyanine green, a dye that emits in the near-infrared spectrum, was coupled with poly (vinyl acetate) to create a light-emitting SMPs web by Torbati et al. When submerged in water, PVAc-webs shrank and displayed noticeably more glowing intensity than films. They came to the conclusion that dye-containing with water-triggered confine could be utilised as antimicrobial medical equipment, temperature sensors, internal suturing, feeding tubes or catheters, and packaging [23, 31].

Enzymatically Triggered SMPs: Certain shape memory proteins (SMPs) may be directly activated by biological stimuli that have cytocompatibility-triggering mechanisms. Biopolymers that occur naturally, such as synthetic polymers (amino acids), are frequently polymers that are sensitive to enzymes. An SMP that can react directly to enzymatic triggers under isothermal cell culture conditions was created by Buffington et al. Shape memory material Pelletane, which is enzymatically fixed, was combined with PCL, which is susceptible to enzymatic degradation, to form the shape-fixing component. When the shape-fixing component degraded due to enzymatic catalysis, the SMP reverted to its initial, predetermined shape [32].

3 Composites of Shape Memory Polymers

Shape memory polymer has limitations and drawbacks that were exposed in the application in the engineering field. In order to improve the characteristics of SMPs, it was doped and reinforced with other materials and the products thus obtained are called as Shape Memory Polymers Composites. By incorporating several additive elements into the SMP framework, SMPCs are created. They have desirable qualities for sophisticated applications, like great mechanical strength and thermal characteristics. The final properties of SMPCs are greatly influenced by additive materials. SMPCs reinforced with fibre, both short and long, offer a wide range of applications. Studies to enhance the SMP features have been conducted in recent decades to increase their usage in a variety of application sectors. The purpose of additive materials is to enhance their characteristics. These additive materials are made up of various particles, fibres, and lengths. Improving SMP characteristics is becoming more and more of an interest, and SMPCs are becoming more well-known. Several additive materials are added to the SMP framework to create SMPCs. They possess desirable qualities for sophisticated applications, like superior mechanical strength and thermal characteristics. The final properties of SMPCs are greatly influenced by additive materials. In the preparation of SMPC, elements derived from fossil fuels are often utilised as reinforcement and matrix materials. Growing environmental awareness has prompted studies into the use of renewable energy sources [33].

One or more of the following carbon materials doping configurations are possible when using a shape memory polymer as a matrix in particles, fibres or vessels, layers or films. Polymers, metals, ceramic particles, etc. are used in SMPC for achieving self-healing, innovative stimulation techniques, reinforcement, and other aims. To create a multipurpose composite material, an SMPC can also be employed as a filler in the form of fibres or similar materials. It is possible to do two-way recovery and other intriguing and practical tasks with a well-designed composite structure. Fillers can be used to design the SMPC. The mechanical and thermal qualities are frequently diminished by excessive filler, which lowers the transition temperature. However, most fillers will lessen a polymer's form memory effect, which is reflected in the recovery and fixation mechanism [33, 34].

SMPs' limited commercial applicability stems from their inferior mechanical characteristics. As such, SMPCs have been created. The continuous fibre typically has the highest reinforcing ability, with short fibres and particles coming in second and third. Due to the high reinforcement, excellent qualities, and general carbon fibre or fabric composition, continuous fibre reinforced SMPCs have dominated the available literature on SMPCs for aerospace applications to date. Although carbon fibres have a maximum elongation ratio of less than 2%, this limits the tensile deformation of SMPCs. These SMPCs with continuous fibre reinforcement were able to withstand significant macro bending deformation. Researchers have worked very hard to investigate the bending process of SMPCs and have come to the conclusion that microbuckling is the cause. When a fibre-reinforced SMPC is heated over T_m/T_g, microbuckling occurs at significant bending ratios. In this case, the SMP matrix lacks the stiffness to support those fibres in compression and has a low shear modulus (of the order of magnitude 10 MPa). As a result of instability, the fibres would have microbuckling. This allows for a comparatively high fibre compression strain since, on a macro scale, a buckled fibre seems shorter than an unbuckled one. The micro buckle's amplitude is set to account for variations in length. Buckled SMPC has a macro strain that exceeds the typical material strain failure limit. Although SMPC is typically less than 2 mm thick, it should be noted that a thicker or larger modulus laminate would have a larger shear force to prevent fibre bending or matrix damage. Based on the type of reinforcements applied SMPCs are generally divided into particles-reinforced and fiber-reinforced composites [35].

3.1 Particles-Reinforced SMPCs

The type of filler used in SMP composites determines whether they are categorised as particle- or fiber-reinforced. Particle-filled or short fiber-filled SMP composites typically exhibit specific properties like high electrical conductivity, magnetic responsiveness, or high microscale stiffness. As a result, this kind of SMP composite is being researched as a material with multiple uses. Particle-reinforced or pure SMPs are not appropriate for use as structural materials; instead, they are typically added to specific particle fillers to act as functional materials. To create electro-active SMPs, materials such as chopped carbon fibres, nickel powder, carbon nanotubes, and electrically conductive carbon black are added to SMPs. SMPs that are magnetised can be induced by ferromagnetic Fe_3O_4-filled SMPs. Additionally, light-induced SMPs might be created by embedding infrared optical fibre within SMPs and using the fibre to transmit infrared light that would heat the SMP [30, 33].

3.2 Fiber-Reinforced Shape Memory Polymers

Fiber-reinforced SMPs have superior mechanical qualities due to their increased strength, stiffness, and resistance to creep and relaxation. These SMPs exhibit promising properties as structural and functional materials for a wide range of cutting-edge applications. As a result, there has been a lot of interest in using fiber-reinforced SMPs in deployable structures including solar arrays, trusses, and antennas. Furthermore, thermoplastic SMP resins like polyurethane SMP are the subject of the majority of investigations on SMP composites. Applications for deployable constructions may benefit from the fiber-reinforced SMP composites' high strain capability and generally strong mechanical qualities. In addition to having superior mechanical and thermomechanical qualities over pure SMPs, the carbon-fiber-reinforced SMP composites may also be turned on by applying a voltage to the carbon fibres that are incorporated into the SMP matrix. As a result, they could be categorised as multipurpose materials in some sense. Shape memory composites that are reinforced with glass or Kevlar fibres are just as significant as those reinforced with carbon fibre [33, 36].

3.3 Applications of SMPCs

SMPCs are smart materials that have applications in anything from space travel to automotive technology. They have recently been designed and qualified, particularly for aeronautical constructions and deployable components. SMPs exhibit promising prospects in biomedicine, intelligent fabrics, self-repairing composite structures, and automotive actuators. Furthermore, a plethora of patents pertaining to SMP applications exist, including those for grippers, intravascular delivery systems, hood/seat assemblies, and adjustable automotive mounts for automobiles. When it comes to conventional aerospace deployable devices, a hinge, or energy devices are used to change the structural structure while in orbit. Traditional deployment devices have some inherent disadvantages, including a difficult-to-assemble procedure, numerous mechanisms, enormous volumes, and unintended consequences during deployment. Conversely, the deployment devices made with SMPs and their composites could be able to get around some fundamental drawbacks [36].

4 Properties of Shape Memory Polymers

Shape memory effect in polymers is depicted in Fig. 6 [8]. The relaxing of polymeric chains brought on by temperature variations is typically responsible for the shape memory effect in polymers. When modifications in the shape memory effect result in the programming and recovery stage's temperature, the phenomenon is referred to as thermally induced SMPs typically consist of two active phases. The "switch phase" is the second, and the fixity phase is the first. While the switch phase permits fixing the temporary shape, the fixity phase allows the polymer to retain its initial shape. Furthermore, the primary shape can be recovered due to the transition temperature T_{trans} present in the switch phase [37].

During polymer synthesis, the polymer chains are arranged in a coil configuration. The beneficial entropic structure is the coil. Chains cannot break through intermolecular or intramolecular bonding during the fixity phase. The form memory effect is enabled through chain switching. Chain movements are prohibited below the switch phase's T_{trans}. The chains can move if the temperature rises above T_{trans}. In this state, the entropy of the chain conformation will decrease if stress is applied because the chains will experience an elastic deformation. When the tension is released, the polymer can return to its initial shape. In contrast, the switching phase will fix the chains while they maintain their temporary structure when the temperature drops below T_{trans}. The stress can be released when the temperature drops below T_{trans}. Ultimately, the polymer will return to its initial shape when the temperature rises over T_{trans} since this shape is more advantageous entropically [37, 38]. Now we discuss the different properties of shape memory effects in polymers.

4.1 Chain Conformation of Linear, Amorphous Polymers

When polymer chains are amorphous, they disperse randomly throughout the matrix, lacking the orderly arrangement of crystallites that characterises semicrystalline polymers. A polymer chain has an internal energy that is the same in all conceivable

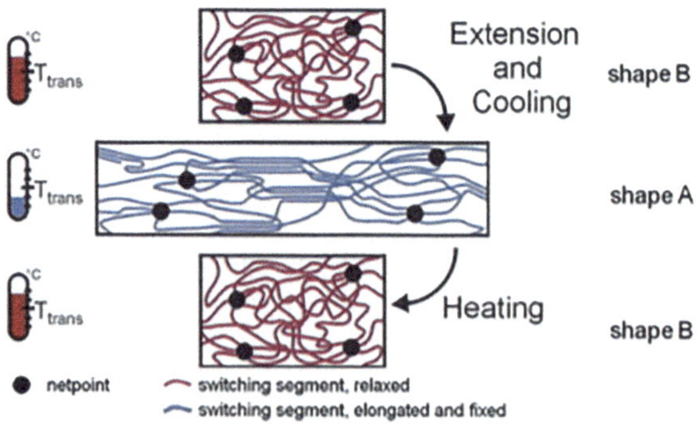

Fig. 6 Shape memory effect in polymers on the application of temperature as stimulus [8]. Copyright 2007. Reproduced with permission from the Elsevier

conformations [39]. The entropy of an amorphous polymer chain is given by

$$S = k \ln W$$

where W denotes the conformation probability.

4.2 Evolution from the Glassy State to the Rubber-Elastic State

In this state, the chains can adopt any of the feasible, energetically comparable conformations without experiencing a considerable degree of disentanglement. Since this conformation is entropically favoured and therefore considerably more likely than a stretched conformation, high number of molecules will form compact coils [40]. When an external force is applied, a polymer with a molecular weight of approximately can stretch in an elastic condition. When the tensile stress is applied for a short span of time, the entanglement of the polymer chains with their immediate neighbours will immediately stop the significant movement of the whole chain. When the external tension is removed, the sample reverts to its initial length. The sample thus demonstrates a form of memory for the non-stretched condition. This recovery, which is commonly referred to as the "memory feature," is predicated on the sample's propensity to revert to the most probable form—its initial, most randomly coiled state [41].

The cross-linkage points keep the chains from separating from one another by using permanent association. Cross-links may be physical or chemical in nature. We refer to those substances as elastomers. Polymers that are cross-linked chemically result in insoluble compounds that swell in suitable solvents. Their cross-linking fixes their shape, and it cannot be altered thereafter. Physical net points are present in thermoplastic elastomers. The presence of a specific material morphology, such as that seen in block copolymers, is necessary for the development of the net points. The phase that forms the hard segments is associated with the highest thermal transition T. Within the thermal transition, the material continues in its permanent shape by using physical net points in the material. Processable straight from the melt, thermoplastic is highly soluble in the solvents. Another feature of these association is the amorphous chain segments that are flexible. The networks will be elastic if these segments have glass transition temperatures lower than the operating temperature. They can be stretched without losing entropy and exhibit entropy elasticity. During stretching, net points get more space between them and acquire orientation. The material regains the entropy it previously lost and takes on its former shape as soon as the external force is removed [42]. As a result this polymer network keeps the mechanical stress in equilibrium. Amazing properties of elastomers include warming up when stretched, increasing elastic modulus when heated, stretched elastomers show negative coefficient of thermal expansion above the glass transition temperature, glass-like behaviour below the T_g value, and contraction upon further cooling. An ideal elastomer cannot be stretched without losing its internal energy [43]. Free energy U estimated using Helmholtz equation is given by

$$U = \frac{1}{2}NkT\left(\lambda_x^2 + \lambda_y^2 + \lambda_z^2 - 3\right)$$

where x, y, and z represent the extension ratios in three dimensions, N represents the count of chain sections between the net points.

4.3 Molecular Mechanism of the Shape Memory Polymers

Elastomers can reveal the shape memory capabilities of the polymer if the material can be sustained in the distorted state within a temperature range for an appropriate application. By employing network chains as a form of molecular switch, this is possible. For this reason, the segments' flexibility ought to depend on the temperature. Chain segments exhibit flexibility at temperatures over T_{trans}. The segment's overall flexibility is restricted when it moves from a rubber-elastic or viscous condition to a glassy state. A portion of the chains remains amorphous since the resulting crystallisation is never complete. The segments are prevented from automatically reverting to their permanent structure, which is estimated by the net points, and from instantly rebuilding the coil-like structure by forming crystallites [44].

Figure 7 shows the diagram of the mechanism of molecules which are thermally stimulated for (a) a copolymer (b) covalently cross-linked polymer and (c) polymer network [4]. The polymer returns to its original shape when the amorphous material deforms quickly due to a sudden drop, withdrawal, or deduction of the external force. These polymers can also display the shape memory effect if the right stimuli method is used. In this instance, the permanent shape can be fixed by using the transient association of the polymer chains, which serve as physical net points. If the amorphous material's glass transition falls within the temperature range that matters for a particular application, this thermal transition can be employed as a switching transition [37].

5 Applications

It is anticipated that future interest in shape memory materials will grow due to their many applications. By breaking them down into subsections, we report the primary uses of SMPs in this section: the primary application is devoted to the biomedical field since this is the most lively sector in terms of industrial advancements and

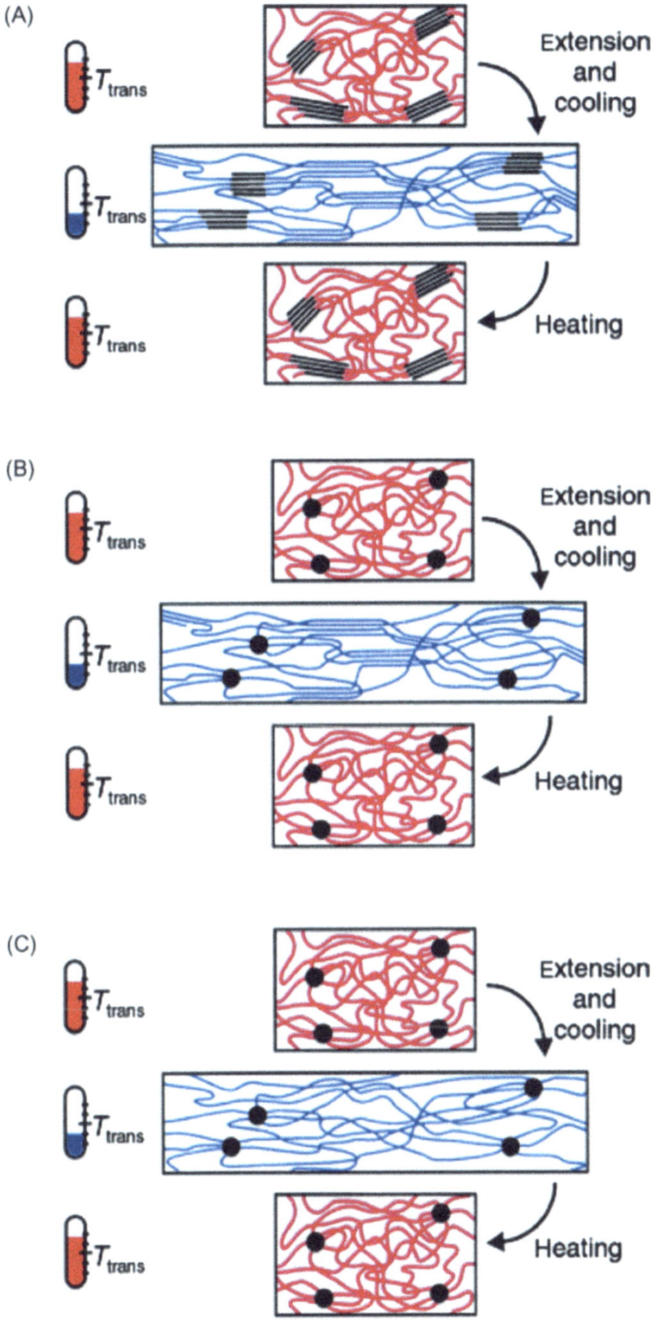

Fig. 7 Diagram of mechanism of molecules which are stimulated thermally for (**a**) a copolymer (**b**) covalently cross-linked polymer and (**c**) polymer network [4]. Copyright 2002. Reproduced with permission from Wiley

current research, whilst the second one encompasses the aerospace uses and the last section compress of the most efficient use of SMP; Printing Technology [45].

5.1 Biomedical Applications

Since the middle of the twentieth century, the biomedical industry has been using polymers more and more. It is imperative to underscore that the use of SMPs for biological applications is subject to specific criteria and constraints. They must be biocompatible, have quick recovery durations, and, depending on their particular uses, additionally biodegradable. Furthermore, in order to prevent tissue damage and enable the use of body temperature as the switching temperature, these SMMs' conversion temperature can be in line with body temperature. Due to their dual properties of biocompatibility and biodegradability, aliphatic polyesters including PLA, PCL, and polydioxanone (PDO) have found widespread use in medicine. The form memory of PLA homopolymer is inherent, as previously stated. Copolymers, polyurethanes, and composites based on PLA have been created to improve their shape memory behaviour. PCL is a semi-crystalline polymer with a liquify point that is close to body temperature. PCL is frequently utilised as a switch phase because of this. A work using glycolic acid to copolymerize PCL was recently published. Lendlein and Langer [46] utilised the biodegradability of polyesters PCL and PDO to create suture wires for tissue healing. The use of vascular stents to prevent vessel obstruction is one of the most researched applications. They are typically based on nitinol, although the benefits of employing SMPs—like their affordability and biodegradability—have raised interest in their use. Additionally, micro-actuators based on SMPs have been created to stop ischemic stroke. According to a 2005 study by Small and colleagues, silicon actuators loaded with carbon nanotubes that are actuated by infrared light are conceivable. In addition to vascular stents and suture wires, other biomedical applications of SMPs include drug delivery systems, ophthalmologic materials, scaffolds for tissue engineering, and self-healing materials for biomimetic devices [47, 46].

5.2 Applications in Aerospace Industry

These days, explosive devices have to be set off in order to deploy the solar wing on space-deployable structures. However, because these explosive devices have temporary effects, the satellites need to pass rigorous mechanical testing. According to NASA, more than 71% of the causes of 88 flight failures were related to the detonation of explosive devices. This figure illustrates how serious this hazard is to space mission completion. As a result, NASA has developed an extensive collection of management procedures. The SMPC is very stiff below and keeps this firmness once it is gradually deployed above, which usually takes a few minutes. When weighed against the explosive device's detonation time (milliseconds), it can prevent severe impact, lower flying risk, and significantly increase aircraft reliability. Hinges, beams, antennae, and space capture equipment are just a few of the applications that have successfully used SMPC-based space-deployable structures, which are gaining increasing attention. Moreover, SMPC's micro-buckling allows it to achieve a higher deformation capacity, raising the deployment/contraction ratio [48].

5.2.1 SMPC Hinge

Usually, the hinge serves as the space-deployable structure's driving power. On the other hand, the driving power of the traditional mechanical hinge is produced by a spring, which vibrates when the hinge is deployed. Such vibration issues can be efficiently avoided by the gradual deployment process of the SMPC. Using two straight SMPC plates, created a thermally equipped hinge and optimised the SMPC cross-section with a high torque to an arc [48, 49].

5.2.2 Deployable Structures of SMPCs

To satisfy the necessities of satellites, a variety of deployable equipment was created, such as the deployable beam, using SMPCs. When the SMPC reaches the designated orbit, it is heated to accomplish deployment, and the composite beam is zigzag-shaped in its contracted condition. The SMPC-crimped beam structure was designed by DSX/Power Sail, and may be frizzled and unfurled to achieve estimated emission and solar cell deployment in specific orbits linked to it. Furthermore, Zhang and colleagues created a removable beam made of SMPC sheet that has a circular cross-section and can be reshaped into a "M" shape. The beam may be deployed in less than a second. A tetrahedral support with springs was created by Shook et al. to improve structural rigidity and decrease vibration. Future aeronautical applications may make use of the cubic deployable truss construction that Leng et al. created, which has three-span beams and a cubic frame. One way to accomplish first- or multidirectional deployment is to regulate the heaters' energization sequence. NASA's Mars exploration mission was made easier by Keller et al.'s deployable, high-precision antenna, which they developed using SMPCs and had good deployment accuracy. Shape memory deployable antennas were the subject of a related study by the Harbin Institute of Technology [48, 50].

5.3 Applications in 3D and 4D Printing

Films, foams, and bulk materials are only a few of the various shapes that SMPs can take. These days, 3D printing offers another option for creating a variety of shapes; this leads to what is known as 4D printing, which denotes that an object is printed flat and then, through heat treatment, is given a three-dimensional shape. This method can be applied, for instance, to controlled sequential folding, which makes it easier to print some structures, or even to make shapes that cannot be printed, or even to create soft robotics. This technique has even been attempted to be used at the microscale or in bioprinting. There are several 3D printing technologies available. Some of them are fused deposition modelling (FDM), selective laser sintering (SLS), selective laser melting (SLM), and stereolithography (SLA). Typically, these 3D printing methods include slicing the model into several layers and are then printed. By doing this, a variety of unique shapes that would be difficult or impossible to make using injection moulding can be produced, opening up new uses for 3D-printed items.

In the past few years, more research and development has been done on integrating the two methods, or 3D printing SMPs. Despite the fact that the first findings on shape memory polymers were published thirty years ago, the number of papers pertaining to this topic is continuously rising quickly. However, there has been a noticeable surge in the study of 3D-printed shape memory polymers, which highlights the significance of this new area of study [49, 51].

Polylactic acid (PLA) is one of the polymers that is most frequently used in FDM 3D printing. This is because of its low toxicity, particularly which is also used rather frequently, as well as it is easy to use, relatively low temperature, and significantly reduced warping than numerous other polymers. However, little is known about its shape memory characteristics. Shape-fixing components are present in PLA. In addition to crystallisation, physical cross-links are attributed to the shape memory capabilities. Pure PLA, on the other hand, cannot be stretched without breaks for longer than about 10% [33]. Because these constraints can be addressed, many uses of PLA require the creation of complex structures, which can be produced by 3D printing. The most current solution to this issue was provided by Langford et al. using origami structures. These origami constructions are designed to allow for folding and unfolding, which allows for the expansion of a small volume into large or vice versa. This feature is useful, for example, for introducing a biomedical platform in the human body in a minimally invasive manner. The PLA filament, they discovered, displayed a consistent shape recovery of roughly 61%, regardless of the applied deformation, but the herringbone tessellated origami tube's recovery rate increased to almost 96% despite some apparent fissures following compression. Following the addition of a porous bone-mimicking infill structure, the healing rate stayed unchanged. Researchers looked into PLA-manufactured cubes with various infills. On the one hand, because the 3D honeycomb structure is totally closed, the interior of the cubes was not as heated as the exterior, which resulted in a poorer recovery in a hot water bath. Examining the gyroid infill structure at varying degrees of infill revealed significantly variable recovery qualities as well as forces required to achieve the intended deformation. Different activation temperatures and printing rates clearly demonstrated time dependencies in the relating angle of the samples, which was evaluated optically. However, the commencement of the unfolding process was not significantly affected by the temperature of the nozzle. The same group also looked into flat PLA origami constructions that could be folded into a pyramid shape with a square base when heated to a temperature close to the glass transition. This structure's concept is an application to release a substance that has been enclosed, grip an object, or close a gap. In a different investigation on parameter optimisation, Wu et al. discovered that the highest recovery occurred at 55 °C for deformation, 70 °C for recovery, 45° for the raster angle during printing, and 0.15 mm for layer thickness. Self-expanding vascular stents printed from PLA were studied by Jia et al. An FDM print of a hexagon-shaped, nested tubular construction with a diameter of 4 mm, length of 40 mm, and wall thickness of 0.4 mm was made. Recovery took place in a water bath at 70 °C following quick cooling and compression at the same temperature. The original shape was regained after 5 s at 70 °C, without any noticeable alterations. For elongated samples, shape recovery rates were examined and found to be roughly 87% after the first cycle and 96% after the second. This was explained by the improved molecular orientation following the first test. In addition to these fundamental research investigations and a few other studies focused on biomedical applications, some authors propose using pure PLA as a shape memory material for smart textiles. However, a lot of research groups mix PLA with different polymers or add fibres, nanoparticles, etc. to PLA [52, 53].

It has been stated that PLA is combined with a variety of materials in various forms to enhance the recovery qualities. Moreover, some printed forms that resembled braids were created using PLA filaments reinforced with carbon nanotubes. Recovery was seen to begin earlier for CNT/PLA specimens than for pure PLA samples and to rely on the printed forms. The temperature was 90 °C, and the recovery rate was 5 K/min. Additionally, silicone elastomer matrix-infused samples consistently demonstrated better form restitution, approaching 97%, when evaluated [54]. PLA can be used with a variety of substances to enhance healing or provide new qualities. Experimenting with various printing parameters revealed that the raster angle, layer thickness, and form memory properties were greatly impacted; however, the shape memory effect was mostly unaffected by the infill density [55]. In recent years, there has been a growing interest in research and development about 3D printing shape memory polymers. Although PLA is the most widely used FDM process in 3D printing polymer, most research organisations use pure PLA; however, a number of publications were found that report on mixing or adding different fillers to PLA. Several academics are actively studying polyurethane, in particular, among the other FDM printable polymers. Different methods exist for preparing SLA printable shape memory polymers in addition to FDM. Nevertheless, very few papers pertaining to the DLP process were located, and there were no publications on shape memory polymers for Polyjet, SLM, or SLS [56].

The field of 4D printing is relatively new and is undergoing rapid change. Since then, a lot of researchers have chosen to focus on 4D printing, and the printed active components (PACs) have been developed and produced. Since then, the industry has seen a tremendous upsurge in advancements, bringing 4D printing to industrial applications. Several 4D printing methods are common, but fused deposition modelling (FDM) is the most widely used. In FDM, the material is extruded one line at a time through a heated nozzle to create a model. Since FDM prints using filaments, it is the most widely used technique due to its comparatively inexpensive cost of printing and material flexibility. The primary distinction between FDM and Direct ink-writing

(DIW) is in the way the material is applied, which is often done using a syringe, and the conditions under which the print can be completed. This holds potential benefits in various biomedical applications involving liquid-based materials. The firmness of printing and the properties of the sample that make it suitable for this process are just two of the many benefits DIW has over other printing techniques. Often called inkjet printing, polyjet printing is a printing technique that uses precisely distributed droplets of polymers on the printing bed to cure them. This allows for the efficient and effective mixing of several materials [57].

Owing to their exceptional qualities and potential uses, SMPs are researched and utilised kind of materials in 4D printing. Well-known SMPs in this area with good shape memory qualities include polylactic acid and polyurethane (PU)-based polymers. These qualities include having a comparatively inexpensive cost of manufacture and being easily enhanced by combining different materials to create composites. Certain composites are designed to alter the kind of stimulus required to activate the SME or to regulate the SMPs' transition temperature. In addition to other materials, SME can be found in semicrystalline and amorphous polymers. Specifically, these SMPs gradually modify their structure as they get closer to a particular temperature where the material transitions from a glassy to a rubbery state, making it simple to mend into a variety of temporary shapes. When the material gets to a particular temperature, the condition of the material changes rapidly in semicrystalline SMPs, in contrast to amorphous SMPs. A force is applied to a material in its rubbery state to bend it while maintaining its new shape. After that, the structure is chilled below its transition temperature so that internal strain allows it to maintain its temporary shape. When the temperature rises over the transition point, the original shape is then restored. In 4D printed structures, diverse material combinations combined with distinct stimuli have made sophisticated, programmable, and varied form modifications possible. These form modifications are meant to demonstrate the potential of using the SME property of the polymer composite in more elaborate proposals and applications [51].

SME can be caused by strain alterations between two constituents, structural extension, electrical current through the heating system, and partial or total print expansion. These modifications to the printed structures can result in a significant form shift where there is a significant variation in strain across the structure's portions. Because it is the most versatile stimulus and can be used with a wide range of materials, ambient heat is the most often employed stimulus in 4D printing. To maintain a transient shape, the pattern is printed in the form of a ball, which is subsequently stretched and chilled. Then through cooling original shape is retrieved. Another common event that causes a form change is swelling in a fluid media. Since the liquid may be removed to reverse the swelling, this procedure is typically easily reversible. Typically, this approach is enhanced by mixing a swellable material polymer (SMP) with the swellable ingredients to maintain the brief shape while the swelling reverses. By activating the SMP to restore its original shape, the original shape is obtained. The heat generated by an electric current is considered to be separate from thermal stimulation. This decision was reached because basic ambient heating does not resemble the targeted propulsion of electrical current flow or the methods used to induce it. Smart materials that are activated by light are another well-liked stimulus that has been investigated. In many circumstances, the targeted or widespread use of light for actuation is highly beneficial. Magnetic field SMP activation is a crucial stimulation approach since it may be done from a distance. Structures that cannot be activated by direct heating can be activated using this technique. Its potential is enormous in a variety of biomedical fields [58].

The actuation required for printed structures is contingent upon the utilisation of 4D printing processes, as some material functional features are specific to the printer. Every printing technique has benefits and drawbacks; these vary depending on the application. FDM is the most widely used printing process since it is an affordable method that allows for material selection freedom. DIW is an additional material extrusion technique that can be used to extrude liquid or hot materials. Although the constituents that can be utilised are restricted to resins, some researchers prefer polyjet and SLA printing because of their accuracy and resolution. Since DLP printing cures an entire layer at a time rather than just one spot at a time, which has an outstanding definition but may have subpar finishing, it is a faster method of light printing. Another well-liked printing technique is polyjet printing, which is employed due to its accuracy and resolution. Despite being more expensive than other printing techniques, this technology has many benefits. Droplets of ink or polymer are applied to a printing tray during polyjet printing, a non-contact printing technique that is often cured by UV radiation. High printing freedom is provided by this method, which is generally preferred for printing multi-material structures. The additive manufacturing process known as polyjet printing features a mechanism that makes it simple to print with many materials. Material grading affects a material's SME response according to its breadth. A polyjet is used to print a graded SMP with sequential shape recovery capabilities, enabling the formation of comparatively intricate recovery shapes. Since heat is what causes the form restitution, submersion in hot water served as a stimulus [58]. Similar to SLA printing, DLP printing is a quick printing process that polymerises an entire design layer using a projector. Since the resin will respond to UV light, any UV projector will work; as a result, it is possible to create an inexpensive printer using just a UV source and a screen. In order to create a polymer with virtuous shape memory and self-healing capabilities, ureido-pyrimidinone units are added to PCL after the printing resin is fixed. Consequently, the developed actuators have desirable qualities and are soft actuators [57]. DLP was used to create an SMP with tunable attributes after fabrication. The reusable nature of the material is made possible by its reconfigurable agreement shape. If a printed object is reconfigured, it can be used for a variety of purposes, offering a broad range of applications. It has also been reported that printable ink with a high transition temperature, excellent form memory, and very high mechanical strength is suitable for industrial use. With DLP, the material can have a very high resolution because it is ink printable.

The usage of 4D printing is highly desired in many sectors, including biomedicine, industrial units, robotics instruments and soft-actuators, and electronics, despite the fact that it is still a relatively young field of study. The latest applications are categorised by looking at the paths that the majority of researchers have taken in the last several years. The application of 4D printing in nutrition manufacturing, energy production, and the usage of fibres in structures are some other areas that have gained attention in the literature. Since 4D printing is a novel manufacturing process with a vast amount of uncharted territory, the field of study is evolving quickly. Recent developments in additive printing have made it possible to prototype highly precise products quickly and at high complexity, opening the door to new developments. There is a lot of interest in 4D printing because scientists have recently found new properties of confirmed materials and complexes. Until a limited years back, the majority of 4D printing literature focused on material and printing method testing for proof of concept, offering solid foundational guidelines for additional study on useful designs. Along with genetic actuators and soft robotics, emerging subfields of 4D printing that are gaining traction include energy absorption and retention of 4D-printed structures. More accurate actuation programming is becoming possible thanks to the development and trying of novel compounds and structural combinations, providing a multitude of attributes for novel concepts to select from. Although it is available, the research on SMPs and the management of their features, in addition to this the degree of control afforded by printing techniques remains largely untapped, despite the abundance of opportunities and prospective advances [51].

6 Conclusions and Future Outlook

In summary, shape memory polymer (SMP) materials represent the most vibrant areas of materials discipline and technology research, offering outstanding chances for scientific advancements. In the high-tech industries of biomedicine, mechanical, mechatronics, and power generation, among others, incredible technological advancements and applications are anticipated in the years to come. To get controlled response materials that are more durable and reliable than those provided by new technologies currently accessible in this industry, however, significant research efforts will be needed [59]. Nearly every aspect of daily life has potential uses for shape memory polymers, including kitchenware, switches, sensors, intelligent packaging, and self-repairing car bodies. Since only a small number of shape memory polymers have been studied yet and even fewer are commercially viable, only a small number of these applications have been put into practice to date. Most of these polymers were not specifically created to be shape memory materials. These advanced printing technologies can also be used in other applications such as, electrochemical energy conversion [60–63], energy storage [64–69], water purification [70–74], biomedical [75], etc.

The intriguing economic potential that shape memory technology has already shown in a short amount of time, significant innovations must be expected. Future advancements will heavily rely on the two-way shape memory effect in addition to other stimuli like light or electromagnetic fields. Modern polymer chemistry and biomaterial science techniques have allowed for the development of polymer systems with shape memory capabilities expressly for medical purposes. This is an illustration of the technology's immense potential. These advancements have led to the creation of intelligent polymers with customised properties. Compared to metal transplants, degradable shape memory polymers provide several intriguing properties. One way to eliminate the need for follow-up surgery to remove the implant is to use minimally invasive treatments that only require a minor incision to introduce the medical product into the body. Patients gain from a more moderate approach in this way, and healthcare expenses can be decreased. Thus, shape memory implant materials could have a significant impact on how pharmaceuticals are made in the future [59].

Together with SMPs, 4D printing technology ushered in a new era by igniting the research domain and presenting a magical world with the easy fabrication of highly intelligent and widely applicable technology. An exponential rise in value can be noted with reference to the use of 4D printed SMPs in biomedical, robotics, biomimetics, origami, and electronics. The harmony of healing, repairability, and origami can improve adaptability in electronics, rockets, satellites, outer space research, biomedical (artificial muscles, organs, bone cement, or medication delivery) [5].

References

1. Meng, H., Li, G.: A review of stimuli-responsive shape memory polymer composites. Polymer **54**(9), 2199–2221 (2013)
2. Dayyoub, T., et al.: Shape memory polymers as smart materials: a review. Polymers **14**(17), 3511 (2022)
3. Lv, H., et al.: Shape-memory polymer in response to solution. Adv. Eng. Mater. **10**(6), 592–595 (2008)
4. Lendlein, A., Kelch, S.: Shape-memory polymers. Angew. Chem. Int. Ed. **41**(12), 2034–2057 (2002)
5. Bhanushali, H., et al.: Shape memory polymer nanocomposite: a review on structure–property relationship. Polym. Bull. **79**(6), 3437–3493 (2022)
6. Hiltz, J.A.: Shape Memory Polymers: Literature Review, vol. 127. Defence R&D Canada, Ottawa, Ontario, Canada, TM (2002)
7. Zhao, Q., Qi, H.J., Xie, T.: Recent progress in shape memory polymer: new behavior, enabling materials, and mechanistic understanding. Prog. Polym. Sci. **49**, 79–120 (2015)
8. Behl, M., Lendlein, A.: Shape-memory polymers. Mater. Today **10**(4), 20–28 (2007)
9. Baghani, M., et al.: A thermodynamically-consistent 3 D constitutive model for shape memory polymers. Int. J. Plast. **35**, 13–30 (2012)
10. Ohki, T., et al.: Mechanical and shape memory behavior of composites with shape memory polymer. Compos. A Appl. Sci. Manuf. **35**(9), 1065–1073 (2004)
11. Hager, M.D., et al.: Shape memory polymers: past, present and future developments. Prog. Polym. Sci. **49**, 3–33 (2015)
12. Kim, B.K., Lee, S.Y., Xu, M.: Polyurethanes having shape memory effects. Polymer **37**(26), 5781–5793 (1996)
13. Koerner, H., et al.: Polymer design for high temperature shape memory: low crosslink density polyimides. Polymer **54**(1), 391–402 (2013)
14. Zhang, H., et al.: A novel type of shape memory polymer blend and the shape memory mechanism. Polymer **50**(6), 1596–1601 (2009)
15. Liu, C., Qin, H., Mather, P.: Review of progress in shape-memory polymers. J. Mater. Chem. **17**(16), 1543–1558 (2007)

16. Tobushi, H., et al.: Thermomechanical constitutive model of shape memory polymer. Mech. Mater. **33**(10), 545–554 (2001)
17. Liu, G., et al.: Novel shape-memory polymer based on hydrogen bonding. Macromol. Rapid Commun. **27**(14), 1100–1104 (2006)
18. Zhu, G., et al.: Shape-memory behaviors of sensitizing radiation-crosslinked polycaprolactone with polyfunctional poly(ester acrylate). J. Appl. Polym. Sci. **95**(3), 634–639 (2005)
19. Jeong, H.M., et al.: Miscibility and shape memory property of poly(vinyl chloride)/ thermoplastic polyurethane blends. J. Mater. Sci. **36**, 5457–5463 (2001)
20. Jeong, H.M., Ahn, B.K., Kim, B.K.: Miscibility and shape memory effect of thermoplastic polyurethane blends with phenoxy resin. Eur. Polymer J. **37**(11), 2245–2252 (2001)
21. Sokolowski, W.M. et al.: Cold hibernated elastic memory (CHEM) self-deployable structures. In: Smart Structures and Materials 1999: Electroactive Polymer Actuators and Devices. SPIE (1999)
22. Ge, Q., et al.: Thermomechanical behavior of a two-way shape memory composite actuator. Smart Mater. Struct. **22**(5), 055009 (2013)
23. Peponi, L., Navarro-Baena, I., Kenny, J.M.: Shape memory polymers: properties, synthesis and applications. In: Smart Polymers and their Applications, pp. 204–236. Elsevier (2014)
24. Meng, Q., Hu, J.: A review of shape memory polymer composites and blends. Compos. A Appl. Sci. Manuf. **40**(11), 1661–1672 (2009)
25. Ikeda, T., et al.: Anisotropic bending and unbending behavior of azobenzene liquid-crystalline gels by light exposure. Adv. Mater. **15**(3), 201–205 (2003)
26. Irie, M., Kunwatchakun, D.: Photoresponsive polymers. 8. Reversible photostimulated dilation of polyacrylamide gels having triphenylmethane leuco derivatives. Macromolecules **19**(10), 2476–2480 (1986)
27. Jiang, H., Kelch, S., Lendlein, A.: Polymers move in response to light. Adv. Mater. **18**(11), 1471–1475 (2006)
28. Gunes, I.S., Jimenez, G.A., Jana, S.C.: Carbonaceous fillers for shape memory actuation of polyurethane composites by resistive heating. Carbon **47**(4), 981–997 (2009)
29. Lee, Y.M., Kim, S.H., Cho, C.S.: Synthesis and swelling characteristics of pH and thermoresponsive interpenetrating polymer network hydrogel composed of poly(vinyl alcohol) and poly(acrylic acid). J. Appl. Polym. Sci. **62**(2), 301–311 (1996)
30. Li, T., et al.: Thermally and near-infrared light-induced shape memory polymers capable of healing mechanical damage and fatigued shape memory function. ACS Appl. Mater. Interfaces **11**(9), 9470–9477 (2019)
31. Yang, B., et al.: On the effects of moisture in a polyurethane shape memory polymer. Smart Mater. Struct. **13**(1), 191 (2003)
32. Fuhrer, R., et al.: Crosslinking metal nanoparticles into the polymer backbone of hydrogels enables preparation of soft, magnetic field-driven actuators with muscle-like flexibility. Small **5**(3), 383–388 (2009)
33. Mu, T., et al.: Shape memory polymers for composites. Compos. Sci. Technol. **160**, 169–198 (2018)
34. Jani, J.M. et al.: A review of shape memory alloy research, applications and opportunities. Mater. Des. (1980–2015) **56**, 1078–1113 (2014)
35. Gao, H., et al.: The research status and challenges of shape memory polymer-based flexible electronics. Mater. Horiz. **6**(5), 931–944 (2019)
36. Leng, J., et al.: Shape-memory polymers and their composites: stimulus methods and applications. Prog. Mater Sci. **56**(7), 1077–1135 (2011)
37. Lendlein, A., Gould, O.E.: Reprogrammable recovery and actuation behaviour of shape-memory polymers. Nat. Rev. Mater. **4**(2), 116–133 (2019)
38. Rousseau, I.A., Mather, P.T.: Shape memory effect exhibited by smectic-C liquid crystalline elastomers. J. Am. Chem. Soc. **125**(50), 15300–15301 (2003)
39. Chung, T., Romo-Uribe, A., Mather, P.T.: Two-way reversible shape memory in a semicrystalline network. Macromolecules **41**(1), 184–192 (2008)

40. Bellin, I., Kelch, S., Lendlein, A.: Dual-shape properties of triple-shape polymer networks with crystallizable network segments and grafted side chains. J. Mater. Chem. **17**(28), 2885–2891 (2007)
41. Nguyen, T.D.: Modeling shape-memory behavior of polymers. Polym. Rev. **53**(1), 130–152 (2013)
42. Li, J., Rodgers, W.R., Xie, T.: Semi-crystalline two-way shape memory elastomer. Polymer **52**(23), 5320–5325 (2011)
43. Jiang, Y., et al.: Influence of compression direction on the shape-memory effect of micro-cylinder arrays prepared from semi-crystalline polymer networks. MRS Adv. **1**(27), 1985–1993 (2016)
44. Neffe, A.T., et al.: Polymer networks combining controlled drug release, biodegradation, and shape memory capability. Adv. Mater. **21**(32–33), 3394–3398 (2009)
45. Pisani, S., et al.: Shape-memory polymers hallmarks and their biomedical applications in the form of nanofibers. Int. J. Mol. Sci. **23**(3), 1290 (2022)
46. Lendlein, A., Robert, L.: Biodegradable, elastic shape-memory polymers for potential biomedical applications. Science **296**(5573), 1673–1676 (2002)
47. Wang, L., et al.: Shape memory polymer fibers: materials, structures, and applications. Adv. Fiber Mater. **4**(1), 5–23 (2022)
48. Mather, P.T., Luo, X., Rousseau, I.A.: Shape memory polymer research. Annu. Rev. Mater. Res. **39**(1), 445–471 (2009)
49. Ehrmann, G., Ehrmann, A.: 3D printing of shape memory polymers. J. Appl. Polym. Sci. **138**(34), 50847 (2021)
50. Alshebly, Y.S., et al.: Review on recent advances in 4D printing of shape memory polymers. Eur. Polym. J. **159**, 110708 (2021)
51. Spiegel, C.A., et al.: 4D printing of shape memory polymers: from macro to micro. Adv. Func. Mater. **32**(51), 2110580 (2022)
52. Subash, A., Kandasubramanian, B.: 4D printing of shape memory polymers. Eur. Polym. J. **134**, 109771 (2020)
53. Pandey, A., et al.: 3D printed biodegradable functional temperature-stimuli shape memory polymer for customized scaffoldings. J. Mech. Behav. Biomed. Mater. **108**, 103781 (2020)
54. Garcia Rosales, C.A., et al.: Characterization of shape memory polymer parts fabricated using material extrusion 3D printing technique. Rapid Prototyp. J. **25**(2), 322–331 (2019)
55. Yeazel, T.R., Becker, M.L.: Advancing toward 3D printing of bioresorbable shape memory polymer stents. Biomacromolecules **21**(10), 3957–3965 (2020)
56. Tang, Z., et al.: 3D printing of a versatile applicability shape memory polymer with high strength and high transition temperature. Chem. Eng. J. **431**, 134211 (2022)
57. Khalid, M.Y., et al.: 4D printing of shape memory polymer composites: a review on fabrication techniques, applications, and future perspectives. J. Manuf. Process. **81**, 759–797 (2022)
58. Bodaghi, M., Damanpack, A., Liao, W.: Triple shape memory polymers by 4D printing. Smart Mater. Struct. **27**(6), 065010 (2018)
59. Panahi-Sarmad, M., et al.: Deep focusing on the role of microstructures in shape memory properties of polymer composites: a critical review. Eur. Polym. J. **117**, 280–303 (2019)
60. Kumar, D., et al.: Nanocarbon assisted green hydrogen production: development and recent trends. Int. J. Hydrogen. Energy **50**, 118–141 (2024)
61. Cherusseri, J., et al.: Polymer-based composite materials: characterizations. In: Composite materials: processing, applications, characterizations, pp. 37–77 (2017)
62. Thomas, S.A., et al.: A comprehensive review on tungsten oxide nanostructures-based electrochromic supercapacitors and machine learning models for design and process parameter optimization. Energy Storage **5**, e499 (2023)
63. Thomas, S.A., et al.: Rapid synthesis of hierarchical tin disulfide (SnS2) nanostructures by a microwave-assisted hydrothermal method for high-performance supercapatteries. ACS Appl. Electron. Mater. **6**, 3346–3361 (2024)
64. Thomas, S.A., Cherusseri, J.: Recent advances in synthesis and properties of zirconium-based MXenes for application in rechargeable batteries. Energy Storage **5**, e475 (2023)

65. Sadanandan, A., et al.: A critical review on two-dimensional $Ti_3C_2T_x$ MXenes for anti-corrosion coatings. Prog. Org. Coat. **183**, 107757 (2023)
66. Thomas, S.A., Cherusseri, J.: Boron carbon nitride (BCN): an emerging two-dimensional nanomaterial for supercapacitors. J. Mater. Chem. A. **11**, 23148–23187 (2023)
67. Pramanik, S., et al.: Metal matrix composites: theory, techniques, and applications. In: Composite materials: processing, applications, characterizations, pp. 369–411 (2017)
68. Cherusseri, J., Kar, K.K.: Recent progress in nanocomposites based on carbon nanomaterials and electronically conducting polymers. In: Polymer nanocomposites based on inorganic and organic nanomaterials, pp. 229–256 (2015)
69. Richard, B., et al.: Minireview on fluid manipulation techniques for the synthesis and energy applications of two-dimensional MXenes: advances, challenges, and perspectives. Energy Fuels **37**, 6999–7013 (2023)
70. Zaed, M.A., et al.: Low-cost synthesis of $Ti_3C_2T_x$ MXene-based sponge for solar steam generation and clean water production. Ceram. Int. **50**, 27910–27922 (2024)
71. Zaed, M.A., et al.: Utilization of recycled materials for low-cost MXene synthesis and fabrication of graphite/MXene composite for enhanced water desalination performance. Sep. Purif. Technol. **354**, 129055 (2025)
72. Zaed, M.A., et al.: Synthesis and characterization of hierarchical $Ti_3C_2T_x$ MXene/graphitic-carbon nitride/activated carbon@ luffa sponge composite for enhanced water desalination. Open Ceram. **19**, 100645 (2024)
73. Zaed, M.A., et al.: Hierarchical $Ti_3C_2T_x$ MXene@ Honeycomb nanocomposite with high energy efficiency for solar water desalination. Chemosphere **366**, 143459 (2024)
74. Zaed, M.A., et al.: Unlocking desalination's potential: harnessing MXene composite for sustainable desalination. Chem. Eng. J. **500**, 156910 (2024)
75. Cherusseri, J., et al.: SARS-CoV-2-on-chip for long COVID management. Biosensors **12**, 890 (2022)

Hydrogels and Hydrogel Nanocomposites for 3D- and 4D-Printed Applications

Rohit Verma, Prashant Hitaishi, Deepshikha Gupta, Sheenam Thatai, and Tejendra Kumar Gupta

Abstract Hydrogels, with their exceptional water retention capabilities and biocompatibility, have emerged as promising materials for a variety of applications in tissue engineering, drug delivery, and soft robotics. The advent of 3D and 4D printing technologies has further expanded the potential uses of hydrogels and hydrogel nanocomposites, enabling the fabrication of complex structures with precise spatial control. This chapter delves into the fundamentals of hydrogels and their nanocomposites, examining their properties, synthesis methods, and suitability for 3D and 4D printing applications. It also highlights recent advancements, discusses challenges, and explores future directions in this rapidly evolving field.

Keywords Hydrogel · Nanocomposites · 3D printer · 4D printers · Click chemistry

1 Introduction

Hydrogels, three-dimensional polymeric networks capable of absorbing and retaining large amounts of water while maintaining their structural integrity, have garnered substantial interest across various scientific disciplines [1, 2]. Hydrogels are made of hydrophilic polymer chains in three-dimensional networks with high swelling capacity in an aqueous solution and without dissolution [3–5]. Hydrogels can absorb water and aqueous solvents, while organogels absorb non-polar and

R. Verma (✉)
Department of Physics, Amity Institute of Applied Sciences, Amity University Uttar Pradesh, Noida, India
e-mail: rverma85@amity.edu

P. Hitaishi
Institute of Experimental and Applied Physics, Kiel University, Kiel, Germany

D. Gupta · S. Thatai · T. K. Gupta (✉)
Department of Chemistry, Amity Institute of Applied Sciences, Amity University Uttar Pradesh, Noida, India
e-mail: tkgupta@amity.edu

S. Moharana et al. (eds.), *Polymer Nanocomposites for 3D, 4D and 5D Printing*, Engineering Materials, https://doi.org/10.1007/978-981-96-4214-4_10

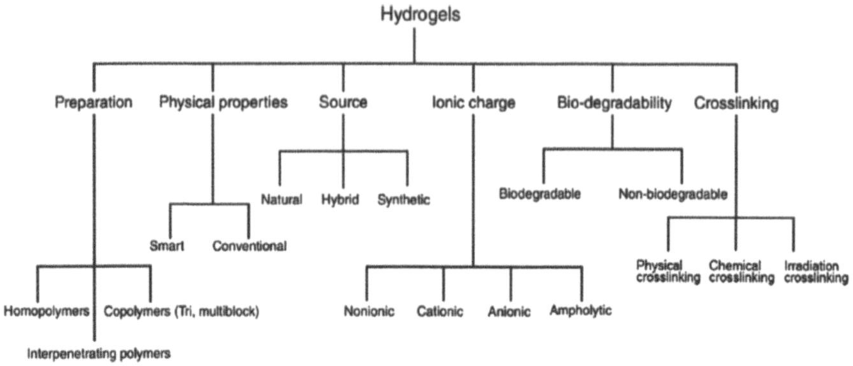

Fig. 1 Classification of polymer hydrogels [15]

organic solvents such as hexane [6]. Hydrogel structure can absorb up to a thousand times its dry weight in aqueous solutions [7]. Their unique properties, including high water content, biocompatibility, and tunable mechanical characteristics, make them versatile materials for applications ranging from biomedical engineering to environmental remediation. The integration of nanotechnology into hydrogels has further expanded their functionalities, paving the way for advanced applications in fields such as drug delivery, tissue engineering, and soft robotics. Previous research has shown that the hydrogel structure and water maintenance are significantly impacted by the kind of polymer, size and ionic charge, polymeric chain flexibility, crosslink type, and density. As a result, many classifications for the hydrogels were offered [8]. Figure 1 presents a summary of the hydrogel categories based on their qualities.

Hydrogels have revolutionized biomedical applications due to their ability to mimic the extracellular matrix's water-rich environment, making them ideal scaffolds for tissue engineering and regenerative medicine [9–11]. These materials offer a supportive framework for cell growth and tissue regeneration, promoting healing and restoration in damaged tissues [12]. Furthermore, their biocompatibility ensures minimal adverse reactions when implanted in living organisms, a critical factor for medical device development [13]. In drug delivery systems, hydrogels serve as effective matrices for controlled release due to their capacity to encapsulate and protect pharmaceutical agents [12]. By modifying the hydrogel's composition and structure, researchers can tailor drug release kinetics to achieve precise therapeutic outcomes, minimizing side effects and optimizing patient compliance [14].

The emergence of hydrogel nanocomposites, integrating nanoscale materials within the hydrogel matrix, represents a significant advancement in material science. Nanoparticles such as carbon nanotubes, graphene, and metallic nanoparticles are commonly incorporated into hydrogels to impart enhanced mechanical strength, electrical conductivity, or stimuli-responsive properties [16]. These enhancements are crucial for applications requiring robust materials that can respond intelligently to external stimuli, such as changes in pH, temperature, or light [17]. Figures 2 and 3 describe hydrogels: classification, characteristics, and applications.

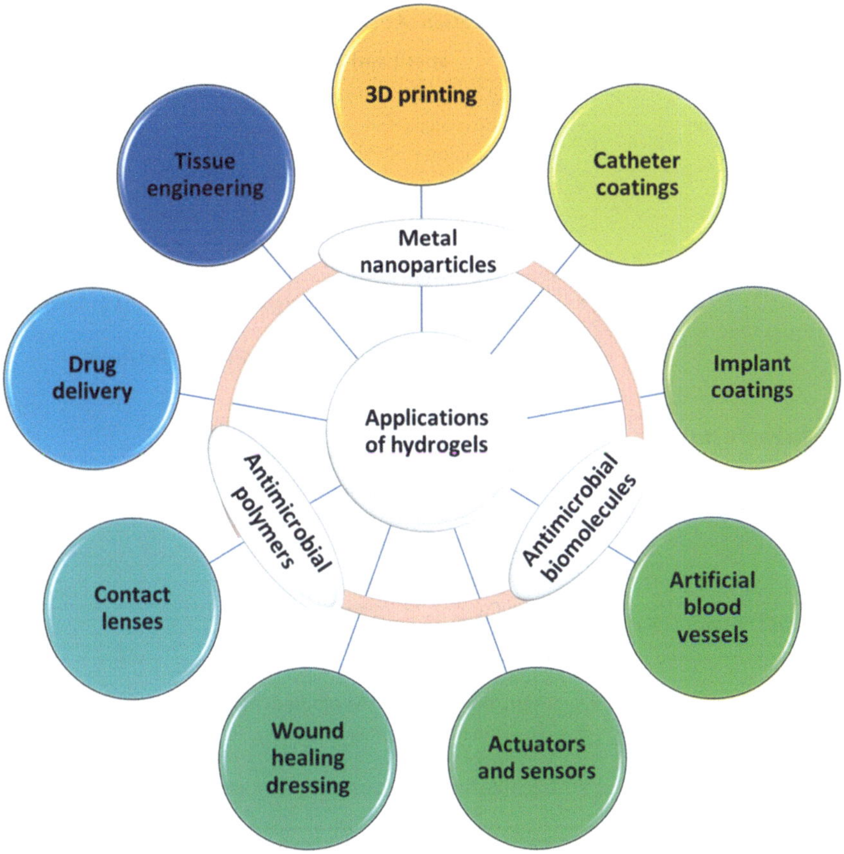

Fig. 2 Hydrogels: classification, characteristics and applications

Nanocomposite hydrogels exhibit superior mechanical properties compared to their pure hydrogel counterparts, making them suitable for load-bearing applications in tissue engineering and orthopedic implants [18]. Moreover, their enhanced electrical conductivity enables the development of bioelectronic devices and sensors that can monitor physiological parameters or deliver electrical stimuli to tissues [19].

The advent of additive manufacturing technologies, particularly 3D printing, has revolutionized the fabrication of hydrogels and nanocomposites [20]. Unlike traditional manufacturing methods, which often rely on molds and manual assembly, 3D printing allows for precise control over the deposition of materials, enabling the creation of complex, patient-specific structures with customizable properties [21].

In 3D printing, hydrogel precursors are extruded, jetted, or cured layer-by-layer to build three-dimensional objects directly from digital models [21]. This layer-by-layer approach facilitates the incorporation of multiple materials and gradients

Classification of hydrogels	
Basis of classification	**Specification and characteristics**
Source	• Natural polymer: Alginate, collagen • Synthetic polymer: poly(vinyl alcohol) • Hybrid polymer: Rubber
Polymer composition	• Homo-polymeric: Poly(N-isopropyl acrylamide) • Hetero-polymeric: Poly(vinyl alcohol)-gelatin • Co-polymeric: Poly(PEGMA-co-monomethyl itaconate • Hybrid: Different polymers or phases • Composites: Organic and inorganic components as fibers and nanoparticles • Interpenetrating polymer network (IPNs): Poly(vinyl alcohol)-alginate
Network structure	• Physical cross-linking: Freeze-thawing • Chemical cross-linking: Glutaraldehyde
Sensitivity to stimuli	• Physical stimulus: Temperature • Chemical stimulus: pH, ionic strength • Bio-chemical stimulus: Antigen, enzyme
Polymer network charge	• Non-ionic: No charge • Ionic: Cationic or anionic • Zwitterion: Both cationic and anionic • Amphoteric: Both acidic and basic groups
Physical aspect	• Micro/nanoparticle: Micro bed/nanogel • Film: Electro spun mat • Matrix: Scaffold • Gel: Injectable drug loaded-hydrogel
Configuration of chains	• Non-crystalline: Random arrangement as amorphous domain • Semi-crystalline: Amorphous + crystalline domains
Pore size	• Nano gel • Micro gel • Macro gel
Interchain interaction	• Physically crosslinked • Chemically crosslinked
Degradability	• Biodegradable • Non-biodegradable

Fig. 3 The hydrogel classification based on their properties

within a single construct, enhancing functionality and mimicking the hierarchical organization found in biological tissues [22].

Building upon the capabilities of 3D printing, 4D printing introduces the dimension of time, enabling structures to transform or self-assemble in response to external stimuli [23]. This dynamic functionality holds tremendous promise for developing smart materials and devices capable of adapting to changing environmental conditions or physiological cues [24]. By harnessing stimuli-responsive hydrogels and nanocomposites, researchers can create self-folding structures, shape-memory materials, and actuators for applications in soft robotics, biomedical implants, and beyond [25].

This chapter explores the synergistic integration of hydrogels and hydrogel nanocomposites with 3D and 4D printing technologies. It begins by elucidating the fundamental properties and synthesis methods of hydrogels, followed by an exploration of the enhancements achieved through nanocomposite formulations. The role of 3D printing in fabricating complex hydrogel structures with tailored properties is then discussed, along with the transformative potential of 4D printing for creating dynamic, responsive materials. Case studies and recent advancements illustrate the practical applications and future directions of these technologies in biomedicine, soft robotics, and beyond.

2 Fundamentals of Hydrogels

2.1 Definition and Properties

Hydrogels are crosslinked polymeric networks that can absorb and retain significant amounts of water. They can be classified based on their origin (natural or synthetic), charge (neutral, anionic, or cationic), and physical structure (amorphous, semi-crystalline, or crystalline). The key properties of hydrogels include:

- **Swelling behavior**: Hydrogels can swell in water due to their hydrophilic nature, which is crucial for their application in drug delivery and tissue engineering [26] Swelling is a result of the interaction between a solvent and a hydrogel matrix. The first step in saving salt or hydrogel into a solvent and extending it to three dimensions is called swelling. The swelling behaviour [27] of a soil hydrogel directly depends upon the structure of the hydrogel, the polymerization process (either grafting or cross-linking), the concentration of the cross-linking agent, and the particle size of the hydrogel. All factors affect the swelling, but the last factor affects the more. If the concentration of the cross-linking agent rises, as a result, high densification of hydrogel and the structure of hydrogels become more congested due to the increased number of cross-links on the polymer chain. Figure 4 shows the factors affecting swelling.
- **Mechanical properties**: The mechanical strength of hydrogels can be tailored by adjusting the crosslinking density and the type of polymer used. The low mechanical strength of hydrogels is usually a consequence of two properties. One cause is that hydrogels have solution-like properties, such as a low density of polymer chains and low friction between polymer chains. The other property is that the structure of the hydrogel network is not homogeneous. Therefore, when the hydrogels are subjected to external forces, the stress is predominantly

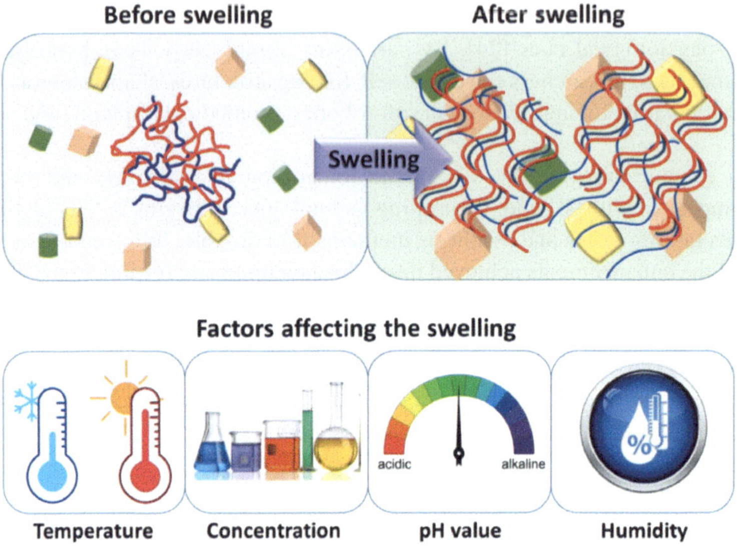

Fig. 4 Factors affecting the swelling of hydrogel

concentrated on short polymer chains, which leads to the failure of the sample under extremely low forces [28].

- **Biocompatibility**: Many hydrogels, especially those derived from natural sources, exhibit excellent biocompatibility, making them suitable for biomedical applications. Because natural polymers have inherent qualities, hydrogels made of them typically have great biocompatibility. Unfortunately, the majority of natural polymeric hydrogels face significant challenges in their use due to their weak mechanical qualities and low toughness. Enzymes have the ability to break down the majority of hydrogels made from natural polymers, enabling cells to proliferate and change their environment. Another necessary component is the rate of biodegradation, which is determined by the molecular weight, crystalline or amorphous, and hydrophilic or hydrophobic behaviors of the polymers [29].
- **Stimuli responsiveness**: Hydrogels can respond to various stimuli, such as pH, temperature, and light, allowing for controlled drug release and smart material applications. For instance, compared to normal tissue (pH 7.4), malignant tissue (pH 6.8) and endosomes/lysosomes (pH 5.5) are more acidic, requiring the use of hydrogels that can change pH to deliver payload drugs [30, 31]. In response to external stimuli, stimulus-responsive hydrogels undergo transitional changes. When temperature, pressure, ionic strength, and insolvent change, the concentration of some biomolecules, such as enzymes, can cause them to enlarge, contract, deteriorate, or go through a sol–gel phase transition. The unusual ability of this kind of hydrogel to carry out certain functions, such as drug release and in situ gel formation, in response to minute changes in surrounding conditions makes it very useful for therapeutic administration.

2.2 Synthesis Methods

Hydrogels can be synthesized using various methods (see Fig. 5) [32–34], described as follows.

- **Free radical polymerization**: A common method where monomers are polymerized in the presence of a crosslinker and an initiator. The most popular and adaptable method for creating hydrogels based on specific monomers, including acrylates and amides, is conventional free radical polymerization [33]. This process is

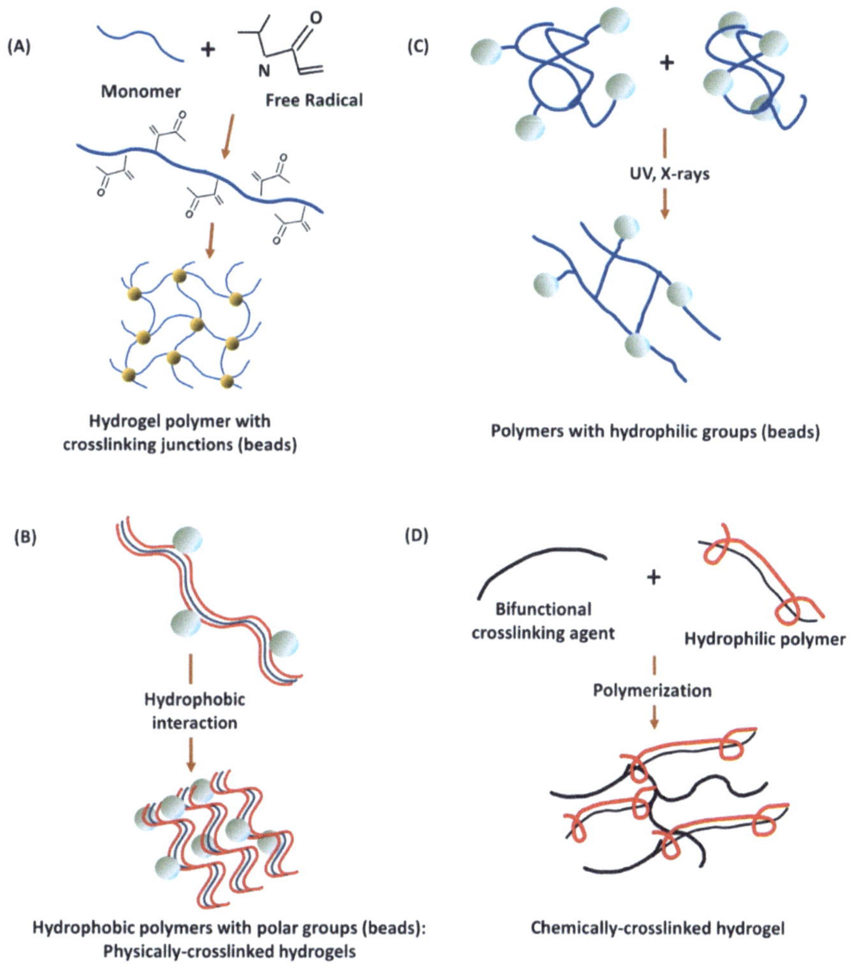

Fig. 5 Synthesis of hydrogels in **a** free radical polymerization, **b** physical cross-linking, **c** irradiation cross-linking, and **d** chemical cross-linking

applied in the creation of hydrogels based on polymers. It is a kind of polymerization that grows in a chain and helps create hydrogels for nanocomposite materials. Using this technique, free radical groups are added one after the other to produce the polymer. These groups serve as templates for the structural build-up (Fig. 5a). This process proceeds through the standard steps of free radical polymerization, which are initiation, propagation, chain transfer, and termination [34].

- **Chemical crosslinking of Hydrogel Polymeric Precursors**: This is one of the important and more often used hydrogel production techniques. Chemical methods for creating crosslinking hydrogels include condensation polymerization reactions, chain-growth polymerization or addition, and the sequential addition of cross-linking agents. A bifunctional crosslinking agent is added to a diluted solution of a hydrophilic polymer with adequate functionality to react with the agent to create hydrogel polymers (Fig. 5d). Both synthetic and natural polymer hydrogels can be made using this technique. For example, formaldehyde or dialdehyde were used as crosslinking agents to create albumin and gelatin-based hydrogels [35].
- **Photopolymerization**: Photopolymerization [36, 37] is a widely used method for the synthesis of hydrogels, characterized by the initiation of polymerization through light exposure, typically UV or visible light. This technique offers spatial and temporal control over the polymerization process, making it highly suitable for creating hydrogels with complex structures and precise properties. Here, we explore the mechanisms, advantages, and applications of photopolymerization in the synthesis of hydrogels. Photopolymerization begins with the absorption of light by photoinitiators, which decompose to form reactive species such as free radicals or cations. These reactive species initiate the polymerization of monomers. The reactive species generated during initiation react with monomer molecules, forming reactive polymer chains. This chain reaction continues, adding monomers to the growing polymer chains, leading to the formation of a crosslinked hydrogel network. The polymerization process concludes when the reactive polymer chains are deactivated or combined, ceasing the growth of polymer chains. This step results in the final crosslinked hydrogel structure [37].
- **Click chemistry**: Click chemistry is a versatile and efficient method for synthesizing hydrogels, characterized by its high specificity, fast reaction rates, and mild reaction conditions [38]. The term "click chemistry" was introduced by K. Barry Sharpless in 2001 to describe a class of biocompatible chemical reactions that are modular, wide in scope, and provide high yields. This method is particularly advantageous for hydrogel synthesis as it allows precise control over the hydrogel network formation and functionalization, making it suitable for various biomedical applications.

The comparison of various hydrogel synthesis methods, their advantages, and disadvantages are collected in Table 1.

Table 1 Comparison of hydrogel synthesis methods

Method	Description	Advantages	Disadvantages	Refs.
Free radical polymerization	Polymerization using a crosslinker and initiator	Simple, widely used	May require toxic initiators	[33, 34]
Chemical crosslinking of hydrogel polymeric precursors	Include condensation polymerization reactions, chain-growth polymerization or addition, and the sequential addition of cross-linking agents	Both synthetic and natural polymer hydrogels can be made using this technique	Limited control over properties	[35]
Photopolymerization	Use of light to initiate polymerization and crosslinking	Precise control, spatial resolution	Requires photosensitive monomers	[36, 37]
Click chemistry	Efficient, versatile method for hydrogel synthesis	High efficiency, tunable	May involve complex chemistry	[38]

3 Hydrogel Nanocomposites

3.1 *Definition and Advantages*

Hydrogel nanocomposites are hybrid materials composed of a hydrogel matrix embedded with nanomaterials, such as nanoparticles, nanofibers, or nanotubes [33]. The incorporation of nanomaterials can significantly enhance the properties of hydrogels, providing advantages such as:

- **Improved mechanical strength**: Nanomaterials can reinforce the hydrogel matrix, making it more robust and durable. Nanocomposites have been increasingly used to enhance the mechanical strength of hydrogels [37–40], which are hydrophilic polymer networks capable of holding a large amount of water. This improvement is due to the unique properties of nanoparticles that can be incorporated into the hydrogel matrix, leading to a synergistic enhancement in the mechanical properties. Nanoparticles, such as silica, clay, graphene, and carbon nanotubes, can be uniformly dispersed within the hydrogel matrix, leading to a homogeneous distribution of stress and preventing localized deformation. The interaction between the polymer chains and nanoparticles can significantly improve the mechanical strength. This interaction can be through hydrogen bonding, covalent bonding, or van der Waals forces, which enhances the load transfer between the hydrogel matrix and the nanoparticles. Nanoparticles can act as additional crosslinking points within the hydrogel network, increasing the crosslinking density and thus the rigidity and strength of the hydrogel. The presence of nanoparticles can enhance the energy dissipation capacity of hydrogels. When subjected to stress, the nanoparticles can absorb and dissipate energy, reducing the likelihood of fracture. Nanoparticles can bridge cracks within the hydrogel matrix, preventing crack propagation and thus enhancing the toughness of the material. Nanoparticles can effectively transfer stress within the hydrogel network, thereby improving the overall mechanical properties. Silica nanoparticles have been widely used to reinforce hydrogels. For instance, the incorporation of silica nanoparticles into a polyacrylamide hydrogel has been shown to significantly improve its mechanical strength and elasticity. The silica nanoparticles act as multifunctional crosslinkers, increasing the crosslinking density and thereby enhancing the mechanical properties of the hydrogel. Graphene oxide (GO) is another nanoparticle that has been used to improve the mechanical strength of hydrogels. GO sheets provide a large surface area for interaction with polymer chains, leading to improved mechanical strength and flexibility. For example, a composite hydrogel made of GO and poly(vinyl alcohol) exhibited enhanced mechanical properties due to the strong hydrogen bonding between the GO sheets and the polymer chains. Clay nanoparticles [40], such as montmorillonite, have also been incorporated into hydrogels to enhance their mechanical properties. These nanoparticles can form an exfoliated structure within the hydrogel matrix, providing a barrier to crack propagation and thereby improving the toughness and strength of the hydrogel.

- **Enhanced thermal stability**: Nanocomposites have shown significant promise in improving the thermal stability of hydrogels. Thermal stability is an essential property for many hydrogel applications, including biomedical devices, sensors, and drug delivery systems, where the material must withstand variations in temperature without degrading. The incorporation of nanoparticles into hydrogels can enhance their thermal properties through various mechanisms. The homogeneous dispersion of nanoparticles within the hydrogel matrix creates a barrier to heat transfer. This barrier effect slows down the thermal degradation of the polymer chains by hindering the movement of heat through the material. Nanoparticles, such as graphene oxide and carbon nanotubes, have high thermal conductivity. When these particles are dispersed within the hydrogel, they can help dissipate heat more efficiently, preventing localized overheating and degradation of the polymer network.
- **Electrical conductivity**: Nanocomposites are a promising solution to enhance the electrical conductivity of hydrogels, which are typically insulating due to their high water content and polymeric structure. The incorporation of conductive nanoparticles into hydrogel matrices can transform them into conductive materials suitable for various applications, including flexible electronics, biosensors, and tissue engineering. Conductive nanoparticles such as graphene, carbon nanotubes (CNTs), and metallic nanoparticles (e.g., gold, silver) can be uniformly dispersed within the hydrogel matrix. This dispersion forms a percolation network that facilitates electron transport throughout the material. Strong interactions between the polymer matrix and nanoparticles, such as π-π stacking between graphene and

aromatic polymer units, enhance the stability and conductivity of the composite hydrogel. These interactions can improve the dispersion and alignment of nanoparticles, optimizing the conductive pathways. Combining different types of nanoparticles (e.g., graphene with CNTs) can create synergistic effects, improving the conductivity more than single nanoparticle types alone. These hybrid structures can take advantage of the unique conductive properties of each nanoparticle type. Graphene, a two-dimensional carbon nanomaterial, has exceptional electrical conductivity. Incorporating graphene into hydrogels can significantly enhance their conductivity. For example, a graphene oxide-polyacrylamide hydrogel exhibited improved electrical properties due to the formation of a conductive network within the hydrogel matrix. Carbon nanotubes (CNTs) are another popular nanomaterial for enhancing hydrogel conductivity. The incorporation of CNTs into hydrogels, such as polyvinyl alcohol (PVA)-CNT hydrogels, has been shown to improve conductivity through the formation of a percolation network of CNTs within the polymer matrix. Metallic nanoparticles, such as gold or silver, can also enhance the conductivity of hydrogels. For instance, a hydrogel incorporating gold nanoparticles (AuNPs) showed increased electrical conductivity and potential applications in bioelectronics and sensing devices.
- **Functionalization**: Nanomaterials have the potential to introduce new functionalities to hydrogels, such as magnetic responsiveness and antibacterial properties. These functionalities arise from the unique physical, chemical, and biological properties of nanoparticles, which can be tailored to meet specific application needs. The incorporation of magnetic nanoparticles such as iron oxide (Fe_3O_4) or magnetite into hydrogels can render the hydrogel magnetically responsive. These nanoparticles align in the presence of an external magnetic field, allowing the hydrogel to move or change shape in response to the field. Magnetic hydrogels can be used for targeted drug delivery. The magnetic field can guide the hydrogel to a specific location within the body, and the release of the drug can be controlled by adjusting the magnetic field intensity and duration. In tissue engineering, magnetic hydrogels can help in the alignment and organization of cells. The magnetic field can be used to create specific patterns or structures, promoting tissue growth and regeneration.

3.2 *Types of Nanomaterials Used*

Various nanomaterials can be incorporated into hydrogels, including:

- **Metallic nanoparticles**: Silver, gold, and platinum nanoparticles can provide antimicrobial properties and enhance mechanical strength [42]. Hydrogels embedded with iron oxide nanoparticles have been widely studied for their magnetic properties. These hydrogels can be manipulated using external magnetic fields, making them useful for applications like remote-controlled drug delivery and magnetic resonance imaging (MRI) contrast enhancement. A study demonstrated that incorporating magnetite nanoparticles into a polyvinyl alcohol

Table 2 Types of nanomaterials used in hydrogel nanocomposites

Nanomaterial type	Properties	Applications	Refs.
Metallic nanoparticles	Antimicrobial, enhanced strength	Wound healing, drug delivery	[41]
Carbon-based nanomaterials	Electrical conductivity, mechanical reinforcement	Bioelectronics, tissue engineering	[39, 41]
Polymeric nanoparticles	Controlled release, biocompatible	Drug delivery, tissue scaffolding	[41]
Silica nanoparticles	Enhanced mechanical properties, thermal stability	Structural reinforcement, sensors	[40, 41]

hydrogel resulted in a magnetically responsive material suitable for remote actuation and controlled drug release.

- **Carbon-based nanomaterials**: Graphene, carbon nanotubes, and fullerenes can impart electrical conductivity and improve mechanical properties.
- **Polymeric nanoparticles**: Polylactic acid (PLA) and poly (lactic-co-glycolic acid) (PLGA) nanoparticles can be used for drug delivery applications.
- **Silica nanoparticles**: These can enhance the mechanical properties and thermal stability of hydrogels.

Types of nanomaterials for the preparation of hydrogels and their uses are given in Table 2.

4 3D Printing of Hydrogels and Hydrogel Nanocomposites

4.1 3D Printing Techniques

3D printing with hydrogels is an emerging field that leverages the unique properties of hydrogels—such as high-water content, biocompatibility, and tunable mechanical properties—to create complex, customizable structures for a variety of applications, including biomedical engineering, tissue engineering, and drug delivery systems. The process of 3D printing using hydrogels is shown in Fig. 6. Several 3D printing techniques are employed to fabricate hydrogel-based structures, each with advantages and limitations. Several techniques are commonly used for printing hydrogels and their nanocomposites [42]:

- **Stereolithography (SLA)**: Stereolithography (SLA) [43, 44] uses a light source, typically a laser or digital light projector, to selectively polymerize a photosensitive hydrogel resin layer by layer. The light causes the hydrogel to solidify in the desired pattern, building the structure from the bottom up. SLA provides high-resolution prints, capable of creating detailed and complex structures. It produces

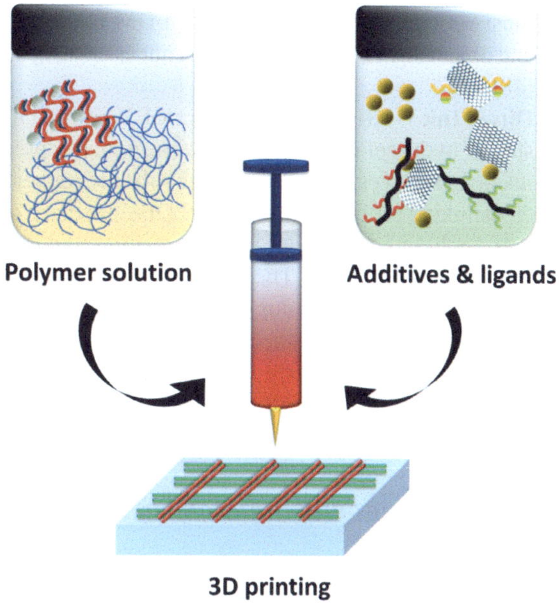

Fig. 6 Process of 3D printing using hydrogels

smooth surfaces, which is beneficial for applications requiring fine detail and precision. One of the limitations of SLA is that the range of hydrogels suitable for SLA is photosensitive [44]. Another limitation is that it requires post-processing to remove uncured resin and further curing to achieve final mechanical properties [43].

- **Inkjet Bioprinting**: Inkjet bioprinting [45, 46] uses a printhead to deposit droplets of hydrogel solutions in a layer-by-layer fashion. The droplets solidify upon deposition, forming the desired structure. Deposits small droplets of hydrogel precursor to form a structure. The advantages of inkjet bioprinting are that this is a gentle deposition process, maintains high cell viability is capable of printing multiple materials simultaneously, and is useful for complex tissue engineering applications. The limitations of this technique are that due to the size of the droplets, the resolution is affected, and the requirement of low-viscosity hydrogels for effective jetting limits material choices [46]. Inkjet bioprinting is extensively used in the creation of cell-laden constructs and complex tissue models, such as skin, cartilage, and liver tissues.
- **Fused deposition modeling (FDM)**: Fused Deposition Modeling (FDM) [47, 48] involves the extrusion of a hydrogel filament through a heated nozzle. The filament is deposited layer by layer onto a build platform, where it solidifies upon cooling. The advantages of this technique are that FDM can use a wide range of hydrogel materials, including those not photosensitive, and are generally more affordable and accessible compared to other 3D printing methods [47]. The limitations of FDM involve lower resolution as compared to SLA, making it less suitable for fine details and FDM is not ideal for hydrogels sensitive to heat, as the material

must be heated for extrusion [48]. FDM is used for fabricating larger structures and objects where high precision is not as critical, such as prosthetics and implants [48].

- **Direct Ink Writing (DIW)**: Direct Ink Writing (DIW) [49, 50] involves the extrusion of hydrogel ink through a nozzle, which is precisely controlled to deposit the ink layer by layer. The hydrogel ink solidifies either through a physical or chemical crosslinking process. The merits of DIW are that it can print a wide range of hydrogels, including those with high viscosities, and is capable of fabricating intricate and porous structures. The demerit of this method is that DIW requires careful optimization of ink rheology to ensure structural stability during and after printing and lower resolution compared to SLA but higher than FDM [49]. DIW is widely used in tissue engineering for creating scaffolds with controlled architecture and porosity, as well as in soft robotics [50].

4.2 Challenges and Solutions of the 3D Printing Using Hydrogels

3D printing techniques with hydrogels offer diverse capabilities for fabricating complex, customizable structures for various applications. Each technique—SLA, FDM, inkjet bioprinting, and DIW—has its unique advantages and limitations, making them suitable for specific applications in biomedical engineering, tissue engineering, and beyond. As technology advances, the development of new hydrogel materials and printing methods will continue to expand the potential of 3D-printed hydrogel structures. Despite the potential of 3D-printed hydrogels, several challenges and their possible solutions [47, 51–57] are described below which need to be addressed.

- **Material Properties**: Hydrogels are inherently soft and contain a high water content, which can make them difficult to print with precision and maintain structural integrity during and after printing. The solution to this may be to use chemical or physical crosslinking agents that can enhance the mechanical strength of hydrogels. For instance, incorporating calcium ions can crosslink alginate-based hydrogels, improving their stability. Developing composite hydrogels that combine natural and synthetic polymers can enhance the mechanical properties of hydrogels while maintaining biocompatibility. For example, blending collagen with synthetic polymers can create hydrogels with improved strength and functionality.
- **Printing Resolution and Accuracy**: Achieving high resolution and accuracy in 3D printing of hydrogels is difficult due to the tendency of the material to spread or sag, leading to lower precision in printed structures. By adjusting printing parameters such as nozzle diameter, extrusion pressure, and printing speed, we can improve resolution, and fine-tuning these parameters for specific hydrogel formulations can result in more precise prints.

- **Structural Integrity and Stability**: Maintaining the structural integrity and stability of printed hydrogel constructs is challenging due to their softness and susceptibility to deformation. Incorporating nanomaterials such as carbon nanotubes or graphene can enhance the mechanical properties of hydrogels without compromising their biocompatibility. These nanocomposite hydrogels exhibit improved tensile strength and elasticity. Applying post-processing techniques such as additional crosslinking, dehydration, or freeze-drying can enhance the mechanical stability of printed hydrogel structures.
- **Biocompatibility and Cell Viability**: Ensuring biocompatibility and maintaining cell viability during and after the printing process is crucial for biomedical applications yet challenging due to the potential cytotoxicity of some printing processes and materials. Using biocompatible and non-toxic hydrogels, such as those based on natural polymers like alginate, gelatin, or hyaluronic acid, can ensure cell viability. These materials support cell attachment, proliferation, and differentiation. Employing mild crosslinking methods, such as using UV light or enzymatic crosslinking, can reduce cytotoxicity and maintain cell viability within printed constructs.
- **Complex Structure Fabrication**: Creating complex, multi-material, and multi-functional structures with hydrogels can be difficult due to the limitations of current printing technologies and material properties. Developing advanced multi-material printing techniques that allow for the simultaneous deposition of different hydrogel formulations can enable the creation of complex structures with varying properties. Utilizing advanced printing technologies such as microfluidic printing or digital light processing (DLP) can enhance the ability to fabricate intricate and heterogeneous structures with high precision.

5 4D Printing of Hydrogels and Hydrogel Nanocomposites

5.1 Concept and Mechanisms

4D printing extends the capabilities of 3D printing by incorporating time as an additional dimension. It involves the creation of structures that can change their shape, properties, or functionality over time in response to external stimuli, such as temperature, pH, or light [51]. This is particularly useful for developing smart materials and devices for biomedical and soft robotics applications. 4D printing is an advanced manufacturing technology that extends 3D printing by incorporating the dimension of time, enabling printed structures to change shape, function, or properties in response to external stimuli such as temperature, pH, light, moisture, or magnetic fields. Hydrogels and hydrogel nanocomposites are particularly well-suited for 4D printing due to their responsive nature and ability to undergo significant deformation and property changes. This technology has promising applications in fields such as biomedical engineering, soft robotics, and smart materials.

5.2 *Materials and Stimuli*

Hydrogels and hydrogel nanocomposites are ideal for 4D printing due to their stimuli-responsive nature. Hydrogels that respond to specific stimuli (e.g., temperature, pH, light) are key to 4D printing. These hydrogels can swell, shrink, or change shape in response to environmental changes. Various stimuli can be used to trigger shape transformations and functional changes, including:

- **Temperature**: Thermo-responsive hydrogels can undergo phase transitions in response to temperature changes. Hydrogels like poly(N-isopropylacrylamide) (PNIPAM) can undergo a reversible phase transition at a specific temperature, leading to shape changes.
- **pH**: pH-sensitive hydrogels can swell or shrink based on the pH of the surrounding environment. Hydrogels made from materials like poly(acrylic acid) can swell or shrink in response to pH changes, enabling controlled deformation.
- **Light**: Photo-responsive hydrogels can change their properties upon exposure to specific wavelengths of light.
- **Magnetic fields**: Magnetic nanoparticles embedded in hydrogels can enable remote control of shape transformations.

5.3 *Applications*

4D printed hydrogels and nanocomposites have numerous potential applications [50], including:

- **Tissue Engineering and Regenerative Medicine**: 4D-printed hydrogels can create dynamic tissue scaffolds that mimic the natural extracellular matrix, promoting cell growth and tissue regeneration [58–60]. These scaffolds can change shape to accommodate growing tissues or respond to biological signals. For example, a 4D-printed hydrogel scaffold that changes shape in response to temperature changes, promoting cell alignment and tissue growth.
- **Drug delivery**: 4D-printed hydrogels can encapsulate drugs and release them in a controlled manner in response to specific stimuli. This allows for targeted and on-demand drug delivery [61]. A pH-responsive hydrogel that releases encapsulated drugs when it encounters the acidic environment of a tumor.
- **Soft robotics**: 4D-printed hydrogels can be used to create soft actuators and sensors that respond to environmental stimuli. These components are essential for the development of flexible and adaptive robots. A soft robotic gripper made from a temperature-responsive hydrogel that can change shape to grasp objects of different sizes and shapes.
- **Smart Materials (Self-Healing Materials)**: 4D-printed hydrogels can be designed to self-heal in response to damage, enhancing the longevity and durability of materials in various applications [62–64]. A hydrogel nanocomposite

that incorporates self-healing polymers and nanoparticles, allowing it to repair itself when exposed to moisture.

6 Recent Advancements and Case Studies

Recent advancements in hydrogels and hydrogel nanocomposites for 3D and 4D printed applications have significantly broadened the scope of these materials in various fields, including biomedical engineering, tissue engineering, drug delivery, and soft robotics. Innovations in material science, printing technologies, and application-specific designs have led to remarkable progress. Here, we discuss some notable advancements and case studies highlighting these developments.

6.1 Advanced Hydrogel Formulations

Multi-Responsive Hydrogels

Multi-responsive hydrogels can respond to multiple stimuli such as temperature, pH, light, and magnetic fields. These hydrogels enable more complex and functional 4D-printed structures. A hydrogel that responds to both temperature and pH changes, enabling precise control over its swelling and mechanical properties. This type of hydrogel can be used in smart drug delivery systems that release drugs in response to specific physiological conditions.

6.2 Nanocomposite Enhancements

Graphene Oxide (GO) Nanocomposites

Incorporating graphene oxide into hydrogels has been shown to improve mechanical strength, electrical conductivity, and biocompatibility. A study demonstrated the use of GO-enhanced hydrogels for neural tissue engineering. The GO provided the electrical conductivity necessary for neural cell function while maintaining the hydrogel's biocompatibility and structural integrity.

Silver Nanoparticles (AgNPs)

Silver nanoparticles are known for their antibacterial properties. Integrating AgNPs into hydrogels can produce materials that prevent bacterial infections, useful for medical implants and wound dressings. AgNP-loaded hydrogels were developed for wound dressings, showing effective antibacterial activity and promoting faster wound healing due to the combined effects of the hydrogel's moisture retention and the nanoparticles' antibacterial properties.

6.3 *Hybrid Printing Techniques*

Combination of 3D and 4D Printing

Integrating both 3D and 4D printing techniques allows for the creation of structures that not only have complex geometries but also dynamic functionalities. Researchers developed a hybrid printing technique that uses 3D printing to create a scaffold and 4D printing to add shape-morphing capabilities. This approach was used to fabricate a vascular stent that can expand in response to body temperature, ensuring proper fit and function within blood vessels.

6.4 *Case Studies*

Several case studies demonstrate the potential of 3D- and 4D-printed hydrogels and hydrogel nanocomposites:

- **Tissue Engineering**: Researchers at Harvard University used a combination of 3D and 4D printing to create hydrogel scaffolds with embedded vascular networks [65, 66]. These scaffolds can mimic the natural extracellular matrix and support the growth and organization of cells into functional tissues. The study demonstrated the fabrication of a hydrogel scaffold that can undergo controlled deformation to facilitate cell alignment and tissue growth. The vascular networks within the scaffold can be perfused with nutrients, promoting cell viability and tissue development.
- **Drug Delivery**: A study published in *Advanced Materials* explored the use of 4D-printed hydrogel-based drug delivery systems that can change shape and release drugs in response to specific physiological triggers [67]. The researchers developed a hydrogel nanocomposite incorporating temperature-responsive and pH-responsive components. This system was tested for its ability to release drugs in a controlled manner, demonstrating significant potential for targeted and on-demand drug delivery.
- **Soft Robotics**: A team from MIT developed self-healing hydrogel-based soft actuators that can repair themselves after being damaged. These actuators are suitable for applications in soft robotics, where durability and longevity are critical. The hydrogel actuators were embedded with microcapsules containing healing agents. Upon damage, the microcapsules ruptured, releasing the agents and initiating the healing process. The actuators regained their functionality within minutes, demonstrating robustness for repetitive use.
- **Environmental Applications-Smart Water Purification Systems**: Researchers at the University of California developed 4D-printed hydrogel nanocomposites for smart water purification systems [68]. These systems can adapt their properties in response to water contaminants, enhancing purification efficiency. These hydrogel nanocomposites incorporated magnetic nanoparticles and temperature-responsive

polymers. In the presence of contaminants, the hydrogels changed their structure to increase adsorption capacity. Magnetic fields were then used to remove the hydrogels along with the adsorbed contaminants from the water.

7 Challenges and Future Directions

7.1 Current Challenges

Despite significant progress, several challenges remain in the field of hydrogel and hydrogel nanocomposite printing:

- **Material limitations**: Developing hydrogels with the desired combination of properties (e.g., mechanical strength, biocompatibility, stimuli responsiveness) remains a challenge.
- **Printing resolution**: Achieving high-resolution and precise control over the printed structures is critical for many applications.
- **Scalability**: Scaling up the production of hydrogel-based structures for commercial applications requires further advancements in printing technologies and material formulations.
- **Long-term stability**: Ensuring the long-term stability and functionality of printed hydrogel structures, especially for biomedical applications, is essential.

7.2 Future Directions

Future research in this field is likely to focus on:

- **Novel materials**: Developing new hydrogel formulations and nanocomposite structures with enhanced and multifunctional properties.
- **Advanced printing techniques**: Innovating printing technologies to improve resolution, efficiency, and scalability.
- **Integration with other technologies**: Combining hydrogel printing with other advanced manufacturing and biomedical technologies to create more sophisticated and functional devices.
- **Clinical translation**: Conducting extensive preclinical and clinical studies to validate the safety and efficacy of hydrogel-based devices for medical applications.

8 Conclusions

Hydrogels and hydrogel nanocomposites hold immense potential for a wide range of applications, particularly in the fields of biomedicine and soft robotics. The advent of 3D and 4D printing technologies has significantly expanded their capabilities, enabling the fabrication of complex, multi-functional structures with precise control over their properties and functionalities. Despite the challenges that remain, ongoing research and technological advancements are likely to drive further innovations in this field, paving the way for new and exciting applications.

Acknowledgements The authors are thankful to the Director, Amity Institute of Applied Sciences (AIAS), and Head, Department of Chemistry, and Head Department of Physics for providing the facilities for the compilation of information. One of the authors, Tejendra K. Gupta, is thankful to the Council of Science and Technology, Lucknow, Uttar Pradesh, India, for the financial support (Vide No. CST/CHEM/D-1431).

References

1. Masuda, T., Akimoto, A.M., Yoshida, R.: 5.1—self-oscillating polymer materials. In: Ebara, M. (ed.) Biomaterials Nanoarchitectonics, pp. 219–236. William Andrew Publishing (2016)
2. Mahdavinia, G.R., Zohuriaan-Mehr, M.J., Pourjavadi, A.: Modified chitosan III, super absorbency, salt- and pH-sensitivity of smart ampholytic hydrogels from chitosan-g-PAN. Polym. Adv. Technol. **15**(4), 173–180 (2004)
3. Sivagangi Reddy, N., Rao, K.S.V.K.: Polymeric hydrogels: recent advances in toxic metal ion removal and anticancer drug delivery applications. Indian J. Adv. Chem. Sci. **4**(2), 214–234 (2016)
4. Gil, E.S., Hudson, S.M.: Stimuli-responsive polymers and their bioconjugates. Prog. Polym. Sci. **29**(12), 1173–1222 (2004)
5. Milosavljevic, N.B. et al.: Preparation and characterization of pH-sensitive hydrogels based on chitosan, itaconic acid and methacrylic acid. Polym. Int. **60**(3), 443–452 (2011)
6. Ibrahim, M.M., Hafez, S.A., Mahdy, M.M.: Organogels, hydrogels and bigels as transdermal delivery systems for diltiazem hydrochloride. Asian J. Pharm. Sci. **8**(1), 48–57 (2013)
7. Argenta, D.F. et al.: Chapter 3—hydrogel nanocomposite systems: physicochemical characterization and application for drug-delivery systems. In: Mohapatra, S.S. et al. (eds.) Nanocarriers for Drug Delivery, pp. 81–131. Elsevier (2019)
8. Baghban Salehi, M. et al.: Superabsorbent sulfonated polyacrylamide/aluminum nitrate hydrogel: swelling, mechanical, thermal and structural properties. Iran. J. Polym. Sci. Technol. **30**(5), 419–433 (2017)
9. Miguel Oliveira, J., Silva-Correia, J., Reis, R.L.: Hydrogels for Tissue Engineering and Regenerative Medicine: From Fundamentals to Applications, 1st edn. (2023, September 18). ISBN: 9780128239483, eBook ISBN: 9780128242254
10. Chaudhary, S., Chakraborty, E.: Hydrogel based tissue engineering and its future applications in personalized disease modeling and regenerative therapy. Beni-Suef Univ. J. Basic Appl. Sci. **11**(1), 1–15 (2022). https://doi.org/10.1186/s43088-021-00172-1
11. Li, L., Scheiger, J.M., Levkin, P.A.: Design and applications of photoresponsive hydrogels. Adv. Mater. **31**(26), 1807333 (2019). https://doi.org/10.1002/adma.201807333
12. Hoffman, A.S.: Hydrogels for biomedical applications. Adv. Drug Deliv. Rev. **64**, 18–23 (2012). https://doi.org/10.1016/j.addr.2012.09.010
13. Peppas, N.A., Bures, P., Leobandung, W., Ichikawa, H.: Hydrogels in pharmaceutical formulations. Eur. J. Pharm. Biopharm. **50**(1), 27–46 (2000). https://doi.org/10.1016/s0939-6411(00)00090-4. PMID: 10840191
14. Slaughter, B.V., Khurshid, S.S., Fisher, O.Z., Khademhosseini, A., Peppas, N.A.: Hydrogels in regenerative medicine. Adv. Mater. **21**(32–33), 3307–3329 (2009). https://doi.org/10.1002/adma.200802106. PMID: 20882499; PMCID: PMC4494665

15. Kumar, A.C., Erothu, H.: Synthetic polymer hydrogels. In: Biomedical Applications of Polymeric Materials and Composites, pp. 141–162. Wiley, Weinheim, Germany (2016)
16. Stevens, K.R., Miller, J.S., Blakely, B.L., Chen, C.S., Bhatia, S.N.: Degradable hydrogels derived from PEG-diacrylamide for hepatic tissue engineering. J. Biomed. Mater. Res., Part A **103**(10), 3331–3338 (2015). https://doi.org/10.1002/jbm.a.35478
17. Zhu, J.: Bioactive modification of poly(ethylene glycol) hydrogels for tissue engineering. Biomaterials **31**(17), 4639–4656 (2010). https://doi.org/10.1016/j.biomaterials.2010.02.044. Epub 2010 Mar 19. PMID: 20303169; PMCID: PMC2907908
18. Gaharwar, A.K., Peppas, N.A., Khademhosseini, A.: Nanocomposite hydrogels for biomedical applications. Biotechnol. Bioeng. **111**(3), 441–453 (2014). https://doi.org/10.1002/bit.25160. Epub 2013 Dec 6. PMID: 24264728; PMCID: PMC3924876
19. Shin, S.R., Jung, S.M., Zalabany, M., Kim, K., Zorlutuna, P., bok Kim, S., Nikkhah, M., Khabiry, M., Azize, M., Kong, J., Wan, K.-T., Palacios, T., Dokmeci, M.R., Bae, H., Tang, X., Khademhosseini, A.: Carbon-nanotube-embedded hydrogel sheets for engineering cardiac constructs and bioactuators. ACS Nano **7**(3), 2369–2380 (2013). https://doi.org/10.1021/nn305559j
20. Mandrycky, C., Wang, Z., Kim, K., Kim, D.H.: 3D bioprinting for engineering complex tissues. Biotechnol. Adv. **34**(4), 422–434 (2016). https://doi.org/10.1016/j.biotechadv.2015.12.011. Epub 2015 Dec 23. PMID: 26724184; PMCID: PMC4879088
21. Gopinathan, J., Noh, I.: Recent trends in bioinks for 3D printing. Biomater. Res. **25**(1), 12 (2021)
22. Kang, H.-W., et al.: A 3D bioprinting system to produce human-scale tissue constructs with structural integrity. Nat. Biotechnol. **34**(3), 312–319 (2016)
23. Ge, Q., et al.: 4D bioprinting: the next-generation technology for biofabrication enabled by stimuli-responsive materials. Biofabrication **12**(4), 042001 (2020)
24. Saroia, J., Yanen, W., Wei, Q., et al.: A review on biocompatibility nature of hydrogels with 3D printing techniques, tissue engineering application and its future prospective. Bio-Des. Manuf. **1**, 265–279 (2018). https://doi.org/10.1007/s42242-018-0029-7
25. Zarek, M., Mansour, N., Shapira, S., Cohn, D.: 4D printing of shape memory-based personalized endoluminal medical devices. Macromol. Rapid Commun. **38**(2), 1600628 (2016). https://doi.org/10.1002/marc.201600628
26. Gupta, N.V., Shivakumar, H.G.: Investigation of swelling behavior and mechanical properties of a pH-sensitive superporous hydrogel composite. Iran. J. Pharm. Res. (IJPR) **11**(2), 481–493 (2011). https://www.ncbi.nlm.nih.gov/pmc/articles/PMC3832170/
27. Rizwan, M., Rubina Gilani, S., Iqbal Durani, A., Naseem, S.: Materials diversity of hydrogel: synthesis, polymerization process and soil conditioning properties in agricultural field. J. Adv. Res. **33**, 15–40 (2021). https://doi.org/10.1016/j.jare.2021.03.007
28. Lin, X., Zhao, X., Xu, C., Wang, L., Xia, Y.: Progress in the mechanical enhancement of hydrogels: fabrication strategies and underlying mechanisms. J. Polym. Sci. **60**(17), 2525–2542 (2022). https://doi.org/10.1002/pol.20220154
29. Ghalia, M.A., Dahman, Y.: Chapter 6—advanced nanobiomaterials in tissue engineering: synthesis, properties, and applications. In: Grumezescu, A.M. (ed.) Nanobiomaterials in soft tissue engineering, pp. 141–172. William Andrew Publishing, Oxford, UK (2016)
30. Qi, X., Wei, W., Li, J., Liu, Y., Hu, X., Zhang, J., et al.: Fabrication and characterization of a novel anticancer drug delivery system: salecan/poly(methacrylic acid) semi-interpenetrating polymer network hydrogel. ACS Biomater. Sci. Eng. **1**(12), 1287–1299 (2015). https://doi.org/10.1021/acsbiomaterials.5b00346
31. Sathian, A., Vijay, N., Joshy, K.S., Bharat Dalvi, Y., Mraiche, F.: Hydrogels: smart materials in drug delivery. In: Hydrogels—From Tradition to Innovative Platforms with Multiple Applications. IntechOpen (2023). https://doi.org/10.5772/intechopen.104804
32. Akhtar, M.F., Hanif, M., Ranjha, N.M.: Methods of synthesis of hydrogels a review. Saudi Pharm. J. **24**(5), 554–559 (2016). https://doi.org/10.1016/j.jsps.2015.03.022
33. Shantha, K.L., Harding, D.R.K.: Synthesis and evaluation of sucrose-containing polymeric hydrogels for oral drug delivery. J. Appl. Polym. Sci. **84**, 2597–2604 (2002)

34. Haraguchi, K., Takehisa, T.: Nanocomposite hydrogels: a unique organic-inorganic network structure with extraordinary mechanical, optical, and swelling/de-swelling properties. Adv. Mater. **14**(16), 1120–1124 (2002)
35. Gyles, D.A., Castro, L.D., Silva, J.O.C., Ribeiro-Costa, R.M.: A review of the designs and prominent biomedical advances of natural and synthetic hydrogel formulations. Eur. Polym. J. **88**, 373–392 (2017)
36. West, J.L., Hubbell, J.A.: Photopolymerized hydrogel materials for drug delivery applications. React. Polym. **25**(2–3), 139–147 (1995)
37. Zhang, H., et al.: Photopolymerizable hydrogels for tissue engineering. Biomed. Mater. **14**(4), 042001 (2019)
38. Qin, G., et al.: In situ click chemistry generation of mucin mimetic poly(ethylene glycol) hydrogels for prevention of cancer cell adhesion. Biomacromolecules **16**(10), 3161–3171 (2015)
39. Xu, Y., et al.: Strong and ductile poly(vinyl alcohol)/graphene oxide composite hydrogels with a layered structure. ACS Nano **4**(7), 3869–3876 (2010)
40. Ahmadi, M., et al.: Clay-based nanocomposite hydrogels: a review on their properties and applications. J. Chem. Soc. Pak. **35**(4), 757–768 (2013)
41. Karchoubi, F., Afshar Ghotli, R., Pahlevani, H., Baghban Salehi, M.: New insights into nanocomposite hydrogels; a review on recent advances in characteristics and applications. Adv. Ind. Eng. Polym. Res. **7**(1), 54–78 (2023). https://doi.org/10.1016/j.aiepr.2023.06.002
42. Baigorria, E., Souza dos Santos, S., De Moura, M., Fraceto, L.: Nanocomposite hydrogels 3D printed for application in water remediation. Mater. Today Chem. **30**, 101559 (2023). https://doi.org/10.1016/j.mtchem.2023.101559
43. Tumbleston, J.R., et al.: Continuous liquid interface production of 3D objects. Science **347**(6228), 1349–1352 (2015)
44. Lim, K.S., et al.: New frontiers in biofabrication with stereolithography. Biomaterials **234**, 119711 (2020)
45. Xu, T., et al.: Inkjet printing of viable mammalian cells. Biomaterials **26**(1), 93–99 (2005)
46. Derby, B.: Printing and prototyping of tissues and scaffolds. Science **338**(6109), 921–926 (2012)
47. Murphy, S.V., Atala, A.: 3D bioprinting of tissues and organs. Nat. Biotechnol. **32**(8), 773–785 (2014)
48. Tao, Y., et al.: 3D printing thermoresponsive polymer materials with improved flexibility and impact resistance. Mater. Des. **134**, 21–29 (2017)
49. Truby, R.L., Lewis, J.A.: Printing soft matter in three dimensions. Nature **540**(7633), 371–378 (2016)
50. Huang, Y., et al.: Three-dimensional printing of biomimetic microstructures for cancer cell migration. ACS Nano **11**(3), 3591–3602 (2017)
51. Pelluau, T., Brossier, T., Habib, M., Sene, S., Félix, G., Larionova, J., Blanquer, S., Guari, Y.: 4D printing nanocomposite hydrogel based on PNIPAM and Prussian blue nanoparticles using stereolithography. Macromol. Mater. Eng. **309**(3), 2300305 (2024). https://doi.org/10.1002/mame.202300305
52. Highley, C.B., Rodell, C.B., Burdick, J.A.: Direct 3D printing of shear-thinning hydrogels into self-healing hydrogels. Adv. Mater. **27**(34), 5075–5079 (2015)
53. Ouyang, L., et al.: The impact of bio-ink properties and printing parameters on 3D bioprinting. J. Biomed. Mater. Res. B Appl. Biomater. **105**(5), 966–976 (2016)
54. Hospodiuk, M., et al.: The bioink: a comprehensive review on bioprintable materials. Biotechnol. Adv. **35**(2), 217–239 (2017)
55. Xu, C., et al.: Advances in 3D printing technology for biomedical applications. Engineering **5**(4), 747–757 (2019)
56. Li, J., et al.: 3D bioprinting of cellular structures with tunable mechanical properties by alginate/graphene oxide bioink. Int. J. Bioprinting **6**(1), 234–245 (2020)
57. Chimene, D., et al.: Advanced bioinks for 3D printing: a materials science perspective. Ann. Biomed. Eng. **48**(1), 200–209 (2020)

58. Miao, S., et al.: 4D printing smart materials for biomedical applications. J. Polym. Sci., Part B: Polym. Phys. **55**(3), 160–167 (2017)
59. Malda, J., et al.: 25th anniversary article: engineering hydrogels for biofabrication. Adv. Mater. **25**(36), 5011–5028 (2013)
60. Jiang, Z., et al.: Magnetic-responsive hydrogels: design, characterization, and biomedical applications. Adv. Polym. Technol. **36**(4), 361–372 (2017)
61. Wan, Z., et al.: 4D-printed shape-memory polymers for remotely controlled drug delivery. ACS Appl. Mater. Interfaces **12**(11), 12445–12452 (2020)
62. Li, V.C.F., Dunn, C.K.: 4D printing: 3D printing meets shape changing materials. Eur. Polymer J. **98**, 362–376 (2018)
63. Ge, Q., et al.: Multimaterial 4D printing with tailorable shape memory polymers. Sci. Rep. **6**, 31110 (2016)
64. Kim, Y.B., et al.: 4D printing of shape memory polymer hydrogels: fabrication and water-triggered shape reconfiguration. Mater. Des. **162**, 278–287 (2019)
65. Shin, S.R., et al.: Graphene-based materials for tissue engineering. Adv. Drug Deliv. Rev. **105**, 255–274 (2016)
66. Khorasani, M.T., et al.: Hybrid 3D and 4D printing of biomimetic scaffolds: an overview. Biomater. Sci. **9**(6), 1881–1902 (2021)
67. Yarali, E., Mirzaali, M.J., Ghalayaniesfahani, A., Accardo, A., Diaz-Payno, P.J., Zadpoor, A.A.: 4D printing for biomedical applications. Adv. Mater. 2402301. https://doi.org/10.1002/adma.202402301
68. Xie, M., et al.: 4D-printed multifunctional hydrogel structures for water purification. J. Mater. Chem. A **7**(17), 9826–9834 (2019)

3D- and 4D-Printed Polymer Nanocomposites for Electronic Applications

Srikanta Moharana

Abstract The integration of 3D and 4D printing technologies with polymer nanocomposites has emerged as a promising approach for fabricating advanced electronic devices. This typical chapter thoroughly investigates the essential elements, processing factors, and different applications of 3D and 4D printed electrically conductive polymer composites. The electrical characteristics of these materials are mainly determined by the establishment of conductive networks within the polymer matrix, which may be meticulously regulated by the judicious selection of filler materials, optimization of processing conditions, and advanced printing techniques. The integration of diverse conductive fillers, such as carbon nanotubes, graphene, and metallic nanoparticles, allows the development of materials with adjustable electrical conductivity, spanning from antistatic to extremely conductive formulations. Furthermore, the advent of 4D printing provides temporal and stimuli-responsive electrical characteristics, broadening the potential applications in intelligent electronics and adaptive systems. Despite substantial advancements in comprehending and regulating the electrical characteristics of printed polymer composites, difficulties persist in attaining uniform performance, enduring stability, and scalable production. This analysis highlights recent advancements, existing problems, and prospective prospects in this swiftly progressing domain, with a specific focus on applications in flexible electronics, biomedical electronic interfaces, and energy storage & conversion.

Keywords Polymer nanocomposites · Smart materials · Conductive polymers · 4D Printing · Electronic applications

S. Moharana (✉)
Department of Chemistry, School of Applied Sciences, Centurion University of Technology and Management, Odisha, India
e-mail: srikantanit@gmail.com

S. Moharana et al. (eds.), *Polymer Nanocomposites for 3D, 4D and 5D Printing*, Engineering Materials, https://doi.org/10.1007/978-981-96-4214-4_11

1 Introduction

The integration of modern manufacturing methods with nanomaterial science has initiated a new epoch in electronic device creation, profoundly altering our methodology for developing complex electronic systems [1]. Three-dimensional (3D) and four-dimensional (4D) printing of polymer nanocomposites represent cutting-edge technologies that combine the versatility of additive manufacturing with the exceptional properties of nanomaterials, enabling unprecedented opportunities for developing next-generation electronic devices [2]. This revolutionary approach addresses critical challenges in traditional electronics manufacturing while opening new avenues for creating adaptive, multifunctional electronic systems. The integration of nanomaterials into polymer matrices has emerged as a transformative strategy for enhancing material properties crucial for electronic applications. Carbon nanotubes, graphene, metallic nanoparticles, and ceramic nanofillers demonstrate remarkable capabilities in modifying the electrical conductivity, thermal stability, and mechanical properties of polymer systems [3]. These enhancements are achieved through complex interactions at the nanoscale, enabling the development of materials with precisely tailored characteristics for specific electronic applications. The synergistic combination of different nanomaterials within polymer matrices creates hybrid systems that exhibit properties surpassing those of individual components [4].

Advanced manufacturing methods, especially 3D printing technology, have transformed the production of polymer nanocomposites by allowing meticulous control of material composition and architecture across various length scales [5]. Conventional manufacturing limitations are progressively being surmounted by new printing techniques, such as fused deposition modelling (FDM), stereolithography (SLA), and selective laser sintering (SLS). These approaches enable the fabrication of intricate geometries and functional structures that were previously unattainable by traditional manufacturing processes [6]. The advent of 4D printing technology signifies a transformative change in material engineering by including time as the fourth dimension via the creation of stimuli-responsive materials [7]. This innovation facilitates the development of intelligent electronic components that may modify their characteristics or configurations in reaction to environmental variations, like temperature, humidity, light, or electrical fields. The amalgamation of shape-memory polymers with responsive nanomaterials has resulted in the creation of adaptive electrical devices with unparalleled functionality [8].

The use of 3D and 4D printed polymer nanocomposites in electronics encompasses a wide range, including flexible sensors, wearable devices, energy storage systems, and biomedical electronics [9]. These materials have distinct benefits for customisation, miniaturisation, and functionality, rendering them especially appropriate for nascent technologies such as Internet of Things (IoT) devices, artificial intelligence hardware, and bio-integrated electronics [10]. Notwithstanding considerable progress, several obstacles persist in the creation and execution of these materials. These include attaining uniform dispersion of nanomaterials, optimizing interfacial interactions, assuring repeatability in manufacturing processes, and preserving constant electrical characteristics across various scales [11]. Moreover, three-dimensional (3D) printing technology has revolutionised fast prototyping by fabricating intricate functional structures layer by layer. While this method facilitates economical bespoke manufacture in several domains, traditional 3D printing with standard polymers produces components that often lack the requisite strength and functionality for sophisticated applications [12]. Researchers have addressed these limits by integrating smart materials such as multi-functional thermally responsive polymers (TRPs) into the printing process, facilitating the efficient manufacture of functional soft systems. In 2013, the Massachusetts Institute of Technology (MIT) made a big advancement in the discipline by introducing 4D printing, which incorporates a temporal "space–time axis" into conventional 3D printing. 4D printing, using similar digital design and production techniques, differentiates itself with the use of smart materials and intelligent design, enabling printed things to adaptively alter form or function in reaction to external stimuli (as seen in Fig. 1).

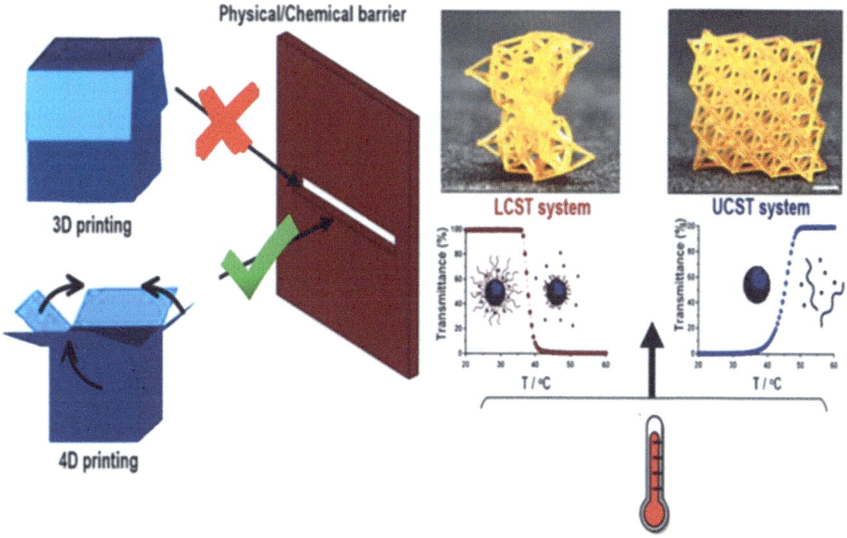

Fig. 1 Fundamental distinctions between 3 and 4D printing [12]. Copyright 2023. Reproduced with permission from Elsevier

2 Nanomaterials in Polymer Composites

2.1 *Types of Nanomaterials*

The use of nanoparticles in polymer matrices has transformed composite material creation, providing exceptional prospects for property improvement and functional capabilities. Carbon-based nanomaterials, such as carbon nanotubes (CNTs), graphene, and fullerenes, constitute a key category of reinforcing agents [13]. These materials have outstanding mechanical capabilities, with Young's modulus values above 1 TPa and intrinsic electrical conductivity akin to those of metals. Multi-walled carbon nanotubes (MWCNTs) and single-walled carbon nanotubes (SWCNTs) have distinct benefits, as MWCNTs deliver improved mechanical reinforcement while SWCNTs exhibit exceptional electrical qualities [14]. Metallic nanomaterials represent a significant group, including noble metal nanoparticles (gold, silver), transition metal oxides, and metallic nanowires. These materials impart distinctive characteristics, including plasmonic effects, catalytic activity, and improved electrical conductivity [15]. Silver nanoparticles are claimed to provide antimicrobial properties and electrical conductivity to polymer composites, making them highly suitable for biomedical applications [16]. Nonetheless, ceramic nanoparticles such as silica, alumina, and titanium dioxide provide improved mechanical characteristics and thermal stability to polymer composites. These compounds often work as multifunctional additives, enhancing mechanical strength, heat resistance, and barrier qualities concurrently [17].

2.2 *Structural Properties*

The structural characteristics of nanomaterials profoundly affect their interaction with polymer matrices and the resultant properties of the composites. Morphological characteristics, including aspect ratio, surface area, and particle size distribution, are essential in defining the ultimate material attributes [18]. High-aspect-ratio nanomaterials, such as carbon nanotubes and graphene nanoplatelets, may establish efficient reinforcing networks at low concentrations, resulting in considerable enhancements in properties [19]. Surface chemistry is a vital structural element, that influences the compatibility between nanomaterials and polymer matrices. Techniques for surface functionalization have been devised to alter nanomaterial surfaces, improving their dispersion and interfacial interactions [20]. Studies indicate that regulated surface modification may enhance mechanical characteristics by as much as 200% relative to unaltered systems [21]. The interaction mechanisms between nanoparticles and polymer matrices transpire over numerous length scales and include diverse physical and chemical phenomena. Interfacial interactions, such as covalent bonding, hydrogen bonding, and van der Waals forces, influence the efficacy of load transmission and the overall performance of composites [22]. Research using sophisticated characterization methods has shown that the development of an interphase zone, with characteristics different from both the bulk polymer and the nanomaterial, considerably affects composite performance [23]. A dual-scale porous polymer based on Mxene was 3D printed by Yu and his coworkers for use in a flexible pressure sensor [24]. Figure 2 demonstrates that the crosslinked network of the porous polymer develops tiny holes as a consequence of phase separation, with an accumulation of haphazardly interlinked spherical particles. The increased phase separation due to higher PEGDA molecular weight and porogen concentration likely results in AP600 exhibiting more interconnected spherical particles and pore architectures compared to AP400 (as shown in Fig. 2a, b). However, the particle dimensions ranging from 1.0 to 1.6 μm were identified via the analysis of images obtained by scanning electron microscopy. The dual-scale porous polymer seen in Fig. 2c was made using a decreased amount of porogen, which is 0.75 times the total mass of PEGDA and PETMP. In contrast to the usual porous polymer structure, the top part has a gel-like shape and a thick surface texture without small holes [24].

Figure 3 shows the structural characteristics and production possibilities of polymers with two scales of porosity. Figure 3a shows an optical image of macroscopic holes with different numbers of them, whereas Fig. 3b shows that they are of the same

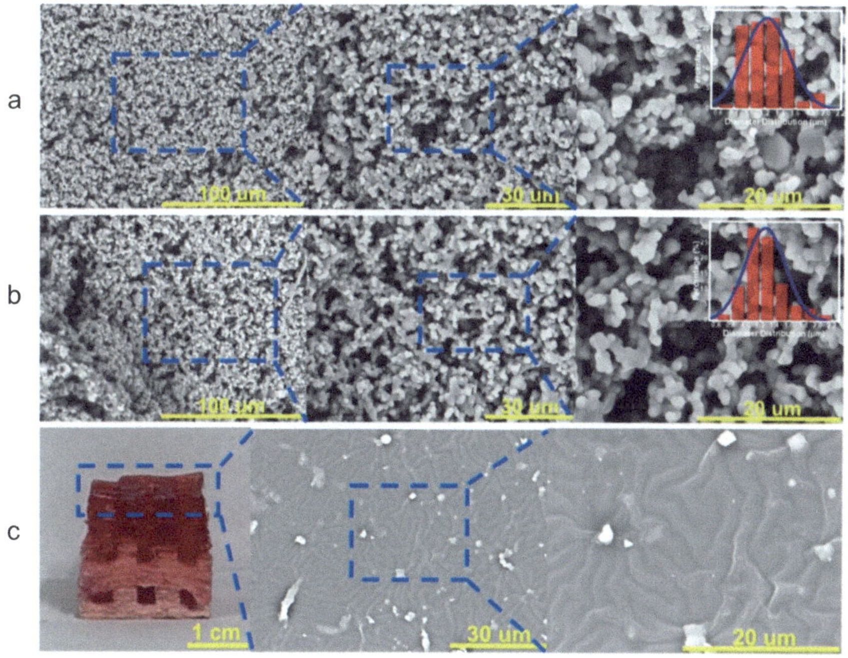

Fig. 2 SEM morphological analysis of phase-separated polymer networks: **a** AP400 porous structure with interconnected spherical domains, **b** AP600 showing increased pore density and particle interconnectivity, and **c** dual-scale morphology resulting from reduced porogen concentration [24]. Copyright 2024. Reproduced with permission from Elsevier

size. In Fig. 3c, we can see that DLP-3D printing produces more precise structures and higher-quality surfaces than the stencil technique, which results in noticeable flaws including fractures, bends, shrinkage, and surface abnormalities. Optical pictures of complicated geometric structures added to DLP-3D printing's adaptability are shown in Fig. 3d, highlighting the technology's benefits in precision production and the creation of elaborate three-dimensional structures.

Bodaghi and his coworkers have developed the 4D printing of polylactic acid (PLA)-polybutylene adipate terephthalate (PBAT) composites using sustainable shape memory polymer [25]. Figure 4 shows scanning electron micrographs (SEM) of PLA-PBAT blends that display significant morphological modifications. The mechanical properties of PLA-PBAT-15% are influenced by interlayer voids dispersed throughout the matrix (Fig. 4a–c). On the other hand, PLA-PBAT-30% shows much less void formation, especially at 50 × magnification, which is probably because the blend's greater volume expansion coefficient improves layer adhesion.

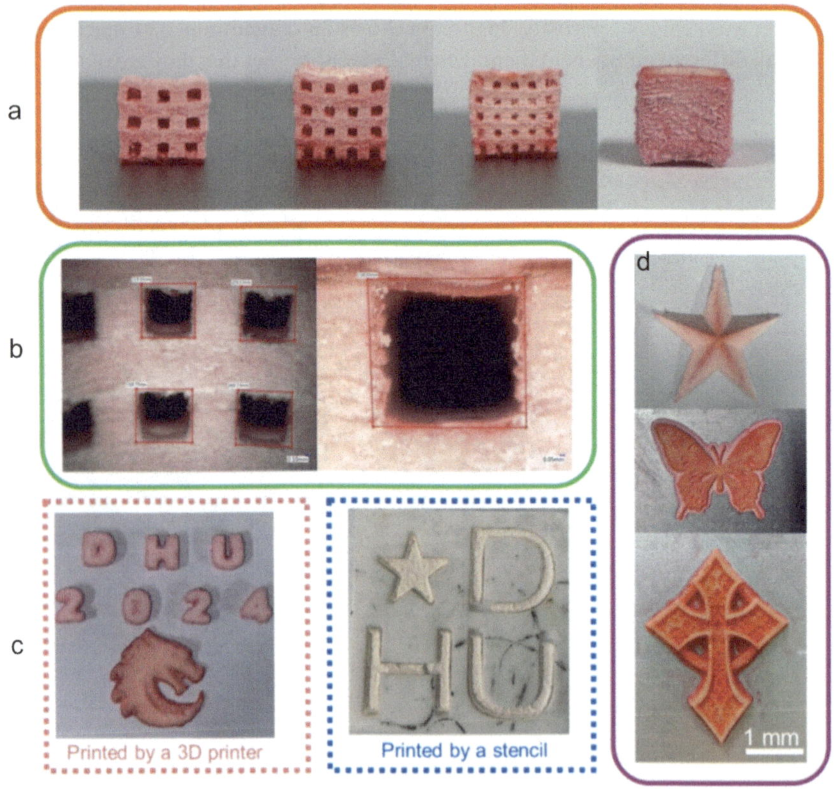

Fig. 3 Porous polymers' structural characterization: **a** different pore densities, **b** study using ultra-field microscopic techniques, **c** comparison of DLP-3D printing with the stencil approach, and **d** complicated geometric designs [24]. Copyright 2024. Reproduced with permission from Elsevier

At 100 × magnification, the presence of smaller but more frequent voids in PLA-PBAT-45% (as shown in Fig. 4 g–i) becomes most apparent, indicating that the PBAT is dispersed finer within the PLA matrix and that phase separation effects are more pronounced [25].

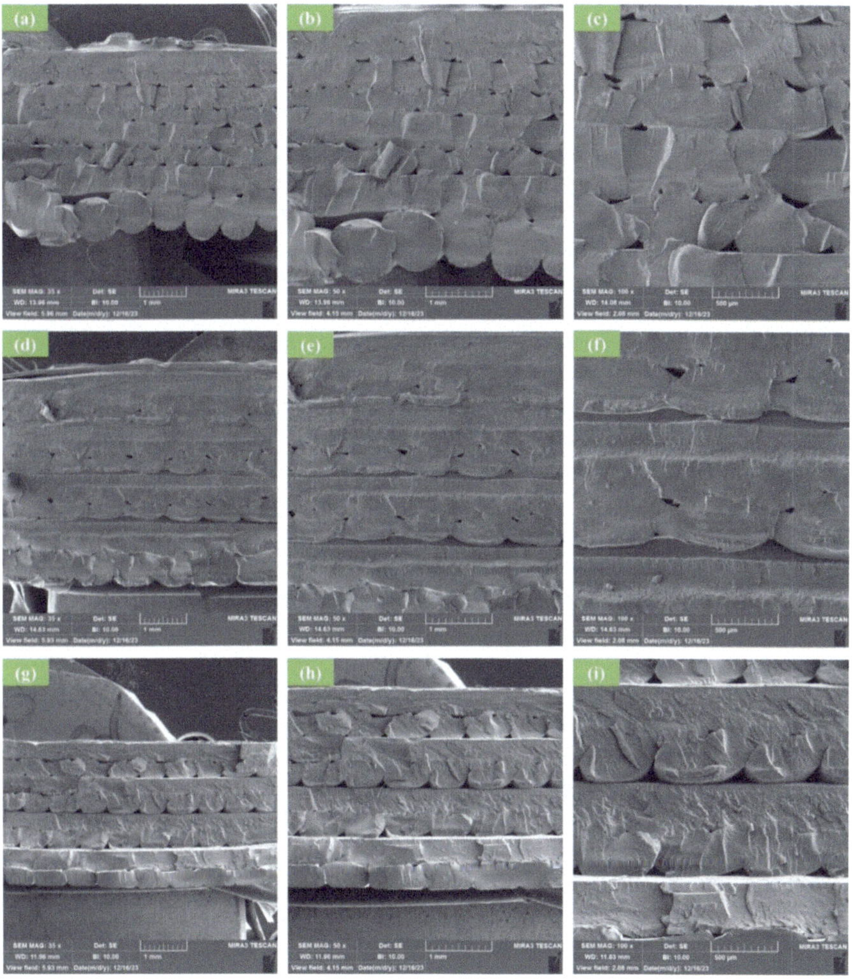

Fig. 4 SEM micrographs of 3D and 4D printed PLA-PBAT blends: **a–c** PLA-PBAT 85/15, **d–f** PLA-PBAT 70/30, and **g–i** PLA-PBAT 55/45 at magnifications of 35 ×, 50 ×, and 100 ×, respectively [25]. Copyright 2024. Reproduced with permission from Elsevier

3 Manufacturing Technologies in Polymer Nanocomposites

3.1 3D Printing Techniques

Fused Deposition Modeling (FDM) has emerged as a versatile technique for fabricating polymer nanocomposites with complex geometries. This process involves the extrusion of thermoplastic nanocomposite filaments through a heated nozzle, enabling layer-by-layer construction of three-dimensional structures [26]. Recent developments in FDM technology have focused on optimizing processing parameters to enhance the alignment of nanomaterials during printing. However, it has been demonstrated that controlled shear forces during extrusion can improve the orientation of carbon nanotubes and graphene platelets, leading to enhanced mechanical and electrical properties in printed components [27]. Stereolithography (SLA) represents a high-resolution printing technique particularly suitable for creating precise structures with photocurable polymer nanocomposites. The process utilizes photopolymerization reactions initiated by focused light to create complex structures with feature sizes down to tens of micrometers [28]. Advanced research has shown that the incorporation of functionalized nanomaterials can enhance both the mechanical properties and cure kinetics of photopolymer systems. Studies have demonstrated that modified ceramic nanoparticles can improve the resolution and surface quality of printed parts while maintaining structural integrity [29]. Selective Laser Sintering (SLS) offers unique advantages for processing polymer nanocomposites, particularly those containing thermally conductive nanomaterials. The technique enables the creation of complex geometries through selective fusion of powder particles using focused laser energy [30]. Recent developments have shown that the incorporation of carbon nanotubes and graphene can significantly improve the thermal conductivity and sintering efficiency of polymer powders, leading to enhanced part quality and reduced processing times [31]. Material extrusion methods have evolved to include direct ink writing and robotic deposition techniques, enabling the fabrication of structures with controlled material placement and orientation. These methods often utilize specially formulated nanocomposite inks with optimized rheological properties [32]. Polymer-based additive manufacturing includes several technologies: FDM, which depends on thermoplastic filament extrusion; SLA/DLP, which employs photopolymerization of liquid resins; DIW, which facilitates direct extrusion of viscous materials; and SLS/SLM, which utilizes powder bed fusion via laser sintering (as shown in Fig. 5a–d). These methods have progressed from basic one-dimensional production to intricate three-dimensional printing, providing exact control over spatial configuration and material characteristics. The advent of 4D printing (as shown in Fig. 5e) enhances these capabilities by integrating time-dependent responsiveness, enabling printed objects to alter their characteristics or configurations in reaction to external stimuli, hence creating new opportunities in smart materials and adaptable structures [33].

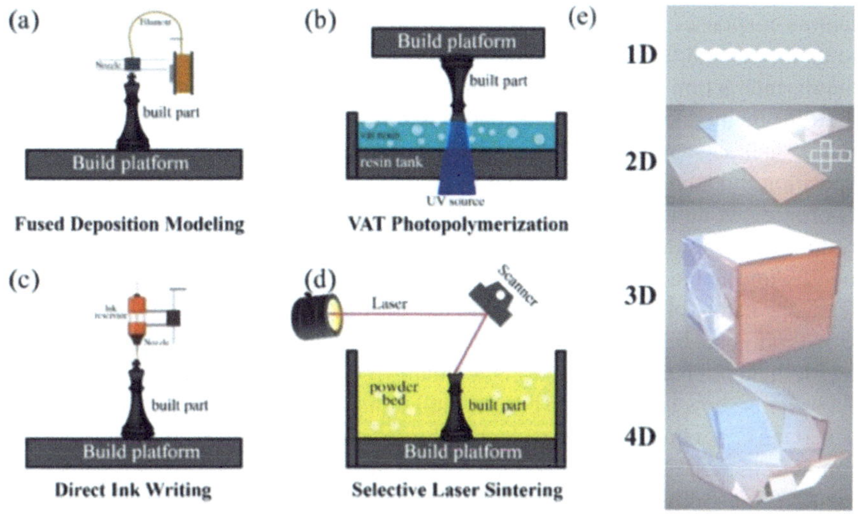

Fig. 5 Overview of polymer-based additive manufacturing: **a** Schematic representation of principal 3D printing techniques: Fused Deposition Modeling (FDM) (**a**), Vat Photopolymerization (SLA/DLP) (**b**), Direct Ink Writing (DIW) (**c**), and Selective Laser Sintering (SLS/SLM) (**d**) and (**e**) Evolution from one-dimensional to four-dimensional printing, highlighting the integration of time-dependent properties in 4D printing [33]. Copyright 2024. Open access. Reproduced with permission from Kufa Journal of Engineering

4 4D Printing Fundamentals

Stimuli-responsive materials represent the foundation of 4D printing technology, enabling printed structures to change their configuration or properties in response to external stimuli. The integration of shape memory polymers with functional nanomaterials has enabled the development of smart composites capable of programmable responses to temperature, light, or electromagnetic fields [34]. Studies have shown that the incorporation of magnetic nanoparticles can enable remote activation of shape changes through applied magnetic fields [35]. Programming techniques for 4D printed structures involve sophisticated methods for encoding desired response behaviors into the printed components. This includes strategic placement of responsive materials, creation of anisotropic structures, and development of multi-material systems with programmed strain distributions [36]. Research has demonstrated that careful control of printing parameters and material composition can achieve complex transformations, including sequential folding and unfolding behaviors [37]. Shape memory mechanisms in 4D printed nanocomposites involve complex interactions between the polymer matrix and incorporated nanomaterials. The presence of nanomaterials can significantly influence the shape memory performance through modifications to the polymer network structure and thermal properties [38]. Advanced characterization techniques have revealed that nanomaterial alignment and distribution play crucial roles in determining the efficiency and repeatability of shape memory responses [39]. 4D printing is largely reliant on five interrelated factors: the additive manufacturing (AM) process, selection of printing materials, applied stimuli, interaction mechanisms, and mathematical modeling. The additive manufacturing method converts digital data directly into printed structures with techniques such as stereolithography (SLA), selective laser sintering (SLS), or fused deposition modeling (FDM) (shown in Fig. 5). These techniques may be used for 4D printing, contingent upon the compatibility of the material with the printer. The printing material must exhibit responsiveness to stimuli while preserving structural integrity during the layer-by-layer production process. These smart materials (SMs) dictate the nature of the needed stimulus and the object's self-transformation capacities. Stimuli may be categorized as physical (temperature, light, moisture, magnetic/electric fields), chemical (pH levels, oxidants/reductants), or biological (enzymes, glucose) (as shown in Fig. 6). These stimuli induce physical or chemical alterations such as tension relaxation, molecule mobility, or phase transitions, resulting in structural deformation. The latter two elements—interaction mechanism and mathematical modeling—direct the transformation process. The interaction method, including mechanical loading or physical manipulation, governs the order of shape alterations, whilst mathematical modeling establishes the time and intensity of stimulus delivery required for the intended transformations [40].

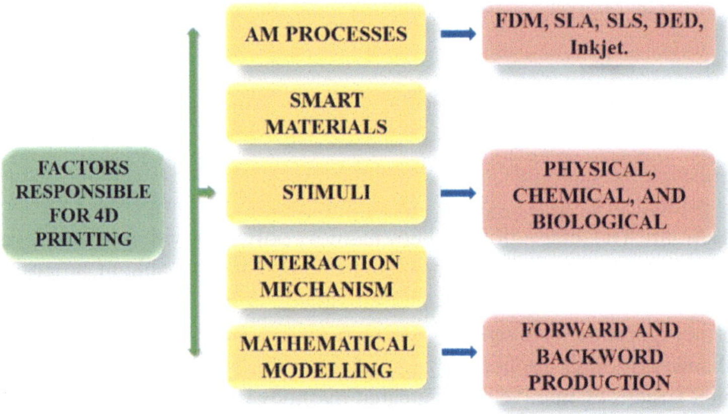

Fig. 6 Diagrammatic illustration of the components involved in 4D printing [40]. Copyright 2021. Reproduced with permission from Elsevier

5 Electrical Properties of 3D and 4D Printing of Polymer Composites

The electrical characteristics of 3D- and 4D-printed polymer composites constitute a vital domain of inquiry and innovation in advanced materials science, with unparalleled prospects for the fabrication of functioning electronic devices and intelligent structures. The integration of electrically conductive fillers into polymer matrices by additive manufacturing has facilitated the creation of materials with adjustable electrical conductivity, leading to novel applications in electronics, sensors, and smart devices [41]. The essential electrical properties of polymer composites are determined by the establishment of conductive networks within the polymer matrix. The incorporation of conductive fillers, such as carbon nanotubes, graphene, or metallic nanoparticles, into the polymer generates linked channels that enhance electron transport across the normally insulating polymer matrix. The efficacy of these conductive networks is fundamentally contingent upon many elements, including filler content, dispersion quality, and the particular processing conditions used throughout the printing process. However, the percolation threshold phenomenon plays a crucial role in determining the electrical behavior of nanocomposite systems. This critical concentration represents the point at which conductive networks form continuous pathways through the polymer matrix, resulting in a dramatic increase in electrical conductivity [42]. Studies have revealed that the percolation threshold can be influenced by multiple factors, including nanomaterial aspect ratio, dispersion quality, and processing conditions. However, it is reported that some nanocomposite systems achieve percolation at concentrations as low as 0.1 wt%, significantly lower than traditional conductive fillers [43].

The advancement of 4D-printed polymer composites adds complexity to the evaluation of electrical properties. These materials, engineered to alter their properties or form in reaction to external stimuli, may demonstrate dynamic electrical characteristics. The capacity to construct structures with programmable electrical characteristics that react to environmental factors like temperature, humidity, or electric fields presents new opportunities for intelligent electronic devices and adaptable systems [44]. The use of these materials in flexible electronics signifies a notably promising avenue. The integration of mechanical flexibility with electrical conductivity facilitates the development of wearable sensors, flexible displays, and adaptable electronic gadgets. The capacity to optimize both mechanical and electrical characteristics via meticulous regulation of material composition and printing settings makes these materials especially appropriate for such applications [45].

6 Thermal and Mechanical Properties

The thermal properties of polymer nanocomposites exhibit intricate relationships with their structure and composition. Thermal stability represents a critical parameter for electronic applications, particularly in high-power devices where heat management is crucial [46]. Nanomaterials such as graphene and ceramic nanoparticles have demonstrated remarkable capabilities in enhancing the thermal degradation temperature of polymer matrices. Research has shown that properly functionalized graphene can increase the thermal decomposition temperature by up to 50 °C while simultaneously improving char formation [47]. Mechanical strength in nanocomposite systems stems from complex interactions between nanomaterials and polymer chains. The incorporation of high-aspect-ratio nanomaterials creates multiscale reinforcement mechanisms that significantly enhance mechanical properties [48]. Carbon nanotubes and graphene, with their exceptional intrinsic strength, contribute to improved tensile strength and modulus through effective load transfer mechanisms. Studies have demonstrated that optimal interfacial bonding can lead to improvements in Young's modulus by up to 300% at relatively low filler concentrations [49]. Thermal conductivity mechanisms in polymer nanocomposites involve sophisticated heat transfer pathways that differ significantly from conventional materials. The creation of thermally conductive networks through strategic nanomaterial alignment and distribution enables efficient heat dissipation crucial for electronic applications [50]. Advanced research has revealed that three-dimensional thermal transport networks can be engineered through careful control of processing parameters and nanomaterial orientation [51]. Fallah and his coworkers investigated the mechanical characteristics of 4D-printed fiber-reinforced PETG composites [52]. Figure 7a displays the tensile strength and modulus results for CCFRP samples over nine trials with different print conditions, including pure PETG as a baseline for comparison. The integration of continuous carbon fiber reinforcement markedly improved the elastic modulus and strength of the PETG matrix. Trials 1 and 7 exhibited exceptional performance, showcasing the superior mechanical characteristics. Trial 1 attained an elastic modulus of 6.89 GPa, indicating a 474% enhancement over unmodified PETG, whilst Trial 7 obtained 6.24 GPa, reflecting a 420% advancement. The tensile strength exhibited comparable improvements, with Trial 1 attaining 129.67 MPa (a 253% increase over pure PETG) and Trial 7 reaching 100.15 MPa (a 173% enhancement). Figure 7b illustrates the tensile stress–strain curves for specimens from Trial 1, Trial 7, and pure PETG. The stress–strain behavior shown by the PETG samples corresponds with the typical response pattern recorded in prior study literature [52].

Kholkhoev and his co-workers [53] have revealed the development of a mechanically robust, high-temperature shape memory polymer by 4D printing for use in the aerospace sector. Figure 8a shows the findings of the thermogravimetric study, which show that the formulated composition allows for the formation of three-dimensional structures with impressive thermal stability. The non-heat-treated samples (MPA-GS) achieved a 10% weight loss temperature ($T_{10\%}$) of 378 °C. As a result of the more thorough conversion of the reactive components, the material shows a little

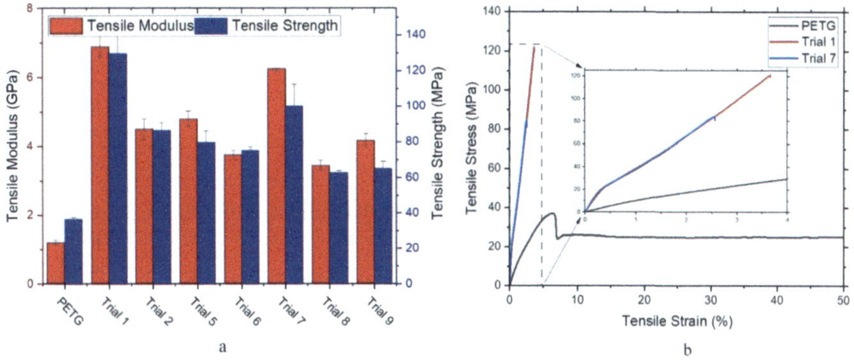

Fig. 7 Mechanical characterization of 4D-printed CCFRP specimens compared to pure PETG: **a** Comparison of tensile strength and elastic modulus; **b** Representative stress–strain curves from optimal specimens [52]. Copyright 2024. Reproduced with permission from Elsevier

improvement in thermal resistance after thermal post-treatment, with T10% climbing to 382 °C. Before and after being exposed to γ-radiation, Fig. 8b shows a comparison of the thermogravimetric behavior of MPA-250 samples. Specimens exposed to radiation had a somewhat lower T10% than the untreated MPA-250, but they nevertheless showed strong thermal stability with thermal decomposition temperatures between 373 and 379 °C.

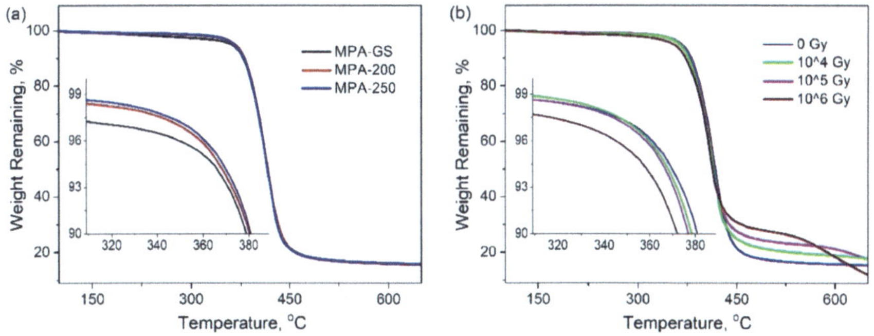

Fig. 8 Thermogravimetric analysis (TGA) results: **a** Thermal decomposition behavior of MPA-derived materials and **b** Effect of γ-radiation exposure on MPA-250 thermal stability [53]. Copyright 2024. Reproduced with permission from Elsevier

7 Electronic Applications of 3D- and 4D-Printed Polymer Nanocomposites

7.1 Flexible Electronics

The domain of flexible electronics signifies a groundbreaking progression in electronic device engineering, especially through the use of 3D- and 4D-printed polymer nanocomposites. Wearable devices have become a principal application area, whereby these materials enable the development of flexible, skin-like electronic systems [54]. The integration of carbon nanotubes and graphene into elastic polymer matrices has enabled the creation of sensors that can monitor physiological parameters while preserving mechanical flexibility and durability [55]. Recent studies have shown that wearable electronics may endure strain tolerances beyond 100% while preserving electrical performance, an essential need for body-mounted applications. Stretchable circuits signify a significant breakthrough in flexible electronics. Researchers have attained significant enhancements in circuit sustainability under mechanical deformation through the strategic design of conductor layouts and the selection of materials [56]. The advancement of serpentine configurations and mesh architectures, coupled with highly conductive polymer nanocomposites, has facilitated the fabrication of electronic circuits that retain operation even under severe tensile circumstances. Research indicates that optimized geometric patterns may attain stretchability of up to 300% while preserving electrical conductivity [57]. Adaptive sensor networks using polymer nanocomposites have transformed environmental and structural monitoring capacities. These networks integrate numerous sensing modalities by strategically combining different nanomaterials, allowing the simultaneous detection of many physical and chemical characteristics [58]. The self-repairing properties of some nanocomposite systems have improved the durability and dependability of these sensor networks, especially under extreme climatic circumstances.

Figure 9a illustrates that in the absence of external pressure, the unique porous structure of the AP porous polymer impedes current passage, leading to elevated electrical resistance. The resistance signal is reduced when the porous network gradually compresses under pressure, resulting in improved conductive pathways. Because it can deform under pressure and completely restore its original shape without permanent deformation when loaded with MXene, the dual-scale porous polymer has remarkable mechanical capabilities; its compression behavior is shown in Fig. 9b. The experimental setup is shown in Fig. 9c, which includes a light bulb, a DC power supply, and an MXene-loaded dual-scale porous polymer. There are various pictures (Fig. 9b, c) showing the changes in light output as a function of various compression loads [24, 59].

Roach and his coworkers created a hybrid additive manufacturing platform that combines several materials and printing methods for advanced 3D-printed structures [60]. The preliminary design components are shown in Fig. 10a, b, depicting a cross-sectional schematic and a CAD representation, respectively. These depict the

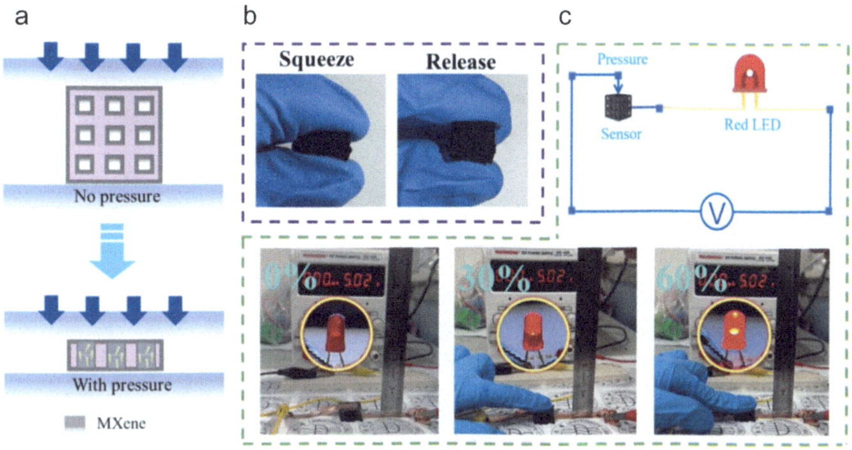

Fig. 9 Compression sensing demonstration and response: **a** Conceptual illustration of the compression sensing mechanism; **b** Visual documentation of the MXene-loaded porous polymer under compression and release cycles; **c** Circuit configuration and real-time LED brightness variation under applied compressive stress [24]. Copyright 2024. Reproduced with permission from Elsevier

three primary components: the PLA framework, silver electrode wires, and the LED chip. The complete fabrication process is documented in Fig. 10c–j. However, the manufacturing process began with FFF printing of the PLA structure, which incorporated channels for silver ink and a designated pocket for the LED chip (Fig. 10c). The platform then relocated to the pick-and-place station, where a 6-axis robotic arm retrieved (Fig. 10d) and positioned (Fig. 10e) the LED chip within the PLA structure. Following this, the platform returned to the extrusion module where conductive silver ink was deposited via DIW printing into the channels, establishing connections for the external power supply (as shown in Fig. 10f). A significant challenge emerged when printing the conductive ink. While previous applications on flat substrates had achieved good conductivity and adhesion, the rough FFF substrate presented complications. The silver ink would seep into gaps between PLA traces, compromising conductivity. This was addressed by applying multiple ink layers until all cracks were filled. The silver traces were then treated using a Xenon photonic curing module, resulting in an estimated resistance of 10 Ω. The concluding production procedures included enclosing the LED with further PLA layers using the FFF printer (Fig. 10g–h). External wires were then put into the silver electrode channels to facilitate power hookup (Fig. 10i). The gadget was effectively lighted upon activation (Fig. 10j), with brightness modifiable via current regulation. The operating conditions were established at 3 V with a maximum current of 1.5 A [60].

Leng and his co-workers have created 4D-printed bioinspired mesh composites exhibiting significant stretchability [61]. The incorporation of 4D printed MCMs (Magnetically-driven Composite Materials) into flexible electronics signifies a notable progression, especially owing to their remarkable stretchability and compatibility with biomaterials in terms of mechanical characteristics. Figure 11

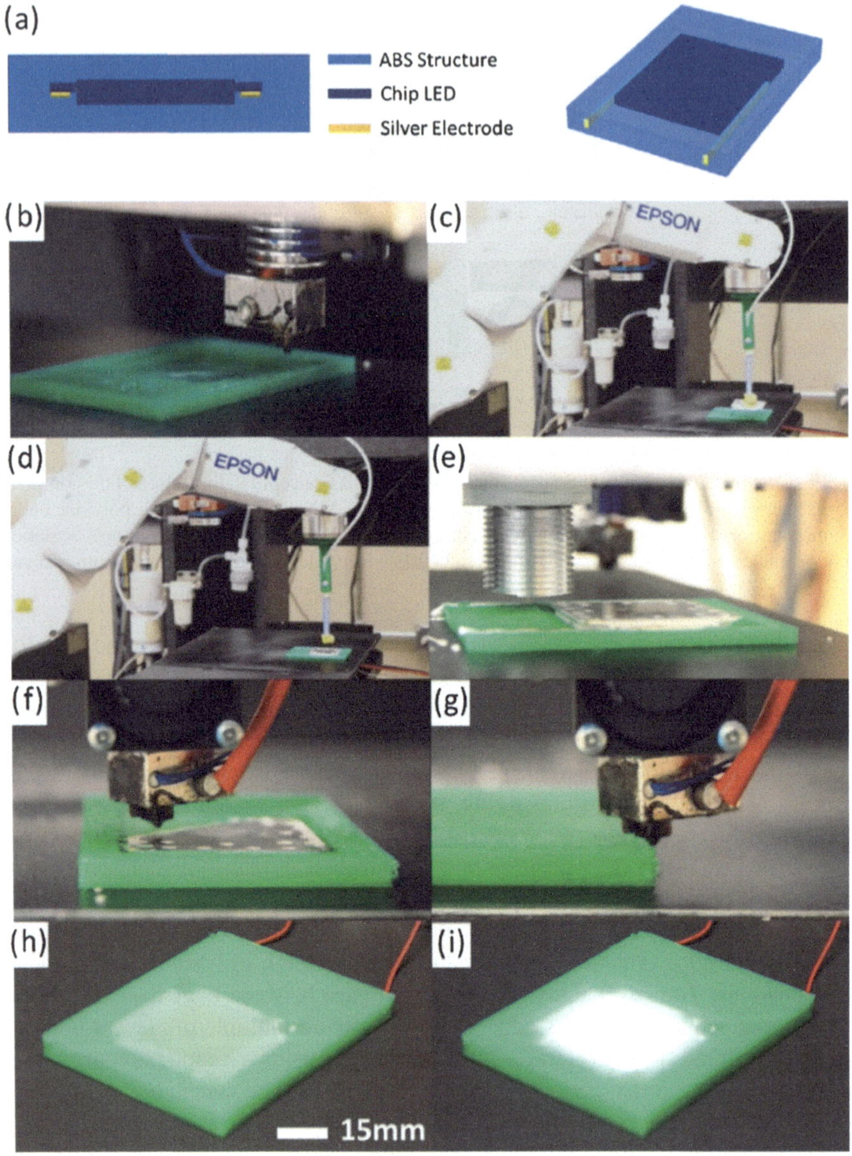
(a)
ABS Structure
Chip LED
Silver Electrode
(b)
(c)
EPSON
(d)
EPSON
(e)
(f)
(g)
(h)
(i)
15mm

◂**Fig. 10** Fabrication sequence of an integrated LED flashlight using hybrid manufacturing **a** Cross-sectional diagram showing the three main components of the digital LED assembly, **b** Three-dimensional isometric view illustrating the complete LED light design, **c** Initial phase: Fused filament fabrication (FFF) process creating the base flashlight housing, **d** Component integration: 6-axis Epson robotic arm initiating the LED pickup sequence, **e** Precise placement of the LED component by the robotic system into the partially printed structure, **f** Application of conductive pathways using direct ink writing (DIW) of silver-based conductive traces, **g**, **h** Final assembly stages showing the encapsulation of the LED component through additional FFF printed layers, **i** Completed LED light assembly after all manufacturing steps and **j** Final product demonstration: Fully assembled flashlight showing successful illumination when connected to power [60]. Copyright 2019. Reproduced with permission from Elsevier

illustrates the practical use of these materials in a stretchy LED electrical gadget. The device design, seen in Fig. 11a, consists of an MCM substrate combined with several LEDs, a cured silver paste creating a conductive layer, and an external DC power source. The system's mechanical integrity is preserved owing to the substantial thickness disparity between the conductive silver film and the MCM substrate, with the former being several orders of magnitude thinner. Moreover, the MCM substrate undergoes little real strain during stretching processes, therefore averting extensive breakdown of the conductive silver coating and guaranteeing device dependability. The performance of the LED electrical gadget was assessed by stretching tests in underpowered settings, concurrently measuring resistance variations. Figure 11b illustrates that the device maintains its performance when elongated from its original condition to a 10% strain, with all three LEDs maintaining their brightness levels. The electrical stability is shown by the steady resistance measurement of around 60 Ω throughout the stretching process, as seen in Fig. 11c. The versatile characteristics of 4D printed MCMs enable the adjustment of mechanical properties to align with biological tissues like human skin, making these devices especially appropriate for human movement monitoring applications, providing benefits over conventional complicated electronic systems (as shown in Fig. 11d). This technology's adaptability surpasses LED circuits, since the exhibited platform may integrate many electronic components often used in stretchy devices and systems [61].

8 Biomedical Electronic Interfaces

The use of polymer nanocomposites in biomedical electronics has established new avenues in healthcare monitoring and therapy. Implantable sensors using these materials demonstrate improved biocompatibility and prolonged durability under physiological conditions [62]. Advanced research has demonstrated the development of glucose sensors utilizing enzyme-functionalized nanocomposites that maintain sensitivity for extended periods while minimizing foreign body responses [63]. Neural interfaces represent a particularly challenging application domain where polymer nanocomposites have shown remarkable promise. The development of flexible neural electrodes with improved signal-to-noise ratios and reduced tissue damage

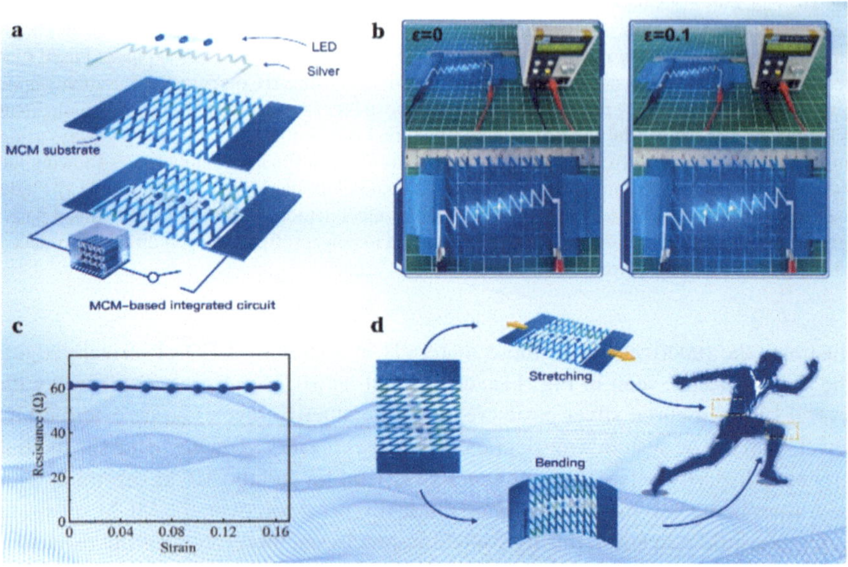

Fig. 11 4D printed MCM-based stretchable electronics: **a** device composition, **b** strain-dependent LED performance, **c** electrical resistance characterization, and **d** wearable sensing demonstration [61]. Copyright 2024. Reproduced with permission from Elsevier

has been achieved through careful materials engineering [64]. Studies have shown that graphene-based nanocomposites can effectively interface with neural tissue while maintaining mechanical compliance matching that of brain tissue, significantly reducing inflammatory responses [65]. Diagnostic platforms utilizing 3D-printed polymer nanocomposites have emerged as powerful tools for rapid medical testing. These platforms integrate multiple functionalities, including sample preparation, analysis, and detection, into single devices [66]. The ability to precisely control material properties and device architecture through 3D printing has enabled the development of highly sensitive and specific diagnostic tools. A 4D-printed shape memory precursor that can be reconfigured and programmed for use in sensing and actuators was developed by Meng and his co-workers [67]. The programmable nature of polymer-derived ceramic (PDC) precursors enables advanced engineering applications beyond sensing capabilities. Figure 12 illustrates the fabrication and implementation of a 4D-printed precursor hinge with integrated resistance circuits for Joule heating-activated shape memory recovery. The fabrication process (Fig. 12a) involves sequential steps: laser-cut polyimide mask preparation (Fig. 12ai–ii), conductive ink patterning (Fig. 12iii), mask removal (Fig. 11aiv), thermal treatment (Fig. 12av), and electrode connection (Fig. 12avi). The thin circuit layer, negligible in thickness and weight compared to the hinge, maintains adaptability during deformation. As shown in Fig. 12c–e, the straight hinge can be programmed to a folded configuration and recover its original shape when activated by a 30-mA current, with uniform heat distribution in the bent zone. The practical application is demonstrated through

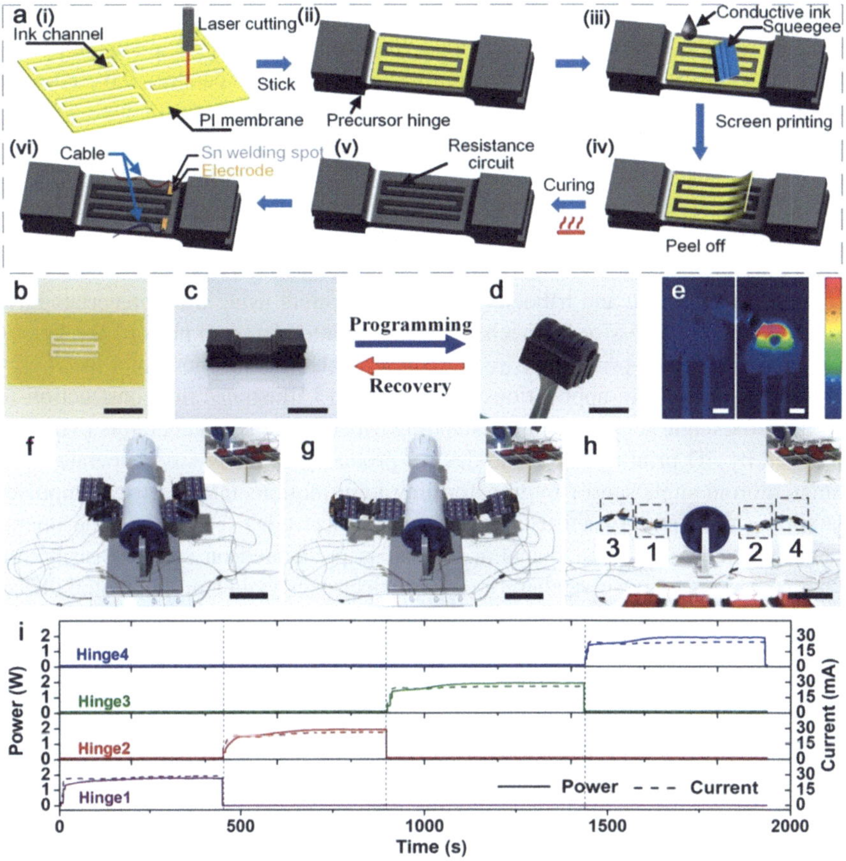

Fig. 12 Shape memory precursor-based deployable system: **a** circuit integration, **b**–**e** hinge programming and activation, **f**–**h** sequential deployment demonstration, and **i** electrical characterization. Scale bars: 1 cm (**b**–**e**), 5 cm (f–h) [67]. Copyright 2023. Reproduced with permission from Elsevier

a deployable solar panel system (Fig. 12f–h), where four independently controlled hinges facilitate sequential panel deployment using approximately 1.8 W power and 30 mA current (Fig. 12i).

9 Energy Storage and Conversion

The domain of energy storage has seen substantial progress due to the use of polymer nanocomposites. Flexible batteries using these materials have shown significant improvements in energy density and mechanical resilience [68]. Recent advancements in structural battery designs indicate the feasibility of incorporating energy storage capabilities into load-bearing elements, resulting in multifunctional systems [69]. Polymer nanocomposite-based supercapacitors have attained outstanding performance metrics by the judicious incorporation of high-surface-area nanoparticles. Studies have shown devices with capacitances over 1000 F/g while preserving mechanical flexibility [70]. The creation of hierarchically porous materials via controlled processing has facilitated rapid charge–discharge capabilities while preserving high energy density. Energy harvesting devices are a burgeoning application area in which polymer nanocomposites have considerable promise. Piezoelectric and triboelectric nanogenerators using these materials have shown effective conversion of mechanical energy into electrical power [71]. Huang et al. developed a 4D-printed self-recovering triboelectric nanogenerator for energy harvesting and sensing applications [72]. Figure 13 illustrates the construction of self-powered angle sensors with freestanding triboelectric nanogenerators (TENGs) produced by 4D printing [72]. The device, produced in many sizes, operates as a human motion angle sensor by transforming joint motions into electrical impulses via sliding contact between tribomaterials and electrodes (Fig. 13b). The operational mechanism, illustrated in Fig. 13c–d, is predicated on charge distribution dynamics: identical positive charges on both electrodes yield a null potential difference (Fig. 13c–i, 13c–iii), whereas total overlap between the friction unit and electrode 2 optimizes the potential difference via charge transfer (Fig. 13c–ii). The six-blade configuration facilitates accurate angle measurement using phase correlation, whereby flexion and extension motions produce unique positive or negative pulse signals. The system exhibits distinct phase-angle correlations: π equates to 30°, 2π to 60°, and 3π to 90°, as seen in Fig. 13d–i to d–iv.

Dong et al. created 3D-printed flexible ceramic-polymer grid composites with a silver covering, using PNN-PZT, for electromechanical energy harvesting [73]. Figure 14 illustrates the manufacturing process and mechanical adaptability of ceramic-polymer grid composites. Figure 14a depicts the layer-by-layer 3D printing technique, whereas Fig. 3b presents the pure PDMS lattice structure. The resultant grid-composite, seen in Fig. 14c–d, demonstrates superior flexibility relative to solid composites. The printing adaptability is shown via several intricate constructions, including biomimetic forms (butterfly, dolphin, bat) and geometric patterns (Fig. 14e). The grid-composite has exceptional mechanical flexibility, as seen by its reaction to various deformation modes such as bending, twisting, and stretching (Fig. 14h–k). The remarkable flexibility and stretchability of the composite render it especially appropriate for wearable applications, as seen in Fig. 14f–g.

10 Challenges and Future Perspectives

The incorporation of nanofillers into polymer matrices for 3D and 4D printing applications signifies a significant advancement in advanced manufacturing technology. This novel method integrates the adaptability of additive manufacturing with the

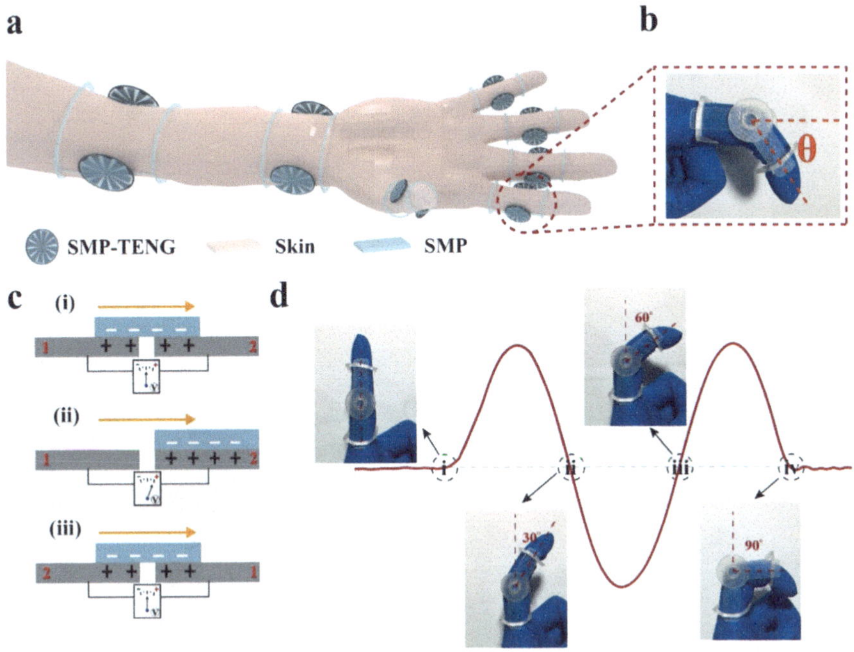

Fig. 13 Self-powered motion sensing system: **a** TENG sensor design, **b** kinematic detection, **c** operational principles, and **d** electromechanical response [72]. Copyright 2021. Reproduced with permission from Elsevier

superior characteristics provided by nanoparticles, generating prospects for transformative applications across several sectors. However, the development and execution of these technologies encounter several difficulties that must be determined to realize their whole potential in industrial applications. One of the primary challenges in developing 3D- and 4D-printed polymer nanocomposites lies in the processing and formulation of materials. The incorporation of nanofillers into polymer matrices presents significant difficulties in achieving uniform dispersion, as nanoparticles naturally tend to agglomerate due to their high surface energy and strong van der Waals forces [9, 73–75]. This agglomeration can lead to inconsistent material properties and potential printing failures -. Furthermore, the rheological properties of nanocomposite inks must be carefully optimized to ensure successful printing, requiring a delicate balance between viscosity and shear-thinning behavior that can accommodate both the printing process and the presence of nanofillers [73–75]. The printing process itself introduces additional complexities that must be carefully managed. Nozzle clogging represents a significant challenge, particularly when working with highly loaded nanocomposites or materials with poor dispersion characteristics [76, 77]. The presence of nanofillers can also affect the heat transfer properties of the printing material, potentially leading to inconsistent layer adhesion and variations in the final product's mechanical properties [77]. These challenges are

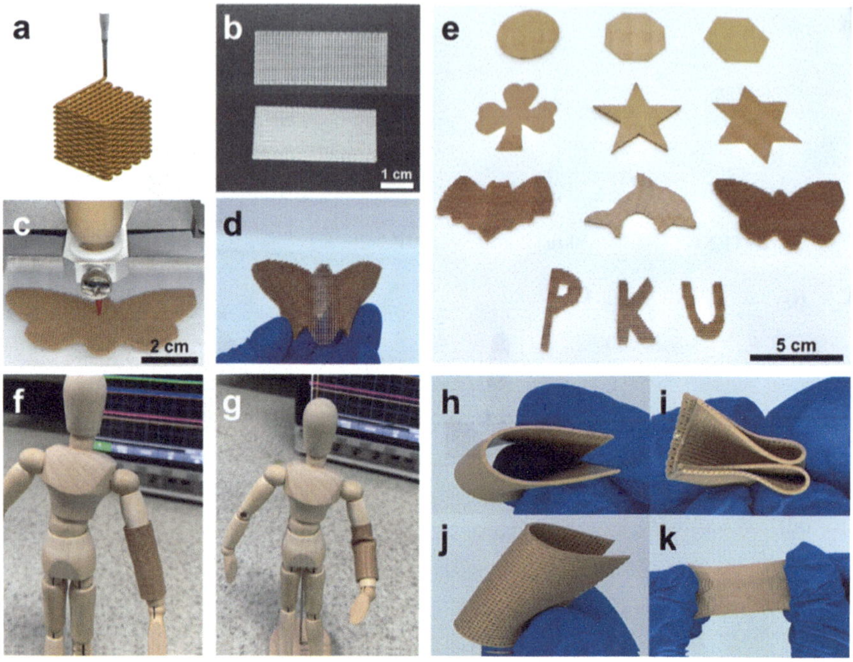

Fig. 14 Flexible ceramic-polymer composites: **a** printing schematic, **b** PDMS template, **c**–**e** composite designs, **f**, **g** wearable application, and **h**–**k** mechanical versatility [73]. Copyright 2020. Reproduced with permission from Elsevier

further complicated by the need to maintain precise control over printing parameters such as temperature, speed, and layer height, all of which can significantly impact the final quality of the printed structure. The achievement of consistent and reproducible properties in printed parts remains a significant hurdle in the field. The layer-by-layer nature of the printing process inherently introduces anisotropic mechanical properties, which can be either advantageous or detrimental depending on the intended application. The orientation and distribution of nanofillers throughout the printed structure can vary significantly, leading to inconsistent performance characteristics. This variability is particularly challenging when attempting to scale up production or when precise property control is required for specific applications. The development of advanced materials concentrates on the creation of smart materials that include integrated sensing and actuation capabilities, alongside bio-inspired nanocomposites including hierarchical structures to enhance the control of nanofiller distribution and orientation. The advancement of self-healing materials with reactive nanofillers signifies a promising avenue, possibly resolving concerns about the long-term durability and dependability of printed structures [76–79].

The future of 3D- and 4D-printed polymer composites with electrical characteristics has disruptive potential across several fields. The future progresses include the development of innovative multi-functional materials that combine electrical conductivity, self-healing, shape-memory, and stimulus responsiveness, advancements in high-resolution printing for microscale electronic circuit fabrication, and the integration of artificial intelligence and machine learning algorithms for improved electrical performance in printing parameters. These advancements have the potential to revolutionize industries such as electronics, aerospace, and healthcare by enabling the production of more efficient and customizable devices. Moreover, the integration of smart polymers with electrical characteristics could lead to the creation of new technologies with enhanced capabilities. For instance, smart polymers could be used to create self-healing electronic devices or sensors that can adapt to changing environments. This innovation has the potential to greatly improve the reliability and longevity of electronic systems in various applications. Overall, the development of smart polymers has the potential to drive innovation and create new opportunities in a wide range of industries. By combining the unique properties of polymers with advanced technology, researchers are paving the way for exciting advancements in materials science.

11 Conclusions

The use of 3D and 4D printing technologies in creating polymer nanocomposites shows great promise for electronic applications. These materials offer improved mechanical, thermal, and electrical properties, making them ideal for a wide range of electronic devices. However, the ability to customize the properties of these materials through precise control of the printing process opens up new possibilities for advanced electronic applications. Overall, the development of 3D- and 4D-printed polymer nanocomposites represents a significant advancement in the field of electronic materials. These innovative materials have the potential to revolutionize the electronics industry by providing enhanced performance and functionality. With ongoing research and development, the future looks bright for the integration of 3D- and 4D-printed polymer nanocomposites in electronic devices. As researchers continue to explore the capabilities of these materials, we can expect to see even more groundbreaking applications emerge in the near future. The versatility and customizable nature of 3D- and 4D-printed polymer nanocomposites make them a promising candidate for next-generation electronic components. These materials offer advantages such as increased durability, flexibility, and conductivity, making them ideal for a wide range of applications. Furthermore, the advancements in structure–property relationships and manufacturing technologies are expected to lead to more advanced applications of electrically conductive printed composites in flexible electronics, wearable devices, biomedical implants, etc. This field, integrating materials science, advanced manufacturing, and computational design, offers solutions to existing challenges and enables fresh capabilities in next-generation electronic devices and intelligent systems.

Acknowledgements The authors gratefully acknowledge the support provided by Centurion University of Technology and Management, Odisha, India, for carrying out the present research work.

References

1. Ninduwezuor-Ehiobu, N., Tula, O.A., Daraojimba, C., Ofonagoro, K.A., Ogunjobi, O.A., Gidi-agba, J.O., Egbokhaebho, B.A., Banso, A.A.: Exploring innovative material integration in modern manufacturing for advancing us competitiveness in sustainable global economy. Eng. Sci. Technol. J. **4**(3), 140–168 (2023)
2. Mahmood, A., Perveen, F., Chen, S., Akram, T., Irfan, A.: Polymer composites in 3D/4D printing: materials, advances, and prospects. Molecules, **29**(2), 319.Rr (2024)
3. Bisht, N., Vishwakarma, J., Jaiswal, S., Kumar, P., Srivastava, A.K., Dhand, C., Dwivedi, N.: Synergizing chemistry: unveiling the potential of hybrid fillers for enhanced performance in shape memory polymers. Adv. Compos. Hybrid Mater. **8**(1), 7 (2025)
4. Pazhamalai, P., Krishnan, V., Mohamed Saleem, M.S., Kim, S.J., Seo, H.W.: Investigating composite electrode materials of metal oxides for advanced energy storage applications. Nano Convergence **11**(1), 30.R (2024)
5. Alarifi, I.M.: Revolutionising fabrication advances and applications of 3D printing with composite materials: a review. Virtual Phys. Prototyping **19**(1), e2390504.R (2024)
6. Lodhi, S.K., Gill, A.Y., Hussain, I.: 3D printing techniques: transforming manufacturing with precision and sustainability. Int. J. Multidiscip. Sci. Arts **3**(3), 129–138.R (2024)
7. Antezana, P.E., Municoy, S., Ostapchuk, G., Catalano, P.N., Hardy, J.G., Evelson, P.A., ... & Desimone, M.F.: 4D printing: The development of responsive materials using 3D-printing technology. Pharmaceutics **15**(12), 2743.R (2023)
8. Stachiv, I., Alarcon, E., Lamac, M.: Shape memory alloys and polymers for MEMS/NEMS applications: review on recent findings and challenges in design, preparation, and characterization. Metals **11**(3), 415 (2021)
9. Kausar, A., Ahmad, I., Zhao, T., Aldaghri, O., Eisa, M.H.: Polymer/graphene nanocomposites via 3D and 4D printing—design and technical potential. Processes **11**(3), 868 (2023)
10. Liu, M., Zhang, Y., Tao, T.H.: Recent progress in bio-integrated intelligent sensing system. Adv. Intell. Syst. **4**(6), 2100280 (2022)
11. Rahman, M., Islam, K.S., Dip, T.M., Chowdhury, M.F., Debnath, S.R., Hasan, S.M., Sakib, M.S., Saha, T., Padhye, R., Houshyar, S.: A review on nanomaterial-based additive manufacturing: dynamics in properties, prospects, and challenges. Progress Additive Manuf. **9**(4), 1197–1224 (2024)
12. Shahbazi, M., Jäger, H., Ettelaie, R., Mohammadi, A., Kashi, P.A.: Multimaterial 3D printing of self-assembling smart thermo-responsive polymers into 4D printed objects: a review. Addit. Manuf. **71**, 103598 (2023)
13. Mishra, R., Militky, J.: Carbon-based nanomaterials. Nanotechnol. Text. Theory Appl., 163–179 (2018)
14. Zhou, W., Bai, X., Wang, E., Xie, S.: Synthesis, structure, and properties of single-walled carbon nanotubes. Adv. Mater. **21**(45), 4565–4583 (2009)
15. Zada, A., Muhammad, P., Ahmad, W., Hussain, Z., Ali, S., Khan, M., Khan, Q., Maqbool, M.: Surface plasmonic-assisted photocatalysis and optoelectronic devices with noble metal nanocrystals: design, synthesis, and applications. Adv. Func. Mater. **30**(7), 1906744 (2020)
16. Kumar, S., Raj, S., Jain, S., Chatterjee, K.: Multifunctional biodegradable polymer nanocomposite incorporating graphene-silver hybrid for biomedical applications. Mater. Des. **108**, 319–332 (2016)

17. Trinh, B.M., Chang, B.P., Mekonnen, T.H.: The barrier properties of sustainable multiphase and multicomponent packaging materials: a review. Prog. Mater Sci. **133**, 101071 (2023)
18. Sung, G., Kim, J.H.: Influence of filler surface characteristics on morphological, physical, acoustic properties of polyurethane composite foams filled with inorganic fillers. Compos. Sci. Technol. **146**, 147–154 (2017)
19. Baig, Z., Mamat, O., Mustapha, M.: Recent progress on the dispersion and the strengthening effect of carbon nanotubes and graphene-reinforced metal nanocomposites: a review. Crit. Rev. Solid State Mater. Sci. **43**(1), 1–46 (2018)
20. Al-Amiery, A.A., Fayad, M.A., Hussain, H.A., Al-Azzawi, W.K., Mohammed, J.K., Majdi, H.S.: Interfacial engineering for advanced functional materials: surfaces, interfaces, and applications. Results Eng., 102125 (2024)
21. Kango, S., Kalia, S., Celli, A., Njuguna, J., Habibi, Y., Kumar, R.: Surface modification of inorganic nanoparticles for development of organic–inorganic nanocomposites—a review. Prog. Polym. Sci. **38**(8), 1232–1261 (2013)
22. Zhou, Y., Fan, M., Chen, L.: Interface and bonding mechanisms of plant fibre composites: an overview. Compos. B Eng. **101**, 31–45 (2016)
23. Fu, S., Sun, Z., Huang, P., Li, Y., Hu, N.: Some basic aspects of polymer nanocomposites: a critical review. Nano Mater. Sci. **1**(1), 2–30 (2019)
24. Zhang, S., Feng, W., Jiang, Y., Li, Y., Liu, Y., Yu, D., Wang, W.: Flexible pressure sensor based on 3D printing MXene@ dual-scale porous polymer. Chem. Eng. J. **498**, 155356 (2024)
25. Rahmatabadi, D., Khajepour, M., Bayati, A., Mirasadi, K., Yousefi, M.A., Shegeft, A., Ghasemi, I., Baniassadi, M., Abrinia, K., Bodaghi, M., Baghani, M.: Advancing sustainable shape memory polymers through 4D printing of polylactic acid-polybutylene adipate terephthalate blends. Eur. Polymer J. **216**, 113289 (2024)
26. Sheikh, T., Behdinan, K.: Fused deposition modelling of thermoplastic polymer nanocomposites: a critical review. C **10**(2), 29 (2024)
27. Ivanov, E., Kotsilkova, R., Xia, H., Chen, Y., Donato, R.K., Donato, K., Godoy, A.P., Di Maio, R., Silvestre, C., Cimmino, S., Angelov, V.: PLA/Graphene/MWCNT composites with improved electrical and thermal properties suitable for FDM 3D printing applications. Appl. Sci. **9**(6), 1209 (2019)
28. Kafle, A., Luis, E., Silwal, R., Pan, H.M., Shrestha, P.L., Bastola, A.K.: 3D/4D printing of polymers: fused deposition modelling (FDM), selective laser sintering (SLS), and stereolithography (SLA). Polymers **13**(18), 3101 (2021)
29. Xu, W., Jambhulkar, S., Ravichandran, D., Zhu, Y., Kakarla, M., Nian, Q., Azeredo, B., Chen, X., Jin, K., Vernon, B., Lott, D.G.: 3D printing-enabled nanoparticle alignment: a review of mechanisms and applications. Small **17**(45), 2100817 (2021)
30. Monfared, V., Bakhsheshi-Rad, H.R., Ramakrishna, S., Razzaghi, M., Berto, F.: A brief review on additive manufacturing of polymeric composites and nanocomposites. Micromachines **12**(6), 704 (2021)
31. Shen, H., Wu, W., Hu, H., Rui, Z., Ye, J., Zhang, C.: Preparation of carbon black/graphene nanosheets/PP composites with 3D separated conductive networks based on selective laser sintering. Polym. Compos. **44**(6), 3522–3534 (2023)
32. Shahbazi, M., Jäger, H., Ettelaie, R., Chen, J., Kashi, P.A., Mohammadi, A.: Dispersion strategies of nanomaterials in polymeric inks for efficient 3D printing of soft and smart 3D structures: a systematic review. Adv. Colloid Interface Sci., 103285 (2024)
33. Hassib, M.A., Islam, M.R., Karim, M.R., Hasan, M.S., Hossain, K.R.: Research progress of 4d printing technology. Kufa J. Eng. **15**(3), 107–133 (2024)
34. Joharji, L., Mishra, R.B., Alam, F., Tytov, S., Al-Modaf, F., El-Atab, N.: 4D printing: a detailed review of materials, techniques, and applications. Microelectron. Eng. **265**, 111874 (2022)
35. Pardo, A., Gómez-Florit, M., Barbosa, S., Taboada, P., Domingues, R.M., Gomes, M.E.: Magnetic nanocomposite hydrogels for tissue engineering: design concepts and remote actuation strategies to control cell fate. ACS Nano **15**(1), 175–209 (2021)
36. Xia, X., Spadaccini, C.M., Greer, J.R.: Responsive materials architected in space and time. Nat. Rev. Mater. **7**(9), 683–701 (2022)

37. Wan, X., Luo, L., Liu, Y., Leng, J.: Direct ink writing based 4D printing of materials and their applications. Adv. Sci. **7**(16), 2001000 (2020)
38. Kouka, M.A., Abbassi, F., Habibi, M., Chabert, F., Zghal, A., Garnier, C.: 4D printing of shape memory polymers, blends, and composites and their advanced applications: a comprehensive literature review. Adv. Eng. Mater. **25**(4), 2200650 (2023)
39. Xia, Y., He, Y., Zhang, F., Liu, Y., Leng, J.: A review of shape memory polymers and composites: mechanisms, materials, and applications. Adv. Mater. **33**(6), 2000713 (2021)
40. Ahmed, A., Arya, S., Gupta, V., Furukawa, H., Khosla, A.: 4D printing: Fundamentals, materials, applications and challenges. Polymer **228**, 123926 (2021)
41. Kumar, S., Duvedi, R.K., Sharma, S.K., Batish, A.: Navigating the frontier: additive Manufacturing's role in synthesizing piezoelectric materials for flexible electronics. J. Thermoplast. Compos. Mater., 08927057241270729 (2024)
42. Sun, X., Huang, C., Wang, L., Liang, L., Cheng, Y., Fei, W., Li, Y.: Recent progress in graphene/polymer nanocomposites. Adv. Mater. **33**(6), 2001105 (2021)
43. Qureshi, N., Dhand, V., Subhani, S., Kumar, R.S., Raghavan, N., Kim, S., Doh, J.: Exploring conductive filler-embedded polymer nanocomposite for electrical percolation via electromagnetic shielding-based additive manufacturing. Adv. Mater. Technol., 2400250 (2024)
44. Ryan, K.R., Down, M.P., Banks, C.E.: Future of additive manufacturing: overview of 4D and 3D printed smart and advanced materials and their applications. Chem. Eng. J. **403**, 126162 (2021)
45. Liu, Y., He, K., Chen, G., Leow, W.R., Chen, X.: Nature-inspired structural materials for flexible electronic devices. Chem. Rev. **117**(20), 12893–12941 (2017)
46. Kwon, Y.J., Park, J.B., Jeon, Y.P., Hong, J.Y., Park, H.S., Lee, J.U.: A review of polymer composites based on carbon fillers for thermal management applications: design, preparation, and properties. Polymers **13**(8), 1312 (2021)
47. Huang, G., Chen, W., Wu, T., Guo, H., Fu, C., Xue, Y., ... Song, P.: Multifunctional graphene-based nano-additives toward high-performance polymer nanocomposites with enhanced mechanical, thermal, flame retardancy and smoke suppressive properties. Chem. Eng. J. **410**, 127590 (2021)
48. Mirabedini, A., Ang, A., Nikzad, M., Fox, B., Lau, K.T., Hameed, N.: Evolving strategies for producing multiscale graphene-enhanced fiber-reinforced polymer composites for smart structural applications. Adv. Sci. **7**(11), 1903501 (2020)
49. Papageorgiou, D.G., Li, Z., Liu, M., Kinloch, I.A., Young, R.J.: Mechanisms of mechanical reinforcement by graphene and carbon nanotubes in polymer nanocomposites. Nanoscale **12**(4), 2228–2267 (2020)
50. Zhang, F., Feng, Y., Feng, W.: Three-dimensional interconnected networks for thermally conductive polymer composites: Design, preparation, properties, and mechanisms. Mater. Sci. Eng. R. Rep. **142**, 100580 (2020)
51. Zhao, H.Y., Yu, M.Y., Liu, J., Li, X., Min, P., Yu, Z.Z.: Efficient preconstruction of three-dimensional graphene networks for thermally conductive polymer composites. Nano-micro letters **14**(1), 129 (2022)
52. Fallah, A., Saleem, Q., Koc, B.: Assessment of mechanical properties and shape memory behavior of 4D printed continuous fiber-reinforced PETG composites. Compos. A Appl. Sci. Manuf. **181**, 108165 (2024)
53. Kholkhoev, B.C., Bardakova, K.N., Nikishina, A.N., Matveev, Z.A., Efremov, Y.M., Frolova, A.A., Akovantseva, A.A., Gorenskaia, E.N., Verlov, N.A., Timashev, P.S., Burdukovskii, V.F.: 4D-printing of mechanically durable high-temperature shape memory polymer with good irradiation resistance. Appl. Mater. Today **1**(36), 102022 (2024)
54. Heng, W., Solomon, S., Gao, W.: Flexible electronics and devices as human–machine interfaces for medical robotics. Adv. Mater. **34**(16), 2107902 (2022)
55. Mishra, S., Saha, B.: Graphene-polymer nanocomposite-based wearable strain sensors for physiological signal Monitoring: Recent progress and challenges. Curr. Opin. Solid State Mater. Sci. **31**, 101174 (2024)

56. Wu, H., Huang, Y., Xu, F., Duan, Y., Yin, Z.: Energy harvesters for wearable and stretchable electronics: from flexibility to stretchability. Adv. Mater. **28**(45), 9881–9919 (2016)
57. Xie, K., Wei, B.: Materials and structures for stretchable energy storage and conversion devices. Adv. Mater. **26**(22), 3592–3617 (2014)
58. Huang, W., Yin, M., Xia, J., Zhang, X.: A review of cross-scale and cross-modal intelligent sensing and detection technology for food quality: mechanism analysis, decoupling strategy and integrated applications. Trends Food Sci. Technol., 104646 (2024)
59. Liu, S., Liu, X., Cui, X., Zhu, M., Lu, S.: 3D printed MXene/rGO hierarchical porous structure based on ultralow-concentration 2D nanomaterial inks for highly sensitive pressure sensing. Chem. Eng. J. **496**, 154250 (2024)
60. Roach, D.J., Hamel, C.M., Dunn, C.K., Johnson, M.V., Kuang, X., Qi, H.J.: The m4 3D printer: a multi-material multi-method additive manufacturing platform for future 3D printed structures. Addit. Manuf. **29**, 100819 (2019)
61. Zeng, C., Liu, L., Xin, X., Zhao, W., Lin, C., Liu, Y., Leng, J.: 4D printed bio-inspired mesh composite materials with high stretchability and reconfigurability. Compos. Sci. Technol. **249**, 110503 (2024)
62. Hu, C., Wang, L., Liu, S., Sheng, X., Yin, L.: Recent development of implantable chemical sensors utilizing flexible and biodegradable materials for biomedical applications. ACS Nano **18**(5), 3969–3995 (2024)
63. Tonelli, D., Gualandi, I., Scavetta, E., Mariani, F.: Focus review on nanomaterial-based electrochemical sensing of glucose for health applications. Nanomaterials **13**(12), 1883 (2023)
64. Lee, J.H., Kim, H., Kim, J.H., Lee, S.H.: Soft implantable microelectrodes for future medicine: prosthetics, neural signal recording and neuromodulation. Lab Chip **16**(6), 959–976 (2016)
65. Sridharan, A., Nguyen, J.K., Capadona, J.R., Muthuswamy, J.: Compliant intracortical implants reduce strains and strain rates in brain tissue in vivo. J. Neural Eng. **12**(3), 036002 (2015)
66. Li, N., Khan, S.B., Chen, S., Aiyiti, W., Zhou, J., Lu, B.: Promising new horizons in medicine: medical advancements with nanocomposite manufacturing via 3D printing. Polymers **15**(20), 4122 (2023)
67. Chen, S., Li, J., Shi, H., Chen, X., Liu, G., Meng, S., Lu, J.: Lightweight and geometrically complex ceramics derived from 4D printed shape memory precursor with reconfigurability and programmability for sensing and actuation applications. Chem. Eng. J. **455**, 140655 (2023)
68. Yang, C., Wei, H., Guan, L., Guo, J., Wang, Y., Yan, X., Zhang, X., Wei, S., Guo, Z.: Polymer nanocomposites for energy storage, energy saving, and anticorrosion. J. Mater. Chem. A **3**(29), 14929–14941 (2015)
69. Nie, B., Lim, J., Liu, T., Kovalenko, I., Guo, K., Liang, J., Zhu, J., Sun, H.: Multifunctional composite designs for structural energy storage. Battery Energy **2**(6), 20230023 (2023)
70. Gong, M., Zhang, L., Wan, P.: Polymer nanocomposite meshes for flexible electronic devices. Prog. Polym. Sci. **107**, 101279 (2020)
71. Jing, Q., Kar-Narayan, S.: Nanostructured polymer-based piezoelectric and triboelectric materials and devices for energy harvesting applications. J. Phys. D Appl. Phys. **51**(30), 303001 (2018)
72. Huang, L.B., Han, J.C., Chen, S., Sun, Z., Dai, X., Ge, P., Zhao, C.H., Zheng, Q.Q., Sun, F.C., Hao, J.: 4D-printed self-recovered triboelectric nanogenerator for energy harvesting and self-powered sensor. Nano Energy **84**, 105873 (2021)
73. Wang, Z., Yuan, X., Yang, J., Huan, Y., Gao, X., Li, Z., Wang, H., Dong, S.: 3D-printed flexible, Ag-coated PNN-PZT ceramic-polymer grid-composite for electromechanical energy conversion. Nano Energy **73**, 104737 (2020)
74. Zhang, X., Li, B.W., Dong, L., Liu, H., Chen, W., Shen, Y., Nan, C.W.: Superior energy storage performances of polymer nanocomposites via modification of filler/polymer interfaces. Adv. Mater. Interfaces **5**(11), 1800096 (2018)
75. Idumah, C.I., Obele, C.M., Ezeani, E.O.: Understanding interfacial dispersions in ecobenign polymer nano-biocomposites. Polymer-Plast. Technol. Mater. **60**(3), 233–252 (2021)
76. Xu, W., Jambhulkar, S., Zhu, Y., Ravichandran, D., Kakarla, M., Vernon, B., Lott, D.G., Cornella, J.L., Shefi, O., Miquelard-Garnier, G., Yang, Y.: 3D printing for polymer/particle-based processing: a review. Compos. B Eng. **223**, 109102 (2021)

77. Siddiqui, S., Surananai, S., Sainath, K., Khan, M.Z., Kuppusamy, R.R.P., Suneetha, Y.K.: Emerging trends in the development and application of 3D printed nanocomposite polymers for sustainable environmental solutions. Europ. Polymer J., 112298 (2023)
78. Uddin, M.N., Hossain, M.T., Mahmud, N., Alam, S., Jobaer, M., Mahedi, S.I., Ali, A.: Research and applications of nanoclays: a review. SPE Polymers **5**(4), 507–535 (2024)
79. Bilisik, K., Akter, M.: Polymer nanocomposites based on graphite nanoplatelets (GNPs): a review on thermal-electrical conductivity, mechanical and barrier properties. J. Mater. Sci. **57**(15), 7425–7480 (2022)

Applications of 3D- and 4D-Printed Polymer Nanocomposites in the Medical and Biomedical Field

Mayank Tiwari and Debabrata Mishra

Abstract This chapter examines the use of polymer nanocomposites in 3D and 4D printing for medical applications. Integrating nanomaterials into polymers improves their electrical, thermal, and magnetic characteristics. 3D printing facilitates the production of personalized medical devices, enhancing their compatibility and their mechanical durability. The chapter focuses on using nanotechnology to improve the process of three-dimensional and four-dimensional printing in biomedical sciences. The addition of shape memory capability offers an additional dimension and introduces a fourth dimension, allowing for adaptive reactions to stimuli. This property is vital for various medical applications, including dental implants, biosensors, occluders, and stents, where accurate and responsive reactions are necessary for optimal performance. This chapter also highlights different types of polymer-nanocomposites used in biomedical printing. Additionally, it examines the developing applications of 4D printing and highlights their potential to transform the manufacturing of medical devices.

Keywords Nanocomposites · 3D printing · Scaffolds · Stents · Shape memory polymers · 4D printing · Shape memory polymer nanocomposites

1 Introduction

Nanomaterials are the fundamental components of nanotechnology and possess dimensions within the nanoscale range, specifically between 1 and 100 nm [51]. Nanomaterials can exist in many different forms, such as nanoparticles (0D), nanowires (1D), and nanosheets (2D). Tiwari et al. [51]. This confinement at the nanoscale within nanomaterials gives rise to distinct electrical, thermal, optical, and magnetic characteristics that can be utilized in different sectors [2]. Polymers on the other hand are macromolecules consisting of recurring monomers that are linked together by covalent bonds [31]. Depending on their molecular structure and the

M. Tiwari · D. Mishra (✉)
Department of Physics and Astrophysics, University of Delhi, New Delhi, Delhi, India
e-mail: dmishra@physics.du.ac.in

S. Moharana et al. (eds.), *Polymer Nanocomposites for 3D, 4D and 5D Printing*, Engineering Materials, https://doi.org/10.1007/978-981-96-4214-4_12

monomers used in their manufacture, polymers might exhibit flexibility or rigidity, transparency or opacity, insulation, or conductivity. Polymers include a variety of features, making them highly versatile and adaptable for use in multiple fields, including but not limited to engineering, medicine, agriculture, and electronics [6].

The integration of nanoparticles into polymers facilitates the development of distinctive materials called polymer nanocomposites [60]. These nanocomposites exhibit specific features or functions, such as enhanced electrical and thermal conductivities, sensitivity to electro-mechanical stimuli, magnetism, and greater mechanical strength [60, 62]. For example, when conducting nanoparticles like gold, graphene, copper, etc. are added to the polymer matrix, it enhances the overall conductivity of the polymer [60, 62]. Introducing magnetic nanoparticles, such as Fe_3O_4, into the polymer results in the formation of a magnetic nanocomposite [39]. This nanocomposite possesses magnetic properties that may be manipulated and regulated by an external magnetic field [60, 62]. Nanomaterials, when used as nanofillers, not merely enhance the characteristics of polymers but also address the challenges associated with the tiny size of nanomaterials during manufacture and manipulation [32]. Furthermore, the nanoparticles can serve as additives to modify the flow properties of polymer materials, thereby rendering them acceptable for specific processing requirements [13].

The development of 3D printing technology has made a major impact on the field of material design [1]. 3D technology is known for its ability to reduce production costs by eliminating the need for tools and molds, minimizing waste products and physical stocks through virtual inventory, and improving the overall quality of the materials [1]. Polymer nanocomposites have the potential to be utilized in 3D printing technology to fabricate devices with distinct features that can be applied in various sectors including engineering, automobile, medicine, etc. [13, 17]. Within the field of medicine, the utilization of polymer nanocomposites in 3D printing allows for the fabrication of customized and precise medical equipment and implantation [60, 62]. These devices can be customized to align with the anatomy and physiology of individual patients, resulting in enhanced outcomes and less likelihood of rejection or problems [13, 60, 62]. One of the most utilized biomedical applications of polymer nanocomposite in 3D printing is scaffold engineering is used to facilitate tissue regeneration and repair in the field of bone tissue engineering [43, 55]. These scaffolds imitate the extracellular matrix and offer a conducive setting for cellular growth, multiplication, and specialization. In addition to their applications in 3D printing, polymer nanocomposites can be engineered to exhibit shape memory behavior which gives them a fourth dimension (4D) to respond to external stimuli [26, 67]. This distinctive characteristic enables the materials to experience deformation in response to external stimuli, such as temperature or light, and subsequently return to their original shape after the stimulus is no longer present. Utilizing this capacity allows for a wide range of creative applications in several domains, such as dental implants, biosensors, occluders, and stent creation [26, 27, 67].

This chapter covers the utilization and implementation of 3D and 4D printed polymer nanocomposites in the medical and biomedical fields. The first section explores the crucial role of nanotechnology in improving 3D printing methods, explaining how nanomaterials help create complex objects with specific features. Following that, it describes different categories of polymer nanocomposites used in 3D printing for medical purposes, emphasizing their distinct features and benefits in producing tailored medical devices. Finally, the chapter explores the growing subject of 4D-printed polymer nanocomposites in biomedicine. It explains how these materials demonstrate shape memory properties and explores their prospective uses in adaptive medical devices such as intelligent implants, scaffolds, and stents.

2 Relationship Between Nanotechnology and 3D-Printing

Nanotechnology and 3D printing have a symbiotic relationship, each enhancing the capabilities and uses of the other [13]. Nanotechnology utilizes the advantages of 3D printing by allowing for precise manipulation of the spatial organization of materials at the nanoscale [10, 13]. This capability enables the creation of sophisticated and versatile structures with customized attributes. One of the most significant benefits that is provided by 3d printing in nanotechnology is 3D patterning. 3D patterning allows for precise manipulation of the arrangement of space of nanoscale materials, promoting the production of complex 3D structures with customized features and capabilities [10, 13]. This capability significantly transforms nanotechnology by enabling the creation and construction of intricate nanostructures and devices that were previously difficult to manufacture. The progress in 3D patterning achieved by 3D printing improves the incorporation and effectiveness of applications at the nanoscale in several domains, including medicine and electronics. Litowczenko et al. demonstrated that the growth of nerve cells was enhanced by the 3D micro/nano-patterns in nerve conduits. The tiniest pattern stimulated the greatest proliferation of neurites, whereas the intermediate-sized patterns of PCL-GO promoted the efficient formation of synapses [28]. Qiao and the group also demonstrated enhanced efficiency in triboelectric nanogenerators through the 3D macro patterning of the device [35].

Moreover, advanced techniques like PolyJet printing can deposit numerous inks at the same time using very small nozzles. This allows for the construction of objects made from different materials, each with its own distinct and configurable features. The use of creative techniques in 3D printing not only improves the ability to design with flexibility but also opens up new opportunities for constructing multifunctional nanostructures and devices.

In contrast, 3D printing utilizes the progress made in nanotechnology to include nanoparticles, nanocomposites, and nanomaterials in printed products, improving their mechanical strength, electrical conductivity, thermal stability, and other useful properties [13, 15]. Nanomaterials can be integrated into the polymer matrix or other materials, improving the final composite's properties. These composites can be employed as the substance (or ink) in 3D printing devices for medical, electronics, and other industrial applications. Gnanasekaran et al. incorporated CNTs into the polymer matrix to form conductive nanocomposites. Integrating Carbon Nanotubes

(CNT) into PBT polymer matrices greatly improved the conductivity of the nanocomposite. PBT/CNT nanocomposites demonstrated higher electrical conductivity and mechanical strength in comparison to conventional polymers such as PLA and ABS. This nanocomposite has the potential to be utilized as a conductive filament for 3D printing [22]. Similarly, the incorporation of magnetic nanomaterials into the polymers can be used to develop magnetic inks for 3d printing. These magnetic nanocomposites can be utilized in developing 3D magnetism-based sensors as well as in hyperthermia. Thus, nanotechnology offers significant potential to 3D printing technology, enhancing its capabilities and versatility.

Hence, the mutually dependent relationship between nanotechnology and 3D printing promotes advancements in several sectors, facilitating the development of modern materials, devices, and systems with superior abilities and performance [13]. The alliance has completely transformed the manufacturing procedures, enabling the creation of complex ideas and practical prototypes that were previously considered unattainable. The incorporation of nanoparticles into 3D printing has significantly transformed various industries, including, auto-motive, healthcare, and consumer electronics, leading to a revolution in innovation [10, 13, 15].

This chapter will focus on the incorporation of different nanomaterials into the polymer matrix to create polymer nanocomposites that are ideal for 3D printing devices used in biomedical applications. The advancements in the synthesis of polymer nanocomposites present a great opportunity to improve the mechanical, thermal, and electrical characteristics of 3D printed parts, hence increasing their versatility and ability to meet specific application needs. Through the utilization of the distinctive characteristics of different nanomaterials, metal nanoparticles, magnetic nanoparticles, metal-oxide nanoparticles, graphene nanosheets, etc. into the polymer matrix, the customization and functionality of polymer-based 3D printed devices can be improved to fulfill the rigorous requirements of biomedical applications.

3 3D Printing Techniques

3D printing leverages layer-by-layer manufacturing technologies to manufacture goods from a digital model. Several widely employed 3D printing methods, such as Fused Deposition Modeling (FDM), Semi-solid extrusion (SSE), Stereolithography (SLA), Selective Laser Sintering (SLS), and Inkjet-based 3D printing, have proven effective in the creation of diverse biomedical materials [62].

3.1 Fused Deposition Modeling (FDM)

FDM is a 3D printing process that uses polymers (thermally melted) to build a 3D object by extrusion. The primary benefits associated with this 3D printing process are the effective printing cost and a large variety of material selection. Low resolution and minimal printing speed are the drawbacks of this technique [62].

3.2 Semi-Solid Extrusion (SSE)

Another kind of extrusion, known as SSE, utilizes a gel or paste to manufacture three-dimensional materials in layers. It uses an ink tank instead of high-temperature polymeric wires. The main advantage of SSE over FDM is that it does not require high-temperature polymers. However, SSE still faces the challenge of resolution and efficiency [62].

3.3 Stereolithography (SLA)

SLA, a printing technology related to photo-polymerization, can convert liquid polymers into solid polymer objects. In general, SLA necessitates the presence of photosensitive monomers in the liquid resin, a source of light, and a shared photo-initiator. The photo-initiator undergoes quick activation when exposed to light, which triggers the polymerization process in the liquid resin, resulting in the creation of 3D objects. This technique provides enhanced accuracy and printing speed but low mechanical characteristics and exposure to UV limit its application [62].

3.4 Selective Laser Sintering (SLS)

SLS utilizes a laser emitter to fuse the powder particles and fabricate three-dimensional materials. The advantage of this technique is in the ability of SLS to produce objects with exceptional strength and rigidity, especially when it comes to fabricating metal-based devices and prostheses that possess exceptional mechanical properties tailored to individual patients. However, the inadequate precision of the lasers and powder unavoidably restricts the level of detail achievable in SLS printed materials [62].

3.5 Inkjet Based

Inkjet-based 3D printing is a non-contact approach that employs heat, electromagnetic, or piezo methods to apply materials or ink onto a substrate to create 3D objects. Inkjet-based 3D printing provides a cost-efficient, accurate, versatile, and compatible approach for fabricating intricate tissues and organs utilized in the fields of tissue engineering and regeneration medicine [62]. Figure 1 illustrates the schematics of the biomedical application of polymer nanocomposites and the different types of 3D printing techniques discussed. Table 1 represents the advantages and disadvantages of different 3D printing techniques.

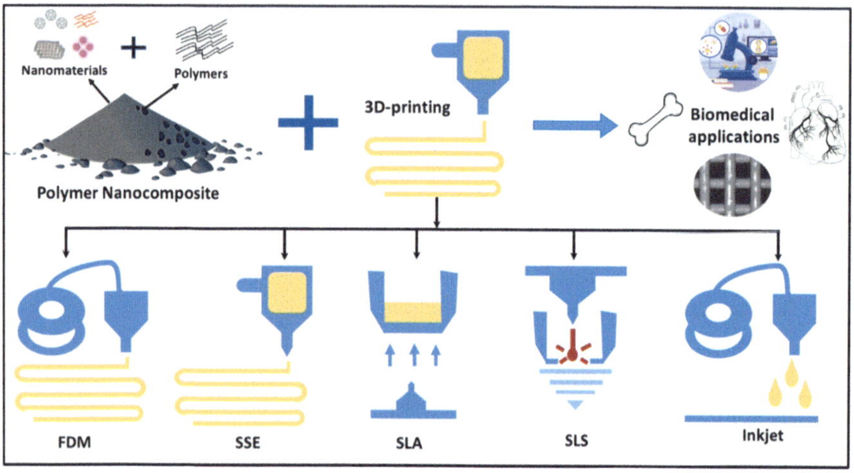

Fig. 1 Schematic diagram illustrating the process of shaping polymer nanocomposites for 3D biomedical applications. Nanomaterials and polymers combine to form a polymer nanocomposite. This composite is used in 3D printing, depicted by a printer nozzle, leading to biomedical applications, symbolized by bone and anatomical illustrations. The diagram also shows five 3D printing techniques: FDM, SSE, SLA, SLS, and Inkjet, each represented by distinct icons

Table 1 Advantages and drawbacks of different 3D printing techniques

3D printing technique	Advantages	Drawbacks
FDM (Fused Deposition Modeling)	Cost-efficient and large material selection	Low speed and resolution
SSE (Syringe Extrusion)	No high-temperature polymers required	Low resolution and efficiency
SLA (Stereolithography)	High printing speed and accuracy	Low mechanical characteristics
SLS (Selective Laser Sintering)	Enhanced strength and rigidity of the manufactured products	Low precision
Inkjet	Cost-efficient and versatile	Requires post-processing

4 3D-Printed Polymer Nanocomposite in the Medical and Biomedical Field

In biomedical science, the incorporation of nanoparticles into polymer matrices represents a new age for materials that have several functions and may be used in a wide range of applications. Metal/metal oxide nanoparticles, graphene, and magnetic materials have potential applications in various fields, including scaffold engineering, bone tissue regeneration, biosensing, and hyperthermia therapy [40, 54, 59]. In the field of scaffold engineering, nanoparticles are utilized to improve the mechanical characteristics of scaffolds. This enhances the overall structural integrity of the scaffolds and promotes the attachment, growth, and specialization of cells, hence facilitating successful tissue regeneration. Incorporating them enhances the effectiveness and compatibility of the scaffold, which is essential for the seamless integration of tissues and good healing [37, 43]. Nanoparticles possess a large surface area and are highly responsive to external stimuli, making them ideal for creating complex biosensors that can detect biological analytes with exceptional accuracy. In addition, integrating magnetic nanoparticles into polymer matrices allows for controlled hyperthermia therapy, a focused method that utilizes external magnetic fields to achieve exact tissue heating [40, 54, 59]. This novel approach has great potential in fighting diseases such as cancer, as it allows for targeted killing of cancerous cells while minimizing harm to healthy tissues.

In the following section, several biomedical applications involving the integration of different types of nanoparticles (NPs) into polymer matrices will be examined. Additionally, advancements in applications such as scaffold engineering, biosensing, and heat-based therapies will be explored.

4.1 Metal/Metal-Oxide-Polymer Nanocomposite in 3D-Printed Scaffold Engineering and Biosensing

Incorporating metal/metal-oxide nanoparticles in the polymer matrix (nanocomposites) has led to significant advancements in various fields, including dental prostheses, biosensors, bone tissue engineering, and antimicrobial applications [11, 36]. These metal/metal-oxide-based nanocomposites are used in bone tissue engineering to enhance mechanical strength, bio-compatibility, and ability to fight against microbes to promote bone regeneration using 3D-printed scaffolds. Moreover, these nanocomposites offer improved sensitivity, stability, and selectivity in 3D-printed biosensors [40, 54]. Furthermore, these nanocomposites are also utilized to fabricate 3D-printed dental prosthetics, enhancing their durability and biocompatibility. Some of the most commonly used metal/metal-oxide nanoparticles with polymers to form nanocomposites discussed in this section are silver nanoparticles (AgNPs), gold nanoparticles (AuNPs), and Titanium dioxide nanoparticles (TiO_2 NPs).

A significant challenge in developing new clinical treatments is the restoration of infected or contaminated bone defects by implanting biomaterials-based 3D-printed orthopedic scaffolds. These scaffolds support tissue regeneration and repair in orthopedic applications, such as bone and cartilage regeneration. However, the 3D-printed scaffolds currently available still face a major challenge of bioactivity and antibacterial efficacy [8]. Silver nanoparticles (AgNPs) are one of the most widely recognized materials in the medical field due to their antimicrobial properties against pathogens like bacteria, fungi, viruses, etc. [48, 49]. AgNPs can be embedded with polymer-based 3D scaffolds thus providing antimicrobial protection and enhancing the regeneration process of the damaged tissue. Merlo et al. in their study utilized AgNPs to improve the quality of 3D-printed scaffolds used in bone-interfacing applications. The scaffolds were covered with PDA (20 nm) using a dip-coating method, and then with AgNPs (70 nm) formed using silver nitrate. The presence of scaffolds coated with a 20 nm layer of PDA and 70 nm AgNPs resulted in a significant decrease in Staphylococcus aureus biofilm formation, reducing it by a factor of 3000–8000 [30]. In another work, Deng et al. fabricated a 3D-printed polyetheretherketone (PEEK) scaffold with AgNPs sandwiched between the first and the second layers of polydopamine (PDA) in the scaffold. Figure 2a demonstrates the schematic representation of the fabrication process of PEEK scaffolds with AgNPs. Detailed process of fabrication can be found elsewhere [12]. The AgNPs incorporated PEEK scaffold exhibits a significant response to bacteria by releasing Ag^+ ions possessing bacteria-killing power. Figure 2a represents the schematics of the fabrication process of AgNPs embedded PEEK scaffold. The Ag^+ ions are released as the scaffold is pH-triggered as shown in Fig. 2b. As the pH drops to 4.5 the Ag ions are released, resulting in protection against the bacteria. These antibacterial-coated scaffolds can reduce post-surgical infections linked to prosthetics, potentially lowering revision surgeries and improving patient outcomes [12]. In recent work, Podstawczyk et al. have developed a filament with antibacterial properties by incorporating silver nanoparticles into polylactic acid (PLA). The method involved mixing silver salts with PLA using both solution-based and solvent-free approaches, which were then extruded to form filaments. The filament exhibited antimicrobial capabilities over Staphylococcus aureus, Escherichia coli, and Pseudomonas aeruginosa. The filament can be employed to fabricate tailor-made antimicrobial 3D objects as required, specifically in scaffold printing applications [34].

Gold nanoparticles (AuNPs) are another type of metal nanoparticle that has been extensively incorporated into polymers. AuNPs have great potential for being incorporated with polymers in 3D-printed devices. AuNPs, renowned for their exceptional conductivity, are highly effective in transferring electrical charges. Moreover, their distinctive surface characteristics provide robust adhesion with a diverse range of enzymes and functional groups, making them highly adaptable for a multitude of applications in cardiovascular tissue engineering, bone-regeneration, and 3D-printed biosensing applications.

Ghaziof et al. developed a 3D scaffold using a nanocomposite made of Poly ϵ-caprolactone (PCL) and AuNPs by FDM technique. Different wt% of AuNPs were

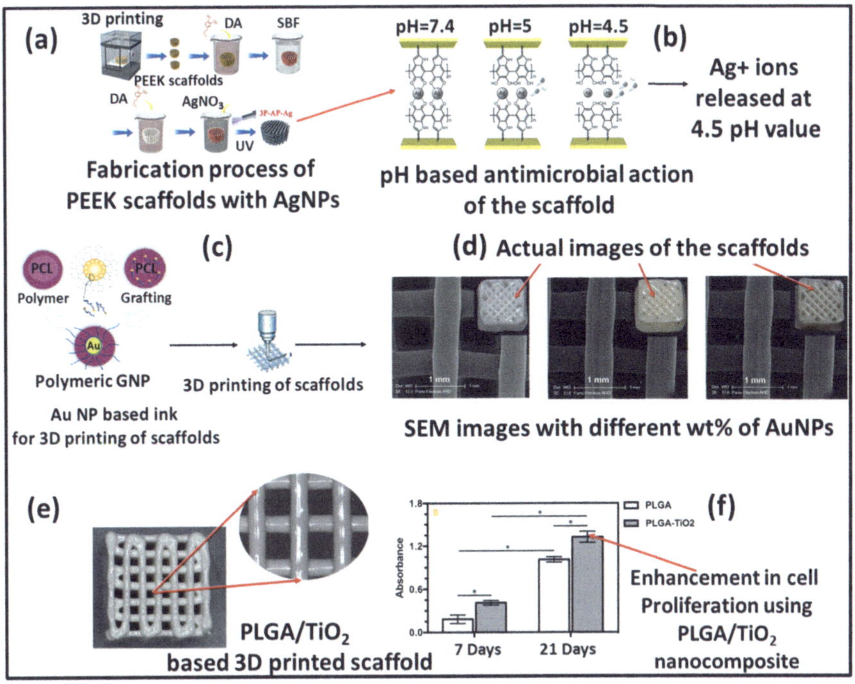

Fig. 2 Schematic representation of **a** the fabrication process of PEEK scaffolds with AgNPs, **b** pH-triggered response of the PEEK/AgNP scaffold. Reprinted from ref Deng et al. [12], Copyright (2020), with permission from the American Chemical Society. **c** Synthesis of PCL/AuNP ink for 3D printing of scaffold. (d) Actual images of the scaffolds with different wt% of AuNPs. Reprinted from ref Ghaziof et al. [21], Copyright (2022), with permission from Elsevier. **e** $PLGA/TiO_2$ based 3D printed scaffold. (f) Enhanced cell proliferation utilizing the $PLGA/TiO_2$ nanocomposite. Reprinted from ref Rasoulianboroujeni et al. [38], Copyright (2019), with permission from Elsevier

embedded into the PCL and it was observed that 0.5 wt% was the optimal concentration of AuNPs. Figure 2c shows the PCL-AuNPs nanocomposite as the base for printing the scaffold. SEM and actual images of the scaffolds with different wt% of AuNPs are shown in Fig. 2d. The electrical conductivity was enhanced by 25% with the incorporation of AuNPs and the compressive strength was also increased by 9% [21]. Hence, AuNPs-based polymer nanocomposites can enhance the electrical signal transmission required in these 3D-printed cardiovascular tissue regeneration scaffolds [21]. Marzo et al. utilized the electrical conductivity and adhesive properties of AuNPs to enhance the electron transfer ability in a 3D-printed enzyme biosensor. The group fabricated a 3D-printed graphene-PLA electrode incubated with AuNPs to study the electron transfer using horseradish peroxidase (HRP) for the detection of hydrogen peroxide (H_2O_2). AuNPs enhanced the process of electron transfer resulting in amplified signal and better performance of the 3D-printed enzyme-based biosensor [29]. Another biomedical application of polymer-AuNP nanocomposites is observed in the field of bone regeneration. Topsakal et al. developed a 3D scaffold as an alternative for bone tissue. The scaffold was 3D printed by utilizing a distinctive PVA ink with a mix of AuNPs and ampicillin (AMP). The purpose of this combination was to enhance the production of bone tissue and also provide antimicrobial qualities. The biocompatibility, osteoinductive, and antibacterial properties of the PVA/AuNP/AMP scaffolds were confirmed using in vitro 3D cell-based assays. The results demonstrated that the 3D-printed PVA/AuNP/AMP composite scaffold is biocompatible, promotes bone formation, and possesses antibacterial characteristics, in contrast to PVA scaffolds [52]. In another study on AuNPs-based polymer nanocomposites, Samadian et al. developed a 3D scaffold using PCL/PLA and embedded a composite of AuNPs and gelatin nanofibres (GNF). The PCL/PLA scaffold was embedded with AuNPs and GNF and the studies were performed on the defect region of the rat's skull. The animal investigation demonstrated that the AuNPs (80 ppm) in PLA/PCL/GNF/AuNPs scaffold resulted in the most significant bone regrowth. The AuNPs with 80 ppm demonstrated the most significant cell proliferation. This study demonstrates that the integration of three-dimensional scaffolds with zero-dimensional (AuNPs) and one-dimensional (GNFs) nanostructures can imitate the natural structure of bone and enhance the process of bone regeneration [42]. Moreover, these studies provide insight into AuNPs incorporation in bone regeneration and implications for managing infections during orthopedic treatments. Polymer-based AuNP-composites play a significant role in enhancing cardiovascular tissue regeneration, bone regeneration, and signal enhancement in 3D-printed biosensors.

In addition to metal NPs, nanocomposites based on metal oxides have also been researched and utilized for biomedical purposes in the regeneration of bone tissues. Titanium dioxide (TiO_2) has gained considerable interest for its biocompatibility, superior mechanical attributes, and rigid structure [4, 56]. Incorporating these traits of TiO_2, 3D-printed structurally compatible scaffolds with mechanical stability and biocompatibility have been fabricated to enhance bone regeneration and help tissue engineering [4, 56]. Rasoulianboroujeni et al. fabricated a 3D-printed porous scaffold utilizing the polylactic-co-glycolic acid (PLGA)/TiO_2 nanocomposite. The image of the PLGA/TiO_2-based 3D printed scaffold is shown in Fig. 2e. The incorporation of TiO_2 NPs enhanced the porosity as well as the compressive modulus of the scaffold by 7.7 and 50 units. Furthermore, the cell proliferation was also enhanced when utilizing the PLGA/TiO_2 nanocomposite as shown in Fig. 2f [38]. Bulut and group studied the bioactive, cytotoxic, and mechanical properties of a 3D printed chitosan/akermanite-TiO_2 (CS/AK/TiO_2) scaffold. A $30 \times 30 \times 3$ mm^3 scaffold was constructed using the 3DP method with different concentrations of TiO_2 (1, 3, 5, and 7%).

Among the various concentrations of TiO_2 NPs, the CS/AK/1Ti scaffold exhibited the highest level of cell proliferation. In addition, for 48 h, increasing TiO_2 concentrations enhanced the cell proliferation and the CS/AK/7TiO_2 scaffold exhibited the greatest cell proliferation. Furthermore, embedding TiO_2 NPs in the scaffold did not result in any cytotoxic effects [3]. In addition to structural stability, TiO_2 also exhibits a broad range of activity against microorganisms like bacteria, fungi, etc. Totu et al. fabricated a 3D-printed denture using PMMA-TiO_2 nanocomposite. It was demonstrated that the addition of TiO_2 to a PMMA polymer matrix inhibits the growth of microbes, particularly Candida species [53]. Therefore, it can be inferred that 3D-printed TiO_2-polymer nanocomposite scaffolds have significant promise as a material for diverse biomedical applications, such as bone tissue regeneration and manufacturing 3D-printed dentures. Moreover, the incorporation of TiO_2 NPs also enhances the osteoconductivity and osteoinductivity of the scaffolds.

4.2 Graphene-Polymer Nanocomposite for Fabricating 3D-Printed Scaffolds, Nerve Guidance Conduits (NGCs), and Biosensors

Graphene consists of a monolayer of carbon atoms organized in a hexagonal lattice. Graphene and its derivatives including graphene oxide (GO) and reduced graphene oxide (rGO) have attracted much attention because of their remarkable characteristics, including outstanding electrical conductivity, mechanical strength, and thermal conductivity [46]. Furthermore, graphene and its derivative-based nanomaterials (GNPs) possess intriguing characteristics including low density, high aspect ratio, 2D shape, strong electrical and thermal conductivity, and mechanical toughness [46, 64]. These unique characteristics are applied in diverse sectors including material science, medicine, electronics, etc.

Zhang et al. [64]. Integration of GNPs in the polymer matrix can enhance the electrical and thermal conductivity, stability, and effective surface area of the GNP-polymer-based nanocomposite which can be utilized in the biomedical field for the development of 3D-printed scaffolds for bone regeneration [46].

Dias et al. fabricated PCL-graphene composite-based 3D scaffolds utilizing the FDM technique for bone tissue engineering. The PCL-graphene composite material obtained a fourfold enhancement in the compressive modulus. Furthermore, the composite scaffolds did not exhibit any cytotoxic effects, and there was a greater proliferation of SaOs-2 cells on the composite scaffolds. Moreover, the 3D-printed PCL/graphene scaffolds demonstrated suitable microstructural and thermal characteristics and positive osteoblastic differentiation ability [14]. Mohammadi and colleagues conducted a study to examine the impact of mechanical properties and biocompatibility by integrating varying weight percentages of graphene oxide (GO) into the 3D-printed scaffolds (2 mm thickness × 15 mm diameter) made from PCL/GO composites. The synthesis procedure of the ink and fabrication of the scaffold with different PCL/GO wt% is shown in Fig. 3a–c. The details of the experimental procedure can be found elsewhere [41]. Integration of GO into the polymer matrix enhanced the pore size as well as the porosity of the scaffolds which can enhance cell multiplication. The biocompatibility and the cell viability were enhanced by the incorporation of GO as shown in Fig. 3d. The excellent cell survival and adhesion of MG-63 osteoblast cells indicate that PCL/GO composite scaffolds possess

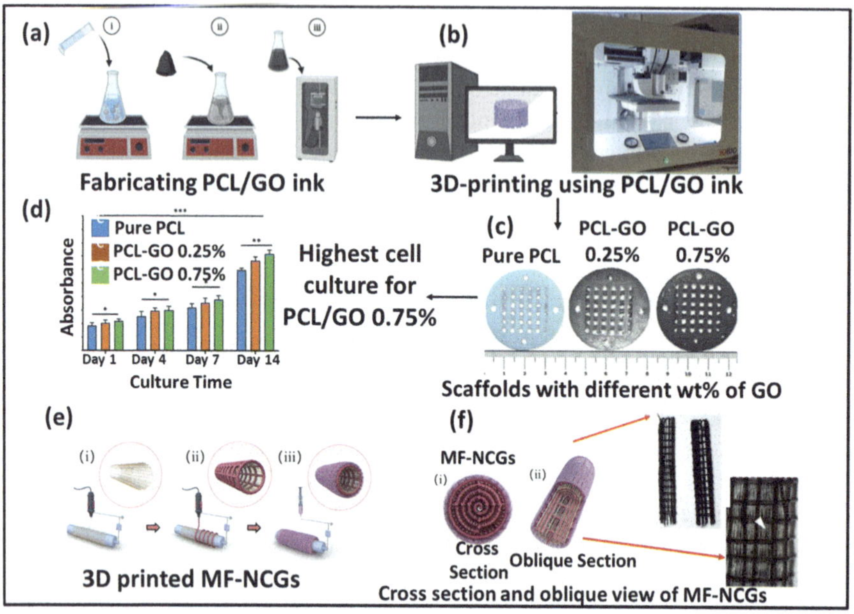

Fig. 3 **a–c** Schematics of the synthesis procedure of the ink and fabrication of the scaffold with different PCL/GO wt%. **d** Enhanced biocompatibility and cell viability by the incorporation of GO. Reprinted from ref Sahafnejad-Mohammadi et al. [41], Copyright (2023), with permission from Wiley. **e** Different components of the 3D printed MF-NCGs. **f** Cross-sectional and oblique view of the MF-NCGs. Reprinted from ref Fang et al. [16], Copyright (2023), with permission from Wiley

highly favorable biocompatibility, making them a highly viable option for bone tissue engineering applications [41]

Peripheral nerves provide connections between these central nervous system structures and the other parts of the body. They have a vital purpose in controlling muscular action, sensory perception, and involuntary activities like heartbeat and digestion. Injuries to peripheral nerves are a prevalent type of trauma affecting millions of individuals globally, making nerve regeneration a crucial area of medical research. Nerve guidance conduits (NGCs) are synthetic substitutes, offering temporary mechanical assistance and a conducive microenvironment for nerve regeneration. The NGCs can be 3D printed and can help in the nerve regeneration process Incorporation of graphene/graphene-derived nanomaterials in the polymer matrix to form 3D-printed NCGs can enhance the conductivity in NGCs which is essential for nerve regeneration because it facilitates the transmission of bioelectrical signals, which are critical for the functional restoration of peripheral nerves These electrical signals play a pivotal role in guiding and promoting the growth of nerve fibers across the injury site, enhancing the speed and effectiveness of nerve regeneration and functional recovery. Fang et al. fabricated conductive 3D-printed NCGs (MF-NCGs) utilizing PCL/RGO microfibres by electrospinning and MEW technique.

Figure 3e shows different parts of the 3D printed MF-NCGs wherein, Fig. 3e-i represents the PCL microfibers, Fig. 3e-ii depicts the microfibers made from MEW-printed reduced graphene oxide (rGO), and polycaprolactone (PCL) (rGO/PCL microfibers). Figure 3e-iii illustrates the process of electrospinning PCL/collagen nanofibers. MF-NCGs have the combination of a backbone made of rGO/PCL microfibers, a coating made of PCL/collagen nanofibers, and an internal structure made of PCL microfibers. Figure 3f displays a cross-sectional and oblique image of the MF-NCGs. The study evaluated the effectiveness of the microfibers in guiding the formation of neurites by growing different cells on MF-NCGs. Through animal research, it was observed that MF-NGCs facilitated the process of neural regeneration and enhanced the functional restoration of both nerve and muscle tissues. The development of conductive MF-NGC offers an exciting and innovative method for treating peripheral nerve damage. Fang et al. [16]. Farzan and his colleagues developed green and conductive composites made of polyurethane/PEGylated graphene oxide (PU/PEG-GO). Stereolithography was employed to 3D print flexible nerve conduits of various shapes, including hollow, permeable, and grooved tubes, utilizing these materials. The composite with 5% PEG-GO had exceptional mechanical characteristics, including a notable tensile strength of 3.51 ± 0.54 MPa and a tensile strain at the point of fracture of roughly 170%. In addition, it exhibited a conductivity value of 1.1×10^{-3} S/cm as well as a minimal contact angle of 72°. These properties were attributed to the strong interfacial interactions between the PEG/GO nanosheets and the PU matrix. Furthermore, the PU/PEG-GO 5% conduit exhibited enhanced compressive strength in comparison to pure PU and experienced appropriate enzymatic breakdown for 6 weeks, indicating its potential long-lasting nature for successful nerve regeneration. Overall, the 3D-printed PU/PEG-GO 5% conduit, characterized by its conductivity, biodegradability, flexibility, and precise geometry, holds promise as a nerve guidance conduit (NGC) for peripheral nerve regeneration [18].

Graphene-based composites have been employed to fabricate 3D-printed electrodes and circuits tailored for sensing applications. Combining graphene derivatives with polymers produces conductive composites that are well-suited for fabricating electrodes and circuits using printing techniques. These can be integrated into complex frameworks, allowing for the electrochemical detection of organic compounds and active biological molecules. Graphene-infused composites can be combined with bio logical components such as enzymes, antibodies, and single-stranded DNA. The 3D-printed graphene biosensors demonstrate consistent and dependable functionality and hold considerable potential for various applications [44]. Silva et al. utilized 3D-printed rGO/PLA electrodes using FDM for biosensing applications. The rGO/PLA electrode demonstrated successful application in detecting serotonin in synthetic urine using differential pulse voltammetry (DPV), exhibiting superior sensing capabilities compared to prior studies. Furthermore, the rGO-PLA surface was utilized to construct a catechol biosensor, which effectively detected the analyte in natural water samples with satisfactory analytical performance. The limit of detection (L.O.D) for serotonin and catechol detection was found to be 0.032 μmol/L and 0.26 μmol/L respectively [47]. In another work on graphene-based composite materials, Cardoso et al. utilized the FDM technique to fabricate biosensing platforms using filaments made of graphene-PLA composite. The biosensor utilizing enzymes was fabricated on the surface of PLA-graphene to detect glucose in blood plasma. The biosensor demonstrated an L.O.D. of μmol/L. Furthermore, the biosensor exhibited improved electrochemical characteristics for directly detecting nitrite and uric acid [5]. The emergence of flexible electronic devices has opened up fresh possibilities for monitoring respiratory function. Humidity sensors, designed with this technical innovation, can detect and track fluctuations near the human body. This capability is extremely advantageous for early detection and treatment of illnesses. Chen and the group created a flexible humidity sensor by using GNPs/MWCNTs (direct ink-printed) as the material for the sensory electrode and PDMS as the bendable substrate. The device demonstrated exceptional performance with an extremely rapid response time of 0.35 s and a recovery time of 2.5 s. It also exhibited high sensitivity, measured at 1.4 RH-1, and excellent repeatability with 500 cycles. Additionally, the capacity of this humidity sensor for detecting moisture without physical contact was investigated. This sensor is expected to be crucial in precisely identifying abnormal breathing patterns, including suffocation, respiratory arrest, and asthma. As a result, it will significantly impact monitoring human health [7].

4.3 3D-Printed Magnetic Polymer Nanocomposite for Hyperthermia Treatment

Cancer is a medical condition where cells divide uncontrollably, leading to the development of abnormal tissue masses called tumors over a long period. Cancer is a prominent cause of death worldwide, resulting in around 9.6 million fatalities in 2018 [20]. Multiple therapeutic interventions, including surgical procedures, radiation therapy, immunotherapy, chemotherapy, and genetic changes, have greatly improved the management of cancer. Nevertheless, several of these methods carry inherent dangers, affecting neighboring healthy cells in the vicinity of the treatment site. Hyperthermia (HT) is a potentially effective alternative to traditional treatments, providing a gentler and non-invasive form of heat therapy [20, 33]. This strategy utilizes the greater susceptibility of cancer cells to heat as a result of their elevated metabolic rates [20]. By exposing cancer cells to temperatures ranging from 41 to 46 °C for prolonged durations, which surpasses the body's usual temperature of 37 °C, it triggers substantial cellular stress. The increase in temperature causes stress within the cells, which results in the disruption of cellular activities. This disruption leads to the breakdown of protein assembly and ultimately results in apoptosis, which is the death of the cell [20].

Magnetic nanoparticles (MNPs) are crucial components utilized in hyperthermia as they possess the capability to transform magnetic energy into thermal energy upon exposure to an external magnetic field. Magnetic hyperthermia (MHT) is a promising therapy for treating cancer. Magnetic nanoparticles (MNPs) including Fe_3O_4, Fe_2O_3,

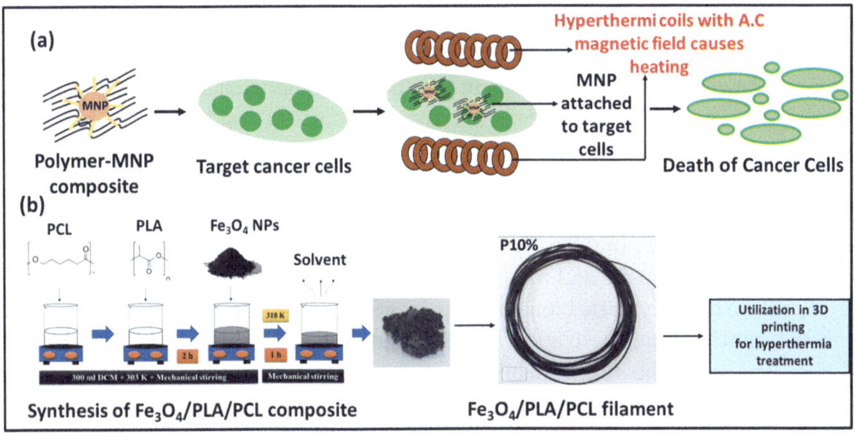

Fig. 4 **a** Schematic representation of utilizing polymer-magnetic-nanocomposite for killing cancer cells by using hyperthermia coils. **b** Fabrication process of Fe_3O_4/PLA/PCL composite and the filament for hyperthermia treatment. Reprinted from ref Galarreta-Rodriguez et al. [19], Copyright (2023), with permission from Springer

Ni, Co, Fe, etc. possess dimensions that allow them to effectively traverse physiological barriers and specifically target metastatic locations. Magnetic nanoparticle-based composites in hyperthermia use the dispersion of magnetic nanoparticles within a matrix material. This enables precise heat therapy, particularly in the treatment of cancer. When subjected to an oscillating magnetic field, these nanoparticles produce concentrated heat, which causes hyperthermia to specifically eliminate or sensitize cancer cells. The magnetic characteristics of the composite allow for precise distribution by utilizing an external magnetic field, hence reducing harm to unaffected tissues. By tailoring the qualities of the composite, it is possible to optimize heat generation, biocompatibility, and stability. Figure 4a shows the schematic illustration of the MHT utilizing magnetic nanoparticles for killing the cancer cells.

Zhang et al. utilized 3D printing technology to fabricate composite scaffolds by integrating magnetic Fe_3O_4 NPs into mesoporous bioactive glass/polycaprolactone (Fe_3O_4/MBG/PCL). The group conducted a comprehensive analysis of the physical and chemical properties, bioactivity in laboratory conditions, ability to deliver anticancer drugs, mechanical strength, magnetic heating capacity, and cellular responses of the Fe_3O_4/MBG/PCL scaffolds. The results showed that the scaffolds had a constant pore size of 400 nm, a porosity that was a high level of 60%, and excellent compressive properties ranging from 13–16 MPa. The magnetic heating capability of the composite with different wt% of Fe_3O_4 was examined and it was observed that the 15Fe_3O_4/MBG/PCL scaffolds increased the temperature from 20 to 43 °C in 2 min using an AC frequency of 409 kHz and a magnetic field of 180 G. Furthermore, a clear correlation was found between the rate at which the temperature was increased and the amount of Fe_3O_4 present in the scaffolds. Furthermore, the Fe_3O_4/MBG/PCL scaffold demonstrated versatile capacities, promoting the process of bone regeneration. Hence, the adaptability of the 3D-printed magnetic Fe_3O_4/MBG/PCL scaffolds demonstrates considerable promise in treating bone deficiencies caused by bone cancers. This potential is ascribed to a synergistic effect resulting from increased bone-forming activity, precisely localized administration of anticancer drugs, and the application of magnetic hyperthermia [61]. In another work, Rodriguez and the group developed low-cost magnetic filaments for FDM 3D printing applications by embedding Fe_3O_4 NPs into the PLA/PCL matrix (Fe_3O_4/PLA/PCL). The fabrication process of the composite and the filament is illustrated in brief in Fig. 4b. Adding Fe_3O_4 NPs to the polymeric blend imparts magnetic characteristics to the composite, with the saturation magnetization being directly proportional to the concentration of NPs. To test the heating capabilities of the composite for MHT application, an assessment was conducted to determine the filament's capacity to generate heat when exposed to low-frequency AC magnetic fields, making use of its magnetic properties. The composite could rapidly increase temperatures when exposed to an alternating current magnetic field (at a frequency of 311 kHz and magnetic field strength of 400 Oe) in different conditions. These magnetic composites-based filaments can be widely utilized in the 3D printing of various materials and scaffolds for magnetic-based medical applications such as magnetic hyperthermia, medication delivery, etc. [19].

Graphene oxide (GO) exhibits exceptional thermal conductivity attributed to its two-dimensional carbon-based structure and the presence of ionized edge groups. GO holds significant potential for advanced thermal management applications. In light of this, Zhang et al. developed a 3D- printed β-tricalcium phosphate bioceramic scaffold was developed, featuring surface enhancements with nanocomposite layers of Fe_3O_4 NPs and GO, termed as βTCP/FeGO nanocomposite. In the βTCP/FeGO scaffolds, after eight coatings, the Fe_3O_4 content remains below 1%. The presence of a low Fe_3O_4 content gives the scaffolds superparamagnetic characteristics and hyperthermia effects. The scaffold temperature can be regulated within the range of 50–80 °C by manipulating the magnetic strength and Fe_3O_4 content while subjecting it to an alternating magnetic field for 15 min. During in vitro experiments, the βTCP/FeGO scaffolds exhibited significant hyperthermia effects, leading to more than 75% mortality in osteosarcoma cells (MG-63). The unique combination of GO and Fe_3O_4 enhances the thermal and magnetic properties respectively in the resulting composite thus ensuring optimal utilization of the composite in killing the cancer/tumor cells [65]. Li and coworkers also examined the synergistic effect of GO and Fe_3O_4 NPs in the composite for tumor treatment. The group fabricated composite hydrogel scaffolds composed of polyvinyl alcohol, sodium alginate, and hydroxyapatite (PVA/SA/HA) using additive manufacturing (AM). These scaffolds were incorporated with different concentrations of GO encapsulated with Fe_3O_4 NPs. The magnetothermal characteristics of the composite scaffolds were studied and it was found that the scaffolds with 15 wt% of magnetic NPs and GO were able to achieve temperatures greater than 45 °C in dry as well as humid environments under the applied magnetic field of 250 Gs. The anti-tumor effect of the composite was verified on mice by utilizing the magneto-thermal effect. The composite scaffold was implanted and an alternating magnetic field (AMF) was applied for 14 days. It was observed that 14 days of treatment with an alternating magnetic field (AMF), successfully inhibited the growth of the tumor. The temperature increase generated by the AMF affected the equilibrium between early and late apoptosis. The observed cytotoxicity at a temperature of 43 °C was primarily attributed to late apoptosis. However, at a temperature of 55 °C, both necrosis and late apoptosis were shown to contribute to the cytotoxic effects. Exposure to temperatures as high as 55 °C is recognized to have the ability to trigger inflammation and promote the dissemination of cancer to other body regions via necrotic cell death. Hence, the group was recommended to employ a treatment temperature of approximately 43 °C for cancer therapy. To achieve the required temperature, one can manipulate the magnetic strength of the implant material and control the intensity of the alternating magnetic field (AMF) [25].

5 4D Printing Technology and Shape Memory Polymer Nanocomposites (SMPNs)

4D printing solutions include integrating a time-dependent component into the 3D printing technique [9]. This enables printed objects to experience alterations in their physical properties, usually in terms of shape or structure, when exposed to specific stimuli. 4D printing is an expansion of 3D printing that includes technological and biomedical aspects [26, 67]. Frequently employed methods encompass light- and extrusion-based methodologies, which expand upon conventional 3D printing systems. These techniques employ diverse printing materials and processes to attain distinct 4D effects, including responsiveness to stimuli, robust mechanical properties, and reversible actuation, customized for specific purposes [45]. By harnessing the capabilities of 4D printing technology, structures with distinct characteristics like self-assembly, multi-functionality, and self-healing may be precisely constructed. These structures can then be utilized across a wide range of industries and applications [26, 67].

Shape memory polymer (SMP) is a substance that retains a temporary shape when exposed to external forces, changes its shape when stimulated, and then returns to its original shape to complete a shape memory cycle. The innovative approach of 4D printing, pioneered by Tibbitis from MIT Tibbits [50], has propelled advancements in various sectors. Shape Memory Polymers (SMPs) find applications in aerospace, additive manufacturing, apparel materials, and biomedical science, owing to their exceptional performance. In biomedical applications, SMPs are crucial for SMP scaffolds, dental equipment, and aneurysm occludes. Moreover, the integration of SMPs into 4D printing techniques has opened up new possibilities in material design and fabrication. By harnessing the principles of shape memory behavior, 4D-printed objects can exhibit dynamic responses to external stimuli, enabling the creation of self-adapting structures and devices [26, 67]. This integration has led to significant

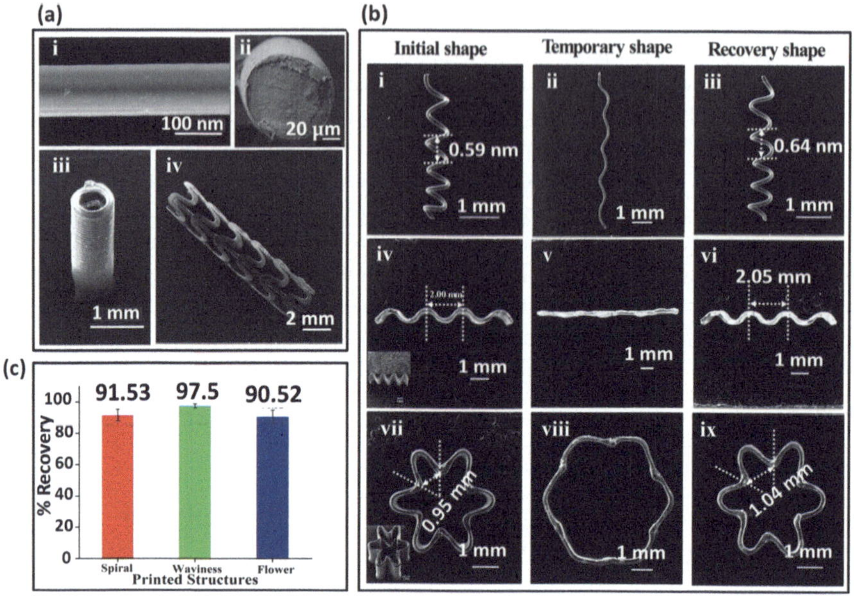

Fig. 5 SEM images and 4D shape-changing behavior of structures printed with c-PLA ink. **a** Examples of printed 4D active structures. **b** Shape-changing behavior: (i–iii) 3D microspiral reverting from 1D shape. (iv–vi) Wavy structure transitioning from 2D sheet. (vii–ix) Flowerlike shape returning from 3D hexagonal shape when heated. **c** Analysis of shape changes, showing deformation and recovery in hot water at approximately 80 °C. Reprinted from ref Wei et al. [58], Copyright (2017), with permission from the American Chemical Society

advancements in additive manufacturing, particularly in the fields of soft robotics and biomedical engineering. Figure 5 depicts scanning electron microscope (SEM) pictures and the dynamic shape-changing behavior of structures fabricated with c-PLA ink [58]. Figure 5a illustrates typical examples of printed 4D active shape-changing architectures. Figure 5b exhibits the shape-altering characteristics of certain printed structures. More precisely, pictures (i–iii) illustrate a 3D microspiral that can store a 1D line-like shape and then return to its original 3D microspiral form when exposed to heat. Images (iv–vi) depict the transformation of a printed waviness-like shape from a temporary 2D sheet form to its original 3D waviness shape when exposed to heat. Images (vii–ix) depict a 3D structure resembling a flower that may be altered into a temporary 3D hexagonal shape and then return to its original flowerlike configuration when subjected to heat. The study offers a quantitative analysis of the morphological changes exhibited by the specific structures. Shape distortion and restoration of the 3D microspiral, 3D waviness-like structure, and 3D flowerlike structure took place in hot water at approximately 80 °C. These structures can be utilized for various biomedical applications which will be discussed in the next section.

In recent years, advancements in nanoscience and nanotechnology have further enhanced the capabilities of SMPs through the development of shape-memory polymeric nanocomposites (SMPNs). These materials, comprising a continuous phase of SMPs and a discontinuous phase of nanoparticles, offer exceptional mechanical, electrical, and optical properties compared to SMPs alone [26, 67]. SMPNs have found widespread applications in aeronautical constructions, medical implants, optical crystals, and soft robots, showcasing their significant potential in numerous industries. The incorporation of SMPNs into 4D printing processes extends the versatility and functionality of printed objects, allowing for precise control over shape-changing behavior and dynamic responses to environmental stimuli. This integration holds promise for the development of advanced materials and devices with tailored properties, driving innovation in fields such as aerospace, healthcare, and beyond. As research in 4D printing and SMPNs continues to advance, we can expect transformative applications in material science, medicine, and engineering [26, 67].

In the upcoming section, this chapter will explore the biomedical applications of integrating nanoparticles (NPs) into shape-memory polymers (SMPs) to create shape-memory polymeric nanocomposites (SMPNs). SMPNs have gained traction in the biomedical field across various sectors. Notably, 4D-printed SMPNs are finding widespread use in scaffold engineering, stent manufacturing, and the development of other biomedical devices. These innovative materials offer tailored mechanical properties, responsiveness to external stimuli, and the ability to adapt to dynamic biological environments. Researchers and practitioners are employing the unique characteristics of SMPNs to tackle issues in tissue engineering, implantable devices, and regenerative medicine. This will lead to the development of innovative solutions that offer improved performance and functionality in biomedical applications.

6 Application of 4D-Printed Shape-Memory Polymeric Nanocomposites (SMPNs) in Medical and Biomedical Field

4D printed shape memory polymeric nanocomposites (SMPNs) offer promising prospects in the biomedical field, particularly for developing occluders, vascular stents, and bone scaffolds. Their inherent features enable the creation of tailored structures that can adjust to physiological conditions providing possibilities for developments in the treatment of cardiovascular disorders and the promotion of bone regeneration.

6.1 4D-Printed SMPNs in Developing Occluders and Stents

Occluders are devices that obstruct or close off unusual pathways or structures, such as congenital heart abnormalities or blood arteries [24]. They remain fixed in position indefinitely, impeding the flow of blood or fluids. On the other hand, stents provide structural support to tubes or vessels, effectively reducing the occurrence of blockages in coronary or peripheral arteries. Stents are inserted in confined or weakened blood vessels to ensure the continuous circulation of blood [58]. SMPNs have been utilized in the fabrication of these stents and occluders to prevent blockages in blood vessels.

Wei and colleagues integrated Fe_3O_4 NPs with ultraviolet (UV)-based PLA inks to create shape-changing smart materials with four dimensions (4D). These materials were designed for use in the fabrication of occluders and stents. The inclusion of iron oxide effectively combines 4D shape-changing objects with swift remotely activated and magnetically controllable characteristics. Figure 6b depicts the shape memory process of the spiral device. When subjected to a confined environment, the spiral structure undergoes form recovery, causing it to expand. This expansion is observed when the spiral is placed within a plastic cylindrical holder with a bigger diameter. The remotely controlled self-expansion process proceeds swiftly by using an alternating magnetic field, resulting in a quick change in the inner diameter from 1 to 2.7 mm within a time frame of 10 s as shown in Fig. 6c. The 4D spiral scaffold exhibits potential as a self-expanding vascular stent, capable of restoring the diameter of confined arteries afflicted by blood clots, hence maintaining unobstructed blood circulation as depicted in Fig. 6d. The inclusion of Fe_3O_4 NPs allows for magnetically directed and remotely controlled actions, hence improving manipulation within the body. The direct-write printing technology enables the creation of adaptable designs, which in turn promotes progress in the development of vascular stents and occluders [58].

Lin et al. also utilized the Fe_3O_4 NPs and PLA matrix to fabricate biodegradable 4D occluders with shape memory behavior. The key features of this occluder are as follows. The occluder can enhance cell adhesion, proliferation, and tissue growth, which helps to facilitate the quick process of endothelialization. Furthermore, the occluder is biodegradable, mitigating the risk of a metallic occluder persisting in the body indefinitely. This occluder can be customized to temporarily assume a linear shape, and the process of deploying it can be controlled remotely. This decreases the scar's size throughout the procedure and increases the likelihood of clinical obstruction. The occluder exhibits strong mechanical reinforcement to effectively seal defects, even after undergoing an additional 16 weeks of degradation in vitro. It effectively and quickly seals all the faults completely. The 4D-printed shape memory occluder is expected to have great potential as an occlusion device, and this research establishes the groundwork for its future development [27]. Wei and the group developed SMPNs using silver-doped carbon nanofibers (Ag/CNF). These hybrid nanocomposites based on Ag/CNF demonstrate rapid and low-voltage-induced shape memory characteristics, which makes it a highly interesting option for printing electroactive devices. Diverse nanocomposite structures with excellent

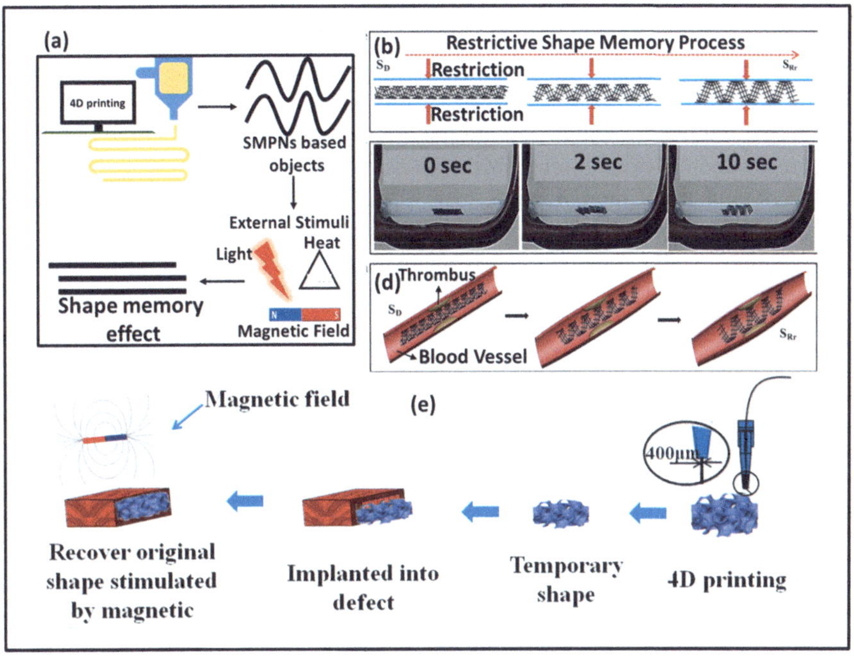

Fig. 6 **a** Utilization of 4D printed shape memory polymeric nanocomposites (SMPNs). **b** Shape memory behavior of the spiral device. **c** The remotely controlled self-expansion process proceeds swiftly by using an alternating magnetic field, resulting in a quick change in the inner diameter from 1 to 2.7 mm within a time frame of 10 s. **d** The 4D spiral scaffold exhibits potential as a self-expanding vascular stent, capable of restoring the diameter of confined arteries afflicted by blood clots, hence maintaining unobstructed blood circulation. Reprinted from ref Wei et al. [58], Copyright (2017), with permission from the American Chemical Society. (e) Schematics representing the shape memory effect for effective deployment of PLA/Fe_3O_4 scaffolds triggered by a magnetic field. Reprinted from ref Zhao et al. [66], Copyright (2021), with permission from Elsevier

conductivity are showcased, such as conductive components that may be printed in ambient conditions, fiber sensors with microstructures, and lightweight scaffolds for diverse applications including scaffold and stent manufacturing. The combination of this groundbreaking method of additive manufacturing and the collaborative impact of the nanocomposite creates new opportunities for the progress of intelligent devices in the biomedical field [57].

6.2 4D-Printed SMPNs in Developing Scaffolds for Bone Tissue Regeneration

SMP nanocomposites exhibit possibilities for addressing the difficulties of bone defect restoration. By combining SMP with nanomaterials, such as nanoparticles or nanofibers, numerous improvements arise. To begin with, nanocomposites improve mechanical strength, which is essential for load-bearing purposes in bone tissue engineering. Additionally, they enhance biocompatibility and facilitate the attachment and growth of cells, which is crucial for the regeneration of tissues. Furthermore, the shape-altering capabilities of the SMPNs can be enhanced and regulated by external elements such as temperature, magnetic field, etc., hence allowing for customization and favorable self-fitted scaffold construction for enhanced bone cell regeneration.

Zhao et al. developed a concept based on shape memory PLA/Fe_3O_4 composite for bone tissue regeneration. These scaffolds are designed to be implanted using minimally invasive surgery and may be activated to return to their functional condition using a magnetic field. Both empirical and micromechanical theoretical investigations support the mechanical integrity of these porous scaffolds. Biological examinations confirm that the scaffolds promote increased cell adhesion and growth, indicating their potential for delivering cells and regenerating bone tissue.

By incorporating Fe_3O_4 nanoparticles into the PLA material, the scaffolds gain the ability to respond to magnetic stimulation. This enables the shape memory effect for effective deployment triggered by a magnetic field as shown in Fig. 6e. This demonstrates the suitability of these scaffolds for less invasive implantation. This approach presents a unique strategy for efficiently healing bone defects by introducing a porous structure into the design of bone tissue scaffolds and combining a shape memory effect [66]. Zhang and group developed a robust nanocomposite, consisting of shape memory polyurethane (SMPU) modified with hydroxyapatite (HA), reduced graphene oxide (rGO), and arginine-glycine-aspartic acid (RGD) for bone repair applications. The SMPU/HA/rGO/RGD nanocomposite exhibits a significant improvement in Young's modulus, with an increase of roughly 200%, as well as a substantial enhancement in tensile strength, with an increase of over 300%, when compared to the original SMPU. These improvements can be attributed to the nanocomposite's intercalated structure and the strengthened internal interactions between the nanofillers and the matrix. The mechanical qualities are further enhanced by chemically immobilizing RGD, which leads to an increase in covalent bonds. RGD promotes cell adhesion, as demonstrated by the enhanced attachment of rabbit osteo mesenchymal stem cells (MSCs), indicating that it could be beneficial for bone regeneration. The nanocomposite's exceptional shape memory performance makes it a highly intriguing option for less-invasive bone healing applications [63]. Guo et al. developed a novel nanocomposite material consisting of a PLA matrix and modified magnesium oxide (MgO) NPs using ferric chloride ($FeCl_3$) and tannic acid (TA), resulting in PLA/$FeCl_3$/TA/MgO nanocomposite. This nanocomposite material is responsive to near-infrared (NIR) light and has shape-memory properties. This design integrates the printability and shape-memory capabilities of PLLA with the

Table 2 Categorization of nanomaterial types and their corresponding biomedical applications within 3D and 4D printed polymer nanocomposites

Biomedical application	Nanomaterials incorporated in polymer matrix	Printing technique	References
Scaffold engineering	Ag, Au, TiO_2, Graphene, GO, rGO	3D	[3, 12, 14, 21, 30, 38, 41, 52]
Biosensing	Au, rGO, Graphene	3D	[5, 29, 47]
NCGs for nerve regeneration	rGO, Graphene, GO	3D	[16, 18]
Hyperthermia treatment	Iron oxide, GO	3D	[19, 25, 61]
Developmentof occluders and stent	Fe_3O_4, Ag	4D	[27, 57, 58]
Bone-tissue regeneration	Fe_3O_4, rGO. MgO	4D	[23, 63, 66]

remarkable biological activity and photothermal properties of MgO nanoparticles that have been modified by $FeCl_3$/TA. The inclusion of elements and the fabrication procedure synergistically improve the mechanical properties and durability of the NIR-enabled shape-memory effect in the composite scaffolds. Additionally, the composite scaffolds acquire antibacterial and antioxidant capabilities. By adding 5 wt% $FeCl_3$/TA/MgO particles, the scaffold's compressive modulus increases to 0.93 MPa, which is four times greater than the original PLA scaffold. This study presents a new composite scaffold made of PLA/$FeCl_3$/TA/MgO that exhibits excellent mechanical properties, as well as a shape-memory effect generated by near-infrared radiation. The material has antibacterial and antioxidant capabilities, making it a highly promising candidate for bone restoration [23] (Table 2) shows the categorization and biomedical applications of different types of nanomaterials reviewed in this chapter.

7 Conclusion, Challenges, and Perspectives

To summarize, the utilization of 3D and 4D printed polymer nanocomposites in the biomedical industry shows great promise for a wide range of applications. The addition of metal/metal oxide nanoparticles such as Ag and Au to a polymer matrix results in significant antimicrobial properties in the composites, which can be employed for 3D printing of scaffolds for bone tissue engineering. Moreover, conductive nanoparticles such as graphene and graphene oxide Au can be employed to augment the sensitivity and signal in three-dimensional printed biosensors. Furthermore, the inclusion of magnetic nanoparticles (MNPs) can lead to the creation of magnetic nanocomposites. These nanocomposites may be manipulated by external magnetic fields and can be used in applications such as hyperthermia treatment. In addition, incorporating the nanocomposite into a polymer matrix can also provide shape memory properties. The printed devices made from shape memory polymeric nanocomposites (SMPNs) can be activated by external stimuli, so introducing a fourth dimension. These SMPNs may be remotely controlled using an external stimulus. They have potential applications in creating stents, occluders, and scaffolds.

Nevertheless, many challenges require consideration. A major issue arises from the restricted selection of materials that can be utilized to produce complex structures with exceptional precision and accuracy. Moreover, several current printing methods exhibit a slow speed, impeding the possibility of extensive output. This highlights the necessity for specialized research that concentrates on creating novel materials or altering current ones to improve their ability to be printed and their level of detail. Furthermore, the problem arises from the inadequate long-term mechanical durability of numerous polymer composites, namely those designed for 4D printing. These materials frequently undergo deterioration problems when exposed to temperature cycles, which undermines their long-term endurance. It is essential to tackle these problems to make progress in the field and fully use the capabilities of 3D and 4D printed polymer nanocomposites in medical applications. The prospects for the utilization of 3D and 4D printed polymer nanocomposites in medical applications appear to be quite favorable since they have the ability to surpass existing constraints through future developments. Ongoing research on innovative materials and advanced printing procedures is anticipated to greatly enhance the resolution, speed, and scalability of these technologies. In addition, the advancement of more durable polymers capable of enduring temperature cycles and mechanical strain will improve the durability and dependability of 4D-printed medical equipment. With the ongoing advancement of these technologies, we may expect a forthcoming era of customized and adaptable medical solutions that will transform the way patients are cared for and improve treatment results. Combining smart materials and biocompatible elements can broaden the range of uses for 4D printing in medical devices capable of dynamic and interactive biomedical applications.

Acknowledgements D.M. acknowledges the financial support from MHRDDST through Grant No. IMP/2018/001668, SERB through Grant No. CRG/2018/004264, IoE, University of Delhi through Grant No. IoE/2021/12/FRP, and UGC through Grant No. F.4-5(202- FRP)/2015(BSR). M.T. acknowledges UGC and CSIR New Delhi for providing financial support.

References

1. Attaran, M.: The rise of 3-d printing: the advantages of additive manufacturing over traditional manufacturing. Bus. Horiz. **60**(5), 677–688 (2017)
2. Bratovcic, A.: Different applications of nanomaterials and their impact on the environment. SSRG Int. J. Mater. Sci. Eng. **5**(1), 1–7 (2019)

3. Bulut, B., Duman, Ş.: Evaluation of mechanical behavior, bioactivity, and cytotoxicity of chitosan/akermanite-tio2 3d-printed scaffolds for bone tissue applications Ceram. Int. **48**(15), 21378–21388 (2022)
4. Carbone, R., Marangi, I., Zanardi, A., Giorgetti, L., Chierici, E., Berlanda, G., et al.: Biocompatibility of cluster-assembled nanostructured tio2 with primary and cancer cells. Biomaterials **27**(17), 3221–3229 (2006)
5. Cardoso, R.M., Silva, P.R., Lima, A.P., Rocha, D.P., Oliveira, T.C., do Prado, T.M., Munoz, R.A.: 3d-printed graphene/polylactic acid electrode for bioanalysis: biosensing of glucose and simultaneous determination of uric acid and nitrite in biological fluids. Sens. Actuators, B Chem. **307**, 127621 (2020)
6. Chanda, M., Roy, S.K.: Industrial Polymers, Specialty Polymers, and Their Applications. CRC Press (2008)
7. Chen, X., Ma, K., Ou, J., Mo, D., Lian, H., Li, X., Luo, Y.: Fast-response non-contact flexible humidity sensor based on direct-writing printing for respiration monitoring. Biosensors **13**(8), 792 (2023)
8. Chen, Z.-Y., Gao, S., Zhang, Y.-W., Zhou, R.-B., Zhou, F.: Antibacterial biomaterials in bone tissue engineering. J. Mater. Chem. B **9**(11), 2594–2612 (2021)
9. Chu, H., Yang, W., Sun, L., Cai, S., Yang, R., Liang, W., Liu, L.: 4d printing: a review on recent progresses. Micromachines **11**(9), 796 (2020)
10. Chung, H., Das, S.: Functionally graded nylon-11/silica nanocomposites produced by selective laser sintering. Mater. Sci. Eng., A **487**(1–2), 251–257 (2008)
11. Correia, T.R., Figueira, D.R., de Sá, K.D., Miguel, S.P., Fradique, R.G., Mendon¸ca, A.G., Correia, I.J.: 3d printed scaffolds with bactericidal activity aimed for bone tissue regeneration. Int. J. Biol. Macromol. **93**, 1432–1445 (2016)
12. Deng, Y., Shi, X., Chen, Y., Yang, W., Ma, Y., Shi, X.-L., Chen, Z.-G.: Bacteria- triggered ph-responsive osteopotentiating coating on 3d-printed polyetheretherketone scaffolds for infective bone defect repair. Ind. Eng. Chem. Res. **59**(26), 12123–12135 (2020)
13. Dermanaki Farahani, R., Dubé, M.: Printing polymer nanocomposites and composites in three dimensions. Adv. Eng. Mater. **20**(2), 1700539 (2018)
14. Dias, D., Vale, A.C., Cunha, E.P., Paiva, C., Reis, R.L., Vaquette, C., Alves, N.M.: 3d-printed cryomilled poly (ε-caprolactone)/graphene composite scaffolds for bone tissue regeneration. J. Biomed. Mater. Res. B Appl. Biomater. **109**(7), 961–972 (2021)
15. Elomaa, L., Kokkari, A., Narhi, T., Seppälä, J.V.: Porous 3d modeled scaffolds of bioactive glass and photocrosslinkable poly (ε-caprolactone) by stereolithography. Compos. Sci. Technol. **74**, 99–106 (2013)
16. Fang, Y., Wang, C., Liu, Z., Ko, J., Chen, L., Zhang, T., Sun, W.: 3d printed conductive multiscale nerve guidance conduit with hierarchical fibers for peripheral nerve regeneration. Adv. Sci. **10**(12), 2205744 (2023)
17. Farahani, R.D., Dubé, M., Therriault, D.: Three-dimensional printing of multifunctional nanocomposites: manufacturing techniques and applications. Adv. Mater. **28**(28), 5794–5821 (2016)
18. Farzan, A., Borandeh, S., Seppälä, J.: Conductive polyurethane/pegylated graphene oxide composite for 3d-printed nerve guidance conduits. Eur. Polymer J. **167**, 111068 (2022)
19. Galarreta-Rodriguez, I., Lopez-Ortega, A., Garayo, E., Beato-López, J.J., La Roca, P., Sanchez-Alarcos, V., Pérez-Landazábal, J.I.: Magnetically activated 3d printable polylactic acid/polycaprolactone/magnetite composites for magnetic induction heating generation. Adv. Compos. Hybrid Mater. **6**(3), 102 (2023)
20. Ganguly, S., Margel, S.: 3d printed magnetic polymer composite hydrogels for hyperthermia and magnetic field driven structural manipulation. Prog. Polym. Sci. **131**, 101574 (2022)
21. Ghaziof, S., Shojaei, S., Mehdikhani, M., Khodaei, M., Nodoushan, M.J.: Electro-conductive 3d printed polycaprolactone/gold nanoparticles nanocomposite scaffolds for myocardial tissue engineering. J. Mech. Behav. Biomed. Mater. **132**, 105271 (2022)
22. Gnanasekaran, K., Heijmans, T., Van Bennekom, S., Woldhuis, H., Wijnia, S., De With, G., Friedrich, H.: 3d printing of cnt-and graphene-based conductive polymer nanocomposites by fused deposition modeling. Appl. Mater. Today **9**, 21–28 (2017)

23. Guo, W., Zhou, B., Zou, Y., Lu, X.: 4d printed poly (l-lactide)/(fecl3–ta/mgo) composite scaffolds with near-infrared light-induced shape-memory effect and antibacterial properties. Adv. Eng. Mater. **26**(2), 2301381 (2024)
24. Huang, Y., Wong, Y.S., Ng, H.C.A., Boey, F.Y., Venkatraman, S.: Translation in cardiovascular stents and occluders: From biostable to fully degradable. Bioengineering & Translational Medicine **2**(2), 156–169 (2017)
25. Li, Y., Huang, L., Tai, G., Yan, F., Cai, L., Xin, C., Al Islam, S.: Graphene oxide-loaded magnetic nanoparticles within 3d hydrogel form high-performance scaffolds for bone regeneration and tumour treatment. Compos. A Appl. Sci. Manuf. **152**, 106672 (2022)
26. Li, Y., Zhang, F., Liu, Y., Leng, J.: 4d printed shape memory polymers and their structures for biomedical applications. Sci. China Technol. Sci. **63**(4), 545–560 (2020)
27. Lin, C., Lv, J., Li, Y., Zhang, F., Li, J., Liu, Y., Leng, J.: 4d-printed biodegradable and remotely controllable shape memory occlusion devices. Adv. Func. Mater. **29**(51), 1906569 (2019)
28. Litowczenko, J., Wychowaniec, J.K., Zaleski, K., Marczak, L., Edwards-Gayle, C.J., Tadyszak, K., Maciejewska, B.M.: Micro/nano-patterns for enhancing differentiation of human neural stem cells and fabrication of nerve conduits via soft lithography and 3d printing. Biomater. Adv. **154**, 213653 (2023)
29. Marzo, A.M.L., Mayorga-Martinez, C.C., Pumera, M.: 3d-printed graphene direct electron transfer enzyme biosensors. Biosens. Bioelectron. **151**, 111980 (2020)
30. Merlo, A., Gonz´alez-Mart´ınez, E., Saad, K., Gomez, M., Grewal, M., Deering, J., et al.: Functionalization of 3d printed scaffolds using polydopamine and silver nanoparticles for bone-interfacing applications. ACS Appl. Bio Mater. **6**(3), 1161–1172 (2023)
31. Milani, P., Fran¸ca, D., Balieiro, A.G., Faez, R.: Polymers and its applications in agriculture. Pol´ımeros **27**, 256–266 (2017)
32. Naskar, A.K., Keum, J.K., Boeman, R.G.: Polymer matrix nanocomposites for automotive structural components. Nat. Nanotechnol. **11**(12), 1026–1030 (2016)
33. Pinel, S., Thomas, N., Boura, C., Barberi-Heyob, M.: Approaches to physical stimulation of metallic nanoparticles for glioblastoma treatment. Adv. Drug Deliv. Rev. **138**, 344–357 (2019)
34. Podstawczyk, D., Skrzypczak, D., Po-lomska, X., Starga-la, A., Witek-Krowiak, A., Guiseppi-Elie, A., Galewski, Z.: Preparation of antimicrobial 3d printing filament: In situ thermal formation of silver nanoparticles during the material extrusion. Polym. Compos. **41**(11), 4692–4705 (2020)
35. Qiao, H., Zhang, Y., Huang, Z., Wang, Y., Li, D., Zhou, H.: 3d printing individualized triboelectric nanogenerator with macro-pattern. Nano Energy **50**, 126–132 (2018)
36. Radhakrishnan, S., Nagarajan, S., Belaid, H., Farha, C., Iatsunskyi, I., Coy, E., et al.: Fabrication of 3d printed antimicrobial polycaprolactone scaffolds for tissue engineering applications. Mater. Sci. Eng., C **118**, 111525 (2021)
37. Rajendran, N.K., Kumar, S.S.D., Houreld, N.N., Abrahamse, H.: A review on nanoparticle based treatment for wound healing. J. Drug Delivery Sci. Technol. **44**, 421–430 (2018)
38. Rasoulianboroujeni, M., Fahimipour, F., Shah, P., Khoshroo, K., Tahriri, M., Eslami, H., Tayebi, L.: Development of 3d-printed plga/tio2 nanocomposite scaffolds for bone tissue engineering applications. Mater. Sci. Eng., C **96**, 105–113 (2019)
39. Rodriguez, A., Coaquira, J., Morales, M., Faria, F., Cunha, R., Santos, J., et al.: Synthesis, characterization and magnetic properties of polymer–fe3o4 nanocomposite. Spectrochim. Acta Part A Mol. Biomol. Spectrosc. **100**, 101–103 (2013)
40. Sabury, S., Kazemi, S.H., Sharif, F.: Graphene–gold nanoparticle composite: application as a good scaffold for construction of glucose oxidase biosensor. Mater. Sci. Eng., C **49**, 297–304 (2015)
41. Sahafnejad-Mohammadi, I., Rahmati, S., Najmoddin, N., Bodaghi, M.: Biomimetic polycaprolactone-graphene oxide composites for 3d printing bone scaffolds. Macromol. Mater. Eng. **308**(5), 2200558 (2023)
42. Samadian, H., Khastar, H., Ehterami, A., Salehi, M.: Bioengineered 3d nanocomposite based on gold nanoparticles and gelatin nanofibers for bone regeneration: In vitro and in vivo study. Sci. Rep. **11**(1), 13877 (2021)

43. Schneider, J., Basak, S., Hou, Y., Koo, J.H., Wardle, B.L., Gadegaard, N., Kumar, S.: 3d- printed polyetheretherketone smart polymer nanocomposite scaffolds: Mechanical, self-sensing, and biological attributes. Adv. Eng. Mater. **26**(4), 2301659 (2024)
44. Sharafeldin, M., Jones, A., Rusling, J.F.: 3d-printed biosensor arrays for medical diagnostics. Micromachines **9**(8), 394 (2018)
45. Shin, D.-G., Kim, T.-H., Kim, D.-E.: Review of 4d printing materials and their properties. Int. J. Precis. Eng. Manuf. Green Technol. **4**, 349–357 (2017)
46. Silva, M., Pinho, I.S., Covas, J.A., Alves, N.M., Paiva, M.C.: 3d printing of graphene-based polymeric nanocomposites for biomedical applications. Funct. Compos. Mater. **2**, 1–21 (2021)
47. Silva, V.A., Fernandes-Junior, W.S., Rocha, D.P., Stefano, J.S., Munoz, R.A., Bonacin, J.A., Janegitz, B.C.: 3d-printed reduced graphene oxide/polylactic acid electrodes: a new prototyped platform for sensing and biosensing applications. Biosens. Bioelectron. **170** 112684 (2020)
48. Skóra, B., Krajewska, U., Nowak, A., Dziedzic, A., Barylyak, A., Kus-Li´skiewicz, M.: Noncytotoxic silver nanoparticles as a new antimicrobial strategy. Sci. Rep. **11**(1), 13451 (2021)
49. Sofi, M.A., Sunitha, S., Sofi, M.A., Pasha, S.K., Choi, D.: An overview of antimicrobial and anticancer potential of silver nanoparticles. J. King Saud Univ. Sci. **34**(2), 101791 (2022)
50. Tibbits, S.: 4d printing: multi-material shape change. Archit. Des. **84**(1), 116–121 (2014)
51. Tiwari, M., Bangruwa, N., Mishra, D.: 0d, 1d, and 2d magnetic nanostructures: classification and their applications in modern biosensors. Talanta Open **8**, 100257 (2023)
52. Topsakal, A., Midha, S., Yuca, E., Tukay, A., Sasmazel, H.T., Kalaskar, D.M., Gunduz, O.: Study on the cytocompatibility, mechanical and antimicrobial properties of 3d printed composite scaffolds based on pva/gold nanoparticles (aunp)/ampicillin (amp) for bone tissue engineering. Mater. Today Commun. **28**, 102458 (2021)
53. Totu, E.E., Nechifor, A.C., Nechifor, G., Aboul-Enein, H.Y., Cristache, C.M.: Poly (methyl methacrylate) with tio2 nanoparticles inclusion for stereolithographic complete denture manufacturing the future in dental care for elderly edentulous patients? J. Dent. **59**, 68–77 (2017)
54. Uddin, M.F., Sun, C.: Strength of unidirectional glass/epoxy composite with silica nanoparticle-enhanced matrix. Compos. Sci. Technol. **68**(7–8), 1637–1643 (2008)
55. Wang, C., Huang, W., Zhou, Y., He, L., He, Z., Chen, Z., et al.: 3d printing of bone tissue engineering scaffolds. Bioactive Mater. **5**(1), 82–91 (2020)
56. Wang, G., Chen, G., Wei, Z., Yu, T., Liu, L., Wang, P., Qi, M.: A comparative study of tio2 and surface-treated tio2 nanoparticles on thermal and mechanical properties of poly (ε-caprolactone) nanocomposites. J. Appl. Polym. Sci. **125**(5), 3871–3879 (2012)
57. Wei, H., Cauchy, X., Navas, I.O., Abderrafai, Y., Chizari, K., Sundararaj, U., Therriault, D.: Direct 3d printing of hybrid nanofiber-based nanocomposites for highly conductive and shape memory applications. ACS Appl. Mater. Interfaces. **11**(27), 24523–24532 (2019)
58. Wei, H., Zhang, Q., Yao, Y., Liu, L., Liu, Y., Leng, J.: Direct-write fabrication of 4d active shape-changing structures based on a shape memory polymer and its nanocomposite. ACS Appl. Mater. Interfaces. **9**(1), 876–883 (2017)
59. Xing, M., Mohapatra, J., Beatty, J., Elkins, J., Pandey, N.K., Chalise, A., Liu, J.P.: Iron- based magnetic nanoparticles for multimodal hyperthermia heating. J. Alloy. Compd. **871**, 159475 (2021)
60. Zarrintaj, P., Vahabi, H., Guti´errez, T.J., Mehrpouya, M., Ganjali, M.R., Saeb, M.R.: Nanocomposite biomaterials made by 3d printing: achievements and challenges. In: Handbook of Polymer Nanocomposites for Industrial Applications, pp. 675–685 (2021)
61. Zhang, J., Zhao, S., Zhu, M., Zhu, Y., Zhang, Y., Liu, Z., Zhang, C.: 3d-printed magnetic fe 3 o 4/mbg/pcl composite scaffolds with multifunctionality of bone regeneration, local anticancer drug delivery and hyperthermia. J. Mater. Chem. B **2**(43), 7583–7595 (2014)
62. Zhang, L., Forgham, H., Shen, A., Wang, J., Zhu, J., Huang, X., Qiao, R.: Nanomaterial integrated 3d printing for biomedical applications. J. Mater. Chem. B **10**(37), 7473–7490 (2022)
63. Zhang, Y., Hu, J., Zhao, X., Xie, R., Qin, T., Ji, F.: Mechanically robust shape memory polyurethane nanocomposites for minimally invasive bone repair. ACS Appl. Bio Mater. **2**(3), 1056–1065 (2019)

64. Zhang, Y., Nayak, T.R., Hong, H., Cai, W.: Graphene: a versatile nanoplatform for biomedical applications. Nanoscale **4**(13), 3833–3842 (2012)
65. Zhang, Y., Zhai, D., Xu, M., Yao, Q., Chang, J., Wu, C.: 3d-printed bioceramic scaffolds with a fe 3 o 4/graphene oxide nanocomposite interface for hyperthermia therapy of bone tumor cells. J. Mater. Chem. B **4**(17), 2874–2886 (2016)
66. Zhao, W., Huang, Z., Liu, L., Wang, W., Leng, J., Liu, Y.: Porous bone tissue scaffold concept based on shape memory pla/fe3o4. Compos. Sci. Technol. **203**, 108563 (2021)
67. Zheng, Y., Du, Y., Chen, L., Mao, W., Pu, Y., Wang, S., Wang, D.: Recent advances on shape memory polymeric nanocomposites for biomedical applications and beyond. Biomater. Sci. (2024)

Elastomeric-Based 3D and 4D Printable Composite Materials for Engineering Applications

Chinmay Shau, Anchit Modi, Pratik Kumar Jagtap, Subhendu Chakroborty, and N. K. Gaur

Abstract This chapter delves into the advancements and applications of elastomeric-based 3D and 4D printable composite materials in engineering. Elastomeric materials, celebrated for their flexibility, resilience, and capacity to revert to their original shape, are increasingly utilized in additive manufacturing for producing highly customizable and high-performance components. The integration of 4D printing technology further enhances these materials by enabling them to undergo shape transformations in response to external stimuli such as light, temperature, or moisture. This capability is particularly advantageous for engineering applications requiring adaptive and responsive solutions. The chapter provides a comprehensive overview of the principles, properties, and manufacturing processes of elastomeric-based composites. It highlights their use in various fields, including aerospace, automotive, and biomedical engineering, where their unique characteristics are leveraged to create soft robotics, adaptive structures, and innovative medical devices. The discussion includes insights into the design and fabrication techniques that enable the creation of these advanced materials and addresses the challenges associated with their adoption, such as material costs, processing complexities, and scalability. By exploring current research and practical applications, the chapter aims to illustrate the transformative potential of elastomeric-based 3D and 4D printable materials in engineering. It underscores their role in advancing the field and suggests future directions for research to overcome existing challenges and expand their utility across diverse engineering domains.

C. Shau (✉) · N. K. Gaur
Department of Physics, Barkatullah University, Bhopal, India
e-mail: chinmaysahu07@gmail.com

A. Modi
Department of Basic Sciences, IITM, IES University, Bhopal, India
e-mail: anchitmodi87@gmail.com

P. K. Jagtap
Faculty of Science and Technology, The ICFAI University, Raipur, Chhattisgarh, India

S. Chakroborty
School of Applied Sciences, Chandigarh University, Unnao, Uttar Pradesh, India

S. Moharana et al. (eds.), *Polymer Nanocomposites for 3D, 4D and 5D Printing*, Engineering Materials, https://doi.org/10.1007/978-981-96-4214-4_13

Keywords Elastomeric composite materials · 3D printing · 4D printing · Engineering applications

1 Introduction

In the rapidly evolving field of material science and engineering, elastomeric-based composite materials stand at the forefront of innovation, offering unparalleled versatility and performance across various applications. These composites, which combine the elastic properties of elastomers with the strength and rigidity of various reinforcements, represent a significant advancement in the development of materials that meet the demanding requirements of modern engineering [1–5]. The genesis of elastomeric composites can be traced back to the intrinsic limitations of conventional elastomers and the subsequent need to enhance their mechanical properties. Elastomers, known for their exceptional elasticity, flexibility, and resilience, are widely used in applications requiring significant deformation under stress and a quick return to their original shape upon removing stress. However, their inherent lack of strength and rigidity often restricts their use in more demanding structural applications [6–8]. The introduction of polymers, particularly elastomers, marked a revolutionary shift, enabling the creation of materials with tailored properties for specific applications. Natural rubber, the earliest form of elastomer, provided the foundation for this shift, offering excellent tensile strength, abrasion resistance, and elasticity [9, 10]. The convergence of elastomeric-based materials with 3D and 4D printing technologies marks a transformative era in engineering, where material science and innovative manufacturing techniques intersect to redefine what is possible. Elastomeric materials, known for their exceptional flexibility, resilience, and ability to return to their original shape after deformation, offer significant advantages when integrated into additive manufacturing processes. This synergy creates composite materials that not only meet the stringent demands of various engineering applications but also open new avenues for design and functionality [11–13].

3D printing, or additive manufacturing, has already established itself as a cornerstone in the production of complex geometries that would be difficult, if not impossible, to achieve with traditional manufacturing methods [14]. By layering materials precisely, 3D printing allows for intricate internal structures, customized designs, and rapid prototyping, significantly reducing the time and cost associated with product development. When elastomeric materials are used in 3D printing, the resulting composites inherit properties such as high elasticity, impact resistance, and durability, which are critical for applications in sectors such as aerospace, automotive, and consumer electronics [15, 16]. 4D printing extends the capabilities of 3D printing by introducing the dimension of time. Materials and structures created with 4D printing can change their shape, properties, or functionality in response to external stimuli such as heat, light, moisture, or magnetic fields. Elastomeric-based 4D printable composites bring forth a new class of materials that can adapt and respond dynamically to their environment [17, 18]. This adaptive behaviour is particularly beneficial in applications requiring self-healing properties, shape memory, or responsive actuation, such as in biomedical devices, soft robotics, and smart textiles [19]. The development of elastomeric-based 3D and 4D printable composite materials is underpinned by significant advancements in material science, including the formulation of novel polymers, the integration of nanomaterials, and the development of hybrid composites. These materials can be engineered at the molecular level to achieve specific performance characteristics, such as enhanced tensile strength, improved thermal stability, and tailored electrical conductivity. Furthermore, the customization of material properties allows for the precise tuning of the composites to meet the specific needs of different applications [20–22].

This chapter delves into the fundamental principles governing elastomeric-based 3D and 4D printable composite materials, exploring the various fabrication techniques and their implications for engineering applications. It examines the mechanical properties, thermal behaviour, and environmental responsiveness of these composites, providing a comprehensive overview of their potential and challenges. Additionally, the chapter highlights case studies and real-world applications that demonstrate the transformative impact of these materials in pushing the boundaries of modern engineering. Through a detailed analysis, this paper aims to shed light on the future directions and innovations in the field of elastomeric-based additive manufacturing, underscoring its significance in shaping the next generation of engineered materials and products.

2 Fundamentals of Elastomeric-Based Composites

2.1 Understanding Elastomers

Elastomers, a class of polymers known for their exceptional elasticity, flexibility, and resilience, are essential to elastomeric-based composites. Despite their ability to deform under stress and return to their original shape, traditional elastomers lack the strength and rigidity required for demanding structural applications. To address this, elastomers are reinforced with materials like glass, carbon, aramid fibers, silica particles, carbon black, and nanoscale fillers such as carbon nanotubes and graphene. This combination retains elastomers' inherent properties while significantly enhancing tensile strength, impact resistance, thermal stability, and chemical resistance [23, 24]. The creation of these composites involves meticulous mixing and compounding techniques to ensure uniform filler distribution. The evolution from natural rubber to advanced synthetic elastomers like SBR, NBR, and silicone rubber has expanded their application range. Understanding the interaction between elastomers and reinforcing agents is crucial for developing advanced materials to meet modern engineering demands in various industries, including automotive, aerospace, healthcare, and consumer goods [25, 26].

2.2 *Types of Elastomers*

Elastomers are versatile polymers known for their high elasticity, flexibility, and resilience, categorized into various types based on their chemical composition and specific properties. Natural Rubber (NR), derived from the latex of rubber trees, offers excellent tensile strength, abrasion resistance, and elasticity, commonly used in tires, footwear, and industrial goods. Styrene-butadiene rubber (SBR), a synthetic elastomer made from styrene and butadiene, provides good abrasion resistance and aging stability, widely employed in automotive tires and conveyor belts. Nitrile Rubber (NBR), also known as Buna-N, made from acrylonitrile and butadiene, exhibits high resistance to oil, fuel, and chemicals, finding use in automotive seals, gaskets, and hoses [27]. Ethylene Propylene Diene Monomer (EPDM), composed of ethylene, propylene, and a diene component, offers excellent weather, ozone, and aging resistance, used in roofing membranes, automotive weather-stripping, and hoses. Silicone Rubber, consisting of silicon, oxygen, carbon, and hydrogen, provides outstanding thermal stability, flexibility, and biocompatibility, commonly used in medical devices, kitchenware, and electrical insulation. Butyl Rubber (IIR), made from isobutylene with a small amount of isoprene, offers excellent impermeability to gases and good chemical resistance, used in inner tubes, pharmaceutical stoppers, and protective clothing. Polyurethane (PU) Rubber, derived from polyols and diisocyanates, provides high abrasion resistance, load-bearing capacity, and versatility when applied in wheels, rollers, and seals. Fluoroelastomers (FKM), made from fluorinated hydrocarbons, offer excellent heat, chemical, and oil resistance, used in aerospace, automotive, and chemical processing industries [28]. Chloroprene Rubber (CR), also known as Neoprene, is made from chloroprene and exhibits good weather, ozone, and chemical resistance, used in wetsuits, hoses, and gaskets. Thermoplastic Elastomers (TPE), a blend of polymers with both thermoplastic and elastomeric properties, are recyclable and easy to process, and they are used in automotive parts, medical devices, and consumer products. These elastomers play integral roles across industries, thanks to their unique properties tailored for diverse applications [29].

3 Composite Materials

Composite materials are engineered materials made from two or more constituent materials with significantly different physical or chemical properties, which, when combined, produce a material with enhanced characteristics compared to its components. These materials typically consist of a matrix material that binds and supports a reinforcement material. The matrix can be a polymer, metal, ceramic, or carbon-based material, while the reinforcement can be fibers, particles, or other structural components. The key advantage of composite materials lies in their ability to tailor properties such as strength, stiffness, weight, and durability to meet specific application requirements [30]. Common types of composite materials include fiber-reinforced polymers (FRP), which utilize fibers such as carbon, glass, or aramid embedded in a polymer matrix; metal matrix composites (MMC), where metal alloys are reinforced with ceramic or metallic fibers or particles; and ceramic matrix composites (CMC), which incorporate ceramic fibers or particles in a ceramic matrix. Composite materials find extensive use in industries ranging from aerospace, automotive, and construction to sports equipment, electronics, and biomedical devices, due to their versatility and ability to offer superior performance characteristics compared to traditional materials. Advances in composite manufacturing technologies, including automated layup processes, additive manufacturing, and nanotechnology, continue to expand the capabilities and applications of composite materials in modern engineering and technology [31].

4 Components of Composite Materials

Matrix: The matrix in composite materials is the continuous phase that binds and supports the reinforcement materials. It can be a polymer, metal, ceramic, or carbon-based material. The choice of matrix depends on the desired properties of the composite, such as strength, stiffness, durability, and resistance to environmental factors like temperature and chemicals. Polymers like epoxy, polyester, and vinyl ester resins are common in fiber-reinforced polymers (FRP), offering versatility and ease of processing. Metal matrices, such as aluminum, magnesium, or titanium alloys, provide strength and thermal conductivity in metal matrix composites (MMC). Ceramic matrices are used in ceramic matrix composites (CMC) for applications requiring high-temperature resistance and wear properties [32].

Reinforcement

The reinforcement component of composite materials enhances specific mechanical properties such as strength, stiffness, and toughness. Reinforcements can be fibers, particles, or structural components embedded within the matrix. Common fiber reinforcements include carbon fibers, glass fibers, and aramid fibers, each offering unique characteristics like high tensile strength (carbon fibers), stiffness (glass fibers), or

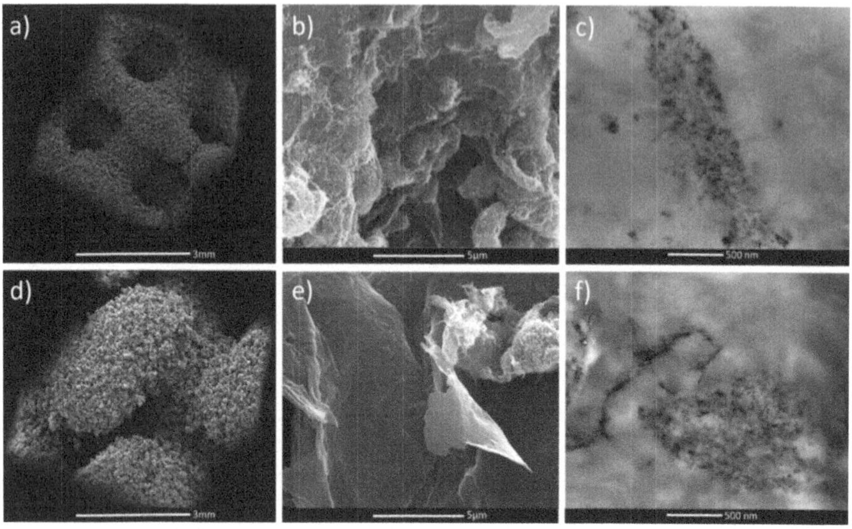

Fig. 1 SEM images of TPU/MWCNTs D60 and G60 samples at 50 × magnification **a**, **d**. SEM **b** and TEM **c** images depict the TPU/MWCNTs G60 sample, while the SEM **e** and TEM **f** images illustrate the porous structures in the TPU/(MWCNTs-GE) G60 sample [33]. Copyright 2020. Reproduced with permission from MDPI

impact resistance (aramid fibers). Particulate reinforcements, such as silica, alumina, or carbon nanotubes, can improve composites' properties like hardness, wear resistance, or thermal conductivity. For instance, thermoplastic polyurethane (TUP) incorporated with fillers like multi-walled carbon nanotubes (MNCNTs) and graphene (GE) with porous structures enhances the functional properties of the 3D-printed elastomeric structures. Figure 1 shows the SEM/TEM images of the TUP/MNCNTs D60 and G60, which demonstrate the porous structures of these types of elastomeric-based composites.

The arrangement and orientation of the reinforcement within the matrix, known as the microstructure, significantly influence the overall performance of the composite material [33]. The synergy between the matrix and reinforcement components allows composite materials to achieve tailored combinations of properties that are superior to those of traditional materials. This capability makes composites suitable for diverse applications across industries, including aerospace, automotive, marine, construction, electronics, and sports equipment, where lightweight, durable, and high-performance materials are essential. Advances in composite manufacturing technologies continue to expand the range of possibilities for creating materials with optimized properties for specific engineering requirements.

5 Elastomeric Composites

Elastomeric composites are advanced materials that combine elastomers with various reinforcing agents to enhance their mechanical and functional properties beyond what traditional elastomers can achieve alone. These composites typically consist of an elastomer matrix, which provides flexibility and resilience, reinforced with materials such as fibers, fillers, or nanoparticles.

The effect of conductive filler geometry is first briefly explained using percolation theory, and then recent efforts related to the design of composites by five different methods are highlighted in Fig. 2, implanting conductive fillers in elastomers, synthesizing metal fillers within elastomers, filling microchannels with liquid metals, infiltrating elastomers in conductive-filler networks, blending conductive fillers with elastic polymers, and implanting conductive fillers in elastomers. The elastomer matrix, often based on polymers like natural rubber or synthetic elastomers (e.g., SBR, NBR, silicone rubber), serves as the continuous phase that embeds and supports the reinforcing agents. The addition of reinforcements significantly improves key characteristics of elastomers, such as tensile strength, tear resistance, impact resistance, thermal stability, and chemical resistance. Common reinforcing materials include fibers like glass, carbon, aramid, and natural fibers, as well as fillers such as silica and carbon black and nanoscale additives like carbon nanotubes and graphene. The choice of reinforcement depends on the desired properties and application requirements of the elastomeric composite [34, 35]. The manufacturing process of elastomeric composites involves meticulous mixing and compounding of the elastomer matrix with the chosen reinforcing agents. Techniques such as two-roll milling, internal mixing using devices like Banbury mixers, and extrusion ensure uniform dispersion of the reinforcements within the elastomer matrix, crucial for achieving consistent mechanical performance. Elastomeric composites find extensive use in industries such as automotive, aerospace, marine, construction, and health care, where materials with high flexibility, resilience, and enhanced mechanical properties are needed. These composites play a vital role in applications ranging from seals, gaskets, and hoses to vibration isolators, conveyor belts, and protective clothing, demonstrating their versatility and utility across diverse industrial sectors [36]. Ongoing research and development in elastomeric composites continue to innovate new formulations and manufacturing techniques, expanding their capabilities and broadening their applications in modern engineering and technology.

6 3D and 4D Printing of Elastomeric-Based Composites

The advancement of additive manufacturing, particularly 3D and 4D printing, has revolutionized the development and application of elastomeric-based composites. These technologies enable the precise fabrication of complex structures with tailored properties that were once unachievable with traditional manufacturing methods [37].

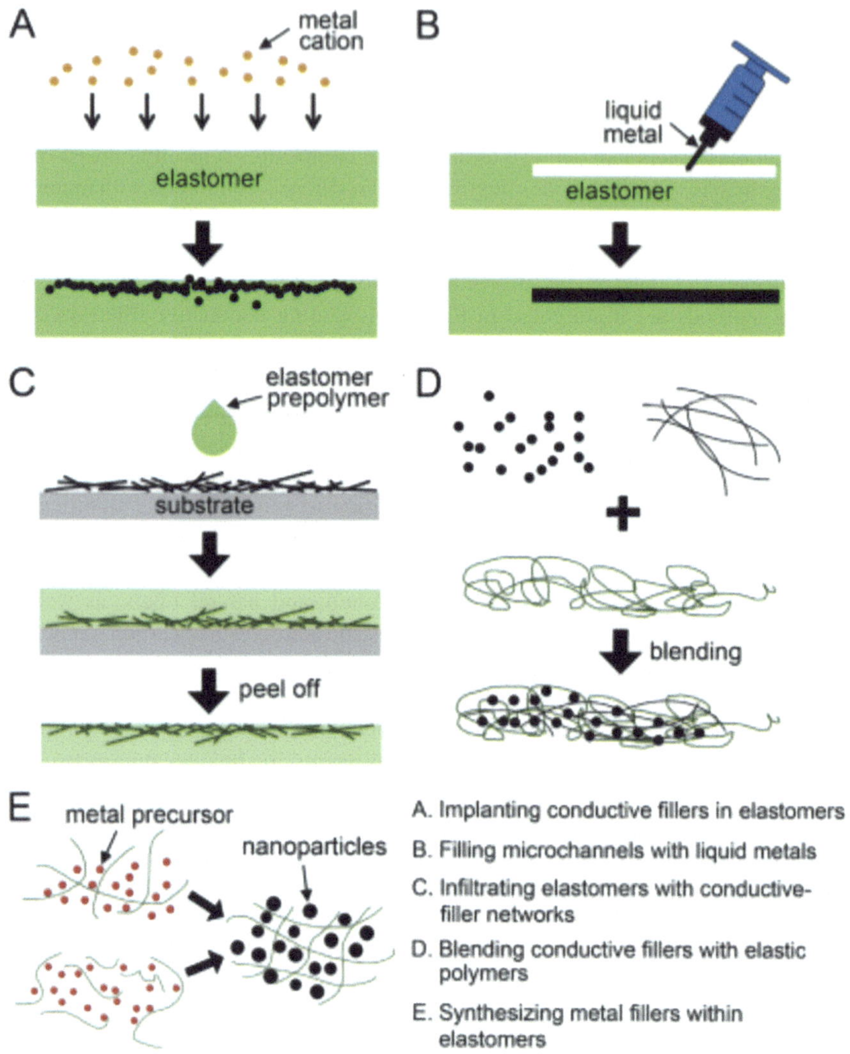

Fig. 2 A schematic representation of five unique approaches for designing conductive composite elastomers [34]. Copyright 2014. Reproduced with permission from Elsevier

Figure 3 shows the progression of 3D printing technology alongside the advancement of polymer materials over time. Customized manufacturing [38]. Elastomeric materials, known for their flexibility, resilience, and capacity to undergo significant deformation before returning to their original shape, form the backbone of these composites. When combined with other materials in composite form, elastomers gain enhanced mechanical properties, such as increased strength, durability, and functionality. This hybridization opens new avenues for creating versatile materials that can

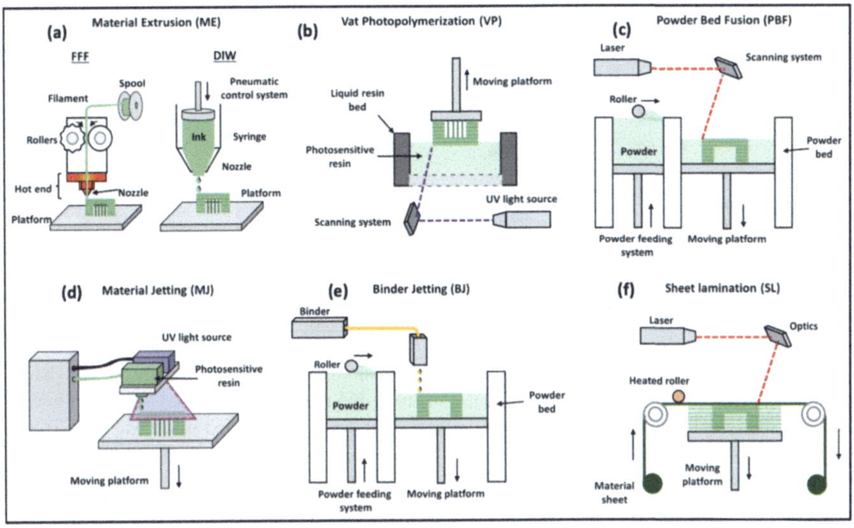

Fig. 3 A depiction of the evolution of 3D printing, highlighting the significant advancements in polymer materials, demonstrates the transition from rapid prototyping to scalable and tailor-made manufacturing [38]. Copyright 2024. Reproduced with permission from MDPI

be used in a wide array of applications, from soft robotics and wearable devices to biomedical implants and aerospace components [37, 38]. In 3D printing, the process of layer-by-layer deposition allows for the fabrication of intricate geometries and customized structures that can be fine-tuned at the micro-scale. This precision is particularly advantageous in the creation of elastomeric-based composites, where the spatial distribution of different materials within the composite can be controlled to achieve desired properties.

For example, researchers have developed multi-material 3D printing techniques that enable the integration of elastomeric polymers with rigid or functional components, resulting in composites that exhibit unique mechanical behaviors, such as graded stiffness or enhanced energy absorption. The ability to customize material properties and geometries on demand has significant implications for industries that require lightweight, flexible, and durable materials [39].

Taking this a step further, 4D printing introduces the dimension of time into the manufacturing process, where 3D-printed structures can change shape or function in response to external stimuli such as temperature, humidity, or light. Elastomeric-based composites are particularly well-suited for 4D printing due to their inherent flexibility and responsiveness to environmental changes. By carefully designing the composition and structure of these materials, it is possible to create smart systems that can self-assemble, self-heal, or adapt to changing conditions over time. For instance, Azobenzene moieties undergo isomerization under UV light, leading to polymer

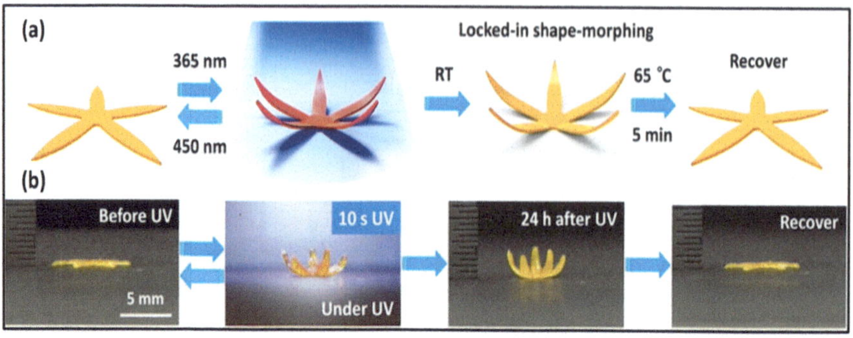

Fig. 4 **a** Illustration and **b** photos showing the light-responsive deformations of a flower-shaped actuator 4D printed with UV-sensitive LCE. Exposure to 365 nm UV light caused the flower to curl, while 450 nm blue light reversed the process. After UV exposure, the curled shape was stabilized at room temperature through UPy crosslink formation. Heating to 65 °C broke the crosslinks, allowing the LCE to return to its original flat shape by switching back to the isotropic state [38]. Copyright 2024. Reproduced with permission from MDPI

matrix expansion, a feature used in 4D printing for creating light-responsive materials. In liquid crystal elastomers (LCEs), these moieties induce stress and misalignment when exposed to UV light, enabling dynamic shape changes. 3D printing of an LCE strip that could lift a weight by contracting under UV light and returning to its original length under blue light. However, stability issues arose due to spontaneous relaxation, which is resolved by adding 2-ureido-4[1H]-pyrimidinone (UPy) crosslinkers, allowing the shape to be stabilized after UV exposure which is shown in Fig. 4 [38–40].

This capability is poised to revolutionize fields like medicine, where 4D-printed implants could dynamically conform to the body's movements, or in aerospace, where materials that adapt to external conditions could enhance the performance and safety of aircraft. The integration of elastomeric materials into 3D and 4D printing is not without challenges. The development of suitable printing techniques, the refinement of material properties, and the optimization of post-processing methods are critical areas of ongoing research. For instance, the rheological properties of elastomers must be carefully controlled to ensure successful extrusion or jetting during the printing process. Additionally, achieving strong interfacial bonding between different materials within a composite is crucial for maintaining the integrity and functionality of the printed structures. Researchers are also exploring the use of advanced simulation tools to predict the ehaviour of elastomeric-based composites under various conditions, aiding in the design of more robust and reliable materials [40].

Despite these challenges, the potential applications of 3D and 4D-printed elastomeric composites are vast and transformative. In the medical field, customized prosthetics, orthotics, and tissue scaffolds can be printed to match the specific anatomical features of individual patients, improving comfort and functionality. In the realm of wearable technology, elastomeric composites can be used to create flexible sensors and actuators that conform to the human body, enabling the development of next-generation smart clothing and accessories. In aerospace, lightweight and adaptive structures can be designed to optimize performance while reducing material waste and energy consumption. As research in this area continues to evolve, the combination of elastomeric materials with advanced printing techniques promises to unlock new possibilities in material science, engineering, and beyond [39, 41].

7 Fabrication Techniques

The fabrication of elastomeric-based composites involves several critical processes to ensure uniformity, strength, and performance. The following sections detail the key fabrication techniques [42, 43].

1. **Mixing and Compounding:** The process begins with mixing the elastomer matrix with reinforcing agents such as fibers, fillers, or nanoparticles. Techniques include two-roll milling, internal mixing using Banbury mixers, or twin-screw extrusion. These methods ensure thorough dispersion of the reinforcements into the elastomer matrix, promoting homogeneity and consistency in material properties.
2. **Molding and Casting:** Once compounded, the elastomeric composite mixture can be molded or cast into desired shapes using techniques like compression molding, injection molding, or transfer molding. These methods allow for the production of complex geometries and precise dimensions while maintaining the integrity of the composite structure.
3. **Calendering:** Calendering involves passing the elastomeric composite through multiple rollers to achieve a uniform thickness and smooth surface finish. This technique is commonly used for producing sheets or films of elastomeric composites used in various applications.
4. **Extrusion:** Extrusion involves forcing the compounded elastomeric composite through a die to create continuous shapes such as profiles, tubes, or hoses. It ensures consistent cross-sectional geometry and mechanical properties along the length of the extrudate.
5. **Curing and Vulcanization:** Curing or vulcanization is a critical step in the fabrication of elastomeric composites where heat and pressure are applied to cross-link the polymer chains within the elastomer matrix. This process enhances the material's strength, durability, and resistance to deformation and aging.
6. **Additive Manufacturing:** Emerging technologies like additive manufacturing (3D printing) are being explored for elastomeric composites. This method enables layer-by-layer deposition of elastomer matrix with reinforcing agents, offering design flexibility and customization in complex geometries.

These fabrication techniques are tailored to optimize the properties and performance of elastomeric composites for specific applications across industries such as automotive, aerospace, healthcare, and consumer goods. Advances in manufacturing technologies continue to drive innovation in elastomeric composite materials, expanding their capabilities and applications in modern engineering and technology.

8 Properties of Elastomeric-Based Composites

8.1 Mechanical Properties

Elastomeric composites exhibit a range of mechanical properties that make them suitable for various engineering applications. These materials are particularly noted for their impressive combination of tensile strength and elasticity, which are crucial for applications that demand both durability and flexibility. The tensile strength of elastomeric composites refers to the maximum stress that the material can withstand while being stretched. This property is essential for ensuring that the material can handle significant loads without experiencing failure or fracture, making it ideal for high-stress environments. Elasticity, another vital property of elastomeric composites, refers to the material's ability to return to its original shape after being deformed.

This characteristic is particularly important in applications where the material is subjected to repeated cycles of stress and strain, as it ensures that the composite can maintain its structural integrity and performance over time. As illustrated in Fig. 5, the PPSe/BM composites demonstrated superior shape memory properties and a quicker degradation rate compared to the pure polyester material. The inherent elasticity of elastomeric composites allows them to absorb and dissipate energy efficiently, reducing the risk of permanent deformation and extending the lifespan of the material. The reinforcement phase within elastomeric composites plays a critical role in enhancing these mechanical properties. By incorporating reinforcing agents such as fibers or particulates, the overall strength and resilience of the composite are significantly improved. These reinforcements enable the material to withstand higher loads and greater deformations than the elastomer matrix alone could endure. As a result, elastomeric composites are able to meet the rigorous demands of various engineering applications, from automotive and aerospace components to industrial seals and gaskets [44].

8.2 Impact Resistance

Elastomeric-based composites are highly valued for their exceptional impact resistance, a critical property for materials used in environments where they must withstand sudden and forceful impacts. Impact resistance refers to the material's capacity to absorb and dissipate energy when subjected to a high-velocity force, thereby

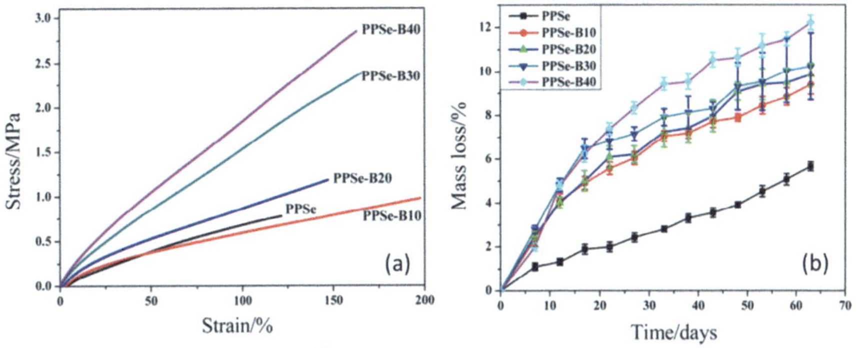

Fig. 5 **a** Mechanical properties of PPSe and PPSe/BM composites, and **b** Degradation of PPSe/BM composites in phosphate-buffered saline at 37 °C, shown by weight loss [18]. Copyright 2016. Reproduced with permission from Elsevier

preventing significant damage and avoiding catastrophic failure. This attribute is particularly important in applications ranging from automotive and aerospace to industrial and consumer products, where durability and safety are paramount. The remarkable impact resistance of elastomeric composites stems from their unique composition. These materials typically consist of a flexible elastomer matrix, which is reinforced with strong, durable fibers or particulates. The elastomer matrix provides inherent flexibility and energy absorption capabilities, allowing the composite to deform under impact and subsequently return to its original shape without cracking or breaking. The reinforcing agents, on the other hand, enhance the overall strength and toughness of the composite, enabling it to withstand higher levels of stress and impact [45].

One of the key advantages of elastomeric-based composites is their ability to maintain structural integrity under repeated impacts. This resilience is particularly beneficial in applications such as automotive bumpers, where the material must endure multiple collisions without losing its protective properties. The energy-absorbing nature of elastomeric composites not only minimizes damage to the material itself but also reduces the force transmitted to other components, thereby enhancing the overall safety of the system. The impact resistance of elastomeric-based composites makes them an indispensable material in various engineering and consumer applications. Their unique combination of flexibility, strength, and energy absorption capabilities ensures that they can effectively protect against high-velocity impacts, maintain structural integrity under stress, and deliver reliable performance in demanding environments. As technology and material science continue to advance, the role of elastomeric composites in enhancing safety, durability, and efficiency across multiple industries is likely to grow even further [24].

8.3 Thermal Properties

Elastomeric-based composites exhibit remarkable thermal properties, making them highly suitable for a diverse array of applications that require thermal stability and resistance. These composites are engineered to withstand significant temperature variations without compromising their mechanical integrity or overall performance. This capability is particularly valuable in industries such as automotive, aerospace, and industrial manufacturing, where materials are frequently exposed to fluctuating and extreme temperatures. The thermal properties of elastomeric composites are primarily derived from their unique composition. The elastomer matrix, which forms the base of these composites, provides inherent flexibility and resilience. This matrix retains its flexibility even under extreme temperatures, ensuring that the composite can absorb and dissipate thermal energy without cracking or losing its shape. This characteristic is crucial for maintaining the performance of components that are subjected to thermal cycling, where rapid temperature changes can lead to material fatigue and failure in less resilient materials [46].

The thermal properties of elastomeric composites also contribute to their longevity and durability. Their resistance to thermal degradation means that these materials can maintain their performance over extended periods, even in harsh thermal environments. This reduces the need for frequent replacements and maintenance, leading to cost savings and increased operational efficiency. Furthermore, the ability to withstand high temperatures without significant deformation or loss of strength enhances the safety and reliability of components made from elastomeric composites. The incorporation of reinforcing agents, such as fibers or particulates, further enhances the thermal stability of elastomeric composites. These reinforcements help to distribute thermal stresses more evenly throughout the material, preventing localized overheating and reducing the risk of thermal degradation. For instance, carbon fibers and silica particles are commonly used to improve the thermal conductivity and heat resistance of the composite. This enhancement allows the material to maintain its structural integrity and mechanical properties over a broad temperature range, making it suitable for high-performance applications [12].

8.4 Chemical Resistance

Elastomeric-based composites are highly esteemed for their exceptional chemical resistance, which makes them suitable for use in a wide range of applications where exposure to harsh chemical environments is a concern. This valuable property is a result of the unique combination of the elastomer matrix and various reinforcing agents, creating a material that can withstand the degrading effects of numerous chemicals, including acids, bases, solvents, and oils. The elastomer matrix in these composites acts as a resilient barrier that is inherently resistant to chemical attack.

This matrix provides a first line of defense, preventing the penetration and subsequent degradation of the material by harmful substances. The inclusion of reinforcing agents such as fibers, particulates, or other additives further enhances this resistance. These reinforcements not only contribute to the mechanical strength and stability of the composite but also improve its ability to repel and withstand chemical exposure, creating a synergistic effect that significantly boosts the overall chemical resistance of the material. The chemical resistance of elastomeric-based composites significantly broadens their range of applications and enhances their durability in challenging environments. This property ensures that these materials can withstand prolonged exposure to aggressive chemicals without losing their structural integrity or functionality. Consequently, elastomeric composites are a versatile and durable choice for industries that demand high-performance materials capable of enduring harsh chemical conditions. As advancements in material science continue, the chemical resistance of elastomeric composites is likely to improve even further, expanding their applicability and reinforcing their role as essential materials in modern engineering and manufacturing [47].

8.5 Electrical Properties

Elastomeric-based composites are distinguished by their diverse and tunable electrical properties, which make them highly valuable in a wide array of electrical and electronic applications. The unique composition of these composites, which combines the inherent flexibility of elastomers with the electrical characteristics of various fillers, allows for a broad spectrum of electrical behaviors, ranging from excellent insulation to enhanced conductivity. This versatility enables the design and engineering of materials that meet specific electrical performance requirements, thereby expanding their applicability in advanced technological fields. One of the key attributes of elastomeric composites is their ability to serve as excellent insulators. The elastomer matrix itself is typically a good insulator, providing high electrical resistance that prevents the flow of current. This property is particularly advantageous in applications such as cable insulation, gaskets, and seals in electrical systems, where preventing electrical leakage and ensuring safety are paramount. The flexibility of the elastomer matrix also allows these composites to be used in dynamic environments where movement and flexibility are essential, without compromising their insulating properties. On the other hand, by incorporating conductive fillers such as carbon nanotubes, graphene, metal particles, or conductive polymers, elastomeric-based composites can be engineered to exhibit significant electrical conductivity. This tunable conductivity is critical for applications in flexible electronics, where materials need to maintain electrical performance while being bent, stretched, or compressed. In wearable electronics, for example, elastomeric composites provide the necessary electrical pathways for sensors and circuits while conforming to the contours of the human body. This ensures both comfort and functionality, making them ideal for smart clothing, health monitors, and other wearable devices [48].

9 Applications in Engineering

9.1 Automotive Industry

The automotive industry has significantly leveraged elastomeric composites across a range of critical components, benefiting from their unique combination of durability, flexibility, and resilience against environmental factors. One of the primary applications is in tires, where elastomeric composites contribute to superior grip, wear resistance, and overall longevity, ensuring optimal performance and safety on the road. Additionally, elastomeric composites play a crucial role in seals and gaskets within automotive systems, where their chemical resistance and flexibility help maintain integrity and prevent leaks even in harsh operating conditions. Hoses and belts also utilize these materials due to their ability to withstand high temperatures, pressures, and mechanical stresses, ensuring reliability in engine and fluid systems. Moreover, elastomeric composites enhance suspension components such as bushings and mounts, effectively absorbing vibrations and improving ride comfort for passengers. These versatile materials continue to drive innovation in automotive engineering, enabling vehicles to achieve higher performance standards while ensuring durability and safety across diverse operating environments [49].

9.2 Aerospace and Defence

In the aerospace and defence sectors, where minimizing weight without compromising performance is crucial, elastomeric composites have emerged as indispensable materials. These composites offer a compelling combination of strength, flexibility, and lightweight properties, making them highly suitable for a diverse range of applications. Aircraft seals, for example, benefit from elastomeric composites due to their ability to maintain excellent sealing performance even under extreme temperatures and pressures, ensuring the integrity of critical compartments and systems within aircraft. Additionally, these materials are employed in vibration dampeners, effectively reducing vibrations and enhancing both passenger comfort and the structural integrity of aircraft components. Elastomeric composites also find application in protective coatings, where they provide a robust defence against corrosion and mechanical damage, extending the lifespan of aerospace structures and equipment. The application features a smart robotic hand in Fig. 6 with fibers orientated at different angles in each finger, enabling precise control over deformation rates, degrees, and forces. Unlike most soft robotic hands, this design allows for sequential actuation. Under UV light, the fingers bend progressively, and the rings attached to them detach in sequence during shape recovery in warm water. This adjustable, fiber-regulated approach shows promise for applications like sequential grasping in soft robotics.

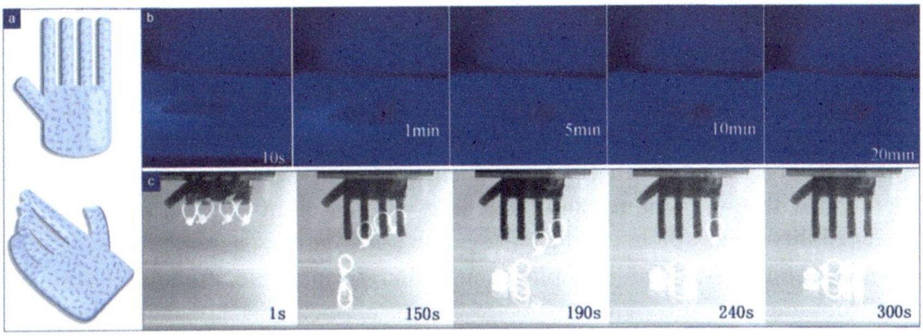

Fig. 6 The step-by-step deployment of a smart robotic hand. **a** Diagram showing the orientation of each finger. **b** Varying bending angles of the fingers after 20 min of exposure to UV light (395 nm wavelength). **c** The fingers returning to their original shape in 60 °C water [50]. Copyright 2019. Reproduced with permission from Elsevier

Furthermore, in the realm of defence, these composites are utilized in advanced protective gear for personnel, offering lightweight yet durable protection against ballistic threats and environmental hazards. Overall, elastomeric composites continue to play a pivotal role in advancing aerospace and defence technologies, enabling enhanced performance, safety, and operational efficiency across various critical applications [51].

9.3 Construction and Civil Engineering

Elastomeric composites play a pivotal role in the construction and civil engineering sectors, where durable and dependable materials are essential for infrastructure projects. These composites are widely utilized in various applications to enhance structural integrity and longevity. For instance, in bridge bearings, elastomeric composites facilitate controlled movement and effective distribution of loads, ensuring the stability and functionality of bridges under varying conditions. Expansion joints, another critical application, rely on these materials to accommodate structural movements caused by thermal expansion and contraction, thereby preventing stress-induced damage and maintaining structural integrity over time. Additionally, elastomeric composites are employed in seismic dampers for buildings and bridges, where they absorb and dissipate seismic energy during earthquakes, minimizing structural damage and improving overall safety. The versatility and resilience of elastomeric composites make them indispensable in construction and civil engineering, contributing to the longevity and performance of infrastructure projects worldwide [52].

9.4 Electronics and Electrical Engineering

The unique electrical properties of elastomeric composites make them indispensable in modern electronics and electrical engineering applications. These materials are particularly valued for their versatility in creating flexible and durable circuits, which can withstand bending and stretching without compromising performance. This capability makes them ideal for use in flexible circuits, enabling innovations in wearable technology, flexible displays, and other portable electronics. Elastomeric composites are also utilized in sensors for various applications, leveraging their flexibility and responsiveness to accurately detect and transmit data which are shown in Fig. 7.

Additionally, composites with conductive fillers excel in electromagnetic interference (EMI) shielding, protecting electronic devices from external electromagnetic signals that could disrupt their operation. In the realm of wearable electronics, these materials offer a combination of comfort, flexibility, and durability, making them suitable for integrating electronic components into clothing, accessories, and health monitoring devices. Overall, elastomeric composites continue to drive advancements in electronics by providing solutions that enhance functionality, reliability, and user experience across diverse applications [53].

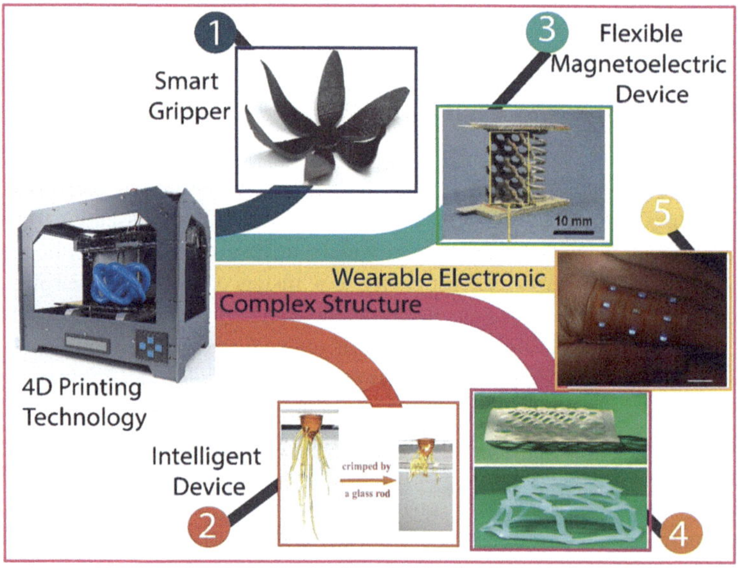

Fig. 7 Recent advancements in 4D printing technology have been utilized in a range of cutting-edge applications, including: (1) smart grippers, (2) intelligent devices, (3) flexible magnetoelectric devices, (4) intricate Kirigami-inspired structures, and (5) wearable electronics [53]. Copyright 2024. Reproduced with permission from Elsevier

9.5 Medical and Health Care

In the medical and healthcare sector, elastomeric composites play a crucial role in various applications where biocompatibility, flexibility, and durability are paramount. These materials are highly valued for their ability to meet the stringent requirements of medical devices and prosthetics.

Thus, a biomedical device can be made with better functionality success rates using freeform manufacturing of the items depicted in Fig. 8, which include wound dressing, surgical gripper systems, and lateral flow sensors for detecting cervical cancer biomarkers. The ability to deliver drugs to specific patients under regulated conditions has been made possible by the development of 3D-printed drug delivery devices with micro and nanoscale structures. In the meantime, biocompatibility and the formation of new tissue are supported by 3D-printed scaffolds containing micro and nanostructures for tissue engineering. Furthermore, blood arteries have been 3D printed in order to modify the structure of the vasculature and add porous elements that let the passage of nutrients and cells. Wearable health monitors also benefit from elastomeric composites, which offer comfort and reliability in devices that continuously monitor vital signs and health metrics. Moreover, these materials can be engineered with antimicrobial properties, enhancing their suitability for use in medical environments where preventing infections is critical. The versatility and performance of elastomeric composites continue to drive innovation in medical technology, contributing to advancements that improve patient care, rehabilitation outcomes, and overall quality of life [54].

10 Future Trends and Challenges

10.1 Advances in Materials Science

Ongoing research in materials science is at the forefront of advancing elastomeric composites with enhanced properties, paving the way for innovations and applications across various industries. One promising area of development is nanocomposites, where nanoscale fillers are integrated into elastomeric matrices to achieve significant improvements in mechanical strength, thermal stability, and electrical conductivity. These nanocomposites offer enhanced performance characteristics compared to traditional composites, making them ideal for high-performance applications in the automotive, aerospace, and electronics sectors. Another significant focus of research is on bio-based elastomers, which are derived from renewable resources such as plant oils and biomass. These materials are designed to reduce dependency on fossil fuels and lower the environmental footprint associated with traditional elastomers. Bio-based elastomers exhibit comparable or superior properties to their petroleum-based counterparts, making them increasingly attractive for use in sustainable products and applications across industries [55].

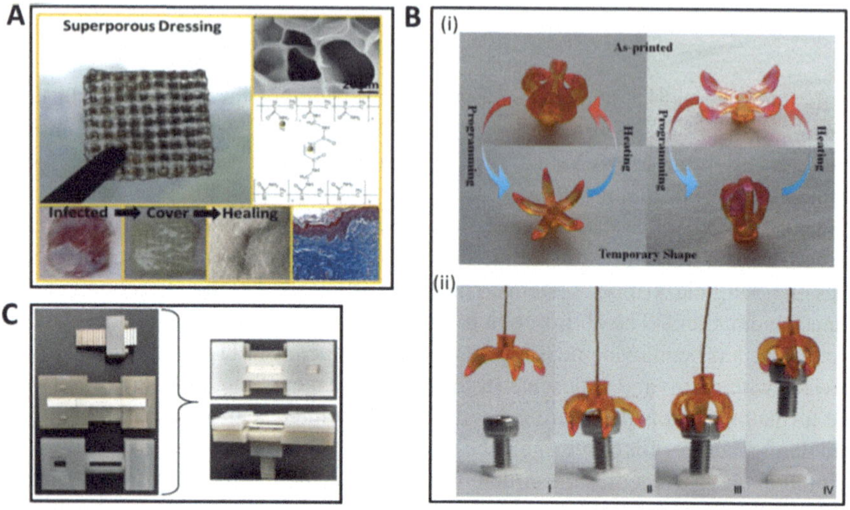

Fig. 8 Recent advances in 3D and 4D printing applied to surgical and diagnostic tools. **a** FDM-based 3D printed AgNP-Pam/HPMC super porous hydrogel used as a wound dressing; **b** 4D printed surgical gripper system: (i) showing the transformation from the printed shape to a temporary shape, (ii) sequential images capturing the gripping of an object; **c** 3D printed device featuring a magnetic focus lateral flow sensor designed for detecting cervical cancer biomarkers [54]. Copyright 2024. Open access. Reproduced with permission from Exploration of Medicine

10.2 Manufacturing Innovations

Advancements in manufacturing technologies, particularly 3D printing, and advanced molding techniques, are anticipated to revolutionize the production landscape of elastomeric composites. These innovations herald a new era of manufacturing capabilities, offering significant advantages over traditional methods. 3D printing, for instance, facilitates the creation of complex shapes and customized components with unprecedented precision and efficiency. The ability to manufacture elastomeric composites through additive manufacturing opens doors to rapid prototyping and iterative design cycles, accelerating innovation and time-to-market for new products. Simultaneously, advancements in advanced molding techniques are enhancing the efficiency and quality of elastomeric composite manufacturing. Innovations in mold design, materials, and processing technologies are improving the consistency and performance of molded components. The integration of these advanced manufacturing technologies not only enhances the capabilities of elastomeric composites but also expands their applicability across diverse industries [56].

10.3 Addressing Challenges

Despite their numerous advantages, elastomeric composites encounter several challenges that must be addressed for broader adoption and optimization in various industries. Key among these challenges are processing difficulties, cost considerations, and the imperative for precise control over material properties. Processing difficulties often stem from achieving uniform dispersion of fillers within the elastomeric matrix, which is crucial for maintaining consistent mechanical, thermal, and electrical properties throughout the composite material. Addressing this challenge requires advancements in processing technologies and formulation techniques to ensure reliable and reproducible results. Cost considerations also pose a significant hurdle, particularly when balancing the desire for enhanced performance with the economic feasibility of large-scale production. Innovations in material sourcing, manufacturing processes, and recycling methods are essential to reduce costs while maintaining or improving the performance of elastomeric composites. Moreover, achieving precise control over material properties such as stiffness, durability, and thermal conductivity remains a critical objective to meet the specific requirements of diverse applications in industries like automotive, aerospace, electronics, and healthcare [57].

11 Conclusion

In conclusion, Elastomeric-based 3D and 4D printable composite materials represent a transformative frontier in engineering, offering new capabilities for creating highly adaptable and durable structures. These materials, characterized by their flexibility, resilience, and ability to return to their original shape after deformation, provide engineers with a unique set of tools to address complex design challenges across various industries. One of the primary advantages of elastomeric composites is their capacity for customization. Through 3D printing, complex geometries and structures can be fabricated with high precision, allowing for the production of components that are tailored to specific applications. This level of customization is particularly valuable in fields such as aerospace, automotive, and medical devices, where the ability to design lightweight, strong, and flexible parts can significantly enhance performance and functionality. The emergence of 4D printing further expands the potential of elastomeric-based materials by introducing the dimension of time into the manufacturing process. 4D printed structures are designed to change shape or function in response to external stimuli such as temperature, light, or moisture. This dynamic behavior is enabled by incorporating stimuli-responsive materials into the elastomeric matrix, allowing the printed object to adapt to its environment after fabrication. In engineering applications, this capability opens up new possibilities for creating smart systems that can respond to changing conditions autonomously.

The widespread adoption of elastomeric-based 3D and 4D printable materials in engineering also presents challenges. Issues such as material cost, processing complexity, and the need for specialized equipment must be addressed to make these technologies more accessible to a broader range of industries. Ongoing research and development are essential to overcoming these hurdles and unlocking the full potential of these materials. As these technologies continue to advance, they will likely play a crucial role in the development of next-generation engineering solutions, offering new ways to address complex challenges across various industries. The future of engineering will undoubtedly be shaped by the continued exploration and application of these versatile materials.

References

1. Papageorgiou, D.G., Kinloch, I.A., Young, R.J.: Carbon **95**, 460–484 (2015). https://doi.org/10.1016/j.carbon.2015.08.055
2. Wang, D., Wang, Z., Ren, S., Xu, J., Wang, C., Hu, P., Fu, J.: Mater. Horiz. **8**(8), 2238–2250 (2021). https://doi.org/10.1039/D1MH00548K
3. Sahoo, B.P., Naskar, K., Tripathy, D.K.: J. Mater. Sci. **47**, 2421–2433 (2012). https://doi.org/10.1007/s10853-011-6065-8
4. Ponnamma, D., Sadasivuni, K.K., Grohens, Y., Guo, Q., Thomas, S.: J. Mater. Chem. C **2**(40), 8446–8485 (2014). https://doi.org/10.1039/C4TC01037J
5. Lu, Y., Liu, Z., Yan, H., Peng, Q., Wang, R., Barkey, M.E., Jeon, J.W., Wujcik, E.K., Appl, A.C.S.: Mater. Interfaces **11**(22), 20453–20464 (2019). https://doi.org/10.1021/acsami.9b05464
6. Alarifi, I.M.: Adv. Ind. Eng. Polym. Res. **6**(4), 451–464 (2023). https://doi.org/10.1016/j.aiepr.2023.05.001
7. Jagadeesh, P., Rangappa, S.M., Siengchin, S., Puttegowda, M., Thiagamani, S.M.K., Rajeshkumar, G., Kumar, M.H., Oladijo, O.P., Fiore, V., Cuadrado, M.M.M.: Polym. Compos. **43**, 5831–5862 (2022). https://doi.org/10.1002/pc.27000
8. Fazli, A., Rodrigue, D.: Mater. **13**(3), 782 (2020). https://doi.org/10.3390/ma13030782
9. Jagadeesh, P., Girijappa, Y.G.T., Puttegowda, M., Rangappa, S.M., Siengchin, S.: J. Nat. Fibers **19**(11), 1–16 (2020). https://doi.org/10.1080/15440478.2020.1854145
10. Sienkiewicz, M., Janik, H., Labuda, K.B., Lipka, J.K.: J. Clean. Prod. **147**, 560–571 (2017). https://doi.org/10.1016/j.jclepro.2017.01.121
11. Sathish, T., Jagadeesh, P., Rangappa, S.M., Siengchin, S.: Polym. Compos. **43**(7), 4297–4305 (2022). https://doi.org/10.1002/pc.26690
12. Zhang, J., Fleuryb, E., Brook, M.A.: Green Chem. **17**, 4647–4656 (2015). https://doi.org/10.1039/C5GC01418B
13. Volpe, V., Davino, D., Sorrentino, L., Gorrasi, G., Pantani, R.: J. Appl. Polym. Sci. **135**(44), 46863 (2018). https://doi.org/10.1002/app.46863
14. Schillfahrt, C., Fauster, E., Schledjewski, R.: Int. J. Mater. Form. **12**, 1–15 (2019). https://doi.org/10.1007/s12289-018-1398-3
15. Kalsoom, U., Nesterenkoab, P.N., Paull, B.: RSC Adv. **6**, 60355–60371 (2016). https://doi.org/10.1039/C6RA11334F
16. Qian, C., Xiao, T., Chen, Y., Wang, N., Li, B., Gao, Y.: Adv. Eng. Mater. **24**(4), 2101068 (2022). https://doi.org/10.1002/adem.202101068
17. Guan, Z., Wang, L., Bae, J.: Mater. Horiz. **9**, 1825–1849 (2022). https://doi.org/10.1039/D2MH00232A

18. Wang, R., Zhang, J., Kang, H., Zhang, L.: Compos. Sci. Technol. **133**, 136–156 (2016). https://doi.org/10.1016/j.compscitech.2016.07.019
19. Ambulo, C., Ford, M., Searles, M., Majidi, C., Ware, T., Appl, A.C.S.: Mater. Interfaces **13**(11), 12805–12813 (2021). https://doi.org/10.1021/acsami.0c19051
20. Kong, J.H., Jang, N.S., Kim, S.H., Kim, J.M.: Carbon **77**, 199–207 (2014). https://doi.org/10.1016/j.carbon.2014.05.022
21. Ryan, K.R., Down, M.P., Banks, C.E.: J. Chem. Eng. **403**, 126162 (2021). https://doi.org/10.1016/j.cej.2020.126162
22. Jiang, Z., Diggle, B., Tan, M.L., Viktorova, J., Bennett, C.W., Connal, L.A.: Adv. Sci. **7**(17), 2001379 (2020). https://doi.org/10.1002/advs.202001379
23. Yang, X., Guo, Y., Luo, X., Zheng, N., Ma, T., Tan, J., Li, C., Zhang, Q., Gu, J.: Compos. Sci. Technol. **164**, 59–64 (2018). https://doi.org/10.1016/j.compscitech.2018.05.038
24. Araby, S., Meng, Q., Zhang, L., Zaman, I., Majewski, P., Ma, J.: Nanotechnology **26**(11), 112001 (2015). https://doi.org/10.1088/0957-4484/26/11/112001
25. Sadasivuni, K.K., Ponnamma, D., Thomas, S., Grohens, Y.: Prog. Polym. Sci. **39**(4), 749–780 (2014). https://doi.org/10.1016/j.progpolymsci.2013.08.003
26. Ma, Z., Li, H., Jing, X., Liu, Y., Mi, H.: Sens. Actuators A: Phys. **329**, 112800 (2021). https://doi.org/10.1016/j.sna.2021.112800
27. Mensah, B., Gupta, K.C., Kim, H., Wang, W., Jeong, K., Nah, C.: Polym. Test. **68**, 160–184 (2018). https://doi.org/10.1016/j.polymertesting.2018.04.009
28. Li, S., Zhang, J., He, J., Liu, W., Wang, Y., Huang, Z., Pang, H., Chen, Y.: Adv. Sci. **10**(34), 2304506 (2023). https://doi.org/10.1002/advs.202304506
29. Pichaiyut, S., Nakason, C., Vennemann, N.: Iran. Polym. J. **21**, 65–79 (2012). https://doi.org/10.1007/s13726-011-0006-x
30. Mathur, V.K.: Constr. Build. Mater. **20**(7), 470–477 (2006). https://doi.org/10.1016/j.conbuildmat.2005.01.031
31. Lunetto, V., Galati, M., Settineri, L., Iuliano, L.: J. Manuf. Process. **85**, 858–874 (2023). https://doi.org/10.1016/j.jmapro.2022.12.020
32. Sharma, A.K., Bhandari, R., Aherwar, A., Rimašauskienė, R.: Mater. Today. Proc. **21**(3), 1559–1562 (2020). https://doi.org/10.1016/j.matpr.2019.11.086
33. Rollo, G., Ronca, A., Cerruti, P., Gan, X.P., Fei, G., Xia, H., Gorokhov, G., Bychanok, D., Kuzhir, P., Lavorgna, M., Ambrosio, M.: Polymers **12**(8), 1841 (2020). https://doi.org/10.3390/polym12081841
34. Park, M., Park, J., Jeong, U.: NanoToday **9**(2), 244–260 (2014). https://doi.org/10.1016/j.nantod.2014.04.009
35. Narayana, K.J., Burela, R.G.: Mater. Today Proc. **5**(2), 5580–5590 (2018). https://doi.org/10.1016/j.matpr.2017.12.149
36. Dolui, T., Natarajan, T.S., Chanda, A.S.J., Ghosh, P., Mukhopadhyay, R., Wießner, S., Heinrich, G., Das, A., Banerjee, S.S.: Adv. Eng. Mater. **25**(20) (2023) 2300584, https://doi.org/10.1002/adem.202300584
37. Barnes, M., Sajadi, S.M., Parekh, S., Rahman, M.M., Ajayan, P.M., Verduzco, R., Appl, A.C.S.: Mater. Interfaces **12**(25), 28692–28699 (2020). https://doi.org/10.1021/acsami.0c07331
38. Mahmood, A., Perveen, F., Chen, S., Akram, T., Irfan, A.: Molecules **29**(2), 319 (2024). https://doi.org/10.3390/molecules29020319
39. Hagaman, D.E., Leist, S., Zhou, J., Ji, H.-F., Appl, A.C.S.: Mater. Interfaces **10**(32), 27308–27315 (2018). https://doi.org/10.1021/acsami.8b08503
40. Lu, X., Ambulo, C.P., Wang, S., Rivera-Tarazona, L.K., Kim, H., Searles, K., Ware, T.H.: Angew. Chem. Int. Ed. **60**(10), 5536–5543 (2021). https://doi.org/10.1002/anie.202012618
41. Khoo, Z.X., Teoh, J.E.M., Liu, Y., Chua, C.K., Yang, S., An, J., Leong, K.F., Yeong, W.Y.: Virtual Phys. Prototyp. **10**, 103–122 (2015). https://doi.org/10.1080/17452759.2015.1097054
42. Malek, P., Walczyk, D.: J. Manuf. Sci. Eng. **138**(2), 021002 (2016). https://doi.org/10.1115/1.4030432
43. Kang, S.S., Choi, K., Nam, J.D., Choi, H.J.: Mater. **13**(20), 4597 (2020). https://doi.org/10.3390/ma13204597

44. Costa, P., Silva, J., Sencadas, V., Simoes, R., Viana, J.C., Méndez, S.L.: J. Mater. Sci. **48**, 1172–1179 (2013). https://doi.org/10.1007/s10853-012-6855-7
45. Mahesh, V., Joladarashi, S., Kulkarni, S.M.: Def. Technol. **17**(1), 161–176 (2021). https://doi.org/10.1016/j.dt.2020.02.013
46. Han, S., Meng, Q., Chand, A., Wang, S., Li, X., Kang, H., Liu, T.: Polym. Test. **80**, 106106 (2019). https://doi.org/10.1016/j.polymertesting.2019.106106
47. Zhang, H., Yan, L., Zhou, S., Zou, H., Chen, Y., Liang, M., Ren, X.: Polym. Eng. Sci. **61**(2), 442–452 (2021). https://doi.org/10.1002/pen.25587
48. Dang, Z., Shehzad, K., Zha, J., Mujahid, A., Hussain, T., Nie, J., Shi, C.: Compos. Sci. Technol. **72**(1), 28–35 (2011). https://doi.org/10.1016/j.compscitech.2011.08.020
49. Barrios, S.U., Verdejo, R., Manchado, M.Á.L., Santana, M.H.: Eur. Polym. J. **190**, 112023 (2023). https://doi.org/10.1016/j.eurpolymj.2023.112023
50. Ren, L., Li, B., Song, Z., Liu, Q., Ren, L., Zhou, X.: Compos. Part B Eng. **164**, 458–466 (2019). https://doi.org/10.1016/j.compositesb.2019.01.061
51. Harussani, M.M., Sapuan, S.M., Nadeem, G., Rafin, T., Kirubaanand, W.: Def. Technol. **18**(8), 1281–1300 (2022). https://doi.org/10.1016/j.dt.2022.03.006
52. Lamour, M., Cecchin, A.: Constr. Build. Mater. **293**(26), 123368 (2021). https://doi.org/10.1016/j.conbuildmat.2021.123368
53. Khalid, M.Y., Arif, Z.U., Tariq, A., Hossain, M., Khan, K.A., Umer, R.: Eur. Polym. J. **205**, 112718 (2024). https://doi.org/10.1016/j.eurpolymj.2023.112718
54. Appuhamillage, G.A., Ambagaspitiya, S.S., Dassanayake, R.S., Wijenayake, A.: Explor Med. **5** 17–47, (2024). https://doi.org/10.37349/emed.2024.00203
55. Romasanta, L.J., Manchado, M.A.L., Verdejo, R.: Prog. Polym. Sci. **51**, 188–211 (2015). https://doi.org/10.1016/j.progpolymsci.2015.08.002
56. Regis, J.E., Renteria, A., Hall, S.E., Hassan, M.S., Marquez, C., Lin, Y.: Mater. **14**(16), 4521 (2021). https://doi.org/10.3390/ma14164521
57. Alghamdi, S.S., John, S., Choudhury, N.R., Dutta, N.K.: Poly. **13**(5), 753 (2021). https://doi.org/10.3390/polym13050753

Fundamentals and Recent Applications of 3D, 4D, and 5D Printing of Polymer and Its Nanocomposites: Challenges and Its Future Prospects

Abhinay Thakur, Valentine Chikaodili Anadebe, Omar Dagdag, Deepak Sharma, and Hari Om

Abstract With the development of 3D, 4D, and 5D printing technology, the field of additive manufacturing has undergone tremendous change. This chapter explores the principles and current developments in these fields with an emphasis on polymers and their nanocomposites. It examines how advances in multi-material and multi-scale 4D printing have transformed a range of industries, including aerospace and healthcare, and highlights how these technologies can be used to build intelligent, shape-shifting objects that are capable of self-assembly. Additionally, the notion of 5D printing is introduced, highlighting its potential for dynamic applications in biomedicine, robotics, and sensors. 5D printing includes time as an additional dimension. Issues with material compatibility, resolution, and post-processing are discussed, in addition to sustainable methods and choices for biodegradable polymers. Incorporating nanoparticles to improve material qualities including strength, conductivity, and thermal stability is also covered in this chapter. Lastly, it offers a futuristic perspective on the widespread industrial use of 3D, 4D, and 5D printing of polymers and their nanocomposites, promoting innovation and sustainability in industries like electronics, personalized medicine, and architecture. For academics, engineers, and innovators working in the ever-changing field of polymer additive manufacturing, this chapter provides a thorough reference.

A. Thakur (✉)
Division of Research and Development, Lovely Professional University, Phagwara, Punjab, India
e-mail: thakurabhinay96@gmail.com

V. C. Anadebe
Department of Chemical Engineering, Alex Ekwueme Federal University Ndufu Alike, Abakakili, Ebonyi State, Nigeria

O. Dagdag
Department of Mechanical Engineering, Gachon University, Seongnam, Republic of Korea

D. Sharma
Deenbandhu Chhotu Ram University of Science & Technology, Murthal, Sonipat, Haryana, India

D. Sharma · H. Om
Department of Chemistry, Bhagat Phool Singh Mahila Vishwavidyalay, Khanpur Kalan, Sonipat, Haryana, India

S. Moharana et al. (eds.), *Polymer Nanocomposites for 3D, 4D and 5D Printing*, Engineering Materials, https://doi.org/10.1007/978-981-96-4214-4_14

Keywords Additive manufacturing · 3D Printing · 4D Printing · 5D Printing · Polymer nanocomposites · Future prospects

1 Introduction

Additive manufacturing (AM), also referred to as 3D printing, has completely changed the fabrication and production sectors by providing a creative and adaptable method for producing intricate parts and structures. AM constructs objects layer by layer from digital models, in contrast to traditional subtractive manufacturing techniques that remove material from a solid block to produce the desired shape [1–4]. This paradigm change has made it possible to produce custom parts that are customized to individual requirements, while simultaneously facilitating quick prototyping and opening up new options for mass customization. AM is significant in a number of industries, including consumer goods, healthcare, automotive, and aerospace, underscoring its revolutionary effect on contemporary production. The inception of additive manufacturing dates back to the 1980s, a time of substantial technological advancement. Stereolithography (SLA) was invented by Charles Hull in 1984, and it was a revolutionary advancement that established the discipline [5–7]. SLA creates complicated and detailed items by layer-curing photopolymer resins using ultraviolet light. This invention served as the model for several other additive manufacturing (AM) processes that have subsequently been created, including digital light processing (DLP), fused deposition modeling (FDM), and selective laser sintering (SLS). These early developments paved the way for AM technologies' quick development and diversification, each of which has its benefits and uses. The first 3D printing technology to be commercialized was stereolithography (SLA), which is still a mainstay in the field [8, 9]. SLA achieves great precision and fine detail by curing liquid resin into hardened plastic with a laser beam, which makes it perfect for producing complex designs and prototypes. One of the most popular AM technologies, fused deposition modeling (FDM), deposits material layer by layer by extruding thermoplastic filaments via a heated nozzle. FDM is a well-liked option for desktop 3D printers because of its accessibility, cost, and variety of available materials. With Selective Laser Sintering (SLS), powdered materials—typically nylon or other thermoplastics—are sinted into solid structures using a laser [10–13]. Layer by layer, the item is constructed by the laser's selective fusion of powder particles. SLS is very useful for creating practical, long-lasting parts with intricate geometry. Similar to SLA, Digital Light Processing (DLP) flashes a single picture of each layer throughout the whole vat of liquid resin using a digital projector screen. This technique is appropriate for intricate and high-quality prints since it offers great resolution and faster print times. A sweeping arm is used in HP's relatively new Multi-Jet Fusion (MJF) technology to apply fusing chemicals over a powder material bed. A heat source then fused the powder into. Offering superior mechanical qualities along with rapid production speeds, MJF is a formidable competitor in the additive manufacturing space. Figure 1 shows the CAD and optical images of 3D-printed objects.

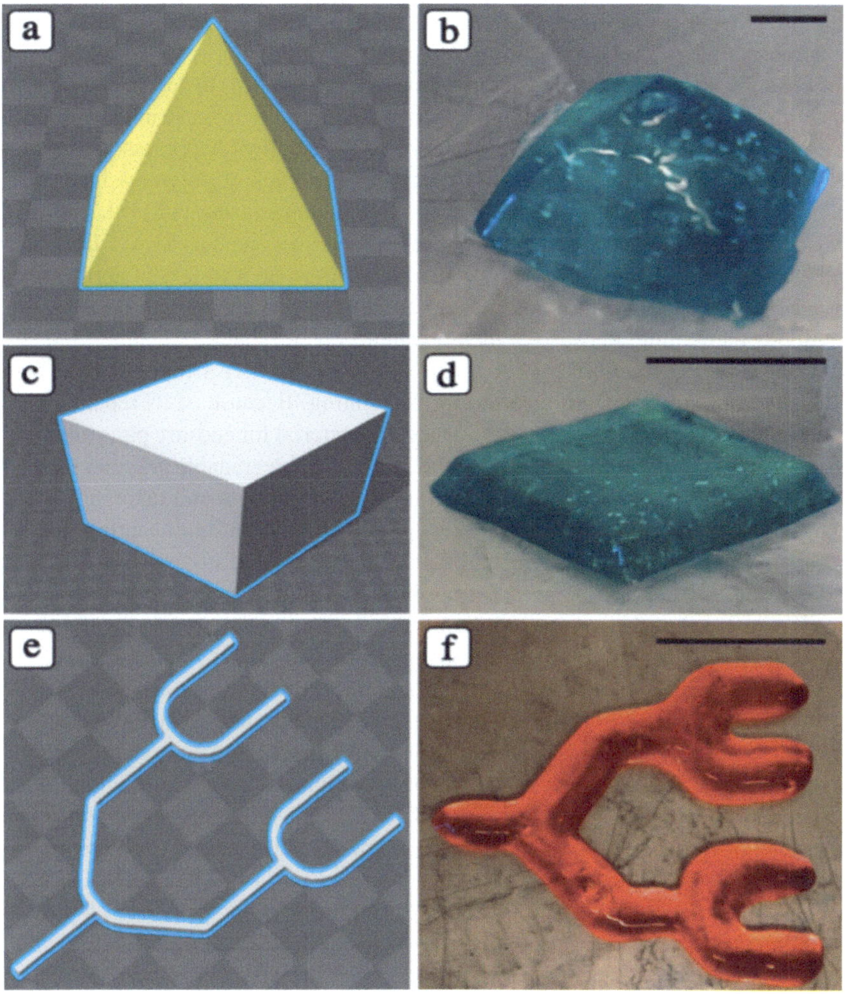

Fig. 1 CAD and optical images of 3D-printed objects. All structures were printed with a layer thickness of 200 μm and a fiber spacing of 500 μm in the X–Y plane. Adapted from Ref. [14] under CCBY 4.0

The adoption of AM technologies has occurred in a number of industries due to their adaptability. AM is used in aircraft to create complicated, lightweight parts that perform better and consume less fuel. AM methods are frequently used to print parts including air ducts, brackets, and turbine blades. Through the creation of bespoke implants, prosthetics, and anatomical models for surgery planning, additive manufacturing (AM) improves the healthcare sector [15, 16]. Another application of AM, known as bioprinting, is being investigated for tissue engineering and regenerative medicine. AM is used in the automobile industry for end-use part manufacture, tooling, and fast prototyping. Customization and small-batch production are especially beneficial for making one-of-a-kind parts and improving vehicle performance. The consumer goods sector uses additive manufacturing (AM) to design and produce customized goods, such as shoes and eyeglasses. The ability to make unique products that fit specific tastes promotes innovation and raises client happiness. The adaptability, ease of processing, and wide range of properties of polymers have made them essential materials in the field of additive manufacturing [17, 18]. Polymers may be utilized in many different 3D printing processes, such as SLA, FDM, SLS, and DLP, because of their versatility. Their adaptability to a variety of applications, from high-strength industrial components to flexible and durable consumer goods, highlights their value. Because of their advantageous mechanical qualities and ease of processing, polymers like PLA (polylactic acid) and ABS (acrylonitrile butadiene styrene) are frequently utilized in 3D printing. Because of its reputation for durability and impact resistance, ABS is a good material for end-use parts and functional prototypes. For desktop 3D printers, PLA—a biodegradable polymer made of renewable resources—is preferred due to its user-friendliness and favorable effects on the environment. The possibilities of 3D printing have been significantly enhanced by the invention of polymer nanocomposites. Properties like strength, conductivity, and thermal stability can be greatly improved by adding nanoparticles to polymer matrices [19, 20]. Additive manufacturing gains a temporal dimension with the use of 4D printing. In contrast to traditional 3D printing, which yields static items, 4D printing produces structures that, in response to external stimuli like moisture, light, or temperature, can gradually change their shape or other characteristics. This dynamic capability is made possible by the employment of responsive polymers or smart materials, which are designed to respond to changes in their surroundings. For instance, a 4D-printed object may be made to self-assemble into a complex shape when damp, or it could be made to fold or expand when heated. The creation of self-adaptive structures and systems that can react to their surroundings is made possible by these materials, and this could result in novel applications in industries including robotics, aircraft, and medical devices [21–24]. The capacity of 4D printing to produce structures that are not only useful in the first place but also able to change or adapt over time to suit changing situations is what gives this technology its transformative promise. With the addition of more degrees of freedom to the additive manufacturing process, 5D printing marks a further advancement. With the help of this sophisticated method, several materials with different qualities may be combined and treated to create complex, multifunctional parts [25–29]. To create items with a variety of dynamic capabilities, 5D printing usually uses complex procedures to combine different elements, such as multi-material extrusion or sophisticated laser techniques. The capacity of 4D printing to produce structures that are not only useful in the first place but also able to change or adapt over time to suit changing situations is what gives this technology its transformative promise. With the addition of more degrees of freedom to the additive manufacturing process, 5D printing marks a further advancement [30, 31]. With the help of this sophisticated method, several materials with different qualities may be combined and treated to create complex, multifunctional parts. To create items with a variety of dynamic capabilities, 5D printing usually uses complex procedures to combine different elements, such as multi-material extrusion or sophisticated laser techniques. Moreover, sustainability issues are addressed by the use of biodegradable polymers in additive manufacturing, which provides environmentally benign substitutes for conventional plastics. These materials can support a circular economy and lessen the negative effects of manufacturing processes on the environment [32]. The range of materials available and their applications are projected to rise as research and development in polymer-based 3D, 4D, and 5D printing continues, spurring innovation and sustainability in additive manufacturing.

2 Fundamentals of 3D, 4D, and 5D Printing

2.1 Basic Principles

A few basic ideas underpin the operations of 3D, 4D, and 5D printing and are shared by several technologies. Digital design is the initial tenet. Using computer-aided design (CAD) software, a digital model is created at the start of the process. After that, the model is transformed into a file in the Standard Tessellation Language (STL) format, which is compatible with 3D, 4D, and 5D printers. The STL file gives the printer the information it needs to build the object layer by layer by describing the surface geometry of the 3D, 4D, and 5D objects as a mesh of triangles. The building method of layer by layer is the second premise. The slicing program divides the prepared digital model into thin horizontal layers [33]. The final object is represented by a cross-section in each slice. Next, layer by layer, the 3D, 4D, and 5D printer constructs the thing by depositing or solidifying material, working upwards from the bottom. Complex shapes that would be challenging or impossible to create with subtractive manufacturing techniques can now be created thanks to this additive manufacturing technique. Material deposition, often known as solidification, is the third principle. Diverse techniques are employed by different 3D, 4D, and 5D printing technologies to deposit or solidify the material [34, 35]. For instance, selective laser sintering (SLS) employs a laser to sinter powdered material, whereas fused deposition modeling (FDM) involves extruding molten thermoplastic through a heated nozzle. Every technology has advantages and disadvantages that affect the material selection, resolution, and mechanical characteristics of the printed object. The most widely used metal 3D printing technology and their uses are displayed in Fig. 2. (a) Shows a 3D printer that uses selective laser melting (SLM). (b) Displays an impeller created with an SLM machine in 3D printing. (c) Shows an example of a 3D printer using fused deposition modeling (FDM). (d) Showcases a sprocket gear made with FDM technology. (e) Showcases a 6-dimensional robotic arm 3D printer using Laser Metal Wire Deposition (LMWD). (f) Displays a rocket body made using the 6-dimensional robotic arm technology of laser metal wire deposition.

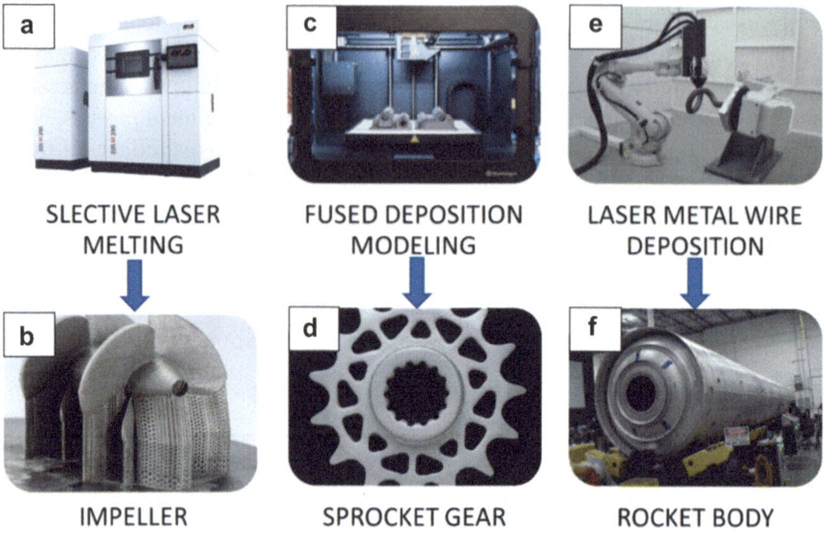

Fig. 2 Leading metal 3D printing technologies and their applications. **a** Selective Laser Melting (SLM) 3D printer. **b** An impeller produced via SLM. **c** Fused Deposition Modeling (FDM) 3D printer. **d** A sprocket gear was created with FDM. **e** Laser Metal Wire Deposition (LMWD) 6-axis robotic arm 3D printer. **f** A rocket body manufactured using LMWD with a 6-axis robotic arm. Adapted from Ref. [36] under CCBY 4.0

2.2 *Key Technologies and Methods*

The fields of 3D, 4D, and 5D printing have seen the emergence of several important technologies and techniques, each with distinct features and uses. It is essential to comprehend these technologies in order to choose the best approach for a certain application.

- Stereolithography (SLA)

The distinction of being the first 3D, 4D, and 5D printing technology to be commercially successful belongs to stereolithography (SLA). Charles Hull developed SLA in 1984, and it revolutionized the production of complex designs and prototypes by providing previously unachievable levels of accuracy and detail. Using a liquid photopolymer resin that solidifies when exposed to ultraviolet (UV) light is the fundamental idea of SLA. A computer model is first created and then divided into thin cross-sectional layers [37–39]. The object is then constructed layer by layer from the bottom up by repeatedly tracing and curing these layers with a laser beam. SLA's operational mechanism is quite simple but incredibly powerful. The first step is to fill a vat with liquid photopolymer resin. After every layer cures, the build platform, which lies slightly below the resin's surface, gradually rises. A laser beam that is guided by galvanometer mirrors carefully transfers the object's cross-sectional pattern onto the resin's surface. The photopolymerization reaction started by the laser's UV light causes the resin to harden exactly where the laser makes contact. A new layer of resin is placed over the previously hardened layer by a recoating blade when a layer has cured and the build platform slides slightly downward. Until the complete item is produced, this process is repeated [40, 41]. SLA's remarkable resolution and fine detail production are two of its main benefits. Much finer than what is possible with many other 3D, 4D, and 5D printing technologies, the laser can attain very high levels of accuracy, often in the range of tens of micrometers. Because of its high resolution, SLA is especially useful for applications like jewelry, dental molds, and precise prototypes that call for complicated geometries and delicate designs. Another benefit of SLA prints over technologies like Fused Deposition Modeling (FDM), which may result in rougher surfaces, is the reduced requirement for considerable post-processing due to their smooth surface quality.

SLA has certain limits that need to be taken into account despite its accuracy and thoroughness. A primary disadvantage is the comparatively sluggish printing speed. Even though laser tracing is quite exact, it can take a while, particularly for larger things. The total print time is increased by the requirement to spread and level a fresh layer of resin after each cure, which necessitates the sequential curing of each layer [42–44]. When time is of the essence in production settings, this reduced pace may operate as a bottleneck. The price of the materials is another factor. Generally speaking, the photopolymer resins used in SLA are more costly than the powders used in Selective Laser Sintering (SLS) or the thermoplastic filaments used in FDM. Viscosity, curing speed, and mechanical strength must all be balanced in these resins, which are specifically made to react to UV radiation. Moreover, post-processing of SLA prints frequently necessitates the use of UV ovens to guarantee complete curing and solvents to remove uncured resin, which raises operational costs. When compared to other 3D, 4D, and 5D printing materials, photopolymer resins may have restricted material properties. Even though there are many distinct types of resins that have unique qualities like hardness, heat resistance, or flexibility, their mechanical performance is usually inferior to that of engineering-grade metals or thermoplastics [45, 46]. This restricts the application of SLA to functional components that have high mechanical or thermal stress requirements.

- Fused Deposition Modeling (FDM)

One of the most popular 3D, 4D, and 5D printing technologies is fused deposition modeling (FDM), especially when it comes to desktop 3D, 4D, and 5D printers. Professionals, educators, and enthusiasts all love FDM because of its ease of use and adaptability. The process of extruding thermoplastic filament through a heated nozzle is the fundamental component of FDM [47–50]. This nozzle builds up the final product by depositing the material layer by layer under the guidance of a computerized design. A spool of thermoplastic filament is inserted into the heated extruder to start the process [51, 52]. The filament is heated by the extruder to a semi-liquid state, which permits it to pass through the nozzle. The nozzle precisely drops the molten plastic onto the build platform as it moves in accordance with the computerized instructions. After extrusion, each layer nearly instantly cools and hardens, creating a firm foundation for the layer that comes after it. Until the entire item is produced, this additive process is continued. FDM is praised for a number of important benefits that support its broad use. It is renowned above all for being user-friendly. Even those with less technical expertise can utilize the technology due to its simple method of feeding filament through a heated nozzle. The printing procedure is made further simpler by the user-friendly interfaces and pre-configured parameters included in many desktop FDM printers. The affordability of FDM is a key component that also greatly adds to its allure. FDM printers and materials are more affordable when compared to other 3D, 4D, and 5D printing technologies like Selective Laser Sintering (SLS) or SLA [53, 54]. Because of its affordability, FDM printing is a desirable choice for small enterprises, educational institutions, and hobbyists who want to experiment with 3D, 4D, and 5D printing without having to make a significant financial commitment.

The variety of materials that are available with FDM is one of its most alluring features. ABS (acrylonitrile butadiene styrene), PLA (polylactic acid), and PETG (polyethylene terephthalate glycol) are common thermoplastics used in FDM. Every material has unique qualities that make it appropriate for a variety of uses. For example, ABS is well-known for its strength and durability, which makes it perfect for end-use parts and functional prototypes. On the other hand, PLA is a popular material for prototypes and educational projects because it is simple to print and biodegradable. PETG offers a well-balanced solution for a range of applications by fusing the strength of ABS with the flexibility of PLA printing [55–57]. Furthermore, specialist filaments like composite, flexible, and heat-resistant materials increase the potential applications of FDM printing. Custom parts, low-volume production runs, and functioning prototypes are especially well-suited applications for FDM. It is a useful tool for engineers and designers who need to test form, fit, and function because of its capacity to produce robust and long-lasting products. Custom parts may be made cheaply and rapidly for both industrial and hobby uses. FDM provides a low-cost substitute for conventional manufacturing techniques in the field of low-volume production, making it possible to produce small batches of parts without the need for pricey molds or tooling. FDM does, however, have certain drawbacks, especially when contrasted with other 3D, 4D, and 5D printing technologies like SLA [33, 58, 59]. A noteworthy disadvantage is the resolution. Although FDM is capable of producing strong, functional things, its resolution is usually lower than SLA's. The surface of the printed product frequently displays the layer lines produced by the extrusion process, which may detract from the fine detail and overall aesthetic appeal. To obtain a smoother surface, these layer lines might need to be treated using post-processing methods including sanding, polishing, or chemical smoothing.

- Selective Laser Sintering (SLS)

A laser is used in Selective Laser Sintering (SLS), a potent and adaptable 3D, 4D, and 5D printing technology, to fuse powdered material into solid structures. While SLS can operate with a wide range of materials, such as metals, ceramics, and composites, it usually uses nylon or other thermoplastics. An object's computer model is divided into thin layers to start the SLS process. These layers direct the laser's selective fusing of the powder particles, constructing the item layer by layer from the bottom up. SLS operates on a highly sophisticated and efficient operating mechanism. A bed of finely ground material is the initial step [60, 61]. This bed contains the build platform, which is initially positioned at its highest point, just below the powder's surface. A laser that is carefully steered by computer-guided optics sweeps across the powder bed's surface. The powder particles merge in a process known as sintering when the laser's energy raises them to a temperature slightly below their melting point. When a layer is finished, the build platform descends a little, and a recoating blade covers the previously sintered layer with a new layer of powder. Until the complete item is produced, this process is repeated [62]. One of SLS's most notable benefits is its capacity to create components with exceptional mechanical qualities that are both useful and long-lasting. Parts made using the sintering process are robust, resilient, and able to bear a great deal of stress and pressure. Because of this, SLS is especially well suited for producing complicated assemblies, functioning prototypes, and end-use components that must function dependably in real-world settings. Strength, flexibility, and heat resistance are all well-balanced in materials used in SLS printing, such as nylon, which expands the practical uses of SLS-printed parts.

The fact that SLS doesn't need support structures sets it apart from other 3D, 4D, and 5D printing technologies like Stereolithography (SLA) and Fused Deposition Modeling (FDM). Support structures are frequently required in SLA and FDM to maintain overhanging elements throughout the printing process. In post-processing, these supports must be manually removed, which can take time and can degrade the finished part's surface quality. In SLS, the part is organically supported during construction by the surrounding unsintered powder [63, 64]. This makes the post-processing phase simpler and does away with the requirement for extra support structures. Furthermore, the lack of support gives designers more creative license to create complex geometries and interconnecting pieces that would be difficult or impossible to make using conventional techniques. One further noteworthy benefit of SLS is its capacity to produce parts with extremely intricate geometries. Detailed and complex designs can be produced thanks to the laser's precision and the powder layers' precise resolution. In sectors like aerospace, automotive, and medical, where sophisticated, lightweight, and customized components are frequently needed, this expertise is very beneficial. Internal cavities, lattice structures, and other complex features that improve performance while using less material and weight can be produced via SLS. SLS has certain drawbacks that need to be taken into account despite all of its benefits. Cost is one of the main disadvantages. Generally speaking, SLS printers cost more than FDM and SLA equipment. The total cost is influenced by the high cost of the laser and the sophisticated optics needed for accurate control [65]. Furthermore, compared to the filaments used in FDM or the resins used in SLA, the powdered ingredients utilized in SLS are typically more expensive. Small enterprises and individual users who wish to use 3D, 4D, and 5D printing technologies may find this higher cost to be a deterrent. Comparing the SLS process to other 3D, 4D, and 5D printing techniques, it can also be quite slow. Print times are increased by the requirement to accurately regulate the laser and guarantee perfect sintering of each layer, particularly for larger objects or parts with fine features. Although post-processing is made easier by the lack of support structures, it can still take a lot of time and work to carefully remove the surrounding unsintered powder and carry out any necessary finishing operations.

- Digital Light Processing (DLP)

Similar to stereolithography (SLA), digital light processing (DLP) is a 3D, 4D, and 5D printing technology that projects a single picture of each layer throughout the whole vat of liquid resin using a digital projector screen. Compared to classic SLA, this method has a number of benefits, especially in terms of speed and resolution. Instead of tracing the layer image point by point with a laser-like SLA, the digital projector in DLP projects the complete layer image at once. The simultaneous curing of each layer in this approach results in much shorter print times. The high resolution possible with DLP is shown by the thousands of small pixels that make up the projected image. The photopolymer resin is selectively cured in the areas exposed to light by the projector's flashing picture, which solidifies the layer [66, 67]. A fresh layer of liquid resin flows over the previously formed layer when each layer has cured, making room for the subsequent picture projection. The build platform then raises slightly. Until the entire object is constructed, this layer-by-layer procedure is continued. The quicker print times of DLP over SLA are one of its main benefits. The time needed to cure each layer is greatly decreased because every layer is simultaneously exposed to light. Because of this, DLP is especially beneficial for generating prints that are detailed and of high quality quickly, which makes it appropriate for applications where accuracy and speed are crucial. For example, DLP is well adapted to meet the demands of businesses like jewelry, dentistry, and high-end consumer products, which frequently require rapid production of small, delicate parts. Another noteworthy advantage of DLP is its great resolution. Depending on the resolution and optics chosen, a digital projector can produce extremely fine details, frequently at the micrometer level [68, 69]. DLP is perfect for producing products with fine features and complex designs that need exact precision because of its high resolution. One additional benefit of using DLP printing to produce visually beautiful parts is that the flawless surface finish it produces often eliminates the need for extensive post-processing.

DLP works very well for generating intricately detailed small to medium-sized parts. Its capacity to project a whole layer at once guarantees consistent curing throughout the layer, producing prints with exceptional accuracy and detail. Because of this, DLP is the material of choice for products where accuracy and fine detail are crucial, like personalized jewelry, hearing aids, and dental aligners. Nonetheless, DLP's main flaw is its constrained construction volume. The maximal construction area is determined by the dimensions of the projector screen and the optics utilized in the system. This limitation may reduce the maximum item size that can be printed, which reduces the suitability of DLP for creating huge parts or components [70, 71]. Even while the technology is excellent at producing small, intricate items, applications needing higher build volumes might not be the greatest fit for it. Pixelation, which can happen if the projector's resolution is too low, is another thing to take into account while using DLP. On curved or inclined surfaces, this might cause a "stair-stepping" effect where individual pixels from the projected picture are visible on the printed object's surface. Although this problem has been much lessened by improvements in projector technology and higher-resolution projectors, it should still be taken into account when choosing DLP for some applications. Comparable to those used in SLA, DLP materials are usually liquid photopolymer resins that solidify in the presence of light. These resins' many qualities, including their toughness, flexibility, and heat resistance, provide a wide range of application possibilities. But much like with SLA, these specialty resins can be expensive in comparison to other materials used in 3D, 4D, and 5D printing. This is something to keep in mind when calculating the total cost of employing DLP technology.

- Multi-Jet Fusion (MJF)

Hewlett-Packard (HP) introduced Multi Jet Fusion (MJF), an innovative 3D, 4D, and 5D printing technology that is becoming more and more well-known for its excellent mechanical qualities and fast production speeds. Applying heat and depositing fusing chemicals are the two steps in the MJF operation process. A fusing agent is first spread over a bed of powder material, usually a thermoplastic polymer like nylon, by a sweeping arm. After that, the powder is selectively heated—usually by an infrared lamp—where the fusing agent has been applied, causing the powder to fuse into solid layers. This process makes it possible to produce intricate geometries quickly and accurately, which makes it ideal for both working parts and prototypes [72, 73]. Because of its effective layer-by-layer processing and lack of need for support structures, which can drastically cut down on post-processing time, MJF is able to make high-quality parts quickly. High throughput and cost-efficiency are two of MJF's main advantages, especially for short to medium-sized production runs. In contrast to conventional approaches, the technology enables batch production, which lowers the total cost per unit by manufacturing many pieces at once. Furthermore, MJF provides the finished parts with strong mechanical properties, such as strength and durability, which are necessary for practical applications. There are restrictions to take into account, though [53, 74]. The initial investment in MJF printers might be substantial due to their high cost. Additionally, compared to other 3D, 4D, and 5D printing technologies like Selective Laser Sintering (SLS) and Fused Deposition Modeling (FDM), the spectrum of materials compatible with MJF is currently more constrained. Because of this, even if MJF works well in some applications, it might not be the best option in every situation, especially if material diversity is important.

3 Applications of 3D, 4D, and 5D Printing Technologies

Many different sectors have adopted 3D, 4D, and 5D printing technologies due to their versatility. These businesses use the unique advantages of additive manufacturing techniques to enhance their processes, products, and overall performance. Figure 3 shows a detailed exploration of how each technology is applied across different sectors.

Fig. 3 Utilization of 3D, 4D, and 5D printing technologies in several fields. Adapted from Ref. [36] Under CCBY 4.0

3.1 3D Printing Applications

- Aerospace

The aircraft sector has seen a significant transformation thanks to 3D printing, especially in the area of producing complex, lightweight components that are difficult to build using standard manufacturing techniques. The employment of sophisticated 3D printing methods like Direct Metal Laser Sintering (DMLS) and Selective Laser Sintering (SLS), which have made major gains in component performance and operating efficiency possible, is an example of this technical progress. The aerospace industry makes substantial use of Direct Metal Laser Sintering (DMLS), a sophisticated additive manufacturing method [75, 76]. It entails layer-by-layer selective fusion of metal powder particles using a powerful laser to produce intricate metallic components. Components with complex geometries that are challenging or impossible to produce with conventional subtractive manufacturing techniques can now be produced thanks to this approach. Turbine blades, for example—essential parts of jet engines—can be made with interior cooling channels that are designed to improve efficiency and thermal management. DMLS can provide a single, integrated component that eliminates the need for assembly and minimizes potential points of failure, in contrast to traditional manufacturing techniques that frequently call for the assembly of several pieces. DMLS has advantages that go beyond intricate geometries. In aircraft applications, the capacity to make lightweight parts is especially valuable. Manufacturers can build components that are substantially lighter than their traditionally made counterparts by utilizing modern metal alloys and optimizing designs through DMLS. Less weight means lighter airplanes use less energy to operate, which improves fuel economy. For instance, DMLS offers lighter weight and better performance for aeronautical parts like fuel nozzles and structural brackets. Because the material qualities were precisely controlled during the printing process, these parts are not only lighter but also show improved strength and durability.

Another significant 3D printing innovation that has had a big impact on the aerospace sector is Selective Laser Sintering (SLS). SLS is the process of layer-by-layer sintering powdered materials, such as metals, ceramics, or polymers, into solid structures using a laser. This process is very helpful for creating intricately detailed items that are difficult to fabricate with traditional methods because of their complicated interior features. SLS is used in the aerospace industry to create parts such as unique tooling, lightweight brackets, and structural supports. SLS has several benefits for the aerospace manufacturing industry [77, 78]. Similar to DMLS, SLS enables the production of lighter parts, increasing total fuel economy and lowering operating expenses. To get the best strength-to-weight ratios, engineers can design and manufacture items with intricate internal geometries, including honeycomb structures or lattice frameworks. SLS is also capable of processing a wide range of materials, such as composite materials and high-performance polymers, which increases its usefulness in the aerospace industry. Beyond simply producing lightweight components, 3D printing technologies like DMLS and SLS have a significant impact on the aircraft sector. Iterative design procedures and quick prototyping are also made easier by these technologies [79–86]. Prototypes of new components may be produced and tested quickly by aerospace industries, which facilitates speedier innovation and design improvements. This feature makes it easier to respond to changing industry requirements and speeds up the development of new technologies. In an experiment, Huang et al. [6] focused on developing a composite material using poly(lactic acid) (PLA) and polycaprolactone (PCL) as the primary components. PLA-g-GMA is created by grafting glycidyl methacrylate (GMA) onto PLA, which serves as a compatibilizer. Melt extrusion is used to create the final polymer substrates. This research is novel in that it uses polypyrrole (PPy) nanoparticles—which have the ability to convert photothermally—into the PLA/PCL matrix to produce shape memory composites that react to light. Furthermore, the functional phase of PPy-coated NdFeB (PPy@NdFeB) is utilized to create self-programmable shape memory polymers. The PLA/PCL/PPy composites exhibit 81.11% and 98.87% shape recovery and shape fixation rates, respectively, under 85.27% strain. The programming conditions are set at 2.14 MPa of stress and 65 °C deformation temperature. Furthermore, the PLA/PCL/PPy@NdFeB composite material demonstrates noteworthy magnetic capabilities in addition to outstanding dual and triple shape memory behavior, which permits self-programmed temporary shapes. By adding a multi-stimulus response mechanism, this study expands the possible uses of temperature-responsive SMPs and addresses their limitations. The results offer useful information for furthering the advancement of SMP development and their application to 3D printing technologies.

- Healthcare

In the field of healthcare, 3D printing has become a game-changer, having a significant impact on the creation and production of anatomical models, implants, prostheses, and medical gadgets. Its capacity to create highly personalized and patient-specific solutions has raised the bar for accuracy and functionality in healthcare, improving patient outcomes and propelling the field of regenerative medicine forward in the process [87, 88]. The capacity to manufacture prosthetics and implants that are specifically designed to meet each patient's unique anatomical needs is one of the most important uses of 3D printing in healthcare. Standardized designs are frequently used in traditional prosthetics and implants, which can result in less-than-ideal fits and functional restrictions. These restrictions are circumvented by 3D printing, which makes it possible to produce devices customized for each patient and created using comprehensive imaging data. For instance, the American business e-NABLE employs 3D printing to create prosthetic limbs for both adults and children. The precise measurements and shapes of a patient's remaining limb are matched by these prosthetics. A computerized model of the patient's limb is created using imaging methods like CT or MRI scans. Then, using this model, a flawlessly fitted prosthetic limb is printed, improving comfort and functionality. By customizing the prosthetic, the patient can benefit from improved mobility and usage as well as better fit and performance. Another example is the creation of 3D-printed prosthetic sockets for amputees in Uganda by a University of Toronto team working with Autodesk Research and CBM Canada [89, 90]. Because each patient's anatomy is unique, these sockets are manufactured to order, greatly enhancing comfort and functionality. In a similar vein, 3D printing makes it possible to design and manufacture implants that exactly fit the recipient's anatomy. This includes intricate constructions like dental or cranial implants, which can be customized to meet unique requirements and blend in perfectly with the patient's body. The capacity to design implants with complex geometries and customized characteristics greatly enhances surgical results and lowers the possibility of problems. For instance, a patient who had lost a sizable piece of his pelvis to cancer received a personalized titanium implant from a Swansea University team using 3D printing technology in the UK. Because the implant was made to exactly fit into the residual bone structure, the patient's quality of life and mobility were enhanced.

In a different instance, medical professionals from China's Shanghai Ninth People's Hospital effectively created a jaw implant for a patient who had a severe bone abnormality using 3D printing. The patient's anatomy was precisely matched in the design of the custom-made implant, which led to an excellent surgical outcome as well as better aesthetic and functional results. In addition to implants and prosthetics, 3D printing is transforming the production of anatomical models for surgery. Surgeons can practice and perfect their procedures on these models of complex surgeries before executing the actual procedure [91]. This improves patient outcomes and surgical precision while lowering the possibility of complications. For example, complex pediatric heart procedures are planned and practiced at Boston Children's Hospital in the United States using 3D-printed heart models. These patient-specific imaging data-derived models offer a precise and comprehensive depiction of the patient's heart, enabling surgeons to better comprehend the anatomy and devise the most successful surgical strategy. Similar to this, the Rady Children's Hospital in San Diego, California, gets ready for intricate spinal procedures by using 3D-printed models. Surgeons may better visualize and perform the process with the aid of these models, which improves surgical results and shortens operating times. Bio-printed tissues and organs are becoming a reality in the field of regenerative medicine thanks to 3D printing. Researchers are investigating the printing of functioning tissues for organ replacement or repair using bio-inks, which are made of biomaterials and living cells. Tobacco plant bio-ink is being used by CollPlant, an Israeli startup, to create 3D-printed scaffolds. These scaffolds may be able to help patients with severe injuries or degenerative disorders since they are made to support the creation of new tissues like bones and tendons. Functional organoids, or tiny replicas of organs that may be used for research and medication testing, are being produced in the United States by researchers at the Wake Forest Institute for Regenerative Medicine via 3D printing. Because they enable the creation of medicines and treatments tailored to each patient individually, these organoids hold the potential to completely transform the field of personalized medicine [92]. Orthopedic and oncological surgical procedures can also be guided by models of complex bone fractures or tumor locations, which can increase accuracy and decrease the possibility of mistakes during the actual treatment. One fascinating area of 3D printing technology application in the healthcare industry is bioprinting. Using bioinks—materials containing living cells—this specialized type of 3D printing produces tissue constructs and organ scaffolds for regenerative medicine. Producing viable biological tissues and organs for transplantation, illness simulation, or medication testing is the aim of bioprinting. The first step in the bioprinting process is to digitally model the tissue or organ that has to be printed. The layer-by-layer deposition of bioinks is then guided by this model. The mixture of cells, growth factors, and biomaterials that promote tissue formation and cell proliferation makes up the majority of bioinks. The printed structures eventually grow into functional tissues that resemble the composition and capabilities of real organs. Applications of bioprinting include the fabrication of organ scaffolds that can serve as a base for the growth of complicated organs like kidneys or livers, as well as tissue patches for the repair of injured organs like skin or heart muscle. Although completely functional organ transplantation is still a ways off, bioprinting has shown to be a useful tool for personalized medicine and therapeutic development by creating tissue models for drug testing and illness study.

- Automotive

3D printing is a game-changing technology that the automotive industry has embraced, drastically transforming the design, testing, and manufacturing of parts. The industry gains from improved rapid prototyping, more effective tooling processes, and the capacity to make end-use and custom parts more quickly and flexibly by utilizing 3D printing. Rapid prototyping is one of the most significant uses of 3D printing in the automotive industry [93]. Historically, the process of developing new car parts has been drawn out and expensive, involving the production of physical prototypes using traditional manufacturing techniques. The testing and improvement of new designs were frequently delayed by the time and expense required to create these prototypes. Rapid prototyping is revolutionized by 3D printing, which makes it possible for engineers and designers to swiftly create physical models of their ideas from digital information. The iteration cycle is greatly accelerated by this approach, enabling a quicker assessment of form, fit, and function. For example, car manufacturers may develop and test several iterations of a component much faster and cheaper than they could via more conventional methods. Teams are able to quickly prototype and develop concepts based on feedback from the actual world, which not only expedites the design process overall but also leads to more creative and adventurous designs. Beyond quick prototyping, 3D printing is being utilized more and more to create tools and production support. Especially for intricate or low-volume products, traditional tooling techniques like injection molds and stamping dies can be costly and time-consuming to manufacture [94, 95]. Lead times and tool production costs can be decreased for car manufacturers by employing 3D printing to build tooling components. For instance, producers are now able to print complex geometries on custom jigs, fixtures, and molds that would be impossible or impractical to create with traditional techniques. By handling or assembling parts on production lines, these machines can increase productivity and precision. Additionally, 3D-printed tooling offers versatility that traditional tooling methods cannot match because it is simple to update or modify as design changes arise.

Significant progress is being made by 3D printing in the automotive industry's end-use part production. Advancements in printing technology and materials have made it possible to produce functioning components that can be installed directly in vehicles, whereas they were previously utilized for tooling and prototyping. This covers everything, from functional items like cooling ducts and housings to interior parts like brackets. The creation of high-performance, lightweight components is one such use. For instance, compared to conventional components, 3D-printed parts consisting of sophisticated polymers or metal alloys can provide weight reductions, which enhances overall fuel efficiency and performance [96, 97]. Lattice structures and integrated assemblies are two examples of complex geometries that can be created using 3D printing but are not possible with traditional production methods. These developments may result in previously unachievable new design solutions and improved part performance. Another significant benefit of 3D printing for the automobile sector is its capacity to manufacture small quantities and customized parts. Historically, there have been significant setup costs and lead times associated with creating low-volume or bespoke products. However, manufacturers can now create customized components quickly and affordably thanks to 3D printing. This capacity is especially useful for creating customized car parts, including unique interior trims or ergonomic adjustments based on personal tastes. Additionally, producers can offer distinctive features and designs without having to worry about substantial retooling or inventory management by using 3D printing to generate parts for limited-edition models or specialized vehicles [98–104].

- Consumer Goods

Due to its ability to create highly customized and individualized products, 3D printing has brought about a revolutionary approach to product design and manufacturing in the consumer goods industry. This technology encourages creativity in the design and manufacture of a broad range of items, from bespoke footwear to custom-fit eyewear, in addition to improving customer happiness. Making custom-fit glasses is one of the most interesting uses of 3D printing in consumer products. In the past, conventional frame forms and sizes were used in the production of eyewear, and manual fits were used to make alterations [105, 106]. This procedure frequently left users with uncomfortable fits. By producing frames that are precisely sized to fit each person's face, 3D printing revolutionizes the eyewear industry. A customer's unique frame design can be created by taking precise measurements of their facial shape using digital scanning technologies. The eyeglass frame is then created using 3D printing to precise measurements, guaranteeing improved comfort and a flawless fit. Furthermore, 3D printing makes it possible to include customized design features, like distinctive patterns, colors, or textures, giving customers a way to show their unique personalities through their eyewear. In addition to enhancing the user experience, this degree of personalization gives eyewear companies a chance to stand out in a crowded market. The footwear business is likewise greatly impacted by 3D printing. In the past, shoe design and manufacturing followed uniform patterns and sizes, which may have caused certain people to feel uncomfortable or have a bad fit. These issues are resolved by 3D printing, which makes it possible to create custom shoes that fit each person's unique foot shape [107, 108]. Customers can obtain shoes that are made using precise measurements and scans of their feet thanks to 3D printing. This personalization goes beyond dimensions and incorporates elements like arch support, cushioning, and general style. As a consequence, you get a pair of shoes that fit you properly and provide customized comfort and support. Innovative design techniques are also made possible by 3D printing, such as the employment of intricate lattice systems to create footwear that is both lightweight and durable. Adaptable parts or replaceable components can be incorporated into the design to provide customers with a high level of customization and flexibility.

Beyond shoes and eyeglasses, 3D printing has become a common feature in many other consumer goods, allowing customized products to be made in a variety of sectors. This covers personalized jewelry, housewares, and even useful products like kitchen appliances or phone covers. Technology makes it possible to design and produce goods that cater to specific needs and tastes, improving the overall. Custom jewelry pieces, such as necklaces or rings, can be crafted with shapes or inscriptions that have special meaning for the wearer. In a similar vein, 3D printing may be utilized to produce one-of-a-kind home décor pieces that satisfy certain design tastes or room constraints, giving customers fresh opportunities to personalize their living environments. Significant innovation in product design is driven by the capacity of 3D printing to design and build bespoke products [109, 110]. Design intricacy and material selections are frequently restricted by traditional manufacturing techniques.

Conversely, designers may now experiment with complex shapes and use cutting-edge materials that were previously unfeasible or prohibitively expensive thanks to 3D printing. This creative freedom in design fosters experimentation and innovation, which results in the creation of novel features and product concepts. For example, the application of sophisticated architecture and cutting-edge materials can provide goods with greater performance qualities or increased durability. Furthermore, rapid prototyping and iteration made possible by 3D printing allow designers to test and improve new concepts more quickly. In the end, 3D printing's capacity to produce unique, customized goods raises customer satisfaction by enabling the creation of goods that are specifically catered to each customer's tastes and requirements. In addition to enhancing a product's fit and functionality, 3D printing customization gives it a feeling of exclusivity and ownership [36]. The potential of 3D printing technology to revolutionize the consumer products industry and provide unprecedented degrees of personalization and customization is expected to grow as it advances, significantly influencing the course of product manufacturing in the future.

3.2 4D Printing Applications

- Smart Materials

By enabling the development of smart materials that may alter their shape or qualities in response to outside stimuli, 4D printing—a technological advancement on the back of 3D printing—brings the dimension of time to additive manufacturing. This technology adds a dynamic component that enables printed items to change or adapt over time, building upon the fundamental capabilities of 3D printing. Developing self-folding structures, morphing parts, and other flexible materials creates new opportunities for robotics, aircraft, and other industries [111, 112]. The creation of self-folding structures is one of the most fascinating uses for 4D printing. These structures have the ability to independently alter their shape in response to environmental cues like light, humidity, and temperature. Usually, the procedure makes use of smart materials, which, under certain circumstances, display programmed deformation characteristics. For example, shape-memory polymers or hydrogels that react to moisture or temperature changes can be used to create self-folding structures. These materials change from a flat or unfolded condition into a predetermined three-dimensional shape when they are exposed to a trigger, like an increase in temperature. This feature is especially useful for developing small, deployable systems that have the potential to grow or change their configuration in response to changes in their surroundings. Self-folding structures have applications in robotics, such as the development of robotic grippers that can adjust and deploy. These systems have the ability to reconfigure themselves to carry out various functions or to fit into limited areas. A robotic arm that folds itself, for instance, may change from a small, portable form to a fully functional arm that can be used in a workspace. This flexibility increases robotic systems' functionality and adaptability, enabling them to perform complicated jobs and navigate a variety of settings [75, 113].

Another application where 4D printing shows promise is in morphing components. These components offer dynamic functionality that is not possible with standard components since they can change their size, shape, or mechanical characteristics in reaction to outside stimuli. The capacity to create components with reconfigurable architecture enables the development of systems with improved performance and flexibility. Aerospace uses morphing components to create sophisticated aerodynamic surfaces and wing structures. For example, morphing wing panels can change shape in midair to maximize aerodynamic efficiency in various scenarios. This feature enhances the aircraft's overall performance, maneuverability, and fuel efficiency. Aerospace technology can become more adaptable and efficient by utilizing morphing components to develop adjustable propulsion systems or adaptive landing gear. Deployable space structures are an innovative use of 4D printing in the aerospace industry [114, 115]. Structures that must be compactly stored for launch and then deployed while in orbit are frequently needed for space missions. These constructions with integrated systems that let them extend or unfold automatically in reaction to external circumstances can be made thanks to 4D printing. For instance, intelligent materials can be printed onto deployable solar arrays or antennas to allow them to expand into their functional configurations after being launched into space. These structures can be made to self-assemble or unfold in reaction to external stimuli like mechanical activation or temperature changes. This capacity is critical for making the most use of the limited space in satellites and spacecraft as well as for guaranteeing the proper deployment of critical equipment once in orbit. Another fascinating use of 4D printing is for components and materials that can mend themselves [116]. Because of their inherent ability to mend themselves, these materials have a longer lifespan and are more dependable. Usually, embedded healing agents or microcapsules that can release repair substances in response to injury are incorporated into the material to facilitate the self-healing process. Self-healing components can be utilized to make parts that can recover from minor wear and tear, decreasing the need for maintenance and repairs in the automotive and aerospace industries. For example, a coating that has the ability to heal itself could restore minor abrasions or scratches on an aircraft surface, maintaining the component's structural integrity and aerodynamic performance. In a similar vein, auto parts made of self-healing materials may enhance the robustness and longevity of parts subjected to demanding operating environments.

- Architecture and Construction

By making it possible to create self-assembling construction modules and adaptable building materials, 4D printing is bringing a new paradigm to the domains of architecture and construction. This cutting-edge technology has a profound effect on the design, construction, and upkeep of buildings by utilizing materials' capacity to alter their form, characteristics, or behavior in response to external stimuli. Architects and builders can achieve lower labor costs, better occupant comfort, and more energy efficiency by incorporating 4D printing into their processes. Making flexible construction materials is one of the most exciting uses of 4D printing in the field of architecture. These materials are made to react quickly to changes in their surroundings, including variations in humidity, light, and temperature [117]. The capacity to adjust to these changes enables more effective building performance and enhanced occupant comfort. For example, facades using adaptive materials can change their characteristics according to the outside temperature. Shape-memory alloys or smart polymers that expand or contract in response to temperature changes could be used in a facade. Expanding the facade during the sweltering summer months would lessen the need for air conditioning and improve energy efficiency by increasing shade and reducing solar heat intake. On the other hand, the facade might shrink throughout the winter to let in more light, which would cut down on the need for heating and increase thermal comfort [118–126]. Adaptive materials can also be utilized for windows and shading systems to maximize thermal insulation and natural light penetration. These systems can automatically alter to improve energy economy and occupant comfort by reacting to changes in temperature or light levels. This lessens the need for artificial lighting and climate control systems [127].

Self-assembling construction modules are another notable development in construction technology made possible by 4D printing. These modules' self-assembling capabilities or ability to change shape while being built ease assembly and minimize the amount of manual effort required. It is possible to create self-assembling modules with integrated smart materials, which, when positioned as intended, allow for automatic assembly. Shape-memory polymer modules, for instance, have the ability to alter their shape in response to mechanical activation or temperature fluctuations. These modules can fold, unfurl, or interlock to create complicated structures under certain circumstances, negating the need for extra assembly steps. By lowering the need for manual labor and lowering the chance of assembly errors, this capacity expedites the building process. Additionally, it makes it possible to quickly and effectively construct intricate structures and geometries that would be difficult to accomplish with more conventional techniques. In architecture, the use of 4D printing to produce deployable and reconfigurable building components is an example of a self-assembling system [128]. These parts can be assembled on-site to create the whole construction after being prefabricated, compacted, and transported. This technology is especially helpful for projects in difficult or isolated areas where using regular construction techniques would not be feasible. The creation of self-assembling façade systems is another illustration. The configuration of these facades can be made to automatically change to maximize building performance. For example, a façade consisting of self-aligning and interlocking modular panels might be utilized to enhance the building's overall utility and efficiency by creating complex architectural features or adapting to changing environmental circumstances.

The efficiency of construction and labor costs can be greatly impacted by the adoption of 4D printing in construction processes. 4D printing can expedite construction operations and reduce labor needs by enabling the automatic adjustment of building components and minimizing the need for human assembly. This effectiveness results in lower costs and quicker project completion timeframes. Moreover, self-assembling modules and adaptable materials might lessen the need for maintenance and repairs as a building age. For instance, facades that can sense and respond to their surroundings automatically can prevent building systems' deterioration, resulting in longer-lasting structures and lower maintenance costs.

- Medical Devices

With 4D printing, medical device technology has advanced significantly as it can now produce devices that can alter shape or function in response to physiological changes. By including a dynamic component that enables gadgets to respond to variations in the environment surrounding the human body, this technology expands upon the ideas of 3D printing. Through adaptable behaviors, 4D printing holds great promise for the medical industry by enhancing the functionality, integration, and general efficacy of medical devices. The creation of adaptable implants is one of the most significant uses of 4D printing in medicine. These implants can alter their size, form, or function in real time because they are made to react to certain physiological cues, such as changes in pH levels, body temperature, or metabolic processes [129, 130]. The functionality and integration of implants within the human body can be greatly improved by this adaptability. Implants composed of shape-memory materials, for example, can enlarge or contract in response to variations in body temperature. Heat-sensitive shape-memory alloys, like Nitinol (nickel-titanium), are well-known for their capacity to revert to a certain shape. By using this characteristic, stents and other vascular implants that expand when they get to the right place in the body can be made, offering the best possible support and enhancing surgery success. Similar to this, orthopedic implants using shape-memory polymers can modify their structure or stiffness in response to changes in body temperature, improving their integration with the surrounding tissues. The use of pH-sensitive polymers in drug-delivery implants is another example. In reaction to variations in the pH levels of their immediate surroundings, these implants have the ability to release therapeutic chemicals. An implanted drug delivery system, for instance, might release medication in response to an acidic or basic environment, enabling the precise and regulated release of medications when and where they are needed.

The development of a self-adjusting prosthesis that can react to variations in the user's physiology or activity level is another promising application of 4D printing. Materials that react to physiological cues, such as changes in temperature, pressure, or movement, can be used to construct prosthetic limbs and gadgets. Prosthetic sockets, for instance, can be designed with materials that alter form and fit in response to variations in body pressure or temperature. By guaranteeing that the prosthetic keeps a snug and comfortable fit during a variety of activities and environmental situations, this adaptability can enhance comfort and functionality [131–133]. These self-adjusting characteristics aid in resolving fit and comfort concerns that prosthesis users frequently encounter. Apart from prostheses and implants, 4D printing is advancing the creation of sensitive surgical instruments. In reaction to certain physiological variables experienced during surgery or recovery, these devices have the ability to modify their design or operation. For example, surgical instruments or equipment that react to changes in mechanical stress or tissue temperature could improve accuracy and lower the chance of problems. Instruments featuring flexible tips or grips that adapt to the properties of the tissue being worked on can increase the efficiency of surgical operations and lower the chance of damaging adjacent tissues. Medical gadgets can work and perform much better when adaptable materials are incorporated into them. 4D printing helps to solve problems with device performance, patient comfort, and overall efficacy by enabling devices to react dynamically to physiological changes. The integration of prostheses and implants with the body can be enhanced by adaptive technologies, such as implants, that change form or function in response to physiological signals. For instance, implants that change size in response to body temperature can offer improved alignment and support, increasing the likelihood of effective results and lowering risks. In addition, medical treatments become more effective and customized with the use of self-adjusting prosthetics and responsive drug delivery systems. Through customization of device functioning to individual physiological parameters, 4D printing contributes to the optimization of medical interventions for the specific requirements of each patient.

3.3 5D Printing Applications

- Aerospace Engineering

By expanding the possibilities of conventional 3D and 4D printing technologies, 5D printing represents the state-of-the-art in additive manufacturing. This sophisticated method allows for the printing of components with integrated, dynamic features by adding new levels of complexity and functionality to the process. 5D printing is changing the game in aerospace engineering, where there are extreme demands on performance, dependability, and resilience [134–136]. This is because it makes it possible to create components with advanced features like embedded sensors, self-healing materials, and other features that are essential for operating in harsh environments. In aerospace engineering, one of the biggest benefits of 5D printing is its capacity to create integrated functioning components that aren't achievable with traditional production techniques. The performance, robustness, and operational capabilities of aircraft components can all be significantly improved by these integrated qualities. One significant use of 5D printing in aerospace is the creation of materials with self-healing properties. Because of their ability to mend themselves when broken, these materials have a far longer lifespan and require less upkeep. Self-repairing components are very useful in aerospace since they are exposed to extreme conditions and high levels of stress. Usually, embedded microcapsules or healing agents are part of the structure of self-healing materials. These microcapsules burst when the material is harmed, releasing the healing agents that interact with the surrounding material to mend the damage. Self-healing composites, for example, can be utilized in airplane components like wings or fuselage sections, where they can take care of small dents or cracks on their own that result from impact or fatigue. This feature lessens the need for frequent inspections and repairs by assisting in maintaining the aircraft's safety and structural integrity.

Direct sensor integration into aircraft components is another revolutionary use of 5D printing. Real-time monitoring of a component's health, performance, and environmental factors is possible using embedded sensors, which can provide insightful information for maintenance. For instance, sensors can be incorporated into fuel tanks or turbine blades to monitor characteristics like vibration, temperature, and pressure. Real-time monitoring lowers the chance of unplanned breakdowns by enabling proactive maintenance and early identification of possible problems. These sensors are included in the components during the printing process, making them intrinsically intelligent [137, 138]. This improves operational efficiency and provides constant feedback. Extreme temperatures, high pressures, and hostile environmental conditions are just a few of the tough environments in which aerospace components must function dependably. By making it possible to create materials and buildings that are specifically made to survive such environments, 5D printing helps to address these issues. For example, sophisticated thermal-resistant materials that maintain their integrity in high-temperature conditions can be used in 5D-printed aerospace parts. Similar to this, materials that can withstand the demands of flight or space travel can be created with great impact resistance and structural stability. These parts are guaranteed to match the exacting specifications of aerospace applications because of the ability to customize material qualities and include functionality directly into the components during the printing process. Additionally, 5D printing makes it possible to create unique designs and intricate geometries that are not possible with standard production techniques. Complex interior structures and shapes are frequently needed in aerospace engineering components in order to maximize performance and save weight. Designers can use 5D printing to create parts with intricate internal lattices, embedded channels, and other intricate features that improve overall efficiency and strength-to-weight ratios. Because of its adaptability, 5D printing makes it possible to modify aircraft parts to satisfy certain design specifications [139, 140]. A component's incorporated cooling channels or aerodynamic features, for instance, can be developed with specific flying conditions or mission profiles in mind. Aerospace systems operate more effectively and efficiently with this degree of customization, which results in higher fuel economy, more cargo capacity, and improved operational capabilities.

- Defense and Military

By making it possible to create multifunctional parts with integrated qualities that are essential for contemporary military applications, 5D printing is revolutionizing the manufacture of innovative materials and components for the defense industry. By adding new functional and complicated dimensions, this cutting-edge technology expands the possibilities of conventional 3D printing, greatly improving the efficacy, resilience, and performance of defense systems and materials. 5D printing makes it possible to create materials with a variety of integrated qualities, meeting different defense application performance criteria [141, 142]. 5D printing enables the simultaneous inclusion of several material properties into a single component, in contrast to conventional manufacturing techniques that usually require separate operations to produce distinct material properties. Materials created as a result of this integration have a variety of properties designed to satisfy certain defense requirements. Impact resistance is a crucial prerequisite for defense applications. It is necessary for military vehicles and equipment to endure severe physical damage from explosions, projectiles, and unforgiving terrain [143, 144]. With 5D printing, impact-absorbing structures or materials may be printed right into the components, allowing for the development of new materials with increased impact resistance. Armor systems, for example, can be made with intricate internal designs that absorb and distribute impact energy, minimizing damage to the underlying structure. This method can be used to improve the protection of persons and equipment in combat scenarios for a variety of defensive components, including ballistic shields, vehicle armor, and protective helmets.

Another essential component of contemporary military operations is stealth technology, which tries to make assets less visible to radar, infrared, and other detection systems. By precisely including elements with certain electromagnetic or thermal properties, 5D printing makes it easier to create materials and components with stealth qualities. Examples of 5D-printed parts that are advantageous for stealth airplanes and marine vessels are those that incorporate thermal insulation or materials that absorb radar directly into their structure [108, 145]. These materials can be embedded into the component to maximize its overall stealth qualities, lowering the chance of detection and increasing platform efficacy. Creating intricate, multipurpose components is one of the main benefits of 5D printing for the defense industry. Multifunctional components reduce the need for separate parts and simplify the overall system design by being built to fulfill numerous roles or offer multiple benefits inside a single unit. 5D printing, for example, can be used to make defense parts with integrated sensors, communication systems, or self-healing capabilities in addition to their structural strength. Systems that are more compact and efficient are the result of this integration's streamlining of the design and production processes. 5D printing may be used to create advanced armor systems with active defense mechanisms, stealth materials, and impact-absorbing layers. For instance, impact-resistant polymers, materials that absorb radar, and embedded sensors for real-time damage and performance monitoring might all be found in a composite armor plate. Drones with integrated features, such as structural reinforcement, stealth coatings, and embedded sensors for communication and navigation, can be produced thanks to 5D printing. These multipurpose drones have improved capabilities and decreased vulnerability, enabling them to carry out a range of tasks like combat, surveillance, and reconnaissance [146, 147]. 5D printing can be used to manufacture impact-resistant materials, lightweight constructions, and integrated communication systems for personal protection equipment, such as helmets and vests. This strategy guarantees that soldiers have access to equipment that offers cutting-edge operational capabilities in addition to physical protection.

There are various operational benefits to the defense industry's adoption of 5D printing. 5D printing simplifies maintenance procedures and supply chains by combining several attributes into a single part. Parts with many functions simplify logistics and eliminate the need for several, independent components, which lowers costs and boosts productivity. Furthermore, the capacity to create and modify sophisticated materials with specific qualities enables quick response to new threats and changing military needs. This flexibility is essential in a defense environment that is changing quickly due to advances in technology and tactics.

- Consumer Electronics

The creation of electronics with embedded sensors and multipurpose characteristics is made possible by 5D printing, a sophisticated development of additive manufacturing that is expanding the possibilities for the construction of electronic devices. By directly integrating electronic circuits, sensors, and components into printed devices, this technology improves the functionality of electronic products while optimizing production procedures. 5D printing solves a number of issues with conventional electronics manufacturing and creates new opportunities for creative electronic applications by integrating electronics right into the printed structures [148, 149]. The capability of 5D printing electronics to incorporate electronic parts like circuits and sensors straight into the printed object is one of its main benefits. Conventional electronic manufacturing frequently uses distinct assembly procedures to join sensors and circuits to prefabricated components. This step can be time-consuming, error-prone, and labor-intensive. On the other hand, 5D printing makes it possible to deposit structural elements and electronic materials at the same time during printing. For instance, conductive materials and inks can be utilized to print complex circuit routes directly onto or into the substrate during the 5D printing process. Production becomes more dependable and effective as a result of this integration, which eliminates the need for extra assembly steps and any misalignments [150, 151]. By integrating several functions into a single component, 5D printing makes it easier to create electronic devices with multiple functions. This feature is especially helpful in applications where there is a shortage of space and a need for multiple operations. For example, 5D printing can be advantageous for wearable electronics, such as fitness trackers or smart clothes. These gadgets become more compact and user-friendly by incorporating communication circuits directly into the fabric and embedding sensors for monitoring health indicators (e.g., temperature, heart rate). Multiple functions can be integrated into a single printed structure, which eliminates the need for additional sensors and connections and improves performance and comfort.

By lowering the number of stages required in the production process, the integration of electronic components during 5D printing streamlines the process. Circuit board assembly, sensor placement, and integration into the finished device are frequently separated processes in traditional electronics production. These components are integrated into a single process with 5D printing, which expedites production and shortens the manufacturing process overall [152, 153]. Furthermore, the risk of errors associated with human assembly is reduced by the direct integration of electronics, increasing the end products' uniformity and reliability. This simplified method may result in lower expenses and more productive output. Because it can create complicated geometries and multifunctional components that are difficult to obtain with other technologies, 5D printing offers unprecedented design flexibility. Electronics can be integrated in ways that maximize the design for particular uses or needs in terms of performance. Device design has been transformed by the incorporation of electrical components onto curved or uneven surfaces, opening the door to the development of more creative and visually beautiful goods. This methodology enables the creation of cutting-edge consumer electronics and personalized smart wearables that can adapt to intricate structures, such as the human body. For instance, in the field of consumer electronics, firms such as Samsung and Apple are incorporating curved and flexible display technology into their most recent smartwatches and smartphones [154, 155]. These designs allow gadgets to fit more easily in the user's hand or on their wrist, which improves ergonomics in addition to aesthetic appeal. The integration of electronic components directly into printed structures presents noteworthy advantages for the automobile and aerospace industries. For example, in the automotive industry, businesses such as Tesla are investigating how to monitor and improve performance by integrating sensors and electronic circuits into car components. Automobile components with embedded electronics can offer real-time information on variables like tire pressure or engine performance, resulting in safer and more effective operation. Because of their increased resilience against environmental factors and vibrations, external wiring and connectors are no longer necessary thanks to integration.

Similar to this, electronics embedded in structural parts of aircraft or spacecraft can significantly increase reliability in aerospace applications. For instance, embedded electronics are used by NASA's Mars rovers and satellites to resist the harsh environments of space flight, such as radiation and temperature changes. These systems benefit from improved lifetime and a lower chance of mechanical failures since sensors and electronics are directly embedded into the structural components. This improves overall performance and lifespan. Furthermore, wearable technology benefits from this embedded method. Fitbit, Under Armour, and other businesses have developed smart clothing and fitness trackers that integrate sensors directly into the fabric. Without sacrificing comfort or style, this integration makes it possible to track physiological data like body temperature or heart rate with ease. The wearables are more useful and durable since these components are integrated during the manufacturing process, shielding the sensors from environmental influences and wear and tear.

4 Polymer-Based Printing Innovations

Polymer-based printing has advanced significantly in additive manufacturing, revolutionizing the manufacture of intricate parts and custom-designed components for a variety of industries. Thanks to advancements in processing methods and materials, polymer-based 3D, 4D, and 5D printing has gained recognition for its exceptional precision in producing functional parts and complex geometries [156, 157]. The creation of sophisticated polymer materials that improve the capabilities of additive manufacturing is one of the major advancements in polymer-based printing.

Because they are readily available and easy to manufacture, traditional thermoplastics like Polylactic Acid (PLA) and Acrylonitrile Butadiene Styrene (ABS) have been utilized for a long time. However new developments have brought about a variety of high-performance polymers with better qualities. For example, the use of polyamide (nylon) and polyether ether ketone (PEEK) is growing. These materials are especially useful in high-stress applications where durability and performance are essential, such as medical implants and aerospace. Polymer-based printing processing methods have also advanced, making it possible to produce parts with improved functionality and precision. Printing with polymers can now be done in novel ways thanks to techniques like Digital Light Processing (DLP) and Direct Ink Writing (DIW). Viscous polymer inks can be deposited using DIW to build complex structures with different material properties, and liquid resins can be cured layer by layer using DLP, which produces high-resolution prints with fine details [158]. These developments make it possible to create complex geometries and functional elements that were previously difficult to accomplish with traditional techniques. The development of polymer-based printing has significantly impacted a number of sectors. Advanced polymers are utilized in aircraft to provide strong, lightweight parts that improve performance and fuel efficiency. Better fit and functioning are provided by tailored implants and prostheses constructed of high-performance polymers in the medical profession, improving patient outcomes. Furthermore, advances in consumer goods have been made possible by the capacity to print intricate geometries and functional components, allowing for the creation of customized items that may be tailored to each customer's preferences and demands.

4.1 Multi-Material and Multi-scale Printing

A major advancement in additive manufacturing has been made possible by the development of multi-material and multi-scale printing, which allows for the construction of intricate structures with a variety of characteristics and uses. Printing with multiple materials at the same time or in a different order is known as multi-material printing. This feature makes it possible to create objects within a single part that have different characteristics, like conductivity, flexibility, and hardness. Multi-material printing is now more widely available because of technologies like Multi-Jet Fusion (MJF) with multiple material jets and Fused Deposition Modeling (FDM) with dual extrusion systems [159, 160]. FDM printers with dual extrusion systems can create complicated items with distinct surface finishes and internal structures by using two different thermoplastics or a thermoplastic and support material. Similar to this, MJF technology uses several material jets to apply various agents and powders selectively, producing objects with variable mechanical and aesthetic qualities. By extending the idea of multi-material printing to several scales, multi-scale printing makes it possible to create structures with a range of feature sizes and resolutions. This method allows items with complex exterior geometries and sophisticated internal characteristics to be produced by integrating micro-scale components with macro-scale structures.

For instance, scientists have created methods for printing microfluidic devices that are integrated into bigger, macro-scale components and feature complex channel networks [161]. This feature is useful for applications where precise fluid control and integration are crucial, like lab-on-a-chip systems. Numerous fields find new applications for the capacity to print at different scales and with numerous materials. Multi-material printing is utilized in the medical field to produce intricate implants and prosthetics that are flexible and hardened to meet each patient's unique requirements. Multi-material parts with integrated functionality, like sensors and actuators, can be developed in the automotive and aerospace industries to improve performance and shorten assembly times. Furthermore, the blending of various materials and scales enables the development of cutting-edge consumer goods with distinctive features and capabilities, such as wearable technology and personalized electronics.

There are still a number of obstacles in the way of multi-material and multi-scale printing breakthroughs. It might be challenging to achieve consistent print quality across different scales and to ensure adequate adhesion between diverse materials. Furthermore, extensive software and control systems are frequently needed for multi-material prints due to their intricacy, which raises the cost and complexity of the printing process. In order to further increase the potential applications of multi-material and multi-scale printing technologies, future research and development activities are concentrated on overcoming these issues, enhancing material compatibility, and improving printing precision.

4.2 Material Compatibility and Challenges

In additive manufacturing, material compatibility plays a crucial role in determining the dependability, performance, and quality of printed parts. Compatibility issues between various materials can affect printed components' robustness, functionality, and longevity. For example, achieving structural integrity and performance in multi-material printing requires effective adhesion between incompatible materials. Incompatible materials might result in lower mechanical qualities, delamination, or weak surfaces [162, 163]. Furthermore, the printing process can be impacted by a material's thermal characteristics, since certain materials need particular temperature profiles to work well. For the production of high-quality products, controlling these heat profiles and guaranteeing uniform material behavior during printing is essential. Difficulties may also arise from the conditions in which various materials are processed. Certain polymers might necessitate particular cooling rates, extrusion temperatures, or post-processing procedures, for instance. It can be difficult to guarantee that these requirements are constantly followed, particularly when dealing with complex geometries or numerous materials. Furthermore, variations in material characteristics, like viscosity and flow characteristics, might have an impact on the printing process's precision and repeatability. To overcome these obstacles and guarantee dependable and consistent outcomes, researchers and producers must create and refine processing parameters. To solve compatibility problems and increase the potential of additive manufacturing, new material development and material improvement are essential [164]. More adaptable and compatible materials are being made possible by developments in polymer science, such as the fabrication of novel resins and composites with specific qualities. Furthermore, advancements in material formulations, including the creation of high-temperature resistant materials or self-healing polymers, provide answers to particular problems and improve the functionality of printed components.

4.3 Sustainable Practices and Biodegradable Options

The desire to minimize environmental effects and encourage resource conservation is driving the growing importance of sustainability as a factor in additive manufacturing. Achieving these objectives largely depends on the development of biodegradable solutions and the adoption of sustainable habits. In additive manufacturing, sustainable practices emphasize resource optimization, waste reduction, and energy conservation. Compared to traditional subtractive manufacturing techniques, additive manufacturing minimizes material waste by producing parts directly from digital models. This is one of its main advantages. Furthermore, a lot of 3D printing technologies only use the material needed to create the part, which reduces waste even further [165, 166]. Aside from producing energy-efficient printers, efforts are also being undertaken to optimize print settings to lower power consumption and increase the energy efficiency of 3D printing processes. A potential remedy for the environmental effects of 3D printing is the use of biodegradable materials. Because they can decompose spontaneously over time and are made of renewable resources, biodegradable polymers—like PLA—have less of an adverse effect on the environment and landfills. For instance, PLA, a more environmentally friendly substitute for conventional polymers derived from petroleum, is manufactured from cornstarch or sugarcane. Additionally, research is being done to create new biodegradable materials with better qualities, such as stronger mechanical performance or compatibility with other printing technologies. In additive manufacturing, recycling and reuse are crucial components of sustainable practices [167, 168]. To lower overall consumption, efforts are being made to create techniques for recycling trash from 3D printing and repurposing materials. For example, in an effort to promote a circular economy and lessen the need for virgin resources, several businesses are looking into using recycled plastics as feedstock for 3D printing. Furthermore, improvements in material recycling technologies—like shredding and repurposing old prints—help to reduce waste and promote sustainability.

5 Nanocomposites in Additive Manufacturing

The understanding that materials at the nanoscale have unique physical and chemical properties because of their high surface area-to-volume ratio gave rise to the idea of nanocomposites. When included in a matrix, the special properties of nanoparticles can improve mechanical strength, electrical conductivity, and thermal stability. Thanks to developments in materials science and nanotechnology, the creation of nanocomposites got underway in the latter half of the twentieth century. Early studies investigated the possible advantages of adding nanoparticles to polymers, and the results showed notable enhancements in a range of material attributes [156, 169]. The matrix type and filler type are the two main factors used to characterize nanocomposites. In polymer nanocomposites, fillers like carbon nanotubes or layered silicates are mixed with a polymer matrix, such as epoxy or polyethylene. Their improved mechanical qualities and resilience to barriers are well-known. Metal matrix nanocomposites mix nanoscale fillers with metal matrices, such as titanium or aluminum, to increase strength and resistance to wear. Ceramics like silicon carbide or alumina are used in ceramic matrix nanocomposites, which have exceptional hardness and thermal stability. Techniques including solution processing, melt processing, in situ polymerization, and mechanical alloying are used in the creation of nanocomposites. The type of nanocomposite being manufactured and the required qualities of the final material determine which of the several methods—each with unique benefits—are selected [170, 171].

The mechanical strength, electrical conductivity, and thermal stability of a matrix material can all be significantly enhanced by the addition of nanoscale fillers. The distinct properties of nanomaterials and their interaction with the matrix are responsible for these advantages. The enhanced mechanical strength of nanocomposites is one of their most noteworthy advantages. Significant gains in tensile strength, hardness, and impact resistance can be achieved by reinforcing the matrix material with nanoscale fillers like graphene or carbon nanotubes. The fillers' efficient load transfer systems are mostly to blame for this improvement. The material's ability to withstand loads is improved by high aspect ratio nanoparticles that have robust interfacial contact with the matrix. Furthermore, these fillers' homogeneous dispersion throughout the matrix inhibits the development of weak areas and lowers the chance of crack propagation [172, 173]. For example, carbon nanotubes are very successful in reinforcing polymer matrices because of their remarkable stiffness and tensile strength. Similar to this, stacked silicates in polymer nanocomposites prevent cracks from spreading, enhancing the material's overall mechanical properties. When conductive nanoscale fillers are added, nanocomposites also show improved electrical conductivity. By forming conductive routes inside the matrix, fillers like carbon nanotubes, graphene, or metallic nanoparticles can greatly enhance the matrix's electrical performance. Reaching the percolation threshold, when the conductive fillers create a continuous network across the matrix, is essential to obtaining high electrical conductivity. The composite material shows significantly improved electrical conductivity once this threshold is reached. For instance, networks of conductive pathways formed by carbon nanotubes improve the electrical characteristics of the composite. In a similar vein, the high electrical conductivity and vast surface area of graphene sheets enable effective charge transfer inside the polymer matrix, improving conductivity overall. Another crucial area where nanocomposites significantly outperform is thermal stability. By adding nanoscale fillers, the matrix's thermal conductivity and resistance to thermal deterioration can both be increased. Because of their excellent thermal conductivity, fillers like graphene and boron nitride are able to transport heat through composite materials more effectively and lower thermal resistance. Furthermore, some fillers have the ability to prevent thermal deterioration, shielding the matrix material from extreme heat. For instance, adding metal oxides to polymer matrices can stop thermal deterioration and increase the material's longevity in hot conditions. Furthermore, by choosing the right fillers and managing their dispersion, the thermal expansion behavior of the composite may be customized, which is essential for applications needing accurate thermal management. In an experiment, Wang et al. [2] focused on the design and synthesis of a hydrogenated glassy polymer, poly(styrene-methyl methacrylate-methylallyl alcohol) (PSMA). This polymer is uniformly integrated into PVDF using a novel voltage-assisted 3D printing-hot pressing technique, resulting in a flexible composite film. The incorporation of PSMA, which is characterized by its high modulus strength and abundant hydrogen bonding, significantly mitigates the ferroelectric phase transition of PVDF. This modification leads to an improvement in charge–discharge efficiency (η). The innovative voltage-assisted 3D printing process also induces an entropy gain mechanism. This mechanism facilitates a more uniform distribution of the two phases within the composite, thereby enhancing interfacial properties and contributing to a higher breakdown strength (Eb). The outcomes of this study reveal a remarkable enhancement in energy storage performance. Specifically, the optimized PVDF/PSMA composites achieve an energy storage density (Ue) of 18.1 J/cm^3 and an efficiency (η) of 80% at an electric field strength of 525 MV/m. These impressive results highlight the potential of voltage-assisted 3D printing technology to significantly improve the energy storage capabilities of PVDF-based dielectric materials. Similarly, Kim et al. [7] introduced a pioneering approach to 3D printing PAI and its functional nanocomposites at room temperature. By implementing a specially designed post-process drying procedure, we achieved precise control over the porosity of the printed structures. This control is crucial as it directly impacts the mechanical properties and overall functionality of the 3D-printed materials. The results from our research are promising. The PAI nanocomposite systems exhibited outstanding performance, with tensile strengths reaching up to 105 MPa and a Tg of 295 °C when reinforced with aramid nanofibers. The integration of carbon nanotubes further enhanced the material properties, imparting an electrical conductivity of 200 S/m at 250 °C. This dual enhancement in mechanical strength and electrical conductivity is a significant achievement in the field of high-performance polymers. The advancements demonstrated in this study underscore the potential of our technique to produce robust, lightweight polymer structures with diverse functionalities. The versatility of these high-performance materials makes them suitable for a wide range of high-tech industries. In aerospace, the lightweight and strong polymer composites can be used to manufacture components that require both mechanical resilience and reduced weight. In air mobility, these materials can contribute to the development of efficient and durable parts for various vehicles. Additionally, the enhanced electrical conductivity and thermal stability of the nanocomposites are particularly advantageous for semiconductor packaging, where materials must withstand high temperatures and maintain electrical performance.

Li et al. [9] engineered a specialized photocurable resin to possess both notable flexibility and substantial stretchability. This was achieved by fine-tuning the proportion of oligomer to acrylate monomer in the resin formulation. By incorporating graphene and silver nanoparticles (AgNPs) into this resin, its electrical conductivity was markedly enhanced. These enhancements are crucial for the resin's application in producing precise electrode structures using photocurable 3D printing technology. The 3D printing process was employed to fabricate electrodes with custom-designed surfaces and line patterns, making them suitable for detecting physiological signals such as electrocardiograms (ECG) and electromyograms (EMG). The resin's performance was further improved by using PIB-POE-PIB, an amphiphilic dispersant, which was added to prevent the aggregation of graphene particles. PIB-POE-PIB helps to stabilize graphene in aqueous solutions by creating non-covalent interactions with the material. Specifically, the lipophilic PIB chain interacts with graphene, while the hydrophilic POE chain enhances the dispersion in water. The photocurable resin, combined with reduced AgNPs, underwent UV-induced photopolymerization, which resulted in AgNPs/graphene/polymer nanocomposite electrodes with exceptional flexibility. These electrodes demonstrated a low surface resistivity of approximately 1×10^2 Ω/sq and were able to endure up to 300% strain before breaking. Even after rigorous washing and perspiration tests, the resistivity of the electrodes remained stable at 6×10^2 Ω/sq, indicating their durability and reliability. Different electrode designs and structures were fabricated using the photocurable 3D printing technique, enabling the effective capture of ECG and EMG signals during various physical activities. This approach highlights the potential of integrating photocurable 3D printing technology with wearable electronics, paving the way for the development of smart textiles that can monitor bioelectric signals effectively. Similarly, Mondal et al. [10] explored the functionalization of calcium-deficient nanohydroxyapatite (nHA) particles with lysine urethane methacrylate (LUM), an amino acid-based compound, which was then reacted with hydroxyethyl methacrylate. These modified nHA particles were incorporated into a resin mixture consisting of methacrylated acrylated epoxidized soybean oil, methacrylated isosorbide, and triethylene glycol dimethacrylate. Using masked stereolithography (mSLA) technology, 3D-printed nanocomposites were produced. The primary hypothesis was that the functionalized nanoparticles would enhance the mechanical properties of the nanocomposites due to improved particle dispersion and stronger interfacial bonding. Various mechanical tests were performed on specimens fabricated with the mSLA printer, including flexural, tensile, compression, and Mode-I fracture toughness tests, following ASTM standards. The results indicated that LUM functionalization of nHA particles improved the dispersion within the resin and increased the viscosity of the uncured nanocomposite ink. Specifically, the flexural strength, yield strength, and Mode-I fracture toughness of the nanocomposites were enhanced by 10%, 30%, and 11%, respectively. The improved mechanical properties are attributed to LUM's ability to create a more robust and stable interface between the matrix and the nanoparticles, reducing debonding and allowing for greater plastic deformation before failure.

6 Challenges and Future Prospects

A key issue in 3D, 4D, and 5D printing is getting process and material compatibility. Specific criteria for material qualities and processing conditions are specific to each printing technology. For example, to create solid structures, 3D printing techniques like Stereolithography (SLA) and Fused Deposition Modeling (FDM) need polymers that are simple to extrude or cure. Conversely, materials used in 4D printing must have certain sensitive polymers in order for them to alter shape in reaction to outside stimuli [174, 175]. The challenge is in choosing or creating materials that meet these various needs while guaranteeing efficient communication with the printing method. When it comes to nanocomposites, it is essential to make sure that the nanoscale fillers are evenly distributed throughout the polymer matrix. Nanoparticle agglomeration can result in uneven material characteristics and reduced functionality. Compatibility of materials and processes is a continuous issue since constant dispersion demands sophisticated methods and specialized tools. For 3D, 4D, and 5D printing to produce high-quality parts and structures, resolution and precision are essential. Capturing intricate features and complicated geometries requires high resolution, and it can be difficult to maintain accuracy when working with various materials and scales. High-resolution printing is possible with technologies like SLA and Multi-Jet Fusion (MJF), but maintaining this accuracy over a wide range of applications necessitates sophisticated process management. Precise control over material properties and printing circumstances is necessary to provide accurate and consistent reactions in 4D printing, where materials are engineered to respond dynamically to stimuli. Similar to this, the demands for resolution and precision are made more challenging by 5D printing, which combines several materials and dynamic functions. The printing process is further complicated by the need to control the interactions between various materials and make sure that their combined qualities fulfill specified performance parameters. Durability and material stability are important considerations for printed component performance. When employed in 3D, 4D, and 5D printing, polymers and nanocomposites must maintain their stability under a variety of circumstances, such as changes in temperature, mechanical strain, and exposure to the environment. Problems with polymer deterioration and filler leaching in 3D printing might affect how well-printed parts function over time. Stability and constant responsiveness are critical in 4D printing, as materials are meant to alter in reaction to external stimuli. Similarly, one of the biggest challenges in 5D printing is guaranteeing the robustness of multi-material constructions and their capacity to consistently carry out various tasks over time. Another issue with 3D, 4D, and 5D printing technologies is their cost and scalability [176, 177]. The use of advanced printing technologies and materials is restricted to specialist industries and small-scale production due to their high cost. Maintaining consistent quality and controlling production efficiency are only two of the additional challenges and expenses that come with scaling up from prototype to mass production. Because of the high expense of nanoscale fillers and the intricate processing needed, nanocomposites are very expensive. Ensuring the economic viability of nanocomposites for wider application requires lowering these expenses without compromising material performance. Concerns over the effects of nanomaterials used in 3D, 4D, and 5D printing on human health, safety, and the environment are becoming more widespread. If nanoparticles are discharged into the atmosphere or become airborne, workers and consumers may be exposed to health concerns. The environmental effects of creating and discarding nanocomposites also need to be taken into account. Creating safer handling protocols, carrying out in-depth risk analyses, and putting sustainable practices into place are all necessary to address these worries and make sure that the advantages of sophisticated printing technologies exceed any possible drawbacks.

Heading ahead, 3D, 4D, and 5D printing are expected to be greatly impacted by developments in material development. New polymers and nanocomposites with improved characteristics that are suited to the needs of these cutting-edge printing technologies are being actively developed by researchers. Future materials for 4D printing should respond more intelligently to outside stimuli including humidity, light, and temperature variations [178, 179]. The spectrum of applications, such as self-healing materials and adaptive structures, will increase with advancements in smart materials and responsive polymers. The development of multifunctional materials with integrated features and capabilities, such as embedded sensors, self-healing mechanisms, and improved mechanical qualities, will be the main emphasis of 5D printing. These developments will expand the capabilities of materials and open up new avenues for additive manufacturing. Printing technology advancements will improve scalability, precision, and resolution. Future advancements in printer design are probably going to include sophisticated control systems and high-resolution print heads, which will make producing complicated structures more accurate and dependable. The precision of 4D and 5D printing will increase with improved software and control algorithms, enabling more complex dynamic responses and multi-material integration. The development of new printing technologies, such as hybrid printing processes and sophisticated laser systems, will help additive manufacturing overcome its present constraints and reach new heights. Scalability and cost-cutting will be essential for 3D, 4D, and 5D printing technologies to be widely used. Future advancements will concentrate on streamlining printing procedures to cut down on material waste and boost output effectiveness. The cost of polymers and nanocomposites will also be reduced by developments in material synthesis and processing. Advances in automation, quality control, and process integration will make it easier to scale up from prototype to mass production and make these technologies more widely available for large-scale manufacturing. The future of additive manufacturing will be heavily focused on sustainability and environmental effects [180]. Environmental pollution and material disposal are challenges that will be addressed by research on eco-friendly nanocomposites and biodegradable polymers. Further lowering additive manufacturing's environmental impact will require the development of recycling and reuse strategies for printed parts. Furthermore, minimizing the negative effects of nanocomposites on human health and the environment can be achieved by using sustainable production and handling procedures. This will guarantee that the advantages of these cutting-edge printing technologies are realized in an ecologically conscious way.

7 Conclusion

The swift advancement of additive manufacturing technologies, which include 3D, 4D, and 5D printing, has resulted in a notable transformation in the handling and use of materials in diverse sectors. These technologies each have special benefits that improve functional capabilities, material customization, and design flexibility, revolutionizing conventional production methods. To fully realize the potential of these cutting-edge printing processes, a complex array of issues involving the integration of polymers and nanocomposites must be addressed. By making it possible to create complex structures and prototypes, 3D printing technologies, such as Fused Deposition Modeling (FDM), Stereolithography (SLA), and Multi-Jet Fusion (MJF), have laid a solid foundation for additive manufacturing. Even with their accomplishment, there are still several obstacles to overcome, including material compatibility, resolution, and precision. Researchers and producers are still focused on finding ways to ensure material stability and durability while maintaining uniform quality and performance over a wide range of materials and scales. With the introduction of 4D printing, additive manufacturing gains a dynamic component thanks to the use of materials that can alter their form or characteristics in reaction to outside stimuli. Although this technology presents new opportunities for the development of responsive systems and smart materials, it also poses problems with the control and responsiveness of materials. The development of polymers that respond to stimuli precisely and consistently while retaining their stability is critical to the evolution of applications involving 4D printing. The capabilities of additive manufacturing are further expanded by 5D printing, which combines various materials and dynamic functionality into a unified process. With the use of this technology, things with sophisticated features like embedded sensors, improved mechanical qualities, and the ability to mend themselves will be created. Significant obstacles, meanwhile, arise from the intricacy of controlling the interactions between various materials and making sure that their overall qualities satisfy performance requirements. Achieving optimal scalability, resolution, and precision in 5D printing will be essential to its effective adoption.

The future of 3D, 4D, and 5D printing technology appears bright, despite these obstacles. It is anticipated that developments in material science will improve the characteristics of polymers and nanocomposites, making them more capable of satisfying the unique requirements of these technologies. Advances in responsive polymers, multifunctional composites, and smart materials will broaden the scope of uses and boost productivity in industries like consumer goods, automotive, aerospace, and healthcare. Advances in printing technology, such as the creation of sophisticated control systems, high-resolution print heads, and novel printing methods, will solve existing issues and expand the potential of additive manufacturing. These technologies will become more affordable and practical for large-scale production as they become more scalable, creating new avenues for their widespread adoption. Future advancements in additive manufacturing will also be heavily reliant on sustainability and environmental effects. Research on recycling techniques, eco-friendly nanocomposites, and biodegradable polymers will solve issues with material disposal and pollution of the environment. By implementing sustainable manufacturing and handling techniques, sophisticated printing technologies can be used to achieve benefits in an environmentally responsible manner.

References

1. Mamatha, G.R., Radha Krushna, B.R., Malleshappa, J., Sharma, S.C., Kumar, S., Krithika, C., et al.: Investigating the influence of mono-, di-, and trivalent co-dopants (Li+, Na+, K+, Ca2+, Bi3+) on the photoluminescent properties and their prospective role in data security applications for SrAl2O4:Tb3+ nanophosphors synthesized via an eco-friendly combu. Mater. Sci. Eng. B [Internet]. 299:117008. (2024). Available from: https://www.sciencedirect.com/science/article/pii/S092151072300750X
2. Wang, J., Peng, B., Zhang, Y., Gong, H., Wang, B., Wu, W., et al.: Voltage-assisted 3D printing of polymer composite dielectric films with low energy loss and high energy storage density. J. Power Sources. **608** 234649. (2024). Available from: https://doi.org/10.1016/j.jpowsour.2024.234649
3. El Aadad, H., El Hamzaoui, H., Brévalle-Wasilewskis, G., Bernard, R., Kinowski, C., Quiquempois, Y., et al.: Solmers: Versatile hybrid resins for nanometric 3D printing of silica-based photonic components. Mater. Today Adv. **22** (2024)
4. Azam, M.U., Belyamani, I., Schiffer, A., Kumar, S., Askar, K.: Progress in selective laser sintering ofmultifunctional polymer composites for strain- and self-sensing applications. J. Mater. Res. Technol. **30** 9625–46. (2024). Available from: https://doi.org/10.1016/j.jmrt.2024.06.024
5. Abdul Hameed, M.M., Thamer, B.M.: Preparation of persistently luminescent polyacrylic acid-based nanocomposite ink for secure encoding. J. Photochem Photobiol A Chem. **448** 115319. (2023). Available from: https://doi.org/10.1016/j.jphotochem.2023.115319
6. Huang, L., Sheng, Y., Mo, Q., Zhang, S., Zheng, Y., Wang, B., et al.: Poly(lactic acid)/polycaprolactone-based self-programmed photothermal responsive shape memory polymers. React. Funct. Polym. **202** (2024)
7. Kim, H.C., Tu, R., Sodano, H.A.: Room temperature 3D printing of high-temperature engineering polymer and its nanocomposites with porosity control for multifunctional structures. Compos. Part B. Eng. **279** 111444. (2024). Available from: https://doi.org/10.1016/j.compositesb.2024.111444

8. Kumar, P., Suryavanshi, P., Kumar Dwivedy, S., Banerjee, S.: Stimuli-responsive materials for 4D printing: mechanical, manufacturing, and biomedical applications. J. Mol. Liq. **410** 125553. (2024). Available from: https://doi.org/10.1016/j.molliq.2024.125553
9. Li, J.W., Chen, H.F., Liu, Y.Z., Wang, J.H., Lu, M.C., Chiu, C.W.: Photocurable 3D-printed AgNPs/Graphene/Polymer nanocomposites with high flexibility and stretchability for ECG and EMG smart clothing. Chem. Eng. J. **484**(2023), 149452. (2024). Available from: https://doi.org/10.1016/j.cej.2024.149452
10. Mondal, D., Willett, T.L.: Functionalization of calcium-deficient nanohydroxyapatite improves the mechanical properties of 3D printed biopolymer nanocomposites. Compos. Sci. Technol. **255**:110707 (2024). Available from: https://doi.org/10.1016/j.compscitech.2024.110707
11. Zhu, J., Gao, H., Zheng, Z., Cao, Y., Wang T.: Photocuring 3D printable flexible strain sensor enhanced by in situ grown silk fibroin nanoparticles. Chem. Eng. J. 154762. (2024). Available from: https://doi.org/10.1016/j.cej.2024.154762
12. Hwang, H., Park, S., Smith, M., Bose, S.T., Peringath, A.R., Zhang, J., et al.: Highly dispersed nanomaterials in polymer matrix via aerosol-jet-based multi-material 3D printing. Nano Energy. **128**(PA):109803. (2024). Available from: https://doi.org/10.1016/j.nanoen.2024.109803
13. Akin, S., Chang, T., Abir, S.S.H., Kim, Y.W., Xu, S., Lim, J., et al.: One-step fabrication of functionalized electrodes on 3D-printed polymers for triboelectric nanogenerators. Nano. Energy. **129**(PB):110082. (2024) Available from: https://doi.org/10.1016/j.nanoen.2024.110082
14. Celikkin, N., Padial, J.S., Costantini, M., Hendrikse, H., Cohn, R., Wilson, C.J., et al.: 3D printing of thermoresponsive polyisocyanide (PIC) hydrogels as bioink and fugitive material for tissue engineering. Polymers (Basel). **10**(5) (2018)
15. Satapathy, Y., Nikitin, V., Hana, J., Venkatesan, K.R., Tran, F., Chen, S., et al.: Multiscale porosity characterization in additively manufactured polymer nanocomposites using micro-computed tomography. Addit. Manuf. **86**:104199. (2024). Available from: https://doi.org/10.1016/j.addma.2024.104199
16. Khodadad Hatkeposhti, J., Kordani, N., Akbarzadeh, Pasha, M., Barari, A.: Fracture load in double keyhole notch PLA-Cu2O nanocomposites manufactured via compression molding and 3D printing: An experimental and numerical study. J. Mech. Behav. Biomed Mater. **153**, 106504 (2024). Available from: https://doi.org/10.1016/j.jmbbm.2024.106504
17. Gad, M.M., Abdullah Alzaki, F., Ahmed Abuwarwar, F., Alhammad, A., Al Hussain, M., Khan, S.Q. et al.: Impact of printing layer thickness on the flexural strength of nanocomposite 3D printed resins: An in vitro comparative study. Saudi. Dent. J. (2024) Available from: https://doi.org/10.1016/j.sdentj.2024.07.009
18. Rodovalho, A.J.R.L., Barbosa, W.T., Vieira, J.L., Oliva, C.A., de, Gonçalves A.P.B., Cardoso, P., da S.M. et al.: Influence of size and crystallinity of nanohydroxyapatite (nHA) particles on the properties of Polylactic Acid/nHA nanocomposite scaffolds produced by 3D printing. J. Mater. Res. Technol. **30**, 3101–11 (2024)
19. Hossain, M.I., Chowdhury, M.A., Mahamud, S., Saha, R.K., Zahid, M.S., Ferdous, J. et al.: Electro-mechanical analysis of nanostructured polymer matrix composite materials for 3D printing using machine learning. Chem. Eng. J. Adv. **19,** 100626 (2024) Available from: https://doi.org/10.1016/j.ceja.2024.100626
20. Zhou, X., Liu, H., Yu, Z., Yu, H., Meng, D., Zhu, L. et al.: Direct 3D printing of triple-responsive nanocomposite hydrogel microneedles for controllable drug delivery. J. Colloid Interface Sci. **670**, 1–11 (2024). Available from: https://doi.org/10.1016/j.jcis.2024.05.045
21. Saenjaiban, A., Thanakkasaranee, S., Jantanasakulwong, K., Punyodom, W., Lee, Y.S., Singjai, P., et al.: Influence of biodegradable polymer films on conformation of polydiacetylene–silver nanocomposites as colorimetric time–temperature indicators. Sensors Actuators A. Phys. **376** 115658 (2024). Available from: https://doi.org/10.1016/j.sna.2024.115658

22. Hanif, M., Zhang, L., Shah, A.H., Chen, Z.: Mechanical analysis and biodegradation of oxides-based magneto-responsive shape memory polymers for material extrusion 3D printing of biomedical scaffolds. Addit. Manuf. **86**, 104174 (2024) Available from: https://doi.org/10.1016/j.addma.2024.104174
23. Kong, D., Guo, A., Wu, H., Li, X., Wu, J., Hu, Y., et al.: Four-dimensional printing of polymer-derived ceramics with high-resolution, reconfigurability, and shape memory effects. Addit. Manuf. **83**, 104050 (2024). Available from: https://doi.org/10.1016/j.addma.2024.104050
24. Yuan, X., Mai, Z., Li, H., Gao, X., Yan, A., Jiang, D., et al.: Enhanced energy density in piezoelectric PVDF-polymer nanocomposite via multiple mechanism synergetic action. Nano Energy [Internet]. **127**, 109734 (2024) Available from: https://doi.org/10.1016/j.nanoen.2024.109734
25. Thakur, R.C., Sharma, R., Kaur, V., Kaur, H.: Molecular interactions of Niacin in binary aqueous solutions of KCl and FeCl 3 at different temperatures : Volumetric Approach. Think India J. **22**(37), 1604–13 (2019)
26. Thakur, R.C., Kaur, R., Kaur, H.: Transport properties of thiaminehydrochlorideín binary aqueous mixtures of galactose. Eur J Mol Clin Med. **7**(07), 2020 (2020)
27. Kaur, H., Thakur, R.C., Pathania, V.S., Sharma, S.: Effect of Choline-based ionic liquid (Cholinium ethanoate) on volumetric and acoustic properties of aliphatic amino acids (glycine and L-alanine). J. Mol. Liq. **346**, 118247 (2022)
28. Nidhi, Kaur, H.: Advancement in field of Energy storage applications by using Ionic Liquids. J. Phys. Conf. Ser. **2267**(1) (2022)
29. Kaur H, Thakur RC, Singh N, Kumar A, Gill R, Thakur A, et al. Influence of 1-butyl-3-propylimidazolium bromide on viscometric and conductometric behavior of proteinogenic amino acids (glycine and L-alanine). AIP Conf Proc [Internet]. 2023;2800(1):20092. Available from: https://doi.org/10.1063/5.0163118
30. Li, Y., Zhang, F., Liu, Y., Leng J.: 4D printed shape memory bismaleimide resin with high storage modulus and low shrinkage rate via second-stage curing. Chem. Eng. J. **491**, 152196 (2024). Available from: https://doi.org/10.1016/j.cej.2024.152196
31. Bernasconi, R., Bellè, U., Brigatti, S., Diamanti, M.V.: 3D printing of photocatalytic nanocomposites containing titania nanoparticles. Addit. Manuf. **79**(2023) (2024)
32. Kasraie, M., Krieg, A.S., Abbott, A.C., Gawde, A., Eisele, T.C., King, J.A., et al.: Carbon nanotube as a conductive rheological modifier for carbon fiber-reinforced epoxy 3D printing inks. Compos Part B Eng. **282**(2023), 111583 (2024). Available from: https://doi.org/10.1016/j.compositesb.2024.111583
33. Nabil, G.M., Althomali, R.H., Mahmoud, M.E.: Decorated gelatin polymer onto copper aluminum layered double hydroxides for superior removal of Congo red: Optimization and adsorption evaluation of kinetics, isotherms, and thermodynamics. J. Mol. Struct. **1319**(P1), 139303 (2024). Available from: https://doi.org/10.1016/j.molstruc.2024.139303
34. Wang, L., Men, J., Feng, J., Jiang, Y., Li, L., Hu, Y., et al.: Extrusion 3D printing of carbon nanotube-assembled carbon aerogel nanocomposites with high electrical conductivity. Nano. Mater. Sci. **6**(3), 312–9 (2024). Available from: https://doi.org/10.1016/j.nanoms.2023.09.002
35. Mondal, D., Willett, T.L.: Dispersion Strategy improves the mechanical properties of 3D-Printed biopolymer nanocomposite. Compos. Part B. Eng. **283**, 111680 (2024). Available from: https://doi.org/10.1016/j.compositesb.2024.111680
36. Iftekar, S.F., Aabid, A., Amir, A., Baig, M.: Advancements and limitations in 3D printing materials and technologies: a critical review. Polymers. **15**(11) (2023)
37. Salimi, S., Graham, A.M., Wu, Y., Song, P., Hart, L.R., Irvine, D.J., et al.: An effective route to the additive manufacturing of a mechanically gradient supramolecular polymer nanocomposite structure. J. Mech. Behav. Biomed Mater. **150**(2023), 106358 (2024). Available from: https://doi.org/10.1016/j.jmbbm.2023.106358
38. Rao, G.N.M., Vakkalagadda, M.R.K.: A review on synthesis, characterization and applications of nanoparticles in polymer nanocomposites. Mater. Today. Proc. **98**(2023), 68–80 (2024) Available from: https://doi.org/10.1016/j.matpr.2023.09.135

39. Korčušková, M., Svatík, J., Tomal, W., Šikyňová, A., Vishakha, V., Petko, F., et al.: Anatase and rutile nanoparticles in photopolymer 3D-printed nanocomposites: Band gap-controlled electron interactions in free-radical and cationic photocuring. React. Funct. Polym. **200** (2024)
40. Appuhamillage, G.A., Ambagaspitiya, S.S., Dassanayake, R.S., Wijenayake, A.: 3D and 4D printing of biomedical materials: current trends, challenges, and future outlook. Explor Med. **5**(1), 17–47 (2024)
41. Sultana, N., Ali, A., Waheed, A., Aqil, M.: 3D Printing in pharmaceutical manufacturing: Current status and future prospects. Mater. Today Commun. **38**, (2024)
42. Zhang, J., Wei, S., Shang, C., Duan, Y., He, Z., An, H., et al.: Modeling and experimental analysis of piezoresistive behavior in conductive porous elastomer under significantly large deformation. Int. J. Eng. Sci. **201**, 104091 (2024). Available from: https://www.sciencedirect.com/science/article/pii/S0020722524000752
43. Sheikh, T., Behdinan, K.: Fused deposition modelling of thermoplastic polymer nanocomposites: a critical review. C-J. Carbon Res. **10**(2) (2024)
44. Siraj, M.T., Alshybani, I., Payel, S.B.: Shahadat MR Bin, Rahman MZ. 12.45—Advances in biocomposite fabrication: Emerging technologies and their potential applications. In: Hashmi, S., (ed.) Compr. Mater. Process. (Second Edition). Second Edi. Oxford: Elsevier. pp 648–71 (2024). Available from: https://www.sciencedirect.com/science/article/pii/B9780323960205000765
45. Gong, L., Zhu, J., Yang, Y., Qiao, S., Ma, L., Wang, H., et al.: Effect of polyethylene glycol on polysaccharides: From molecular modification, composite matrixes, synergetic properties to embeddable application in food fields. Carbohydr Polym. **327**, 121647 (2024). Available from: https://www.sciencedirect.com/science/article/pii/S0144861723011128
46. Kaur, S., Kaur, H., Rao, A.S., Prakash, G.V.: A review on photoluminescence phosphors for biomedical, temperature sensing, photovoltaic cell, anti-counterfeiting and white LED applications. Phys. B Condens. Matter. **690**: 416224 (2024). Available from: https://www.sciencedirect.com/science/article/pii/S0921452624005659
47. Nidhi, Kaur, H., Thakur, R.C., Kumar, A.: Effect of temperature on conductance studies of 1,4-dimethyl-4H-1,2,4-triazolium iodide in binary aqueous solutions of benzylamine and benzamide. AIP Conf. Proc. **2800**(1), 20229 (2023). Available from: https://doi.org/10.1063/5.0163221
48. Pathania, V., Sharma, S., Vermani, B.K., Kaur, H., Thakur, R.C.: Investigating the solvation behavior of some perchlorates in non-aqueous medium and predicting their ion-solvation thermodynamics using Born formula. J. Mol. Liq. **371**, 121067 (2023). Available from: https://doi.org/10.1016/j.molliq.2022.121067
49. Rana, S., Thakur, R.C., Kaur, H., Sharma, A., Sharma, R., Thakur, V., et al.: Investigating the solvation behavior of some lithium salts in binary aqueous mixtures of 1-Ethyl-3-methylimidazolium Tetrafluoroborate ([EMIM][BF4]) at equidistant temperatures (T = 298.15, 303.15, 308.15, 313.15, 318.15) K. J. Chem. Eng. Data. **68**(6), 1291–304 (2023) Available from: https://doi.org/10.1021/acs.jced.3c00037
50. Thakur, R.C., Sharma, A., Sharma, R., Kaur, H.: A comparative analysis of volumetric, viscometric and conductometric properties of Triethylmethylammonium Tetrafluoroborate (TEMABF4) and Tetraethylammonium Tetrafluoroborate (TEABF4) in pure propylene carbonate (PC) and binary aqueous propylene carbonate. J. Mol. Liq. **374**, 121244 (2023). Available from: https://doi.org/10.1016/j.molliq.2023.121244
51. Yang, X.C., Wang, X.X., Wang, C.Y., Zheng, H.L., Yin, M., Chen, K.Z., et al.: Silk-based intelligent fibers and textiles: structures, properties, and applications. Chem. Commun. **60**(61), (2024). 7801–23. Available from: https://www.sciencedirect.com/science/article/pii/S135973452401348X
52. Naghib, S.M., Ahmadi, B., Kangarshahi, B.M., Mozafari, M.R.: Chitosan-based smart stimuli-responsive nanoparticles for gene delivery and gene therapy: Recent progresses on cancer therapy. Int. J. Biol. Macromol. 134542 (2024) Available from: https://www.sciencedirect.com/science/article/pii/S0141813024053479

53. Makurat-Kasprolewicz, B., Ipakchi, H., Rajaee, P., Ossowska, A., Hejna, A., Farokhi, M., et al.: Green engineered biomaterials for bone repair and regeneration: Printing technologies and fracture analysis. Chem. Eng. J. **494**, 152703 (2024). Available from: https://www.sciencedirect.com/science/article/pii/S1385894724041901
54. Navya, N., Krushna, B.R.R., Sharma, S.C., Srinivasan, A.R., George, A., Sunder mohapatra, S., et al.: In-situ embedding of carbon dots in perovskite GdAlO3:Eu3+/Li+ composite: Boosting phosphor stability, optical thermometry, flexible displays and data protection applications. Mater. Today Sustain. **27**, 100840 (2024). Available from: https://www.sciencedirect.com/science/article/pii/S2589234724001763
55. Sadeghzade, S., Hooshiar, M.H., Akbari, H., Tajer, M.H.M., Sahneh, K.K., Ziaei, S.Y., et al.: Recent advances in Organ-on-a-Chip models: How precision engineering integrates cutting edge technologies in fabrication and characterization. Appl. Mater. Today. **38**, 102231 (2024). Available from: https://www.sciencedirect.com/science/article/pii/S2352940724001768
56. Li, L., Chen, X., Feng, Z., Jiang, Y., Dai, L., Zhu, J., et al.: Recent advances and perspectives of practical modifications of vanadium redox flow battery electrodes. Green Chem. **26**(11), 6339–60 (2024). Available from: https://www.sciencedirect.com/science/article/pii/S1463926224004825
57. Pakeeza, Draz, M.U., Yaqub, A., Jafry, A.T., Khan, M., Ajab, H.: Electrochemical sensing of B-complex vitamins: current challenges and future prospects with microfluidic integration. RSC Adv. **14**(15), 10331–47 (2024). Available from: https://www.sciencedirect.com/science/article/pii/S2046206924008866
58. Camargo, J.R., Crapnell, R.D., Bernalte, E., Cunliffe, A.J., Redfern, J., Janegitz, B.C., et al.: Conductive recycled PETg additive manufacturing filament for sterilisable electroanalytical healthcare sensors. Appl. Mater. Today. **39** (2024)
59. Zhao, J., Zhang, M., Guo, Y., Liu, W., Li, Y., Yang, Y.: Colorful cholesteric liquid crystal polymer network patterns prepared using cholesteric liquid crystal inks. Giant. **17**, 1–7 (2024)
60. Galván-Romero, V., Gonzalez-Salazar, F., Vargas-Berrones, K., Alcantara-Quintana, L.E., Martinez-Gutierrez, F., Zarazua-Guzman, S., et al.: Development and evaluation of ciprofloxacin local controlled release materials based on molecularly imprinted polymers. Eur. J. Pharm. Biopharm. **195**(2024)
61. Mahjoubnia, A., Cai, D., Wu, Y., King, S.D., Torkian, P., Chen, A.C., et al.: Digital light 4D printing of bioresorbable shape memory elastomers for personalized biomedical implantation. Acta Biomater. **177**, 165–77 (2024)
62. Xie, J., Wang, X., Wang, Y., Li, J., Cao, C., Jin, M.: Effect of substituents on the performance of N-carbazolyl double benzylidene ketones in visible/NIR light one/two-photon polymerization. Eur. Polym. J. **202**(2023) (2024)
63. Xie, Z., Chen, S., Zhang, S., Pei, Y., Li, L., Wu, P.: Enhanced performance of fiber-based perovskite solar cell achieved by multi-functional high polymer doping strategy. J. Colloid Interface Sci. **663**(January), 309–28 (2024)
64. Ramalho, A., Freitas, D., Almeida, H.: Evaluation of the structural strength of anisotropic PLA components manufactured by 3D printing. Procedia. Struct. Integr. **53**(2022), 81–8. (2024). Available from: https://doi.org/10.1016/j.prostr.2024.01.011
65. Qu, S., Yuan, Y., Ye, X., Xu, W.: High sensitivity artificial synapses using printed high-transmittance ITO fibers for neuromorphic computing. Chinese Chem. Lett. 110030 (2024)
66. Furia, F., Roels, E., Terryn, S., Lutz, A., Vanderborght, B., Van Assche, G., et al.: Investigation of the rheological behaviour for improved extrusion-based additive manufacturing of polymer composites based on dissociative dynamic covalent bonds. Addit. Manuf. **91**, 104342 (2024). Available from: https://doi.org/10.1016/j.addma.2024.104342
67. Ouyang, Q., Yao, C., Chen, H., Song, L., Zhang, T., Chen, D., et al.: Machine learning-coupled tactile recognition with high spatiotemporal resolution based on cross-striped nanocarbon piezoresistive sensor array. Biosens Bio. **246**(2023), 115873 (2024). Available from: https://doi.org/10.1016/j.bios.2023.115873

68. Zhong, W., Wang, Q., Shen, X.: Quinoa protein/polysaccharide electrostatic complex stabilized vegan high internal phase emulsions for 3D printing: Role of complex state and gelling-type polysaccharides. Food Chem. **434**(2023), 137447 (2024). Available from: https://doi.org/10.1016/j.foodchem.2023.137447
69. Wang, M., Bao, Y., Li, D., Bian, Y., Si, X., Gao, N., et al.: Scientometrics and visualized analysis of 3D food printing: A boost to future food customized development. Food Biosci. **58**, (2024).
70. Li, Z.J., Zhang, T.X., Dai, H.L., Yao, Y., Xie, P.H.: Thermal-based efficient modeling for mechanical properties of thermoplastic polymers in material extrusion. Int. J. Heat Mass. Transf. **226**, 125526 (2024). Available from: https://doi.org/10.1016/j.ijheatmasstransfer.2024.125526
71. El Aadad, H., El Hamzaoui, H., Brévalle-Wasilewskis, G., Bernard, R., Kinowski, C., Quiquempois, Y., et al.: Solmers: Versatile hybrid resins for nanometric 3D printing of silica-based photonic components. Mater. Today Adv. **22**, (2024)
72. Yin, L., Huang, J.B., Yue, T.C., Wang, L.L., Wang, D.Z.: Two coordination polymers based on imidazopyridine carboxylic acid ligand for efficient catalytic cycloaddition of CO2 with epoxides and cyanosilylation. J. Mol. Struct. 1309, 138165 (2024). Available from: https://doi.org/10.1016/j.molstruc.2024.138165
73. Kang, J., Yoon, J., Lee, B., Jung, H., Kim, J., Nam, W., et al.: Ultrathin, solvent-resistant dielectric for monolithic fabrication of low-power, intrinsically stretchable active-matrix electronic skin. Device. **2**(6), 100426 (2024). Available from: https://doi.org/10.1016/j.device.2024.100426
74. Zhang, J., Zhou, W., Yang, L., Qu, W., Yao, Y., Lu, T., et al.: Visible-light excitable thermally activated delayed fluorescence from hydrogen-bonded D-A aromatic N-heterocycle/polymer material and the application for anti-counterfeiting. Dye Pigment. **223**(2023), 111927 (2024). Available from: https://doi.org/10.1016/j.dyepig.2023.111927
75. Zhang, H., Huang, S., Sheng, J., Wang, C., Zhang, J., Zhu, M., et al.: 4D printing of shape memory polymer nanocomposites for enhanced performances and tunable response behavior. Eur. Polym. J. **201**, 112568 (2023). Available from: https://www.sciencedirect.com/science/article/pii/S0014305723007516
76. Khater, A.Z., Saadi, M.A.S.R., Bhattacharyya, S., Kutana, A., Tripathi, M., Kamble, M., et al.: Processing dynamics of carbon nanotube-epoxy nanocomposites during 3D printing. Cell Reports Phys. Sci. **4**(10), 101617 (2023). Available from: https://doi.org/10.1016/j.xcrp.2023.101617
77. Liu. Y., Cao, L., Wang, P., Pang, X., Li, L., Mei, Q., et al.: Multi-Color Room temperature phosphorescent Silicon-Nanodot-Based nanocomposites with silane tuning and applications to 5D information encryption. Chem. Eng. J. **470** (2023)
78. Verma, P., Ubaid, J., Alam, F., Deveci, S., Kumar, S.: Multifunctional characteristics of 3D printed polymer nanocomposites under monotonic and cyclic compression. Def Technol. **30**, 13–22 (2023). Available from: https://doi.org/10.1016/j.dt.2023.05.017
79. Thakur, A., Kumar, A.: Unraveling the multifaceted mechanisms and untapped potential of activated carbon in remediation of emerging pollutants: A comprehensive review and critical appraisal of advanced techniques. Chemosphere. **346**(2023), 140608 (2024). Available from: https://doi.org/10.1016/j.chemosphere.2023.140608
80. Thakur, A., Dagdag, O., Berisha, A., Ebenso, E., Kumar, A., Sharma, S.: Experimental accompanied with computational (atomic / electronic)- level simulation investigations of Polygonum cuspidatum root extract as sustainable corrosion inhibitor for mild steel in aggressive corrosive media. Environ. Sci. Pollut. Res. (0123456789) (2024). Available from: https://doi.org/10.1007/s11356-024-34141-9
81. Amaka Chikaodili, N., Sunday Ume, C., Chukwudi Nnaji, P., Bartholomew Iroha, N., Dagdag, O., Okechukwu Ezeugo, J., et al.: Physisorption or chemisorption: Insight from AI computing model based on DFT, MC/MD-simulation for prediction of MOF-based inhibitor adsorption on Cu in brine solution. Comput. Theor. Chem. **1238**, 114730 (2024). Available from: https://doi.org/10.1016/j.comptc.2024.114730

82. Chikaodili, V., Abeng, F.E., Thakur, A., Ebenso, E.: Surface interaction and inhibition mechanism prediction of Aciclovir molecule on Fe (110) using computational model based on DFT , RDF and MD simulation. Comput. Theor. Chem. **1238**, 114702 (2024). Available from: https://doi.org/10.1016/j.comptc.2024.114702
83. Thakur, A., Kumar, A., Dagdag, O., Kim, H., Berisha, A., Sharma, D., et al.: Unraveling the corrosion inhibition behavior of prinivil drug on mild steel in 1M HCl corrosive solution : insights from density functional theory , molecular dynamics , and experimental approaches. Front Chem. 1–22 (2024)
84. Bensalah. J., Thakur, A., Kumar, A.: Investigating the adsorption processes of polymer resins for the removal of micropollutants: a comprehensive review in the field of environmental remediation. Environ. Res. **254**, 119128 (2024). Available from: https://doi.org/10.1016/j.envres.2024.119128
85. Thakur, A., Kumar, A.: Ecotoxicity analysis and risk assessment of nanomaterials for the environmental remediation. Macromol. Symp. **410**(1), 1–23 (2023)
86. Thakur, A., Savaş, K., Kumar, A.: Recent trends in the characterization and application progress of nano-modified coatings in corrosion mitigation of metals and alloys. Appl. Sci. 13:730 (2023)
87. Mahmood, A., Akram, T., Shenggui, C., Chen, H.: Revolutionizing manufacturing: A review of 4D printing materials, stimuli, and cutting-edge applications. Compos. Part B Eng. **266**, 110952 (2023). Available from: https://doi.org/10.1016/j.compositesb.2023.110952
88. Patil, S.L., Pawar, O.Y., Patil, H.S., Sutar, S.S., Kamble, G.U., Kim, D. kee, et al.: The g-C3N4-TiO2 nanocomposite for non-volatile memory and artificial synaptic device applications. J. Alloys Compd. **962**, 171024 (2023). Available from: https://doi.org/10.1016/j.jallcom.2023.171024
89. Wu, T., Huan, X., Zhang, H., Wu, L., Sui, G., Yang, X.: The orientation and inhomogeneous distribution of carbon nanofibers and distinctive internal structure in polymer composites induced by 3D-printing enabling electromagnetic shielding regulation. J. Colloid Interface Sci. **638**, 392–402 (2023). Available from: https://doi.org/10.1016/j.jcis.2023.02.014
90. Al-Qahtani, S.D., Al-Bonayan, A.M., Alenazi, N.A., Sallam, S., Abualnaja, M.M., Ashour, G.R.S., et al.: Recycled poly(propylene-co-methy l methacrylate) as printing binder for photoluminescent textiles. Mater. Chem. Phys. **307**, 128099 (2023). Available from: https://doi.org/10.1016/j.matchemphys.2023.128099
91. Vidakis, N., Petousis, M., Velidakis, E., Mountakis, N., Grammatikos S, Tzounis L. Multifunctional medical grade Polyamide12/Carbon black nanocomposites in material extrusion 3D printing. Compos. Struct. **311**(2021), 116788 (2023). Available from: https://doi.org/10.1016/j.compstruct.2023.116788
92. Narasimhamurthy, K.N., Krushna, B.R.R., Manjunatha, K., Chiu, H.H., Subramanian, B., Wu, S.Y., et al.: Novel intense blue emitting Bi2Zr2O7:Ce3+ nanocomposites insertion into flexible polymer films for anti-counterfeiting, long-term storage fingerprints, and display device applications. Mater. Today Commun. **37**, 106883 (2023). Available from: https://doi.org/10.1016/j.mtcomm.2023.106883
93. kabir, A., Chowdhury. M.A., Hossain, N., Islam, M.A., Aoyon, H., Masud Rana, M.: Graphene reinforced 3D printed polymer nanocomposites for diverse applications. Results Chem. **6**, 101197 (2023). Available from: https://doi.org/10.1016/j.rechem.2023.101197
94. Petousis, M., Ninikas, K., Vidakis, N., Mountakis, N., Kechagias, J.D.: Multifunctional PLA/CNTs nanocomposites hybrid 3D printing integrating material extrusion and CO_2 laser cutting. J. Manuf. Process. **86**(2022), 237–52 (2023). Available from: https://doi.org/10.1016/j.jmapro.2022.12.060
95. Li, J.W., Lee, J.C.M., Chuang, K.C., Chiu, C.W.: Photocured, highly flexible, and stretchable 3D-printed graphene/polymer nanocomposites for electrocardiography and electromyography smart clothing. Prog. Org. Coatings. **176**(2022), 107378 (2023). Available from: https://doi.org/10.1016/j.porgcoat.2022.107378
96. Aufa, A.N., Ismail, Z., Zaki Hassan, M.: Emerging trends in 4d printing of hydrogels in the biomedical field: A review. Mater Today Proc. (xxxx), 1–6 (2023) Available from: https://doi.org/10.1016/j.matpr.2023.01.101

97. Dey, A., Varagnolo, S., Power, N.P., Vangapally, N., Elias, Y., Damptey, L., et al.: Doped MXenes—A new paradigm in 2D systems: Synthesis, properties and applications. Prog. Mater Sci. **139**, 101166 (2023). Available from: https://www.sciencedirect.com/science/article/pii/S0079642523000981
98. Kaya, S., Lgaz, H., Thakkur, A., Kumar, A., Özbakır, D., Karakuş, N. et al.: Molecular insights into the corrosion inhibition mechanism of omeprazole and tinidazole : a theoretical investigation. Mol. Simul. 1–15 (2023)
99. Sharma, D., Thakur, A., Sharma, M.K., Sharma, R., Kumar, S., Sihmar, A, et al.: Effective corrosion inhibition of mild steel using novel 1,3,4-oxadiazole-pyridine hybrids: Synthesis, electrochemical, morphological, and computational insights. Environ. Res. **234**, 116555 (2023). Available from: https://doi.org/10.1016/j.envres.2023.116555
100. Arifa Farzana, B., Mujafarkani, N., Thakur, A., Kumar, A., Mushira Banu, A., Shifana, M.: Evaluating (p-Semidine-Guanidine-Formaldehyde) terpolymer resin efficiency as anti-corrosive agent for mild steel in 1 M H2SO4: An experimental and computational approach. Inorg Chem. Commun. **158**(P1), 111572 (2023). Available from: https://doi.org/10.1016/j.inoche.2023.111572
101. Sharma, D., Thakur, A., Sharma Kumar, M., Jakhar, K., Kumar, A., Sharma Kumar, A., et al.: Synthesis, Electrochemical, Morphological, Computational and Corrosion Inhibition Studies of 3-(5-Naphthalen-2-yl-[1,3,4]oxadiazol-2-yl)-pyridine against Mild Steel in 1 M HCl. Asian J. Chem. **35**(5), 1079–88 (2023)
102. Dhonchak, C., Agnihotri, N.: Computational Insights in the Spectrophotometrically 4H-chromen-4-one Complex using DFT Method. Biointerface Res. Appl. Chem. **13**(4), 357 (2023)
103. Thakur, A., Kumar, A.: Exploring the potential of ionic liquid-based electrochemical biosensors for real-time biomolecule monitoring in pharmaceutical applications : From lab to life. Results Eng. **20**, 101533 (2023). Available from: https://doi.org/10.1016/j.rineng.2023.101533
104. Thakur, A., Kumar, A., Kaya, S., Benhiba, F., Sharma, S.: Electrochemical and computational investigations of the Thysanolaena latifolia leaves extract : An eco-benign solution for the corrosion mitigation of mild steel. Results Chem. **6**, 101147 (2023). Available from: https://doi.org/10.1016/j.rechem.2023.101147
105. Khandan Barani, A., Roudini, G., Barahuie, F., Binti Masuri, S.U.: Design of hydrophobic polyurethane–magnetite iron oxide-titanium dioxide nanocomposites for oil-water separation. Heliyon. **9**(5), e15580 (2023). Available from: https://doi.org/10.1016/j.heliyon.2023.e15580
106. Mahato, S.K., Gupta, V.K., Kumar, A., Kumar, N., Trehan, R., Singh, R.P.: Material selection considerations for 3D printing of electrically conductive polymer composites: A review. Mater. Today Proc. (xxxx) (2023). Available from: https://doi.org/10.1016/j.matpr.2023.07.185
107. Gu, T., Ji, T., Bi, H., Ding, K.J., Sun, H., Zhai, W., et al.: 4D printed and multi-stimulus responsive shape memory polymer nanocomposites developed on hydrogen bonding–metal-phenolic sacrificial network: Application for hazardous chemical operations soft robots. Appl. Mater. Today. **35**, 102009 (2023). Available from: https://doi.org/10.1016/j.apmt.2023.102009
108. Yao, J., Morsali, M., Moreno, A., Sipponen, M.H., Hakkarainen, M.: Lignin nanoparticle-enhanced biobased resins for digital light processing 3D printing: Towards high resolution and tunable mechanical properties. Eur. Polym. J. **194**, 112146 (2023). Available from: https://doi.org/10.1016/j.eurpolymj.2023.112146
109. Gu, T., Bi, H., Sun, H., Tang, J.F., Ren, Z., Zhou, X., et al.: Design and development of 4D-printed cellulose nanofibers reinforced shape memory polymer composites: Application for self-deforming plant bionic soft grippers. Addit. Manuf. **70**(2022), 103544 (2023). Available from: https://doi.org/10.1016/j.addma.2023.103544
110. Nafady, A., Alothman, A.A., Shaikh, S.F.: Fabrication of photoluminescent electrically conductive and flame-retardant cellulose fabric incorporating polyaniline/strontium aluminate nanocomposite for a plethora of useful applications. Int. J. Biol. Macromol. **244**, 125384 2023. Available from: https://doi.org/10.1016/j.ijbiomac.2023.125384

111. Zhou, Y., Chen, J., Liu, X., Xu, J.: Three/four-dimensional printed pla nano/microstructures: crystallization principles and practical applications. Int. J. Mol. Sci. **24**(18) (2023)
112. Tao, Y., Lin, L., Ren, X., Wang, X., Cao, X., Gu, H., et al.: Four-Dimensional Micro/ Nanorobots via Laser Photochemical Synthesis towards the Molecular Scale. Micro. **14**(9) (2023).
113. Jia, L., Hao, J., Feng, Q., Li, H., Liu, K.: A multifunctional integrated carbon nanotubes/ polyphenylene sulfide composite: preparation, properties and applications††Electronic supplementary information (ESI) available. See https://doi.org/10.1039/d2na00855f. Nanoscale Adv. **5**(6), 1740–9 (2023). Available from: https://www.sciencedirect.com/science/article/pii/S2516023023023833
114. Chen, Y., Zhao, L., kang, Zhou, J., long, Bian, Y., hua, Gao, X., wen, Chen, H., et al.: Advances in the use of carbonaceous scaffolds for constructing stable composite Li metal anodes. New Carbon Mater. **38**(4), 698–718 (2023). Available from: https://www.sciencedirect.com/science/article/pii/S1872580523607346
115. Qiu, Z. han, Zhao, G. yu, Sun, Y., Wang, X. zhen, Zhao Z. bin, Qiu J. shan.: Carbon-based photothermal materials for the simultaneous generation of water vapor and electricity. New Carbon Mater. **38**(6), 997–1017 (2023). Available from: https://www.sciencedirect.com/science/article/pii/S1872580523607851
116. Chen, D., Zhao, Z., Chen, G., Li, T., Chen, J., Ye, Z., et al.: Metal selenides for energy storage and conversion: A comprehensive review. Coord Chem. Rev. 479, 214984 (2023). Available from: https://www.sciencedirect.com/science/article/pii/S0010854522005793
117. Zhang, Y., An, Q.: Self-powered assemblies for tissue engineering applications. Supramol Mater. **2**, 100036 (2023). Available from: https://www.sciencedirect.com/science/article/pii/S2667240523000065
118. Thakur, A., Kumar, A.: Recent trends in nanostructured carbon-based electrochemical sensors for the detection and remediation of persistent toxic substances in real-time analysis. Mater. Res. Express. **10**(3) (2023)
119. Verma, C., Thakur, A., Ganjoo, R., Sharma, S., Assad, H., Kumar, A., et al.: Coordination bonding and corrosion inhibition potential of nitrogen-rich heterocycles: Azoles and triazines as specific examples. Coord Chem. Rev. **488**, 215177 (2023). Available from: https://doi.org/10.1016/j.ccr.2023.215177
120. Thakur, A., Kumar, A., Singh, A.: Adsorptive removal of heavy metals, dyes, and pharmaceuticals: Carbon-based nanomaterials in focus. Carbon, N.Y. **217**(2023), 118621 (2023). Available from: https://doi.org/10.1016/j.carbon.2023.118621
121. Kaya, S., Thakur, A., Kumar, A.: The role of in Silico/DFT investigations in analyzing dye molecules for enhanced solar cell efficiency and reduced toxicity. J. Mol. Graph. Model. **124**, (2023).
122. Thakur, A., Kumar, A., Sharma, S., Ganjoo, R., Assad, H.: Computational and experimental studies on the efficiency of Sonchus arvensis as green corrosion inhibitor for mild steel in 0.5 M HCl solution. Mater. Today Proc. **66**(xxxx), 609–21 (2022). Available from: https://doi.org/10.1016/j.matpr.2022.06.479
123. Thakur, A., Sharma, S., Ganjoo, R., Assad, H., Kumar, A.: Anti-Corrosive Potential of the Sustainable Corrosion Inhibitors Based on Biomass Waste: A Review on Preceding and Perspective Research. J. Phys. Conf. Ser. **2267**(1) (2022)
124. Thakur, A., Sharma, S., Ganjoo, R., Assad, H., Kumar, A.: Anti-Corrosive potential of the sustainable corrosion inhibitors based on biomass waste: a review on preceding and perspective research. J. Phys. Conf. Ser. **2267**(1), 012079 (2022)
125. Thakur, A., Kaya, S., Kumar, A.: Recent innovations in nano container-based self-healing coatings in the construction industry. Curr. Nanosci. **18**(2), 203–16 (2021)
126. Kumar, A., Thakur, A.: Encapsulated nanoparticles in organic polymers for corrosion inhibition. Corrosion Protection at the Nanoscale. Elsevier Inc. pp. 345–362 (2020). Available from: https://doi.org/10.1016/B978-0-12-819359-4.00018-0
127. Tang. S., Lu, X., Geng, P., Chen, D., Shi, Y., Su, J., et al.: A non-covalent supramolecular dual-network polyelectrolyte evaporator based on direct-ink-writing for stable solar thermal

evaporation††Electronic supplementary information (ESI) available. See https://doi.org/10.1039/d2ma00927g. Mater. Adv. **4**(1), 223–30 (2023). Available from: https://www.sciencedirect.com/science/article/pii/S2633540923017619
128. Li, X., Ding, Z., Kong, L., Fan, X., Li, Y., Zhao, J., et al.: Recent progress in organic-based radiative cooling materials: fabrication methods and thermal management properties. Mater. Adv. **4**(3), 804–22 (2023). Available from: https://www.sciencedirect.com/science/article/pii/S2633540923017929
129. Feng, X., Hu, Y., Cao, L., Liu, L.: Artificial trachea design, construction, and application: Materials, cells, and growth factors. Appl. Mater. Today. **35**, 101968 (2023). Available from: https://www.sciencedirect.com/science/article/pii/S235294072300238X
130. Jia, Z., Xu, X., Zhu, D., Zheng, Y.: Design, printing, and engineering of regenerative biomaterials for personalized bone healthcare. Prog. Mater. Sci. **134**, 101072 (2023). Available from: https://www.sciencedirect.com/science/article/pii/S007964252300004X
131. Thijssen, Q., Toombs, J., Li, C.C., Taylor, H., Van Vlierberghe, S.: From pixels to voxels: A mechanistic perspective on volumetric 3D-printing. Prog. Polym. Sci. 147, 101755 (2023). Available from: https://www.sciencedirect.com/science/article/pii/S0079670023001090
132. Ying, X., Zhou, L., Fu, W., Wang, Y., Su, B.: Electrochemiluminescence devices for point-of-care testing. Sens. Diagn. **2**(3), 480–91 (2023). Available from: https://www.sciencedirect.com/science/article/pii/S2635099823000638
133. Mohd Yusoff, N.H., Chong, C.H., Wan, Y.K., Cheah, K.H., Wong, V.L.: Optimization strategies and emerging application of functionalized 3D-printed materials in water treatment: A review. J. Water Process Eng. **51**, 103410 (2023). Available from: https://www.sciencedirect.com/science/article/pii/S2214714422008546
134. Liu, Z., Zhong, Y., Hu, Z., Zhang, W., Zhang, X., Ji, X., et al.: Modification of ZIF-8 nanocomposite by a Gd atom doped TiO2 for high efficiency photocatalytic degradation of neutral red dye: An experimental and theoretical study. J. Mol. Liq. **380**, 121729 (2023). Available from: https://www.sciencedirect.com/science/article/pii/S0167732223005329
135. Yin, H., Zhang, W., Zhu, L., Meng, F., Liu, J., Wen, G.: Review on lattice structures for energy absorption properties. Compos. Struct. **304**, 116397 (2023). Available from: https://www.sciencedirect.com/science/article/pii/S0263822322011291
136. Wang, S., Zhang, L., Ma, R., Yu, J., Zhang, X., Shi, C., et al.: A novel one-pot strategy to construct 3D-printable cellulose nanofiber/poly(deep eutectic solvent) conductive elastomers. Chem. Eng. J. **454**, 140022 (2023). Available from: https://www.sciencedirect.com/science/article/pii/S1385894722055024
137. Zhang, Q., Wang, Q., Cui, J., Zhao, S., Zhang, G., Gao, A., et al.: Structural design and preparation of Ti3C2Tx MXene/polymer composites for absorption-dominated electromagnetic interference shielding. Nanoscale Adv. **5**(14), (2023) 3549–74. Available from: https://www.sciencedirect.com/science/article/pii/S2516023023023043
138. Patadiya, J., Naebe, M., Wang, X., Joshi, G., Kandasubramanian, B.: Emerging 4D printing strategies for on-demand local actuation & micro printing of soft materials. Eur. Polym. J. **184**, 111778 (2023). Available from: https://www.sciencedirect.com/science/article/pii/S0014305722007820
139. Li, B., Wang, S., Chen, J., Li, Z., Shan, W., Wang, X., et al.: 3D Printing of LuAG:Ce transparent ceramics for laser-driven lighting. Ceram Int. **49**(23), 38708–16 (2023). Available from: https://doi.org/10.1016/j.ceramint.2023.09.207
140. Jiang, G., Li, Z., Zhang, Y., Chen, R., Ma, Y., He, W., et al.: A facile and green method to prepare highly luminescent and mechanically robust polylactide-based composites for three-dimensional printing. Ind. Crops Prod. **202**, 117094 (2023). Available from: https://doi.org/10.1016/j.indcrop.2023.117094
141. Hogan, K.J., Perez, M.R., Öztatlı, H., Si, S., Wang, Z., Jiang, E.Y., et al.: Development of 3D-printing composite inks based on photoreactive cartilage extracellular matrix and gelatin nanoparticles. Bio. **36**, (2023).
142. Jin, H., Yeong Park, G., Kyong Kim, M., Cha, J., Seok Ham, D., Kim, M.: Eco-friendly Solvent-Processible and highly luminescent perovskite nanocrystals with polymer zwitterions

for Air-Stable optoelectronics. Chem. Eng. J. **459**(2022), 141531 (2023). Available from: https://doi.org/10.1016/j.cej.2023.141531
143. Jiao, C., Wang, Q., Li, L., Chen, W., Liu, J., Xu, Y., et al.: In situ 3D printing of liquid metal-hydrogel hybrid for multifunctional soft bioelectronics and devices. Cell Reports Phys. Sci. **4**(11), 101640 (2023). Available from: https://doi.org/10.1016/j.xcrp.2023.101640
144. Kurowska, A., Nikodem, A., Jabłoński, A., Janusz, J., Szczygieł, P., Ziąbka, M., et al.: Layered PCL scaffolds modified with bioactive additives fabricated by electrospinning and 3D-printing for the nasal bone and cartilage defects. Mater. Des. **233**, (2023)
145. Sharifi, A.M., Kwon, D.J., Shah, S.Z.H., Lee, J.: Modeling of frontal polymerization of carbon fiber and dicyclopentadiene woven composites with stochastic material uncertainty. Compos. Struct. **326**, 117582 (2023). Available from: https://doi.org/10.1016/j.compstruct.2023.117582
146. Liu, J., Cai, X., Wang, Y., Ren, G., Zhou, L., Mahmoud, M., et al.: Moisture-induced electrical power generation with waste activated sludge. Chem. Eng. J. **472**, 144868 (2023). Available from: https://doi.org/10.1016/j.cej.2023.144868
147. Kurth, D., Ristok, S., Rühle, S., Verl, A., Giessen, H.: Multi-axis two photon polymerization machine and software concept for the manufacturing of aspheric lenses on non-planar substrates. Procedia CIRP. **118**, 682–7 (2023). Available from: https://doi.org/10.1016/j.procir.2023.06.117
148. Liu, Y., Cao, L., Wang, P., Pang, X., Li, L., Mei, Q., et al.: Multi-Color Room temperature phosphorescent Silicon-Nanodot-Based nanocomposites with silane tuning and applications to 5D information encryption. Chem. Eng. J. **470**, (2023)
149. Zhang, T.T., Yang, X.N., Hu, J.H., Luo, Y., Liu, H.J., Tao, Z., et al.: New dual-mode orthogonal tunable fluorescence systems based on cucurbit[8]uril: White light, 3D printing, and anti-counterfeit applications. Chem. Eng. J. **452**(P1), 138960 (2023). Available from: https://doi.org/10.1016/j.cej.2022.138960
150. Li, Z., Dong, J., Zhang, S., Li, Y., Deng, C., Li, D., et al.: Phase transition reversible 3D printing of magnetic thixotropic fluid. Appl. Mater. Today. **34**, 101920 (2023). Available from: https://doi.org/10.1016/j.apmt.2023.101920
151. Ahmad, N.D., Kusmono, Wildan, M.W., Herianto.: Preparation and properties of cellulose nanocrystals-reinforced Poly (lactic acid) composite filaments for 3D printing applications. Results Eng. **17**(2022) (2023)
152. Al-Qahtani, S.D., Al-Bonayan, A.M., Alenazi, N.A., Sallam, S., Abualnaja, M.M., Ashour, G.R.S., et al.: Recycled poly(propylene-co-methyl methacrylate) as printing binder for photoluminescent textiles. Mater. Chem. Phys. **307**, (2023).
153. Sun, G., Wang, J., Wang, K., Baghani, M., Peng, Y., Rao, Y.: Repeatable compressive functionality of 3D printed shape-memory thin-walled corrugated structures. Int. J. Mech. Sci. **257**, 108552 (2023). Available from: https://doi.org/10.1016/j.ijmecsci.2023.108552
154. Reggio, D., Mirabile, A., Lazzari, M.: Sensing soluble molecules through SERS substrates in one-step procedure: Unrevealing the meiji woodblock printing materials. Talanta. **254**(2022), 124177 (2023). Available from: https://doi.org/10.1016/j.talanta.2022.124177
155. Parker, M., Ezeokeke, N., Matsuzaki, R., Arola, D.: Strength and its variability in 3D printing of polymer composites with continuous fibers. Mater. Des. **225**, 111505 (2023). Available from: https://doi.org/10.1016/j.matdes.2022.111505
156. Mohd Yusoff, N.H., Irene, Teo, L.R., Phang, S.J., Wong, V.L., Cheah, K.H., Lim, S.S.: Recent advances in Polymer-based 3D printing for wastewater treatment application: an overview. Chem. Eng. J. **429**, 132311 (2022). Available from: https://www.sciencedirect.com/science/article/pii/S1385894721038900
157. Moola, A.R., Santo, J., Penumakala, P.K.: Multiscale analysis for predicting elastic properties of 3D printed polymer-graphene nanocomposites. Mater. Today Proc. **62**, 4025–9 (2022). Available from: https://doi.org/10.1016/j.matpr.2022.04.600
158. Ibarhiam, S.F., Alshareef, H.F., Alqarni, S.A., Shah, R., Al-Qahtani, S.D., Almehmadi, S.J., et al.: Novel nanocomposite film developed via screen-printing of viologen polymer for anti-counterfeiting applications: Photochromism, thermochromism and vapochromic. React.

Funct. Polym. **172**(2021), 105186 (2022). Available from: https://doi.org/10.1016/j.reactfunctpolym.2022.105186
159. Lee, H.U., Jin, J.H., Chung, B.G.: High-capacity three-dimensional solar rechargeable micro-supercapacitor using MnO_2/V2O5-based binary metal oxide nanocomposite ink. J. Ind. Eng. Chem. **115**, 544–53 (2022). Available from: https://doi.org/10.1016/j.jiec.2022.08.042
160. Choi, K., Son, H., Park, J., Han, I., Han, B., Youn, B., et al.: Demonstration of durable electronic textiles via mechanically assisted highly adhesive printing of carbon nanotube-polymer composites on commercial fabrics. J. Ind. Eng. Chem. **108**, 508–13 (2022). Available from: https://doi.org/10.1016/j.jiec.2022.01.031
161. Younes, H., Kuang, X., Lou, D., DeVries, B., Rahman, M.M., Hong, H.: Magnetic-field-assisted DLP stereolithography for controlled production of highly aligned 3D printed polymer-Fe3O4@graphene nanocomposites. Mater. Res. Bull. **154**, 111938 (2022). Available from: https://doi.org/10.1016/j.materresbull.2022.111938
162. Wan, X., He, Y., Liu, Y., Leng, J.: 4D printing of multiple shape memory polymer and nanocomposites with biocompatible, programmable and selectively actuated properties. Addit. Manuf. **53**(2021), 102689 (2022). Available from: https://doi.org/10.1016/j.addma.2022.102689
163. Li, H., Lim, S.: Screen printing of surface-modified barium titanate/polyvinylidene fluoride nanocomposites for high-performance flexible piezoelectric nanogenerators. Nanomaterials. **12**(17), (2022)
164. Vidakis, N., Petousis, M., Mountakis, N., Maravelakis, E., Zaoutsos, S., Kechagias, J.D.: Mechanical response assessment of antibacterial PA12/TiO2 3D printed parts: parameters optimization through artificial neural networks modeling. Int. J. Adv. Manuf. Technol. **121**(1–2), 785–803 (2022). Available from: https://doi.org/10.1007/s00170-022-09376-w
165. Bianhong, L., Wei, Q., Qiong, W.: Research progress of carbon materials in the field of three-dimensional printing polymer nanocomposites. Nanotechnol. Rev. **11**(1), 1193–208 (2022)
166. Wu, Y., Lu, Y., Zhao, M., Bosiakov, S., Li, L.: A critical review of additive manufacturing techniques and associated biomaterials used in bone tissue engineering. Polymers (Basel). **14**(10), 1–24 (2022)
167. Sonatkar, J., Kandasubramanian, B., Ismail, S.O.: 4D printing: Pragmatic progression in biofabrication. Eur. Polym. J. **169**, 1–33 (2022)
168. Park, Y.G., Yun, I., Chung, W.G., Park, W., Lee, D.H., Park, J.U.: High-Resolution 3D printing for electronics. Adv. Sci. **9**(8), (2022)
169. Zong, W., Ouyang, Y., Miao, Y.E., Liu, T., Lai, F.: Recent advances and perspectives of 3D printed micro-supercapacitors: from design to smart integrated devices. Chem. Commun. **58**(13), 2075–95 (2022). Available from: https://www.sciencedirect.com/science/article/pii/S1359734522005079
170. Zhao, Y. rong, Liu, C. cong, Lu, Q. qiong, Ahmad, O., Pan, X. jun, Daria, M.: Recent progress on freestanding carbon electrodes for flexible supercapacitors. New Carbon Mater. **37**(5), 875–97 (2022). Available from: https://www.sciencedirect.com/science/article/pii/S1872580522606371
171. He, S. jiao, Zhang, K. qiang, Zou, Y. jun, Tian, Z. hong.: Three-dimensional printed carbon-based microbatteries: progress on technologies, materials and applications. New Carbon Mater. **37**(5), 898–917 (2022). Available from: https://www.sciencedirect.com/science/article/pii/S1872580522606346
172. Khan, K., Tareen, A.K., Iqbal, M., Zhang, Y., Mahmood, A., Mahmood, N., et al.: Recent advance in MXenes: New horizons in electrocatalysis and environmental remediation technologies. Prog. Solid State Chem. **68**, 100370 (2022). Available from: https://www.sciencedirect.com/science/article/pii/S0079678622000243
173. Yao, Z., Coatsworth, P., Shi, X., Zhi, J., Hu, L., Yan, R., et al.: Paper-based sensors for diagnostics, human activity monitoring, food safety and environmental detection. Sens. Diagn. **1**(3), 312–42 (2022). Available from: https://www.sciencedirect.com/science/article/pii/S263509982300081X

174. Wijesundara, Y.H., Herbert, F.C., Trashi, O., Trashi, I., Brohlin, O.R., Kumari, S., et al.: Carrier gas triggered controlled biolistic delivery of DNA and protein therapeutics from metal–organic frameworks††Electronic supplementary information (ESI) available. See https://doi.org/10.1039/d2sc04982a. Chem Sci. **13**(46), 13803–14 (2022). Available from: https://www.sciencedirect.com/science/article/pii/S2041652023005953
175. Khalid, M.Y., Arif, Z.U., Noroozi, R., Zolfagharian, A., Bodaghi, M.: 4D printing of shape memory polymer composites: A review on fabrication techniques, applications, and future perspectives. J. Manuf. Process. **81**, 759–97 (2022). Available from: https://www.sciencedirect.com/science/article/pii/S1526612522005035
176. Ansari, A.A., Muthumareeswaran, M.R., Lv, R.: Coordination chemistry of the host matrices with dopant luminescent Ln3+ ion and their impact on luminescent properties. Coord Chem. Rev. **466**, 214584 (2022). Available from: https://www.sciencedirect.com/science/article/pii/S0010854522001795
177. Padil, V.V.T., Akshay Kumar, K.P., Murugesan, S., Torres-Mendieta, R., Wacławek, S., Cheong, J.Y., et al.: Sustainable and safer nanoclay composites for multifaceted applications. Green Chem. **24**(8), 3081–114 (2022). Available from: https://www.sciencedirect.com/science/article/pii/S1463926222002965
178. Livi, S., Baudoux, J., Gérard, J.F., Duchet-Rumeau, J.: Ionic liquids: a versatile platform for the design of a multifunctional epoxy networks 2.0 generation. Prog. Polym. Sci. **132** 101581 (2022). Available from: https://www.sciencedirect.com/science/article/pii/S007967002200079X
179. Yuan, G., Chen, Z.Q., Zhang, C., Xie, Z.Y., Liu, S.Y., Meng, X.H., et al.: 2D lanthanide coordination polymers as multi-responsive luminescence sensors for selective and sensitive recognition of Cr(VI)/MnO_4—anions and broad-spectrum detection of antibiotics. J. Solid State Chem. **315**, 123442 (2022). Available from: https://doi.org/10.1016/j.jssc.2022.123442
180. Liu, H., Zhang, Y., Liu, X., Duan, W., Li, M., Zhou, Q., et al.: Additive manufacturing of nanocellulose/polyborosilazane derived CNFs-SiBCN ceramic metamaterials for ultra-broadband electromagnetic absorption. Chem. Eng. J. **433**(P3), 133743 (2022). Available from: https://doi.org/10.1016/j.cej.2021.133743

3D-Printed Polymer Nanocomposites for Water Treatment

Mohammad Harun-Ur-Rashid, Md. Abu Bin Hasan Susan, and Abu Bin Imran

Abstract 3D-printed polymer nanocomposites are transforming water treatment technology. Materials of this kind allow sophisticated water filtration systems to be designed and built with remarkable accuracy using additive manufacturing. Nanomaterials in polymer matrices give these composites outstanding mechanical, thermal, and chemical capabilities, making them efficient removers of water contaminants. Nanoscale additions, including metal oxides, carbon-based compounds, and organic–inorganic hybrids, improve the adsorption, catalytic, and antibacterial capabilities of the filtration system. This chapter covers the latest 3D-printed polymer nanocomposites for water treatment, including design, fabrication, and functionalization. It explores how nanoparticles improve 3D-printed structures for water treatment applications such as filtration, adsorption, and photocatalytic degradation. Material compatibility, structural integrity, and cost-effectiveness issues have been addressed when scaling these technologies for real-world applications. The recent case studies showing how these modern water treatment methods are implemented and benefit the environment have been discussed. Finally, the potential of 3D-printed polymer nanocomposites to revolutionize water purification and develop more sustainable and efficient systems has been explored.

Keywords 3D printed polymer nanocomposites · Water treatment · Additive manufacturing · Nanomaterial functionalization · Removal efficiency · Sustainable purification

---

M. Harun-Ur-Rashid
Department of Chemistry, International University of Business Agriculture and Technology (IUBAT), Dhaka, Bangladesh

Md. A. B. H. Susan (✉)
Department of Chemistry, University of Dhaka, Dhaka, Bangladesh
e-mail: susan@du.ac.bd

Dhaka University Nanotechnology Center (DUNC), University of Dhaka, Dhaka, Bangladesh

A. B. Imran (✉)
Department of Chemistry, Bangladesh University of Engineering and Technology (BUET), Dhaka, Bangladesh
e-mail: abimran@chem.buet.ac.bd

S. Moharana et al. (eds.), *Polymer Nanocomposites for 3D, 4D and 5D Printing*, Engineering Materials, https://doi.org/10.1007/978-981-96-4214-4_15

1 Introduction

In the rapidly changing field of environmental technology, sustainable water treatment methods are crucial. 3D-printed polymer nanocomposites represent a significant advancement in water purification, combining the precision of additive manufacturing (AM) with the functional benefits of nanotechnology. These innovations address critical global challenges like water scarcity and pollution, which are intensified by industrial growth, population rise, and environmental harm. 3D printed polymer nanocomposites, which integrate nanoscale materials into a polymer base using 3D printing, offer tailored, effective, and scalable solutions for water treatment. This approach leverages the ability of nanotechnology to manipulate matter at an atomic level, producing materials with superior properties due to their nanostructures. These include nanoparticles, nanotubes, and quantum dots, which exhibit unique physical and chemical characteristics distinct from their bulk counterparts. This is mainly due to their high surface-to-volume ratio and nanoscale effects. Nanocomposites combine multiple materials, including at least one nanometric, into a single matrix, which can be polymer, metal, or ceramic and is enhanced with nanoparticles or fibers. This enhancement significantly improves the mechanical, thermal, and electrical properties of materials. Adding clay nanoparticles to polymer nanocomposites can substantially increase tensile strength and heat resistance. This synergy between nanotechnology and 3D printing opens new possibilities for creating more efficient water treatment systems, marking a step forward in tackling one of the most urgent environmental challenges of our era. The application of nanomaterials, such as nanocomposite gels [21, 36, 46, 47] and films [19, 39, 40], structural colored nanomaterials [25, 37, 76], organo-metallic nanomaterials [22], molecular machines [41] have advanced the capabilities of intelligent materials. In the medical field, nanotechnology has led to groundbreaking developments like nanovaccines [32] and nanobiomedicine [31], enhancing treatment personalization and effectiveness. The diagnostics and sensing sector has also benefitted from nanobiosensors [20], offering enhanced sensitivity and precision. These nanomaterials extend beyond laboratory settings, revolutionizing industrial manufacturing [23, 24, 26–28, 34], aiding environmental pollution control [29, 30, 35, 38], and contributing to cleaner, more efficient energy solutions [33, 42, 43, 49].

Nanomaterials are transforming water treatment with their immense surface area-to-volume ratio and distinct chemical and physical properties. When merged with polymer matrices using AM, the resulting 3D-printed polymer nanocomposites harness the functional benefits of nanomaterials and the structural flexibility of 3D printing. These composites offer unmatched structural variety and functional versatility, allowing customization to specifically target contaminants such as heavy metals, organic pollutants, and microorganisms with high precision. This capability arises from the enhanced adsorption, catalytic, and antimicrobial properties of nanomaterials. Advancements in 3D printing technology enable the development of intricate structures optimized for maximum water treatment efficiency. These advancements contribute to the development of devices that excel in performance while prioritizing environmental sustainability. 3D-printed polymer nanocomposites align with green chemistry and sustainable engineering principles by designing systems that reduce waste and energy use. Exploring 3D-printed polymer nanocomposites for water treatment is vital in addressing global water purity challenges. It encourages an interdisciplinary approach among researchers, policymakers, and practitioners, emphasizing the necessity for continued research and development, supportive policy frameworks for innovation, and education to equip future professionals. Engaging with these materials could lead to eco-friendly purification systems, marking a significant advancement in combining technological innovation with environmental stewardship.

2 Types of 3D Printing Technologies

The 3D printing process is classified into: binder jetting (BJT), directed energy deposition (DED), material extrusion (MEX), material jetting (MJT), powder bed fusion (PBF), sheet lamination (SHL), vat photopolymerization (VPP), and hybrid techniques [45].

2.1 Binder Jetting Technology

BJT uses a liquid bonding agent sprayed onto a powder layer to selectively bind materials. This process is similar to inkjet printing and can include color printing by adding colored ink. After printing, models undergo thermal treatment for solidification, and excess powder is removed. Finishing procedures such as sintering or applying an epoxy layer may follow. BJT is used with materials like metal powders, quartz sand, gypsum, and polymer powders. It gives high print quality, precise dimensions, and process efficiency, capable of up to 40 kg/h, facilitated by the compact arrangement of models and the ability to print multiple models simultaneously. However, the quality is dependent on the quality and composition of the powders, with fine-grained powders offering superior performance despite challenges like caking. The layer height varies based on the material used, and the quality of the binder also affects the final quality of the model. BJT faces limitations, such as the porosity of metal components, which ranges from 3 to 10% after finishing, making them unsuitable for high-performance applications [70].

2.2 *Directed Energy Deposition*

DED is a 3D printing process using concentrated thermal energy—from laser beams, electron beams, or plasma arcs—to melt materials during deposition [75]. This technique is primarily used with metals like titanium, inconel, tantalum, among others, in wire or powder form. DED enables the creation of parts with varying compositions or hybrid structures incorporating multiple materials. It builds or repairs components layer by layer, melting the material as it's deposited through a nozzle attached to a multi-axis arm. DED is particularly efficient in fabricating and restoring large industrial metal frameworks, notably in the aerospace sector, where it is used to prepare durable and lightweight composite parts. Several variations of DED technology exist, including laser-engineered net shaping (LENS), direct metal deposition (DMD), electron beam additive manufacturing (EBAM), and wire arc additive manufacturing (WAAM). AREVO has developed a variant called polymer DED for producing composite components using a carbon fiber wire or filament within a thermoplastic polymer matrix, which is notable for not requiring a specialized environment. Unlike other 3D printing methods, DED is best suited for large-scale components that require less precision but benefit from increased productivity and speed. The fabrication rate varies with the energy source and material deposition rate, with LENS at about 0.5 kg/h, EBAM up to 9 kg/h, and WAAM reaching up to 10 kg/h. The process typically results in components that need minimal tooling and post-processing.

2.3 *Material Extrusion*

MEX in 3D printing involves stacking thermoplastics through nozzle extrusion, with fused deposition modeling (FDM) and fused filament fabrication (FFF) being common methods [67]. Emerging technologies like fused granulate fabrication (FGF), fused pellet fabrication (FPF), powder melt extrusion (PME), and direct ink writing (DIW) are gaining traction. FDM/FFF operates by extruding thermoplastic filaments from a spool through an extruder and printer, where it solidifies upon reaching the workstation, with printhead temperatures ranging from 190 to 280 °C and nozzle diameter varying by material and desired print quality. The work table moves vertically for layer deposition. FGF/FPF techniques utilize thermoplastic pellets instead of filaments, heated and ejected through a nozzle, while PME uses powder. Continuous advancements and new material introductions make extrusion techniques highly popular due to their ease of use, affordability, and the variety of durable components they can produce, including those from materials like polyetherimide (PEI) resins ULTEM and PEEK. FFF involves converting polymers into filaments before printing, which may alter characteristics and cost. FGF allows the direct use of unprocessed pellets, enabling material mixing during printing and the use of materials unsuitable for filament synthesis. FGF is suited for larger, simpler pieces due to higher productivity but coarser resolution, whereas FFF is preferred for smaller, intricate details. DIW centers on extruding liquid ink layer by layer for 3D models, which is ideal for detailed designs in sectors like ceramics, aerospace, and architecture. It is also pivotal for 4D printing in biomedicine and robotics. The print quality of DIW depends on ink properties, offering material flexibility, the ability to print fine details, and cost-effectiveness due to the simplicity of the machine design.

2.4 Material Jetting

MJT is a 3D printing method where three-dimensional models are created by precisely depositing liquid materials, such as photopolymer resins or wax, onto a build platform via specialized nozzles [72]. After deposition, these droplets rapidly solidify, with the build platform moving downward after each layer to construct the model. In the case of polymers, solidification is achieved through UV light from a lamp integrated within the printhead, drawing parallels to traditional inkjet printing. MJT has evolved into several variations, including PolyJet matrix technology by Objet, multi-jet printing (MJP) by 3D Systems using thermoset photopolymers, drop-on-demand (DOD) by Solidscape utilizing wax, and nanoparticle jetting (NPJ) by XJet for ceramics. These methods produce models with outstanding surface quality, accurate dimensions, and robust mechanical and thermal properties, allowing for the use of diverse materials. This includes a variety of resins—such as tough, durable, clear, rubber-like, heat-resistant, and ceramic-filled—alongside dental, castable resins, and wax, highlighting versatility of MJT. MJT employs a continuous deposition process with multiple printheads, enabling the creation of prints that blend multiple materials and colors and offer various finishes like glossy or matte. Complex models necessitate support structures, made simultaneously with the primary print using a soluble material for easy removal post-printing. The layer thickness in MJT varies with the material, generally ranging from 16 to 32 μm, with printing settings determined by the complex droplet formation process and primarily dictated by the equipment manufacturer.

2.5 Powder Bed Fusion

PBF is a 3D printing technique that involves the selective fusion of powder particles using various thermal energy sources, including lasers, electron beams, heated printheads, or intense light with masks [63]. This technique includes methods like selective laser sintering (SLS), selective laser melting (SLM), among others, and uses a range of polymer powders, such as polyamides (PA12, PA11), polystyrene, thermoplastic elastomers, and polyaryletherketones. SLS, a prominent PBF method, uses a laser to sinter polymer powders without fully melting them, creating objects layer by layer without needing external support structures. The powder itself supports the item being printed. SLS can process multiple materials simultaneously, each with different thermal and optical properties, by controlled heating and liquefaction with a laser and micromirrors. Selective heat sintering (SHS), introduced in 2011, offers an alternative to SLS by using a thermal head instead of a laser for powder sintering. Despite potentially lower efficiency, this makes it more scalable and suited for thermoplastic prototyping. Multi-jet fusion (MJF) of HP, patented in 2015, combines PBF with 2D inkjet printing technology. It involves layering polymer powder, applying fusing and detailing agents via inkjet heads, and then melting the powder in targeted areas with an infrared heater. MJF stands out for its efficiency, speed, and ability to produce multicolor prints, significantly speeding up prototype production and improving powder reuse.

2.6 Sheet Lamination

SHL is a 3D printing method that constructs models by stacking and bonding layers of material, including laminated object manufacturing (LOM), computer-aided manufacturing of laminated engineering materials (CAM-LEM), and other processes [7]. LOM works with a variety of materials, such as composites, paper, textiles, and plastics, while CAM-LEM primarily uses polymer tapes combined with ceramic or metal foils. Selective deposition lamination (SDL) involves bonding paper sheets with adhesive applied only in specific areas, and selective lamination composite object manufacturing (SLCOM) utilizes fiber composites within a thermoplastic matrix. Ultrasonic additive manufacturing (UAM) fuses metal sheets through ultrasonic welding. In LOM, sheets are layered onto a work table, bonded, and then cut into shape, a method known as "bond-then-cut" for its precision and efficiency in minimizing errors by using extra material for support. CAM-LEM shapes the sheets before bonding, applying adhesive as needed and using heat or pressure to set it. Each repetition of the process lowers the table by the thickness of a sheet. SHL techniques exhibit material versatility and cost-effectiveness, especially with paper, producing durable models that can be easily modified (painted, drilled, or sanded). However, the method has its drawbacks, including a "stepping" effect that may reduce surface quality, potential inaccuracies from laser cutting, the inability to create objects with internal spaces, material wastage during cutting, and high energy consumption, which limits its applications.

2.7 Vat Photopolymerization

VPP is a 3D printing technique that solidifies liquid photopolymer resin into solid structures using UV or visible light [65]. This method encompasses several technologies, including stereolithography (SLA), continuous digital light processing (CDLP), and digital light processing (DLP), which utilize various light sources like lasers, digital projections, LEDs, and oxygen for curing. The resins used in these processes range from epoxy to polyurethane and acrylic, often requiring support structures for complex designs. SLA, the pioneer among VPP and AM processes, creates objects by solidifying liquid resin layer by layer with a computer-controlled laser. The process is known for its high precision and smooth finish, although it comes with higher costs, the need for support structures, and limited resin options. The size of the laser beam plays a critical role in determining the quality of the final product. DLP operates similarly to SLA but uses a digital light projector instead of a laser, which can impact accuracy, based on the resolution of the projector and may result in pixelation effects. DLP is more suitable for smaller projects due to its spot irradiation method, which helps prevent resin shrinkage. CDLP, or CLIP/DLS, stands out by employing digital projection in a continuous, rather than layer-by-layer, manufacturing process facilitated by an oxygen-permeable window. This approach ensures a steady fluid level and enhances print quality, speed, and resolution, enabling faster prototype production compared to SLA and DLP. Other VPP techniques, like two-photon lithography, aim to create micro- and nanostructures through nonlinear polymerization rates when exposed to light, pushing the limits of conventional 3D printing to achieve exceptionally detailed structures.

3 Types of 3D Printing Materials

3D printing technology is rapidly evolving, which makes it increasingly relevant for industrial applications. However, employing this technology introduces safety concerns due to the release of volatile organic compounds and ultrafine particles. Adopting natural polymers for 3D printing could significantly address these safety issues. The use of biodegradable, environmentally friendly, and recyclable materials presents benefits in terms of easy disposal or repurposing at the end of their useful life. Furthermore, adopting recyclable and biodegradable materials can reduce the costs associated with the 3D printing process. Figure 1 summarizes the wide variety of materials used in 3D printing, including plastics, metals and alloys, composites, and other pertinent materials [29]

Plastic is used in 3D printing for its versatility, cost-effectiveness, and easy post-processing. Key materials like polylactic acid (PLA), acrylonitrile butadiene styrene (ABS), polyethylene terephthalate glycol-modified (PETG), and thermoplastic polyurethane (TPU) are used in a variety of applications, including prototyping and consumer products. Despite their broad utility, challenges such as environmental impact and toxic emissions during printing are concerns. Advances towards sustainable materials, like bioplastics and composites, are promoting eco-friendliness. Thermoplastics, used in material extrusion and powder bed fusion, vary in their softening temperatures and require support structures for complex designs. Amorphous thermoplastics like ABS and PLA are favored for their melting properties. Semisolid crystalline polymers like nylon are used in powder bed fusion for their ability to fuse at controlled temperatures, offering support for overhangs and

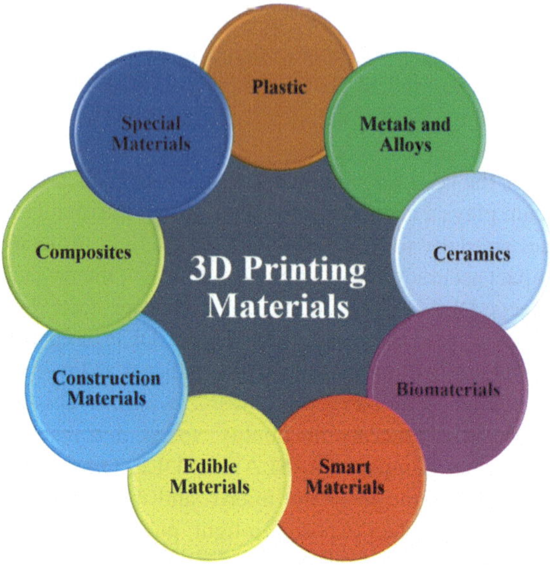

Fig. 1 Different types of materials in the 3D printing manufacturing process. Reproduced from Ref. [29] with permission from Elsevier, Copyright 2024 by Elsevier

high-density outputs despite the complexity of feedstock preparation and potential surface quality issues. Thermosets in 3D printing, particularly in vat photopolymerization, involve photopolymers like epoxies and acrylates. Epoxies have reduced shrinkage and warping compared to acrylates, benefiting from cationic polymerization that minimizes volume change. Commercial resins often blend acrylates for speed and epoxies for strength, creating durable structures with less energy requirement. However, resins can degrade over time, influencing the longevity of printed objects. Innovative methods like laser additive manufacturing process (LAMP) use photopolymer composites for specialized applications, highlighting ongoing developments toward improving mechanical properties and product longevity in 3D printing.

3D metal printers use laser or electron beam technologies to melt metallic materials, either wire or powder, layering them based on a 3D CAD model to create physical objects. This method excels at producing complex designs challenging for traditional manufacturing, making it highly valuable for producing small quantities in sectors like automotive, aerospace, medical, and defense. Common metals used include aluminum, cobalt, nickel alloys, stainless steels, and titanium alloys. Cobalt alloys are favored in dental applications for their strength and heat resistance, nickel alloys in aerospace for their corrosion and high-temperature endurance, and titanium alloys in aerospace and medical fields for their mechanical properties and corrosion resistance. The process involves various techniques like powder bed fusion and material jetting but faces challenges such as dimensional inaccuracies and the need for post-printing finishing.

Ceramics are celebrated for their exceptional hardness, heat resistance, and corrosion resistance but are marked by brittleness. Advances in 3D printing have enabled the production of ceramic objects without cracks or significant porosity, enhancing their mechanical strength. These advances make ceramics suitable for intricate designs, potentially revolutionizing architecture and construction. Alumina, bioactive glasses, and zirconia are used in 3D printing in high-tech, dentistry, aeronautics, and nuclear reactors owing to their characteristics. Challenges remain, such as a limited range of ceramic materials suitable for 3D printing, measurement inaccuracies, and surface finish issues. Biomaterials in 3D printing are transforming healthcare by allowing customized medical solutions with improved precision and cost-effectiveness. This innovation supports the development of medical devices, cell printing, and implants, extending into organ model fabrication, dentistry, and tissue engineering. Despite the progress, designing organs with functional blood vessels remains a challenge.

Smart materials in 3D printing can change their properties—like color, shape, and size—under external stimuli, creating soft robotics and self-evolving structures. Shape memory polymers and alloys, such as Ni–Ti, are notable for their application in various fields, facilitated by 3D printing to produce complex shapes with precision and fine surface quality. Edible printing materials have brought innovation to the food industry, allowing for customization and reduced waste. This technology supports the creation of food items with diverse materials, textures, and designs, promoting commercialization in both small-scale and industrial food production. Construction materials, especially concrete, have seen advancements in 3D printing, offering the potential for constructing buildings and infrastructure with reduced waste, cost, and labor. Challenges include ensuring strong layer adhesion in concrete structures to maintain integrity and durability.

Composite production in 3D printing involves meticulous selection and preparation of materials, aiming for consistent quality and desired properties. This technology has advanced to produce crack-free and low-porosity ceramic artifacts, enhancing their mechanical robustness. Polymer composites benefit from material extrusion methods for layered fabrication, incorporating additives to achieve optimal viscosity and structural integrity. Innovations have led to the creation of composites with dispersed nanotubes and fiber-reinforced materials, enhancing conductivity and mechanical properties. Metal composites in 3D printing cover a range of types, including particulate, fibrous, and laminates, with SLM and laser metal deposition (LMD) being preferred methods. These composites find applications in aerospace due to their graded properties, and innovative techniques like laser peening and ultrasonic consolidation have improved consolidation and accuracy. Ceramic composites, influenced by biomaterials, allow for the immediate use of bioceramic scaffolds post-fabrication, bypassing traditional sintering. Techniques like selective laser gelation and material jetting have pushed the boundaries in creating high-resolution structures with precise electrical performance. However, challenges remain in achieving uniform phase distribution and mechanical properties comparable to conventionally pressed samples, highlighting the need for further research in composite material optimization in 3D printing.

3D printing utilizes a diverse array of materials beyond the conventional ones, including photopolymers, wood, glass, and cellulose compounds. Glass is printed using the fused deposition modeling technique, while cellulose, a sustainable biopolymer, is versatile in applications ranging from healthcare and electronics to textiles and construction. Various 3D printing techniques such as SHL, DED, PBF, MEX, BJT, and MJT make products from these materials. Photopolymers react to light and are used in vat photopolymerization processes. Wood printing, executed through fused filament manufacturing, seeks innovative methods to repurpose discarded furniture into wood-based 3D printed items, indicating a push towards more sustainable manufacturing practices.

4 Fundamentals of 3D-Printed Polymer Nanocomposites

3D-printed polymer nanocomposites for water treatment combine advanced AM techniques with nanotechnology to prepare materials tailored for purifying water. These composites comprise a polymer base integrated with nanoparticles, such as titanium dioxide, silver, and graphene oxide, which impart special properties for enhanced water treatment. The design flexibility offered by 3D printing allows for producing structures with complex geometries, high surface areas, and customizable porosity, crucial for effective contaminant removal. The selection of polymers and nanoparticles is key, focusing on non-toxic, chemically resistant materials that prevent harmful leaching into the water while offering high mechanical strength and durability against the physical and chemical stresses of water treatment processes. Furthermore, these nanocomposites can be engineered to target specific pollutants through selective adsorption, photocatalytic degradation, or antimicrobial activity, highlighting their adaptability to various water treatment needs. Sustainability is also a cornerstone of their design, with a preference for biodegradable or recyclable materials to minimize environmental impact post-use. In essence, 3D-printed polymer nanocomposites embody a promising avenue for enhancing water treatment technologies, combining effectiveness, customization, and environmental sustainability.

4.1 Polymer Matrices

3D-printed polymer nanocomposites use polymer matrices to disperse nanoparticles to generate water treatment products with improved characteristics. The selection of an appropriate polymer matrix is crucial, as it determines the mechanical strength, chemical resistance, processability, and environmental impact of the composite. Various polymer matrices are commonly used to fabricate 3D-printed polymer nanocomposites for water treatment, highlighting their unique attributes and contributions to the field.

4.1.1 Polylactic Acid

PLA is a favored choice due to its biodegradability, which originates from renewable resources like corn starch or sugarcane. Its widespread use in 3D printing can be attributed to its ease of processing and relatively low melting temperature, making it suitable for various printing technologies. In water treatment applications, PLA-based nanocomposites can be engineered to degrade safely, reducing environmental impact after their lifecycle [10].

4.1.2 Acrylonitrile Butadiene Styrene

ABS is known for its strength, making it excellent for applications needing long-lasting components. Its excellent mechanical properties and resistance to chemical degradation enable the production of water treatment devices that can withstand harsh environments and repetitive use. However, ABS is not biodegradable, necessitating issues for recycling and appropriate waste management techniques [18].

4.1.3 Polyethylene Terephthalate Glycol-Modified

PETG combines PLA and ABS characteristics, including chemical resistance, durability, and ease of manufacturing. It exhibits superior moisture resistance compared to PLA, making it particularly suitable for water treatment environments where prolonged exposure to water is expected. The recyclability of PETG further enhances its appeal as a sustainable option for 3D-printed applications [55].

4.1.4 Thermoplastic Polyurethane

TPU is distinguished by its flexibility and elasticity, allowing the creation of 3D-printed components with unique mechanical properties such as bendability and shock absorption. These characteristics are advantageous in designing water treatment devices that require flexibility or need to withstand vibration and impact. Resistance of TPU to oil and grease contributes to its utility in applications involving oily water separation [3].

4.1.5 Polyetherimide

PEI, with its high strength and thermal stability, is suitable for high-performance applications. Its ability to endure high temperatures and resist chemical wear makes it an excellent candidate for water treatment processes that involve heat and corrosive substances. Despite its advanced properties, the Processing of PEI requires higher temperatures, often necessitating specialized 3D printing equipment [2].

4.1.6 Polycaprolactone

Polycaprolactone (PCL) is a biodegradable polyester with a low melting point, making it amenable to 3D printing processes. Its degradation rate can be adjusted by blending with other polymers or incorporating specific nanoparticles, tailoring the lifespan of water treatment devices according to their application needs. This feature is particularly relevant for temporary installations or single-use applications in water treatment [1].

Incorporating nanoparticles into these polymer matrices enhances the functionality of composite materials, enabling the targeted removal of contaminants, antibacterial activity, and improved mechanical properties. The choice of polymer matrix and nanoparticles is guided by the specific requirements of the water treatment process, including the types of pollutants to be removed, the operational environment, and sustainability considerations. Future advancements in polymer science and nanotechnology are expected to expand the capabilities of 3D-printed polymer nanocomposites in addressing global water treatment challenges. Table 1 serves as a guide for selecting polymer matrices for 3D-printed polymer nanocomposites based on the specific requirements of water treatment applications, considering factors such as durability, chemical resistance, and environmental impact.

4.2 Nanomaterials

Nanomaterials enhance the properties of 3D-printed polymer nanocomposites for water treatment applications. Their integration into polymer matrices leads to materials that exhibit superior mechanical strength, chemical resistance, and specific functionalities such as antimicrobial properties and pollutant adsorption. Various nanomaterials are unique in their contributions to the effectiveness of 3D-printed nanocomposites in water purification.

4.2.1 Titanium Dioxide

Titanium dioxide (TiO_2) nanoparticles are known for photocatalyzing organic pollutants under UV light. This feature is particularly beneficial for water treatment, as it can lead to harmful chemicals and bacteria breakdown, rendering the water safe for consumption or release into the environment. The incorporation of TiO_2 into polymer matrices for 3D printing allows for the creation of complex structures with high surface area-to-volume ratios, enhancing the photocatalytic efficiency of the composite [11].

Table 1 Comparison of polymer matrices for 3D printed polymer nanocomposites in water treatment: Assessing biodegradability, mechanical properties, chemical resistance, and sustainability

Polymer matrix	Biodegradability	Mechanical properties	Chemical resistance	Applications in water treatment	Sustainability notes
PLA	Biodegradable	Good strength, but brittle without modification	Moderate	Filtration systems, biodegradable meshes	Derived from renewable resources; compostable under industrial conditions
ABS	Non-biodegradable	High strength and toughness	High	Durable components for filtration and water processing units	Recyclable, but derived from petrochemicals
PETG	Non-biodegradable (recyclable)	High toughness, impact resistance	High	Reservoirs, piping, and connectors exposed to moisture	Recyclable; offers a balance between performance and environmental impact
TPU	Non-biodegradable (recyclable in some cases)	Flexible, elastic, and durable	Good resistance to oil and grease	Flexible hoses, seals, and membranes for water treatment	Recycling depends on the availability of specific facilities
PEI	Non-biodegradable	High mechanical strength, excellent thermal stability	Very high	High-temperature filtration, components resistant to harsh chemicals	High performance but challenging to recycle
PCL	Biodegradable	Low strength, highly flexible	Moderate	Temporary water treatment solutions, dissolvable supports	Biodegradable with a slower degradation rate than PLA

4.2.2 Silver Nanoparticles

Silver nanoparticles (AgNPs) exhibit antimicrobial efficacy, making them ideal for water treatment applications aimed at disinfection. AgNPs can be embedded within polymer matrices to produce nanocomposites with surfaces inhospitable to bacteria, viruses, and other microorganisms. This antimicrobial action, combined with the physical filtering capabilities of the polymer structure, provides a dual-function approach to water purification [50].

4.2.3 Carbon-Based Nanomaterials

Graphene oxide (GO) and carbon nanotubes (CNTs) are carbon-based nanomaterials known for their exceptional mechanical properties and high surface areas. These characteristics make them suitable for adsorption of heavy metals and other pollutants from water. Furthermore, their incorporation into polymer matrices can significantly enhance the mechanical strength and durability of the resulting nanocomposite, making them suitable for applications in water treatment systems that require robustness and longevity [60].

4.2.4 Zeolites

Zeolites are microporous, aluminosilicate minerals with a well-defined pore structure, making them excellent for adsorbing contaminants such as heavy metals and ammonia from water. By integrating zeolites into 3D-printed polymer matrices, it is possible to produce composites that combine the mechanical properties of the polymer with the adsorption capabilities of the zeolites, leading to the efficient removal of specific pollutants from water [73].

4.2.5 Bioactive Glasses

Bioactive glasses can release ions in a controlled manner, which can be beneficial for neutralizing harmful substances in water. Their integration into 3D-printed polymer nanocomposites could be explored for applications where ion exchange or the release of beneficial ions into the water is desired, potentially opening up new avenues for water treatment [13].

4.2.6 Magnetic Nanoparticles

Magnetic nanoparticles, such as iron oxides, offer the unique advantage of being easily separable from water using a magnetic field. This property is valuable for water treatment processes where the nanocomposite acts as an adsorbent for pollutants and is subsequently removed from the water stream magnetically, ensuring no secondary pollution. Embedding magnetic nanoparticles in 3D-printed structures could facilitate the development of reusable water treatment devices [78].

Integrating these nanomaterials into 3D-printed polymer matrices for water treatment combines the advantages of AM and nanotechnology, leading to innovative solutions for purifying water. Future research and development in this area are expected to focus on optimizing the dispersion of nanoparticles within the polymer matrix, enhancing the interfacial interaction between the nanoparticles and the polymer, and exploring the environmental impact of these nanocomposites. Table 2 provides a snapshot of the diversity of nanomaterials available for enhancing the

Table 2 Comprehensive overview of nanomaterials used in 3D printed polymer nanocomposites for water treatment, highlighting key properties, applications, and environmental impact

Nanomaterial	Key properties	Applications in water treatment	Environmental impact
TiO_2	Photocatalytic, UV-reactive	Degradation of organic pollutants, disinfection	Generally low toxicity, but potential impact requires careful management
AgNPs	Antimicrobial	Disinfection, bacterial growth inhibition	Concerns over toxicity to aquatic life; requires responsible use
CNTs	High surface area, mechanical strength	Adsorption of heavy metals, organic pollutants	Potential environmental persistence; research into safe disposal needed
GO	Large surface area, adsorptive	Removal of heavy metals, dyes, and other pollutants	Biodegradability varies; concerns over long-term environmental impact
Zeolites	Microporous structure	Adsorption of heavy metals, ammonia, and other contaminants	Generally regarded as environmentally benign
Bioactive glasses	Controlled ion release	Neutralizing harmful substances, ion exchange	Biocompatible and generally considered safe for the environment
Magnetic nanoparticles (iron oxides)	Magnetic properties	Facilitated removal of pollutants through magnetic separation	Low toxicity, but environmental dispersion should be monitored

efficacy of 3D-printed polymer nanocomposites in water treatment applications, highlighting their specific roles and potential environmental considerations.

5 Desirable Properties of Polymer Nanocomposites

5.1 Photocatalytic Activity

Photocatalytic activity is an essential feature in certain nanocomposites, especially those incorporating semiconducting nanoparticles like TiO_2 or zinc oxide (ZnO). These materials harness UV light to activate photocatalytic reactions that degrade organic pollutants into harmless byproducts. The effectiveness of this process is amplified by the large surface area provided by the nanoscale additives, enhancing

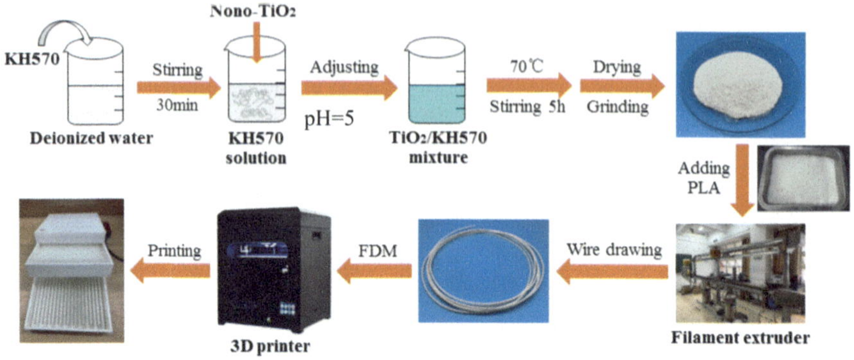

Fig. 2 Illustration of the procedure for preparing, producing materials, and the final printing process of modified TiO_2. KH-570 is a methacryl-functional silane. Reproduced from Ref. [56], [https://doi.org/10.3390/polym13132196], under the terms of the CC BY 4.0 license [https://creativecommons.org/licenses/by/4.0/]

the exposure of photocatalytic sites to pollutants. This property contributes to the breakdown of stubborn contaminants and works under ambient conditions, offering a sustainable approach to water purification.

Mai et al. have pioneered the integration of photocatalytic technology into 3D printing materials to fabricate water treatment facilities [56]. Their innovative approach involved the preparation of a novel 3D printing material, endowed with photocatalytic activity, used to print water treatment equipment. This equipment was subsequently enhanced by applying a silver-loaded TiO_2 coating to boost its catalytic degradation performance. The fabrication process utilized melt blending to produce composite filaments, comprising 10 wt% modified TiO_2 and 90 wt% PLA, with a precise diameter of 1.75 ± 0.05 mm. The silver-loaded TiO_2 was applied using a UV-curing method to achieve a uniform coating, ensuring an even dispersion of the modified particles within the PLA matrix. Incorporating 10 wt% TiO_2 into the PLA effectively yielded stable printing filaments, while adding 5 wt% silver loading significantly enhanced the photocatalytic degradation capabilities of the material. This novel 3D printed facility shows the potential to efficiently degrade organic compounds in wastewater and exhibits a bactericidal effect, marking a significant step forward in the application of 3D printing technology for future drinking water treatment solutions. Figure 2 illustrates the procedure for preparing, and producing materials, and the final printing process of modified TiO_2.

5.2 *Antimicrobial Properties*

Integrating metal nanoparticles such as Ag and Cu endows polymer nanocomposites with potent antimicrobial properties. These metals release toxic ions to a wide range of microorganisms, effectively disinfecting water without additional chemical agents.

The antimicrobial action of these composites is broad-spectrum, targeting bacteria, viruses, and fungi, thereby significantly reducing the microbial load in treated water. This feature is particularly critical in preventing the spread of waterborne diseases in vulnerable communities.

Bayraktar et al. have successfully demonstrated the preparation and 3D printing of silver nanowire (Ag NW) loaded PLA nanocomposites [5]. Through simple solution mixing methods followed by 3D printing, they have developed nanocomposites that exhibit excellent dispersion of Ag NWs within the PLA matrix. The antibacterial effects of these nanocomposites against common bacterial strains such as *Staphylococcus aureus* and *Escherichia coli* were significant, even a 4 wt% Ag NW loading can achieve a 100% kill rate against these bacteria within two hours. The presence of Ag NWs not only enhances the antibacterial properties but also influences the thermal behavior of the PLA matrix, delaying degradation and increasing crystallinity. These findings suggest the potential of Ag NW/PLA nanocomposites, fabricated via 3D printing, as effective antibacterial materials for public health and wastewater treatment applications, particularly in areas with high exposure to bacterial contamination. Figure 3 showcases SEM images illustrating the effectiveness of 4 wt% Ag NW/PLA nanocomposites in killing bacteria. Figure 3a displays *S. aureus* colonies after an 8 h exposure, with damaged bacteria highlighted in red and the few survivors in white, confirming the 100% kill rate of the nanocomposites up to 8 h. Figure 3b reveals the aftermath of a 2 h incubation, showing dead *E. coli* colonies on the sample surfaces, thus demonstrating the lethal effect of nanowires on these bacteria. The analysis suggests that Ag NWs exhibit a higher antibacterial performance against Gram-negative bacteria like *E. coli*, indicating that the 3D-printed nanocomposites possess potent and enduring antibacterial activity against such pathogens.

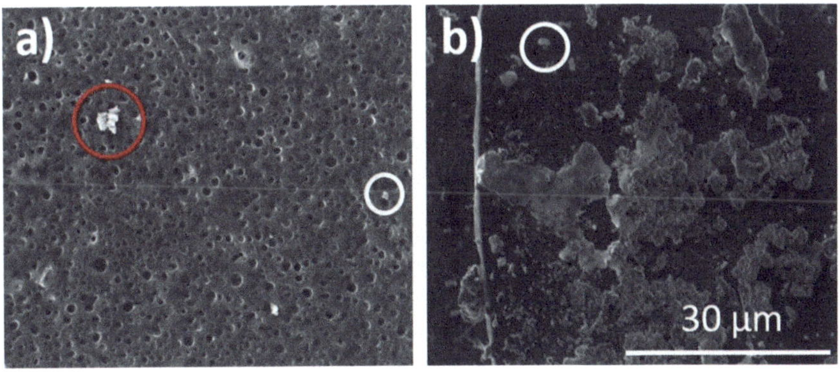

Fig. 3 SEM images display **a** *S. aureus* bacteria following 8 h of exposure and **b** *E. coli* bacteria after 2 h of exposure to 4 wt% Ag NW/PLA nanocomposite disks, with both images presented at the same scale. Reproduced from Ref. [5] with permission from Elsevier, Copyright 2019

5.3 Adsorption Capacity

Incorporating carbon-based nanomaterials like activated carbon, CNTs, and GO into polymer matrices significantly boosts the adsorption capacity of the resulting composites. These materials have vast surface areas and porous structures, ideal for trapping heavy metals, organic chemicals, and other contaminants. Their high adsorption efficiency is attributed to the myriad active sites on their surfaces, which bind pollutants through physical and chemical interactions. Optimizing the pore size and surface chemistry of these nanomaterials can further enhance selectivity and capacity for specific contaminants. Baigorria et al. manufactured hybrid polymer nanocomposite hydrogels with high adsorption capability using 3D printing technology for wastewater treatment [4]. Figure 4 illustrates the development of nanocomposite polymer hydrogel for wastewater treatment by adsorption method.

5.4 Mechanical Strength and Durability

The structural integrity of water treatment systems is paramount, necessitating materials that exhibit high mechanical strength and durability. Nanocomposites leverage the reinforcing potential of nanoparticles to improve the tensile strength, impact resistance, and longevity of polymer matrices. These enhanced mechanical properties ensure that the nanocomposites can withstand the rigors of water treatment operations, including high pressures, flow rates, and cyclic loading, without succumbing to wear or deformation [16].

5.5 Chemical Resistance

Water treatment environments are chemically harsh, exposing materials to a spectrum of disinfectants and contaminants. Polymer nanocomposites destined for such applications must exhibit intrinsic chemical resistance to maintain their structure and functionality over time. This resistance is crucial for preventing the leaching of polymer constituents or nanoparticles into the water, ensuring the safety and purity of the treated water. Chemical resistance also extends the service life of the materials, reducing the need for frequent replacements [68].

5.6 Biodegradability and Environmental Sustainability

The environmental impact of water treatment materials is an increasing concern. Biodegradable polymer nanocomposites offer a solution by breaking down into

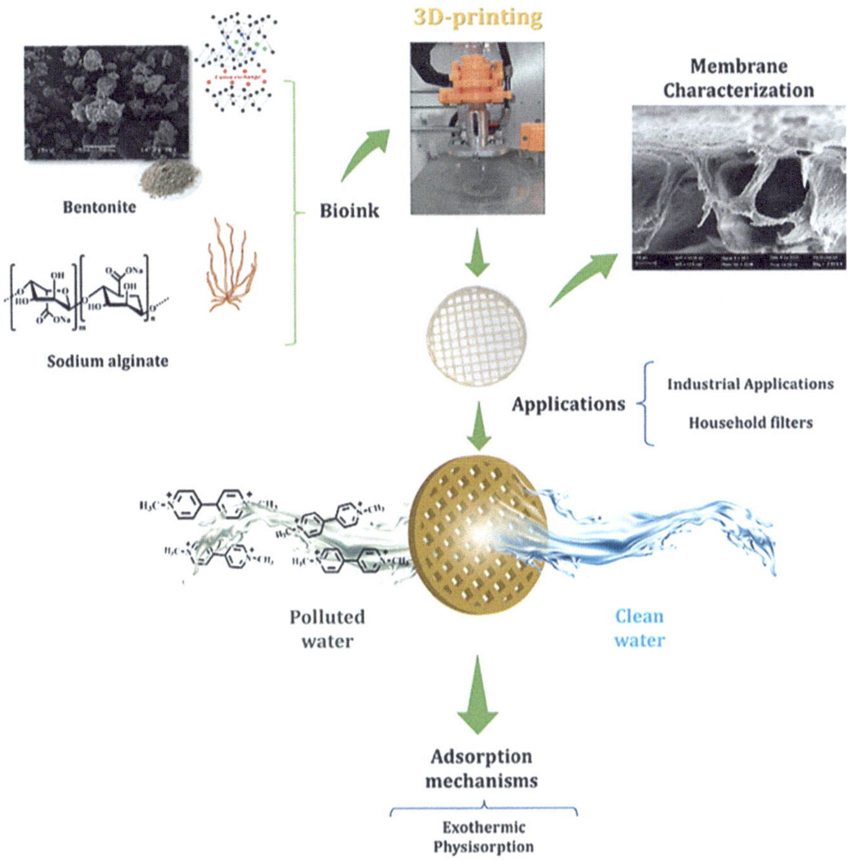

Fig. 4 Graphical representation of the entire process of fabricating nanocomposite polymer hydrogels for wastewater treatment through the adsorption technique. Reproduced from Ref. [4] with permission from Elsevier, Copyright 2024

environmentally benign byproducts at the end of their lifecycle. PLA and PCL are polymers that can be engineered with nanoparticles to retain functional efficacy while ensuring biodegradability. This property aligns with sustainable water treatment practices, minimizing waste and facilitating eco-friendly disposal.

5.7 Selective Pollutant Targeting

Customizing nanocomposites for selective pollutant targeting leverages the specificity of molecular interactions between contaminants and the composite surface. By functionalizing the polymer or nanoparticle components, these materials can preferentially adsorb or react with specific pollutants, such as heavy metals, organic compounds, or pathogens. This selective targeting enhances the efficiency of water treatment processes, removing hazardous substances without affecting the essential qualities of water [71].

3D-printed polymer nanocomposites for water treatment exemplify a harmonious blend of material science and environmental engineering. Researchers and engineers can tackle the multifaceted water purification challenges by carefully designing these composites to exhibit properties such as photocatalytic activity, antimicrobial efficacy, and selective pollutant targeting. Future advancements will likely focus on enhancing these properties, exploring new nanomaterials, and addressing sustainability concerns to meet the global demand for clean water. Table 3 encapsulates the critical properties that 3D printed polymer nanocomposites must possess to be effective in water treatment applications, highlighting how each property contributes to improving water quality and ensuring the sustainability of treatment processes.

6 Design and Fabrication Techniques

The design and fabrication of 3D-printed polymer nanocomposites for water treatment applications involve intricate processes that blend material science, nanotechnology, and AM techniques. This inclusive approach ensures the production of materials with desired properties for specific water treatment needs, such as adsorption, photocatalytic activity, and antimicrobial effects. The various aspects of these techniques, emphasizing their significance in enhancing water purification systems have been discussed. Figure 5 represents different steps for manufacturing 3D-printed polymer nanocomposites for water treatment applications.

6.1 Conceptual Design and Simulation

The conceptual design and simulation phase are the cornerstone in developing 3D-printed polymer nanocomposites for water treatment. At this stage, leveraging advanced software tools enables the theoretical construction and fine-tuning of nanocomposite architectures even before any physical prototype is produced. This proactive approach allows for the meticulous planning of the morphology of the nanocomposite, including its porosity, surface area, and the distribution of nanoparticles within the polymer matrix. Computational fluid dynamics (CFD) simulations stand at the forefront of this phase, providing a powerful computational technique that models the flow of water and the dynamic interactions with dissolved or suspended pollutants. Through CFD, designers can visualize the flow patterns around and through the nanocomposite structures, identifying areas where pollutants might accumulate or flow too rapidly for effective treatment. This insight is invaluable for tailoring the design of the nanocomposite to enhance contact time with water,

Table 3 Desirable properties of 3D-printed polymer nanocomposites for water treatment and their impact on purification processes

Property	Significance	Potential impact on water treatment
Photocatalytic activity	Enables degradation of organic pollutants under light exposure	Facilitates the removal of harmful chemicals and pathogens, reducing water toxicity
Antimicrobial properties	Provides bactericidal and viricidal effects to prevent microbial growth	Ensures water safety by eliminating disease-causing microorganisms
Adsorption capacity	High surface area for trapping contaminants like heavy metals and organic pollutants	Enhances water purity by effectively removing a wide range of pollutants
Mechanical strength and durability	Withstands physical stresses from water flow and filtering processes	Guarantees longevity and reliability of water treatment systems, reducing maintenance needs
Chemical resistance	Resists degradation by chemicals used in water treatment processes	Preserves material integrity and effectiveness over time, ensuring consistent water quality
Biodegradability/ recyclability	Minimizes environmental impact post-use through degradation or recycling	Supports sustainable water treatment practices by reducing waste and promoting material reuse
Selective pollutant targeting	Tailored to adsorb or react with specific contaminants	Increases efficiency of contaminant removal, allowing for targeted water purification strategies
High surface area	Enhances the interaction between water and the nanocomposite material	Improves the efficiency of photocatalytic activity, adsorption, and microbial disinfection
Thermal stability	Maintains performance under varying thermal conditions	Ensures consistent water treatment efficacy in environments with fluctuating temperatures
Cost-effectiveness	Affordable production and operation costs	Makes advanced water treatment solutions accessible to a wider range of communities and industries

ensuring that the maximum amount of pollutants can be adsorbed, degraded, or otherwise neutralized.

Optimization algorithms can be applied in conjunction with CFD simulations to refine the design iteratively. These algorithms adjust parameters such as the shape and size of the nanocomposite, the distribution of functional nanoparticles, and the

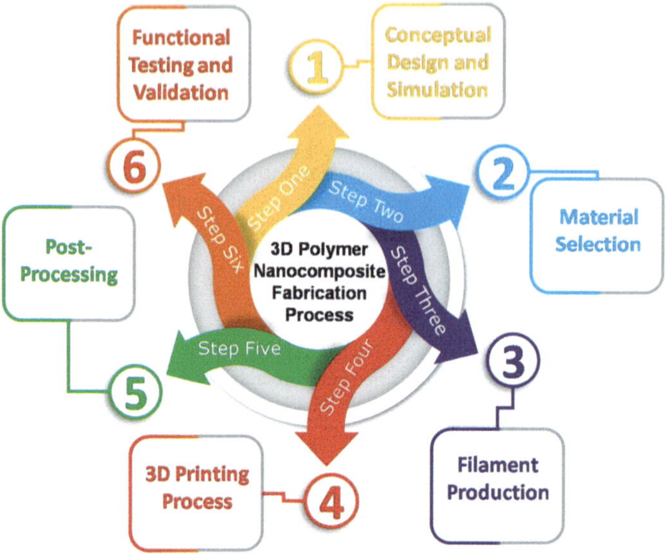

Fig. 5 Six-step cycle of designing and fabricating 3D printed polymer nanocomposites for water treatment (from conceptualization to validation)

overall structure of the water treatment device to find the optimal configuration that yields the highest purification efficiency. This might involve tweaking the design to increase surface roughness in areas critical for pollutant capture or modifying the internal geometry to promote turbulent flow that enhances photocatalytic reactions. The predictive power of CFD simulations also extends to evaluating the mechanical strength and durability of the proposed designs under realistic water flow conditions. This ensures that the nanocomposites are effective in purifying water and resilient enough to withstand the physical stresses encountered in real-world water treatment settings. By integrating CFD simulations with optimization techniques, the conceptual design and simulation phase establishes a robust foundation for developing 3D-printed polymer nanocomposites. This systematic approach accelerates the R&D process by minimizing the need for trial and error in physical testing and ensures that the final product is both effective in treating water and economically viable to produce. The culmination of this phase is a well-defined blueprint for a nanocomposite primed for fabrication and poised to address specific water treatment challenges with unprecedented efficiency.

6.2 *Material Selection and Nanoparticle Functionalization*

Choosing the optimal polymer and nanoparticles is fundamental in creating effective 3D-printed polymer nanocomposites for water treatment. The process is further refined by functionalizing the nanoparticles tailoring their surfaces to improve interactions with target pollutants. This surface modification can enhance the ability of the nanocomposites to capture, adsorb, or decompose specific contaminants, thus significantly increasing their water treatment capabilities. Techniques such as the grafting of functional groups or coating with specific agents can be employed to modify the nanoparticle surfaces. These adjustments not only bolster the affinity of the nanoparticles for particular pollutants but also can impart additional functionalities, like antimicrobial or anti-fouling properties. Such strategically functionalized nanoparticles within the polymer matrix create a composite that is more effective in purifying water and versatile across various applications, from industrial wastewater treatment to potable water purification [8].

6.3 Nanocomposite Preparation and Filament Production

After choosing materials, creating the nanocomposite entails integrating the functionalized nanoparticles with the polymer base until a uniform mixture is achieved. The well-dispersed blend ensures that the nanoparticles are evenly distributed, which is crucial for the consistency of the properties of the nanocomposite. This mixture is subsequently shaped into filaments for FDM or formulated into a resin appropriate for SLA and SLS printing methods. This step is critical to produce a consistent filament or resin that maintains the desired properties throughout the volume of the material, leading to high-quality and functionally reliable 3D printed structures.

6.4 3D Printing Process

The 3D printing process involves choosing an appropriate AM technique based on the design requirements and material properties. FDM is the go-to technique for ease of use and straightforward access. In this process, filaments consisting of polymer-nanomaterial blends are extruded through a precisely heated nozzle to construct the object incrementally. Critical parameters like the temperature of the nozzle and printing speed are finely tuned to the materials used, ensuring the final structure boasts integrity and the intended functionality. SLA harnesses the power of UV light to solidify a liquid resin that has been combined with nanoparticles, producing highly accurate parts with smooth finishes. Its precision makes it ideal for crafting intricate devices for water treatment that demand exact specifications and close-fitting components. SLS, utilizes a concentrated laser to meld nanoparticle-infused polymer powders into solid forms, enabling sturdy items with sophisticated geometries. Crucially, this method facilitates the production of items with porous structures, which can significantly improve both water flow and pollutant removal efficiency. The technique of BJT builds parts by depositing a binding agent onto a bed of polymer and nanomaterial powders, one meticulous layer at a time. Once the printing is done, these parts undergo a curing or sintering phase to reach their full strength and durability. This technique is particularly adept at effectively creating large-scale and complex shapes.

6.5 Post-processing and Performance Enhancement

Post-processing, encompassing drying, curing, or sintering, is essential in solidifying the mechanical properties and ensuring the stability of the nanocomposites. These steps are critical in translating the precision of 3D printing into functional resilience. Moreover, applying performance enhancement treatments, such as protective surface coatings or targeted chemical modifications, further refine the nanocomposites, tailoring them for specific water treatment capabilities. Such enhancements improve the efficacy of contaminant removal and can impart additional functionalities like antimicrobial action, contributing to the overall performance optimization of the water treatment process [12, 64].

6.6 Functional Testing and Validation

The conclusive development phase for 3D-printed nanocomposites entails thorough testing to authenticate their effectiveness in water treatment scenarios. This rigorous evaluation encompasses assessing their adsorptive strength, their capability to catalyze the degradation of pollutants under light, and their potency against microbial contamination. It is crucial to verify the resilience and endurance of these materials against mechanical stresses and chemical aggressors, replicating controlled lab conditions and variable real-world environments. Such comprehensive validation ensures the nanocomposites meet the stringent demands of water treatment applications, confirming their readiness for deployment in the field [59].

By systematically following these six steps, researchers and engineers can design, fabricate, and implement 3D-printed polymer nanocomposites that are structurally sophisticated and functionally optimized for advanced water treatment solutions. This comprehensive approach ensures the development of materials that effectively address specific water purification challenges, contributing to the sustainability and efficiency of water treatment processes. Table 4 provides a structured overview of the systematic approach taken from conceptualization to validation, ensuring the development of effective and reliable 3D-printed polymer nanocomposites for water treatment applications.

Table 4 Comprehensive stages of design and fabrication techniques for 3D-printed polymer nanocomposites in water treatment (from concept to validation)

Step	Description	Purpose	Impact
Conceptual design and simulation	Utilizing software for the design and fluid dynamics simulation of the nanocomposite structure	To optimize the structure for maximum pollutant interaction and removal efficiency	Ensures the final product is designed for optimal functionality in water treatment applications
Material selection and nanoparticle functionalization	Choosing appropriate polymers and functionalizing nanoparticles to target specific pollutants	To ensure the composite material has the desired properties for pollutant adsorption and degradation	Tailors the nanocomposite's capabilities for specific water treatment needs
Nanocomposite preparation and filament production	Creating a homogeneous mixture of nanoparticles within the polymer matrix and extruding into filaments	To produce consistent quality material for FDM or as a resin for SLA and SLS printing	Allows for the reliable creation of nanocomposite structures with uniform properties
3D printing process	Employing additive manufacturing techniques like FDM, SLA, SLS, or BJ to create the nanocomposite structure	To fabricate the designed structure with precision and accuracy	Results in the production of complex, functional water treatment devices
Post-processing and performance enhancement	Conducting steps such as drying, curing, sintering, and applying surface coatings	To improve mechanical strength, stability, and water treatment functionality	Enhances the efficiency and durability of the nanocomposite in water treatment applications
Functional testing and validation	Rigorous testing for adsorption capacity, photocatalytic activity, antimicrobial efficiency, and durability	To validate the performance of the nanocomposite in simulated and real-world water treatment scenarios	Confirms that the nanocomposite meets the necessary standards for effective water treatment

7 Materials in 3D Printing

Understanding material intricacies and their associated challenges is essential for improving 3D printing processes. Central to this is grasping the binding mechanisms that affect material adhesion during printing [9]. These mechanisms span a wide range, from secondary phase-assisted techniques like adhesive additives and liquid phase sintering to chemically induced methods involving reactive techniques and polymerization. Through low-viscosity flow and melting, processes such as solid-state sintering and liquid fusion also play critical roles. However, while promising, these mechanisms can introduce defects such as balling, porosity, cracks, distortion, and chemical degradation in the printed components. Addressing these material-related issues is essential to ensure the structural integrity and longevity of 3D-printed products.

7.1 Binding Mechanisms

In 3D printing, materials bond and solidification process impact the structural integrity of the final product. These mechanisms range from secondary phase-assisted binding, which uses additional substances to enhance cohesion, to chemically induced binding reliant on reactions or polymerization. Physical alteration methods like liquid fusion and solid-state sintering are also pivotal, offering diverse approaches to material binding. Despite their potential, these mechanisms can introduce flaws such as porosity, balling, cracks, and distortion, necessitating careful management to ensure product quality. Secondary phase-assisted binding employs supportive substances or phases for improved cohesiveness, which is vital for quality in printed structures. This includes sticky additives for adhesion and processes like liquid phase sintering for fusing material particles, scalable and versatile for various 3D printing applications. Chemical binding, where materials bond through chemical reactions, is essential for creating strong, durable connections between materials. This includes reactive binding for initiating chemical reactions during printing and polymerization for converting monomers into polymer chains, offering precision and adaptability in printed structures. Solid-state sintering involves particle coalescence at high temperatures without melting, minimizing grain growth, and preserving material properties. Liquid fusion relies on melting to merge particles, advantageous for creating smooth, unified structures and filling voids for compact, evenly distributed components, highlighting its importance in 3D printing for achieving high-strength and resilient parts.

7.2 Defects in 3D-Printed Materials

Defects in 3D printed materials significantly affect their quality and functionality. These defects include porosity, cracks, distortion, and issues with surface finish, all of which can compromise the mechanical properties and aesthetics of the final product. The balling phenomenon, particularly in SLM processes, leads to irregular sintering and affects the consistency of metal prints. Porosity, categorized into irregular and spherical pores, arises from shrinkage, inadequate binding, or gas entrapment, affecting material integrity. Cracks develop due to thermal stresses, rapid cooling, or compositional issues, especially in materials with low thermal shock resistance. Distortion and delamination occur from uneven cooling and contraction, undermining geometric accuracy and layer adhesion of printed objects. Poor surface finish can result from layering, material aging, or improper deposition sizes, often necessitating post-production treatments to meet quality standards. Chemical degradation and oxidation, influenced by environmental conditions and process parameters, can further degrade material properties. Addressing these defects is crucial for enhancing the reliability and application scope of 3D printing technologies.

8 Applications in Water Treatment

Integrating 3D-printed polymer nanocomposites into water treatment technologies presents a transformative approach to addressing global water purity challenges. These materials are applied within various segments of water treatment, namely filtration systems, adsorption processes, and photocatalytic degradation, elucidating their design, efficiency, and mechanism of action.

8.1 Filtration Systems

Incorporating 3D-printed polymer nanocomposites has significantly enhanced the design and efficiency of filtration systems. These materials, characterized by their high surface area, tailored pore size distribution, and functionalized surfaces, offer superior performance over traditional filtration media. The design process involves computational modeling to optimize the geometry and surface characteristics of the filters, ensuring maximum contaminant removal while minimizing pressure drop across the filter. Efficiency is further improved by incorporating specific nanomaterials within the polymer matrix, which can target and remove specific contaminants, including heavy metals, organic pollutants, and pathogens, from water [51, 61, 77]. Kozior et al. have made significant advancements in creating mechanically stable air and water purification filters by combining electrospinning and 3D printing technologies [52]. Their research addresses the mechanical fragility of electrospun nanofiber mats by 3D printing a rigid scaffold, and then electrospinning the nanofiber mat onto this scaffold. While adhesion between the nanofiber mats and common rigid printing polymers was insufficient, using TPU significantly improved the bond between the two materials. This method not only enhances the durability of the filters but also maintains the beneficial properties of the nanofiber mats, such as the potential for photodegradation and antimicrobial activity. Figure 6 presents a novel approach to producing durable and efficient filtration systems, showcasing the promising integration of electrospinning with 3D-printed structures.

Fijoł et al. developed PLA nanocomposite-based filters which are suitable for wastewater treatment [15]. PLA composites reinforced with 2,2,6,6-tetramethyl-1-piperidinyloxy (TEMPO)-oxidized cellulose nanofibers (TCNF) or chitin nanofibers

Fig. 6 3D-printed polymer nanocomposite-based filter for water treatment. Reproduced from Ref. [52], [https://doi.org/10.3390/polym11122034], under the terms of the CC BY 4.0 license [https://creativecommons.org/licenses/by/4.0/]

(ChNF) were developed through a four-step process, including melt blending, thermally induced phase separation (TIPS) pelletization, freeze-drying, and extrusion into 3D printing filaments. These biocomposite filters, printed in various geometries with multiscale pores, demonstrated superior mechanical strength and water throughput compared to pure PLA filters. The enhanced adsorption capacity of filters was notable for copper ions, with TCNF and ChNF filters achieving up to 234 and 208 mg/gNF, respectively, significantly outperforming the 4 mg/g of pure PLA filters. They also showed improved microplastic removal from laundry effluent, with 54% and 35% separation efficiencies for TCNF/PLA and ChNF/PLA filters, respectively, compared to 26% for PLA filters. This advancement illustrates the potential of combining eco-friendly materials with 3D printing for creating efficient, customizable water filtration systems, particularly beneficial in regions lacking access to clean water.

Ogazi and Osifo have contributed significantly to developing water filtration technologies by studying the efficacy of 3D-printed chitosan (CS) membranes modified with GO and AgNPs [62]. They primarily focused on characterizing CS-GO-AgNP composite membranes and assessing the influence of dimethylacetamide (DMAc), used as a co-solvent, on the performance of the nanocomposite. Increasing DMAc concentration while reducing the CS mole fraction in the ink formulation led to a broader distribution of AgNPs across the membrane surface. This modification enhanced the mechanical strength and surface hydrophilicity of the CS membranes and significantly improved their antibacterial efficiency. The augmented GO-AgNPs concentration within the CS matrix markedly inhibited the proliferation of fecal coliforms, total coliforms, and *E. coli* in contaminated surface water. Specifically, the sample labeled B-12, which comprised 79% CS, 21% DMAc, and 1.2 mL of GO-AgNP, showed exceptional performance by inhibiting over 95% of fecal and total coliforms and completely inactivating 99.9% of *E. coli* cells. This demonstrates the potential of 3D-printed modified CS membranes in effectively treating water, highlighting a significant advancement in nanocomposite materials for microbial contamination control. A foundation for future research on scalable and efficient water treatment methods has thus been established, leveraging the antimicrobial properties of nanomaterials integrated into 3D printed structures. Table 5 represents the multifaceted benefits and design considerations of 3D-printed polymer nanocomposites in water treatment applications. These filters offer a tailored approach to addressing a wide array of water pollutants, making them a promising technology for future advancements in water purification.

8.2 Adsorption Processes

3D-printed polymer nanocomposites are effective adsorbents for removing a wide range of contaminants from water in adsorption processes. The versatility of 3D printing technology allows for fabricating adsorbents with complex shapes and structures that maximize contact with water contaminants, enhancing adsorption efficiency. These nanocomposites can be engineered to include functional groups that specifically bind to certain pollutants, including pharmaceuticals, dyes, and heavy metals, facilitating their removal from water [14, 53]. The reusability and ease of regeneration of these adsorbents make them economically attractive for water treatment applications.

Masud et al. have innovated water purification technology by creating graphene-biopolymer aerogels through a direct ink writing (DIW) 3D printing and freeze-drying process [58]. They incorporated polydopamine (PDA) and bovine serum albumin (BSA) into the graphene ink to enhance printability and improve the contaminant removal efficiency of the material. This method significantly advanced the scalability and application potential of graphene aerogels in removing a wide range of contaminants, including heavy metals, organic dyes, and solvents, from water. The aerogels demonstrated exceptional removal performance of contaminants and notable regeneration capability and were effectively applied as bottle-cap filters for dye removal. Masud et al. have emphasized the versatility and efficiency of 3D-printed graphene-biopolymer aerogels in water treatment, which is a significant step towards scalable and practical purification technologies.

Pei et al. have advanced the field of water purification by integrating metal–organic frameworks (MOFs) with biocompatible polymers through 3D printing, creating highly effective adsorbents for organic dye removal [66]. Their innovative approach, employing direct ink writing, mitigates the limitations associated with powdery MOF materials, such as instability and high-pressure drop, and enhances mechanical properties via calcium alginate and gelatin (CA–GE) blend. The study highlights the

Table 5 Overview of 3D printed polymer nanocomposites-based filters for water treatment: design features, functions, target pollutants, and key advantages

Feature	Design aspects	Function	Target pollutants	Advantages
Filtration efficiency	Customizable pore size and distribution through 3D printing	Physical removal of contaminants	Particulates, bacteria, microorganisms	High precision in contaminant removal, tailored to specific water quality needs
Surface functionalization	Integration of functional materials (e.g., activated carbon, TiO_2) during the printing process	Enhanced adsorption and photocatalytic degradation	Heavy metals, organic compounds, dyes	Increased removal efficiency for a broad range of pollutants through combined physical and chemical processes
Mechanical stability	Reinforced structures using polymer nanocomposites	Withstands high pressure and flow rates	N/A	Durable and suitable for continuous operation in harsh environmental conditions
Antimicrobial activity	Incorporation of antimicrobial nanoparticles (e.g., AgNPs)	Prevents biofilm formation and microbial contamination	Bacteria, viruses, and other pathogens	Ensures the production of microbiologically safe water, minimizing health risks
Photocatalytic properties	Embedding photocatalysts (e.g., ZnO, TiO_2 nanoparticles) within the matrix	Sunlight-driven degradation of pollutants	Organic pollutants, pharmaceuticals, pesticides	Enables self-cleaning capabilities and degradation of complex contaminants without additional energy or chemical inputs
Regeneration and reusability	Design for easy cleaning and replacement of filter segments	Extendable lifecycle of the filter	N/A	Cost-effective and environmentally friendly by reducing the need for frequent filter replacements
Customizability	Modular design enabling scalability and adaptability to various water treatment systems	Adaptable to different scales and needs of treatment	N/A	Flexibility in application, from small-scale point-of-use systems to large-scale water treatment facilities

significant impact of printing geometry, MOF loading, and MOF size on adsorption efficiency, with the MOF/CA–GE composite achieving up to 99.8% removal of methylene blue within 20 min. This performance is attributed to the synergistic effect of the microporous structure of the MOF and the meso/macroporous configuration of the 3D matrix. The printed adsorbents demonstrated superior reusability, withstanding at least 7 regeneration cycles without efficiency loss, vastly outperforming traditional powdery MOFs. Pei et al. have developed more stable, efficient, and reusable adsorbents for water treatment applications.

Ivanov et al. explored the electrical and thermal properties of PLA composites enhanced with graphene nanoplates (GNP) and multi-walled carbon nanotubes (MWCNT) for potential applications in FDM 3D printing, focusing on water treatment [44]. These composites exhibited improved filler dispersion, resulting in higher electrical and thermal conductivity. This enhanced performance, along with uniform nanoparticle distribution, suggests suitability for developing multifunctional filtration devices for water treatment. The study underscores the potential of the composites for efficient pollutant removal, making them valuable for sustainable water purification systems. Achieving a homogeneous dispersion of MWCNT and GNP in polymer melts is crucial for creating enhanced filler-based filaments suitable for fused deposition modeling (FDM). Heterogeneous dispersions can cause nozzle blockages and flux instability during printing. Current research aims to determine the optimal concentration of carbon nanofillers that exceed the percolation threshold while maintaining printability. SEM images in Fig. 7 demonstrate the distribution of carbon nanotubes and GNPs in PLA composites, while TEM images in Fig. 8 reveal the formation of an interconnected network in bi-filler composites, indicating a synergistic improvement in electrical and mechanical properties.

Frimodig and Haukka have explored the potential of 3D-printed polymers in removing estrogens, a group of endocrine-disrupting chemicals (EDCs), from aqueous solutions [17]. Their research demonstrates that filters made from PA, TPU, polypropylene (PP), and polystyrene (PS) can effectively adsorb estrogen hormones, with PA and TPU, polymers rich in functional groups, showing superior adsorption efficiency compared to the simpler PP and PS. The kinetic models developed for polyamide filters further underscore their effectiveness in capturing estrone, 17β-estradiol, and 17α-ethinylestradiol, highlighting the critical role of the powder-bed fusion 3D printing technique in fabricating filters with optimal porosity and flow characteristics. This contributes significantly to developing new filtration technologies using 3D printing to mitigate the environmental impact of EDCs. Table 6 highlights the versatility and effectiveness of 3D-printed polymer nanocomposites in adsorption processes for water treatment. By leveraging the advanced capabilities of 3D printing, these materials can be precisely engineered to enhance adsorption capacity, specificity, and durability, offering innovative solutions for improving water quality.

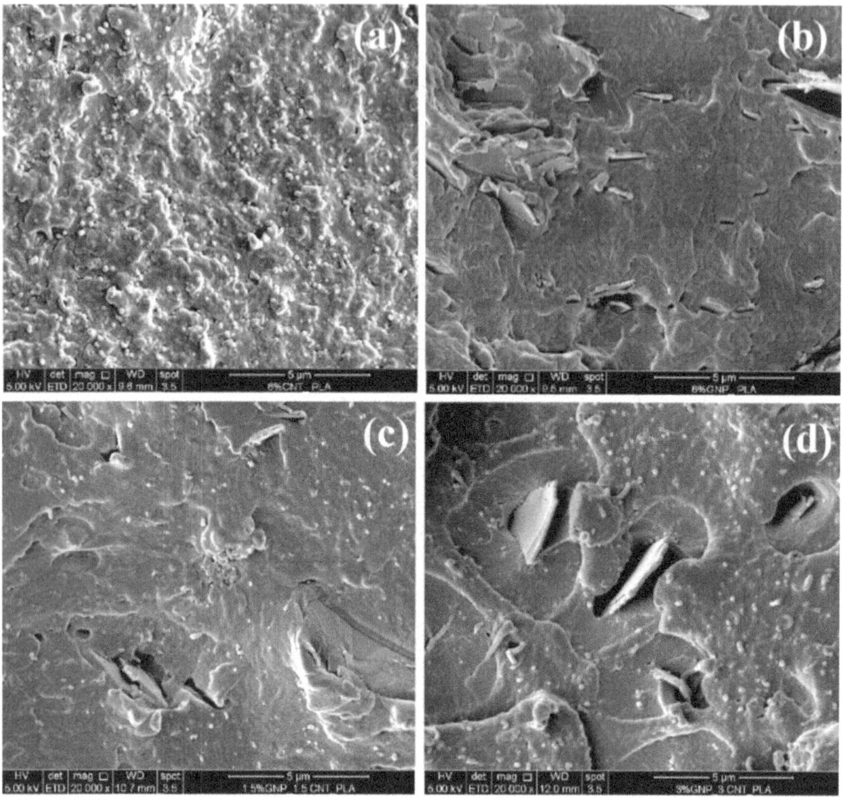

Fig. 7 SEM images of cryo-fractured surfaces of mono-filler composites with **a** 6 wt% multi-walled carbon nanotubes (MWCNT) and **b** 6 wt% graphene nanoplates (GNP), as well as bi-filler composites containing **c** 1.5% GNP/1.5% MWCNT and **d** 3% GNP/3% MWCNT, respectively. All images are magnified at 20,000×. Reproduced from Ref. [44], [https://doi.org/10.3390/app9061209], under the terms of the CC BY 4.0 license [https://creativecommons.org/licenses/by/4.0/]

8.3 *Photocatalytic Degradation*

The application of 3D printed polymer nanocomposites in photocatalytic degradation leverages the power of sunlight to drive the degradation of pollutants such as organic pollutants [54, 57, 74], drug residue [6, 69], algal toxins [48], etc. in water. These materials are engineered to include photocatalytic nanoparticles, such as TiO_2 or ZnO, within the polymer matrix. Upon exposure to sunlight, these nanoparticles generate reactive oxygen species (ROS) that can oxidize and break down a wide range of organic pollutants, including pesticides, pharmaceuticals, and endocrine-disrupting chemicals, into less harmful substances. The design of these photocatalytic systems focuses on maximizing the surface area exposed to sunlight and ensuring efficient separation and recovery of the photocatalyst after treatment.

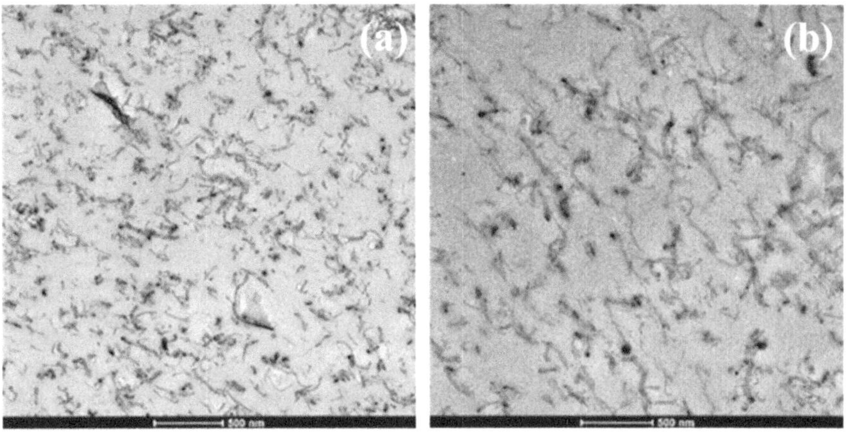

Fig. 8 TEM images of bi-filler composites containing **a** polylactic acid (PLA) with 1.5 wt% graphene nanoplates (GNP) and 4.5 wt% MWCNT, and **b** PLA with 3 wt% GNP and 3 wt% MWCNT, respectively. Reproduced from Ref. [44], [https://doi.org/10.3390/app9061209], under the terms of the CC BY 4.0 license [https://creativecommons.org/licenses/by/4.0/]

Li et al. have innovated in the photocatalytic degradation of organic pollutants by ZnO and TiO_2 and iron-based MOF (Fe-MOF) nanoparticles with 3D-printed fractal polymer surfaces [54]. Through a novel application of core–shell synthesis, 3D printing, and plasma grafting, they successfully attached these nanoparticles onto the substrates, enhancing the degradation of pollutants like Rhodamine B dye and Ciprofloxacin antibiotic. Using poly(vinylphosphonic acid) as an intermediary layer significantly improves the grafting process, suggesting a versatile method for functionalizing polymeric surfaces with inorganic compounds. Pre-treating metal oxide nanoparticles with organic coupling agents before plasma grafting can increase photodegradation efficiency by up to 20%. The feasibility of employing SLA for creating high-surface-area supports that facilitate the immobilization of catalysts underscores the potential of integrating advanced manufacturing techniques with photocatalysis.

Bergamonti and coworkers have developed an innovative approach for wastewater treatment through the creation of TiO_2-supported chitosan scaffolds (TiO_2/CS), utilizing 3D printing technology for the photocatalytic degradation of amoxicillin (AMX) under UV/Vis irradiation [6]. TiO_2/CS showed necessary potential and promises as a material for removing pharmaceutical compounds from wastewater. By incorporating commercial P25-TiO_2 into chitosan at specified concentrations and ensuring the homogeneous distribution of TiO_2 particles within the scaffold, they have demonstrated that these 3D printed scaffolds not only retain their photocatalytic activity after repeated use but also possess the mechanical durability required for practical applications. The efficiency of AMX degradation was systematically evaluated, showing high photodegradation rates, which surpass those of direct photolysis. The photocatalytic degradation follows a zero-order kinetic model, differing from

Table 6 Comprehensive overview of 3D printed polymer nanocomposites in water treatment through adsorption processes, highlighting design aspects, functions, targeted pollutants, and advantages

Feature	Design aspects	Function	Target pollutants	Advantages
Adsorption capacity	High surface area and porosity tailored through 3D printing techniques	Adsorption of contaminants	Heavy metals, organic pollutants, dyes, pharmaceuticals	Enhanced efficiency in removing various pollutants due to optimized surface characteristics
Surface functionalization	Incorporation of selective adsorption sites (e.g., ion exchange materials, activated carbon)	Targeted contaminant removal	Specific pollutants based on functionalization	High specificity and selectivity towards targeted contaminants, improving water quality significantly
Reusability and regeneration	Design for easy desorption of adsorbed materials and regeneration of the adsorbent	Sustainable use of filter material	N/A	Cost-effective and environmentally friendly, reducing the need for frequent replacements and waste generation
Mechanical stability	Reinforced polymer matrix integrating nanofillers or fibers to withstand operational conditions	Durable under various water flows and pressures	N/A	Ensures long-term operation without degradation, maintaining consistent adsorption performance
Modularity and scalability	Modular design allowing for scalability and customization to fit specific treatment systems	Adaptable to different treatment capacities	N/A	Flexibility to cater to both small-scale household and large-scale industrial water treatment needs

(continued)

Table 6 (continued)

Feature	Design aspects	Function	Target pollutants	Advantages
Chemical stability	Selection of chemically resistant polymers and composites to prevent leaching and degradation	Safe for long-term water treatment	N/A	Ensures that the filter does not introduce new contaminants into the water, preserving water quality
Integration with other processes	Compatibility with additional treatment stages (e.g., UV degradation, biological treatment) for comprehensive purification	Enhanced overall water purification	Multiple types of pollutants in a sequential treatment setup	Offers a holistic approach to water treatment by efficiently removing a wide range of pollutants through combined methodologies

the pseudo-first-order kinetics observed with water-dispersed TiO_2. The breakdown of AMX results in environmentally less harmful compounds. This has been a significant advancement in photocatalytic water treatment, offering a novel, efficient, and reusable solution for mitigating antibiotic pollution.

Kennedy et al. have developed a groundbreaking approach for the degradation of microcystin, a harmful algal bloom toxin, utilizing 3D-printed polymer composites embedded with photocatalytic TiO_2 [48]. By integrating biocompatible PLA with TiO_2 at various loadings, they have created customizable, high-surface area structures capable of specific kinetic reactions against algal toxins. Their research demonstrates that these PLA-TiO_2 composites, particularly those with a high TiO_2 content (34% w/w), are effective in photocatalytically reducing microcystin under UV and solar-simulated light, achieving faster degradation rates than photolysis alone. The innovative use of 3D printing allows for the design of structures like lattices that are deployable and retrievable in natural water bodies and overcome the challenges of poor recoverability of TiO_2. A significant foundation for the optimization of photocatalytic composites for water treatment was provided, marking a notable advancement in the application of 3D printing and nanotechnology in environmental remediation. Table 7 showcases how 3D-printed polymer nanocomposites can be effectively used in photocatalytic degradation processes for water treatment, emphasizing their design features, functionality, target pollutants, and unique advantages. By capitalizing on the capabilities of 3D printing technology, these materials present innovative solutions for efficient and sustainable water purification.

3D-printed polymer nanocomposites represent a promising class of materials for water treatment, offering customizable solutions to a wide range of contamination challenges. These materials provide effective, efficient, and economically viable options for improving water quality worldwide through the intelligent design of filtration systems, adsorption processes, and photocatalytic degradation mechanisms.

Table 7 Comprehensive applications of 3D printed polymer nanocomposites in water treatment through photocatalytic degradation—design features, functions, target pollutants, and advantages

Feature	Design aspects	Function	Target pollutants	Advantages
Photocatalytic efficiency	Integration of photocatalytic nanoparticles (e.g., TiO_2, ZnO) within the polymer matrix	Degradation of organic pollutants under sunlight	Dyes, pharmaceuticals, pesticides, organic pollutants	High degradation rates of complex organic compounds, transforming them into harmless substances with minimal secondary waste
Surface area optimization	High surface-to-volume ratio structures designed for maximum exposure to light	Maximizes the photocatalytic reaction sites	N/A	Increased reaction efficiency due to optimized nanocomposite structures that enhance light absorption and pollutant interaction
Durability in water treatment	Material stability under UV exposure and in aqueous environments	Long-term operational stability	N/A	Enhanced material longevity and consistent performance, reducing the need for frequent replacement or maintenance
Reusability	Easy recovery and reuse of photocatalytic materials without significant loss of activity	Cost-effective and environmentally friendly purification	N/A	Decreases operational costs and environmental impact by minimizing the need for new materials

(continued)

Table 7 (continued)

Feature	Design aspects	Function	Target pollutants	Advantages
Selective contaminant targeting	Customizable nanocomposite formulations for specific pollutant degradation	Tailored degradation pathways	Specific organic pollutants tailored by photocatalyst	Allows for the targeted removal of particular pollutants, enhancing water treatment efficiency for specific applications
Integration with other processes	Compatibility with pre- and post-treatment processes (e.g., filtration, adsorption) for comprehensive treatment	Enhanced overall water quality	Broad range of pollutants in a multi-step treatment system	Offers a holistic water treatment solution by combining photocatalytic degradation with other purification techniques for thorough pollutant removal

Their adaptability and performance in various water treatment applications highlight the potential of combining nanotechnology with 3D printing to address the pressing need for clean and safe water.

9 Challenges and Future Perspectives

Integrating 3D-printed nanocomposites into water treatment processes represents a significant step forward in addressing global water purity challenges. However, this promising technology also faces several hurdles that must be overcome to realize its full potential. The challenges are associated with scalability and cost-effectiveness to assess the environmental impact and sustainability of these materials and explore future directions for their application in water treatment.

9.1 Scalability and Cost-Effectiveness

Scaling up the production and use of 3D printed nanocomposites presents a complex challenge, primarily due to the intricate processes involved in nanocomposite fabrication and the cost implications of 3D printing technologies. Currently, the high cost of raw materials, especially high-quality nanofillers, and the energy-intensive nature of 3D printing processes limit the widespread adoption of this technology. While beneficial for customization, the bespoke nature of 3D printed objects poses difficulties in mass production, as the printing time and costs increase significantly with the complexity and size of the printed structures.

To address these challenges, future research and development efforts must focus on optimizing the production processes to reduce material wastage and energy consumption. Advancements in 3D printing technologies that allow for faster printing speeds and more cost-effective yet high-performance materials will be crucial. Moreover, developing standardized design templates for typical water treatment applications could streamline production, reducing costs and facilitating scalability.

9.2 Environmental Impact and Sustainability

The lifecycle and environmental footprint of nanocomposite materials are critical considerations, especially given the growing emphasis on sustainable development. The production of nanocomposites involves synthesizing and manipulating nanoparticles, which can pose environmental and health risks if not properly managed. Furthermore, the end-of-life disposal of these materials, particularly the non-biodegradable components, presents significant environmental challenges. Efforts to minimize the environmental impact of 3D printed nanocomposites should focus on the entire lifecycle, from raw material extraction and processing to end-of-life disposal or recycling. The development of biodegradable nanocomposites, coupled with green synthesis methods for nanoparticles, could significantly reduce the environmental footprint of these materials. Additionally, establishing recycling protocols for nanocomposite filters and exploring the possibility of repurposing or composting biodegradable components could enhance the sustainability of this technology.

9.3 Challenges and Author Viewpoints

Integrating 3D-printed polymer nanocomposites into water treatment systems presents several challenges, notably scalability, cost-effectiveness, and environmental impact. While these challenges are significant, they are surmountable with continued interdisciplinary research and innovation. Optimizing production processes and advancing 3D printing technologies are crucial for reducing material wastage and energy consumption, thus making production more cost-effective. Developing standardized design templates for water treatment applications could streamline production, facilitating scalability and reducing costs.

In terms of environmental sustainability, the development of biodegradable nanocomposites and green synthesis methods for nanoparticles is advocated. Establishing recycling protocols and exploring the repurposing or composting of biodegradable components could significantly enhance the sustainability of these technologies. There is also potential for incorporating smart materials and sensors into nanocomposite structures, leading to adaptive filters that respond to changing water quality conditions or detect specific contaminants in real time. Additionally, decentralized water treatment systems using 3D-printed nanocomposites, especially in remote or underdeveloped regions, show promise. Such systems could provide sustainable solutions to the global water crisis by enabling local production and maintenance of water treatment devices tailored to specific community needs. Overall, addressing these challenges requires a holistic approach that combines technological innovation with sustainable practices and supportive policy frameworks. This integrated strategy will be crucial in realizing the full potential of 3D-printed polymer nanocomposites for water treatment applications.

9.4 Future Directions

Emerging trends in water treatment suggest a bright future for 3D-printed nanocomposites, with potential applications extending beyond current capabilities. The integration of smart materials and sensors into nanocomposite structures could lead to the development of filters that can adapt dynamically to changing water quality conditions or detect specific contaminants in real time. In addition, advances in nanotechnology could enable the design of nanocomposites with enhanced photocatalytic, adsorptive, and antimicrobial properties, further improving the efficiency and range of applications in water treatment. Exploring the use of 3D-printed nanocomposites in decentralized water treatment systems, particularly in remote or underdeveloped regions, presents an exciting opportunity. Such systems could provide a sustainable solution to the global water crisis by enabling local production and maintenance of water treatment devices tailored to the specific needs of each community. While challenges in scalability, cost-effectiveness, and environmental impact must be addressed, the future of 3D-printed nanocomposites in water treatment looks promising. Ongoing research and innovation in materials science, 3D printing technology, and environmental engineering will be key to unlocking the full potential of this technology for sustainable water purification solutions worldwide.

10 Conclusion

The exploration of 3D-printed polymer nanocomposites for water treatment underscores a significant stride towards innovative and sustainable solutions. The versatility of 3D printing technologies and the advanced properties of polymer nanocomposites offer promising avenues for addressing global water purification needs. A comprehensive understanding emerges through a detailed analysis of various 3D printing technologies, materials, and their applications in water treatment. Challenges such as scalability, cost-effectiveness, and environmental sustainability highlight the need for ongoing research and development. However, the potential to revolutionize water treatment processes through custom-designed, efficient, and environmentally friendly systems is profound. Future directions point towards harnessing emerging trends and optimizing the use of nanocomposites in water treatment, aiming for broader application scopes and minimizing the environmental footprint. The integration of smart materials and the development of biodegradable composites further exemplify the dynamic nature of this field. The convergence of nanotechnology, 3D printing, and environmental science holds the promise of developing next-generation water treatment solutions that are both effective and sustainable, marking a pivotal shift in how we approach water purification challenges.

Acknowledgements The authors acknowledge the Ministry of Science and Technology for special allocation for a Project related to the work. The authors also acknowledge the Bose Centre for Advanced Research in Natural Science for financial support.

Conflicts of Interest There are no conflicts to declare.

References

1. Abbasi Geravand, M.H., Saljoughi, E., Mousavi, S.M., Kiani, S.: Biodegradable polycaprolactone/MXene nanocomposite nanofiltration membranes for the treatment of dye solutions. J. Taiwan Inst. Chem. Eng. **128**, 124–139 (2021). https://doi.org/10.1016/j.jtice.2021.08.048
2. Al-Ghafri, B., Lau, W.-J., Al-Abri, M., Goh, P.-S., Ismail, A.F.: Titanium dioxide-modified polyetherimide nanofiber membrane for water treatment. J. Water Process Eng. **32**, 100970 (2019). https://doi.org/10.1016/j.jwpe.2019.100970
3. Allami, T., Alamiery, A., Nassir, M.H., Kadhum, A.H.: Investigating physio-thermo-mechanical properties of polyurethane and thermoplastics nanocomposite in various applications. Polymers **13**(15) (2021). https://doi.org/10.3390/polym13152467
4. Baigorria, E., Souza dos Santos, S., de Moura, M.R., Fraceto, L.F.: Nanocomposite hydrogels 3D printed for application in water remediation. Mater. Today Chem. **30**, 101559 (2023). https://doi.org/10.1016/j.mtchem.2023.101559
5. Bayraktar, I., Doganay, D., Coskun, S., Kaynak, C., Akca, G., Unalan, H.E.: 3D printed antibacterial silver nanowire/polylactide nanocomposites. Compos. B Eng. **172**, 671–678 (2019). https://doi.org/10.1016/j.compositesb.2019.05.059
6. Bergamonti, L., Bergonzi, C., Graiff, C., Lottici, P.P., Bettini, R., Elviri, L.: 3D printed chitosan scaffolds: a new TiO_2 support for the photocatalytic degradation of amoxicillin in water. Water Res. **163**, 114841 (2019). https://doi.org/10.1016/j.watres.2019.07.008
7. Bhatt, P.M., Kabir, A.M., Peralta, M., Bruck, H.A., Gupta, S.K.: A robotic cell for performing sheet lamination-based additive manufacturing. Addit. Manuf. **27**, 278–289 (2019). https://doi.org/10.1016/j.addma.2019.02.002
8. Bortoli, L.S.D., Farias, R.D., Mezalira, D.Z., Schabbach, L.M., Fredel, M.C.: Functionalized carbon nanotubes for 3D-printed PLA-nanocomposites: effects on thermal and mechanical properties. Mater. Today Commun. **31**, 103402 (2022). https://doi.org/10.1016/j.mtcomm.2022.103402
9. Bourell, D., Kruth, J.P., Leu, M., Levy, G., Rosen, D., Beese, A.M., Clare, A.: Materials for additive manufacturing. CIRP Ann. **66**(2), 659–681 (2017). https://doi.org/10.1016/j.cirp.2017.05.009

10. Cairone, S., Hegab, H.M., Khalil, H., Nassar, L., Wadi, V.S., Naddeo, V., Hasan, S.W.: Novel eco-friendly polylactic acid nanocomposite integrated membrane system for sustainable wastewater treatment: performance evaluation and antifouling analysis. Sci. Total Environ. **912**, 168715 (2024). https://doi.org/10.1016/j.scitotenv.2023.168715
11. Dharma, H.N., Jaafar, J., Widiastuti, N., Matsuyama, H., Rajabsadeh, S., Othman, M.H. et al.: A review of titanium dioxide (TiO_2)-based photocatalyst for oilfield-produced water treatment. Membranes **12**(3) (2022). https://doi.org/10.3390/membranes12030345
12. Dizon, J.R.C., Gache, C.C.L., Cascolan, H.M.S., Cancino, L.T., Advincula, R.C.: Post-processing of 3D-printed polymers. Technologies **9**(3) (2021). https://doi.org/10.3390/technologies9030061
13. El-Sayyad, G.S., Abdel Maksoud, M.I.A., Fahim, R.A., Bekhit, M., Kassem, S.M., Awed, A.S., et al.: Gamma radiation induced synthesis of novel chitosan/gold/bioactive glass nanocomposite for promising antimicrobial, and antibiofilm activities. J. Cluster Sci. **34**(4), 1877–1891 (2023). https://doi.org/10.1007/s10876-022-02357-9
14. Fijoł, N., Aguilar-Sánchez, A., Mathew, A.P.: 3D-printable biopolymer-based materials for water treatment: a review. Chem. Eng. J. **430**, 132964 (2022). https://doi.org/10.1016/j.cej.2021.132964
15. Fijoł, N., Aguilar-Sánchez, A., Ruiz-Caldas, M.-X., Redlinger-Pohn, J., Mautner, A., Mathew, A.P.: 3D printed polylactic acid (PLA) filters reinforced with polysaccharide nanofibers for metal ions capture and microplastics separation from water. Chem. Eng. J. **457**, 141153 (2023). https://doi.org/10.1016/j.cej.2022.141153
16. Fijoł, N., Mathew, A.P.: Accelerated ageing of 3D printed water purification filters based on PLA reinforced with green nanofibers. Polym. Test. **129**, 108270 (2023). https://doi.org/10.1016/j.polymertesting.2023.108270
17. Frimodig, J., Haukka, M.: Removal of estrogens from aqueous solutions using 3D-printed polymers. Environ. Sci.: Adv. **2**(12), 1739–1745 (2023). https://doi.org/10.1039/D3VA00159H
18. Han, L., Shen, L., Lin, H., Huang, Z., Xu, Y., Li, R., et al.: 3D printing titanium dioxide-acrylonitrile-butadiene-styrene (TiO_2-ABS) composite membrane for efficient oil/water separation. Chemosphere **315**, 137791 (2023). https://doi.org/10.1016/j.chemosphere.2023.137791
19. Harun-Ur-Rashid, M., Foyez, T., Imran, A.B.: Fabrication of stretchable composite thin film for superconductor applications. In: Pal, K. (ed.) Sensors for Stretchable Electronics in Nanotechnology, 1st edn., pp. 63–78. CRC Press, Boca Raton (2021)
20. Harun-Ur-Rashid, M., Foyez, T., Jahan, I., Pal, K., Imran, A.B.: Rapid diagnosis of COVID-19 via nano-biosensor-implemented biomedical utilization: a systematic review. RSC Adv. **12**(15), 9445–9465 (2022). https://doi.org/10.1039/D2RA01293F
21. Harun-Ur-Rashid, M., Imran, A.B.: Superabsorbent hydrogels from carboxymethyl cellulose. In Mondal, M.I.H. (ed.) Carboxymethyl Cellulose, vol. 1, pp. 159. Nova Science Publisher, Inc., New York, USA (2019)
22. Harun-Ur-Rashid, M., Imran, A.B.: Organometallic nanomaterials synthesis and sustainable green nanotechnology applications. In: Pal, K. (ed.) Green Nanoarchitectonics, 1st edn., pp. 249–270. Jenny Stanford Publishing, New York (2022)
23. Harun-Ur-Rashid, M., Imran, A.B.: Engineered nanomaterials for energy conversion cells. In: Singh, N.B., Susan, M.A.B.H., Chaudhary, R.G. (eds.) Applications of Emerging Nanomaterials and Nanotechnology, vol. Materials Research Foundations, pp. 103–126. Materials Research Forum LLC, Millersville, USA (2023a)
24. Harun-Ur-Rashid, M., Imran, A.B.: Nanomaterials in the automobile sector. In: Singh, N.B., Susan, M.A.B.H., Chaudhary, R.G. (eds.) Emerging Applications of Nanomaterials, vol. 141, pp. 124–150. Materials Research Foundations, Millersville, USA (2023b)
25. Harun-Ur-Rashid, M., Imran, A.B., Seki, T., Takeoka, Y., Ishii, M., Nakamura, H.: Template synthesis for stimuli-responsive angle independent structural colored smart materials. Trans. Mater. Res. Soc. Jpn. **34**(2), 333–337 (2009a). https://doi.org/10.14723/tmrsj.34.333
26. Harun-Ur-Rashid, M., Imran, A.B., Susan, M.: Prospective nanomaterials for food packaging and safety. In: Singh, N.B., Susan, M.A.B.H., Chaudhary, R.G. (eds.) Emerging Applications of Nanomaterials, Materials Research Forum LLC, vol. 141, pp. 327–352. Materials Research Foundations, USA (2023)

27. Harun-Ur-Rashid, M., Imran, A.B., Susan, M., Hasan, A.B.: Green polymer nanocomposites in automotive and packaging industries. Curr. Pharm. Biotechnol. **24**(1), 145–163 (2023). https://doi.org/10.2174/1389201023666220506111027
28. Harun-Ur-Rashid, M., Imran, A.B., Susan, M.A.B.H.: Fire-resistant polymeric foams and their applications. In: Gupta, R.K. (ed.) Polymeric Foams: Applications of Polymeric Foams, vol. ACS Symposium Series, pp. pp 97–121. ACS Publications, USA (2023)
29. Harun-Ur-Rashid, M., Imran, A.B., Susan, M.A.B.H.: Emerging 3D printed polymers and composites for water quality preservation. In: Reference Module in Materials Science and Materials Engineering. Elsevier (2024a)
30. Harun-Ur-Rashid, M., Imran, A.B., Susan, M.A.B.H.: Magnetic polymer nanocomposites for removal of dyes and metals from wastewater. In: Reference Module in Materials Science and Materials Engineering. Elsevier (2024b)
31. Harun-Ur-Rashid, M., Jahan, I., Foyez, T., Imran, A.B.: Bio-inspired nanomaterials for micro/nanodevices: a new era in biomedical applications. Micromachines **14**(9) (2023). https://doi.org/10.3390/mi14091786
32. Harun-Ur-Rashid, M., Jahan, I., Imran, A.B.: Recent advancement of nanostructured materials for clinical challenges in vaccinology. In: Pal, K. (ed.) Nanovaccinology: Clinical Application of Nanostructured Materials Research to Translational Medicine, pp. 135–160. Springer International Publishing, Cham (2023)
33. Harun-Ur-Rashid, M., Jahan, I., Islam, M.J., Kumer, A., Huda, M.N., Imran, A.B., et al.: Global advances and smart innovations in supramolecular polymers. J. Mol. Struct. **1304**, 137665 (2024). https://doi.org/10.1016/j.molstruc.2024.137665
34. Harun-Ur-Rashid, M., Naidu Krishna, S.B., Golla, N., Bin Imran, A.: Polyphenol oxidases: the future toward global sustainability. In: Verma, P., Agrawal, K., Shah, M.P. (eds.) Function, Wastewater Remediation, and Biosensors, pp. 309–344. De Gruyter (2024)
35. Harun-Ur-Rashid, M., Pal, K., Imran, A.B.: Hybrid nanocomposite fabrication of nanocatalyst with enhanced and stable photocatalytic activity. Top. Catal. (2023). https://doi.org/10.1007/s11244-023-01809-4
36. Harun-Ur-Rashid, M., Seki, T., Takeoka, Y.: Structural colored gels for tunable soft photonic crystals. Chem. Rec. **9**(2), 87–105 (2009). https://doi.org/10.1002/tcr.20169
37. Harun-Ur-Rashid, M., Bin Imran, A., Seki, T., Ishii, M., Nakamura, H., Takeoka, Y.: Angle-independent structural color in colloidal amorphous arrays. ChemPhysChem **11**(3), 579–583 (2010). https://doi.org/10.1002/cphc.200900869
38. Hassan, M.K., Karim, M.T., Biswas, P., Howlader, D., Harun-Ur-Rashid, M., Kumer, A.: Computational investigation for tetragonal crystals of $Zn(GaS_2)_2$, $Zn(GaSe_2)_2$, and $Zn(GaTe_2)_2$ photocatalysts for wastewater treatment: first principle approaches. Adv. J. Chem.-Sect. B: Nat. Prod. Med. Chem. **6**(1), 46–66 (2024). https://doi.org/10.48309/ajcb.2024.421434.1204
39. Huda, M.N., Seki, T., Suzuki, H., ANM, H.K., Harunur-Rashid, M., Takeoka, Y.: Characteristics of high-density poly(N-isopropylacrylamide) (PNIPA) brushes on silicon surface by atom transfer radical polymerization. Trans. Mater. Res. Soc. Jpn. **35**(4), 845–848 (2010). https://doi.org/10.14723/tmrsj.35.845
40. Huda, N., Rashid, M., Harun-Ur-Rashid, M.: A review on stimuli-responsive grafted membranes based on facile synthesis process and extensive applications. Int. J. Innov. Appl. Stud. **8**(3), 1296–1312 (2014)
41. Imran, A.B., Harun-Ur-Rashid, M., Takeoka, Y.: Polyrotaxane actuators. In: Asaka, K., Okuzaki, H. (eds.) Soft Actuators: Materials, Modeling, Applications, and Future Perspectives, 2nd edn., pp. 81–147. Springer Nature, Singapore (2019)
42. Islam, M.J., Pal, K., Abdullah, M.M., Harun-Ur-Rashid, M.: Threshold bandgap of lead halide hybrid perovskites for making stable single crystal by NIR laser trapping. J. Mol. Struct. **1301**, 137322 (2024). https://doi.org/10.1016/j.molstruc.2023.137322
43. Islam, M.J., Pal, K., Harun-Ur-Rashid, M., Kumar, A., Hossain, M.F., Asthana, N., et al.: Optical maneuvering of photofunctioning hybrid perovskite for future photonics potential application. J. Mol. Liq. **408**, 125343 (2024). https://doi.org/10.1016/j.molliq.2024.125343

44. Ivanov, E., Kotsilkova, R., Xia, H., Chen, Y., Donato, R.K., Donato, K. et al.: PLA/graphene/MWCNT composites with improved electrical and thermal properties suitable for FDM 3D printing applications. Appl. Sci. **9**(6) (2019). https://doi.org/10.3390/app9061209
45. Izdebska-Podsiadły, J.: Chapter 3—classification of 3D printing methods. In: Izdebska-Podsiadły, J. (ed.) Polymers for 3D Printing, pp. 23–34. William Andrew Publishing (2022)
46. Karim, M.R., Harun-Ur-Rashid, M., Imran, A.B.: Effect of sizes of vinyl modified narrow-dispersed silica cross-linker on the mechanical properties of acrylamide based hydrogel. Sci. Rep. **13**(1), 5089 (2023). https://doi.org/10.1038/s41598-023-32185-4
47. Karim, M.R., Harun-Ur-Rashid, M., Imran, A.B.: Highly stretchable hydrogel using vinyl modified narrow dispersed silica particles as cross-linker. ChemistrySelect **5**(34), 10556–10561 (2020). https://doi.org/10.1002/slct.202003044
48. Kennedy, A.J., McQueen, A.D., Ballentine, M.L., May, L.R., Fernando, B.M., Das, A., et al.: Degradation of microcystin algal toxin by 3D printable polymer immobilized photocatalytic TiO_2. Chem. Eng. J. **455**, 140866 (2023). https://doi.org/10.1016/j.cej.2022.140866
49. Khair, A., Dey, N.K., Harun-Ur-Rashid, M., Alim, M.A., Bahadur, N.M., Mahamud, S., Ahmed, S.: Diffusimetry renounces Graham's law, achieves diffusive convection, concentration gradient induced diffusion, heat and mass transfer. Defect Diffus. Forum **407**, 173–184 (2021). https://doi.org/10.4028/www.scientific.net/DDF.407.173
50. Khan, S.A., Jain, M., Pandey, A., Pant, K.K., Ziora, Z.M., Blaskovich, M.A.T., et al.: Leveraging the potential of silver nanoparticles-based materials towards sustainable water treatment. J. Environ. Manag. **319**, 115675 (2022). https://doi.org/10.1016/j.jenvman.2022.115675
51. Khan, S.B., Irfan, S., Lam, S.S., Sun, X., Chen, S.: 3D printed nanofiltration membrane technology for waste water distillation. J. Water Process Eng. **49**, 102958 (2022). https://doi.org/10.1016/j.jwpe.2022.102958
52. Kozior, T., Mamun, A., Trabelsi, M., Wortmann, M., Lilia, S., Ehrmann, A.: Electrospinning on 3D printed polymers for mechanically stabilized filter composites. Polymers **11**(12) (2019). https://doi.org/10.3390/polym11122034
53. Kumar, N., Gusain, R., Pandey, S., Ray, S.S.: Hydrogel nanocomposite adsorbents and photocatalysts for sustainable water purification. Adv. Mater. Interfaces **10**(2), 2201375 (2023). https://doi.org/10.1002/admi.202201375
54. Li, K., de Rancourt de Mimérand, Y., Jin, X., Yi, J., Guo, J.: Metal oxide (ZnO and TiO_2) and Fe-based metal–organic-framework nanoparticles on 3D-printed fractal polymer surfaces for photocatalytic degradation of organic pollutants. ACS Appl. Nano Mater. **3**(3), 2830–2845 (2020). https://doi.org/10.1021/acsanm.0c00096
55. Mahesh, V.: ANN based prediction of the absorption characteristics of additive manufactured glycol-modified polyethylene terephthalate nanocomposites reinforced with short-carbon fibers and nanoclay fillers. Polym. Compos. **44**(12), 8223–8239 (2023). https://doi.org/10.1002/pc.27337
56. Mai, Z., Liu, D., Chen, Z., Lin, D., Zheng, W., Dong, X. et al.: Fabrication and application of photocatalytic composites and water treatment facility based on 3D printing technology. Polymers **13**(13) (2021). https://doi.org/10.3390/polym13132196
57. Martín de Vidales, M.J., Nieto-Márquez, A., Morcuende, D., Atanes, E., Blaya, F., Soriano, E., Fernández-Martínez, F.: 3D printed floating photocatalysts for wastewater treatment. Catal. Today **328**, 157–163 (2019). https://doi.org/10.1016/j.cattod.2019.01.074
58. Masud, A., Zhou, C., Aich, N.: Emerging investigator series: 3D printed graphene-biopolymer aerogels for water contaminant removal: a proof of concept. Environ. Sci. Nano **8**(2), 399–414 (2021). https://doi.org/10.1039/D0EN00953A
59. Mohd Yusoff, N.H., Chong, C.H., Wan, Y.K., Cheah, K.H., Wong, V.L.: Optimization strategies and emerging application of functionalized 3D-printed materials in water treatment: a review. J. Water Process Eng. **51**, 103410 (2023). https://doi.org/10.1016/j.jwpe.2022.103410
60. Nasrollahzadeh, M., Sajjadi, M., Iravani, S., Varma, R.S.: Carbon-based sustainable nanomaterials for water treatment: state-of-art and future perspectives. Chemosphere **263**, 128005 (2021). https://doi.org/10.1016/j.chemosphere.2020.128005

61. Ogazi, A.C., Osifo, P.O.: Effects of dimethylacetamide on chitosan/AgNP/GO fluid properties for 3D printing of water filtration membranes. Polym. Adv. Technol. **34**(4), 1220–1230 (2023). https://doi.org/10.1002/pat.5964
62. Ogazi, A.C., Osifo, P.O.: Inhibition of coliforms and Escherichia coli bacterial strains in water by 3D printed CS/GO/AgNP filtration membranes. J. Polym. Environ. **31**(10), 4448–4467 (2023). https://doi.org/10.1007/s10924-023-02889-9
63. Pal, R., Basak, A.: Linking powder properties, printing parameters, post-processing methods, and fatigue properties in additive manufacturing of $AlSi_{10}Mg$. Alloys **1**(2), 149–179 (2022). https://doi.org/10.3390/alloys1020010
64. Paz, R., Moriche, R., Monzón, M., García, J.: Influence of manufacturing parameters and post processing on the electrical conductivity of extrusion-based 3D printed nanocomposite parts. Polymers **12**(4) (2020). https://doi.org/10.3390/polym12040733
65. Pazhamannil, R.V., Govindan, P.: Current state and future scope of additive manufacturing technologies via vat photopolymerization. Mater. Today: Proc. **43**, 130–136 (2021). https://doi.org/10.1016/j.matpr.2020.11.225
66. Pei, R., Fan, L., Zhao, F., Xiao, J., Yang, Y., Lai, A., et al.: 3D-Printed metal-organic frameworks within biocompatible polymers as excellent adsorbents for organic dyes removal. J. Hazard. Mater. **384**, 121418 (2020). https://doi.org/10.1016/j.jhazmat.2019.121418
67. Pierre, J., Iervolino, F., Farahani, R.D., Piccirelli, N., Lévesque, M., Therriault, D.: Material extrusion additive manufacturing of multifunctional sandwich panels with load-bearing and acoustic capabilities for aerospace applications. Addit. Manuf. **61**, 103344 (2023). https://doi.org/10.1016/j.addma.2022.103344
68. Rubio, L.R., Teijido, R., Veloso-Fernández, A., Pérez-Yáñez, S., Vilas-Vilela, J.L.: Polymeric nanocomposite membranes for water remediation: from classic approaches to 3D printing. In: Shalan, A.E., Hamdy Makhlouf, A.S., Lanceros-Méndez, S. (eds.) Advances in Nanocomposite Materials for Environmental and Energy Harvesting Applications, pp. 191–243. Springer International Publishing, Cham (2022)
69. Sevastaki, M., Suchea, M.P., Kenanakis, G.: 3D Printed fully recycled TiO_2-polystyrene nanocomposite photocatalysts for use against drug residues. Nanomaterials **10**(11) (2020). https://doi.org/10.3390/nano10112144
70. Shahrubudin, N., Lee, T.C., Ramlan, R.: An overview on 3D printing technology: technological, materials, and applications. Procedia Manuf. **35**, 1286–1296 (2019). https://doi.org/10.1016/j.promfg.2019.06.089
71. Siddiqui, S., Surananai, S., Sainath, K., Zubair Khan, M., Raja Pandiyan Kuppusamy, R., Kempaiah Suneetha, Y.: Emerging trends in development and application of 3D printed nanocomposite polymers for sustainable environmental solutions. Eur. Polym. J. **196**, 112298 (2023). https://doi.org/10.1016/j.eurpolymj.2023.112298
72. Sireesha, M., Lee, J., Kranthi Kiran, A.S., Babu, V.J., Kee, B.B.T., Ramakrishna, S.: A review on additive manufacturing and its way into the oil and gas industry. RSC Adv. **8**(40), 22460–22468 (2018). https://doi.org/10.1039/C8RA03194K
73. Sodha, V., Shahabuddin, S., Gaur, R., Ahmad, I., Bandyopadhyay, R., Sridewi, N.: Comprehensive review on zeolite-based nanocomposites for treatment of effluents from wastewater. Nanomaterials **12**(18) (2022). https://doi.org/10.3390/nano12183199
74. Sreedhar, N., Kumar, M., Al Jitan, S., Thomas, N., Palmisano, G., Arafat, H.A.: 3D printed photocatalytic feed spacers functionalized with β-FeOOH nanorods inducing pollutant degradation and membrane cleaning capabilities in water treatment. Appl. Catal. B **300**, 120318 (2022). https://doi.org/10.1016/j.apcatb.2021.120318
75. Svetlizky, D., Das, M., Zheng, B., Vyatskikh, A.L., Bose, S., Bandyopadhyay, A., et al.: Directed energy deposition (DED) additive manufacturing: physical characteristics, defects, challenges and applications. Mater. Today **49**, 271–295 (2021). https://doi.org/10.1016/j.mattod.2021.03.020
76. Takeoka, Y., Yoshioka, S., Teshima, M., Takano, A., Harun-Ur-Rashid, M., Seki, T.: Structurally coloured secondary particles composed of black and white colloidal particles. Sci. Rep. **3**(1), 2371 (2013). https://doi.org/10.1038/srep02371

77. Tijing, L.D., Dizon, J.R.C., Ibrahim, I., Nisay, A.R.N., Shon, H.K., Advincula, R.C.: 3D printing for membrane separation, desalination and water treatment. Appl. Mater. Today **18**, 100486 (2020). https://doi.org/10.1016/j.apmt.2019.100486
78. Vinosha, P.A., Manikandan, A., Preetha, A.C., Dinesh, A., Slimani, Y., Almessiere, M.A., et al.: Review on recent advances of synthesis, magnetic properties, and water treatment applications of cobalt ferrite nanoparticles and nanocomposites. J. Supercond. Novel Magn. **34**(4), 995–1018 (2021). https://doi.org/10.1007/s10948-021-05854-6

5D-Printed Polymer Nanocomposites for Healthcare Applications

Mohammad Hossain and Md. Abu Bin Hasan Susan

Abstract Modern civilization has experienced significant advancements in material technologies to facilitate the use of various innovative materials in a range of applications. The versatile functions and associated beneficial properties of polymer nanocomposites (PNCs), such as their high elastic modulus, large surface area to volume ratio, large flame retardancy, and improved density among the many novel hybrid biomaterials have revolutionized the field. Due to the structuring that occurs during the processing of the polymer at the nanoscale, the processing of nanocomposite (NC) polymers is the most crucial step in bridging the technology between the composition of the NCs and their ultimate properties and derived applications. PNCs can be produced using a variety of methods. The goal of these methods is to create NC materials that are dispersed uniformly and do not aggregate. The most widely used methods for creating PNCs include melting and mixing, in-situ polymerization, electrospinning, selective laser sintering, and additive manufacturing (AM) methods like 3D, 4D, and 5D technologies. Stereolithographic processes are the foundation of AM or 3D printing (3DP). It is based on the x, y, and z axes, but 4D printing (4DP) adds a new dimension: time. Compared with 3DP and 4DP, 5D printing (5DP) can be used to manufacture more complex structures or designs that need more strength or have curved layers. Two more axes are involved in 5DP, namely the rotation of the extruder head and the print bed. This chapter provides a quick overview of 5DP technology and its various uses. The presentation includes a thorough review of the various healthcare applications, including drug delivery, dentistry, orthopaedics, bioprinting, tissue engineering, nanomedicine, and medical and biological areas.

Keywords 5D printing · Polymer nanocomposites · Additive manufacturing · Complex structure · Health care applications

M. Hossain
Department of Chemistry, Jashore University of Science and Technology, Jashore, Bangladesh
e-mail: mohammadhossainh59@gmail.com

Md. A. B. H. Susan (✉)
Department of Chemistry, University of Dhaka, Dhaka, Bangladesh
e-mail: susan@du.ac.bd

Dhaka University Nanotechnology Center (DUNC), University of Dhaka, Dhaka, Bangladesh

S. Moharana et al. (eds.), *Polymer Nanocomposites for 3D, 4D and 5D Printing*, Engineering Materials, https://doi.org/10.1007/978-981-96-4214-4_16

Abbreviations

PNC	Polymer nanocomposite
NC	Nanocomposite
AM	Additive manufacturing
CAD	Computer-aided design
3DP	3D printing
4DP	4D printing
5DP	5D printing
MRI	Magnetic resonance imaging
rBMSC	Rat bone marrow mesenchymal stem cells
TGF-β	Transforming growth factor beta
AA	Ascorbic acid
SMC	Smooth cell muscle
PDLLA-co-TMC	Poly (D, L-lactide-co-trimethylene carbonate)
GelMA	Gelatin methacryloyl
CT	Computed tomography
FDG	^{18}F-fluorodeoxyglucose
PET	Positron emission tomography
PPE	Personal protective equipment

1 Introduction

An increasing number of polymer materials with diverse structures, functions, and properties are needed as material science moves closer to greater advancements. Since polymer nanocomposites (PNCs) are innovative materials with programmable tasks and remarkable properties, they deserve a careful examination [38, 57]. Currently, nanotechnology is thriving across all scientific and technical fields. Applications for nanomaterials and nanocomposites (NCs) have been presented in a variety of fields, such as tissue engineering, agriculture, food packaging, medicine, the aerospace sector, textiles, optoelectronics, and automotive sectors. NCs are widely used in academia and industry because of their peculiar qualities, inventive design capabilities, quick manufacture, eco-friendliness, and affordability. In essence, NCs are solid materials having multiple phases made of one substance embedded in another material with a reinforcing phase that can take many different shapes, such as particles, sheets, or threads [48]. As PNCs, among other NCs have an immense potential to develop the engineering sector, they have drawn substantial interest from several researchers in the healthcare industry. Moreover, the improved biocompatibility and biodegradability have made the PNCs appealing for tissue engineering, particularly in the 3D scaffolds. Since PNCs have improved mechanical properties, high-performance materials like silk and bone can be replaced with them [41]. Certain discoveries point to methods for modeling biological tissues with both soft and rigid components. Bioinspired materials can be created with an amazing combination of a hard nanostructure and a soft polymer matrix. Consequently, the process of creating these NCs involves disbanding hard nanostructures onto a soft polymer matrix [49, 73]. Biotechnology and healthcare have shown a considerable deal of interest in these NCs [23]. These biomaterials, which are used in healthcare applications, are the focus of research and unite the collaborative efforts of biological and nanotechnological science.

It is possible to create polymer matrix NCs mechanically or chemically. Numerous attempts have been made by researchers to use chemical processes, complex polymerization events, or surface modification of filler materials to disseminate nanofillers evenly and homogeneously throughout the polymer matrix. These include additive manufacturing (AM), sol–gel, intercalation, mixing of polymer and nanofillers, and in situ methods. A number of AM techniques have been applied to rapid prototyping. Typically, computer-aided design (CAD) models are used with AM technology. The purpose of AM has grown in recent years in a variety of sectors, with the pharmaceutical and medical sectors placing a greater emphasis on the creation of drug items and medical equipment [50]. As a type of AM, 3DP is extensively used in many sectors, together with the design of jewelry, electronics, textile industries, robotics, automobiles, clothing, scaffolding, and digital filaments. Another AM technology known as 4DP has become a breakthrough. The 4D principle opens up new avenues for AM research that may lead to greater applications for 3DP and 4DP technologies [60]. A 5-axis printer is used to print a 3D object in 5DP technology [26]. As a result, products produced with this technique have three to five times the strength of those produced by a typical 3D printer, leading to innovations in manufacturing, medicine, and other disciplines.

2 5D Printing

One recent development in AM is 5DP. Having the degrees of freedom of five for both printable items and the print head, this technology can produce curved and contoured structures in place of flat ones. As the print head prints, the printable item rotates. Therefore, unlike 3D printers, which move via a straight layer, printing traces the contour of the item being created. The main advantage of this technique is to create objects with curved layers that have enhanced properties like color, porosity, or strength [26]. This five-axis printing technology generates items with multiple dimensions where three axes are there in 3DP. Apart from the x-axis, y-axis, and z-axis of 3DP technologies, the printable item in this 5-axis printing moves back and forth which adds two more rotational axes. As a result, this technology may produce more intricate and inventive items than those produced with traditional 3DP. A portion of the polytopic roadmap for 5DP that has been produced is shown in Fig. 1. Having 3D as the initial frame, the 3D, 4D, and 5D frames are displayed.

In contrast to 3DP, which uses only three axes, 5DP allows things to be created from five axes [61]. Apart from the three print head axes utilized in 3DP technologies,

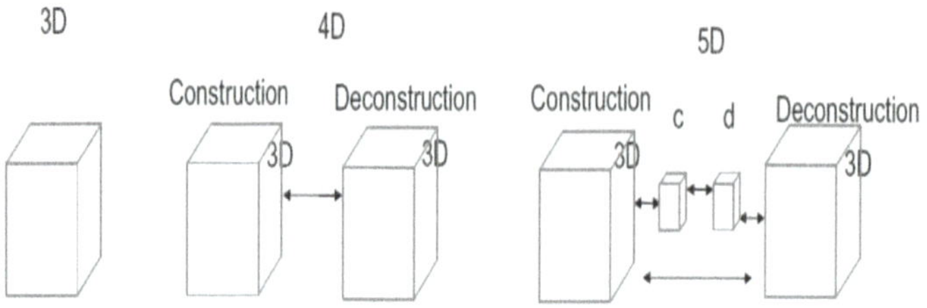

Fig. 1 Polytopic roadmap developed for 5DP. Reproduced from [29] with permission from Springer International Publishing, Copyright [2023]

Fig. 2 All the 5 axes of a 5D printer. Reproduced from [19] with permission from Food Science and Nutrition, Copyright [2023]

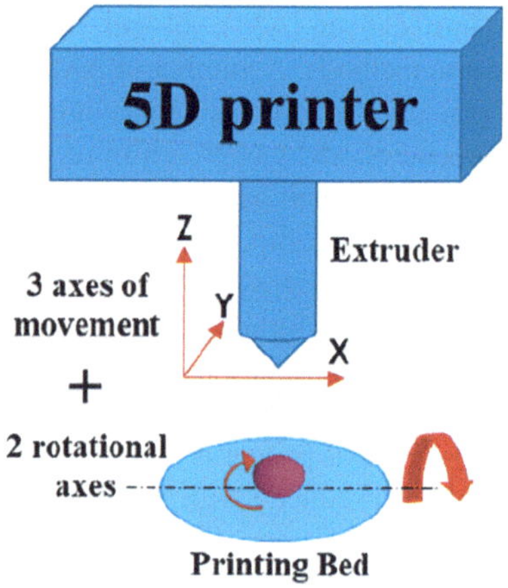

the printable bed moves on two additional rotational axes to produce the five axes of 5DP (Fig. 2).

The print head may travel in different directions, enabling it to print any necessary layers, planes, or curves. This is not achievable with 3DP. Because of its better movements, the print head can now trace the shapes and contours of items. Furthermore, when the print head follows a linear route on a static printable item, it can produce curved shapes as opposed to flat ones. The produced curved layers give strength to complexly designed items. The "Best of Innovation" award for the 3DP category was given out by the C.E.S. 2018 to the Bengaluru, India-made Ethereal Halo 5D printer, which has demonstrated innovation in AM technology (Fig. 3c). To construct intricate designs for practical or decorative uses, the machine can spin and swivel the pieces (Fig. 15.3a). In addition to cutting soft metals and other materials, this printer may also be used to cut jewelry and auto parts. The fundamental advantage of 5DP technology is its capacity to produce robust items with numerous dimensions by printing in five different axes, which enhances the strength of any necessary layers, planes, or curves (Fig. 3b). In addition, the most intricate constructions that were previously impossible to 3D print because of the requirement of many fillers and support can now be made. Cardiovascular stents, made of metal or polymeric mesh, can be used to treat restricted or choked thrombus in the heart. These stents can be extended or shrunk in accordance with the blood flow (Fig. 3d). Sequenced computed tomography (CT), magnetic resonance imaging (MRI), and positron emission tomography (PET) scan data of a patient can be used to generate a 5D anatomic model. Figure 3e shows the printed full-size anatomical models of the remaining tumor [21]. Since the 5D printer could print curved layers, it has been possible to manufacture items that 3DP failed (Fig. 3). 3D-printed products are claimed not to be nearly as strong as 5D-printed objects. This is due to the horizontal layers seen in 3D-printed products, which lead to weak spots. During the 5DP process, a variety of forces move layers in various directions that are contrary to the creating directions, such as through, up, and down. A stronger product is produced as a result of these forces and the durability of these materials [61]. The fact that 5DP needs 25% less amount of material than 3DP is another significant benefit of this technology [61].

3 Polymer Nanocomposites in Healthcare Applications

PNCs are extensively employed in the energy sector, food packaging, military systems, safety, automobiles, various types of sensors, catalysis, and information industries. They are significant materials for both industrial and research reasons [2, 9, 67]. PNCs have become more prevalent as a result of the ongoing development of this field. Owing to their enhanced mechanical, thermal, electrical, optical, and magnetic qualities, PNCs are extensively employed in numerous domains and have gradually become an indispensable aspect of our existence. NC systems have several promising uses, ranging from the creation of novel materials to the improvement of previously established functionality of the devices, such as drug delivery, bioimplants, the detection and treatment of diabetes and cancer, bioimaging, etc. The probable uses of PNCs are listed in Table 1.

4 5D Printing Technology in Healthcare Applications

The healthcare industry is home to one notable application. Better dental care, orthopaedic surgery, prostheses, and surgical instruments are all made possible by 5DP technology, which opens the door to more advanced medical procedures and patient care. 5DP helps the production of extremely strong medical tool manufacturing equipment. Any curved surface needed for automotive applications can be produced using it. It is employed in the production of surgical instruments such as debakey forceps, monopolar diathermy, and mosquito forceps. Prosthetics are devices that are used to replace lost body parts in accidents or to help disabled individuals from birth. Artificial hands, legs, lower jaws, and teeth with intricately shaped, highly durable implants are produced by 5DP. Furthermore, the combination of machine learning and artificial intelligence algorithms promises to improve

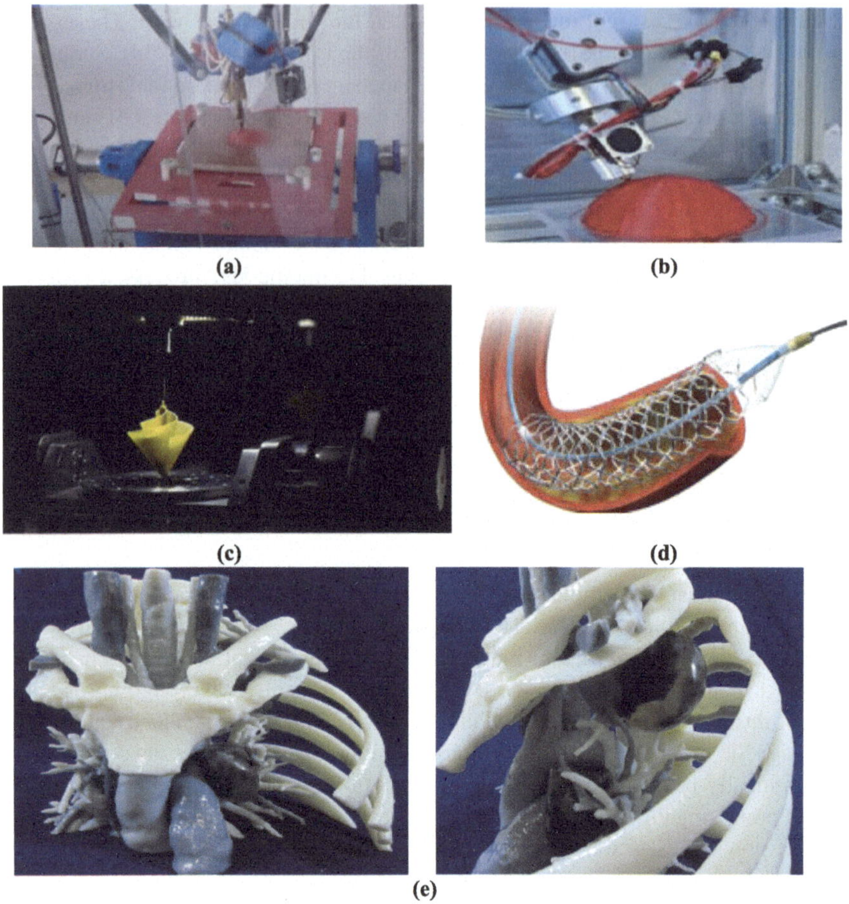

Fig. 3 **a** Moveable printhead and print bed (Reproduced from [6] with permission from Select Proceedings of ICAMMM 2021, Copyright [2022].), and printing of **b** curved cap (Reproduced from [55] with permission from Journal of Agriculture and Food Research, Copyright [2022].), **c** complex structure (Reproduced from [6] with permission from Select Proceedings of ICAMMM 2021, Copyright [2022].), **d** cardiac stent (Reproduced from [6] with permission from Select Proceedings of ICAMMM 2021, Copyright [2022].), and **e** 5D model demonstrating fused pre and posttreatment superior sulcus lung cancer and level 5 lymph nodes with a clear representation of the relationship of the tumor to surrounding structures (Reproduced from [6] with permission from Select Proceedings of ICAMMM 2021, Copyright [2022].)

predictive modeling, optimize material utilization, and streamline the design process. The possible uses of 5DP-related technology are summarized in (Table 2).

Table 1 The possible uses of PNCs

Industry/area	Particular uses of PNCs
Cancer diagnosis and treatment	Identifying tumor cells and selectively irradiating specific tumor areas have been two ways that PNCs have been applied for the diagnosis and treatment of cancer over the past decades [42, 44, 58]. PNCs of magnetic nanoparticles have also been investigated for many promising applications [18, 43].
Diabetes diagnosis and treatment	The use of NCs in the control of diabetes has recently gained recognition in several ways. These initiatives have made a significant impact on the development of nanosensors, catalysts, and sensor surface area [1, 8]. There have been reports of glucose detection using sensors that are made of carbon nanotube-reinforced PNCs. These are very selective, sensitive, stable, and accountable [64, 81].
Ophthalmic drug delivery systems	As PNCs are biocompatible, mucoadhesive, nontoxic, and biodegradable, they are used more frequently than other nanomaterials as ophthalmic drug delivery systems [33, 53]. These NCs provide improved bioavailability, a lower dose requirement, improved cellular absorption of pharmaceuticals that are poorly absorbed by cells, good therapeutic efficacy, and a decreased risk of side effects like blurred vision and/or immunological reactions.
3D bioimplants	Because of their biodegradable kinetics and biocompatibility, PNCs are a subject of intense attention in tissue engineering research [78]. The PNCs-based scaffolds are porous, biocompatible, and biodegradable. The production of these scaffolds has been approached in recent years in a number of ways, including temperature-induced phase separation [54], particulate leaching [65], rapid prototyping [17], electrospinning [56], polymer foaming, and emulsion freeze-drying [5].
Fluorescence bioimaging	A clinical method called "bioimaging" is used to identify if organs and tissues are operating normally or abnormally. As a new type of fluorescent probes in bioimaging, PNCs have surfaced recently [10, 46].

Table 2 The possible uses for technology related to 5DP

Industry/area	Particular uses for 5DP
Bioprinting	In contrast to 3DP, 5DP can create intricate, contoured, and strong structures that may preserve their unique structural and mechanical qualities after printing.
Medical	It is capable of printing intricate orthopaedic implants with ease, even those that need to be much strong to support a patient during the surgery. It can also be used for surgical education, learning, and planning [22, 25]. It enables the creation of complexly shaped artificial body parts, including hands, lower jaws, and legs, but it also demands a great degree of strength [63].
Surgical tools	Instruments for surgery such as mosquito forceps, DeBakey forceps, and monopolar diathermy.

5 Healthcare Applications of 5D-Printed Polymer Nanocomposites

5.1 Dentistry

The claim that 5D-printed things are far stronger than 3D-printed ones is proven [25]. Hence, by developing any complicated curved dental product with significantly increased strength that seems dependable for surgery, the life of a patient can be improved. Different difficulties that are not addressed by 3DP and 4DP technologies can be readily taken up by 5DP. The majority of dental implants, dentures, and other devices have curved shapes rather than being flat. With this method, a complicated curved shape with greater strength might be produced [26]. These can enhance post-operative patient outcomes and implant dependability. Missing teeth of a person can be produced with ease if the necessary surface polish and dependability are met. As a result, it offers extremely precise dental-surgical guidelines. It is useful to create orthodontic models and aligners with the necessary strength. This technology will be useful for planning, teaching, manufacturing complex shapes, and creating dependable operations. The primary drawback of this innovation is that it is expensive for the movable print head and bed. To build and manage the 5D printer, highly qualified human resources are also required. Over time, this technology will greatly increase reliability and minimize adverse effects, making the job of a surgeon easier. It works effectively for printing dental implants composed of several biocompatible materials. 5D-printed materials will emerge as the ideal answer for any curved product, including bones, dentures, stents, and other dental instruments and equipment. By utilizing digital data, this technique advances dentistry to the point of mass customization, contributing to the development of digital dentistry. The goal is to lower implant costs in comparison to 3DP technology, 5DP uses reduced material waste. It has enormous potential to address a number of dental issues, including enhancing current dental procedures and meeting specific needs for intricate shapes.

5.2 Orthopaedics

The utilization of 5DP in orthopaedics is anticipated. As the bones in the human body are made of curved surfaces rather than flat ones, 5D-printed artificial bones can be useful in surgery. Since the primary goal is to get beyond the drawbacks of 4DP and 3DP which include their inability to produce curved layer objects and their lack of maximum strength, for the purpose of creating prosthetic bones, 5DP will be a very useful tool with exceptional strength [84]. Artificial bones are needed for bone implantation in orthopaedics, and 5DP can be a valuable resource for this [25]. The study indicates that the artificial bones or objects made with 5DP have five times more strength than the products printed with 3DP. The caps used in orthopaedic surgery have been shown to withstand four times the MPa of caps manufactured using 3DP.

As 5D printed materials are stronger, they require fewer materials to manufacture artificial bone implants that are just as robust as those made with previous printing technologies (Mpofu et al.). Using information from various sources like CT and MRI, 4D printed materials can be effectively utilized in orthopaedics to construct implants for a difficult surgical scenario. Then, utilizing a variety of readily available programs, such as Vesalius, Mimics, Magic, 3D doctors, 3D slicer, and OsiriX Imaging software, this data can be simply changed to an appropriate 3D CAD format. Compared to 3D or 4D printed products, printed objects can withstand pressure four to five times higher, which is why orthopaedic surgeons prefer to use 5DP. Artificial bone caps with complex or curved shapes may be manufactured with ease using 5DP, making it a valuable tool for the healthcare sector. Furthermore, it is possible to make good use of a mix of smart materials [13]. The fundamental advantage of 5DP above other AM technologies is its ability to print surgical complex implants in accordance with patient procedures. Similar to dental surgery, orthopaedics has the same 5DP limitations. There is a need for highly competent and efficient workers for the upkeep of a 5D printer [31].

5.3 Bioprinting

With the use of the 5DP technology, things that are subject to specific environmental conditions can be printed which can change their shape and size with time. Several sectors, including bioprinting, can benefit from the application of this technology. Bioprinting is the process of printing 3D objects with biomaterials such as body tissue and living cells. By controlling the chemical and physical characteristics of the printed objects, it offers an effective way to construct various intricate structures. The utilization of this technology facilitates the printing of complex microstructure biopolymers, living cells, and bioinks, hence permitting the creation of highly structurally intact functional tissue constructs. It is possible to create intricate structures with particular mechanical and biological characteristics by adjusting the physical and chemical qualities of printed materials.

Organoids, prosthetic limbs, and tissue architectures for scientific study are all created with this technology. The application of 5DP to bioprinting technology may pave the way for novel developments of reformative drugs by producing intricate structures made of living tissues. Additionally, 5DP can make 3D scaffolds, which enable more precise and controlled development of tissue structures. Personalized medical therapies can be developed through the use of 5DP to generate tissue structures tailored to individual patients [27]. Since it can print layered structures of distinct characteristics in one step, 5DP has prospects for a complete transformation of bioprinting [24]. 5DP gives more precision to shape, size, and intricacy than typical 3DP. It may be possible to print complex vascular networks and add different types of cells to print materials using 5DP. Moreover, with the use of this technology, patients could receive individualized care that catered to their specific needs [14].

5.4 Tissue Engineering

The interdisciplinary area that merges clinical medicine, engineering, and biology is tissue engineering. It offers a vital, exciting, and promising way to replace damaged or diseased human body tissues or organs with artificial ones, keeping or even improving tissue and organ functions. The use of printing technologies has undoubtedly greatly enhanced tissue engineering, a crucial component of biomedical engineering. A wide range of tissue engineering scaffolds with customized structures and properties are created with 3DP utilizing different biomaterials, like polymers, ceramics, metals, and their composites [75, 83]. Printing technology has improved to 4DP within the recent years, adding time as a fourth dimension to the 3Ds of space, to overcome various constraints of 3DP and bioprinting. Under the right conditions, the form, characteristics, and functioning of structures that are 4D printed can change [36]. Since information is so important in the modern world, it might be considered the fifth dimension that will help 4DP evolve into 5DP and lead to a new dimension of AM [76, 77]. In contrast to 4DP, which utilizes time to increase 3D space, and 3DP, which uses space alone, 5DP uses information in addition to time and space. To accomplish the goals of 5DP, information—the fifth dimension—may take the shape of species that, when released into the environment, alter the surrounding conditions. In order to achieve the greatest therapeutic results with tissue engineering, 5DP will make it possible to fabricate items with information implanted and disseminated in them as well as objects that may change shape. To enhance the development of new tissues, biomolecules are frequently employed in tissue engineering. Consequently, biomolecules of tissue engineering can hasten the regeneration of desired tissues by encasing bioactive biomolecules. Innovative tissue engineering items with the capacity to change shape and deliver information in a regulated manner could be created using 5DP in order to regenerate complicated bodily tissues. One crucial path for creating biological replacements for tissue preservation, repair, and enhancement is scaffold-based tissue engineering. Here, the research and creation of innovative, multi-layered, cell-filled structures for the purpose of producing tubular tissues (such as blood arteries) in the body serves as an example of 5DP. This particular kind of 5D printed object may change its shape and convey encoded information in a controlled manner (biomolecules in this case). The procedure for creating cell-laden constructions through 5DP is shown in Fig. 4. These consist of two layers: one that changes shape and contains rat bone marrow mesenchymal stem cells (rBMSC) and another that delivers biomolecules.

5.5 Nanomedicine

"Nanomedicine," or the application of nanotechnology to medicine, is at the vanguard of revolution in modern healthcare and presents a viable means of delivering medication [11, 20, 34]. Accordingly, the U.S. Food and Drug Administration recently

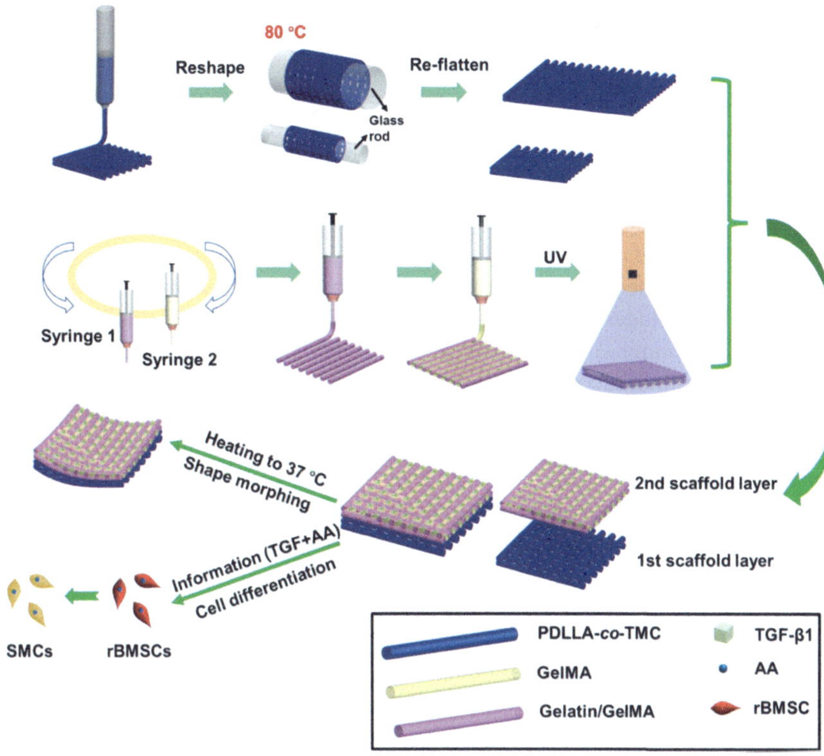

Fig. 4 Diagrammatic representation of the 5D bioprinting process for bilayer cell-scaffold structures that may change shape (first scaffold layer) and deliver information in-situ (second scaffold layer); TGF-β: Transforming growth factor beta; AA: Ascorbic acid; SMCs: Smooth cell muscles; PDLLA-co-TMC: Poly(D, L-lactide-co-trimethylene carbonate); GelMA: Gelatin methacryloyl. Reproduced from [40] with permission from the Journal of Materials Research, Copyright [2023]

approved a number of first-generation nanoproducts for clinical use. More clinical and preclinical studies are being conducted on the nanosystems polymer and liposomal drug conjugate [79]. The applications of printing technology in medicine and related toxicity/ vitality investigations show the great potential of interactive processes [32]. The materials used to create bio-printed scaffolds are primarily either from synthetic molecules (polyethylene glycol) [28] or naturally occurring polymers (gelatine, collagen, alginate, chitosan, and hyaluronic acid). This allows for precise control [72] over the chemical and physical properties of materials [15]. 5DP combines physiological activity data for individualized treatment with 3D model creation data [21].

5.6 Thoracic Surgery

Anatomical models printed in 3D for complicated surgical situations enhance resident and patient education, facilitate operative team preparation, and direct the procedure. For surgical planning, a 5DP technology has also been developed. In patients with complex thoracic pathology, 5D-printed models help in surgical planning, margin selection for resection, predicting possible challenges, training learners, and patient education. The model is more useful when data regarding physiological activity (the 5th dimension) and the therapy for tumors with time (the 4th dimension) are included. Therefore, 5DP includes the three x, y, and z axes as height, breadth, and depth and the other two as tumor size (determined by CT and change in physiology (change in avidity of 18F-fluorodeoxyglucose, FDG determined by PET). Surgical planning is complicated in complex thoracic surgical patients due to deformed tissue of the tumor tissues and other neighboring tissues. In particular, two distinct advantages come from using 5DP to create a material depiction of the efficacy of treatment found in FDG-PET imaging. First, there may be a chance to improve the R0 resection rate by employing the control of residual FDG activity and the previous tumor site. Second, because of prior radiation exposure, the team can access spare parts of structures that were never clinically affected but may appear strange following excision. Further developments in surgical anatomy and physiology may result in the addition of new dimensions of information that can be printed and integrated in the future, further improving the assimilation of complicated data required for oncologic and surgical planning.

5.7 Medical

In 2016, American colleges introduced the concept of 5DP for the first time [26]. Building a strong portion of a curved layer is the main benefit of this approach [16]. Etherea has experience with 3DP technology. They constructed a 5D printer and won the Consumer Electronics Show award for outstanding innovation [19]. Bone scaffolds that are specifically tailored for use in surgery could be made using a 5DP prototype. Artificial bones composed of 5DP are necessary to provide these bone implants exceptional strength because human bones are curved rather than flat. There is a lot of promise for this technology to meet this essential demand. As 3DP employs linearly, this is not suitable for creating intricately curved orthopaedic implants. The test findings indicate that 5DP products are three to five times more robust than 3D-printed items. Studies have revealed that a cap made from 5DP can tolerate four times the pressure and is four times stronger than that made by 3DP [26].

After printing, curved layers get stronger because of the material force present in 5DP [61]. Complex, robust, curved-surface implants are necessary for orthopaedic surgery. In addition to being helpful for creating these intrusive medical implants

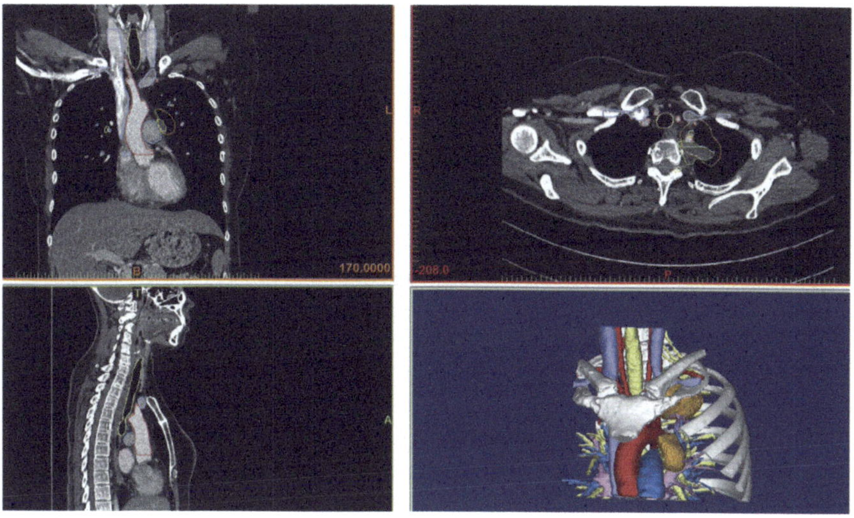

Fig. 16.5 Following image segmentation using computed tomography images, a three-dimensional anatomic model is created. The completed segmentation is shown in the image at the bottom right. Reproduced from [21] with permission from The Annals of Thoracic Surgery, Copyright [2016]

depending on the real surgery of the patients, 5DP is also helpful for preoperative care, education, and learning. Because of this, 5DP can readily produce intricate, curved structures that need to be strong significantly. The primary distinction between 3 and 5DP is that the former yields a flat surface, while the latter generates stronger sections with a contoured layer. For the most part, both procedures use the same technologies, including 3D scanners, 3DP materials, and 3D CAD file input.

High-resolution axial chest CT scans with intravenous contrast and FDG PET are carried out before therapy in surgical planning. Using sequenced imaging data from the CT, MRI, and PET scans of the patient, a 5D anatomic model is created [21]. Using color-coding for FDG activity, the 5D anatomic models integrate pretreatment and posttreatment CT and PET images to produce a composite image that shows the tumor before treatment and the residual tumor after therapy. A 3D virtual anatomic model is created using the segmented data (Fig. 5).

A few difficult facets of the field are the developing hotspots related to vascularization and bioprinting. Current hot topics in bio-fabrication research include multi-material printing for potential studies in the near future.

5.8 *Biomedical*

The medical items made with AM help in the healthcare sector by mitigating the interruptions caused by fluctuations in worldwide supply and demand. The use of 5DP in medicine, dentistry, and related professions is widespread. With the aid of this cutting-edge technology, challenging and intricately designed items, prosthetics, medical devices, and bioimplants can be created [16]. In addition, bioprinted devices and biomimetic objects are created using 5DP processes. The improvement of nanosystem 5DP equipment has received backing from both the Food and Drug Administration and the European Medicines Agency. Utilizing vascular smooth muscle cells in a solution, these devices were utilized to create biomaterial scaffolds, or scaffolds loaded with nanoparticles, which were then investigated in the rat vena cava and showed promise in regulating the release of a medication gradually [16]. It is necessary to use robust and intricate contoured objects in orthopaedics and dentistry. Using CAD, 5DP technology is becoming more and more common in the creation of orthodontic braces, dental implants, and aligners [16].

AM technology is used in the production of sanitary personal protective equipment (PPE), such as surgical masks, valves, ventilators, face shields, test swabs, safety goggles, visors, N95 respirators, and ear protectors [69]. During COVID-19, the healthcare industry made significant progress in producing biomedical products, demonstrating the superiority of biomedical applications over other industries. Due to local and worldwide supply restrictions, technology has helped manufacture a variety of medical gadgets, such as PPE [85]. Therefore, printing technology has the ability to significantly improve emergency situations by enabling mass customization in both design and complexity [39]. During the pandemic, a number of firms, including Lonati, 3D systems, Materialize, Airwolf3D, and Isinnova, assisted the healthcare sector in producing ventilators [68]. A wide range of polymeric materials were used to manufacture PPE [7]. During the epidemic, face shields, masks, ventilators, and swabs were the most important medical supplies which indicates a great contribution of printed technology. Also, by creating recyclable items like respirators and filters, 5DP technology can help reduce environmental risks associated with medical waste from disposable equipment [30]. It is anticipated that biomedical implants will enhance physiological, tribological, and biomechanical performance [12]. For example, biomedical titanium alloys are extensively utilized in the orthopaedic and orthodontic fields to create various artificial implants [35].

Graphene-based PNCs (Fig. 6) are among the most significant NCs [82] because they exhibit desirable properties like increased flexibility, elasticity, and the capacity to adapt to both flat and irregularly shaped surfaces [66]. They also play remarkable roles in continuous proliferation, controlled adhesion, and improved differentiation. After 10 days of use, a membrane with chitosan blended over the surface of polyvinyl-containing graphene completely heals skin wounds, this effect was not observed in the control groups where graphene was absent [45].

A procedure that uses less energy and is more sustainable must be identified, as AM for biological applications is developing daily. 5DP can be the most suitable candidate in this regard.

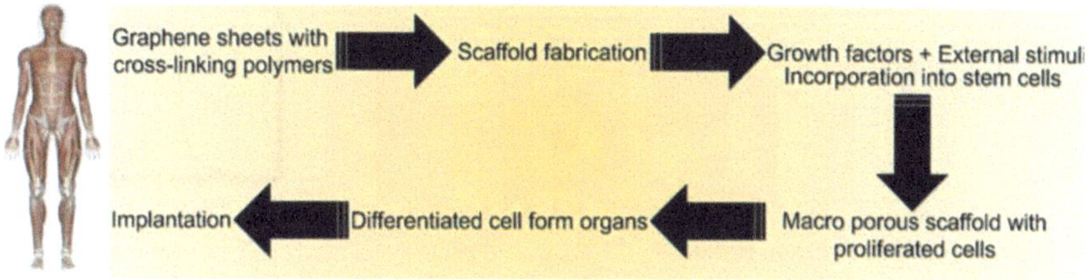

Fig. 6 PNCs based on graphene for biomedical applications. Reproduced from [66] with permission from Elsevier, Copyright [2023]

5.9 Drug Delivery

Pharmaceutical drugs are essential for the treatment of various diseases. A collection of technologies known as "drug delivery" is employed to securely introduce a pharmaceutical substance for a particular therapeutic purpose [71]. From tablets to drug-eluting implants, drug delivery techniques have advanced dramatically in the last few years. The concept behind this delivery system is that a drug is incorporated, its active chemicals are released, and the drugs go to the active sites. The ailment, the intended outcome, the effectiveness of the drug-effective materials, and occasionally the preferences of the patient play a role in the drug delivery method selection process.

5DP has changed the pharmaceutical distribution industry to offer more personalized services. The route for drug distribution by additive manufacturing is shown in Fig. 7. A laboratory test that comes before a 5DP drug delivery system may include an MRI or CT scan. This offers important details about the condition of the patient in addition to the design parameters intended to provide a patient-centered dosage form and a scheduled release profile. The next step is to create a treatment plan that may include the recommended drugs and the best release profile for them. 5DP technology not only offers customized treatment but also programmable medication release. A digital model is then created using CAD. Finally, the drug-loaded delivery system is made by 5DP and distributed through channels [50]. Delivering a drug delivery system to patients is thought to provide a barrier due to patient variability. Pharmaceutical manufacturing procedures including capsule filling and tableting have progressed, especially when it comes to managing drug-releasing profiles with the help of tools like biomimetic particles, functionalized liposomes, and nanoparticles. Enhancing drug delivery systems employing these methods, nevertheless, would require funding for related research and scaling up.

The drug distribution sector has undergone a transformation because of AM, allowing for personalized medication. A laboratory test, which may involve MRI or CT, is the first step in a conventional AM drug delivery system. This provides helpful information about the health of a patient and the specifications for the design that aims to provide a dosage form that is centered around the patient and a scheduled release kinetics. A treatment plan is developed, which may include the recommended drugs

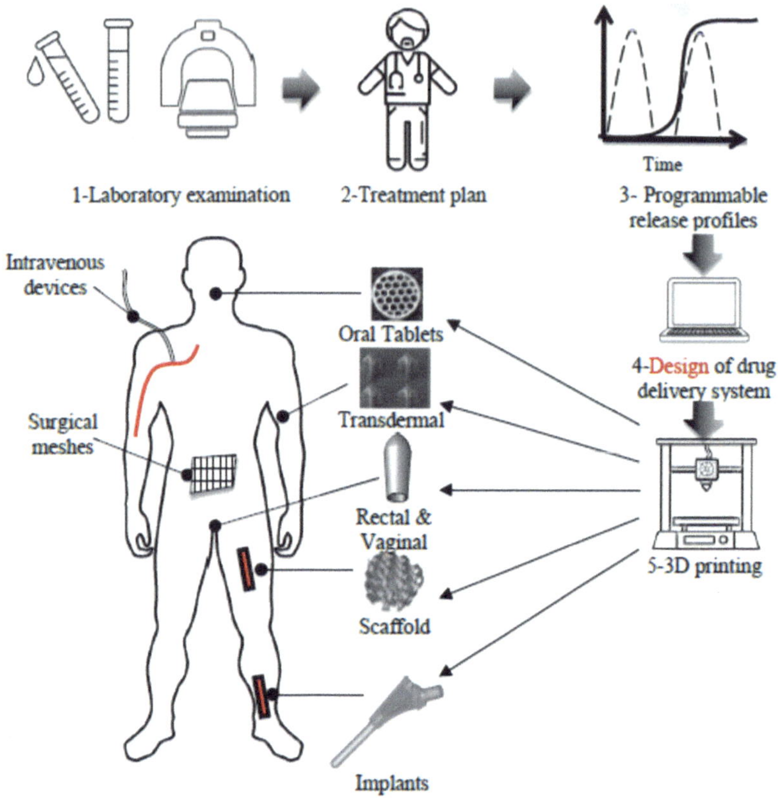

Fig. 7 The procedure and available body paths of 5D printed drug delivery systems. Reproduced from [50] with permission from the International Journal of Pharmaceutics, Copyright [2020]

and their optimal release kinetics. Not only can AM technology offer customized treatment, but it can also administer drugs with programmable release. The next step involves using CAD to construct a digital model. Last but not least, the drug-loaded delivery system is created by AM and can be delivered via scaffolds for drug delivery, oral solid pharmaceuticals, intravenous devices, rectal delivery, transdermal patches, and biomedical implants [50]. Variations in the release profiles of dosage forms have been documented in the literature, contingent upon the filament type and the printing method. Variations in the release profiles of dosage forms have been documented in the literature, contingent upon the filament type and the printing method. Masood demonstrated that selected FDM process parameters, including the raster gap and raster angle, had a significant effect on the drug release characteristics in addition to the printed matrix [47].

The surface/mass ratio can be modified with various printing methods of solid dosage forms of distinct shapes and sizes, which ultimately affects the release of the active pharmaceutical ingredient [59].

6 Challenges

Although PNCs have many advantages, they have several disadvantages too. As these materials are of high viscous nature, non-uniform distribution of dispersed phase remains throughout the PNCs and agglomeration occurs. Moreover, taking into account the financial ramifications, the production of polymer/metal NCs industrially could have a positive impact on the lives of people and communities. However, uniformity of the synthesis pathway and manufacturing process is crucial for creating high-quality goods on a big scale [80]. There are still a number of obstacles that are required to be cleared in order to get a complete realization of the potential of polymer-based NCs for bio applications, despite several attempts to do so. Biocompatibility and biodegradability, together with other aspects that need to be carefully examined with appropriate preclinical investigations, are necessary conditions for using these NCs as biomaterials in order to prevent unrealistic expectations.

5DP is a new technology that has a number of drawbacks and difficulties that should be taken care of by the scientific community. The majority of the few solutions available for 5DP are either intended for industrial use only or come at a high cost. The fact that 3DP is more widely supported than 5DP could be another barrier. Because there may not yet be established support networks, users may need to rely on their own independent education of the technology or search for resources specifically related to 5DP. Slicing software has not kept up with these advances, therefore creating and slicing models for 5DP can also be highly complicated. Few slicing applications allow for 5-axis printing, much less real 3DP. Therefore, creating complex structures and controlling the extra axes in 5DP call for a higher degree of skill. Lastly, since 5DP is still in its infancy, there are few qualified resources available to create and maintain 5D printers. This could result in a more complicated and expensive 5DP solution implementation.

The primary constraint on this technique is the additional expense required for two additional axes. Accurate hardware and software are needed to operate the customized computer. In order to build and manage the 5D printer, highly qualified human resources are also needed. The formulation, development, and processing of bio-inks present a challenge in achieving the desired qualities of printed products. A thorough understanding of cells, printing techniques, biomaterials, and the in vivo biological milieu is required. For the purpose of healthcare applications, a multidisciplinary effort is required to comprehend the structural, chemical, physical, mechanical, rheological, biological, and immunological aspects. Another serious issue arises which is increased exposure to potentially harmful plastics and plastic inhalation will result in the body of the operator accumulating carcinogenic substances. Despite the rapid advancement of bioprinting technology, emissions have gradually decreased.

7 Prospects

Real-time cost estimation can be possible using 5DP technology. Subsequent developments in this technology can concentrate on employing complex algorithms to estimate costs in real time [4, 37]. These algorithms will provide precise and up-to-date cost estimates during the building process by taking into account variables including material pricing, labor costs, equipment utilization, and project schedules. It can make exact project scheduling possible [70]. Experts are able to make thorough timetables, monitor work in real time, and spot possible delays. Schedules will be optimized by advanced analytics and AI algorithms, guaranteeing effective resource use and on-time project completion. To improve financial accountability and transparency, blockchain technology can be included in 5DP procedures [74].

At this point, the devices are temporarily dependable for the intended use. Future directions in the use of bioprinting techniques in biomaterials engineering are thought to be examined from a need-based standpoint. In order to achieve this, a complex, multipurpose printing device must incorporate the capabilities of 4D and 5D printers. Meanwhile, more information may help offer useful advice for dealing with penalties and solving temporary issues. One potential use for 5DP is cancer decoding, which makes it easier to plan surgical procedures, choose accurate resection margins, anticipate challenges or hazards, and provide students with more easily available teaching [21]. Furthermore, the tumor and its microenvironment can be extensively studied using the anatomical models of malignant tissue that are provided by CT, MRI, and PET scans both before and after the treatment. Other aspects that are very important to surgical planning with 5D printers include tracking the anatomical aberration of the tumor tissues and neoadjuvant treatments [3]. However, segmenting images from an anatomic model is necessary for advanced bioprinting using 5D printers. Future developments in surgical anatomy, body physiology, organ responses to malignant cells, and, most importantly, the dynamic modeling of growing organs through data assimilation may all demonstrate the potency of the recently invented printing machines [51]. Printing multifunctional scaffolds for clinical transplantation can be accomplished by combining tissue engineering and vascularized osteogenesis using 5DP technology [76, 77]. With considerably more dependability and fewer adverse effects, this technique will eventually make the job of a surgeon easier. It works effectively for printing dental implants composed of several biocompatible materials. Stents, bones, dentures, and other curved-shaped dental instruments and gadgets will all be perfectly solved by 5DP technology. This will be useful for complex cases and has the potential to be successfully used in dentistry in the next years. It has enormous potential to address a number of dental issues, including enhancing current dental procedures and meeting specific needs for intricate shapes. This technology has the potential to bring about revolutionary innovation in orthopaedics in the future, similar to 3DP and 4DP. It will open up countless opportunities and offer top-notch care to preserve the lives of the patients. It appears that this technique can create intricately designed implants for orthopaedics and medicine. Using these technologies also opens up the possibility of organ manufacture, tissue scaffolds, manufacturing biomimetic materials as well as splint and stent production, bio-printing, and the improvement of orthopaedic instruments for human growth. Because it prints robust structures made of curved layers efficiently, it will soon be an essential tool for various sophisticated applications. Notwithstanding a number of drawbacks, 5DP holds enormous promise to transform the world of materials in ways that are currently unimaginable. In certain respects, it is not strange to state that science and science fiction overlap in this field. This raises high expectations, requiring scientists to do more research and use their newfound knowledge in a variety of healthcare applications. Figure 8 shows the rising flashpoints in bioprinting, vascularization, and other difficult issues.

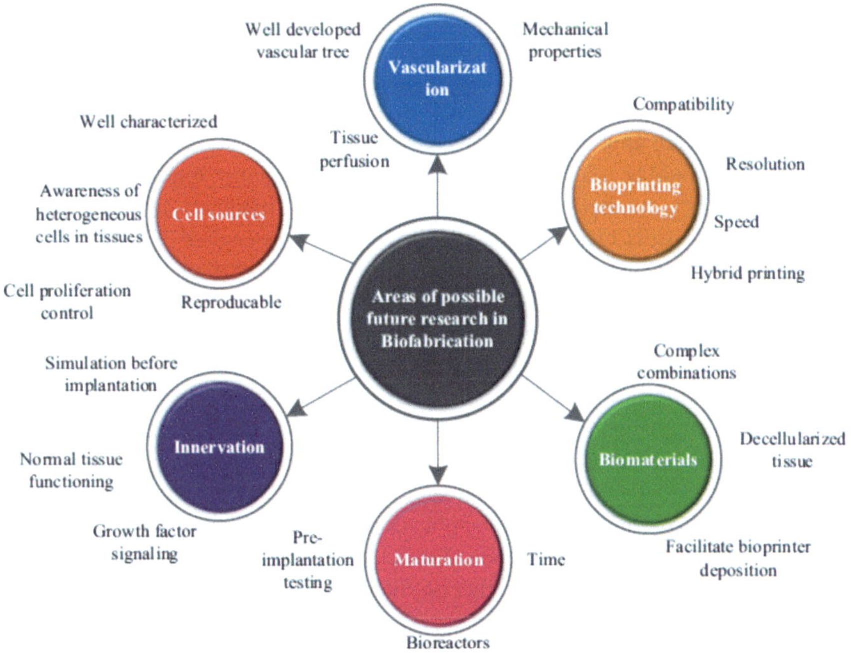

Fig. 8 An example to illustrate many directions for future research and development in biomedical AM. Reproduced from [62] with permission from Progress in Additive Manufacturing, Copyright [2023]

8 Conclusion

Significant advancements in the manufacturing of polymer-based NCs have been made possible by the special properties of nanostructured reinforcing elements. The synthesis and functionalization of nanofillers open up new avenues for research into how they might be used to create polymer-based NCs for possible medical applications. Due to the significantly superior properties of these PNCs over virgin polymers, significant advancements have been made in the medical field. This chapter presents a summary of the numerous healthcare applications of PNCs. PNCs have extensive applications, for example, dental and medical implants, drug and gene delivery, biosensors, bioimaging, tissue engineering, and wound healing. These show considerable potential for the identification and treatment of disorders. In cancer treatment, the distribution of various therapeutic agents, or medications, has been addressed. Along with orthopaedics, tissue regeneration, and bioprinting are also covered. The development of biomedical toolkits has also been explored with PNCs because of their hybrid features and synergistic effects. Therefore, in the near future, PNCs have a great deal of promise as the primary therapeutic and diagnostic agent for use in healthcare applications. 5DP enables the fabrication of incredibly precise structures that would be tough to achieve using 3DP technology. It has huge potential for utilization in a number of sectors, one of its most important qualities may be its extraordinary stress-bearing capacity—five times greater than 3DP, which may make it far more effective in healthcare applications. By simulating the physiological actions of illnesses, medications, and other treatments, 5DP has proven to be an unwavering benefit to the medical field, saving countless lives. Although there hasn't been much research carried out on cutting-edge techniques like 5DP, these studies may be able to address some of the biomimicry problems that crop up in traditional 3D bioprinting. As more and more biomaterials are found, it is important to consider both their biocompatibility and possible human toxicity. It is necessary to create new strategies to address the time consumption problems. 5DP opens up new possibilities for the growth of AM and healthcare applications, resulting in the creation of unique tissue engineering scaffolds, biomedical tool kits, dental implants, artificial bones, individualized nanomedicine, and other products with great clinical performance.

Acknowledgements The authors acknowledge the Ministry of Science and Technology for special allocation for a project related to the work. The authors also acknowledge the Bose Centre for Advanced Research in Natural Science for financial support.

Conflicts of Interest There are no conflicts to declare.

References

1. Al-Mokaram, A.M.A.A.A., et al.: The development of non-enzymatic glucose biosensors based on electrochemically prepared polypyrrole–chitosan–titanium dioxide nanocomposite films. Nanomaterials **7**(6), 129 (2017)

2. Alateyah, A.I., Dhakal, H.N., Zhang, Z.Y.: Processing, properties, and applications of polymer nanocomposites based on layer silicates: a review. Adv. Polym. Technol. **32**(4), 21368 (2013)
3. Anas, S., et al.: Concept of 5D printing technology and its applicability in the healthcare industry. Mater. Today Proceed. **56**, 1726–1732 (2022)
4. André, J.C.: From Additive Manufacturing to 3D/4D Printing 2: Current Techniques, Improvements and their Limitations. John Wiley & Sons (2017)
5. Aranaz, I., et al.: Preparation of chitosan nanocompositeswith a macroporous structure by unidirectional freezing and subsequent freeze-drying. Mar. Drugs **12**(11), 5619–5642 (2014)
6. Aruna, R. et al.: Evolution of 5D printing and its vast applications: a review. In: Recent Advances in Materials and Modern Manufacturing: Select Proceedings of ICAMMM 2021 pp. 691–714 (2022)
7. Brighenti, R., et al.: Laser-based additively manufactured polymers: a review on processes and mechanical models. J. Mater. Sci. **56**, 961–998 (2021)
8. Cash, K.J., Clark, H.A.: Nanosensors and nanomaterials for monitoring glucose in diabetes. Trends Mol. Med. **16**(12), 584–593 (2010)
9. Chow, W.S., Mohd Ishak, Z.A.: Smart polymer nanocomposites: a review. Express Polym. Lett. **14**(5), 416 (2020)
10. Corr, S.A., Rakovich, Y.P., Gun'ko, Y.K.: Multifunctional magnetic-fluorescent nanocomposites for biomedical applications. Nanoscale Res. Lett. **3**, 87–104 (2008)
11. Davis, M.E., et al.: Evidence of RNAi in humans from systemically administered siRNA via targeted nanoparticles. Nature **464**(7291), 1067–1070 (2010)
12. Davoodi, E., et al.: Additively manufactured metallic biomaterials. Bioact. Mater. **15**, 214–249 (2022)
13. Dawood, A., et al.: 3D printing in dentistry. Br. Dent. J. **219**(11), 521–529 (2015)
14. Dickey, M.D.: Shaped after print. Nat. Mater. **15**(4), 379–380 (2016)
15. Elviri, L., et al.: 3D-printed polylactic acid supports for enhanced ionization efficiency in desorption electrospray mass spectrometry analysis of liquid and gel samples. Talanta **155**, 321–328 (2016)
16. Foresti, R., et al.: In-vivo vascular application via ultra-fast bioprinting for future 5D personalised nanomedicine. Sci. Rep. **10**(1), 3205 (2020)
17. Fradique, R., et al.: Production of new 3D scaffolds for bone tissue regeneration by rapid prototyping. J. Mater. Sci. Mater. Med. **27**, 1–14 (2016)
18. Gaharwar, A.K., Peppas, N.A., Khademhosseini, A.: Nanocomposite hydrogels for biomedical applications. Biotechnol. Bioeng. **111**(3), 441–453 (2014)
19. Ghazal, A.F., et al.: Progress in 4D/5D/6D printing of foods: applications and R&D opportunities. Crit. Rev. Food Sci. Nutr. **63**(25), 7399–7422 (2023)
20. Giardiello, M., et al.: Accelerated oral nanomedicine discovery from miniaturized screening to clinical production exemplified by paediatric HIV nanotherapies. Nat. Commun. **7**(1), 13184 (2016)
21. Gillaspie, E.A., et al.: From 3-dimensional printing to 5-dimensional printing: enhancing thoracic surgical planning and resection of complex tumors. Ann. Thorac. Surg. **101**(5), 1958–1962 (2016)
22. Godoi, F.C., Prakash, S., Bhandari, B.R.: 3d printing technologies applied for food design: status and prospects. J. Food Eng. **179**, 44–54 (2016)
23. Goenka, S., Sant, V., Sant, S.: Graphene-based nanomaterials for drug delivery and tissue engineering. J. Control. Release **173**, 75–88 (2014)
24. Guo, C., Zhang, M., Devahastin, S.: Color/aroma changes of 3D-Printed buckwheat dough with yellow flesh peach as triggered by microwave heating of gelatin-gum Arabic complex coacervates. Food Hydrocoll. **112**, 106358 (2021)
25. Haleem, A., Javaid, M.: Future applications of five-dimensional printing in dentistry. Curr. Med. Res. Pract. **9**(2), 85–86 (2019)
26. Haleem, A., Javaid, M., Vaishya, R.: 5D printing and its expected applications in Orthopaedics. J. Clin. Orthop. Trauma **10**(4), 809 (2019)

27. He, C., Zhang, M., Devahastin, S.: Investigation on spontaneous shape change of 4D printed starch-based purees from purple sweet potatoes as induced by microwave dehydration. ACS Appl. Mater. Interfaces. **12**(34), 37896–37905 (2020)
28. Hunt, N.C., Grover, L.M.: Cell encapsulation using biopolymer gels for regenerative medicine. Biotech. Lett. **32**, 733–742 (2010)
29. Iordache, O.: Prospective. In: Iordache, O. (ed.) Roads to Higher Dimensional Polytopic Projects, pp. 177–202. Springer International Publishing, Cham (2023)
30. Irfan Ul Haq, M. et al.: 3D printing for development of medical equipment amidst coronavirus (COVID-19) pandemic—review and advancements. Res. Biomed. Eng. **38**, 305–315 (2020)
31. Javaid, M., Haleem, A.: Significant advancements of 4D printing in the field of orthopaedics. J. Clin. Orthop. Trauma **11**(4), S485–S490 (2020)
32. Kang, H.-W., et al.: A 3D bioprinting system to produce human-scale tissue constructs with structural integrity. Nat. Biotechnol. **34**(3), 312–319 (2016)
33. Kaur, H., et al.: Synthesis, characterization and evaluation of thiolated tamarind seed polysaccharide as a mucoadhesive polymer. Carbohyd. Polym. **90**(4), 1543–1549 (2012)
34. Kim, B.Y., Rutka, J.T., Chan, W.C.: Effect of Coumarate 3-zhydroxylase Down regulation on lignin structure. Nanomed New England J Med **363**, 2434–2443 (2010)
35. Kolli, R.P., Devaraj, A.: A review of metastable beta titanium alloys. Metals **8**(7), 506 (2018)
36. Kuang, X., et al.: Advances in 4D printing: materials and applications. Adv. Func. Mater. **29**(2), 1805290 (2019)
37. Kumar, A., Mittal, R.K., Haleem, A.: Advances in Additive Manufacturing: Artificial Intelligence, Nature-Inspired, and Biomanufacturing. Elsevier (2022)
38. Kumar, S., et al.: Recent advances and remaining challenges for polymeric nanocomposites in healthcare applications. Prog. Polym. Sci. **80**, 1–38 (2018)
39. Kunovjanek, M., Wankmüller, C.: An analysis of the global additive manufacturing response to the COVID-19 pandemic. J. Manuf. Technol. Manag. **32**(9), 75–100 (2021)
40. Lai, J., Wang, M.: Developments of additive manufacturing and 5D printing in tissue engineering. J. Mater. Res. **38**(21), 4692–4725 (2023)
41. Lee, C., et al.: Measurement of the elastic properties and intrinsic strength of monolayer graphene. Science **321**(5887), 385–388 (2008)
42. Liao, J., et al.: Polymer hybrid magnetic nanocapsules encapsulating IR820 and PTX for external magnetic field-guided tumor targeting and multifunctional theranostics. Nanoscale **9**(7), 2479–2491 (2017)
43. Lim, E., et al.: PH-triggered drug-releasing magnetic nanoparticles for cancer therapy guided by molecular imaging by MRI. Adv. Mater. **23**(21), 2436 (2011)
44. Liu, M., et al.: Recent advance in research on halloysite nanotubes-polymer nanocomposite. Prog. Polym. Sci. **39**(8), 1498–1525 (2014)
45. Lu, B., et al.: Graphene-based composite materials beneficial to wound healing. Nanoscale **4**(9), 2978–2982 (2012)
46. Ma, P., et al.: Rational design of multifunctional upconversion nanocrystals/polymer nanocomposites for cisplatin (IV) delivery and biomedical imaging. Adv. Mater. **25**(35), 4898–4905 (2013)
47. Masood, S.H., Singh, S.: (2007) Characterization of micro-features in polymeric drug delivery devices using FDM. In: ASME International Mechanical Engineering Congress and Exposition, pp. 463–470
48. Mezzenga, R., Ruokolainen, J.: Nanoparticles in the right place. Nat. Mater. **8**(12), 926–928 (2009)
49. Mitragotri, S., Lahann, J.: Physical approaches to biomaterial design. Nat. Mater. **8**(1), 15–23 (2009)
50. Mohammed, A., et al.: Additive manufacturing technologies for drug delivery applications. Int. J. Pharm. **580**, 119245 (2020)
51. Mota, F.B., et al.: Future of bioprinted tissues and organs: a two-wave global survey. Foresight STI Gov. **16**(1), 6–20 (2022)

52. Mpofu, T.P., Mawere, C., Mukosera, M.: The impact and application of 3D printing technology. Int. J. Sci. Res. (IJSR) **3**(6), 2148–2152 (2014)
53. Nagarwal, R.C., et al.: Polymeric nanoparticulate system: a potential approach for ocular drug delivery. J. Control. Release **136**(1), 2–13 (2009)
54. Nam, Y.S., Park, T.G.: Biodegradable polymeric microcellular foams by modified thermally induced phase separation method. Biomaterials **20**(19), 1783–1790 (1999)
55. Nida, S., Moses, J.A., Anandharamakrishnan, C.: Emerging applications of 5D and 6D printing in the food industry. J. Agric. Food Res. **10**, 100392
56. Park, S.H., et al.: Development of dual scale scaffolds via direct polymer melt deposition and electrospinning for applications in tissue regeneration. Acta Biomater. **4**(5), 1198–1207 (2008)
57. Paul, D.R., Robeson, L.M.: Polymer nanotechnology: nanocomposites. Polymer **49**(15), 3187–3204 (2008)
58. Peng, H., et al.: Polymeric multifunctional nanomaterials for theranostics. J. Mater. Chem. B **3**(34), 6856–6870 (2015)
59. Pietrzak, K., Isreb, A., Alhnan, M.A.: A flexible-dose dispenser for immediate and extended release 3D printed tablets. Eur. J. Pharm. Biopharm. **96**, 380–387 (2015)
60. Quanjin, M., et al.: Recent 3D and 4D intelligent printing technologies: a comparative review and future perspective. Proced. Comput. Sci. **167**, 1210–1219 (2020)
61. Reddy, P.R., Devi, P.A.: Review on the advancements of additive manufacturing-4D and 5D printing. Int. J. Mech. Prod. Eng. Res. Dev. **8**(4), 397–402 (2018)
62. Rehman, M., et al.: Additive manufacturing for biomedical applications: a review on classification, energy consumption, and its appreciable role since COVID-19 pandemic. Prog. Addit. Manuf. **8**(5), 1007–1041 (2023)
63. Sadiq, H.A.J., Patil, P.P.: Review on 4D and 5D printing technology. Int. Res. J. Eng. Technol. **7**(6), 744–751 (2020)
64. Salavagione, H.J., et al.: Chemical sensors based on polymer composites with carbon nanotubes and graphene: the role of the polymer. J. Mater. Chem. A **2**(35), 14289–14328 (2014)
65. Sin, D., et al.: Polyurethane (PU) scaffolds prepared by solvent casting/particulate leaching (SCPL) combined with centrifugation. Mater. Sci. Eng. C **30**(1), 78–85 (2010)
66. Shahbaz, A. et al.: Polymer nanocomposites for biomedical applications. In: Smart Polymer Nanocomposites, pp. 379–394. Elsevier (2023)
67. Sun, X., et al.: Recent progress in graphene/polymer nanocomposites. Adv. Mater. **33**(6), 2001105 (2021)
68. Tareq, M.S., et al.: Additive manufacturing and the COVID-19 challenges: an in-depth study. J. Manuf. Syst. **60**, 787–798 (2021)
69. Tarfaoui, M., et al.: 3D printing to support the shortage in personal protective equipment caused by COVID-19 pandemic. Materials **13**(15), 3339 (2020)
70. Titotto, S.: From sketches and installations to bioinspired 5D printing models: representation interactions for smart cities. In: Handbook of Research on Developing Smart Cities Based on Digital Twins, pp. 365–387. IGI Global (2021)
71. Tiwari, G., et al.: Drug delivery systems: an updated review. Int. J. Pharm. Investig. **2**(1), 2 (2012)
72. Tonelli, A., et al.: The geek and the chemist: antioxidant capacity measurements by DPPH assay in beverages using open source tools, consumer electronics and 3D printing. Sens. Actuators B Chem. **282**, 559–566 (2019)
73. Vaia, R., Baur, J.: Adaptive composites. Science **319**(5862), 420–421 (2008)
74. Varriale, V. et al. (2023) Industry 5.0 and triple bottom line approach in supply chain management: the state-of-the-art. Sustainability **15**(7), 5712
75. Wang, C., et al.: 3D printing of bone tissue engineering scaffolds. Bioact. Mater. **5**(1), 82–91 (2020)
76. Wang, M.: 5D printing and its application in tissue engineering. In: Materials Research Society (MRS) Spring Meeting. Materials Research Society (2021)
77. Wang, P., et al.: 3D printing of tissue engineering scaffolds: a focus on vascular regeneration. Bio-Design Manuf. **4**(2), 344–378 (2021)

78. Wang, X., et al.: 3D printing of polymer matrix composites: a review and prospective. Compos. B Eng. **110**, 442–458 (2017)
79. Weissig, V., Pettinger, T.K., Murdock, N.: Nanopharmaceuticals (part 1): products on the market. Int. J. Nanomed. **9**, 4357–4373 (2014)
80. Zare, Y., Shabani, I.: Polymer/metal nanocomposites for biomedical applications. Mater. Sci. Eng., C **60**, 195–203 (2016)
81. Zhang, Q., et al.: Fabrication of polymeric ionic liquid/graphene nanocomposite for glucose oxidase immobilization and direct electrochemistry. Biosens. Bioelectron. **26**(5), 2632–2637 (2011)
82. Zhang, Y., et al.: Graphene: a versatile nanoplatform for biomedical applications. Nanoscale **4**(13), 3833–3842 (2012)
83. Zhao, S., et al.: A review of magnesium corrosion in bio-applications: mechanism, classification, modeling, in-vitro, and in-vivo experimental testing, and tailoring Mg corrosion rate. J. Mater. Sci. **58**(30), 12158–12181 (2023)
84. Zhou, L., Fu, J., He, Y.: A review of 3D printing technologies for soft polymer materials. Adv. Func. Mater. **30**(28), 2000187 (2020)
85. Zmuda Trzebiatowski, P., et al.: Additive production management in COVID 19 pandemic. Eur. Res. Stud. J. **25**(1), 352–365 (2022)

Multifunctional Polymer Nanocomposites for 3D and 4D Printing: Idea of 6D Printing

Akshita, Devang Sukhla, Rohit Verma, Sheenam Thatai, S. P. Singh, and Tejendra K. Gupta

Abstract The field of 3D and 4D printing has significantly evolved with the advent of multifunctional polymer nanocomposites, which offer enhanced properties and capabilities for advanced manufacturing applications. This abstract explores the concept of integrating these advanced materials into 6D printing, a futuristic paradigm that extends beyond traditional 3D and 4D techniques. Multifunctional polymer nanocomposites, which combine polymers with nanomaterials, provide unique attributes such as improved mechanical strength, electrical conductivity, and responsiveness to environmental stimuli. These properties are crucial for 3D printing, which constructs objects layer by layer, and 4D printing, which introduces the dimension of time by enabling printed structures to change shape or properties in response to external stimuli. The proposed 6D printing concept aims to incorporate an additional dimension—dynamic adaptability—to the traditional three spatial dimensions and one temporal dimension. This involves developing nanocomposites that not only respond to environmental changes but also interact with dynamic systems, such as real-time data inputs or active external controls. By integrating sensors, actuators, and adaptive materials into the printing process, 6D printing envisions creating objects that can autonomously adjust their properties and functionalities in response to continuous external influences. This chapter outlines the potential benefits of 6D printing, including the creation of highly adaptive and interactive materials for applications in fields such as aerospace, healthcare, and robotics. It also addresses the challenges in material design, process control, and system integration required to realize this advanced manufacturing paradigm. Ultimately, 6D printing represents a significant leap forward in additive manufacturing, promising unprecedented levels of adaptability and functionality in future products and systems.

Akshita · D. Sukhla · S. Thatai · T. K. Gupta (✉)
Department of Chemistry, Amity Institute of Applied Sciences, Amity University Uttar Pradesh, Noida, India
e-mail: tkgupta@amity.edu

R. Verma (✉) · S. P. Singh
Department of Physics, Amity Institute of Applied Sciences, Amity University Uttar Pradesh, Noida, India
e-mail: rverma85@amity.edu

S. Moharana et al. (eds.), *Polymer Nanocomposites for 3D, 4D and 5D Printing*, Engineering Materials, https://doi.org/10.1007/978-981-96-4214-4_17

Keywords 3D printers · 4d printers · 6D printers · Polymer nanocomposites

1 Introduction

Multifunctional structures have emerged because of the merger of science and engineering, opening the door for multidisciplinary study targeted at improving effectiveness, economy, and quality of life. Multifunctional materials, also known as advanced materials, have sparked research into the creation and design of new goods. Designing structural elements with multifunctionality in mind entails giving them the ability to carry out duties beyond their main mechanical purpose. Energy harvesting and storage, self-healing, improved electrical and thermal conductivity, integrated sensors and actuators, and adaptive response (morphing) processes are some of these extra features. A possible approach for fiber-reinforced polymer composites for use in next-generation transportation vehicles is their multi-functionalization. To prove the competence and dependability of multifunctional materials, it is crucial to assess their attributes in terms of their mechanical, electrical, and other functional qualities. Multifunctional materials help cars and airplanes reduce weight by incorporating energy storage into their structural components [1]. The development of novel manufacturing techniques is just as important to the growth of scientific advances as the creation of sophisticated materials. It is anticipated that the development of new manufacturing processes will turn traditional procedures into sophisticated processing platforms that combine several local functions into a worldwide system. Multifunctionality may be included through the use of additive manufacturing as, unlike traditional techniques, it does not necessitate the actual inclusion of functional layers or laminas. Instead, it promotes customizable or arbitrary geometries during creation. Functional category-based characterization is acknowledged as important for the design and construction of multifunctional buildings. The reliability of future goods also depends on the integration of these materials with mechanical systems, therefore improving sustainable production processes for advanced materials is not the only issue [2]. In order to create the multidisciplinary multifunctional material system that is needed, theoretical, computational, and experimental methodologies must be integrated methodically. This is an example of an interdisciplinary technique that uses smart materials effectively and efficiently, requiring knowledge from the fields of mechanical, electrical, and materials science. A collaborative structure like this makes it easier to design intelligent or useful materials and gadgets. A polylactic acid matrix is applied at the nozzle to continuous carbon or jute fibers prior to three-dimensional (3D) printing. Comparing the fiber form to bulk 3D printed materials, experimental results show improved physical attributes such as increased tensile strength and elastic modulus [3]. This development has made way for the subsequent wave of composite constructions that are produced in three dimensions. These multipurpose materials may be printed in three dimensions, which offers several benefits for large-scale manufacturing: rapid fabrication, low processing costs, and the capacity to print intricate models. One of the printing process's main advantages is its adaptability, but there are still issues that need to be resolved in order to print intelligent or useful materials and structures.

With respect to 3D printing technologies, polymer-based smart materials are becoming more and more popular because of their many benefits, which include great cost-effectiveness, adaptability, and versatility. The research and development of multifunctional materials and related 3D printing techniques are receiving a lot of attention in an effort to address the growing demand for precise and effective production of complex multifunctional structures [4]. Graphene has proven to be an effective nanocarbon nanofiller for polymers [5]. It is utilized in the production of numerous rubbers, thermosets, thermoplastics, and conductive polymeric matrices. It has been found that the presence of trace amounts of graphene enhances the characteristics of these nanocomposites [6]. Many straightforward methods have been employed to produce polymer–graphene nanocomposites with fine nanoparticle dispersion and enhanced physical properties [7].

Nanomaterials combining polymer and graphene have shown promise in technological domains like energy, electronics, biomedicine, and engineering [8]. 3D and 4D printing methods have been employed to produce polymer–graphene nanoparticles [9, 10]. 3D printing is the process of producing three-dimensional objects using various instruments, materials, and procedures, including selective laser sintering, stereolithography, and fused deposition modeling. While 3D printing and 4D printing share similar technologies, 4D printing uses more advanced materials that respond to moisture, light, and heat. It has been found that graphene-based nanocomposites can be produced using both 3D and 4D printing techniques [11, 12]. The three- and four-dimensional polymer-graphene nanoparticles have shown remarkable characteristics and unique nanostructures. 3D and 4D printing technologies have led to a few useful advancements in the field of polymer-graphene nanoparticles [13, 14].

2 Characteristics of 3D and 4D Printing Methods

3D printing is recognized as an efficient manufacturing technique for producing 3D-printed goods when utilized appropriately [15]. 3D printing has become a major technological advancement in recent decades [16]. Effective 3D printing requires the controlled deposition of printable material. The layered deposition method, or additive manufacturing, is an essential 3D process [17]. A number of factors influence the effectiveness of 3D printing technology, including the choice of printable material, morphology, cost, structural complexity, and manufacturing process selection [18, 19]. Figure 1 depicts the various 3D printing technologies.

Moreover, in additive manufacturing, the complexity of the process determines the structural geometry [20]. Subtractive manufacturing has also been effectively used in conjunction with additive manufacturing to create 3D-printed goods [21]. Selective laser sintering, fused deposition modeling, stereolithography, selective layer melting, electron beam melting, and laminated object manufacturing are some of the techniques used to create 3D-printed objects. Every printing method has its benefits and

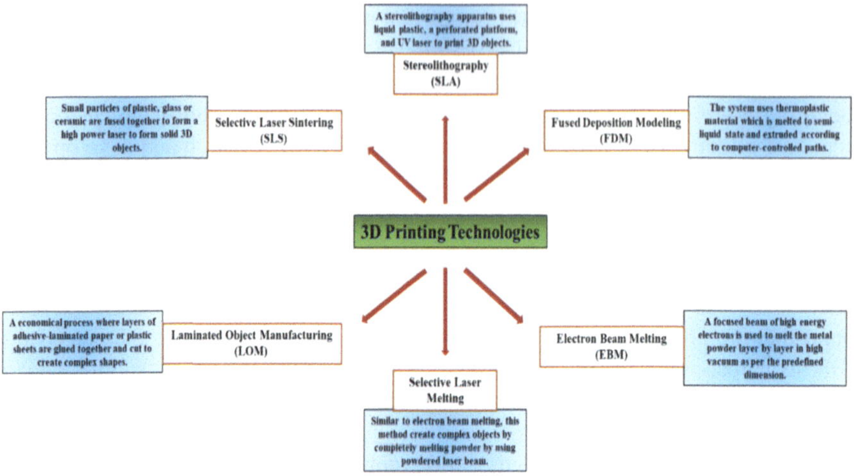

Fig. 1 Schematic diagram of 3D printing technologies

drawbacks when it comes to creating materials. The cost and time efficiency of the commercially available 3D printing technologies.

4D printing was made possible by developments in printing technology illustrated in Fig. 2 [22]. In contrast to 3D printing, 4D printing uses printed materials that self-assemble in reaction to outside stimuli [23]. Stimulus-responsive programming makes 4D printing methods novel [24]. 4D printing incorporates temporal dimensions and monitoring, in contrast to 3D printing. The 4D printing process has been used to create multifunctional next-generation sensors, actuators, robots, self-assembling structures, and other dynamic devices [25]. Applications for 3D and 4D printing may be found in the automotive, aerospace, robotics, electronics, and medical implant industries [26]. Controlling the production parameters necessary for these printing processes also means controlling the mechanical and geometric qualities of the printed products [27]. In addition to polymers and nanocomposites, 3D and 4D printing techniques have also been used to print metal-based materials [28]. As such, these sophisticated printing methods have the potential to be used in the design and production of sophisticated materials in the future [29].

3 Multifunctional Structures

A variety of secondary functional members are integrated with the load-bearing component (substrate), which acts as the primary functional member of an intelligent multifunctional structure. In this case, the system's multifunctionality is enhanced by flexible electronic components, photovoltaics (PVs), batteries, and piezoelectrics.

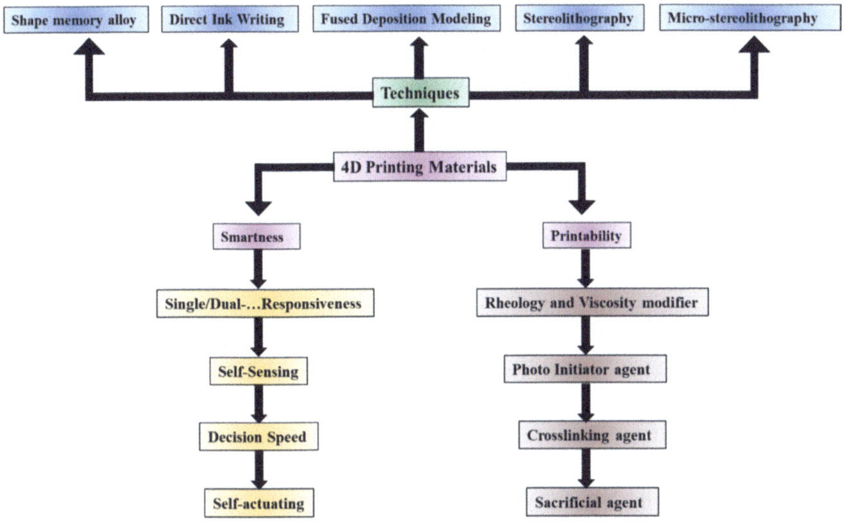

Fig. 2 Flowchart of 4D printing technologies

It is possible that some of these secondary functional elements are not polymeric. However, our goal is for all polymer-based products to meet the demand soon.

3.1 *Piezoelectric Structures*

In the 1880s, Pierre and Jacques Curie made the first observations of the electromechanical conversion phenomenon [30]. A property known as piezoelectricity, or electromechanical coupling, causes some crystals (like quartz, tourmaline, barium titanate, Rochelle salt, and so forth), polymers, and ceramics to respond mechanically to an electrical input and vice versa. Piezoelectric materials have enormous potential for application as actuators and sensors in electrical and energy-harvesting systems [31, 32]. Braille displays tactile devices, like vibrating actuators, are made of electroactive polymers like cellulose acetate and graphene oxide [33]. When an electric field is provided, these materials elongate (strain) in actuator mode; when mechanical strain is applied, they create charge or voltage in sensor mode. Due to their fragile nature, piezoceramics are usually produced as tiny, simple-shaped pieces. Piezopolymers, on the other hand, are easily cut into a variety of forms and sizes due to their flexibility and thin sheet form. Piezopolymers have better sensing capabilities because they are more sensitive to mechanical loads under different loading conditions than piezoceramics. It is believed that piezopolymer-based additive manufacturing, or 3D printing, is a viable technology for producing useful devices [32]. Bodkhe et al. did not use traditional attachment techniques; instead, they directly printed sensors on substrates (such as shoe surfaces) [34] The sensing capacity of

PVDF with 10 weight percent of $BaTiO_3$ nanoparticles was increased by 300% when the two processes were carried out simultaneously. This increase occurred when the PVDF was exposed to electric fields up to 1 MV/m. Comparing this efficiency to non-in situ poling procedures, it was higher. Hoejin et al. used a fused deposition method 3D printing technique to create a filament with a diameter of 2.7 mm [35]. In contrast to the conventional solvent-based method, the extruded and spooled filament produced a threefold increase in piezoelectric response and showed a consistent dispersion of BTO (barium titanate) nanoparticles in PVDF (polyvinylidene fluoride). The electromechanical performance of $BaTiO_3$/PVDF was greatly improved by using a thermal poling technique. Moreover, Hoejin et al. used a corona poling technique to activate 3D-printed PVDF sheets as piezoelectric-sensitive materials [35]. Using a state-of-the-art process called "Integrated 3D Printing and Corona Poling Process (IPC)," this was achieved. The result was piezoelectric (BTO/PVDF) composites, which are perfect for sensor and energy-harvesting applications and are simple, dependable, and reasonably priced to produce. Because of the IPC process, the final tensile strength of the created piezoelectric composites increased and they could generate current up to 0.106 nA, or 12 kV of electric power. Higher bed temperatures were shown to decrease piezoelectricity, though. By distributing piezonanoparticles in a photolabile polymer solution, a microscale digital projection printing technique was used to create three-dimensional piezoelectric materials. The superior performance of 3D-printed BTO/PVDF nanocomposite films over solvent-based films has been shown by several experimental characterizations [36]. Results from scanning electron microscopy (SEM) showed that, in comparison to solvent-based films, 3D printed films exhibit reduced aggregation, porosity, and fractures. In a comparable manner, X-ray diffraction (XRD) spectra showed a uniform distribution of BTO nanoparticles, while results from Fourier transform infrared spectroscopy (FTIR) suggested a higher β-phase concentration in 3D printed films. Additionally, it was found that this method generated more current [35].

3.2 Battery Fiber Structures

Due to their promise, multifunctional materials with uses in flexible electronics and energy storage have drawn a lot of research attention [37]. For the use of cutting-edge additive manufacturing processes like liquid deposition modeling (LDM), stereolithography equipment (SLA), fused deposition modeling (FDM), and electron beam manufacturing (EBM) these materials must be developed. In contrast to conventional subtractive approaches, these techniques add material by consecutive layers to form 3D structures [38]. Electrochemical devices, such as batteries, directly transform chemical energy into electrical energy. They are made up of many voltaic cells with two half-cells in each. One half-cell's positive electrode draws in cations, whereas the other half-cell's negative electrode draws in anions. A conductive electrolyte containing the appropriate metal cations connects these half-cells in series.

The rapid development of electric vehicles and portable electronics has accelerated the search for novel energy storage technologies with higher power and energy densities [39]. The present state of lithium-ion intercalation technology is dominated by its extended lifespan, energy densities, lacking effects of memory and high power [40]. Applying a voltage during battery charging encourages electron mobility and electrode reactions. Battery technology advancements often aim to minimize volume growth while increasing power density, energy density, and cycle ability. This may be accomplished by creating innovative electrode materials. Because activated carbon is readily available and reasonably priced, it has been extensively utilized for electrochemical energy storage. However, because of their improved mechanical, electrical, and catalytic qualities, carbon nanomaterials including graphene, carbon nanotubes (CNTs), and ordered mesoporous carbon (OMC) may be better substitutes [41]. Novel materials are being developed and fabricated for flexible energy storage devices, including polystyrene sulfonic acid and polyvinyl alcohol reinforced mix nanocomposites. These nanocomposites contain carbon black nanoparticles for reinforcement [42]. The size and form of batteries have a big influence on how gadgets are made and how long they last. The volume that batteries take up in laptops and mobile phones has a direct impact on these devices' thinness and battery life. To maximize energy storage and minimize the weight and size of batteries, new methods for quick design, prototyping, and mass production in a variety of desirable shapes must be developed. Unconventional form factors have been studied in the past, such as flexible wire-shaped batteries, printable solid-state batteries, thin paper batteries, and spray-painted batteries [43]. 2005 witnessed the publication of one of the earliest reports on a polymer battery in the shape of a flexible fiber. The battery employed polypyrrole-hexafluorophosphate as the active cathode material and polypyrrole-polystyrene sulfonate as the anode material electrodeposited on a platinum wire substrate. The battery was a major development since it established a new form factor, even if its capacity was only about 10 mAh/g. These substances are capable of providing batteries for multiple uses, which creates opportunities for their application as textile materials and structural elements. Structural batteries with many functions can potentially boost the system's overall power and energy densities. One instance of this is the application of reinforced electrode composite, which has the potential to be used as body panels for unmanned aerial aircraft and electric cars. Moreover, lithium-ion batteries (LIBs) that are readily accessible for purchase can be included in a panel's cavity and used as a structural element of a satellite or spacecraft. Because of their intrinsic multifunctionality, these batteries have the potential to be utilized in a wide range of electric vehicles in the future, including those for air, sea, and road transportation. This is because they can provide nearly "massless" energy storage [44]. Unlike typical manufacturing methods, the manufacture of these multifunctional materials provides a demanding study topic. Novel additive manufacturing techniques have emerged as next-generation technologies for creating complex multifunctional materials. Examples of these techniques are 4D and 3D printing. With the aid of additive manufacturing, intricate three-dimensional structures may be produced quickly and with no need for significant adjustments to the manufacturing method already in use. Additionally, the incorporation of functional conductive materials into the 3D printing process enables the production of structural components that are linked to electronics. For instance, adding conductive nanoparticles to the additive manufacturing process has greatly streamlined the microfabrication of microscale electrical circuits [45]. Additionally, the blend of additive manufacturing using smart materials has given rise to new area of study in 4D printing [46]. While the field of advanced functional materials additive manufacturing is still in its infancy, breakthroughs in this area are anticipated to propel research across several disciplines. As was previously noted, conductive features in multifunctional materials are often achieved through the employment of lithium ions and carbon-based materials in the additive manufacturing process. In order to give polymer-based materials electrical and sensing properties, carbon-based components such as graphene and carbon nanofiber are frequently incorporated [47]. Microbatteries are commonly made using conductive materials such as graphene, zinc, and lithium ions. It may be possible to boost the energy density of additively built microbatteries by making efficient use of the restricted 3D area [48]. The most successful, effective, and efficient forms of 3D printing are believed to be extrusion-based processes. These methods have been widely applied to the development of 3D systems for conventional energy storage applications in a fast and reliable manner. In an application of energy storage systems, Wang et al. [49] demonstrated the 3D printing of all-fiber lithium-ion batteries. These batteries are used as energy storage devices and are made using an extrusion-based 3D printing technique. The printing inks, which were composed of CNTs, lithium iron oxide, and lithium titanium dioxide, were used to print lithium iron phosphate fiber cathodes and lithium titanium oxide fiber anodes. The substantial flexibility and electrochemical capabilities of the manufactured fiber electrodes were used to highlight their multifunctional qualities. Mechanical tensile tests were used to verify the 3D-printed fibers' strength and flexibility. The electrochemical properties of 3D-printed fiber electrodes were evaluated using coin cell designs and charge–discharge curves. For electronic wearables, the manufacturing of these fiber-shaped batteries presents a feasible future. Reyes et al. [43] provided evidence of the creation of fused filament manufacturing technique-based 3D-printed lithium-ion batteries. They also demonstrated how cutting-edge materials are used to create wearable electrical devices that are 3D printed and have built-in batteries. An electrical conductivity source meter from the Keithley 2400 series was used to calibrate it. Conductivity, filler loading, and charge storage capacity were shown to be related. The combination of graphene nanoplates and lithium titanate material in the anode and carbon nanotubes and lithium manganese oxide in the cathode produced the maximum charge storage capacity, according to the data. Structural batteries are multipurpose devices that improve the system's overall power and energy density. For instance, the body panels of electric and unmanned aerial aircraft are composed of structural batteries made of electrodes reinforced with composite materials [50]. Also, by incorporating readily available commercial lithium-ion batteries into a cavity panel, a multifunctional structural device utilized in aerospace applications is created [51]. The capacity of three-dimensional printing to create any kind of three-dimensional geometric component is a skill that is heavily employed in the production of wearable electronics. The ability to 3D print wearable electronics with built-in batteries has been shown. In one instance, the system powers an LCD panel that darkens the screen of 3D-printed, battery-operated LCD sunglasses. It does this by applying electricity to a switch. These batteries have a series connection. The installation of the battery components is made easier by the shape of the sunglasses' frame. The battery's separator, anode, and cathode are printed individually and filled with electrolytes before integration. A bracelet that emits light and has an integrated battery. The cathode and anode are curved electrodes that are 3D printed to match the contour of the bangle's design. Utilizing a Celgard polymer separator, the printed device enhances battery performance and facilitates the integration of 3D battery printing technology with readily available materials. LED can run for 60 s on the assembled bangle battery. Global reliance on energy is gradually moving away from finite fossil fuels and toward sustainable and renewable energy sources. As a result, conventional power sources are beginning to give way to energy storage technologies like lithium-ion batteries and supercapacitors. These energy storage devices still have performance challenges to be solved before they can attain high energy densities, long cycle lives, and flexible features. Next-generation energy storage devices could be produced using 3D printing techniques because they enable accurate functional element supply and mixing during the process [52]. Materials designed for 3D printing that have more functions than conventional materials (pure resins) can satisfy targeted multifunctional structural applications. For conventional 3D printing materials, nanoparticles can provide, among other multifunctionalities, mechanical, thermal, actuation, sensing, conductive, and magnetic capabilities [53]. The addition of nanocomposite materials into 3D printing techniques opens up significant possibilities for the production of materials with optimized and multifunctional properties. Along with improving the functionality (mechanical strength and electrical conductivity) of manufactured goods, this integration makes 3D printing processes more printable. To fully benefit from the integration of nanotechnology into 3D printing technologies, it is imperative to understand the various types of functional nanocomposite materials and their processing techniques. Developing suitable printing methods is required to produce 3D-designed structures. Furthermore, a few problems (such as printing and dependability) need to be fixed in order to fully realize the promise of 3D printing with multifunctional nanocomposites. One example of an issue with the extrusion-based 3D printing method could be that the addition of nanomaterials causes the printing materials to become more viscous, and this could lead to their aggregation and block the nozzle. Therefore, it is necessary to employ suitable and effective material mixing techniques. Materials used in unmanned aerial vehicles, satellites, and flexible electronics require high specific weight and strength as well as high specific energy densities. The multifunctional properties of CNTs are well known to most people. Carbon nanotubes (CNTs) must be integrated as fillers into polymer composites in order to achieve the required properties. Careful engineering of the chemical attachment at the interface between CNTs and the polymer matrix is necessary to ensure mechanical load transmission to individual nanotubes. The mechanical properties are affected by the volume fractions of CNTs incorporated into the polymer matrix. Li-ion batteries and supercapacitors are suitable, effective electrochemical storage devices with the required functional characteristics. Compared to Li-ion batteries, which have a moderate power density but a slightly higher energy density, supercapacitors may deliver 20–100 times more power per unit volume. Benson et al. have demonstrated that the combination of electrodes posing electrically conductive polyaniline (PANI) polymer onto carbon nanotube (CNT) fabric is a unique material that can be used to create desired functional features [54]. PANI was selected due to its superior mechanical properties, elevated conductivity, and robust resistance to environmental changes. The resulting flexible composites showed a specific capacitance that was significantly higher than commercially available carbon nanopowders when evaluated for supercapacitors. These features offer good energy and power densities along with a long cycle life. One common 3D printing technique for producing 3D battery and supercapacitor electrodes is direct ink writing. This method is used in the 3D printing of energy devices because it is low-cost, adaptable to different types of materials, and capable of producing complex three-dimensional structures without requiring expensive secondary machining processes. The 3D printing technology allows the functional inks to be precisely engineered for a range of complex geometries in filament form. For instance, 3D printing can be used to create helical fluidic microchannels and microstructured fibers using the thermoplastic polymer polylactide (PAL). The greater energy and power density made possible by the larger electrode surface area are the primary advantages of this method. Printing processes and ink preparation must be understood in order to design the material system. These techniques include the use of CNT-, graphene-, and Zn-based electrode materials to fabricate supercapacitors and energy storage devices. This method includes a micro-nozzle, dispenser, and basic desktop-based 3D printer for a few hundred dollars upfront. Furthermore, "3D direct-ink writing" describes a manufacturing process that lets a simple desktop computer oversee the production of complex 3D objects. Methodically, volatile solutions of carbon fibers, polymers, solvents, and other additive materials are mixed to create carbon-based inks. In order to achieve the required level of printability, more polymers are required to control the viscosity of inks derived from carbon-based materials. When it comes to active materials for electrode fabrication, graphene-based materials are commonly used due to their excellent chemical stability, large specific surface area, and high electrical conductivity. Many synthetic methods, such as chemical vapor deposition (CVD) and ethanol-based CVD, are used to create graphene-based materials [55]. The increasing demand for efficient and sustainable energy storage devices can be satisfied by using additive manufacturing techniques to create electrodes and electrical energy storage systems with the required 3D geometries while preserving the multifunctional properties of processing materials. One-step 3D printing of materials is now possible thanks to the development of water-based thermoresponsive inks. Examples include the 3D printing of graphene-based electrodes for lithium-ion batteries and the multi-material printing of graphene and copper [56]. Graphene's unique properties make it valuable for a wide range of applications, including multifunctional composites and Li-ion batteries. This material's strong microstructure, large surface area, and superior electrical conductivity enable improvements to be made to the features and operation of current electrical energy devices. Graphene's introduction into additive manufacturing, specifically in the 3D printing of electrodes and electrical components for energy storage, presents a promising path for practical applications. There are limitations on the ways in which graphene can be produced while preserving its special properties and how it can be mixed with other practical materials. Controlling graphene at both macro and micro scales is desirable for a wide range of potential applications. In this sense, printing graphene-based 3D functional materials seems like a promising application for additive manufacturing methods. With the direct-ink writing methodology, an extrusion printing technique that deposits a liquid ink substance that quickly hardens upon extrusion, building three-dimensional structures layer by layer is made easier. This multimaterial printing technique offers many advantages for performing various tasks in an advanced 3D printing format.

Jakus et al. [57] reported creating graphene composite material at room temperature through the use of liquid ink and extrusion-based 3D printing. The 3D-printed graphene composite, which is composed of graphene and polylactide-co-glycolide, exhibits exceptional properties for application in energy storage and functional composites. Compared to 3D-printed carbon materials, the final 3D-printed material is more flexible, stronger, and electrically conductible. The direct ink is prepared by standard mixing of readily available ingredients (dissolved polymer and graphene), which is then allowed to evaporate at room temperature. To facilitate proper dispersion and mixing with the graphene material, polylactide-co-glycolide is first dissolved in dichloromethane. The ready ink is stored in a glass container that is tightly sealed and kept in the refrigerator. The ink can then be extruded using a nozzle to produce stacked three-dimensional objects. Graphene polymer composite's mechanical strength and flexibility enable the construction of complex thin-structure geometries that are sliced, rolled, and folded. The 3D-printed composites made of graphene can withstand stresses of more than 210% before breaking. By altering the composition of the initial inks, it is feasible to tailor the mechanical and electrical properties of the 3D-printed graphene material structure for specific applications. Electrical conductivity is measured by connecting a triggered LED in series with the assembled 3D open-mesh cylinders to verify resistance. The methods used in the processing of graphene inks intended for 3D printing. The procedure entails mixing the dissolved polymer and dispersing graphene powder in a graded solvent. The 3D graphene (3DG) inks are manufactured and then used to quickly construct unique 3D structures that can be used for a range of applications, as mentioned in Fig. 3. A multifunctional property-based 3D liquid sensor composed of PLA (polylactic acid)/multi-walled CNTs solutions is produced by a solvent-based 3D printing technique.

The high electrical conductivity of the manufactured helix provides exceptional sensitivity for sensing applications. The 3D-printed liquid sensor is a great option for liquid sensing applications because of its many uses, including its electrical conductivity, structural functions like liquid trapping, and lightweight design. To make the ink, PLA is dissolved in melted dichloromethane, a low-boiling-point solvent. The nanocomposite solutions are then continuously forced through a micro-nozzle using a syringe. Rapid solvent evaporation causes a quick phase transition from fluid to solid, resulting in the formation of very effective and efficient features. The process

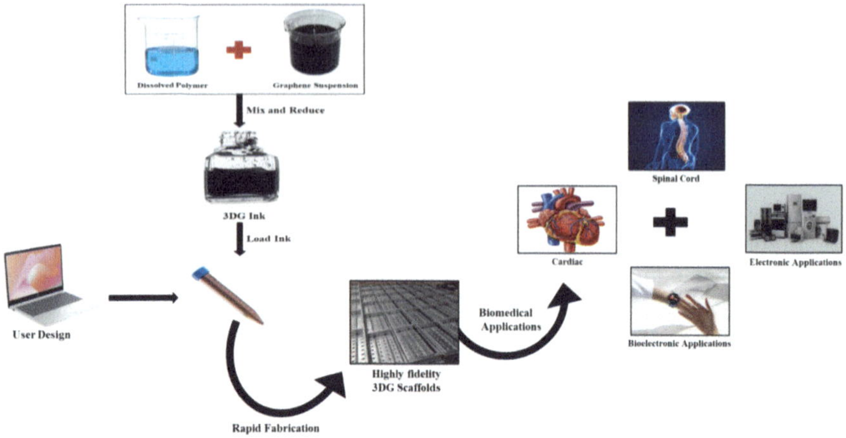

Fig. 3 The processing method of synthesis of 3D structures, and possible uses of 3D printable graphene (3DG) inks [59]. Copyright 2015. Reproduced with permission from the American Chemical Society

can be altered by appropriately altering the printing parameters (temperature, pressure, and flow rate) and ink properties [58]. A brief demonstration of piezoelectric output current measurement is mentioned in Fig. 4.

3.3 PV Structures

By using the photovoltaic effect, a solar cell, sometimes referred to as a photovoltaic (PV) cell, transforms sunlight into electrical energy. A solar module or photovoltaic panel is made up of many of these solar cells. PV panels that are linked in series have accumulating voltages and steady currents. On the other hand, when they are joined in parallel, the voltage stays constant, but their currents increase. PV cells are divided into two types:

(1) Inorganic semiconductor materials, such as crystalline, polycrystalline, amorphous, and microcrystalline silicon, are used to create inorganic solar cells or PV cells.
(2) Organic solar cells, or OSCs, use conductive organic polymers or molecules to capture solar radiation and use the photovoltaic effect to generate electricity.

Compared to inorganic solar cells, organic solar cells (OSCs) have a few benefits, such as the capacity to be produced as thin films with high sunlight absorption coefficients, low specific weight, and biodegradability. Additionally, they are very transparent and flexible, which makes them suitable for usage on a variety of substrates, including window glass panels. Solar energy is captured and stored by organic photovoltaics (OPVs), and gel-based solar cells can be made through additive

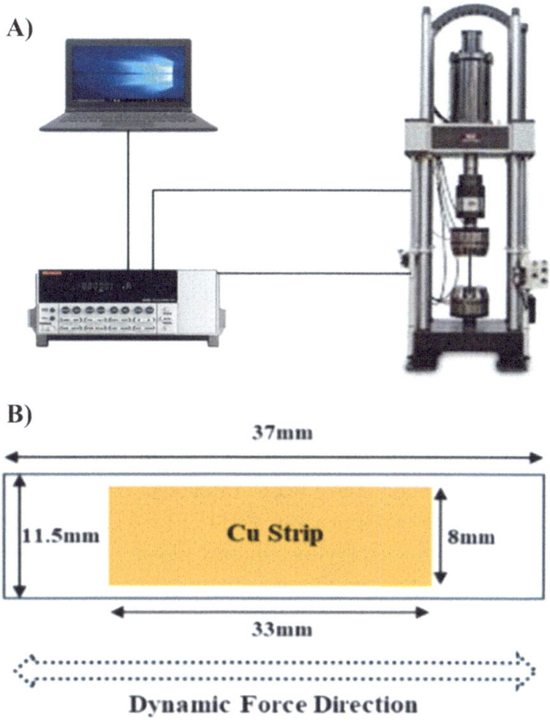

Fig. 4 Schematic diagram of (A) Experimental setup for piezoelectric output current measurement, and (B) Sample and electrode design [60]. Copyright 2017. Reproduced with permission from Sage Publications

manufacturing, or 3D printing, which is a promising technique [61]. Additionally, due to this method's scalability and ease of transferability, devices on various substrates with precise sizes can be created, resulting in the production of solution-processed solar cells from lab to fab and roll-to-roll (R2R) [62]. Both thin solid films and large layers (100 μm) may be produced by 3D printers [63]. Because inkjet printing technology allows for the accurate application of conductive ink at a reduced cost, it is a viable option for producing 3D-printed bulk heterojunction solar cells. Thin-film transistors, photodetectors, and organic light-emitting diodes (OLEDs) may all be produced affordably with inkjet printing. The energy conversion efficiency of inkjet-printed organic photovoltaic devices is on par with those obtained by traditional spin coating techniques. The shape and microstructure of polymer-fullerene films need to be regulated in order to increase the performance efficiencies of inkjet-printed solar cells [64]. Slot die 3D printing is the method of choice for OSCs, even though FDM is more widely used and inkjet printing is more efficient. In order to maximize cost-effectiveness and reduce material waste, slot die printing uses a solution dispenser that is synced with existing hardware to achieve consistent stripes in both width and thickness. [63] Power conversion efficiency (PCE) of about 4–5% was obtained by this approach, which overcomes the limits of nozzle-based printing in producing the requisite low thickness. After thermal annealing, the PCE climbed to 6%. While the PCE of perovskite solar cells (PCE ≈23%) is higher than that of organic photovoltaics (OPVs), perovskite cells are not as popular as OPVs because of the toxicity of the chemicals used in them and the need for continued research to enhance their stability and environmental effect. 3D printing can also be used to produce perovskite solar cells, while problems with scalability and cell shape still exist. These solar cells have been mass-produced using a slot die coater and an R2R base in order to overcome scaling issues. IV curves (current vs. voltage) are used to characterize the performance of solar cells. In a scientific environment, a solar simulator is used to mimic the sun's power output using several kinds of lamps, most frequently Xe lamps. This makes it possible to expose materials and electronics to irradiance and to characterize solar cells electrically [65]. Krebs et al. [66] have performed mechanical fatigue tests on rolled 3D-printed OSCs under various bending and torsion loading situations. In addition, failure areas under cyclic loading circumstances were identified using finite element analysis. These findings provided strategies to extend the life of 3D-printed solar cells. Studies comparing OPVs for two sets of polymers (PBDTTTz-4 and P3HT) that were created by spin coating and R2R methods have been conducted. Validation experiments showed that when employing the R2R process, PBDTTTz-4 is more efficient than P3HT. Additionally, it was shown that drying kinetics plays a critical role in the slot-die-coated R2R process [67]. One major obstacle to the growth of renewable energy is its high cost. Solar gadgets are still pricey and out of reach for many people. Using 3D printing as a production technique might save costs and improve product quality. This method permits design changes to be made as the product is being printed, as well as the prudent use of raw resources. While searching for the perfect material for solar cells, 3D printing is a good production solution. It takes a lot of testing and prototyping to create new solar panels with cutting-edge materials and technological characteristics. Scientists at CSIRO have produced A3-sized solar cells using 3D printing that can be rolled up and applied to a variety of surfaces, including windows and buildings. PV ink distributed on flexible plastic strips is used in the production of these useful and effective solar panels. Screen printing, gravure coating, and slot-die coating are some of the manufacturing techniques used. These solar cells can now be precisely 3D printed thanks to additive manufacturing, which presents a huge opportunity for mass customization in this industry. Customized 3D-printed solar panels with a desired shape and size can be produced. Researchers in Mexico have used additive manufacturing to optimize light collection by reshaping absorbers, leading to a 20% increase in plant efficiency. Consequently, even though producing solar panels may seem challenging, it is entirely feasible to create 3D-printed solar panels using simpler and quicker processes [68].

3.4 Flexible Electronics Structures

There are always fresh opportunities being presented by the research and manufacturing of wearable electronics and embedded electronics. Conventional manufacturing methods are based on assembling discrete pieces that are externally manufactured using many machining operations; these parts are limited in terms of both functionality and design. Unified manufacturing processes may be used to produce functional materials thanks to the progress made in material interface engineering. As a result, multidimensional control over embedded and wearable electronics is made possible by additive manufacturing techniques, allowing for the quick production of numerous functional pieces composed of various materials. To maximize the benefits of 3D printing, new fabrication techniques and improved material qualities are needed. For instance, high-performance multifunctional nanocomposites reinforced with graphene that exhibit enhanced thermal conductivity and dielectric characteristics are much sought after for use in integrated electronics applications [69]. The creation of conductive graphene nanocomposites—a blend of flexible conductive film and graphene oxide—offers several benefits to electronic-based touchscreen device development, including ultrahigh flexibility, chemical stability, and conductivity [70]. The synthesis of nanostructured ZnO by biological processes and combustion offers additional advantages for the economical and environmental benefits of sensor device production [71]. Hybrid zinc ferrite nanostructures prepared by spin-coating have been used to create chemical sensors capable of identifying liquefied petroleum gas at low temperatures [72]. Additionally, chemical sensors that can detect liquid petroleum gases at room temperature have been created using white graphene-reinforced mix nanocomposites made of polyvinyl alcohol and polypyrrole [73]. Strong sensitivity and resistance to high temperatures are attributes of oil sensors made possible by PANI-based polymer composite films [74]. Using blending techniques to improve the dielectric characteristics of cellulose acetate and aluminium nanoparticles will improve their performance and make them useful for embedded electronics design [75]. Self-transforming electronic structures are becoming more and more important in flexible electronics because of their extraordinary capacity to take on preprogrammed complicated configurations for certain capabilities. Sundaram et al. [76] presented a technique for the instant folding of 3D-printed functional composites with integrated electronics. The composite material was printed using multimaterial printing without the need for any additional processing processes. The composite material uses the leftover pressures from the polymer-based swelling process to self-shape during the layer-by-layer manufacturing process. These devices that change shape offer a way to adapt planar electronics with intricate geometries into nonplanar electronics. Self-transforming electronics and 3D printing may be used to allow a variety of applications, including robots and sensors. Using 3D printing, Sampada et al. [77] created coextruded multifunctional piezoelectric sensors in a single step. This technique was created to create conformal sensors without using an industrial poling procedure. These sensors are used in wearable electronics to track the aeroelastic stability of 3D-printed wings and to study breathing and knee motion. Fiber encapsulation additive manufacturing (FEAM) has been used to 3D print customized, complex electromechanical, and electrical equipment using a simple 3D printer [78]. The electrical performance, printability, and material compatibility of three solder, solder paste, and conductive polymer composite electrical connectors have all been assessed. Combining the benefits of FEAM and thermoplastic elastomer additive manufacturing (TEAM) processes, Cohen et al. [79] developed a force sensor. Mechanical and electrical characterizations were done in order to perform sensitivity analysis and assess the performance under various loading conditions.

3.5 4D Structures

A specific class of functional materials that alter their characteristics over time in response to various stimuli (such as thermal, mechanical, chemical, magnetic, electric, and photochromic stimuli), either separately or in combination, enable 4D structures—dynamic, perceptive 3D structures [80]. Additive manufacturing techniques combined with multifunctional materials can create multi-stimulation objects or devices. Cao et al. [80] examined the effects of both numerous stimuli and single stimuli, including chemical and physical changes brought on by heat, light, and pressure. Complex functional components with topological optimization have been created and produced for use in industries such as robotics, electronics, smart clothing, aerospace, and textiles [81]. The deformation of multimaterial grippers has been shown using a dependable and simple method. These shape-memory polymer (SMP)-based 4D printing techniques can be applied to medical equipment, solar panels, and aircraft parts.

4 3D and 4D Printed Polymeric–Graphene Nanocomposite

Notable advancements in the domains of 3D and 4D printing technologies have occurred during the last few decades [82]. Thermoplastic polymers, metals, alloys, organic and inorganic nanomaterials, and other materials have been treated with these methods. Studies on graphene nanomaterials that are 3D- and 4D-printed have concentrated on their enhanced properties, processing techniques, technological promise, and difficulties [83]. Direct ink writing [84], inkjet 3D printing [85], stereolithography (SLA) [86], fused deposition modeling (FDM) [87], extrusion printing [88], binder-jet printing [89], and other 3D printing techniques are utilized to create graphene nanomaterials. Direct 3D printing works by extruding a viscous material from a pressure-filled syringe to create three-dimensional objects. The syringe head moves in three dimensions while the platform remains stable to build the item layer by layer. The controlled extrusion of thermoplastic filaments powers the FDM printer. Layer by layer, the semi-liquid filaments in FDM are extruded onto the platform where they solidify into the final pieces after melting at the nozzle. By altering the printing settings, the printed product's quality may be managed. In stereolithography, a resin reservoir is exposed to a regulated UV laser to facilitate photocuring and the creation of a patterned layer. Acrylic and epoxy resins are two examples of typical polymers used in SLA. As both methods rely on powder processing, selective laser sintering, and SLA are comparable processes. On the other hand, liquid binder is not used in SLS; instead, a laser beam is used for controlled route scanning and printing. Molecule diffusion is used to fuse powders using high-power lasers. Typically, variables like scan speed, laser power, and powder particle size affect the final product's resolution. To obtain the desired nanomaterial with excellent qualities, the right printing technique must be chosen. Therefore, the performance of printed material is affected by a number of factors, including the polymer matrix, graphene dispersion, graphene functionality, matrix-nanofiller interaction, interface formation, and printing parameters like process type, printing direction, and ink viscosity. The mechanical, electrical, thermal, and other crucial physical characteristics of 3D-printed polymer-graphene nanomaterials have been the subject of numerous investigations [90]. In this instance, 3D printing has mostly made use of thermoplastic materials. But 3D printing has also been used to produce certain thermosets and stretchy polymers. For example, in the 3D direct ink writing technique, a pressured syringe is used to extrude the polymer-graphene nanocomposite [91]. Before choosing a particular printing method, one must weigh the relative benefits and drawbacks of each of these 3D printing approaches [92, 93]. The 3D direct ink printing method was utilized by Qian and colleagues [94] to create poly(lactic acid)–graphene oxide nanocomposites. Comparing the poly(lactic acid)–graphene oxide nanocomposite to pure poly(lactic acid), differential scanning calorimetry (DSC), and thermogravimetric analysis (TGA) revealed higher melting and thermal degradation temperatures. The higher melting temperature of the 3D-printed poly(lactic acid)–graphene oxide nanocomposite was caused by the increased plasticity capability and dispersion of nanofillers. Moreover, the compatibility between the matrix and nanofiller was enhanced by the hydrogen bonds between poly(lactic acid) and graphene oxide, which led to a high level of thermal stability for the nanocomposites relative to the pure polymer. Mechanical testing showed that the 3D-printed material with 1.6 weight percent graphene oxide had a tensile strength improvement of 67.4% over the plain printed polymer. The homogeneous distribution of nanoparticles in the polymer resulting from the printing process enhanced the strength of the nanocomposite by acting as fine reinforcement. Consequently, research has focused on the 3D direct ink printing technique as a viable means of creating high-performing nanocomposites. The processing of nanocomposite materials has been studied using 4D printing techniques [95]. Micro-stereolithography, FDM, SLA, and direct ink writing are popular 4D printing techniques. The reaction of the 4D-printed complex nanostructures to various environmental stimuli has been investigated. As was previously said, 4D printing is an improved version of 3D printing that includes stimuli-responsive elements and time limits. Smart shape memory items have been designed using 4D-printed nanocomposites. Smart nanomaterials such as graphene nanocomposites combined with shape programming typically result in unique processes for shape-changing when printing objects with 4D printing. Wei et al. [96] created graphene-derived, poly(lactic acid), and acrylonitrile–butadiene–styrene nanocomposites using FDM-based 4D printing. For computer-based models, the 4D-printed materials effectively included 5.6 weight percent graphene components. The ultimate nanocomposite displayed a linear thermal coefficient of 75 ppm$^{\circ}C^{-1}$, suggesting that the chosen printing process enhanced the material's ability to expand appropriately at elevated temperatures. The SLA-based 4D printing method was used by Zarek and associates [97] to create printed polymer–graphene nanostructures. The behaviour of shape memory showed that the 3D technique may be used in several types of printed models. Digital models were produced using commercial computer-aided design (CAD) software (Autodesk Inventor, Mill Valley, CA, USA) that had been modified using the printer software. At room temperature, the printed objects had a hard, wax-like surface. The structures become elastomeric and flexible above the melting temperature (Tm), which for polycaprolactone is 55 °C. The structures were cooled below Tm to rectify their deformation. The constructions regained their original printed shape upon heating. With the use of a heat gun, the temperature was raised to 70 °C to initiate the form memory procedure. Thermoset shape memory vascular stents with submillimetre thicknesses and plenty of voids were produced. By using this method, high-resolution and intricate models were produced, along with a scaled Eiffel Tower and a bird that were printed. For technological applications, pyrolytic carbon microelectrodes have been acknowledged as a potential substitute for their traditional metallic counterparts [98]. Graphene-like materials have been created from graphene oxide or pyrolytic carbon via laser printing technology. Highly oriented pyrolytic graphite and graphene-based nanomaterials have been used to create screen printing electrodes and other carbon electrode materials [99]. A multiple pulsed laser approach was utilized by Chyan et al. [100] to create laser-induced graphene (LIG) on a range of substrates. To start the creation of graphene on biodegradable substrates including paper, fabric, coconut shells, cork, and potato skins, a CO_2 laser beam was used. The generated nanomaterial is now appropriate for flexible micro-supercapacitors on renewable materials thanks to the use of LIG on these substrates. Moreover, high-temperature engineering polymers like poly(ether imide), Kevlar, Kapton, and polysulfones have been used to create LIG. Crosslinked polystyrene has also been used as a substrate for LIG synthesis. By utilizing a continuous wave semiconductor-diode laser to selectively pyrolyze SU-8 photoresist, Ludvigsen et al. [101] created pyrolytic carbon microelectrodes. The SU-8 precursor was effectively converted into pyrolytic carbon by exposure to laser light. It was suggested that the resulting extremely porous materials be used as electrodes in electrochemical energy storage and biosensing.

5 Idea of 6D Printing

The benefits of 3D printing are widely known and include cost-effectiveness, creative possibilities, accelerated production, accessibility, tangible design, design freedom, part consolidation, removal of tooling requirements, accelerated production, and the capacity to create an infinite number of shapes and geometries using a range of raw materials. Furthermore, 3D printing lowers manufacturing risks and wastes production. The practical advantages seem extensive when viewed through the lens of engineering. Given these advantages, one might assume that 3D printing would have gained widespread traction as a manufacturing technique by now [102]. This is untrue, as the industry has a difficult time incorporating this technology into regular production. At the moment, less than 2% of the manufacturing market is made up of additive manufacturing. This slow adoption is caused by several issues, such as limitations to conventional material components, lengthy production times, the requirement for extensive post-processing, and imperfect final products. These difficulties prevent 3D printing from being widely used in industrial settings. Advances in 4D printing technology can be broadly classified into two categories, with current constraints in the 3D printing process having a direct impact on the latter. First, there is the manufacturing process itself; second, there are the features, functions, and capacities of the final smart structures that are produced [102, 103]. The need for support materials and the subsequent requirement to remove them after printing present major challenges in terms of manufacturing limitations. Furthermore, the lack of extra degrees of freedom during printing requires lengthier processing routes, which leads to longer production times. Additionally, this restriction limits the structural complexity and coherence that can be achieved, both of which are essential for smart material (SM) strength. While more complex programming can be used to create intelligent components with complex orientations using 4D printing, this method lacks the high manufacturing flexibility offered by 6D printing, which enables layer-by-layer orientation. This talk presents the novel idea of six-dimensional (6D) printing in order to improve and address some aspects of additive manufacturing. It also seeks to provide a scientific basis for the use of 6D printing and its possible advantages [104].

By fusing the ideas of 4D and 5D printing processes, the authors present the idea of 6D printing for the first time. Using five degrees of freedom during the printing process, this innovative method creates final goods with intelligent characteristics that allow them to change shape in response to outside stimuli. This approach's main benefit is that it can create structures with distinct "response to stimulus" characteristics because of their unique structural features. It is anticipated that the installation of two additional printing axes will make it easier to produce a new line of smart materials (SMs) with tighter specifications and tolerances [104]. Furthermore, 6D printing seeks to reduce processing time while improving product strength and quality. Numerous industries, including manufacturing, aerospace, construction, and medicine, may use this technology. Advanced on-demand goods, self-adjusting systems, and material- and energy-efficient prosthetics can all be made with it. The capacity of 6D printing to accept both single-phase and nano-reinforced polymers as raw materials can expand its versatility [104, 105]. The final smart materials (SMs) made using this method should have specialized properties due to the incorporation of appropriate nano-inclusions and nano-reinforcements [106, 107]. However, the design of the method becomes more complex when nanomaterials are added to the polymer base. The reason for this complexity is that accurate printer calibration, modifications to the entire printing setup, and thorough computational modeling and simulation predictions are necessary to guarantee the correct design of the manufacturing process and the finished product [105–107].

6 Conclusions

Multifunctional materials are used in many products today, even though their manufacturing process is still in its infancy. It is expected that multifunctional fiber-reinforced polymer composites will be easier to integrate into daily life as research into their advances. Many polymers have shown in recent years to be capable of multifunctional characteristics. The printing methods of multifunctional materials have seen significant advancements in research, which have broadened their potential applications in fields including electronics, construction, aerospace, and energy [108]. Additive manufacturing has enabled the creation of incredibly complex multifunctional structures [109] in a range of industries, including aerospace and biomedicine, by carefully balancing structural and non-structural elements. Biomimetic devices like the lotus effect for super-hydrophobicity, crab claw movement to incorporate electromechanical connection, and shark skin for decreased drag have all been made possible by 3D printing [109]. As a result, multifunctional composite structures are engineered to fulfill basic structural responsibilities such as strengthening, stiffening, and stabilizing, in addition to a range of secondary purposes such as sensing, actuation, and energy storage/harvesting. Improving the functional characteristics of materials is essential for creating highly dependable goods; piezoelectric fillers, for instance, can increase efficiency [110]. High surface resolution must be attained throughout the digital manufacturing process of 3D-printed objects of all sizes, from continuum to nano [109]. Furthermore, innovative design is necessary in addition to understanding multifunctional materials and printing methods [109]. Additionally, a reliable mathematical/computational model [111] with multiscale analysis will be developed in order to enhance understanding, provide quantitative descriptions, and aid in the best possible design of multifunctional composite structures involving mechanical, electrical, chemical, thermal, and other multi-physics interactions. The founders and pilots of Solar Impulse, Bertrand Piccard and André Borschberg, have created aviation history by flying 40 k kilometers around the world [112]. Building on this success, the idea of integrating solar panels in a monolithic fashion on carbon fiber-based composite structural members, together with the ability to store and harvest energy using piezoelectric or battery fibers/laminates that have integrated 3D printed flexible electronics, will enable the production of lightweight, extremely dependable, energy-efficient aircraft as well as useful products. Thus, in the not-too-distant future, these multipurpose 3D-printed constructions are going to be the most sustainable, eco-friendly, and regenerative items available.

Acknowledgements The authors are thankful to Director, Amity Institute of Applied Sciences (AIAS), and Head, Department of Chemistry, and Head Department of Physics for providing the facilities for the compilation of information. One of the authors, Tejendra K Gupta is thankful to the Council of Science & Technology, Lucknow, Uttar Pradesh, India, for the financial support (Vide No. CST/CHEM/D-1431).

References

1. Adam, T.J. et al.: Multifunctional composites for future energy storage in aerospace structures. Energies (Basel) **11**(2) (2018). https://doi.org/10.3390/en11020335
2. Lendlein, A., Trask, R.S.: Multifunctional materials: concepts, function-structure relationships, knowledge-based design, translational materials research. Multifunct. Mater. **1**(1) (2018). https://doi.org/10.1088/2399-7532/aada7b
3. Matsuzaki, R., et al.: Three-dimensional printing of continuous-fiber composites by in-nozzle impregnation. Sci. Rep. **6**(1), 23058 (2016). https://doi.org/10.1038/srep23058
4. Oliveira, J., Correia, V., Castro, H., Martins, P., Lanceros-Mendez, S.: Polymer-based smart materials by printing technologies: improving application and integration. Addit. Manuf. **21**, 269–283 (2018). https://doi.org/10.1016/j.addma.2018.03.012
5. Kausar, A.: Shape memory polymer/graphene nanocomposites-state-of-the-art. E-Polym. (2022). https://doi.org/10.1515/epoly-2022-0024
6. Gopal, J., Muthu, M., Sivanesan, I.: A Comprehensive compilation of graphene/fullerene polymer nanocomposites for electrochemical energy storage (2023). https://doi.org/10.3390/polym15030701
7. De Angelis, F. et al.: In vitro mechanical properties of a novel graphene-reinforced pmma-based dental restorative material (2023). https://doi.org/10.3390/polym15030622
8. Kausar, A., Bocchetta, P.: Polymer/graphene nanocomposite membranes: status and emerging prospects. J. Compos. Sci. **6**, 76 (2022). https://doi.org/10.3390/jcs6030076
9. Morales, M.A., Maranon, A., Hernandez, C., Michaud, V., Porras, A.: Colombian sustainability perspective on fused deposition modeling technology: opportunity to develop recycled and biobased 3D printing filaments (2023). https://doi.org/10.3390/polym15030528
10. Xu, D., Shi, J., Qiu, R., Lei, W., Yu, W.: Comparative investigations on properties of three kinds of FDM 3D-printed natural plant powder/poly(lactic acid) biocomposites (2023). https://doi.org/10.3390/polym15030557
11. Chen, Z., et al.: 3D printing of ceramics: a review. J. Eur. Ceram. Soc. **39**(4), 661–687 (2019). https://doi.org/10.1016/j.jeurceramsoc.2018.11.013
12. Zhou, W., et al.: 4D-printed dynamic materials in biomedical applications: chemistry, challenges, and their future perspectives in the clinical sector. J. Med. Chem. **63**(15), 8003–8024 (2020). https://doi.org/10.1021/acs.jmedchem.9b02115
13. Wang, Y., Zhou, W., Cao, K., Hu, X., Gao, L., Lu, Y.: Architectured graphene and its composites: manufacturing and structural applications. Compos. Part A Appl. Sci. Manuf. **140**, 106177 (2021). https://doi.org/10.1016/j.compositesa.2020.106177
14. Hartings, M., Ahmed, Z.: Chemistry from 3D printed objects. Nat. Rev. Chem. **3** (2019). https://doi.org/10.1038/s41570-019-0097-z

15. Joharji, L., King Fahd University of Petroleum and Minerals Mishra, R.B., Alam, F. King Abdullah University of Science and Technology Tytov, S.: 4D printing: a detailed review of materials, techniques, and applications. Microelectron Eng. (2022)
16. Costa, L.A., Rangel, B., Lino Alves, J., Marques, A.T., BM da Silva, A.F., Esfandiari, P., MG da Silva, J.F., Rita Silva, A., Parente, M.P.L.: 4D structures for the short-time building of emergency shelters. Proceed. Inst. Mech. Eng. Part L J. Mater. Des. Appl. (2022)
17. Shen, B., Erol, O., Fang, L., Kang, S.H. search by orcid Shen.: Programming the time into 3D printing: current advances and future directions in 4D printing. Multifunct. Mater. **3** (2020)
18. Rayate, A., Jain, P.K.: A review on 4D printing material composites and their applications (2018). www.sciencedirect.comwww.materialstoday.com/proceedings2214-7853
19. Li, X., Shang, J., Wang, Z.: Intelligent materials: a review of applications in 4D printing. Assem. Autom. **37**(2), 170–185 (2017). https://doi.org/10.1108/AA-11-2015-093
20. Raj Agarwal, S.M.V.G.V.J.: The application of three-dimensional printing on foot fractures and deformities: a mini-review. Annal. 3D Print. Med. **5** (2022)
21. Muhammad Yasir Khalid, Z.U.A.W.A.: 4D printing: technological and manufacturing renaissance. Macromol. Mater. Eng. (2022)
22. Teunis van Manen*, V.M.D.M.C.S.M.J.M.A.A.Z.: Theoretical stiffness limits of 4D printed self-folding metamaterials. Commun. Mater. **3**(1) (2022)
23. †ORCID and Cuan-Urquizo, E. 1, 2, *, ‡ORCID by Tejada-Ortigoza, V. 1.: Towards the development of 3D-printed food: a rheological and mechanical approach. Food Eng. Technol. (2022)
24. Khorsandi, D. et al.: 3D and 4D printing in dentistry and maxillofacial surgery: printing techniques, materials, and applications. Acta Materialia Inc (2021). https://doi.org/10.1016/j.actbio.2020.12.044
25. Quanjin, M., Rejab, M.R.M., Idris, M.S., Kumar, N.M., Abdullah, M.H., Reddy, G.R.: Recent 3D and 4D intelligent printing technologies: a comparative review and future perspective. In: Procedia Computer Science, pp. 1210–1219. Elsevier B.V. (2020). https://doi.org/10.1016/j.procs.2020.03.434
26. Wu, J.J., Huang, L.M., Zhao, Q., Xie, T.: 4D printing: history and recent progress. Springer Verlag (2018). https://doi.org/10.1007/s10118-018-2089-8
27. Yang, D. et al.: 3D/4D printed tunable electrical metamaterials with more sophisticated structures. Royal Soc. Chem. (2021). https://doi.org/10.1039/d1tc02588k
28. Wan, X., Luo, L., Liu, Y., Leng, J.: Direct ink writing based 4D printing of materials and their applications. Adv. Sci. **7**(16) (2020). https://doi.org/10.1002/advs.202001000
29. Guo, Y., Patanwala, H.S., Bognet, B., Ma, A.W.K.: Inkjet and inkjet-based 3D printing: connecting fluid properties and printing performance. Rapid Prototyp. J. **23**(3), 562–576 (2017). https://doi.org/10.1108/RPJ-05-2016-0076
30. Krucińska, I., Cybula, M., Rambausek, L., van Langenhove, L.: Piezoelectric textiles: state of the art. Mater. Technol. **25**(2), 93–100 (2010). https://doi.org/10.1179/175355510X12716725525717
31. Lee, C., Tarbutton, J.: Electric poling-assisted additive manufacturing process for lead-free piezoelectric device fabrication. Proced. Manuf. **1**, 320–326 (2015). https://doi.org/10.1016/j.promfg.2015.09.035
32. Nadgorny, M., Ameli, A.: Functional polymers and nanocomposites for 3D printing of smart structures and devices. ACS Appl. Mater. Interfaces **10**(21), 17489–17507 (2018). https://doi.org/10.1021/acsami.8b01786
33. Mohiuddin, M., Sadasivuni, K.K., Mun, S., Kim, J.: Flexible cellulose acetate/graphene blueprints for vibrotactile actuator. RSC Adv. **5**(43), 34432–34438 (2015). https://doi.org/10.1039/C5RA03043A
34. Bodkhe, S., Rajesh, P.S.M., Gosselin, F., Daniel, T.: Simultaneous 3D printing and poling of PVDF and its nanocomposites. ACS Appl. Energy Mater. **1** (2018). https://doi.org/10.1021/acsaem.7b00337
35. Kim, H., Fernando, T., Li, M., Lin, Y., Tseng, B.: Fabrication and characterization of 3D printed BaTiO 3 /PVDF nanocomposites. J. Compos. Mater. **52**, 002199831770470 (2017). https://doi.org/10.1177/0021998317704709

36. Kim, K., et al.: 3D optical printing of piezoelectric nanoparticle-polymer composite materials. ACS Nano **8**(10), 9799–9806 (2014). https://doi.org/10.1021/nn503268f
37. Chang, J. et al.: Advanced material strategies for next-generation additive manufacturing (2018). https://doi.org/10.3390/ma11010166
38. Maurel, A., et al.: Highly loaded graphite-polylactic acid composite-based filaments for lithium-ion battery three-dimensional printing. Chem. Mater. **30**, 7484–7493 (2018). https://doi.org/10.1021/acs.chemmater.8b02062
39. Schmidt-Rohr, K.: How batteries store and release energy: explaining basic electrochemistry. J. Chem. Educ. **95** (2018). https://doi.org/10.1021/acs.jchemed.8b00479
40. Wang, Q., et al.: A flame retardant ionic conductor additive for safety-reinforced liquid electrolyte of lithium batteries. J. Electrochem. Soc. **164**(7), A1559 (2017). https://doi.org/10.1149/2.1111707jes
41. Yang, Z. et al. (2015). Recent advancement of nanostructured carbon for energy applications. Chem. Rev. **115** (2015). https://doi.org/10.1021/cr5006217
42. Mohanapriya, M. et al.: Polyvinyl alcohol (PVA)/polystyrene sulfonic acid (PSSA)/carbon black nanocomposite for flexible energy storage device applications. J. Mater. Sci. Mater. Electron. **28** (2017). https://doi.org/10.1007/s10854-016-6287-2
43. Reyes, C. et al.: 3D Printing a complete lithium ion battery with fused filament fabrication. ACS Appl. Energy Mater. **1** (2018). https://doi.org/10.1021/acsaem.8b00885
44. Fredi, G., et al.: Graphitic microstructure and performance of carbon fibre Li-ion structural battery electrodes. Multifunct. Mater. **1**, 15003 (2018). https://doi.org/10.1088/2399-7532/aab707
45. Ahn, B.Y., Lorang, D.J., Lewis, J.A.: Transparent conductive grids via direct writing of silver nanoparticle inks. Nanoscale **3**(7), 2700–2702 (2011). https://doi.org/10.1039/C1NR10048C
46. Momeni, F., Mehdi Hassani, S.M., Liu, N.X., Ni, J.: A review of 4D printing. Mater. Des. **122**, 42–79 (2017). https://doi.org/10.1016/j.matdes.2017.02.068.
47. Kamyshny, A., Magdassi, S.: Conductive nanomaterials for printed electronics. Small **10** (2014). https://doi.org/10.1002/smll.201303000
48. Fu, K. et al.: Graphene oxide-based electrode inks for 3D-printed lithium-ion batteries. Adv. Mater. **28**, p. n/a-n/a (2016). https://doi.org/10.1002/adma.201505391
49. Wang, Y. et al.: 3D-printed all-fiber Li-ion battery toward wearable energy storage. Adv. Funct. Mater. **27**, 1703140-n/a (2017). https://doi.org/10.1002/adfm.201703140
50. Deka, B., Hazarika, A., Kim, J., Park, Y.-B., Park, H.: Recent development and challenges of multifunctional structural supercapacitors for automotive industries: review on multifunctional structural supercapacitors. Int. J. Energy Res. **41** (2017). https://doi.org/10.1002/er.3707
51. Hellqvist Kjell, M., Jacques, E., Zenkert, D., Behm, M., Lindbergh, G.: PAN-based carbon fibers for structural lithium-ion batteries. In: ECCM 2012-Composites at Venice, Proceedings of the 15th European Conference on Composite Materials (2012)
52. Zhu, C. et al. (2017). 3D printed functional nanomaterials for electrochemical energy storage. Nano Today **15** (2017). https://doi.org/10.1016/j.nantod.2017.06.007
53. Dermanaki Farahani, R., Dubé, M., Daniel, T.: Three-dimensional printing of multifunctional nanocomposites: manufacturing techniques and applications. Adv. Mater. **28** (2016). https://doi.org/10.1002/adma.201506215
54. Benson, J., Kovalenko, I., Boukhalfa, S., Lashmore, D., Sanghadasa, M., Yushin, G.: Multifunctional CNT-polymer composites for ultra-tough structural supercapacitors and desalination devices. Adv. Mater. **25**(45), 6625–6632 (2013). https://doi.org/10.1002/adma.201301317
55. Wei, M., Zhang, F., Wang, W., Alexandridis, P., Zhou, C., Wu, G.: 3D direct writing fabrication of electrodes for electrochemical storage devices. J. Power Sour. **354** (2017). https://doi.org/10.1016/j.jpowsour.2017.04.042
56. García Rocha, V. et al.: Multimaterial 3D printing of graphene-based electrodes for electrochemical energy storage using thermoresponsive inks. ACS Appl. Mater. Interfaces **9** (2017). https://doi.org/10.1021/acsami.7b10285

57. Jakus, A., Secor, E., Rutz, A., Jordan, S., Hersam, M., Shah, R.: Three-dimensional printing of high-content graphene scaffolds for electronic and biomedical applications. ACS Nano **9** (2015). https://doi.org/10.1021/acsnano.5b01179
58. Guo, S., Yang, X., Heuzey, M.-C., Daniel, T.: 3D Printing of multifunctional nanocomposite helical liquid sensor. Nanoscale **7** (2015). https://doi.org/10.1039/C5NR00278H
59. Jakus, A.E., Secor, E.B., Rutz, A.L., Jordan, S.W., Hersam, M.C., Shah, R.N.: Three-dimensional printing of high-content graphene scaffolds for electronic and biomedical applications. ACS Nano **9**(4), 4636–4648 (2015). https://doi.org/10.1021/acsnano.5b01179
60. Kim, H., Fernando, T., Li, M., Lin, Y., Tseng, T.L.B.: Fabrication and characterization of 3D printed BaTiO3/PVDF nanocomposites. J. Compos. Mater. **52**(2), 197–206 (2018). https://doi.org/10.1177/0021998317704709
61. Choi, S., Potscavage, W., Kippelen, B.: Area-scaling of organic solar cells. J. Appl. Phys. **106**, 54507 (2009). https://doi.org/10.1063/1.3211850
62. Huang, C.Y.: Extrusion-based 3D printing and characterization of edible materials, pp. 8–25 (2018)
63. Vak, D. et al.: Solar cells: 3D printer based slot-die coater as a lab-to-fab translation tool for solution-processed solar cells (Adv. Energy Mater. 4/2015). Adv. Energy Mater. **5** (2014). https://doi.org/10.1002/aenm.201401539
64. Jung, S. et al.: Organic solar cells: all-inkjet-printed, all-air-processed solar cells (Adv. Energy Mater. 14/2014). Adv. Energy Mater. **4**, p. n/a-n/a (2014). https://doi.org/10.1002/aenm.201470072
65. Hamadani, B., Dougherty, B.: Solar cell characterization. Springer Ser. Mater. Sci. **218**, 229–245 (2016). https://doi.org/10.1007/978-3-319-20331-7_8
66. Iii, M.F. et al.: Version of record: https://www.sciencedirect.com/science/article/pii/S0927024817304531. pp. 1–29
67. Rossander, L.H., et al.: In-line, roll-to-roll morphology analysis of organic solar cell active layers. Energy Environ. Sci. **10**(11), 2411–2419 (2017). https://doi.org/10.1039/C7EE01900A
68. Yang, Z., Chueh, C.-C., Zuo, F., jong h Kim, Liang, P., Jen, A.: High-performance fully printable perovskite solar cells via blade-coating technique under the ambient condition. Adv. Energy Mater. **5** (2015). https://doi.org/10.1002/aenm.201500328.
69. Deshmukh, K., et al.: Solution-processed white graphene-reinforced ferroelectric polymer nanocomposites with improved thermal conductivity and dielectric properties for electronic encapsulation. J. Polym. Res. **24**(2), 27 (2017). https://doi.org/10.1007/s10965-017-1189-4
70. Ponnamma, D., kumar Sadasivuni, K., Cabibihan, J.-J., Yoon, J., Kumar, B.: Reduced graphene oxide filled poly(dimethyl siloxane) based transparent stretchable, and touch-responsive sensors. Appl. Phys. Lett. **108** (2016). https://doi.org/10.1063/1.4947595
71. Goutham, S. et al.: Nanostructured ZnO gas sensors obtained by green method and combustion technique. Mater. Sci. Semicond. Process **57** (2017). https://doi.org/10.1016/j.mssp.2016.09.037
72. Goutham, S., Kumar, D., kumar Sadasivuni, K., Cabibihan, J.-J., Rao, K.: Nanostructure ZnFe2O4 with bacillus subtilis for detection of LPG at low temperature. J. Electron. Mater. **46** (2017). https://doi.org/10.1007/s11664-016-5273-z
73. Sadasivuni, K.K., Deshmukh, K., M.A.-A.T.A.-T.T.- Al-Maadeed.: 3D and 4D printing of polymer nanocomposite materials : processes, applications, and challenges. Elsevier, Amsterdam (2020). LK-https://worldcat.org/title/1124925813
74. Sadasivuni, K.K., Ponnamma, D., Kasak, P., Krupa, I., Ali M., Al-Maadeed, S.A.: Designing dual phase sensing materials from polyaniline filled styrene–isoprene–styrene composites. Mater. Chem. Phys. **147**(3):1029–1036 (2014). https://doi.org/10.1016/j.matchemphys.2014.06.055
75. Deshmukh, K. et al.: Newly developed biodegradable polymer nanocomposites of cellulose acetate and Al2O3 nanoparticles with enhanced dielectric performance for embedded passive applications. J. Mater. Sci. Mater. Electron. **28** (2017). https://doi.org/10.1007/s10854-016-5616-9

76. Science, C., Science, C., Matusik, W., Science, C., Supervisor, T., Kolodziejski, L.: 3D-printing form and function (2018)
77. Bodkhe, S., Noonan, C., Gosselin, F., Daniel, T.: Coextrusion of multifunctional smart sensors. Adv. Eng. Mater. **20** (2018). https://doi.org/10.1002/adem.201800206
78. Saari, M., Cox, B., Richer, E., Krueger, P., Cohen, A.: Fiber encapsulation additive manufacturing: an enabling technology for 3D printing of electromechanical devices and robotic components. 3D Print Addit. Manuf. **2** (2015). https://doi.org/10.1089/3dp.2015.0003
79. Saari, M., Xia, B., Cox, B., Krueger, P., Cohen, A., Richer, E.: Fabrication and analysis of a composite 3D printed capacitive force sensor. 3D Print Addit. Manuf. **3**, 136–141 (2016). https://doi.org/10.1089/3dp.2016.0021
80. Boydston, A.J., et al.: Additive manufacturing with stimuli-responsive materials. J. Mater. Chem. A. Mater. **6**(42), 20621–20645 (2018). https://doi.org/10.1039/C8TA07716A
81. Gardan, J.: Smart materials in additive manufacturing: state of the art and trends. Virtual Phys. Prototyp. (2018). https://doi.org/10.1080/17452759.2018.1518016
82. Wu, J.-J., Huang, L.-M., Zhao, Q., Xie, T.: 4D printing: history and recent progress. Chin. J. Polym. Sci. **36**, 1–13 (2017). https://doi.org/10.1007/s10118-018-2089-8
83. Yang, D. et al.: 3D/4D printed tunable electrical metamaterials with more sophisticated structures. J. Mater. Chem. C Mater. **9** (2021). https://doi.org/10.1039/D1TC02588K
84. Wan, X., Luo, L., Liu, Y., Jinsong, L.: Direct ink writing based 4D printing of materials and their applications. Adv. Sci. **7** (2020). https://doi.org/10.1002/advs.202001000
85. Guo, Y., Patanwala, H., Bognet, B., Ma, A.: Inkjet and inkjet-based 3D printing: Connecting fluid properties and printing performance. Rapid Prototyp. J. **23** (2017). https://doi.org/10.1108/RPJ-05-2016-0076
86. Shirazi, S. et al.: A review on powder-based additive manufacturing for tissue engineering: selective laser sintering and inkjet 3D printing. Sci. Technol. Adv. Mater. (2015)
87. Kafle, A., Luis, E., Silwal, R., Pan, H.M., Shrestha, P., Bastola, A.: 3D/4D printing of polymers: fused deposition modelling (FDM), selective laser sintering (SLS), and stereolithography (SLA). Polym. (Basel) **13**, 3101 (2021). https://doi.org/10.3390/polym13183101
88. Ponnamma, D., Yin, Y., Salim, N., Parameswaranpillai, J., Thomas, S., Hameed, N.: Recent progress and multifunctional applications of 3D printed graphene nanocomposites. Compos. B Eng. **204**, 108493 (2021). https://doi.org/10.1016/j.compositesb.2020.108493
89. Mostafaei, A., et al.: Binder jet 3D printing–process parameters, materials, properties, and challenges. Prog. Mater. Sci. **119**, 100707 (2020). https://doi.org/10.1016/j.pmatsci.2020.100707
90. Singh, S., Ramakrishna, S., Berto, F.: 3D printing of polymer composites: a short review. Mater. Des. Process. Commun. **2** (2019). https://doi.org/10.1002/mdp2.97
91. Wang, X., Jiang, M., Zhou, Z., Gou, J., Hui, D.: 3D printing of polymer matrix composites: a review and prospective. Compos. B Eng. **110**, 442–458 (2017). https://doi.org/10.1016/j.compositesb.2016.11.034
92. Zhang, Z., Corrigan, N., Bagheri, A., Jin, J., Boyer, C.: A versatile 3D and 4D printing system through photocontrolled RAFT polymerization. Angewandte Chemie Intern. Edit. **58** (2019). https://doi.org/10.1002/anie.201912608
93. Pan, H.M.: Advanced materials in 3D/4D printing technology (2022). https://doi.org/10.3390/polym14163255
94. Qian, Y., Li, C., Qi, Y., Zhong, J.: 3D printing of graphene oxide composites with well controlled alignment. Carbon N Y **171**, 777–784 (2021). https://doi.org/10.1016/j.carbon.2020.08.077
95. Joshi, S., et al.: 4D printing of materials for the future: opportunities and challenges. Appl. Mater. Today **18**, 100490 (2020). https://doi.org/10.1016/j.apmt.2019.100490
96. Wei, X., et al.: 3D printable graphene composite. Sci. Rep. **5**, 11181 (2015). https://doi.org/10.1038/srep11181
97. Zarek, M., Layani, M., Cooperstein, I., Sachyani, E., Cohn, D., Magdassi, S.: 3D printing: 3D printing of shape memory polymers for flexible electronic devices (Adv. Mater. 22/2016). Adv. Mater. **28**, 4166 (2016). https://doi.org/10.1002/adma.201670148

98. Chang, Y., Cao, Q., Venton, B.J.: 3D printing for customized carbon electrodes. Curr. Opin. Electrochem. **38**, 101228 (2023). https://doi.org/10.1016/j.coelec.2023.101228
99. Zhang, W., Zhu, S., Luque, R., Han, S., Hu, L., Xu, G.: Recent development of carbon electrode materials and their bioanalytical and environmental applications. Chem. Soc. Rev. **45** (2015). https://doi.org/10.1039/c5cs00297d
100. Chyan, Y., Ye, R., Li, Y., Singh, S., Arnusch, C., Tour, J.: Laser-induced graphene by multiple lasing: toward electronics on cloth, paper, and food. ACS Nano **12**, 2176–2183 (2018). https://doi.org/10.1021/acsnano.7b08539
101. Ludvigsen, E. et al.: Selective direct laser writing of pyrolytic carbon microelectrodes in absorber-modified SU-8. Micromachines (Basel) **12**, 564 (2021). https://doi.org/10.3390/mi12050564
102. El-Sayegh, S., Romdhane, L., Manjikian, S.: A critical review of 3D printing in construction: benefits, challenges, and risks. Springer (2020). https://doi.org/10.1007/s43452-020-00038-w
103. Reddy, R., Pinninti Chaitanya Bharathi Institute of Technology Devi, P.A.: Review on the advancements to additive manufacturing-4D and 5D printing. Int. J. Mech. Prod. Eng. Res. Dev. (2018)
104. Tzou, H.S., Lee, H.J., Arnold, S.M.: Smart materials, precision sensors/actuators, smart structures, and structronic systems. Mech. Adv. Mater. Struct. **11**(4–5), 367–393 (2004). https://doi.org/10.1080/15376490490451552
105. Huang, S.H., Liu, P., Mokasdar, A., Hou, L.: Additive manufacturing and its societal impact: a literature review (2013). https://doi.org/10.1007/s00170-012-4558-5
106. Kaleem, M.A.: Manufacturing graphene based polymer matrix composites (GPMCs) via 3D printing (additive manufacturing): a review
107. Yu, X., Cheng, H., Zhang, M., Zhao, Y., Qu, L., Shi, G.: Graphene-based smart materials. Nature Publishing Group (2017). https://doi.org/10.1038/natrevmats.2017.46
108. Biswas, K., et al.: Additive manufacturing integrated energy—enabling innovative solutions for buildings of the future. J. Sol. Energy Eng. **139**, 15001 (2017)
109. Lin, D. et al.: Three-dimensional printing of complex structures: man made or toward nature?. ACS Nano. **8** (2014). https://doi.org/10.1021/nn504894j
110. Kim, H., Torres, F., Wu, Y., Villagran, D., Lin, Y., Tseng, B.: Integrated 3D printing and corona poling process of PVDF piezoelectric films for pressure sensor application. Smart Mater. Struct. **26**, 9 (2017). https://doi.org/10.1088/1361-665X/aa738e
111. Gupta, R., Harursampath, D.: Dielectric elastomers: asymptotically-correct three-dimensional displacement field. Int. J. Eng. Sci. **87**, 1–12 (2015). https://doi.org/10.1016/j.ijengsci.2014.10.006
112. Abbe, G., Smith, H.: Technological development trends in solar-powered aircraft systems. Renew. Sustain. Energy Rev. **60**, 770–783 (2016). https://doi.org/10.1016/j.rser.2016.01.053

4D-Printed Polymer Nanocomposites for Smart Textiles

Biswajeet Acharya, Srikanta Moharana, and Amulyaratna Behera

Abstract The emergence of 4D printing technology is revolutionizing textile engineering by integrating traditional materials with advanced manufacturing techniques to produce smart textiles that can adapt dynamically to environmental stimuli. Unlike conventional 3D printing, which creates static objects, 4D printing allows textiles to change their shape, properties, and functions over time, responding to temperature, humidity, light, and magnetic fields. This adaptability is achieved through the use of responsive polymers, polymer nanocomposites, and nanotechnology, enabling fabrics to self-assemble, self-heal, and modify their characteristics in real time. Key 4D printing techniques—including fused deposition modeling, stereolithography, and multi-material printing—are utilized to fabricate these smart textiles, which find applications in healthcare, sports, defense, and fashion. Although challenges related to material complexity, durability, and production scalability exist, ongoing research aims to enhance the functionality and affordability of 4D-printed textiles, paving the way for innovative applications in various industries.

Keywords 4D printing · Smart textiles · Polymer nanocomposites · Responsive materials · Adaptive fabrics

B. Acharya
School of Pharmacy and Life Sciences, Centurion University of Technology and Management, R.Sitapur, Paralakhemundi, Odisha, India

S. Moharana
Department of Chemistry, School of Applied Sciences, Centurion University of Technology and Management, R.Sitapur, Paralakhemundi, Odisha, India

A. Behera (✉)
School of Pharmacy, DRIEMS University, Tangi, Cuttack, Odisha, India
e-mail: dr.amulyaratnabehera@gmail.com

S. Moharana et al. (eds.), *Polymer Nanocomposites for 3D, 4D and 5D Printing*, Engineering Materials, https://doi.org/10.1007/978-981-96-4214-4_18

1 Introduction

The advent of 4D printing has introduced transformative capabilities in textile engineering, merging traditional materials with advanced technologies to create smart textiles that not only exhibit flexibility but also adapt over time in response to specific stimuli. This convergence allows for textiles that respond to environmental changes, bringing dynamic and programmable properties to fabrics [1]. Unlike traditional 3D printing, which produces static objects, 4D printing enables textiles to alter their shape, properties, or functions over time when exposed to certain external conditions such as temperature, humidity, light, or magnetic fields. This adaptability relies on combining materials science, responsive polymers, and nanotechnology to create fabrics that can self-assemble, self-heal, or adjust their properties in real time based on the user's needs or surrounding environment [2]. Shape-memory polymers, hydrogels, and other programmable materials play a critical role in enabling these dynamic capabilities, making it possible to fine-tune textile characteristics for a variety of applications, including sportswear, medical devices, and adaptive furnishings.

A central element of 4D-printed textiles is polymer nanocomposites, which enhance the mechanical, thermal, and responsive properties of these materials [3]. These nanocomposites, made of polymers combined with nanoscale particles like carbon nanotubes, graphene, or metallic nanoparticles, provide unique qualities that significantly improve the flexibility, durability, and environmental responsiveness of textiles. For example, carbon nanotube-polymer composites greatly increase a textile's electrical conductivity and elasticity, making them ideal for wearable electronics [4]. Furthermore, nanocomposites enable textiles to respond to multiple stimuli simultaneously, such as heat and light, expanding the range of possible applications. They also offer the advantage of creating lightweight yet robust materials that withstand varied environmental conditions, essential in fields like aerospace, automotive, and defense [5]. To fabricate these smart textiles, various 4D printing techniques are utilized, each suited to different applications. Key techniques include fused deposition modeling (FDM), stereolithography (SLA), selective laser sintering (SLS), direct ink writing (DIW), and multi-material printing. SLA, for instance, enables high-resolution printing, while FDM is prized for scalability and cost-effectiveness [6]. Multi-material printing is particularly important for 4D printing in textiles, as it allows the integration of different materials with distinct properties into a single fabric. This capability enables textiles to perform complex, multi-step reactions to environmental stimuli, further expanding the possibilities for creating advanced, multi-functional materials. These smart functionalities transform 4D-printed textiles into adaptive systems that interact with the wearer and the environment [7]. For example, shape-memory polymers enable textiles to change shape in response to temperature, making them ideal for wearables that provide ventilation or warmth as needed. Self-healing materials can repair minor damages, extending the lifespan and usefulness of garments. Additionally, conductive nanomaterials allow textiles to function as sensors, capable of monitoring vital signs or environmental changes and transmitting data to electronic devices. Such responsive properties have far-reaching applications in healthcare, sports, fashion, and defense, where textiles are expected to enhance user experience and adapt to various conditions [8].

In practice, 4D-printed smart textiles have found applications in diverse sectors. In healthcare, they are used in wearable devices for health monitoring, compression therapy, and localized treatments. Sports and fitness industries leverage them for smart apparel that adjusts fit, regulates temperature, or provides biomechanical feedback, enhancing athletic performance [9]. In defense, adaptive camouflage and responsive armor are possible applications, improving soldier comfort and safety in challenging environments. Meanwhile, fashion designers are exploring the aesthetic and functional possibilities of 4D-printed textiles, designing garments that change shape or color with movement or environmental changes. Despite the significant potential, 4D-printed textiles face challenges, notably the complexity and cost of materials, particularly polymer nanocomposites and other responsive materials, which can limit large-scale production [10]. Durability is another issue; textiles that frequently change shape or function may wear out more quickly. Moreover, technical limitations in printing resolution and material precision must be addressed to ensure reliable performance in various conditions. However, the future of 4D-printed textiles is promising. Research is focused on developing more affordable materials, refining printing techniques, and enhancing the smart functionalities of textiles. With advances in nanotechnology, materials science, and engineering, 4D printing has the potential to revolutionize multiple industries, creating textiles that are not just functional clothing but integral parts of smart, interactive systems [11]. This technology could redefine what textiles can achieve, opening a new era of intelligent fabrics tailored to meet the dynamic needs of a rapidly evolving world.

2 Fundamentals of 4D Printing for Textiles

4D printing is a cutting-edge evolution of traditional 3D printing, enhancing the technology by enabling materials and structures to change shape or functionality over time when exposed to specific external stimuli such as temperature, humidity, light, or magnetic fields. While 3D printing focuses on creating static objects by layering materials in three dimensions, 4D printing incorporates a "fourth dimension"—time—which allows these printed structures to morph and adapt post-fabrication [12]. This transformative potential is particularly valuable in textiles, where adaptability, responsiveness, and interactivity are becoming central to innovation. The introduction of additive manufacturing (AM) techniques in the 1980s marked a pivotal moment for the exploration and development of shape-memory polymers (SMP). This innovative approach provided researchers with the tools to fabricate complex geometries and tailored properties, expanding the potential applications of SMPs across various fields [13]. Since this breakthrough, SMPs have attracted considerable interest from the scientific community, leading to ongoing research and advancements in their synthesis, characterization, and application. As a result, SMPs

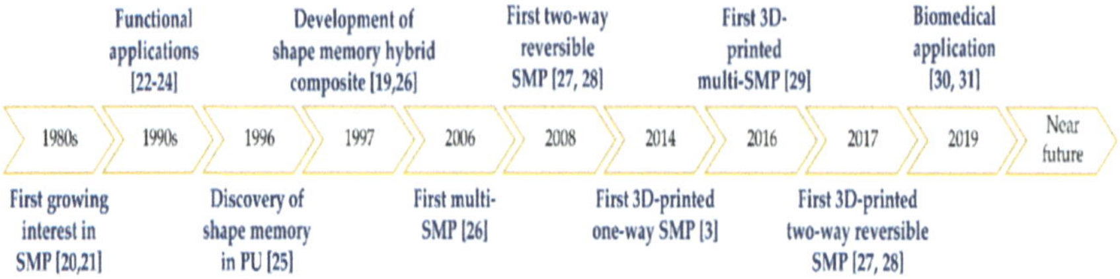

Fig. 1 Progress and advancement of additive manufacturing techniques [13]. Copyright 2019. Reproduced with permission from MDPI

have become a focal point of study for developing responsive materials capable of transforming their shape and functionality under specific stimuli (as shown in Fig. 1).

A flow diagram illustrating the evolution from 3D printing (3DP) to 4D printing (4DP) technology would showcase the advancements in material science, design capabilities, and functional outcomes over time. Starting with traditional 3DP, which focuses on creating static, three-dimensional objects from materials like plastics, metals, and resins, the diagram would highlight innovations such as the incorporation of smart materials and responsive polymers that respond to external stimuli (e.g., heat, light, moisture). This transition to 4DP emphasizes the added "dimension" of time, where structures can undergo programmed transformations or adapt their shapes post-printing. Key stages in this evolution would include the introduction of shape-memory polymers, self-healing materials, and the use of nano-fillers to modify properties like flexibility, conductivity, and durability, ultimately enabling applications in fields such as biomedical devices, aerospace, and robotics (as shown in Fig. 2) [10].

In textile engineering, the success of 4D printing relies on several essential principles, with each contributing to the material's ability to respond and adapt to external stimuli. Material selection is a critical initial step, as the chosen materials must be responsive to environmental triggers such as temperature, moisture, light, or magnetic fields [14]. Shape-memory polymers (SMPs), for instance, are commonly used due to their ability to change shape when exposed to heat, while hydrogels respond dynamically to moisture by expanding or contracting. Each material type possesses its own unique reaction mechanisms. For example, SMPs have a "programmed" shape memory that activates upon heating, allowing them to return to a predetermined form, which is particularly useful for applications that require adaptable structure and fit. Hydrogels, on the other hand, respond to hydration by altering their volume, making them suitable for fabrics that need to manage moisture levels for user comfort [15].

An equally important principle in 4D-printed textiles is printing technology. The printing process must be advanced enough to place and layer these responsive materials with high precision to ensure the desired functionality. This precision enables specific segments of the textile to activate only under certain conditions, giving designers the flexibility to create complex, multi-functional fabrics [16]. The goal of incorporating 4D printing in textiles is to produce materials that are not only adaptive but can serve both practical and aesthetic purposes by shifting in response

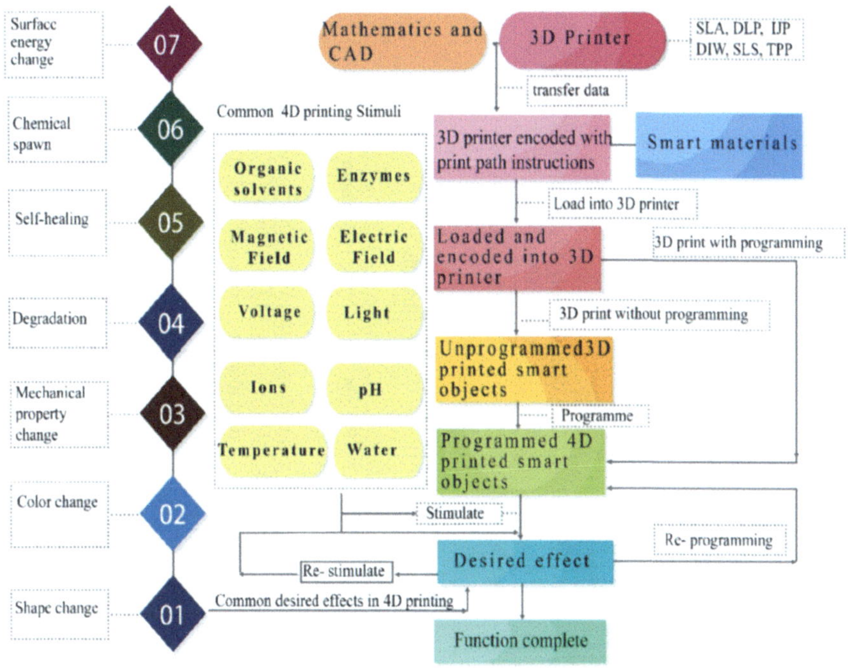

Fig. 2 Flow diagram illustrating the evolution of 3DP technology to 4DP technology [10]. Copyright 2022. Reproduced with permission from Elsevier

to real-time conditions. Achieving this adaptability requires in-depth knowledge of advanced material science, especially in areas like nanocomposites and responsive polymers [17]. Through this knowledge, researchers can embed programmable properties directly into textile fibers, which allows for remarkable functions—self-assembly, shape-shifting, and even self-repair—that go far beyond the limitations of conventional fabrics. By integrating these advanced materials, 4D-printed textiles can offer a new level of responsiveness and functionality, with applications ranging from fashion and healthcare to sports and military gear, marking a transformative shift in textile design and use [18]. Below is a table that summarizes some of the key aspects related to the fundamentals of 4D printing for textiles, including types of materials, stimuli, key properties, applications, and advantages (Table 1).

Rodriguez utilized carbon nano-fillers to modify the mechanical properties of the resin, aiming to increase the glass transition temperature (Tg) and enhance conductivity. The addition of these nano-fillers raised the resin's viscosity and introduced shear-thinning behavior, allowing printing at lower pressures as the ink undergoes shear inside the nozzle. Ensuring that the storage modulus is lower than the loss modulus is crucial; this ensures the ink remains flowable under shear, preventing it from solidifying immediately after leaving the nozzle and enabling effective shape

Table 1 Overview of responsive materials for 4D-printed smart textiles: stimuli, mechanisms, properties, and applications

Material type	Stimulus	Mechanism of response	Key properties	Applications	Advantages	Limitations	Examples	References
Shape-memory polymers	Temperature, light	Phase transition	Shape recovery, durability	Adaptive clothing, medical devices	High responsiveness, reusability	Cost, durability concerns	Polyurethane, PLA	[19]
Hydrogels	Humidity, pH	Absorption, swelling	Flexibility, softness	Wearables, soft robotics	Biocompatibility	Dehydration risk	Polyvinyl alcohol (PVA)	[20]
Electroactive polymers	Electric field	Electric activation	Conductivity, flexibility	Wearable electronics, sensors	High flexibility	Power requirement	Polyaniline, PVDF	[21]
Carbon nanotubes	Heat, electric field	Conductivity enhancement	High strength, lightweight	Military fabrics, sportswear	High strength, elasticity	High cost	CNT-polymer blends	[22]
Shape-memory alloys	Heat	Shape transformation	High strength, elasticity	Medical implants, responsive fabrics	Long-term stability	Weight, expensive	Nickel-Titanium alloys	[23]
Liquid crystal polymers	Light, temperature	Molecular rearrangement	Thermal response	Smart curtains, temperature-controlled textiles	Rapid response	UV degradation potential	LC-Polymer composites	[24]
Thermoplastic elastomers	Temperature	Phase change	Flexibility, durability	Sportswear, fashion	Ease of processing	Limited shape memory	TPU, SBS (Styrene-Butadiene)	[25]

fixation at room temperature. Rodriguez observed a corresponding increase in electrical conductivity with added carbon nanotubes (CNT), with concentrations ranging from 0.6 to 5.6 vol% boosting conductivity from 0.0001 to 0.4 S/cm. This enhanced conductivity was sufficient to activate low-power electronics. In one experiment, a 5.6 vol% SMP composite was connected to a 9 V power source without physical contact; when placed in an 85 °C oven, the SMP returned to its original shape, completing the circuit and illuminating an LED, as demonstrated in Fig. 3 [26].

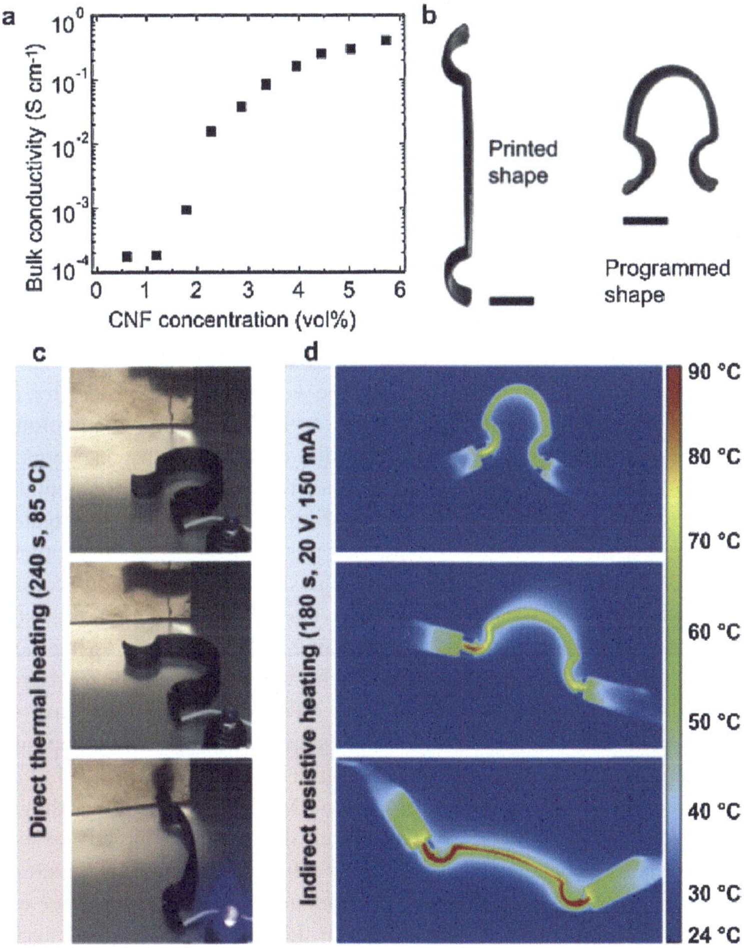

Fig. 3 Electrical conductivity of the composite SMPs [26]. Copyright 2016. Reproduced with permission from Springer Nature

3 Key Materials in 4D Printing of Textiles

In 4D printing, stimuli-responsive materials form the foundation of creating adaptive, interactive textiles, each type offering distinct properties suitable for various applications in wearable technology. A principal category of these materials is Shape-Memory Polymers (SMPs), which are designed to change shape when exposed to specific stimuli such as heat, light, or pH variations. SMPs are especially valuable in developing textiles that can alter their fit, structure, or support based on environmental conditions [26]. For instance, fabrics embedded with SMP fibers can adjust tightness in response to body temperature changes, which enhances comfort and functionality. Such textiles might be useful in sportswear, where a garment that adapts to body heat could improve the athlete's performance and experience by creating a second-skin feel. SMPs could also be applied in medical textiles, where adaptive support in compression garments, braces, or casts can better meet therapeutic needs by changing pressure or rigidity in specific areas as required [27]. Hydrogels are another group of stimuli-responsive materials that play a critical role in the creation of 4D-printed textiles. These materials are particularly sensitive to moisture and humidity, responding by expanding, contracting, or modifying their porosity. This characteristic is especially useful in textiles intended to maintain or regulate body temperature by managing moisture levels. In practice, hydrogel-based fabrics can wick away moisture, which keeps the wearer comfortable and dry, making them suitable for sports and outdoor clothing. As hydrogels change in volume with humidity, they naturally allow for the production of moisture-responsive garments, providing an added layer of utility that responds to changing environmental conditions [28]. Ding et al. explored a direct 4D printing approach to investigate shape memory effects using a PolyJet technique that jets multi-material ink droplets layer by layer, curing them with ultraviolet light. The study utilized various commercial materials, including TangoBlack + and Tango + as elastomers and VeroClear as the shape-memory polymer (SMP). When heated, the SMP softens, allowing the object to transition into a new permanent shape, which can subsequently be reprogrammed into multiple subsequent forms. This process effectively consolidates five distinct printing and programming steps into a single step. The 3D-printed component can quickly change its shape upon exposure to heat (as shown in Fig. 4) [29].

Thermo-responsive polymers are similar to SMPs but differ in that they do not rely on a "memory" shape to function; instead, they respond directly to temperature fluctuations by expanding, contracting, or altering their internal structure [30]. This responsiveness can be leveraged in garments that are designed for climate-specific insulation, as textiles embedded with thermo-responsive polymers could help regulate body temperature by providing insulation or cooling effects depending on the wearer's surroundings [31]. For instance, in colder temperatures, thermo-responsive materials could expand to increase fabric density and enhance warmth, while in warmer temperatures, the fabric could contract, allowing heat to dissipate more easily. Another innovative class of materials for 4D-printed textiles is photo-responsive materials, which are engineered to react to light, particularly ultraviolet

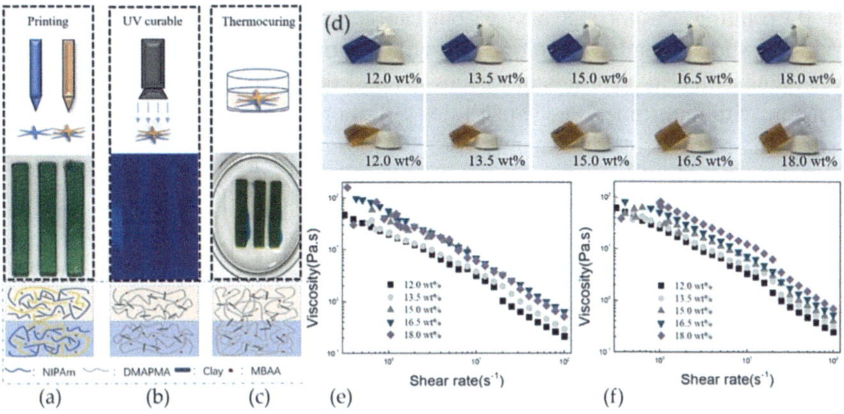

Fig. 4 **a** Photographs illustrating the printing process and the final printed results. **b** Image showing the state during photocuring. **c** Depiction of the thermo-curing process. **d** Tilted photographs showcasing the two printing fluids with varying clay content. **e**, **f** Graphs displaying the viscosity as a function of shear rate for the top and bottom layers [29]. Copyright 2023. Reproduced with permission from MDPI

(UV) or visible light. These materials offer unique functionalities for textiles, such as color-changing or opacity-shifting abilities, which add both practical and aesthetic value [32]. For example, UV-sensitive fabrics could serve as wearables that indicate UV exposure levels by changing color when the user has been exposed to a certain amount of sunlight. This type of visual feedback can be especially useful in environments where UV exposure needs to be monitored for health reasons, such as in sun-protective clothing for outdoor activities or in settings, where UV intensity is high. Additionally, photo-responsive textiles can contribute to fashion-forward applications by offering dynamic, interactive designs that shift colors with light exposure, introducing a new layer of creativity to garment design [33].

Conductive nanocomposites represent another class of materials that are integral to 4D-printed textiles, especially those that are considered "smart" or "connected" wearables. By embedding nanoparticles like carbon nanotubes or metallic particles within textile fibers, these materials can conduct electricity or detect various environmental factors, such as pressure or strain [34]. Conductive textiles are increasingly used in applications that require real-time monitoring, such as smart wearables that track biometrics like heart rate, body temperature, or muscle movement. The integration of conductive nanocomposites allows textiles to not only sense environmental or bodily changes but also communicate data to external devices, such as smartphones or medical monitoring systems. The presence of these materials within the fabric enables a higher level of interactivity and functionality, making it possible to produce clothing that not only fits well but also serves as an interface between the wearer and digital devices [35]. Each of these materials is integrated into textile fibers during the 4D printing process, with specific layers or fiber segments carefully designed to respond to unique stimuli. This approach ensures that textiles produced via 4D printing are not just passive but can adjust characteristics like thickness, flexibility, support, and even color based on real-time environmental conditions. By embedding different materials in various parts of the textile, it is possible to create garments with multifunctional capabilities [36]. For instance, a jacket could contain areas with SMPs for fit adjustment, hydrogels for moisture-wicking, and conductive nanocomposites for biometric monitoring, all within a single, cohesive fabric. This level of customization in 4D-printed textiles presents an innovative leap in wearable technology, with applications spanning fashion, healthcare, sports, and even military gear, where adaptive, high-performance materials are increasingly necessary. As material science continues to advance, the possibilities for 4D-printed textiles grow, offering the potential for garments that not only serve as clothing but act as extensions of the wearer's environment, responding dynamically to create a more comfortable, protective, and interactive experience [37]. Figure 5 illustrates the various methods for the preparation of polymer-based inorganic nanoparticle composites. Each method is represented visually, showing the steps involved in the preparation process. Arrows and labels indicate how nanoparticles are incorporated into the polymer matrix through different techniques. The diagram might depict a sequence for each method, starting from the initial components (polymers and nanoparticles) through the processing steps, and culminating in the final composite material. Visual representations such as flowcharts or schematic diagrams may illustrate the processes of intercalation, polymerization, or mixing, highlighting the interactions between the polymer and nanoparticles. Each method may be accompanied by a brief description or key features, emphasizing the advantages or specific applications of that particular technique. Overall, Fig. 5 serves as a comprehensive overview of the methodologies available for creating polymer-based inorganic nanoparticle composites, providing clarity on the procedural differences and potential applications of each method [38].

The textile value chain encompasses various forms of textiles and textile products, beginning with the fiber stage. Fibers can be derived from natural sources, such as cotton and wool, or synthesized from polymers through spinning processes. These individual fibers are then spun together to create yarn. The unique interlacing of these yarns results in the formation of textile fabrics, which serve as the foundational materials for a wide range of textile applications and products (as shown in Fig. 6) [39].

4 Polymer Nanocomposites in 4D-Printed Textiles

Polymer nanocomposites are foundational to advancing 4D-printed textiles, enhancing mechanical, thermal, and responsive properties that enable textiles to change dynamically under specific stimuli [40]. These nanocomposites are formed by embedding nanoparticles—such as carbon nanotubes, graphene, metallic nanoparticles, or silica—into polymer matrices, which amplifies the polymer's intrinsic characteristics. When used in 4D printing, polymer nanocomposites significantly improve durability, flexibility, conductivity, and responsiveness to environmental changes.

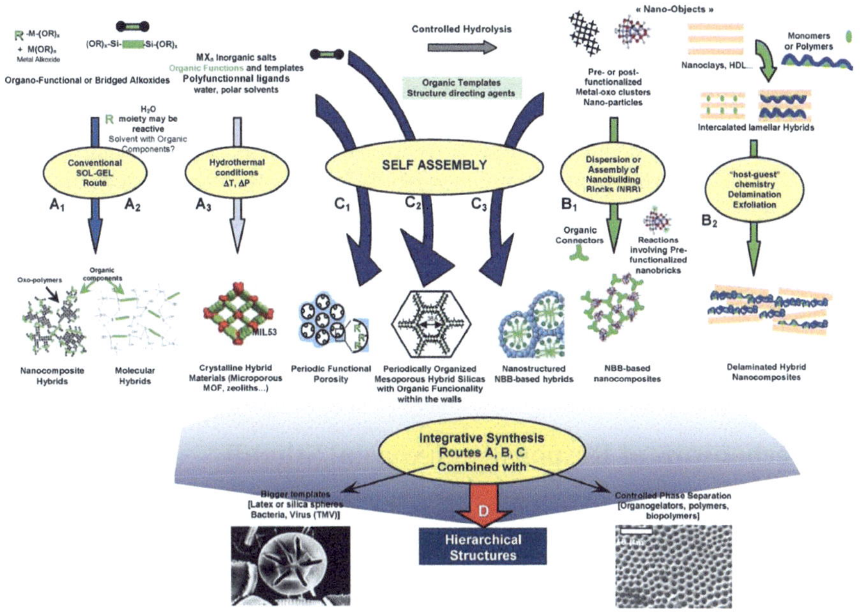

Fig. 5 Schematic representation illustrating the primary chemical pathways for synthesizing polymer-inorganic nanocomposites. Path A: Sol–gel process; Path B: Assembly or dispersion techniques; Path C: Self-assembly methods; Path D: Integrative synthesis approaches [38]. Copyright 2005. Reproduced with permission from the Royal Society of Chemistry

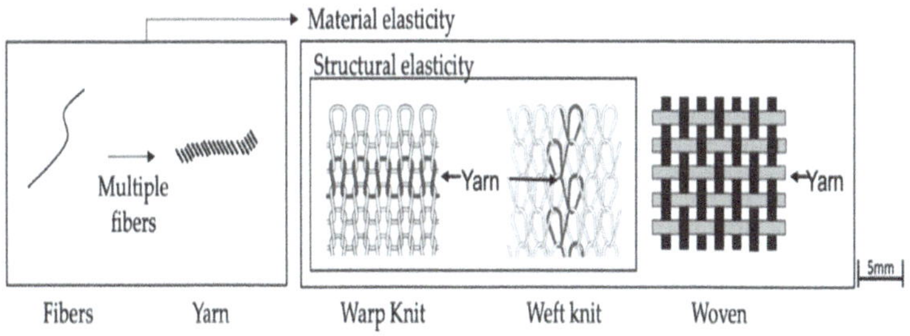

Fig. 6 The textile production chain from fiber to yarn and selected textile structures [39]. Copyright 2021. Reproduced with permission from MDPI

This makes them highly suitable for applications that demand high adaptability, such as smart wearable textiles, healthcare devices, and military fabrics [13, 41]. In the context of 4D printing, these nanocomposites respond to stimuli such as heat, light, pH, and moisture, which allows them to alter their structure, shape, or function in real time. For example, carbon nanotube (CNT)-reinforced polymers provide enhanced electrical conductivity and elasticity, ideal for wearable electronics that must flex and conform to the human body without losing functionality [42]. Similarly, incorporating graphene into polymers can improve mechanical strength while maintaining lightweight properties, making the material ideal for protective clothing and responsive garments. The versatility of polymer nanocomposites in 4D-printed textiles also allows for multi-stimuli responsiveness, where a textile can react to more than one external factor simultaneously [43]. This feature broadens application potential, enabling the creation of garments or fabrics that can, for example, self-ventilate in response to temperature increases or change color based on UV exposure. Additionally, polymer nanocomposites allow for strength, flexibility, and responsiveness customization, enabling textiles to meet the specific needs of different industries, including sports, fashion, healthcare, and defense (Table 2).

5 Mechanisms of Response to External Stimuli

The underlying principle of 4D printing in textiles centers on how materials respond to external stimuli, influencing their practical applications across various domains. One significant mechanism is thermal responsiveness, where materials such as shape-memory polymers (SMPs) and certain specialized polymers are engineered to react to temperature changes [51]. This capability is particularly beneficial for garments that need to adapt to body contours, as these fabrics can "remember" a specific shape and revert to it when exposed to predetermined temperatures. For example, SMPs facilitate shape-shifting textiles, allowing for applications like adaptive compression garments or temperature-sensitive clothing that adjusts fit based on the wearer's activity level [52]. Another critical response mechanism is moisture responsiveness, which involves hydrogels and other hydrophilic materials that react to humidity and moisture levels by altering their volume and porosity. This characteristic is essential in performance fabrics, including sportswear and military gear, where moisture-wicking capabilities and breathability are vital for optimal wearer comfort. A fabric designed to sense and respond to sweat levels, for instance, could increase airflow when moisture is detected, significantly enhancing comfort during extended physical exertion [53].

Light responsiveness is another fascinating aspect of 4D-printed textiles, wherein photo-responsive materials can change color or opacity based on light exposure. This quality allows for innovative applications, such as outdoor clothing made from UV-sensitive materials that shift colors when subjected to high levels of ultraviolet radiation. Such garments can serve as visual indicators, alerting the wearer to take necessary protective measures against sun exposure, thus promoting health and safety in outdoor activities [54]. Lastly, incorporating electrical and magnetic responsiveness into textiles, achieved through embedding conductive materials like carbon nanotubes or metallic particles, enables fabrics to react to electrical or magnetic fields. This capability has significant implications for health monitoring, allowing sensors embedded in the fabric to track various physiological variables, including

Table 2 Nanoparticle-enhanced polymer matrices for smart textile applications: Stimuli, benefits, and challenges

Nanoparticle	Polymer matrix	Stimulus response	Key benefits	Applications	Challenges	References
Carbon nanotubes (CNTs)	Polyurethane, polyethylene	Electric field, temperature	High conductivity, elasticity	Wearable electronics, smart textiles	High-cost, complex synthesis	[44]
Graphene	Polyvinyl alcohol (PVA), PLA	Light, mechanical stress	Strength, lightweight	Protective wear, sensors	Limited scalability	[45]
Metallic nanoparticles	Polyvinyl chloride (PVC), TPU	Heat, magnetic field	Durability, flexibility	Medical wearables, sports textiles	Corrosion, stability issues	[46]
Silica	Polyimide, silicone	Heat, chemical exposure	Thermal stability, hardness	Fire-resistant fabrics, medical wear	Brittle, processing difficulties	[47]
Silver nanoparticles	PLA, polyethylene glycol (PEG)	Antimicrobial response	Antibacterial, anti-fungal	Healthcare wear, hygiene products	Toxicity, regulatory concerns	[48]
Cellulose nanocrystals	PCL, PVA	Moisture, temperature	Biocompatibility, sustainability	Biodegradable wearables, soft robotics	Lower conductivity	[49]
Nanoclays	Epoxy, PLA	Mechanical, chemical resistance	Enhanced strength, stability	Defense wear, industrial textiles	Processing uniformity issues	[50]

body temperature, heart rate, and movement [55]. Such textiles have promising applications in wearable technology, facilitating real-time data collection and interaction with external devices, ultimately paving the way for more sophisticated and responsive clothing that enhances both comfort and functionality. This convergence of material science and technology within 4D printing represents a transformative step in the evolution of textiles, allowing for garments that serve traditional purposes and engage dynamically with the wearer's environment [56].

Razzaq et al. reported that polylactic acid (PLA) is a widely used thermoplastic material in fused filament fabrication (FFF) that exhibits a thermo-responsive shape memory effect (SME). To enhance the shape memory properties, PLA is often blended with other thermoplastics. For example, mixtures of PLA and thermoplastic polyurethane (TPU) were prepared at weight ratios of 90/10, 80/20, 70/30, and 60/40 using a twin-screw extruder (as shown in Fig. 7a–e). Scanning electron microscopy (SEM) revealed that these PLA/TPU mixtures were immiscible, with TPU droplets transforming into fibrils as the TPU content increased from 10 to 40% in the printed specimens due to shear elongation. All samples were capable of being fixed into temporary shapes at room temperature without the need for additional heating or cooling cycles. Notably, the blend containing 80% PLA exhibited the highest fixity ratio (Rf), achieving values of 82% in tension, 91% in compression, and 84% in bending at room temperature. However, this blend also displayed brittle behavior, becoming increasingly brittle with repeated loading cycles. In this context, the crystalline domains of PLA acted as crosslink points, retaining the original shape, while the TPU phase accommodated strain, preventing PLA from breaking under high deformations. When heated above their glass transition temperature (Tg) of 60 °C, the mobility of PLA molecules increased, allowing them to release stored stresses and return to their original shapes. In another study, 80/20 TPU/PLA mixtures were compounded via twin-screw extrusion, converted into filaments, and processed using FFF. However, the high TPU content rendered the filaments very flexible and difficult to print, necessitating the use of a screw-based material extrusion (MEX) system. It was also observed that PLA tended to fibrillate under certain extrusion conditions, particularly when the viscosity ratio of PLA to TPU was close to 1. This fibrillation occurred at extrusion temperatures below 190 °C in the 3D printer. The PLA fibrils were visualized by dissolving the TPU in dimethylformamide (DMF) and imaging the remaining material using SEM, revealing that PLA fibrils with large aspect ratios formed in the printer nozzle, oriented in the direction of deposition. The degree of crystallinity of the printed specimens was significantly higher than that of samples produced by compression molding, attributed to different heating and cooling rates, as well as levels of polymer chain orientation. This increased crystallinity resulted in enhanced tensile strength and Young's modulus for the printed specimens. Further evaluation of the shape memory capabilities of these TPU/PLA polymer composites revealed that as the PLA content increased from 0 to 30%, the fixity ratio (Rf) improved while the recovery ratio (Rr) decreased. The optimal overall shape memory behavior was achieved with the addition of 10% PLA to TPU, yielding Rf values of 91% and Rr values of 71%. The orientation of the infill also influenced the obtained shape memory ratios. When the infill was aligned parallel

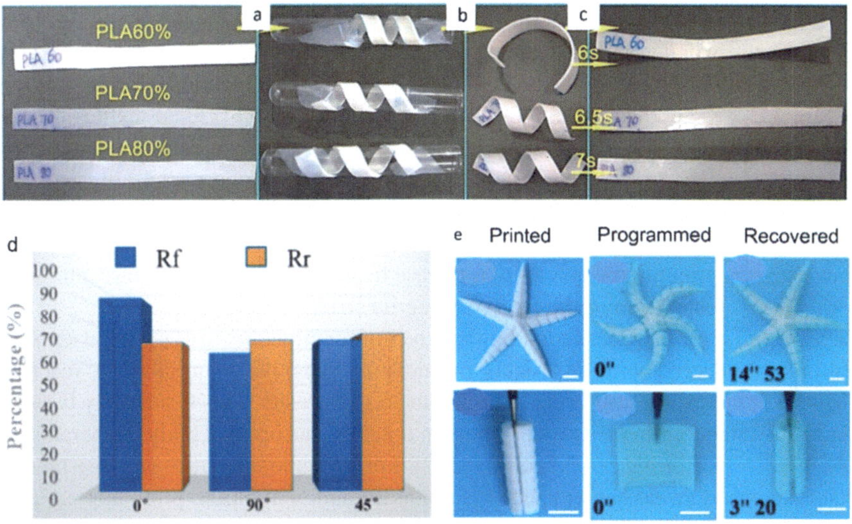

Fig. 7 **a** Strips of PLA/TPU polymer–polymer composites, **b** room temperature shape programming of PLA/TPU composites, **c** shape recovery of PLA/TPU composites after heating to 70 °C, **d** fixity ratio (Rf) and recovery ratio (Rr) for TPU/PLA composites printed with different infill orientations (0°, 90°, and 45°), **e** FFF produced PBS/PLA composite specimens after printing, programmed in water at 85 °C, and shape recovery in water at 85 °C (scale bar 10 mm) [57]. Copyright 2022. Reproduced with permission from MDPI

to the direction of the applied force (0°), the highest fixity ratio (Rf) of 90% was recorded. This was followed by an infill orientation at 45° to the loading direction, which resulted in an Rf of 72%. Conversely, an infill pattern perpendicular to the applied force (90°) yielded the lowest Rf value of 65%. In contrast, the recovery ratio (Rr) remained relatively consistent across all specimens, with values ranging from 71 to 75% (as shown in Fig. 7) [57].

6 Smart Functionalities in 4D-Printed Textiles

Smart functionalities in 4D-printed textiles signify a revolutionary advancement in material technology, enabling textiles to perform dynamic, adaptive actions in response to various stimuli. Unlike traditional fabrics that remain static, 4D-printed textiles integrate capabilities that allow them to change properties, shapes, and behaviors, significantly enhancing their utility across various applications, including fashion, healthcare, sports, and military use [58]. One of the primary smart functionalities is shape memory, which allows textiles to alter their form based on external conditions such as temperature, light, or moisture. For instance, garments embedded with shape-memory polymers (SMPs) can conform to the wearer's body or shift between shapes depending on their environment. This adaptability improves comfort and optimizes fit and support during physical activities [59]. In medical applications, shape-memory materials can design compression garments that adjust pressure levels according to the wearer's needs, enhancing therapeutic effectiveness. Self-healing capabilities are another groundbreaking functionality in 4D-printed textiles. These textiles can repair minor cuts or abrasions autonomously by incorporating specific polymers that can reform their structure when damaged, extending their lifespan and reducing waste [60]. This feature is particularly valuable in rugged environments, such as military or outdoor settings, where durability is crucial. Self-healing materials often employ microcapsules that release healing agents upon rupture, allowing for seamless restoration of the fabric's integrity [61].

Moisture responsiveness adds another layer of functionality to 4D-printed textiles. Fabrics engineered with hydrogels or moisture-sensitive materials can react to changes in humidity by altering their thickness or porosity. This technology is especially advantageous in activewear, where maintaining an optimal body temperature is essential for performance and comfort [62]. For example, moisture-wicking fabrics that respond to sweat can enhance breathability and ventilation, providing athletes with a more comfortable experience during intense physical exertion. Color-changing capabilities represent an innovative approach to aesthetics in 4D printed textiles. Certain photo-responsive materials can alter their color or patterns in response to UV light or temperature fluctuations [63]. This feature adds a dynamic visual element to garments and can serve functional purposes, such as indicating sun exposure or temperature changes. In fashion, these color-changing textiles create opportunities for personalized and interactive designs, allowing garments to adapt their appearance according to the wearer's preferences or environmental conditions [64].

Moreover, the integration of conductive materials into 4D printed textiles enables the development of smart fabrics that can sense environmental variables and communicate with external devices. Textiles embedded with sensors can monitor biometric data, such as heart rate or body temperature, and transmit this information to smartphones or healthcare devices for real-time monitoring [65]. This capability paves the way for advanced wearable technology, where garments not only serve aesthetic or protective functions but also provide critical health insights, enhancing the wearer's overall well-being. Additionally, 4D printed textiles can exhibit shape-shifting properties that allow them to change from flat structures into three-dimensional forms, enabling applications in adaptive architecture and fashion design [66]. This technology can be used to create garments that transform during use, adapting to different functions such as casual wear, formal attire, or protective clothing based on the context (Table 3).

The 4D-printed polymer-textile composites (PTCs) are comprised of i) a shape memory polymer (SMP), ii) thermoplastic polyurethane or elastomer (TPU/TPE), and iii) a nylon textile, as shown in Fig. 8a, b, and c. Initial experiments revealed that the SMP had difficulty adhering to the textile substrate during the printing process, leading to nozzle clogging issues. To address this, flexible filaments, known to adhere well to nylon substrates in prior studies, were used to bond the materials. Previous research has indicated that the Z-axis height plays a critical role in adhesion quality between the polymer and textile layers. To improve surface finish and dimensional

Table 3 Smart functionalities in textiles: Materials, mechanisms, and applications

Smart functionality	Description	Materials used	Response mechanism	Applications	Benefits	Examples	References
Shape memory	Changes shape based on external conditions	Shape-memory polymers (SMPs)	Temperature, light, or moisture	Adaptive clothing, compression garments	Enhanced fit and comfort	Garments that adjust fit with body temp	[67]
Self-healing	Repairs minor damages autonomously	Healing polymers, Microcapsules	Chemical reaction	Military gear, outdoor apparel	Increased durability and lifespan	Fabrics that seal cuts and abrasions	[68]
Moisture responsiveness	Alters thickness or porosity with humidity	Hydrogels, Moisture-sensitive materials	Humidity	Activewear, sports gear	Improved breathability and comfort	Fabrics that wick away sweat	[69]
Color-changing	Changes color or pattern in response to light	Photo-responsive materials	UV light, temperature	Fashion, outdoor wear	Interactive and personalized designs	UV-sensitive shirts that change color	[70]
Conductive textiles	Senses environmental factors and communicates	Conductive polymers, Nanocomposites	Electrical fields	Wearable technology, health monitoring	Real-time health insights	Smart fabrics that monitor heart rate	[71]
Shape-shifting	Transforms from flat to three-dimensional forms	Responsive polymers	Mechanical stimuli	Fashion design, adaptive architecture	Versatile use for different contexts	Clothing that adapts to various styles	[72]
Temperature regulation	Adjusts insulation or cooling based on temperature	Thermo-responsive polymers	Temperature	Healthcare, outdoor clothing	Maintains optimal body temperature	Jackets that provide warmth or cooling	[73]

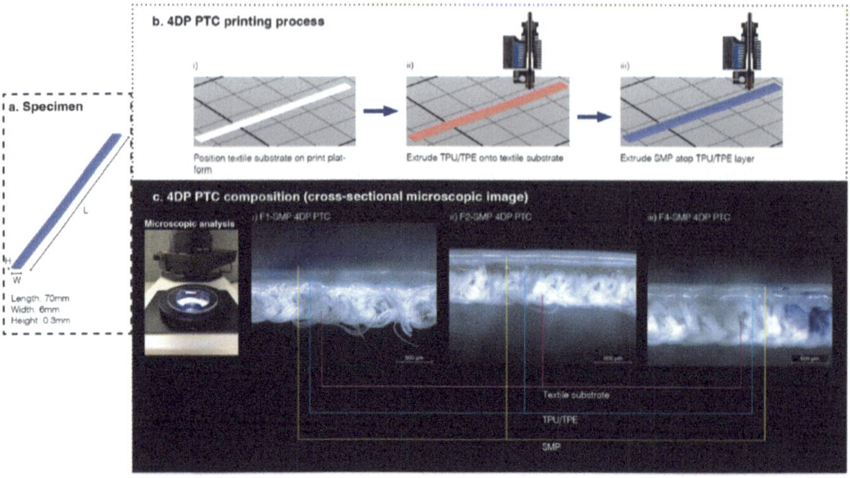

Fig. 8 Schematic illustration of the 4D printing process using PTC (Phase Transition Composite) material (**a**, **b**) and **c** Cross-sectional SEM image (500 μm) [5]. Copyright 2024. Reproduced with permission from Elsevier

accuracy, polymer layers were printed at a 0.15 mm height. A preliminary calibration of the Z-axis was conducted to determine the optimal distance between the nozzle and print surface, ensuring the polymer adhered smoothly to the substrate without gaps or ridges between lines. Through a process of trial and error, the ideal Z-axis height when printing on the platform was established at approximately 1.600 mm, though this may vary depending on the specific machine. For printing onto a textile substrate with a thickness of 0.50 mm, the Z-axis height was adjusted to around 1.100 mm to account for the textile's thickness. Setting an inaccurate height may lead to under-extrusion, filament retraction, and nozzle clogging issues (as shown in Fig. 8) [5].

7 Applications of 4D-Printed Smart Textiles

4D-printed smart textiles represent a transformative advancement in textile technology, merging fashion, functionality, and technology. These textiles, which respond dynamically to environmental stimuli, pave the way for innovative applications across various sectors. Fused Deposition Modeling (FDM) is a widely used extrusion-based additive manufacturing (AM) technique, as shown in Fig. 9a. In this process, a filament is fed into a nozzle by a motor-driven pinch roller, where it is then heated to a molten state. With applied mechanical force, the molten filament is extruded onto a platform to create the first 2D layer, followed by additional layers to form a 3D structure. Once deposited, each polymer layer solidifies through mechanisms like crystallization, chain alignment, non-covalent bond recovery, or chemical cross-linking. FDM is particularly popular for 3D printing due to its simplicity and low cost and is increasingly being applied in 4D printing for various polymers and composites. Commonly used materials in FDM-based 4D printing include polylactic acid (PLA) and polyurethane (PU), both exhibiting shape-memory properties. PLA, a thermoplastic and semi-crystalline material, can undergo shape transformation. For example, Liu et al. demonstrated that adding carbon nanotubes (CNTs) to PLA (creating CNT/PLA composites) enabled earlier activation of the shape-memory response, enhancing recovery force by 144%, and improved both flexural strength and modulus, as shown in Fig. 9a. Another approach by Yu et al. involved a bi-layer composite of PLA and carbon fiber-reinforced PLA (CF/PLA) with an actuator block and a constraint block, each printed with different directional paths, as illustrated in Fig. 9b. Finite element analysis (FEA) allowed precise control of shape changes in these 4D-printed parts. Under heating at 80 °C, differential shrinkage between the two blocks resulted in a bending effect [15]. To enable remote actuation and avoid the need for external heaters, researchers have explored conductive CNT/PLA composites. For instance, Liu et al. examined 4D printing with CNT/PLA filaments, shown in Fig. 9b, where the structures were activated by current-induced differential heating through the conductive layers. Joule heating created temperature gradients that triggered shape-memory responses, with optimized parameters—such as slower printing speeds, thicker layers, and 0° raster angle—enabling faster electro-activated recovery. These advancements highlight the potential of 4D-printed devices in applications requiring responsive, programmable shape transformations.

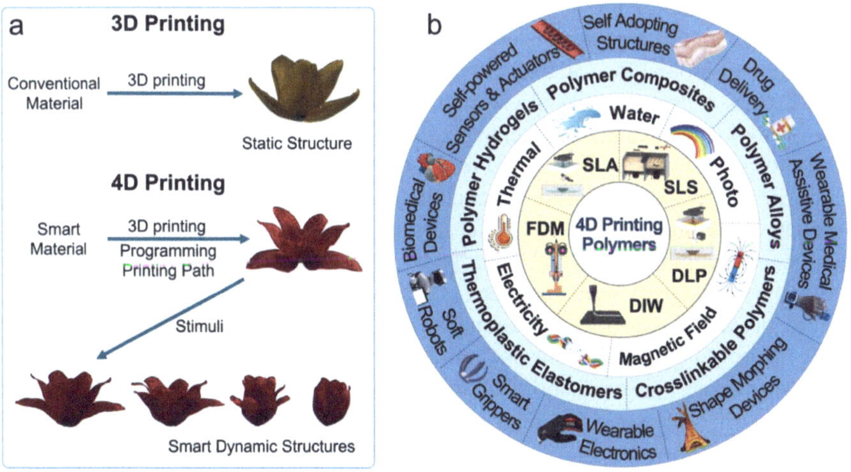

Fig. 9 **a** Brief comparison of 3D and 4D printing. **b** Representative printing methods, external stimuli, printable polymeric materials, and applications of 4D printing [15]. Copyright 2022. Reproduced with permission from Elsevier

8 Fashion and Wearable Technology

The fashion industry is undergoing a remarkable transformation with the integration of 4D printed textiles, particularly those employing polymer nanocomposites. These sophisticated materials can dynamically modify their color, shape, or texture in response to a range of environmental stimuli, including temperature fluctuations, humidity variations, and changes in light conditions [74]. Polymer nanocomposites, which combine polymers with nanoscale fillers—such as carbon nanotubes, graphene, or metallic nanoparticles—exhibit enhanced mechanical, thermal, and electrical properties compared to traditional textiles. This unique combination of features empowers designers to create garments that are aesthetically pleasing and functionally versatile, catering to the specific needs and preferences of individual users [75]. The application of photo-responsive nanocomposite fibers in garment design exemplifies the innovative potential of these materials. For instance, photochromic nanocomposites can change their color or opacity when exposed to varying light intensities, allowing for the creation of garments that visually adapt to different environments. Such textiles can serve practical purposes, such as providing wearers with protection against UV radiation by darkening in response to sunlight, thereby reducing the risk of sunburn and skin damage [76]. Additionally, the aesthetic versatility afforded by these materials enables designers to craft clothing that can shift in appearance throughout the day, offering a customizable wardrobe experience that aligns with the wearer's activities and preferences.

Thermo-responsive nanocomposites further illustrate the potential of 4D-printed textiles in fashion. These materials can respond to changes in body temperature or ambient temperature, allowing garments to alter their fit and comfort dynamically [77]. For example, a running shirt composed of thermochromic nanocomposites could tighten around the torso to provide support when the wearer's body temperature increases during physical exertion, subsequently relaxing as the body cools down. This capability not only enhances comfort but also improves performance by allowing the garment to respond to the wearer's physiological state, potentially leading to better endurance and reduced fatigue during athletic activities [78]. Moreover, the integration of embedded sensors in nanocomposite-based smart textiles significantly amplifies the functionality of garments. These sensors can monitor various physiological parameters, such as heart rate, temperature, and moisture levels, thereby providing real-time data to the wearer. For instance, a fitness-oriented shirt could incorporate moisture sensors that detect sweat levels, triggering the fabric to change its porosity or breathability to improve comfort and ventilation during intense workouts. This interconnectedness between the garment and the wearer enables a feedback loop, allowing for personalized adjustments that enhance the overall user experience [79].

In addition to enhancing individual performance and comfort, the development of 4D printed textiles utilizing polymer nanocomposites also holds implications for sustainability in the fashion industry. By creating garments that can adapt to different environmental conditions, the need for multiple items in a wardrobe may be reduced, promoting a more sustainable approach to clothing consumption [80]. Additionally, the incorporation of eco-friendly nanomaterials in textile production can minimize the environmental impact associated with traditional textile manufacturing processes. For example, the use of biodegradable polymers or sustainably sourced nanofillers can contribute to the development of eco-conscious fashion, addressing growing consumer concerns regarding sustainability and ethical production practices. The interplay between nanotechnology and fashion also paves the way for innovative applications beyond aesthetic and functional enhancements [81]. For instance, smart textiles equipped with sensors could facilitate enhanced user interaction through mobile applications, allowing wearers to customize their garments in real time based on their preferences or environmental conditions. This level of interactivity could extend to the integration of augmented reality (AR) technologies, where users can visualize and select garment alterations using AR interfaces, further enhancing the personalized experience [82].

Shie et al. reported that Fig. 10A illustrates the process by which digital light from a commercial printer is directed onto a printable solution positioned between two slides and separated by a spacer. This setup enables the light to attenuate along a gradient due to variations in thickness, allowing for the creation of different curing conditions. Once the unreacted chemical agent is removed, a polymer film with various structural shapes is produced. The differences in thickness introduce built-in stress within the films, facilitating the transformation of the flat film into a three-dimensional structure upon the release of this stress. Figure 10B demonstrates how the choice of planar light patterns enables the formation of complex permanent shapes that can be altered into temporary forms and recovered upon reheating. This capability stems from the unique properties of shape memory polymer (SMP) hydrogels, which can undergo structural rearrangements in response to changes in their degree of swelling when exposed to specific stimuli. Additionally, research has shown that both ordinary and auxetic structures can exhibit reversible shape changes using 4D printing techniques [83].

9 Sportswear and Activewear

Athletes are increasingly turning to 4D-printed textiles as a means to enhance performance and overall experience during physical activities. These advanced smart fabrics are engineered to dynamically respond to the wearer's physiological conditions, particularly body temperature and moisture levels, thereby optimizing comfort and functionality during athletic performance. One key aspect of these 4D-printed textiles is their ability to incorporate materials with moisture-wicking capabilities [84]. Hydrophilic fibers in the fabric can rapidly absorb and transport sweat away from the skin, promoting evaporative cooling and maintaining optimal thermal comfort. This feature is particularly crucial in sportswear, as it helps to regulate body temperature during intense physical exertion, reducing the risk of overheating and enhancing the athlete's endurance. Additionally, the incorporation of shape-memory

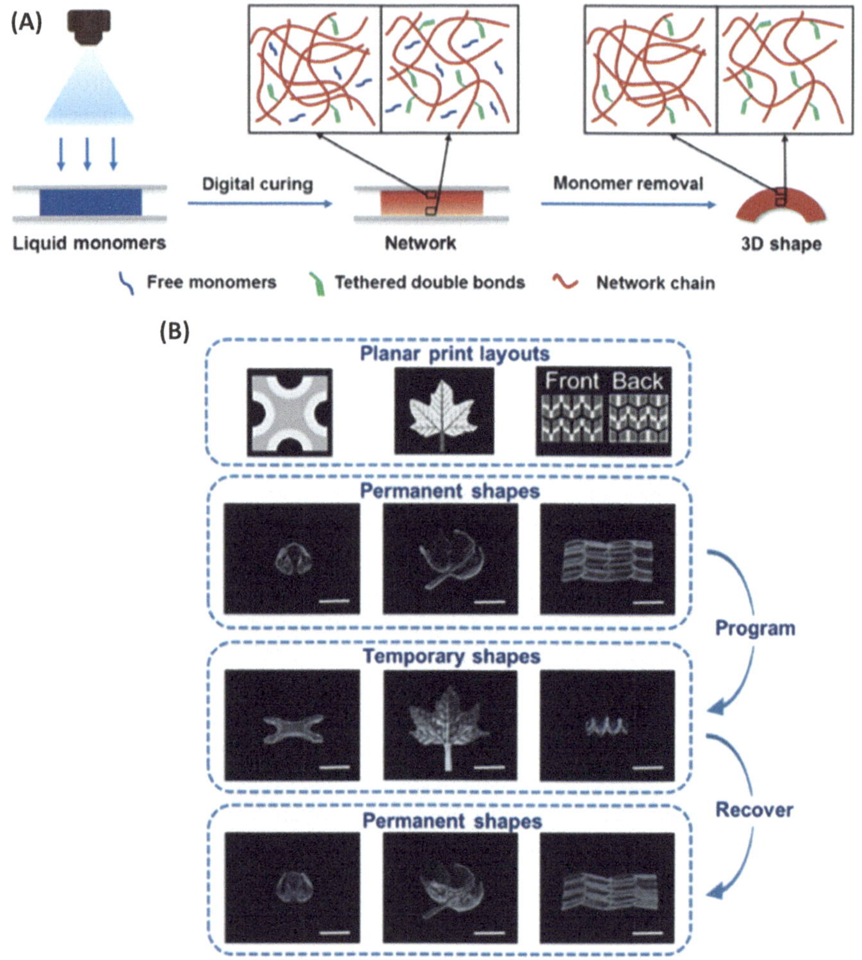

Fig. 10 **a** Schematic representation of the digital shape-memory polymer (SMP) fabrication process. **b** Digital fabrication of intricate permanent shapes, demonstrating their shape memory behavior. In the planar print layouts, the black background indicates areas of no light exposure, while the light and dark regions represent light exposure times of 14 s and 30 s, respectively [13]. Copyright 2019. Reproduced with permission from MDPI

polymers (SMPs) in 4D-printed activewear allows the garment to conform closely to the athlete's body shape and movement patterns. These materials are programmed to respond to temperature changes, enabling them to provide targeted support where needed [85]. For example, as an athlete's body temperature rises during exercise, the SMPs can tighten around key muscle groups, offering compression that enhances blood circulation and reduces muscle vibration. Once the body cools down, these materials can revert to their original, relaxed state, providing freedom of movement without compromising support [86]. Breathability is another essential characteristic of 4D-printed sportswear, which can be achieved through the strategic design and layering of nanocomposite materials. By integrating porous structures within the textile, designers can facilitate airflow and moisture transfer, further enhancing the fabric's ability to keep the athlete dry and comfortable. This breathable architecture helps maintain a balanced microclimate next to the skin, allowing athletes to perform optimally without being distracted by discomfort [87].

Figure 11a illustrates the chemical structures of epoxidized soybean oil and bisphenol F diglycidyl ether (BFDGE), as detailed in Rodriguez's work (Fig. 11b). In this study, carbon nanotubes were incorporated into the resin, creating non-Newtonian inks (Fig. 11c). Rodriguez employed two distinct post-printing curing routes. The first method, shown in Fig. 11d, e, involves thermal curing at 80 °C for 16 h followed by a post-cure at 150 °C for 2 h. In the second method (Fig. 11f), the printed part is folded, then initially cured at 80 °C for 4 h, followed by a full cure at 80 °C for 12 h, and finally post-cured at 150 °C for 2 h (Fig. 11g). Both routes converge at Fig. 11h, where the printed structure is heated to 80–100 °C for 5 min. At this stage, an external force is applied to shape the structure, which is maintained as the temperature cools to 20 °C. The structure is then able to return to its original shape when reheated above its glass transition temperature (Tg). This '4D' printing technique, using shape memory polymers, enables the fabrication of complex structures with multi-functional capabilities and diverse thermal shape recovery profiles. Such materials could be used in applications like electrical interconnects and thermal sensors. Rodriguez highlights challenges in achieving complete resin curing through photochemical methods, which may impair longevity, as well as in creating inks with sufficient pot life, printability, stiffness, conductivity, and shape-memory properties [26].

10 Healthcare and Medical Textiles

The integration of 4D printed textiles into healthcare applications holds transformative potential, particularly in the development of adaptive medical garments designed to meet the dynamic needs of patients. These innovative textiles can be engineered to provide personalized compression therapy, a crucial aspect of managing various medical conditions such as venous insufficiency, lymphedema, and post-surgical recovery [88]. By utilizing shape-memory polymers (SMPs) and other responsive materials, these garments can autonomously adjust their pressure levels in response to the wearer's movements or changes in body temperature. This adaptability enhances comfort and promotes optimal circulation, thereby facilitating healing processes and reducing the risk of complications associated with static compression.

For example, adaptive compression garments made from 4D-printed textiles can respond to physical activity by dynamically varying the compression levels applied to different body segments [88]. As a patient engages in movement, the garment could tighten in areas requiring additional support—such as around the ankle or knee—while relaxing in regions where less compression is needed. This targeted approach

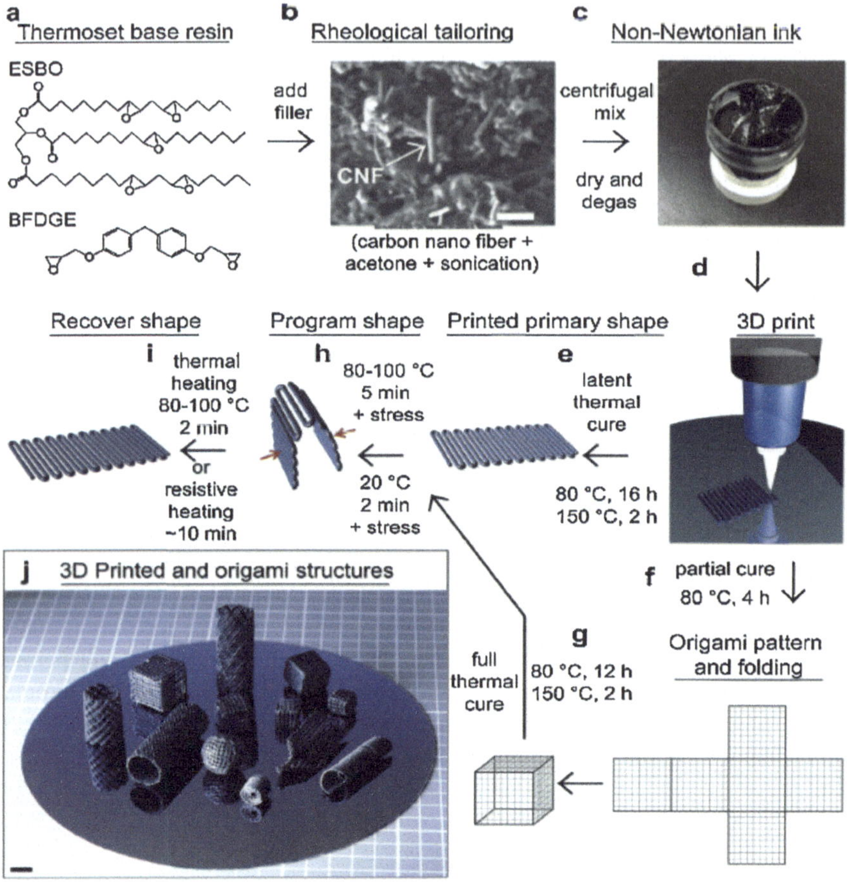

Fig. 11 Ink development and manufacturing process [26]. Copyright 2016. Reproduced with permission from Springer Nature

not only improves blood flow and lymphatic drainage but also helps to mitigate discomfort associated with traditional, non-adaptive compression garments, which may feel constricting during movement [89]. Moreover, incorporating sensors into 4D-printed medical textiles allows for real-time monitoring of vital signs, such as heart rate, respiratory rate, and temperature. These smart textiles can be designed to collect and transmit health data wirelessly to healthcare professionals, enabling continuous patient monitoring without the need for intrusive devices. For instance, garments embedded with conductive fibers and nanosensors can track physiological parameters and alert medical personnel in case of anomalies, facilitating timely interventions [90]. This capability is particularly beneficial in settings such as post-operative care, chronic disease management, and rehabilitation, where early detection of health issues can significantly improve patient outcomes.

In addition to their applications in compression therapy and vital sign monitoring, 4D-printed medical textiles can also be designed to enhance wound care. Textiles that respond to changes in humidity or temperature can provide a controlled environment for wound healing, promoting optimal conditions for tissue regeneration [90]. For example, hydrogels integrated into the fabric can absorb excess exudate from wounds while maintaining a moist environment, thereby accelerating the healing process. Furthermore, some advanced textiles can release therapeutic agents in response to specific triggers, such as the presence of bacteria, thereby providing localized treatment and minimizing infection risk. The development of 4D-printed adaptive medical textiles is a significant step toward personalized healthcare solutions [91]. By leveraging advances in material science and engineering, these textiles can be tailored to the unique needs of individual patients, offering a level of customization that traditional medical garments cannot achieve. The interplay of responsive materials, integrated sensors, and smart technologies in healthcare textiles not only enhances patient comfort and mobility but also empowers healthcare providers with real-time data that can inform clinical decisions. As research in this field continues to evolve, the potential applications of 4D-printed textiles in healthcare are vast, promising to redefine the standards of care and patient experience [92, 93]. The illustration provides a visual comparison between two textile concepts: a shape memory composite integrated into the traditional fabric and a dedicated shape memory smart textile. In Fig. 12a, the diagram showcases a conventional fabric that incorporates shape memory materials (SMMs). This integration allows the fabric to exhibit unique properties, such as the ability to revert to a pre-defined shape when exposed to specific stimuli (like heat). The traditional fabric serves as the base, while the SMMs enhance its functionality, enabling applications that require adaptability and responsiveness without compromising the fabric's original look or feel. Figure 12b, the focus shifts to a shape-memory smart textile specifically designed for dynamic functionality. This textile is engineered from the ground up with shape memory polymers, allowing for greater flexibility in design and performance. The smart textile is illustrated to highlight its capability to change shape or structure autonomously in response to environmental factors such as temperature, humidity, or light. This enhanced interactivity and adaptability open new possibilities for applications in fashion, sportswear, and healthcare, making the fabric not only functional but also responsive to the wearer's needs (as shown in Fig. 12) [3, 93].

Recently, a novel shape memory polymer (SMP) has been developed for the treatment of cranio-maxillofacial defects (as shown in Fig. 13a–c). The process involves briefly treating a PCL diacrylate with warm saline, followed by UV polymerization. This is then combined with a solvent-cast particulate leaching method to produce porous scaffolds. The unique "self-fitting" behavior of the SMP enables it to adapt to irregular defects, allowing for the creation of customizable shapes and structures tailored to meet the specific needs of individual patients [13, 94].

Similar to FDM, DIW 4D printing is gaining considerable interest due to its versatile applications. Lin et al. [95] developed a composite ink made from magnetic Fe_3O_4 and PLA and used the DIW technique to create shape memory occluders. These occluders serve as temporary platforms for atrial septal defect (ASD) repair,

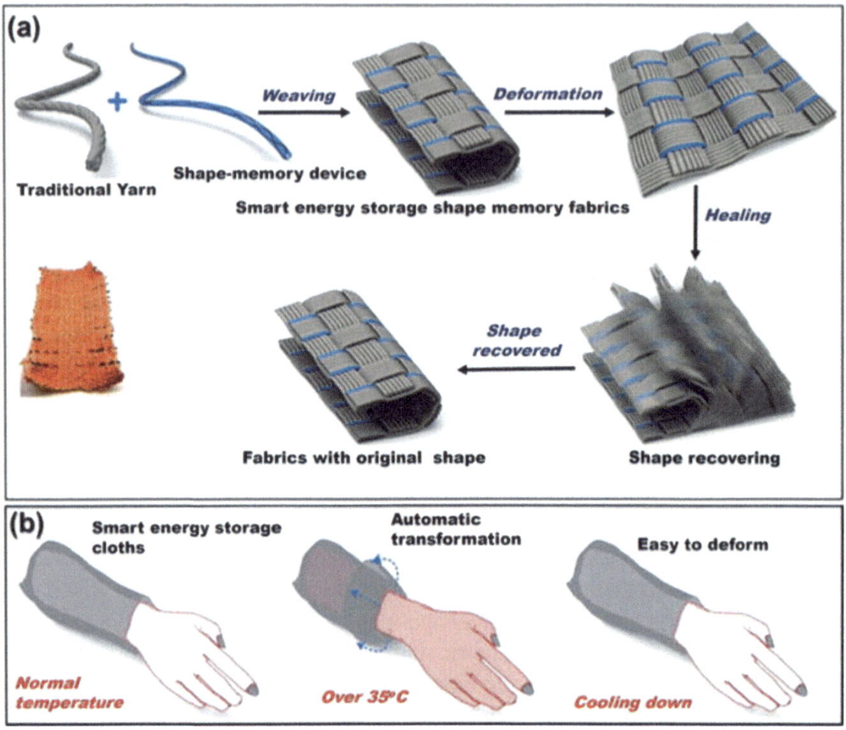

Fig. 12 Schematic illustration of **a** a shape memory composite integrated into the traditional fabric and **b** a shape memory smart textile [3]. Copyright 2024. Reproduced with permission from MDPI

as shown in Fig. 14a, supporting tissue growth before full endothelialization occurs. The occluder design includes a support frame printed from the composite ink and thin PLA membranes fabricated through electrospinning. To minimize scarring from interventional surgery, the occluder can be reshaped from its permanent double-disk form into a temporary linear configuration using stimuli such as heat or a magnetic field (as shown in Fig. 14b–c). Once implanted, the device reverts to its original shape under a remote stimulus, facilitated by the shape memory effect from Fe_3O_4 particles in the ink. The occluder demonstrated strong cell adhesion, enhanced proliferation, and promoted tissue development, aiding in rapid endothelialization, and offering a promising alternative to metal-based occlusion devices [95].

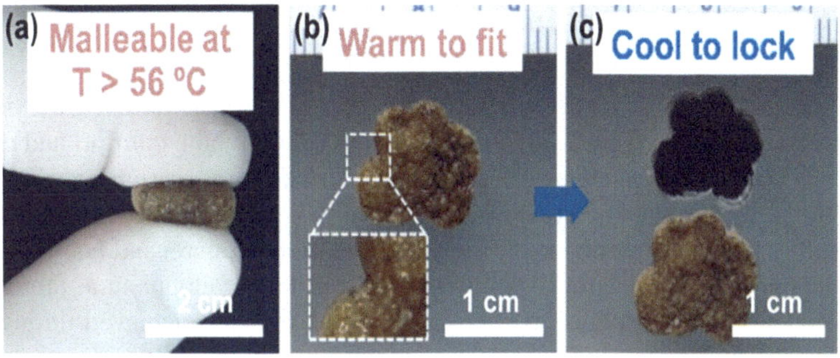

Fig. 13 Images showcasing the self-fitting behavior of a polydopamine-coated PCL shape memory polymer (SMP) scaffold. When the scaffold is heated above the melting temperature (Tm) of PCL, the cylindrical support softens, allowing it to conform to irregular shapes effectively. Upon cooling, the scaffold retains the newly formed temporary shape even after the bracket is removed from the defect [94]. Copyright 2014. Reproduced with permission from Elsevier

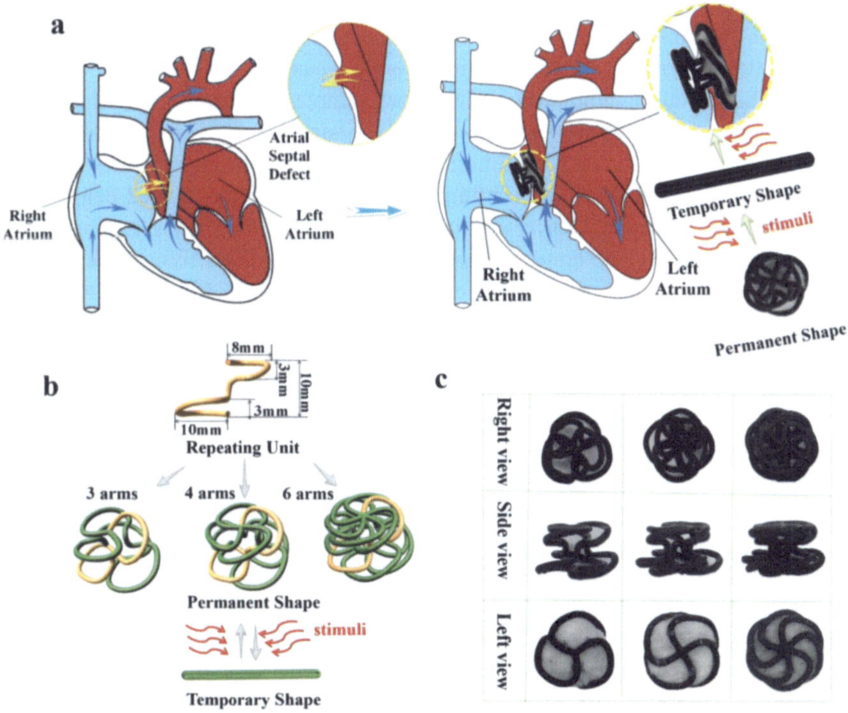

Fig. 14 (A) Diagram showing the atrial septal defect (ASD) prototype before and after occlusion therapy using an occlude (**a**), **b** Design overview of three occluder frames with varied arm structures, and **c** Detailed view of the printed occluder structures [95]. Copyright 2019. Reproduced with permission from Wiley & Sons

11 Home Textiles and Interior Design

The emergence of smart textiles in home furnishings is revolutionizing the field of interior design by offering adaptive solutions that enhance both comfort and energy efficiency. These innovative textiles can be integrated into various applications, including curtains, upholstery, and bedding, providing functionalities that respond dynamically to environmental stimuli such as light, temperature, and humidity [96]. One of the notable advancements in home textiles is the development of fabrics that alter their opacity in response to sunlight. This capability is achieved through the incorporation of photo-responsive materials, which can change their transparency or color when exposed to varying levels of UV radiation [97]. For instance, smart curtains made from these advanced textiles can transition from transparent to opaque as sunlight intensity increases, effectively regulating indoor lighting and reducing the need for artificial illumination. This not only enhances the aesthetic appeal of a space but also contributes to energy savings by minimizing reliance on electrical lighting, thereby reducing overall energy consumption [98]. Moreover, smart textiles that respond to humidity play a crucial role in regulating indoor climate. Hydrophilic materials embedded in these textiles can absorb or release moisture based on the surrounding humidity levels. For example, upholstery fabrics designed with moisture-responsive properties can help maintain a comfortable indoor environment by adjusting to changes in humidity. When humidity levels rise, these textiles can absorb excess moisture, preventing the buildup of dampness that can lead to mold and mildew growth. Conversely, in drier conditions, they can release moisture to enhance air quality and comfort, thereby creating a healthier living environment [99]. In addition to their functional benefits, smart textiles also offer opportunities for enhanced user interaction and engagement. Many of these textiles can be integrated with sensors and smart home systems, enabling them to respond to environmental changes and user preferences. For example, a smart bedding system could adjust its thermal properties based on the sleeper's body temperature, ensuring optimal nighttime comfort. Similarly, smart curtains could be programmed to open or close automatically based on the time of day or user-defined preferences, further enhancing convenience and energy efficiency [100].

The integration of nanotechnology into home textiles further elevates their capabilities. Nanocomposite materials can enhance the durability, stain resistance, and performance of fabrics while enabling them to maintain their adaptive properties over time. For instance, textiles embedded with nanoparticles can provide antimicrobial properties, ensuring that upholstery remains hygienic and resistant to bacteria and odors, thus extending the lifespan of home furnishings [101]. From an environmental perspective, the use of smart textiles in interior design also aligns with sustainability goals. By effectively managing indoor lighting and climate, these textiles contribute to reducing energy consumption and greenhouse gas emissions associated with heating, cooling, and lighting homes. Furthermore, the potential for incorporating recycled or sustainably sourced materials in the production of smart textiles

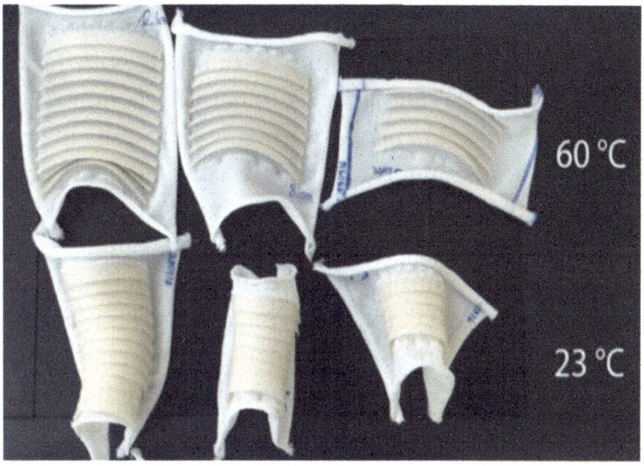

Fig. 15 Thermal-responsive 4D textile [103]. Copyright 2018. Reproduced with permission from Springer Nature

can enhance their eco-friendliness, promoting a more sustainable approach to interior design [102].

Shape-shifting textiles can be developed by a novel method that integrates 3D printing with pre-strained textile substrates, wherein the textile serves as an energy storage medium. These hybrid 4D textiles combine conventional fabric with printed polymer elements to attain dynamic functionality. At the Institut für Textiltechnik (ITA) at RWTH Aachen University, Schmelzeisen and coworkers have innovated the creation of time-dependent microstructures with adjustable mechanical characteristics through this textile-based 3D printing technique. Their approach utilizes the adaptability of 3D printing—encompassing material selection, process settings, and dimensional precision—while employing pre-stretched hyperelastic fabrics for energy storage [103, 104]. This combination facilitates enhanced shape-changing effects in the resultant 4D-printed structures, as illustrated in Fig. 15.

11.1 Environmental Monitoring and Protection

There are fascinating prospects for developing interactive and immersive experiences that enthrall audiences when 4D-printed textiles are used in the entertainment sector. These textiles can transform costumes and stage designs by changing their appearance in real time, thus enhancing storytelling and engagement during performances. For instance, costumes made from shape-memory polymers (SMPs) can be programmed to alter their form or color based on specific cues, such as lighting changes or choreography, allowing performers to embody different characters or emotions throughout the show [105]. This dynamic capability adds a layer of visual intrigue and deepens the narrative experience by allowing the fabric to act as a part of the performance. Additionally, textiles with interactive features can respond to audience engagement, making the viewing experience more participatory. Fabrics embedded with sensors can react to movement, sound, or even the audience's emotional responses, fostering a more interactive environment. For example, in a theatrical setting, seats could be outfitted with fabrics that pulse or change color in response to the audience's collective reactions, thereby creating a feedback loop between performers and viewers that heightens the overall atmosphere [106]. In the realm of gaming, interactive textiles can enhance tactile feedback, providing players with physical sensations that correspond to in-game actions. This level of immersion not only enriches the gaming experience but also opens up new avenues for storytelling by allowing players to feel the game environment more viscerally. Beyond the entertainment sector, 4D-printed smart textiles hold significant promise for environmental monitoring and protection. By integrating advanced sensing technologies into fabrics, these textiles can continuously monitor environmental conditions and alert users to hazardous situations [107]. For example, clothing for outdoor activities could be equipped with sensors that detect UV radiation levels, providing real-time feedback to the wearer regarding their sun exposure. Such functionality encourages proactive measures to prevent overexposure to harmful rays, thereby reducing the risk of skin damage and related health issues.

Moreover, smart textiles can monitor air quality and pollution levels. Fabrics embedded with chemical sensors can detect harmful pollutants such as volatile organic compounds (VOCs) or particulate matter in the environment [108]. When pollution levels exceed safe thresholds, the textiles can signal the wearer through color changes or notifications via connected devices. This capability empowers individuals to take timely action, such as seeking shelter or wearing additional protective gear in polluted areas, thereby enhancing public health and safety. The application of smart textiles in environmental protection extends to specialized gear for first responders or outdoor workers [109]. For instance, uniforms made from these advanced textiles can include sensors that monitor exposure to toxic substances or hazardous conditions, ensuring that workers are alerted to potential dangers in real time. This enhances personal safety and allows for better compliance with occupational health standards, ultimately protecting the well-being of individuals in high-risk environments.

12 Future Prospects

The future of 4D printed textiles appears promising, with ongoing advancements in materials science and printing technologies poised to overcome current limitations. Research is focusing on developing cost-effective, durable polymer nanocomposites that can withstand the rigors of everyday use while maintaining their dynamic properties. Innovations in printing techniques will likely enhance the resolution and precision of 4D-printed textiles, enabling the creation of more intricate designs and functionalities. Additionally, the integration of artificial intelligence and machine learning into textile design and production processes may facilitate the development of truly adaptive textiles that respond intelligently to user needs and environmental changes. As consumer demand for smart textiles grows, we can expect to see increased collaboration between researchers, manufacturers, and designers to explore new applications in healthcare, fashion, and beyond.

13 Conclusions

In conclusion, 4D printing is set to transform the textile industry by introducing intelligent fabrics that exhibit unprecedented adaptability and functionality. By merging advanced technologies with traditional materials, 4D printed textiles offer exciting possibilities for enhancing user experience across various sectors, from healthcare and sports to defense and fashion. Despite existing challenges such as material costs and durability concerns, the continuous evolution of 4D printing technologies holds the potential to redefine the capabilities of textiles. As research progresses, we anticipate a future where smart textiles become integral components of interactive systems, catering to the dynamic requirements of a rapidly changing world. This innovative technology not only paves the way for practical applications but also challenges the conventional understanding of what textiles can achieve, heralding a new era of intelligent fabrics.

Acknowledgements The authors gratefully acknowledge the support provided by Centurion University of Technology and Management, Odisha, India, for carrying out the present research work.

References

1. Falahati, M., Ahmadvand, P., Safaee, S., Chang, Y.C., Lyu, Z., Chen, R., Li, L., Lin, Y.: Smart polymers and nanocomposites for 3D and 4D printing. Mater. Today **1**(40), 215–45 (2020)
2. Kausar, A., Ahmad, I., Zhao, T., Aldaghri, O., Eisa, M.H.: Polymer/graphene nanocomposites via 3D and 4D printing—design and technical potential. Processes **11**(3), 868 (2023)
3. Franco Urquiza, E.A.: Advances in additive manufacturing of polymer-fused deposition modeling on textiles: from 3D printing to innovative 4D printing—a review. Polymers **16**(5), 700 (2024)
4. Eutionnat-Diffo, P.A., Cayla, A., Chen, Y., Guan, J., Nierstrasz, V., Campagne, C.: Development of flexible and conductive immiscible thermoplastic/elastomer monofilament for smart textiles applications using 3D printing. Polymers **12**(10), 2300 (2020)
5. Cheung, T.C., Choi, S.Y., Jiang, S.: Influence of thermoplastic polyurethane and elastomer polymers on self-folding behaviour of 4D-printed polymer-textile composites. Addit. Manuf. **5**(79), 103909 (2024)
6. Biswas, M.C., Chakraborty, S., Bhattacharjee, A., Mohammed, Z.: 4D printing of shape memory materials for textiles: mechanism, mathematical modeling, and challenges. Adv. Func. Mater. **31**(19), 2100257 (2021)
7. Mahmood, A., Perveen, F., Chen, S., Akram, T., Irfan, A.: Polymer composites in 3D/4D printing: materials, advances, and prospects. Molecules **29**(2), 319 (2024)
8. Burela, R.G., Kamineni, J.N., Harursampath, D.: Multifunctional polymer composites for 3D and 4D printing. 3D and 4D Print. Polym. Nanocomposite Mater **1**, 231–57 (2020)

9. Sajjad, R., Chauhdary, S.T., Anwar, M.T., Zahid, A., Khosa, A.A., Imran, M., Sajjad, M.H.: A review of 4D printing–technologies, shape shifting, smart polymer based materials, and biomedical applications. Adv. Ind. Eng. Polym. Res. **7**(1), 20–36 (2024)
10. Khalid, M.Y., Arif, Z.U., Noroozi, R., Zolfagharian, A., Bodaghi, M.: 4D printing of shape memory polymer composites: a review on fabrication techniques, applications, and future perspectives. J. Manuf. Process. **1**(81), 759–97 (2022)
11. Manaia, J.P., Cerejo, F., Duarte, J.: Revolutionising textile manufacturing: a comprehensive review on 3D and 4D printing technologies. Fash. Textiles. **10**(1), 20 (2023)
12. Li, J.W., Chen, H.F., Liu, Y.Z., Wang, J.H., Lu, M.C., Chiu, C.W.: Photocurable 3D-printed AgNPs/Graphene/Polymer nanocomposites with high flexibility and stretchability for ECG and EMG smart clothing. Chem. Eng. J. **15**(484), 149452 (2024)
13. Shie, M.Y., Shen, Y.F., Astuti, S.D., Lee, A.K., Lin, S.H., Dwijaksara, N.L., Chen, Y.W.: Review of polymeric materials in 4D printing biomedical applications. Polymers **11**(11), 1864 (2019)
14. Chen, A.Y., Pegg, E., Chen, A., Jin, Z., Gu, G.X.: 4D printing of electroactive materials. Adv. Intell. Syst. **3**(12), 2100019 (2021)
15. Fu, P., Li, H., Gong, J., Fan, Z., Smith, A.T., Shen, K., Khalfalla, T.O., Huang, H., Qian, X., McCutcheon, J.R., Sun, L.: 4D printing of polymers: techniques, materials, and prospects. Prog. Polym. Sci. **1**(126), 101506 (2022)
16. Vera, M., Mella, C., Urbano, B.F.: Smart polymer nanocomposites: recent advances and perspectives. J. Chil. Chem. Soc. **65**(4), 4973–81 (2020)
17. Li, J.W., Lee, J.C., Chuang, K.C., Chiu, C.W.: Photocured, highly flexible, and stretchable 3D-printed graphene/polymer nanocomposites for electrocardiography and electromyography smart clothing. Prog. Org. Coat. **1**(176), 107378 (2023)
18. Zhang, W., Zhang, F., Lan, X., Leng, J., Wu, A.S., Bryson, T.M., Cotton, C., Gu, B., Sun, B., Chou, T.W.: Shape memory behavior and recovery force of 4D printed textile functional composites. Compos. Sci. Technol. **26**(160), 224–30 (2018)
19. Nadgorny, M., Ameli, A.: Functional polymers and nanocomposites for 3D printing of smart structures and devices. ACS Appl. Mater. Interfaces **10**(21), 17489–507 (2018)
20. Subash, A., Kandasubramanian, B.: 4D printing of shape memory polymers. Eur. Polymer J. **5**(134), 109771 (2020)
21. Etawy, M.S., Nassar, G.E., Mohammed, N., Nawar, S.H., Hassabo, A.G.: 4D printing of stimuli-responsive materials. J. Text. Color. Polym. Sci. **21**(2), 241–58 (2024)
22. Yan, S., Zhang, F., Luo, L., Wang, L., Liu, Y., Leng, J.: Shape memory polymer composites: 4d printing, smart structures, and applications. Research **7**(6), 0234 (2023)
23. Rahmatabadi, D., Mirasadi, K., Bayati, A., Khajepour, M., Ghasemi, I., Baniassadi, M., Abrinia, K., Bodaghi, M., Baghani, M.: 4D printing thermo-magneto-responsive PETG-Fe3O4 nanocomposites with enhanced shape memory effects. Appl. Mater. Today **1**(40), 102361 (2024)
24. Zhang, K., Huang, X., Cui, Z., Panahi-Sarmad, M., Xiao, X.: 4D printed Electroactive carbon fiber and carbon nanotube synergistically reinforced composites for supporting personalized insole. Compos. Commun. **1**(51), 102049 (2024)
25. Khoo, Z.X., Teoh, J.E., Liu, Y., Chua, C.K., Yang, S., An, J., Leong, K.F., Yeong, W.Y.: 3D printing of smart materials: a review on recent progresses in 4D printing. Virtual Phys. Prototyp. **10**(3), 103–22 (2015)
26. Rodriguez, J.N., Zhu, C., Duoss, E.B., Wilson, T.S., Spadaccini, C.M., Lewicki, J.P.: Shape-morphing composites with designed micro-architectures. Sci. Rep. **6**(1), 27933 (2016)
27. Cheung, T.C., Choi, S.Y., Jiang, S.X.: Influence of thermoplastic polyurethane polymers on self-folding behaviour of 4d-printed polymer-textile composites. SSRN 4530257
28. Kouka, M.A., Abbassi, F., Habibi, M., Chabert, F., Zghal, A., Garnier, C.: 4D printing of shape memory polymers, blends, and composites and their advanced applications: a comprehensive literature review. Adv. Eng. Mater. **25**(4), 2200650 (2023)
29. Alsaadi, M., Hinchy, E.P., McCarthy, C.T., Moritz, V.T., Zhuo, S., Fuenmayor, E., Devine, D.M.: Liquid-based 4D printing of shape memory nanocomposites: a review. J. Manuf. Mater. Process. **7**(1), 35 (2023). https://doi.org/10.3390/jmmp7010035

30. Tamay, D.G., Dursun Usal, T., Alagoz, A.S., Yucel, D., Hasirci, N., Hasirci, V.: 3D and 4D printing of polymers for tissue engineering applications. Front. Bioeng. Biotechnol. **9**(7), 164 (2019)
31. Ponnamma, D., Parangusan, H., Tanvir, A., AlMa'adeed, M.A.: Smart and robust electrospun fabrics of piezoelectric polymer nanocomposite for self-powering electronic textiles. Mater. Des. **15**(184), 108176 (2019)
32. Chizari, K., Arjmand, M., Liu, Z., Sundararaj, U., Therriault, D.: Three-dimensional printing of highly conductive polymer nanocomposites for EMI shielding applications. Mater. Today Commun. **1**(11), 112–8 (2017)
33. Tariq, A., Arif, Z.U., Khalid, M.Y., Hossain, M., Rasool, P.I., Umer, R., Ramakrishna, S.: Recent advances in the additive manufacturing of stimuli-responsive soft polymers. Adv. Eng. Mater. **25**(21), 2301074 (2023)
34. Alsaadi, M., Hinchy, E.P., McCarthy, C.T., Moritz, V.F., Zhuo, S., Fuenmayor, E., Devine, D.M.: Liquid-based 4D printing of shape memory nanocomposites: a review. J. Manuf. Mater. Process. **7**(1), 35 (2023)
35. Salimon, A.I., Senatov, F.S., Kalyaev, V., Korsunsky, A.M.: Shape memory polymer blends and composites for 3D and 4D printing applications. In 3D and 4D Printing of Polymer Nanocomposite Materials 2020 Jan 1, pp. 161–189. Elsevier
36. Rayate, A., Jain, P.K.: A review on 4D printing material composites and their applications. Mater. Today Proceed. **5**(9), 20474–84 (2018)
37. Sanatgar, R.H., Campagne, C., Nierstrasz, V.: Investigation of the adhesion properties of direct 3D printing of polymers and nanocomposites on textiles: effect of FDM printing process parameters. Appl. Surf. Sci. **1**(403), 551–63 (2017)
38. Sanchez, C., Julián, B., Belleville, P., Popall, M.: Applications of hybrid organic–inorganic nanocomposites. J. Mater. Chem. **15**(35–36), 3559–3592 (2005)
39. Koch, H.C., Schmelzeisen, D., Gries, T.: 4D textiles made by additive manufacturing on pre-stressed textiles—an overview. Actuators **10**(2), 31 (2021). https://doi.org/10.3390/act10020031
40. Zhang, Z., Demir, K.G., Gu, G.X.: Developments in 4D-printing: a review on current smart materials, technologies, and applications. Int. J. Smart Nano Mater. **10**(3), 205–24 (2019)
41. McLellan, K., Sun, Y.C., Naguib, H.E.: A review of 4D printing: Materials, structures, and designs towards the printing of biomedical wearable devices. Bioprinting **1**(27), e00217 (2022)
42. Chow, W.S., Mohd Ishak, Z.A.: Smart polymer nanocomposites: a review. Express Polym. Lett. 2020 1, 14(5)
43. Grancarić, A.M., Jerković, I., Koncar, V., Cochrane, C., Kelly, F.M., Soulat, D., Legrand, X.: Conductive polymers for smart textile applications. J. Ind. Text. **48**(3), 612–42 (2018)
44. Bora, L.V., Vadaliya, K.S., Bora, N.V.: Sustainable feedstocks for 4D printing: biodegradable polymers and natural resources. Green Mater. **40**(XXXX), 1–7 (2023)
45. Khalid, M.Y., Arif, Z.U., Noroozi, R., Hossain, M., Ramakrishna, S., Umer, R.: 3D/4D printing of cellulose nanocrystals-based biomaterials: additives for sustainable applications. Int. J. Biol. Macromol. **11**, 126287 (2023)
46. Deshmukh, K., Muzaffar, A., Kovářík, T., Křenek, T., Ahamed, M.B., Pasha, S.K.: Fundamentals and applications of 3D and 4D printing of polymers: challenges in polymer processing and prospects of future research. In: 3D and 4D Printing of Polymer Nanocomposite Materials 2020 Jan 1, pp. 527–560. Elsevier
47. Khalid, M.Y., Arif, Z.U., Ahmed, W., Umer, R., Zolfagharian, A., Bodaghi, M.: 4D printing: technological developments in robotics applications. Sens. Actuators A **16**(343), 113670 (2022)
48. Razzaq, M.Y., Gonzalez-Gutierrez, J., Mertz, G., Ruch, D., Schmidt, D.F., Westermann, S.: 4D printing of multicomponent shape-memory polymer formulations. Appl. Sci. **12**(15), 7880 (2022)
49. Shi, J., Liu, S., Zhang, L., Yang, B., Shu, L., Yang, Y., Ren, M., Wang, Y., Chen, J., Chen, W., Chai, Y.: Smart textile-integrated microelectronic systems for wearable applications. Adv. Mater. **32**(5), 1901958 (2020)

50. Gnanasekaran, K., Heijmans, T., Van Bennekom, S., Woldhuis, H., Wijnia, S., De With, G., Friedrich, H.: 3D printing of CNT-and graphene-based conductive polymer nanocomposites by fused deposition modeling. Appl. Mater. Today **1**(9), 21–8 (2017)
51. Mallakpour, S., Tabesh, F., Hussain, C.M.: 3D and 4D printing: from innovation to evolution. Adv. Coll. Interface. Sci. **1**(294), 102482 (2021)
52. Peng, S., Yu, Y., Wu, S., Wang, C.H.: Conductive polymer nanocomposites for stretchable electronics: material selection, design, and applications. ACS Appl. Mater. Interfaces **13**(37), 43831–54 (2021)
53. Huang, X., Panahi-Sarmad, M., Dong, K., Li, R., Chen, T., Xiao, X.: Tracing evolutions in electro-activated shape memory polymer composites with 4D printing strategies: a systematic review. Compos. A Appl. Sci. Manuf. **1**(147), 106444 (2021)
54. Haq, M.I., Raina, A., Ghazali, M.J., Javaid, M., Haleem, A.: Potential of 3D printing technologies in developing applications of polymeric nanocomposites. Tribol. Polym. Polym. Compos. Ind. **4**, 193–210 (2021)
55. Koch, H.C., Schmelzeisen, D., Gries, T.: 4D textiles made by additive manufacturing on pre-stressed textiles—an overview. In: Actuators 2021 Feb 10, Vol. 10, No. 2, p. 31. MDPI
56. Bhanushali, H., Amrutkar, S., Mestry, S., Mhaske, S.T.: Shape memory polymer nanocomposite: a review on structure–property relationship. Polym. Bull. **79**(6), 3437–93 (2022)
57. Razzaq, M.Y., Gonzalez-Gutierrez, J., Mertz, G., Ruch, D., Schmidt, D.F., Westermann, S.: 4D printing of multicomponent shape-memory polymer formulations. Appl. Sci. **12**(15), 7880 (2022). https://doi.org/10.3390/app12157880
58. Manapat, J.Z., Chen, Q., Ye, P., Advincula, R.C.: 3D printing of polymer nanocomposites via stereolithography. Macromol. Mater. Eng. **302**(9), 1600553 (2017)
59. Ma, S., Zhang, Y., Wang, M., Liang, Y., Ren, L., Ren, L.: Recent progress in 4D printing of stimuli-responsive polymeric materials. Sci. China Technol. Sci. **63**(4), 532–44 (2020)
60. Wu, H., Fahy, W.P., Kim, S., Kim, H., Zhao, N., Pilato, L., Kafi, A., Bateman, S., Koo, J.H.: Recent developments in polymers/polymer nanocomposites for additive manufacturing. Prog. Mater. Sci. **1**(111), 100638 (2020)
61. Park, S., Shou, W., Makatura, L., Matusik, W., Fu, K.K.: 3D printing of polymer composites: materials, processes, and applications. Matter **5**(1), 43–76 (2022)
62. Malekmohammadi, S., Sedghi Aminabad, N., Sabzi, A., Zarebkohan, A., Razavi, M., Vosough, M., Bodaghi, M., Maleki, H.: Smart and biomimetic 3D and 4D printed composite hydrogels: opportunities for different biomedical applications. Biomedicines **9**(11), 1537 (2021)
63. Lyu, Z., Wang, J., Chen, Y.: 4D printing: interdisciplinary integration of smart materials, structural design, and new functionality. Int. J. Extreme Manuf. **5**(3), 032011 (2023)
64. Zhong, J., Zhou, G.X., He, P.G., Yang, Z.H., Jia, D.C.: 3D printing strong and conductive geo-polymer nanocomposite structures modified by graphene oxide. Carbon **1**(117), 421–6 (2017)
65. Farham, B., Baltazar, L.: A review of smart materials in 4D printing for hygrothermal rehabilitation: innovative insights for sustainable building stock management. Sustainability. **16**(10), 4067 (2024)
66. Mondal, K., Tripathy, P.K.: Preparation of smart materials by additive manufacturing technologies: a review. Materials **14**(21), 6442 (2021)
67. Liu, T., Zhou, T., Yao, Y., Zhang, F., Liu, L., Liu, Y., Leng, J.: Stimulus methods of multi-functional shape memory polymer nanocomposites: a review. Compos. A Appl. Sci. Manuf. **1**(100), 20–30 (2017)
68. Wan, X., Zhang, F., Liu, Y., Leng, J.: CNT-based electro-responsive shape memory functionalized 3D printed nanocomposites for liquid sensors. Carbon **1**(155), 77–87 (2019)
69. Mahmood, A., Akram, T., Shenggui, C., Chen, H.: Revolutionizing manufacturing: a review of 4D printing materials, stimuli, and cutting-edge applications. Compos. B Eng. **6**, 110952 (2023)
70. Li, Z., Yang, F., Yin, Y.: Smart materials by nanoscale magnetic assembly. Adv. Func. Mater. **30**(2), 1903467 (2020)

71. Massaro, A.: Intelligent materials and nanomaterials improving physical properties and control oriented on electronic implementations. Electronics **12**(18), 3772 (2023)
72. Shah, M.A., Pirzada, B.M., Price, G., Shibiru, A.L., Qurashi, A.: Applications of nanotechnology in smart textile industry: a critical review. J. Adv. Res. **1**(38), 55–75 (2022)
73. Zhou, W., Qiao, Z., Nazarzadeh Zare, E., Huang, J., Zheng, X., Sun, X., Shao, M., Wang, H., Wang, X., Chen, D., Zheng, J.: 4D-printed dynamic materials in biomedical applications: chemistry, challenges, and their future perspectives in the clinical sector. J. Med. Chem. **63**(15), 8003–24 (2020)
74. Liang, C., Rogers, C.A., Malafeew, E.: Investigation of shape memory polymers and their hybrid composites. J. Intell. Mater. Syst. Struct. **8**(4), 380–6 (1997)
75. Sydney Gladman, A., Matsumoto, E.A., Nuzzo, R.G., Mahadevan, L., Lewis, J.A.: Biomimetic 4D printing. Nat. Mater. **15**(4), 413–8 (2016)
76. Al Rashid, A., Khan, S.A., Al-Ghamdi, S.G., Koç, M.: Additive manufacturing of polymer nanocomposites: needs and challenges in materials, processes, and applications. J. Market. Res. **1**(14), 910–41 (2021)
77. Aufa, A.N., Ismail, Z., Hassan, M.Z.: Emerging trends in 4d printing of hydrogels in the biomedical field: A review. Mater. Today Proceed. (2023)
78. Vatani, M., Lu, Y., Engeberg, E.D., Choi, J.W.: Combined 3D printing technologies and material for fabrication of tactile sensors. Int. J. Precis. Eng. Manuf. **16**, 1375–83 (2015)
79. Yao, D., Cui, H., Hensleigh, R., Smith, P., Alford, S., Bernero, D., Bush, S., Mann, K., Wu, H.F., Chin-Nieh, M., Youmans, G.: Achieving the upper bound of piezoelectric response in tunable, wearable 3D printed nanocomposites. Adv. Func. Mater. **29**(42), 1903866 (2019)
80. Zhang, M., Zhao, M., Jian, M., Wang, C., Yu, A., Yin, Z., Liang, X., Wang, H., Xia, K., Liang, X., Zhai, J.: Printable smart pattern for multifunctional energy-management E-textile. Matter. **1**(1), 168–79 (2019)
81. Fantino, E., Chiappone, A., Calignano, F., Fontana, M., Pirri, F., Roppolo, I.: In situ thermal generation of silver nanoparticles in 3D printed polymeric structures. Materials **9**(7), 589 (2016)
82. Maurya, M.R., Sadasivuni, K.K., Cabibihan, J., Ahmad, S., Kazim, S.: Shape Memory Composites Based on Polymers and Metals for 4D Printing. Springer International Publishing AG, Switzerland (2022)
83. Kuang, X., Yue, L., Qi, H.J.: Introduction to 4D printing: concepts and material systems. Addit. Manuf. Technol. Design Optim. Model. **28**, 1–42 (2023)
84. Chizari, K., Daoud, M.A., Ravindran, A.R., Therriault, D.: 3D printing of highly conductive nanocomposites for the functional optimization of liquid sensors. Small **12**(44), 6076–82 (2016)
85. Madgula, K., Puli, V.S.: Recent progress in synthesis methods of shape-memory polymer nanocomposites. Shape Memory Compos. Based Polym. Metals 4D Print. Process. Appl. Chall. **19**, 173–212 (2022)
86. Chen, Q., Mangadlao, J.D., Wallat, J., De Leon, A., Pokorski, J.K., Advincula, R.C.: 3D printing biocompatible polyurethane/poly (lactic acid)/graphene oxide nanocomposites: anisotropic properties. ACS Appl. Mater. Interfaces **9**(4), 4015–23 (2017)
87. Llewellyn-Jones, T.M., Drinkwater, B.W., Trask, R.S.: 3D printed components with ultrasonically arranged microscale structure. Smart Mater. Struct. **25**(2), 02LT01 (2016)
88. Eutionnat-Diffo, P.A., Chen, Y., Guan, J., Cayla, A., Campagne, C., Zeng, X., Nierstrasz, V.: Stress, strain and deformation of poly-lactic acid filament deposited onto polyethylene terephthalate woven fabric through 3D printing process. Sci. Rep. **9**(1), 14333 (2019)
89. Rahmatabadi, D., Bayati, A., Khajepour, M., Mirasadi, K., Ghasemi, I., Baniassadi, M., Abrinia, K., Bodaghi, M., Baghani, M.: Poly (ethylene terephthalate) glycol/carbon black composites for 4D printing. Mater. Chem. Phys. **1**(325), 129737 (2024)
90. Bekas, D.G., Hou, Y., Liu, Y., Panesar, A.: 3D printing to enable multifunctionality in polymer-based composites: a review. Compos. B Eng. **15**(179), 107540 (2019)
91. Schmelzeisen, D., Koch, H., Pastore, C., Gries, T.: 4D textiles: hybrid textile structures that can change structural form with time by 3D printing. Narrow Smart Text 189–201 (2018)

92. Wei, H., Cauchy, X., Navas, I.O., Abderrafai, Y., Chizari, K., Sundararaj, U., Liu, Y., Leng, J., Therriault, D.: Direct 3D printing of hybrid nanofiber-based nanocomposites for highly conductive and shape memory applications. ACS Appl. Mater. Interfaces **11**(27), 24523–32 (2019)
93. Huang, Y., Zhu, M., Pei, Z., Xue, Q., Huang, Y., Zhi, C.: A shape memory supercapacitor and its application in smart energy storage textiles. J. Mater. Chem. A **4**(4), 1290–1297 (2016)
94. Zhang, D., George, O.J., Petersen, K.M., Jimenez-Vergara, A.C., Hahn, M.S., Grunlan, M.A.: A bioactive "self-fitting" shape memory polymer scaffold with potential to treat cranio-maxillo facial bone defects. Acta Biomater. **10**(11), 4597–4605 (2014)
95. Lin, C., Lv, J., Li, Y., Zhang, F., Li, J., Liu, Y., Liu, L., Leng, J.: 4D-printed biodegradable and remotely controllable shape memory occlusion devices. Adv. Func. Mater. **29**(51), 1906569 (2019)
96. Khalid, M.Y., Arif, Z.U., Ahmed, W.: 4D printing: technological and manufacturing renaissance. Macromol. Mater. Eng. **307**(8), 2200003 (2022)
97. Kouhi, M., Sajadi-Javan, Z.S., Abedi, N.: 3D printing of bionanocomposites and their broad spectrum of applications. In: Advances in Bionanocomposites 2024 Jan 1, pp. 247–276. Elsevier
98. Bodkhe, S., Turcot, G., Gosselin, F.P., Therriault, D.: One-step solvent evaporation-assisted 3D printing of piezoelectric PVDF nanocomposite structures. ACS Appl. Mater. Interfaces **9**(24), 20833–42 (2017)
99. Lund, A., van der Velden, N.M., Persson, N.K., Hamedi, M.M., Müller, C.: Electrically conducting fibres for e-textiles: an open playground for conjugated polymers and carbon nanomaterials. Mater. Sci. Eng. R. Rep. **1**(126), 1–29 (2018)
100. Sondagar, H.D., Kumar, S., Sharma, V.S.: A state-of-the-art review of 4D printing techniques, stimuli activated materials and future perspectives. Mater. Today Proceed. (2024)
101. Li, Y., Zhang, F., Liu, Y., Leng, J.: 4D printed shape memory polymers and their structures for biomedical applications. Sci. China Technol. Sci. **63**(4), 545–60 (2020)
102. Gardan, J.: Smart materials in additive manufacturing: state of the art and trends. Virtual Phys. Prototyp. **14**(1), 1–8 (2019)
103. Schmelzeisen, D., Koch, H., Pastore, C., Gries, T.: 4D textiles: hybrid textile structures that can change structural form with time by 3D printing. Narrow Smart Text 189–201 (2018)
104. Neuß, J., Kreuziger, M., Grimmelsmann, N., Korger, M., Ehrmann, A.: Interaction between 3D deformation of textile fabrics and imprinted lamellae. In Proceedings of the Aachen-Dresden-Denkendorf International Textile Conference, Dresden, Germany, pp. 24–25 (2016)
105. Saska, S., Pilatti, L., Blay, A., Shibli, J.A.: Bioresorbable polymers: advanced materials and 4D printing for tissue engineering. Polymers **13**(4), 563 (2021)
106. Arif, Z.U., Khalid, M.Y., Zolfagharian, A., Bodaghi, M.: 4D bioprinting of smart polymers for biomedical applications: recent progress, challenges, and future perspectives. React. Funct. Polym. **1**(179), 105374 (2022)
107. Bodaghi, M., Zolfagharian, A.: 4D printing principles and manufacturing. Insmart materials in additive manufacturing 2022 Jan 1, pp. 1–17. Elsevier
108. Wang, X., He, Y., Liu, Y., Leng, J.: Advances in shape memory polymers: remote actuation, multi-stimuli control, 4D printing and prospective applications. Mater. Sci. Eng. R. Rep. **1**(151), 100702 (2022)
109. Kuang, X., Roach, D.J., Wu, J., Hamel, C.M., Ding, Z., Wang, T., Dunn, M.L., Qi, H.J.: Advances in 4D printing: materials and applications. Adv. Func. Mater. **29**(2), 1805290 (2019)

Smart Materials for Next-Generation Manufacturing of 3D, 4D, and 5D Printing Technologies

Rohit Verma, Alka Nair, Akshita, Deepshikha Gupta, and Tejendra K. Gupta

Abstract Shape-Memory Polymers (SMPs) are a class of smart materials that have garnered significant attention due to their ability to revert to a predetermined shape upon exposure to external stimuli such as heat, light, or moisture. This unique property makes SMPs highly valuable in advanced manufacturing technologies like 3D, 4D, and 5D printing, where precision, adaptability, and multifunctionality are essential. In 3D printing, SMPs allow the creation of complex, customizable structures with potential applications in biomedical devices, aerospace components, and flexible electronics. Moving beyond 3D, 4D printing integrates the dimension of time, enabling printed objects to change shape or function after fabrication. SMPs play a critical role here, allowing for the design of self-assembling structures, adaptive textiles, and dynamic medical implants that respond to environmental changes. The emerging field of 5D printing further extends these capabilities by adding additional directional complexity and multifunctionality to printed objects, potentially revolutionizing industries like automotive engineering, biomedical industries, and robotics. This chapter explores the versatile applications of SMPs in these cutting-edge printing technologies, highlighting their transformative impact on manufacturing, material science, and engineering.

Keywords Shape Memory Polymers (SMPs) · 3D-printed SMPs · 4D-printed SMPs · 5D-printed SMPs · Multiple SMPs · Thermoplastic and thermoset SMPs

R. Verma (✉) · A. Nair
Department of Physics, Amity Institute of Applied Sciences, Amity University Uttar Pradesh, Noida, India
e-mail: rverma85@amity.edu

Akshita · D. Gupta · T. K. Gupta (✉)
Department of Chemistry, Amity Institute of Applied Sciences, Amity University Uttar Pradesh, Noida, India
e-mail: tkgupta@amity.edu

S. Moharana et al. (eds.), *Polymer Nanocomposites for 3D, 4D and 5D Printing*, Engineering Materials, https://doi.org/10.1007/978-981-96-4214-4_19

1 Introduction

We all have come across the word ‘polymer’. A polymer is a large molecule formed by the repetition of many subunits called ‘mer’ units and is a type of macromolecule. It can be made naturally or synthetically. With the development of technology, scientists have derived several types of polymers that have found their uses in numerous fields. One such class of polymers is called shape memory, shape-shifting, or shape-changing polymers. As the name suggests, shape memory polymers are those materials that possess the ability to regain their shape after the applied external stimulus is removed. Just like other materials, these polymers can modify their shape, size, stiffness, and strain responses when external stimuli or forces such as heat, light, and electric or magnetic field are applied to it. However, after the removal of the external force such materials regain the original structure, thus they are memorizing their original shape and size before exposure to force. SMPs are being rigorously investigated for their biomedical applications due to various properties that it possesses which are favorable in mimicking the extracellular matrix (ECM). Also, since such materials have an extraordinarily small amount of toxicity and minimally invasive procedures, it is of immense importance in biomedicine. Due to the versatility of such materials, it has a broad range of applications such as in the manufacturing of sensors, textiles, robotics, aerospace engineering, and so on [1].

2 Internal Structure of Shape-Memory Polymers

As discussed previously, polymers are macromolecules comprising long chains. The property of shape memory polymers is not an intrinsic effect and hence not all polymers acquire such properties. Shape-shifting polymers are manufactured by specific morphological processes such as extruding and injection molding. Crystalline or liquid crystalline, and amorphous phases are introduced into the polymer chains in specific places called switching segments. These switching segments are soft points that have more flexibility and elasticity than other parts of the polymer. The hard parts which do not participate in making the macromolecule flexible are called net points. This accounts for the fixed shape of the fiber of the polymer. The liquid crystalline or amorphous phase in the switching segments helps in the fixity and recovery of the shape when the external stimulus is applied or removed. At the net points, chemical crosslinks, which are non-flexible portions of the polymer, chain entanglements, interpenetrating networks, or interlocked supramolecular complexes are present which provides rigidity to the polymer. The crosslinking at the net points is obtained in a polymer that consists of a morphology that has at least two segregated domains. Such segregated domains are widely found in copolymers. The domain that possesses the highest thermal transitional temperature (T_{perm}) is called net points and the domains that have the second highest transitional temperature (T_{trans}) are called switching segments.

3 Classification of Shape-Memory Polymers

Based on the process of several types, SMPs are classified into the following types:

(1) Shape-memory blocks,
(2) Shape-memory foams,
(3) Shape-memory fibers,
(4) Shape-memory films.

Based on the shape memory effect, SMPs can be classified as follows:

3.1 One-Way Shape-Memory Effect (OWSMEs)

Such polymer substances require another step to induce temporary shape when the material loses its original structure on the application of an external stimulus.

3.2 Two-Way Shape-Memory Effect (TWSMEs)

Such materials possess the ability to regain their original shape without applying any force for reshaping the material. It can also be called the reverse shape-memory effect (reversible SMEs) and the polymers that have this property are known as reversible SMPs.

3.3 Multiple SMPs

Such materials can take several temporary shapes in addition to their original shape. The conversion from their original shape to the temporary shape is achieved by applying external stimuli and the transition from the temporary shape to the original shape is done by an additional stimulus provided [2]. These different kinds of SMPs are shown in Fig. 1.

The type of stimulus-induced on a shape memory polymer also helps in differentiating them into chemically induced and physically induced SMPs. There are certain SMPs that can be subjected to more than one type of stimuli at the same time or a combination of two or more stimuli simultaneously. Such type of SMPs is called multi-stimuli responsive SMPs. Physically induced SMPs include thermally induced SMPs, light-induced SMPs, electric or magnetic-induced SMPs. Chemically induced SMPs include pH-induced SMPs, water-induced SMPs, and enzymatically triggered SMPs.

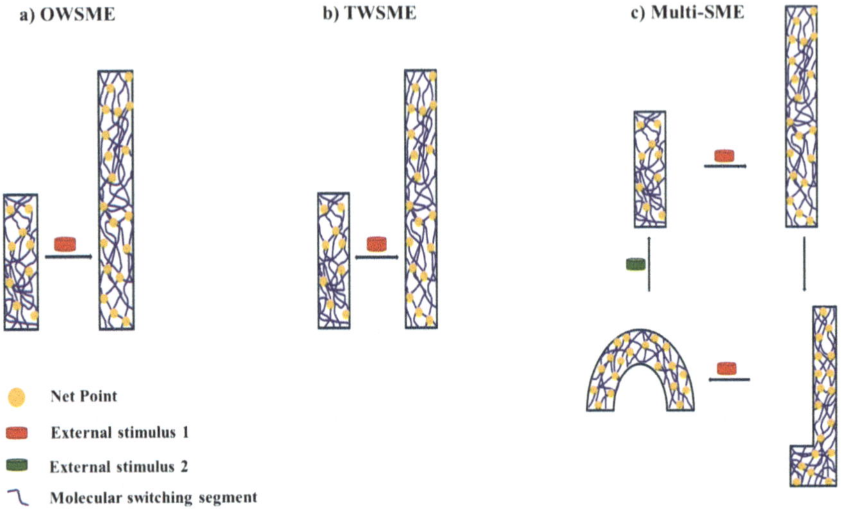

Fig. 1 The figure shows the distinct types of SMPs based on the shape memory effect [3]. Copyright 2023. Reproduced with permission from MDPI

4 Thermoset and Thermoplastic SMPs

We know that thermosets are those polymers that have a fixed shape and size at each temperature and can only be deformed when the temperature changes. Thermoset shape memory polymers show restricted shape transformation and inferior reformation ability. This is because the polymer chains in its structure are chemically cross-linked and require more thermal energy to break the chemical bonds. Thermoset SMPs show a stronger structure at the microlevel than thermoplastic SMPs. They have a high shape fixity rate and recovery rate but have low deformation ability and are more brittle. The methods in which the thermoset SMPs can be deformed are very restricted. Some thermoset SMPs include epoxy resin, cyanate resin, etc.

In the case of thermoplastic SMPs, the polymer chains are physically cross-linked. It consists of a two-phase morphology wherein one phase is responsible for physical cross-linking between the polymer chains and the other phase behaves as a molecular switch. Thermoplastic SMPs also have high shape fixity and recovery rates but are also relatively soft materials that do not require remarkably high thermal energy to regain their original shape. The deformation that can be applied to such materials is also quite high. Some examples of thermoplastic SMPs are polyurethanes and polyether esters [4].

5 Reinforcements and Fillers in Shape-Memory Polymer Composites

Shape memory polymers are composite materials that are constructed by injecting various crystalline materials to enhance the elastic and shape-reforming properties of the material. Such materials that are introduced into the SMPs are called reinforcements and fillers. Composite fillers and reinforcements have been widely utilized to change and enhance the physical and mechanical properties of SMPs. Fillers are used to modify or improve properties such as thermal conductivity, electrical resistivity, friction, wear resistance, and flame resistance and to lower the cost of the material. Fillers are of two types, namely, conductive fillers and extender fillers. Conductive fillers are used to enhance the electrical and thermal conductivity of the SMP composite, and the extender filler is used to provide more material to the matrix of the composite. Fiber reinforcement has a very significant effect on the properties of the composites to which they are added. Reinforcements are specific particulates, fibers, or fabrics that are used to strengthen or toughen SMP composites.

Studies are constantly being done to improve the applicability of SMPs in various areas. One such observation includes the effect of the use of additives on the properties of SMPs. Additive materials comprise short and long fibers that are introduced into the molecular structure which gives the desired properties such as mechanical strength, thermal properties and widens the applications. Fossil-based materials are used as matrix and reinforcement materials.

6 3D-Printed Shape-Memory Polymers

3D printing is a new-age process that has been developed to create 3D models by adding layers of materials repeatedly. This is the opposite of subtractive manufacturing processes which include cutting a 3D material into various layers. This technique is usually computer-controlled and helps in converting graphic models into physical models. In science, 3D printing has a wide variety of applications which include the following.

6.1 Rapid Prototyping

3D printing allows physicists to convert graphic models into physical prototypes without any time consumption. This helps in reducing the time consumed in the manufacturing of models and gives a more accurate picture of the physical model.

6.2 Tissue Engineering

It can be used to create functional tissues by deposition of layers of cells, which is of utmost importance in tissue culture and biological regeneration of organs and drug testing.

6.3 Medical Devices

3D printing technology can be used to create prosthetics and implants which can be implanted in the patient's body as a replacement for some existing organ. This technology also helps in creating biological models which helps in the study of anatomical structures and pathological models.

SMPs can be used in 3D printing with the help of a fused deposition model (FDM). An FDM 3D printer runs by melted filament material deposition over a build platform layer by layer until a completed part is formed. FDM makes use of digitally designed files that are uploaded to the computer, and they get translated into physical dimensions. FDM-based materials include polymers such as ABS (acrylonitrile–butadiene–styrene), PLA (polylactic acid), PETG (polyethylene terephthalate glycol) and PEI (polyethyl enimine), which get fend by the machine as threads or fibers through a heated nozzle. The FDM machine is utilized by loading a spool of thermoplastic filament of SPM into the printer. Once the desired temperature is reached by the nozzle, the printer releases the filament through the nozzle.

This nozzle is connected to a three-axis system which allows it to move across the X, Y, and Z axes. The printer expels melted material in the shape of thin strands and these strands get deposited on the layers along a particular path that is already fixed by the design. Once the material gets deposited, it cools and solidifies. There are some important factors that should be taken into consideration which are build size and layer height. The build size of a desktop 3D printer is 200 × 200 × 200 mm and of an industrial 3D printer it can be 1,000 × 1,000 × 1,000 mm. The layer height that is provided by an FDM machine ranges between 50 and 400 microns. When shorter layers are printed, we get more accurate details of the model, and the exact geometry of the model created on the computer can be retained. Various shape memory polymers such as polylactide (PLA), polytetrafluoroethylene (PFTE), and ethylene–vinyl acetate (EVA) can be used for layer deposition in 3D printing technology [5].

Origami structures have recently been used to create expandable biomedical scaffolds that can be compressed into a small volume for minimally invasive introduction into the human body, and then expanded to a larger volume once inside. This property of folding and unfolding to change volume is advantageous for applications like delivering scaffolds for tissue engineering. Langford et al. compared two origami patterns for this application:

1. Waterbomb tessellation,

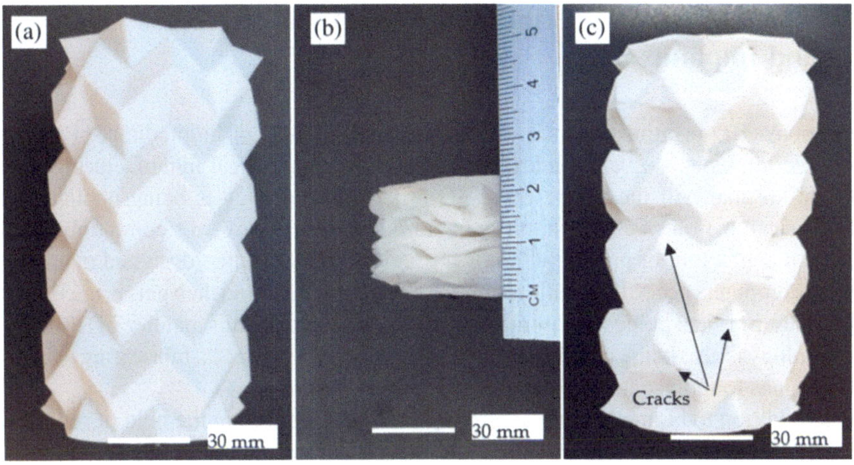

Fig. 2 Herringbone tessellated tube **a** after 3D printing; **b** after compression; **c** after recovery. Figure taken from the Ref. [6] under CC-BY license

2. Herringbone tessellation.

They found that the waterbomb pattern deformed in an unreliable way, while the herringbone tessellated tubes could be strongly compressed and then mostly restored to their original shape as shown in Fig. 2. PLA filament origami tubes showed a constant shape recovery of about 61%, regardless of the amount of deformation applied. Herringbone origami tubes had an increased shape recovery of approximately 96%, and only a few cracks were seen [6].

7 Tailoring Mechanical Properties Through Composite Formulation

Composite materials have provided researchers with opportunities for instilling specific mechanical properties to the structures and materials to meet load requirements which can be induced by changing environments. This leads to their increased use in structural applications. Using composite materials requires a crucial step of selecting a fiber-resin combination, which will match all the requirements of the SMPs, while making the choice of a material. This process involves first choosing the fiber, and the resin, and then deciding the proportion of these two constituents in the composite material. The process of selection of such materials possessing the required qualities is called laminate selection. This is the task that involves choosing a fiber and resin combination that supports the technical and economic factors. Material tailoring involves manipulation of the constitution of the composite material to get desired properties and it is the selection of fiber and resin.

8 Application of 3D Printing in Biomedical Devices and Implants

Nowadays 3D printing is being widely used in the field of biomedicine. Patient-specific implants, surgical guides, bioprinting, pharmaceutical manufacturing, and so on, are many of the fields where 3D printing technology is being applied. As the basic function of 3D printing is creating models, it is being commonly used in the manufacture of prosthetic limbs for patients which is first designed manually using computers according to the requirement of the patient and then gets printed with accuracy. Surgical guides and implants can also be created in this manner. 3D printing technology is also being utilized in tailoring drug delivery devices, fabricating tissues, and organs to replace natural but faulty organs in patients.

In addition to biomedicine, 3D printing is also being used in the field of robotics. This facilitates prototyping, tooling, and manufacturing which lowers the costs of the machines so manufactured, and time consumed and assists robotics engineers to achieve their targets & goals on the pre-decided time. They can customize designs based on applications of the robot according to consumption requirements and optimize designs based on research and testing with the help of quick prototyping that has been accomplished by 3D printing. 3D printing is suitable for quickly prototyping robotic design. It requires a lot of trial and error to design robots. This requires first creating a framework and several criteria of architecture are utilized to follow a planned application or routine. Using a 3D printing process, prototypes can be produced easily, inexpensively, and with low rates of faults or errors. A 3D-printed prototype can be generated in less amount of time than it takes for its machine counterpart. The reduced processing time and the simplicity of low-volume production associated with rapid prototyping provide more versatility to the manufacture of robots. 3D printing also helps in the digital fabrication of pneumatic silicon actuators which can further be programmed to create bioinspired motions. The actuator is an elastomeric body with its surface comprising reinforced, well-defined lead strips. Like the architecture of muscle fibers found in the human body, these lead angles can be altered to achieve elongation, contraction, twisting, and other motions that can be exhibited by the muscles. Since SMPs have the property to regain their original shape, these materials can be used for the manufacture of actuators which will pave the way for the manufacturing of humanoid robots [7]. The effect of temperature on the shapeshift of the SMPs, and the shape memory effect (SME) is shown in Fig. 3.

9 Challenges in 3D Printing

3D printing technology has especially important benefits over traditional manufacturing methods but also like every other technological advancement it comes with several challenges that must be solved to facilitate its widespread use in the medical sector. One of the key challenges of 3D printing technology in the medical sector is

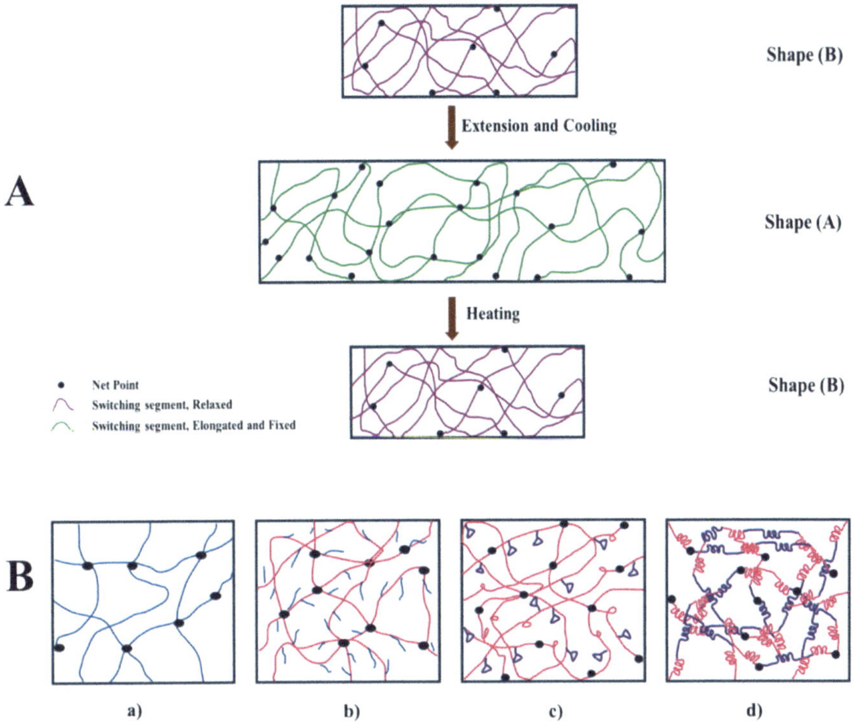

Fig. 19.3 Schematic of the (A) Shapeshift in shape memory polymers because of temperature, (B) Polymer network showing SME (red = molecular switches and gray = net points) [8]. Copyright 2022. Reproduced with permission from MDPI

material selection. The materials used in printing are required to be biocompatible, sterilizable, and safe for human use. Along with that, the material's properties must match the intended application, such as strength, flexibility, and wear resistance. Therefore, it is crucial to identify and develop new biocompatible, sterilizable materials that have the capability to withstand the enzymatic, unstable environment of the human body.

Accuracy and precision are very important factors in 3D printing for biomedical applications. The medical industry requires high accuracy and precision for printed parts, and technology has made commendable developments in achieving these requirements. However, problems persist in ensuring consistent quality in large-scale production. The tolerance for the medical printed parts must be very tight, and any divergence from the required specifications can result in significant consequences for the patient. Achieving a specific level of accuracy and precision can be challenging and requires cautious monitoring of the printing process, materials, design, and adhering to strict quality control standards. Regulatory compliance is another important challenge that must be dealt with in 3D printing for medical applications. Medical devices created using 3D printing must match strict regulatory standards to make sure that the safety and effectiveness of the model when used remains intact. These standards may change depending on the country or region where the device is being used. Any design or production process deflection must be carefully evaluated and documented to ensure that the final product matches all regulatory requirements. Costs are another crucial challenge for medical applications that must be considered in 3D printing. While 3D printing can significantly reduce the cost of production of medical devices, there are still some costs that add up with the use of technology. These costs, which include the equipment, materials, and time required to produce the device and software, can be a prominent issue for many medical facilities.

9.1 4D Printing Technology

4D printing technology was introduced as an advancement to the 3D printing technology with the addition of time as a fourth dimension. The 4D-printed models include live or time-varying 3D models which can change their shape or size when an external stimulus is applied to it over a specific period. Such models include the use of hydrogel resins, active polymers, and even the use of live tissues. In this process, the 3D model is printed by using the same methods as explained previously but only with the addition that the model evolves with time when it is brought into contact with external factors such as temperature, humidity, light, pressure, electric or magnetic field, and so on. 4D printing makes it possible for a model to bend, repair, assemble, or even disintegrate itself. It acquires a new shape or functionality on its own by reacting to the environment. Thus, 4D printing helps in the creation of three-dimensional objects that can adapt to external environmental circumstances without the intervention of humans, robots, or computers.

The shapeshifting smart materials that constitute the 4D-printed items respond to environmental factors such as heat, water, light, wind, or electricity based on a set of instructions fended into their geometric coding. They may elongate, bend, wrinkle, fold, twist, or even disintegrate once they are made to get activated. Anything from wood to rubber can form layers with stimuli-responsive materials. The purpose of this invention is to discover new properties and manipulate objects, independent of human or mechanical intervention [7].

Figure 4 shows the color and shape change in 3D-printed structures, which are composed of SMPs. The as-printed structure with upright grids (**left**) functions as a structural color filter that transmits only specific wavelengths of visible light. Deformation of the structures at elevated temperature flattens the nanostructures (**right**), rendering the part colorless; it remains in an invisible state after cooling to room temperature. Heating recovers both the original geometry and color of nanostructures, leading to a submicron demonstration of 4D printing [9].

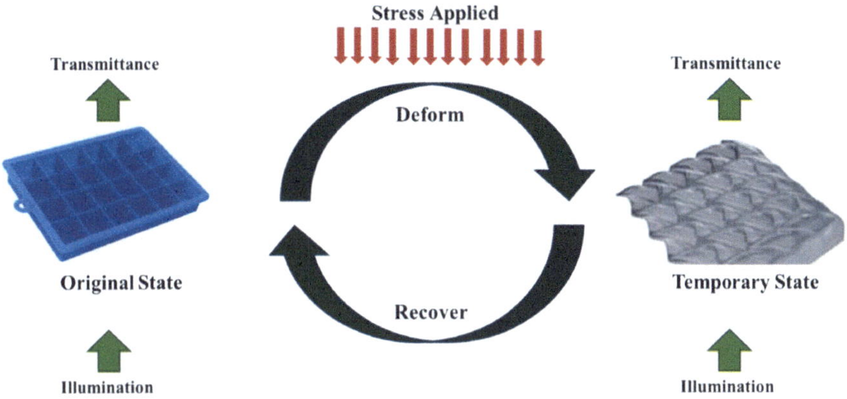

Fig. 4 Schematic of color and shape change of a constituent nanostructured element of a "disappearing" 3D-printed structure composed of SMP

10 Properties of the Materials Used in 4D Printing

The following are the properties that are required by the material to be suitable for utilization in 4D printing [10]:

- ***Hydrogels***: These materials are receptive to moisture. These materials show change when they get exposed to electric energy.
- ***Piezoelectric***: These materials change according to applied mechanical stress applied to them, such as pressure or latent heat.
- ***Thermo-reactive***: Heat or temperature fluctuations can lead to creating physical transformation in these materials
- ***Photo-reactive***: These materials are quick to react to light.
- ***Magento-reactive***: Materials that transform when subjected to magnetic energy.
- ***pH-reactive***: Materials that are sensitive to pH levels.

4D printing technology can be developed at the macroscale as well as the microscale. Digital light processing (DLP) and direct laser writing (DLW) are selected as suitable 3D printing technologies to cover both regimes. Firstly, a system based on monofunctional isobornyl acrylate and two crosslinkers consisting of a soft and a hard diacrylate are tested and selected to be compatible with both printing techniques. Employing DLP, a large variety of structures that show distinct complexity can be printed. These structures can be converted from simple frames to more arduous 3D geometries such as double platform structures, infinity rings, or cubic grids. The shape memory effect is exhibited by all the 3D geometries. Excellent shape fixity as well as recovery and repeatability can be observed in so-formed structures. Further, the foundation is adapted for fast 4D printing at the microscale using DLW. Specifically, the 4D-printed microstructures display remarkable shape memory properties. The possibility of trapping and releasing micro-objects, such as microspheres, is demonstrated by designing smart box-like 4D microstructures that are capable of being thermally actuated which provides the required evidence for the versatility and potential of the reported system.

The role of shape-memory polymers in the evolution of 3D printing technology to 4D printing is vital. One of the main drawbacks of 3D printing was that the model thus created by it was static and inanimate. However, with the addition of temporal dimension to the 3D-printed model, 4D printing has revolutionized its applications. 4D printing helped with the fabrication of complex models with the use of shape memory polymers which are now being applied as smart implants, biomedical sensing, electronics, robotics, and many such fields. The materials used for 4D printing should have higher stiffness, faster response rate, facile fabrication, etc., and shape-memory polymers are the best candidates for such materials.

10.1 Self-Healing Property of SMPs

Shape-memory effect is intricately linked to the shape-healing effect since the former is believed to help close the fractures so that the polymer can heal. However, there are not many reports of electroactive polymer composites studying both effects. Some of these reports use reversible tie linkages that are noncovalent, and the shape-healing effect is the aftermath of the rearrangement of the thermoplastic polymers in the material. This technique has been tested to have satisfactory performance. However, the lack of covalent tie linkages makes these materials nonresistant to solvents and may limit their thermosets and elastomeric characteristics. Self-healing of thermosetting shape-memory polymer (SMP) composites takes place as per the close-then-heal (CTH) healing strategy. Three types of composite structures are constructed, including a thermosetting SMP-based syntactic foam-cored sandwich, a grid-stiffened thermosetting SMP-based syntactic foam-cored sandwich, and a 3-D woven fabric-reinforced thermosetting SMP-based syntactic foam composite. The prepared fiber-reinforced composite panels are constantly damaged with the help of scratches of a few mm or by causing the material to suffer an impact externally and get healed per the CTH scheme. The healing efficiency is evaluated as per applicable mechanical strength tests. Damage sensing is monitored by a percolated carbon nanotube network and the healing process of the material is also scrutinized. The comprehensive experimental program asserts that the CTH scheme can heal structural-length scale cracks repeatedly, efficiently, timely, and molecularly, and it contains the potential to heal the material without any external intervention [3]. The shape-memory-assisted self-healing process of PCL based on disulfide bond and SME is shown in Fig. 5.

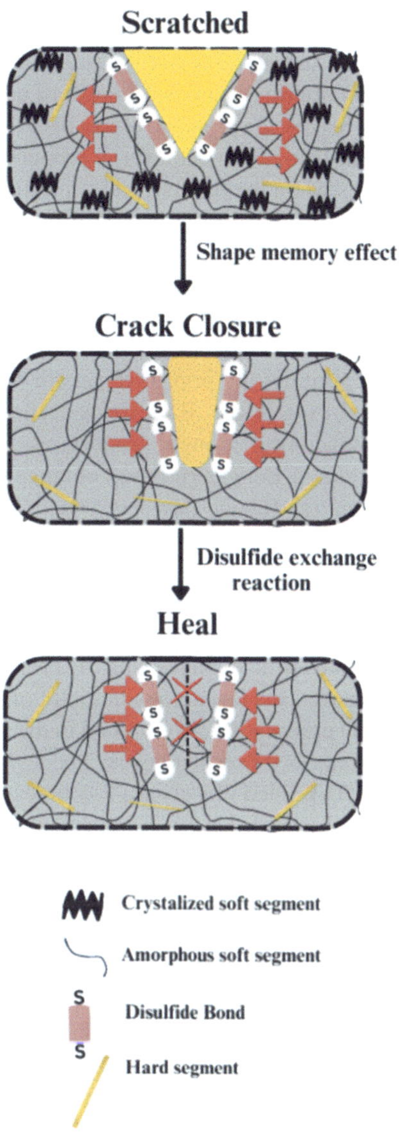

Fig. 5 Shape-memory-assisted self-healing process based on disulfide bond and SME of PCL [11]. Copyright 2018. Reproduced with permission from Springer Nature

10.2 5D Printing Possibilities

In the previous years, 3D printing has explored its potential applications in various fields of engineering, medical, dentistry, aerospace, and many other areas. The concept of 4D printing was introduced in 2014, and the possibility of using and inspecting its applications in different fields is now a major topic of research. Some American universities have given the concept of 5D printing in 2016. Today, it has been started to get implemented by Mitsubishi Electric Research Labs (MERL) with the assistance of a senior research scientist, William Yerazunis.

5D printing is a novel branch of additive manufacturing. In this technology, the print head & the printable object have five degrees of freedom which directly indicates that instead of the flat layer, it produces curved layers. In this process, the print part moves while the printer head is printing. So, printing undertakes the curved path of the part being printed instead of operating through a straight layer as in the case of 3D printers. The most prominent advantage of this technology is to create a part with a curved layer with improved strength.

Instead of 3 axes used in 3D printing, 5D printing technologies use a five-axis printing technique which gives rise to objects in multiple dimensions. In this five-axis printing, the print bed can stir back and forth on two axes besides the X, Y, and Z axes of the 3D printing technologies. Thus, this technology is capable of manufacturing stronger products as compared to parts made through 3D printing. In India, a Bengaluru-based company has worked in the business of 3D printing technology for the past few years. They are now trying to develop a 5D printer, which if they could produce successfully, would help them win the CES (Consumer Technology Association) award for best innovation.

A 5D-printed model gives the potential to fabricate artificial bone for surgery. Because human bones are not flat and have a curved surface, there is a requirement to create artificial bones with 5D printing to provide excellent strength to these bone implants. This technology has a lot of potential to fulfill this primary requirement.

3D printing is not very good in the manufacturing of complex curved Orthopedic implants as it uses flat layers. The test results of some experiments conducted show that 5D-printed objects are 3–5 times stronger than a 3D-printed object. It has also been observed that 5D-printed caps are also fortified and handle four times more megapascals of pressure than 3D-printed caps. There is the force exerted by the material during the curving of layers that makes printable objects stronger. Therefore, there is a lower requirement for raw material as compared to 3D printing for creating implants of the same strength [12]. Comparison of 3D, 4D, and 5D printing technologies, their key features are given in Table 1.

Thus, some of the advantages of 5D printing over 3D printing can be given as:

Table 1 Comparison between 3D, 4D, and 5D printing technology [5, 10, 12]

Type	Dimensions	Key features	Example
3D printing	3 Axes (X, Y, Z)	Using a fixed plateau, print flat layers	Common industrial and desktop 3D printers
4D printing	3 Axes + Smart materials	Print objects that alter shape over time in response to outside stimuli	Self-assembling buildings and smart textiles
5D printing	5 Axes (X, Y, Z, Rotation of X and Y)	Features a movable plateau, curved layer printing, and more durable 5D-printed parts	Intricate constructions with curved layers and housing parts

- **Fewer supports**: Thanks to being able to rotate the printhead or bed in the X- and Y-axes, it can be ensured that the printer is almost always printing on a solid part of the model, eliminating the need for excessive supports, and enabling more intricate and complex models to be printed, where supports would have to be provided and the removal of the support would have been impossible.
- **Stronger prints**: Standard 3D prints are always weakest across their layer lines. With 5-axis printing, it is not limited to stacking flat layers in the same way and so much stronger prints can be produced.
- **Less material used**: Being able to print at a variety of angles reduces the need for infill and supports, so the amount of material used can be significantly reduced.
- **Less post-processing**: By not needing so much support and not being limited to flat layers, it is possible to produce much smoother, higher-quality prints with less need for post-processing.

However, along with advantages, there are certain disadvantages in 5D printing which are as follows:

- **Few options**: It is not possible to find a whole lot of options out there, and the options that do exist tend to be intended for industrial use or are expensive, making 5-axis printing inaccessible for most.
- **Limited support**: There is extremely limited support available for these printers compared with something like standard 3D printing, where there are huge communities of users. This is obviously something that will gradually improve with time, but as of now, the industrial prospect of using this technology is still far-fetched.
- **Complex slicing**: Creating and slicing models for 5-axis printing can be very tough. 5-axis printing requires a vastly different approach, so a lot of changes must be made in the traditional way of manufacture and the structure of the machinery used to break when it comes to designing or selecting models to print. Furthermore, most slicers do not even support non-planar (true 3D) printing, so 5-axis printing is still out of the question. In other words, if a model is to be sliced and printed using a 5-axis machine, we have limited and mostly paid options, along with a steep learning curve.

10.3 Use of Nanoparticles, Fibers, and Fillers in SMPs

SMPs can be manufactured in many forms and sizes from the nanoscale to the macroscale, including nanofibers, nanoparticles, thin films, micro foams, and bulk devices. The introduction of nanostructure into SMPs can lead to enhancement in mechanical properties, unique structural color, specific surface area, and multiple functions. It is essential to enhance the current understanding of the various nano or microstructures of SMPs and their fabrication, and to find suitable approaches for producing SMP-based nano or microstructures for different applications. Various types of hybrids or fillers can be used to enhance the properties of shape memory polymers. Carbon nanotubes when integrated with SMP help in quick shape recovery with the help of an electronic flow controlled by mild warming. This can help the shape memory polymers to elaborate their usage to tech-initiated sensors which are crucial parts of actuators and small-scale ethereal vehicle control. The surface of the SMP is changed with a solvent containing nitric acid and sulfuric acid and multi-walled carbon nanotubes which are integrated between the polymeric materials and permeable additives. Similarly, with the introduction of electromagnetic fillers, SMPs can be enhanced to activate complex shape developments in the presence of an external electromagnetic field. Ni chain or fiber can also be integrated into the SMP structure to enhance the shape recovery process of the SMP. For the modification of SMP fibers, there are many fiber formation processes which include melt spinning, dry spinning, wet spinning, gel spinning, emulsion spinning, shear spinning and electrospinning. Among these, the electrospinning fibers fabrication technique is the most repeatedly used technique due to its ability to produce fibers ranging from micrometer to nanometer scale. Electrospinning shows numerous advantages including easy processing and continuous fiber manufacturing. Electro-spun polymer fibers show high surface area per volume unit, high porosity, small diameter, low density, desirable fiber orientation, and nanoarchitecture that can imitate native Extra Cellular matrix (ECM). Due to these reasons, electro-spun nanofibers are extensively used in different fields as nanosensors, filters, biomedical and tissue engineering supports (i.e., as drug delivery systems, hemostatic gauze, artificial vessels, and scaffolds) [13].

11 Processing Techniques for SMP Composites

11.1 Fused Deposition Model (FDM)

Fused deposition modeling is an additive manufacturing technology that makes 3D components using a constant thermoplastic or composite material thread in filament form. An extruder feeds the plastic filament with the help of an extruding nozzle, which is melted and then selectively deposited layer by layer onto the build platform in a predetermined automated path. The first few steps are like any other additive manufacturing technology, which starts with building preparation software. The initial stage is to import the design file and select options for the build, such as layer height, orientation, and infill percentage. The software then computes sections and slices the part into numerous layers. The program then makes extruder paths and creates instructions based on the sectioning data to drive the extrusion heads.

11.2 Stereolithography (SLA)

Stereolithography is one of the most useful methods in the manufacturing of shape memory polymers, especially utilized in 3D printing techniques. This is done by curing layers of photopolymer (which is a light-sensitive liquid plastic) to make a model. The precise layering is steered by a CAD file which is converted into an STL file which helps in digitally slicing a 3D object into layers. Each layer of photopolymer is subjected to an ultraviolet laser that simultaneously hardens it. Subsequent layers are added on top of the previous layer until a complete 3D model has been generated. It produces prototypes that are small, precise, and faster to generate in comparison to standard machinery.

11.3 Digital Light Processing (DLP)

Digital light processing (DLP) is another type of SLA. The major difference between DLP and laser-based SLA is that DLP uses the technique of shining ultraviolet (UV) light (or visible light) from a digital projector to flash a single image of the layer across the entire resin at once. One of the key components of DLP is a digital micromirror device (DMD) chip. The DMD chip consists of an array or a group of reflective aluminum micromirrors that deflect incoming light from the UV source to project an image of a particular pattern or a layer of a 3D CAD model. The projection of an image is constituted of small square pixels called "voxels" which passes through an optical lens and cures the photo-curable polymer resin. To fabricate a high-resolution structure, setting up parameters such as the curing time of each layer, layer thickness, and intensity of the UV light is crucial. These parameters are dependent on the constitution and types of photopolymer and photo-initiator cast for the resin. SMPs are attractively used in biomedical fields, such as laser-induced devices for the removal of blood clots. The device is inserted into the blood vessel by nominally invasive surgical treatment, and the laser is applied, the SMPs coils into their permanent shape, and allow the removal of the blood clots from the body [14].

11.4 Selective Laser Sintering (SLS)

Selective laser sintering (SLS) is a powder-based additive manufacturing technology that uses energy imparted by the laser to melt and fuse the powders and then stack layers to form a printed part based on 3D model data. The powder quality significantly affects the performance of SLS sintered parts and thus the powder design and production are the core technology for SLS.

11.5 Powder Bed Fusion (PBF)

Powder bed fusion (PBF) methods use either a laser or an electron beam to melt and fuse material powder together. The electron beam melting method requires a vacuum but can be used with metals and alloys in the production of functional parts. All PBF processes involve the outspread of the powder material over previous layers. There are many mechanisms to enable this, including a roller or a blade.

12 Conclusions

It is evident that SMPs have attained great progress in architecture, triggering new methods and biomedical applications. A substantial amount of work has been directed to identifying network design principles that can tune polymer chemistries for a variety of applications, with increasing effort towards understanding SME and biological response to these materials. However, many challenges also come up with the development of any innovative technology. Considering various shape-memory effects, TWSME is mostly preferred over OWSME, so reversible micro or nanostructure SMPs should be further investigated to accomplish better performance and more durable scaffoldings. Thermo-responsive SMP fibers are the most investigated due to their wide applications, but it is still necessary to program a more precise transition temperature range for biomedical applications as different temperatures are registered during inflammation, infection, or other pathological conditions. There are various regulatory measures that need to be undertaken when applying the technology of shape memory polymers in biomedical applications as these include the matter of life of a human being. Regulatory engineering is a discipline that consists of the development and application of engineering methods, tools, approaches, and other relevant processes derived from various engineering disciplines used to support regulatory and other policy objectives. With the introduction of innovative technology in the market, there is a lot of skepticism that gets involved. The general masses are not completely ready to accept the new idea and hence certain statutory bodies have been formed to address the grievances and doubts of the people. The prosthetic limbs, implants, stents, and other surgical equipment that are developed with the help of shape memory polymers are first passed through many tests, which ensure the safety and overall functioning of the model before it is evaluated on human beings. All these are taken care of by regulatory bodies that manage the variety of new models that are being introduced to the biomedical world. There are abundant approaches to implementing the shape memory effect of polymers which includes the application of its electro or thermal response and structural changes. With the incorporation of nanoparticles, fibers, and fillers, the effect on the shape memory polymers can be vividly studied and their range of applications can hence be widened based on the findings.

Acknowledgements The authors are thankful to Director, Amity Institute of Applied Sciences (AIAS), and Head, Department of Chemistry, and Head Department of Physics for providing the facilities for the compilation of information. One of the authors, Tejendra K, Gupta, is thankful to the Council of Science & Technology, Lucknow, Uttar Pradesh, India, for the financial support (Vide No. CST/CHEM/D-1431).

References

1. Zare, M., Prabhakaran, M.P., Parvin, N., Ramakrishna, S.: Thermally-induced two-way shape memory polymers: mechanisms, structures, and applications. Elsevier B.V (2019). https://doi.org/10.1016/j.cej.2019.05.167
2. Zhao, W., Liu, L., Zhang, F., Leng, J., Liu, Y.: Shape memory polymers and their composites in biomedical applications. Elsevier Ltd (2019). https://doi.org/10.1016/j.mscc.2018.12.054
3. Rokaya, D. et al.: Shape memory polymeric materials for biomedical applications: an update. MDPI (2023). https://doi.org/10.3390/jcs7010024
4. Mora, P., Schäfer, H., Jubsilp, C., Rimdusit, S., Koschek, K.: Thermosetting shape memory polymers and composites based on polybenzoxazine blends, alloys and copolymers. John Wiley and Sons Ltd (2019). https://doi.org/10.1002/asia.201900969.
5. Gong, J. a c, Cao, P. a c, Tao, L. a c, Pei, X. a c, Wang, T. a c, Zhang, Y. a c, Wang, Q. a c, Zhang, J. b, Zhangzhang Tang a b.: 3D printing of a versatile applicability shape memory polymer with high strength and high transition temperature. Chem. Eng. J. **431** (2022)
6. Ehrmann, G., Ehrmann, A.: 3D printing of shape memory polymers. John Wiley and Sons Inc (2021). https://doi.org/10.1002/app.50847
7. Alam, F., Ubaid, J., Butt, H., El-Atab, N.: Swift 4D printing of thermoresponsive shape-memory polymers using vat photopolymerization. NPG Asia Mater **15**(1) (2023). https://doi.org/10.1038/s41427-023-00511-x
8. Pisani, S., Genta, I, Modena, T., Dorati, R., Benazzo, M., Conti, B.: Shape-memory polymers hallmarks and their biomedical applications in the form of nanofibers (2022). MDPI. https://doi.org/10.3390/ijms23031290
9. Razzaq, M.Y., Gonzalez-Gutierrez, J., Mertz, G., Ruch, D., Schmidt, D.F., Westermann, S.: 4D printing of multicomponent shape-memory polymer formulations. MDPI (2022). https://doi.org/10.3390/app12157880
10. Ahmed, A., Arya, S., Gupta, V., Furukawa, H., Khosla, A.: 4D printing: fundamentals, materials, applications and challenges. Elsevier Ltd (2021). https://doi.org/10.1016/j.polymer.2021.123926.
11. Xu, Y., Chen, D.: Shape memory-assisted self-healing polyurethane inspired by a suture technique. J. Mater. Sci. **53**(14), 10582–10592 (2018). https://doi.org/10.1007/s10853-018-2346-9
12. Khan, M.Y. b, Rafey, M. c, Faheem, K. d, Anas, S. a.: Concept of 5D printing technology and its applicability in the healthcare industry. Mater. Today Proc. **56**, 1726–1732 (2022)
13. Chavali, M., Nikolova, M.P., Tripati, G.K., Subbaroy Sarma, G.V.S.: Effect of nano and hybrid fillers on shape-memory polymers properties. Springer, pp. 213–251 (2022)
14. Behl, M., Lendlein, A.: Shape-memory polymers. Mater. Today **10**(4), 20–28 (2007). https://doi.org/10.1016/S1369-7021(07)70047-0

The manufacturer's authorised representative in the EU is Springer Nature Customer Service Centre GmbH, Europaplatz 3, 69115 Heidelberg, Germany. If you have any concerns regarding our products, please contact ProductSafety@springernature.com

Printed and bound by CPI Group (UK) Ltd, Croydon, CR0 4YY
01/10/2026
02225929-0019